D1761375

Linear Electronics

Linear Electronics

Theodore F. Bogart, Jr.

University of Southern Mississippi

MERRILL, AN IMPRINT OF
MACMILLAN PUBLISHING COMPANY
New York

MAXWELL MACMILLAN CANADA
Toronto

MAXWELL MACMILLAN INTERNATIONAL
New York   Oxford   Singapore   Sydney

Editor: Dave Garza
Development Editor: Carol Robison
Production Supervisor: bookworks
Production Manager: Aliza Greenblatt
Text Designer: Susan E. Frankenberry
Cover Designer: Thomas Mack
Cover Photograph: Copyright © Tom Tracy/Photo Network
Illustrations: Precision Graphics

This book was set in Times Roman by Bi-Comp, Inc.,
printed and bound by Arcata Martinsburg.
The cover was printed by Lehigh Press, Inc.

Macmillan Publishing Company
866 Third Avenue, New York, New York 10022

Macmillan Publishing Company is
part of the Maxwell Communication
Group of Companies.

Maxwell Macmillan Canada, Inc.
1200 Eglinton Avenue East
Suite 200
Don Mills, Ontario M3C 3N1

Library of Congress Cataloging-in-Publication Data

Bogart, Theodore F.
    Linear electronics / Theodore F. Bogart, Jr.
        p.   cm.—(Merrill's international series in engineering
    technology)
    Includes index.
    ISBN 0-02-311601-3
        1. Linear integrated circuits—Design.   2. Operational amplifiers.
    3. Electronic circuit design.   4. Semiconductors.   I. Title.
    II. Series.
    TK7874.B63   1994
    621.3815—dc20
                                                        93-13493
                                                          CIP

Printing: 1 2 3 4 5 6 7   Year: 4 5 6 7 8 9 0

MERRILL'S INTERNATIONAL SERIES IN ENGINEERING TECHNOLOGY

Microcomputer Servicing

Adamson, *Microcomputer Repair*, 0-02-300825-3

Asser, Stigliano, & Bahrenburg, *Microcomputer Servicing: Practical Systems and Troubleshooting, 2nd Edition*, 0-02-304241-9

Asser, Stigliano, & Bahrenburg, *Microcomputer Theory and Servicing, 2nd Edition*, 0-02-304231-1

Programming

Adamson, *Applied Pascal for Technology*, 0-675-20771-1

Adamson, *Structured BASIC Applied to Technology, 2nd Edition*, 0-02-300827-X

Adamson, *Structured C for Technology*, 0-675-20993-5

Adamson, *Structured C for Technology (with disk)*, 0-675-21289-8

Nashelsky & Boylestad, *BASIC Applied to Circuit Analysis*, 0-675-20161-6

Instrumentation and Measurement

Berlin & Getz, *Principles of Electronic Instrumentation and Measurement*, 0-675-20449-6

Buchla & McLachlan, *Applied Electronic Instrumentation and Measurement*, 0-675-21162-X

Gillies, *Instrumentation and Measurements for Electronic Technicians, 2nd Edition*, 0-02-343051-6

Transform Analysis

Kulathinal, *Transform Analysis and Electronic Networks with Applications*, 0-675-20765-7

Biomedical Equipment Technology

Aston, *Principles of Biomedical Instrumentation and Measurement*, 0-675-20943-9

Mathematics

Monaco, *Essential Mathematics for Electronics Technicians*, 0-675-21172-7

Davis, *Technical Mathematics*, 0-675-20338-4

Davis, *Technical Mathematics with Calculus*, 0-675-20965-X

INDUSTRIAL ELECTRONICS/INDUSTRIAL TECHNOLOGY

Bateson, *Introduction to Control System Technology, 4th Edition*, 0-02-306463-3

Fuller, *Robotics: Introduction, Programming, and Projects*, 0-675-21078-X

Goetsch, *Industrial Safety and Health: In the Age of High Technology*, 0-02-344207-7

Goetsch, *Industrial Supervision: In the Age of High Technology*, 0-675-22137-4

Goetsch & Davis, *Introduction to Total Quality: Quality, Productivity, and Competitiveness*, 0-02-344221-2

Horath, *Computer Numerical Control Programming of Machines*, 0-02-357201-9

Hubert, *Electric Machines: Theory, Operation, Applications, Adjustment, and Control*, 0-675-20765-7

Humphries, *Motors and Controls*, 0-675-20235-3

Hutchins, *Introduction to Quality: Management, Assurance, and Control*, 0-675-20896-3

Laviana, *Basic Computer Numerical Control Programming*, 0-675-21298-7

Pond, *Fundamentals of Statistical Quality Control*, 0-02-396034-5

Reis, *Electronic Project Design and Fabrication, 2nd Edition*, 0-02-399230-1

Rosenblatt & Friedman, *Direct and Alternating Current Machinery, 2nd Edition*, 0-675-20160-8

Smith, *Statistical Process Control and Quality Improvement*, 0-675-21160-3

Webb, *Programmable Logic Controllers: Principles and Applications, 2nd Edition*, 0-02-424970-X

Webb & Greshock, *Industrial Control Electronics, 2nd Edition*, 0-02-424864-9

MECHANICAL/CIVIL TECHNOLOGY

Dalton, *The Technology of Metallurgy*, 0-02-326900-6

Keyser, *Materials Science in Engineering, 4th Edition*, 0-675-20401-1

Kokernak, *Fluid Power Technology*, 0-02-305705-X

Kraut, *Fluid Mechanics for Technicians*, 0-675-21330-4

Mott, *Applied Fluid Mechanics, 4th Edition*, 0-02-384231-8

Mott, *Machine Elements in Mechanical Design, 2nd Edition*, 0-675-22289-3

Rolle, *Thermodynamics and Heat Power, 4th Edition*, 0-02-403201-8

Spiegel & Limbrunner, *Applied Statics and Strength of Materials, 2nd Edition*, 0-02-414961-6

Spiegel & Limbrunner, *Applied Strength of Materials*, 0-02-414970-5

Wolansky & Akers, *Modern Hydraulics: The Basics at Work*, 0-675-20987-0

Wolf, *Statics and Strength of Materials: A Parallel Approach to Understanding Structures*, 0-675-20622-7

DRAFTING TECHNOLOGY

Cooper, *Introduction to VersaCAD*, 0-675-21164-6

Ethier, *AutoCAD in 3 Dimensions*, 0-02-334232-3

Goetsch & Rickman, *Computer-Aided Drafting with AutoCAD*, 0-675-20915-3

Kirkpatrick & Kirkpatrick, *AutoCAD for Interior Design and Space Planning*, 0-02-364455-9

Kirkpatrick & Kirkpatrick, *AutoCAD for Interior Design and Space Planning, Release 12 Version, 2nd Edition*, 0-02-364471-0

Kirkpatrick, *The AutoCAD Book: Drawing, Modeling, and Applications, 2nd Edition*, 0-675-22288-5

Kirkpatrick, *The AutoCAD Book: Drawing, Modeling, and Applications, Including Release 12, 3rd Edition*, 0-02-364440-0

Lamit & Lloyd, *Drafting for Electronics, 2nd Edition*, 0-02-367342-7

Lamit & Paige, *Computer-Aided Design and Drafting*, 0-675-20475-5

Maruggi, *Technical Graphics: Electronics Worktext, 2nd Edition*, 0-675-21378-9

Maruggi, *The Technology of Drafting*, 0-675-20762-2

Sell, *Basic Technical Drawing*, 0-675-21001-1

TECHNICAL WRITING

Croft, *Getting a Job: Resume Writing, Job Application Letters, and Interview Strategies*, 0-675-20917-X

Panares, *A Handbook of English for Technical Students*, 0-675-20650-2

Pfeiffer, *Proposal Writing: The Art of Friendly Persuausion*, 0-675-20988-9

Pfeiffer, *Technical Writing: A Practical Approach, 2nd Edition*, 0-02-395111-7

Roze, *Technical Communications: The Practical Craft, 2nd Edition*, 0-02-404171-8

Weisman, *Basic Technical Writing, 6th Edition*, 0-675-21256-1

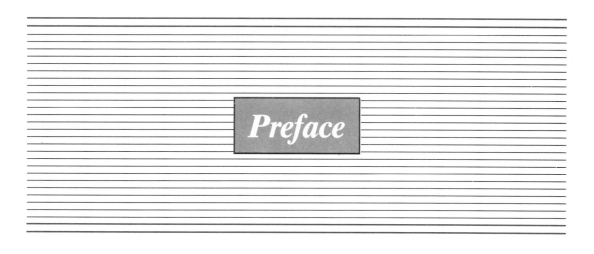

Preface

Linear Electronics is designed to support the traditional two-semester introductory course in electronic circuit theory at the engineering technology level. Although semiconductor device theory is treated in several chapters, the emphasis in this book is more on system *concepts* than on transistor circuit analysis. Toward that end, topics such as gain, frequency response, feedback, filtering, waveshaping, and oscillation theory are treated from a generalized block-diagram approach. In many instances, the principal block in the treatment is an operational amplifier. This approach stands in contrast to the traditional pedagogy in which each principle is taught in the context of a circuit containing a particular semiconductor device. It is justified by at least three important considerations:

1. Integrated circuits have made most modern electronic circuits and systems modular in nature. In such cases, an in-depth study of the peculiarities of individual semiconductor devices is not relevant to practical design and troubleshooting activities.
2. Although students should have an understanding of fundamental device theory, too much emphasis on the analysis of specific transistor circuits tends to engross them in a maze of "formulas" and computational tasks that have limited scope. Formula-driven electronics distracts from the understanding of more important concepts and the broad principles underlying the theory.
3. When the emphasis on discrete circuit analysis is reduced, more time is available to treat topics for which little or no room was available in traditional courses. For example, *Linear Electronics* includes coverage of phase-locked loops, logarithmic amplifiers, transconductance amplifiers, A/D and D/A converters, modulators, and waveform generators, among others. (Since most of these topics are independent of one another, instructors can elect to include or omit them at their own discretion.)

Notwithstanding the foregoing remarks, the author recognizes that discrete circuit theory must be included in a text of this nature. There are, of course, numerous applications involving the control of heavy currents, high voltages, and large power dissipation that can only be served by discrete devices. Indeed, a survey of potential users of the book revealed that there is considerable interest in retaining coverage of such topics, so the original scope of the work was expanded to include chapters on power supplies, voltage regulators, and power amplifiers. These chapters are provided in addition to early chapters on diode and transistor theory and single-transistor amplifiers. Besides providing students with a knowledge of semiconductor fundamentals, discrete device theory exposes them to such important concepts as families of characteristic curves, load lines, quiescent values, operating limits, and temperature effects. Although the author believes these chapters should be included in an introductory electronics course, the book is organized so that coverage of discrete devices could be omitted without loss of continuity, if so desired.

Much of the material emphasizing broad principles and concepts as well as some device theory covered in *Linear Electronics* is excerpted from the author's *Electronic Devices and Circuits* (Macmillan/Merrill, 3rd edition). In *Linear Electronics* the treatment of semiconductor theory has been condensed, discrete circuit theory reduced, systems concepts expanded, and the scope has been broadened.

Many chapters are accompanied by SPICE and PSpice examples and exercises. Appendix A contains SPICE instructional material that parallels the sequence in which electronics theory is introduced in the book, so new simulation skills can be developed at the same pace that new theory is taught.

The book contains a large number of end-of-chapter exercises, with answers provided for odd-numbered exercises. One reason for the large quantity of exercises is that every effort was made to include at least one odd-numbered exercise and one even-numbered exercise for every important topic or concept discussed in the book. Thus, instructors who prefer to assign exercises with answers and those who prefer exercises without answers will both find an ample selection.

ACKNOWLEDGMENTS

I wish to thank my colleagues who have been most helpful in reviewing the manuscript. They have provided numerous suggestions for the development of this first edition: Andrew Anderson, Brookdale Community College; Rick Ackerman, ITT Technical Institute; Ambrose Barry, University of North Carolina-Charlotte; Harold Broberg, Indiana University-Purdue University; Don Custer, Western Iowa Technical Community College; Hobart McWilliams, Montana State University; John Myers, Montana State University; Prof. Akbar Nouhi, California State Polytechnic University, Pomona; Malcolm Skipper, Midlands Technical College; and John Slough, DeVry Institute of Technology-Irving.

Thanks also to my editorial group at Macmillan Publishing Company, especially Dave Garza, Carol Robison, and Aliza Greenblatt.

Table of Contents

13   POWER SUPPLIES AND VOLTAGE REGULATORS   515

14   POWER AMPLIFIERS   577

Linear Electronics

Basic Concepts in Linear Circuits

1

1.1 LINEARITY

Since the subject of this book is *linear* electronics, we begin our study with a review of the definition of linearity, the practical implications of that concept, and some of the terminology associated with it. In the process, we will learn that the scope of linear electronics is not necessarily limited to electronic devices that adhere strictly to the definition of linearity.

A passive device is said to be linear if the current through it is directly proportional to the voltage across it. Thus, for example, if we double the voltage across a linear device, the current through it will also double. The name *linear* is derived from the fact that a plot of the current through such a device versus the voltage across it is a straight line. A very familiar example of a passive linear device is a resistor. We know that a resistor having resistance R is linear because Ohm's law expresses the fact that the current I through it is directly proportional to the voltage V across it:

$$I = \frac{V}{R} \qquad (1.1)$$

Note that I and V in equation 1.1 are variables and we may regard the quantity $1/R$ as a *constant of proportionality:*

$$I = \left(\frac{1}{R}\right) V \qquad (1.2)$$

This constant is of course the conductance ($G = 1/R$) of the resistor. Suppose $R = 50\ \Omega$ and the voltage across R is 50 V. Then, $I = 50\ \text{V}/50\ \Omega = 1$ A. If we double the voltage to $V = 100$ V, then the current also doubles: $100\ \text{V}/50\ \Omega = 2$ A. Similarly, increasing or decreasing V by any factor increases or decreases I by the same factor.

When plotting the current through and voltage across an electronic device, it is conventional to plot current along the vertical axis and voltage along the horizontal axis. Such a plot is called the current–voltage *characteristic* of the device. Figure 1.1 shows the current–voltage (I–V) characteristic of a 1-kΩ resistor. We see that the characteristic of this linear device is indeed a straight line.

1

Figure 1.1
I–V characteristic of a 1-kΩ resistor.

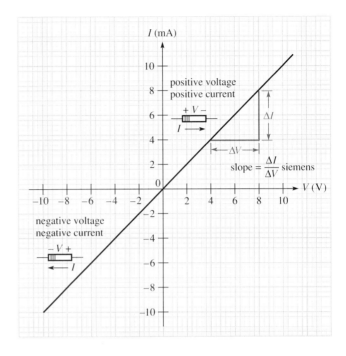

In any investigation of current and voltage relationships, we assume, or de-fine, one voltage polarity to be positive and one direction of current flow to be positive. A negative voltage is one having a polarity opposite that assumed to be positive and a negative current is one that flows in a direction opposite that assumed to be positive. We know that reversing the polarity of the voltage across a resistor reverses the direction of the current through it, and Figure 1.1 shows that the linearity property of the 1-kΩ resistor also applies in that situation: The straight line extends into the third quadrant where both voltage and current are negative.

The *slope* of an *I–V* characteristic is the *change* in current divided by the *change* in voltage that caused it. A change in current is denoted by ΔI and a change in voltage by ΔV. See Figure 1.1. Since ΔI has the units of amperes (A) and ΔV has the units of volts (V), the units of the slope are A/V, i.e., the same as the units of conductance: siemens (S).

$$\text{slope} = \frac{\Delta I}{\Delta V} \qquad \text{siemens} \qquad (1.3)$$

It is important to note that the slope of the *I–V* characteristic of a linear device is the same no matter where along the characteristic it is calculated. Furthermore, the value of the slope is the same as the value of $1/R$ in equation 1.1. In other words, the value of conductance calculated from $G = I/V$ at any single point on the linear characteristic is the same as the slope $\Delta V/\Delta I$ calculated anywhere along the characteristic. These ideas are illustrated in the next example.

Example 1.1

1. Find the slope of the *I–V* characteristic shown in Figure 1.2 when the current through the device having that characteristic changes from 2 mA to 4 mA.

Figure 1.2
(Example 1.1) Values for part (a) are
shown.

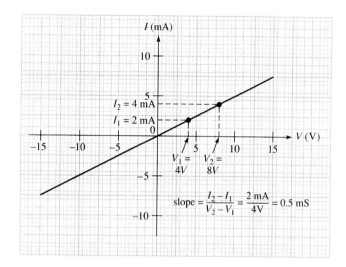

2. Repeat (1) when the current changes from 6 mA to −1 mA.
3. Repeat (1) when the voltage changes from 5 V to 10 V.
4. Repeat (1) when the voltage changes from −4 V to 0 V.
5. What is the resistance of the device?

Solution.   The slope is calculated using the relationship

$$\text{slope} = \frac{\Delta I}{\Delta V} = \frac{I_2 - I_1}{V_2 - V_1} \tag{1.4}$$

where I_1 and V_1 are values at one point (point 1) on the characteristic and I_2 and V_2 are values at a second point (point 2). It does not matter which point is designated 1 and which is designated 2. We will arbitrarily choose the second value cited in each part of this example as point 2. Using values obtained from the I–V characteristic, we find:

1. The voltage changes from 4 V to 8 V when the current changes from 2 mA to 4 mA. Thus,

$$\text{slope} = \frac{I_2 - I_1}{V_2 - V_1} = \frac{4 \text{ mA} - 2 \text{ mA}}{8 \text{ V} - 4 \text{ V}} = \frac{2 \text{ mA}}{4 \text{ V}} = 0.5 \text{ mS}$$

2.

$$\text{slope} = \frac{I_2 - I_1}{V_2 - V_1} = \frac{-1 \text{ mA} - 6 \text{ mA}}{-2 \text{ V} - 12 \text{ V}} = \frac{-7 \text{ mA}}{-14 \text{ V}} = 0.5 \text{ mS}$$

3.

$$\text{slope} = \frac{I_2 - I_1}{V_2 - V_1} = \frac{5 \text{ mA} - 2.5 \text{ mA}}{10 \text{ V} - 5 \text{ V}} = \frac{2.5 \text{ mA}}{5 \text{ V}} = 0.5 \text{ mS}$$

4.

$$\text{slope} = \frac{I_2 - I_1}{V_2 - V_1} = \frac{0 \text{ mA} - (-2 \text{ mA})}{0 \text{ V} - (-4 \text{ V})} = \frac{2 \text{ mA}}{4 \text{ V}} = 0.5 \text{ mS}$$

We see that the slope has the same value (0.5 mS) no matter where along the characteristic it is calculated and no matter how large or how small the increments in current and voltage (ΔI and ΔV).

5. Since the device is linear, the slope of the $I-V$ characteristic equals its conductance G. Thus,

$$G = \frac{1}{R} = 0.5 \text{ mS}$$

and

$$R = \frac{1}{G} = \frac{1}{0.5 \times 10^{-3} \text{ S}} = 2 \times 10^3 \ \Omega = 2 \text{ k}\Omega$$

Verify that the ratio of I to V at any *single* point on the linear $I-V$ characteristic is $G = 0.5$ mS. It follows that the resistance, $R = V/I = 2$ kΩ, is the same at every point on the characteristic.

Inductors and capacitors are further examples of passive linear devices. In these cases, we interpret the definition of linearity in terms of the magnitudes of *ac* voltages and currents and assume that the frequency is constant. Since the frequency is constant, the magnitudes of the inductive and capacitive reactances are constant:

$$|X_L| = 2\pi f L \quad \text{and} \quad |X_C| = \frac{1}{2\pi f C}$$

The magnitude of the ac voltage, $|v|$, and of the ac current, $|i|$, are then related by

$$|i| = \frac{|v|}{|X_L|} \quad \text{and} \quad |i| = \frac{|v|}{|X_C|} \tag{1.5}$$

and we see again that current is directly proportional to voltage.

Static (dc) and Dynamic (ac) Resistance

When we use an $I-V$ characteristic of a device to find the ratio of the voltage V to current I at any single point on the characteristic, we have found the *static*, or dc, resistance of the device:

$$R_{dc} = \frac{V}{I} \quad \text{ohms} \tag{1.6}$$

We know that R_{dc} has the same value at every point on the $I-V$ characteristic of a linear device such as a resistor. Imagine now that we connect an ac voltage source across the device and thus change the voltage across it between some minimum and maximum values. This change in voltage, ΔV, causes a change in current, ΔI, through the device. The ratio of ΔV to ΔI is called the *dynamic*, or ac, resistance of the device:

$$r_{ac} = \frac{\Delta V}{\Delta I} \quad \text{ohms} \tag{1.7}$$

(Note the use of lower-case r, in accordance with the convention for representing ac quantities.) It is clear that r_{ac} is the reciprocal of the slope of the $I-V$ characteristic, as defined by equation 1.3. We therefore know that a linear device, such as a resistor, has the same value of r_{ac} no matter where along its $I-V$ characteristic it is calculated and that $r_{ac} = R_{dc}$. As we shall eventually see, such is not the case for many practical electronic devices that do not fit our strict definition of linearity.

1.2 A BROADER DEFINITION OF LINEARITY

We can think of the I–V characteristic of a device as a plot of experimental data obtained by adjusting the voltage to different values and observing the values of current that result. In other words, the voltage is the *stimulus*, or input, to the device and the current is the *response*, or output, of the device. In many practical electronic devices and components, such as amplifiers, the input may be either a voltage or a current and the output may be either a voltage or a current. For example, a voltage amplifier is a device whose input is a voltage and whose output is a voltage (an amplified version of the input voltage). In general, such a device is said to be linear if the output is directly proportional to the input, i.e., if a plot of output versus input is a straight line.

Figure 1.3 shows a plot of output voltage versus input voltage for a linear voltage amplifier. Note that output voltage, V_o, is plotted along the vertical axis and input voltage, V_{in}, along the horizontal axis. The ratio of an output voltage to an input voltage, V_o/V_{in}, is called the voltage *gain*, and is designated A_v. Since the plot is linear, the voltage gain has the same value no matter at which point the ratio is calculated. Furthermore, the slope, $\Delta V_o/\Delta V_{in}$, is the same no matter where along the plot it is calculated, and the slope has the same value as the ratio V_o/V_{in} at any single point on the plot:

$$A_v = \frac{V_o}{V_{in}} = \frac{\Delta V_o}{\Delta V_{in}} \tag{1.8}$$

Figure 1.3
A plot of output voltage (V_o) versus input voltage (V_{in}) for a linear voltage amplifier.

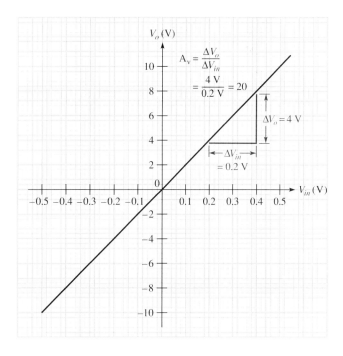

Figure 1.4
(Example 1.2)

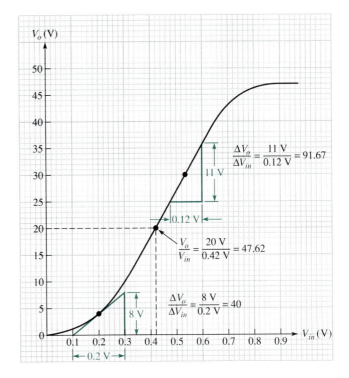

Since the units of V_o, ΔV_o, V_{in}, and ΔV_{in}, are all volts, the units of A_v are V/V, which is also considered dimensionless. As can be seen in the figure, the value of A_v in this example is 20.

Figure 1.4 shows a plot of V_o versus V_{in} for a typical voltage amplifier that does not satisfy the definition of linearity. The next example demonstrates that V_o/V_{in} and $\Delta V_o/\Delta V_{in}$ in this case have different values, depending on where along the plot they are calculated.

Example 1.2

For the amplifier whose output voltage is plotted versus input voltage in Figure 1.4, find

1. $\Delta V_o/\Delta V_{in}$ when $V_{in} = 0.2$ V;
2. $\Delta V_o/\Delta V_{in}$ when $V_o = 30$ V;
3. V_o/V_{in} when $V_o = 20$ V.

Solution

1. As shown in Figure 1.4, we construct a right triangle whose hypotenuse is a line tangent to the curve at the point where $V_{in} = 0.2$ V. Then

$$\frac{\Delta V_o}{\Delta V_{in}} = \frac{8 \text{ V} - 0 \text{ V}}{0.3 \text{ V} - 0.1 \text{ V}} = \frac{8 \text{ V}}{0.2 \text{ V}} = 40$$

2. Repeating the procedure at the point where $V_o = 30$ V, we find

$$\frac{\Delta V_o}{\Delta V_{in}} = \frac{36 \text{ V} - 25 \text{ V}}{0.6 \text{ V} - 0.48 \text{ V}} = \frac{11 \text{ V}}{0.12 \text{ V}} = 91.67$$

3. At the point where $V_o = 20$ V,

$$\frac{V_o}{V_{in}} = \frac{20 \text{ V}}{0.42 \text{ V}} = 47.62$$

Notice that all three values calculated in this example are different. Also observe that the slope of the curve is *zero* at values of V_{in} greater than about 0.9 V. In both of these regions the curve is horizontal, meaning that the *change* in V_o, (ΔV_o), is zero, and so therefore is $\Delta V_o / \Delta V_{in}$.

In the context of a voltage amplifier, we interpret the results of this example as follows: A dc input voltage of 0.42 V will produce a dc output voltage of 20 V; an ac input voltage that has a peak-to-peak value of 0.12 V (and centered on a dc value of about 0.49 V) will produce an ac output voltage having a peak-to-peak value of 11 V (centered on a dc value of 30 V). Any ac voltage that is centered on a dc voltage less than 0 V or greater than 0.9 V will produce zero ac output voltage.

1.3   THE SCOPE OF LINEAR ELECTRONICS

As might be expected, the field of study we call linear electronics includes the study of devices that adhere strictly to the definition of linearity. It also includes devices that have a linear *range,* such as the amplifier in Example 1.4. Furthermore, it may include devices whose plots of output versus input are nonlinear over their entire range. (We often *idealize* such devices and assume that they are at least approximately linear over some small range.)

The study of linear electronics is not so much the study of devices that adhere (more or less) to the definition of linearity as it is the study of devices that are operated (used) over some *continuous* range of input and output voltages. For example, if the amplifier of Example 1.2 were used to amplify any or all voltages in the range from, say, $V_{in} = 0.5$ V to $V_{in} = 0.7$ V, we would say that it is being used in a linear application. On the other hand, if the input voltages were always either 0 V or 0.9 V, we would say it is being used in a *digital* application. In other words, it is conventional to use the word *linear* to distinguish an application from *digital,* rather than to mean the opposite of nonlinear. In short, linear electronics means nondigital electronics. A more accurate term that is often used to mean linear as opposed to digital is the word *analog.* Thus, analog devices are those that are intended to be operated over some continuous range of inputs and outputs and that therefore fall within the scope of *linear* electronics.

EXERCISES

Section 1.1   Linearity

1.1 What is the slope of the I–V characteristic of a 500-kΩ resistor?

1.2 For the device whose I–V characteristic is shown in Figure 1.5,
  a. Find the slope when the current changes from 8 μA to 24 μA.
  b. Find the slope when the voltage changes from −8 V to +4 V.
  c. What is the resistance of this device?

1.3 What is the slope of the I–V characteristic of a 25-mH inductor when the ac current through it has a frequency of 10 kHz?

1.4 The ac voltage across a 0.1-μF capacitor is 9 V rms and has a frequency of 1.5 kHz. If the frequency is held constant and the voltage is increased by 10%, what is the rms value of the current at the larger voltage?

Figure 1.5
(Exercises 1.2 and 1.5)

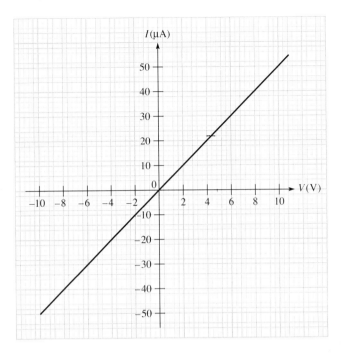

1.5   For the device whose *I–V* characteristic is shown in Figure 1.5,
    a. Find r_{ac} at the point where $V = 6$ V.
    b. Find R_{dc} at the point where $I = -32\ \mu A$.
    c. What is the conductance of the device?

1.6   When the voltage across a certain linear device is increased from -2 V to $+7$ V, it is found that the current increases from -4 mA to 14 mA.
    a. What is the value of r_{ac}?
    b. What is the dc current in the device when the dc voltage is 15 V?

Section 1.2   A Broader Definition of Linearity

1.7   For the voltage amplifier whose plot of V_o versus V_{in} is shown in Figure 1.6,
    a. Find $\Delta V_o/\Delta V_{in}$ when $V_{in} = 30$ mV.

    b. Find the voltage gain when $V_{in} = 30$ mV.
    c. Find the output voltage when the input voltage is -10 mV dc.

1.8   For the voltage amplifier whose plot of V_o versus V_{in} is shown in Figure 1.7,
    a. Find the voltage gain when $V_{in} = 0.2$ V.
    b. Find the voltage gain when $V_{in} = 0.4$ V.
    c. Find the voltage gain when $V_{in} = -0.24$ V.
    d. Over what range of input voltages is the amplifier linear?

1.9   The input to the amplifier whose plot of V_o versus V_{in} is shown in Figure 1.7 is $v_{in}(t) = 0.1 + 0.1 \sin(\omega t)$ V.
    a. What is the peak-to-peak output voltage?
    b. Write a mathematical expression for the output voltage, $v_o(t)$.

Figure 1.6
(Exercise 1.7)

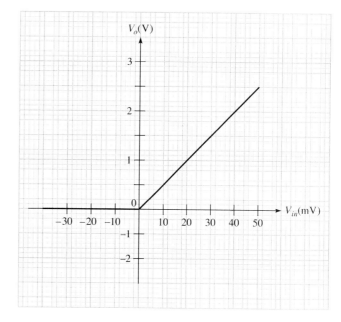

Figure 1.7
(Exercises 1.8 and 1.9)

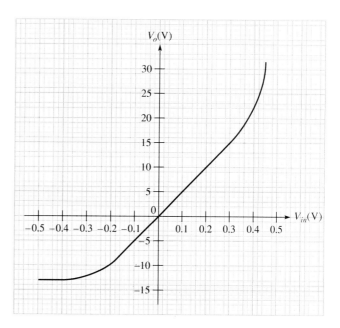

Semiconductor Theory and Junction Diodes

2

2.1 SEMICONDUCTOR THEORY

Semiconductor Materials

Modern electronic devices—diodes, transistors, integrated circuits—are constructed from a special class of materials called *semiconductors*. As the name implies, a semiconductor is neither a good conductor of electrical current nor an insulator. However, it is not the ability or lack of ability to conduct current that makes semiconductors useful. Rather, it is their ability to form *crystals* having special electrical properties that makes them so valuable. The element *silicon* (Si) is the most widely used semiconductor material, followed by *germanium* (Ge).

Figure 2.1(a) is a diagram of the atomic structure of the silicon atom. Notice that its nucleus contains 14 positively charged protons (designated 14P) and 14 neutrons (14N), which have no electrical charge. When the atom is electrically neutral, as in the figure, it contains 14 negatively charged electrons distributed among four electron shells surrounding the nucleus. The outermost shell contains four electrons, called the *valence* electrons, which are the ones most responsible for the electrical properties of the material. Figure 2.1(b) is an abbreviated diagram of the atom in which only the four valence electrons and the nucleus are shown. We will use this type of diagram in all future discussions. Germanium atoms also have four valence electrons and will be diagrammed the same way.

Since neither silicon nor germanium occur naturally in a state suitable for use as a semiconductor material, they must be subjected to a complex manufacturing process in which crystals are "grown" from a batch of melted, highly purified material. Figure 2.2 is a diagram of the atomic structure of a semiconductor crystal. Notice that each atom *shares* valence electrons with four of its neighbors, as indicated by the ovals enclosing electrons around each nucleus. Shared electrons tend to stay bound to their "parent" atoms in *covalent bonds,* which are symbolized in the figure by the ovals. Although the interlocking structure of the crystal is highly stable, it is possible for a valence electron to acquire enough energy (usually heat energy) to overcome the covalent bond and thus escape its parent atom. We say that a covalent bond has been *ruptured*. The escaped electron becomes a *free* electron that can wander about in the material. There is enough heat energy at room temperature to free a large number of electrons,

Figure 2.1
Structure of the silicon atom.

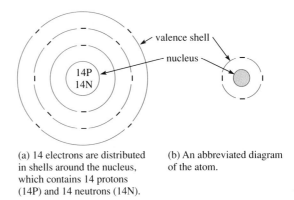

(a) 14 electrons are distributed
in shells around the nucleus,
which contains 14 protons
(14P) and 14 neutrons (14N).

(b) An abbreviated diagram
of the atom.

which are then available for the conduction of electrical current through the crystal, just as the flow of electrons through a conductor constitutes current. (However, there are vastly more free electrons in a conductor than in a semiconductor.)

When an electron escapes a covalent bond, it leaves behind a *hole* in the crystal structure. Since electrically neutral atoms have as many positively charged protons in their nucleus as they do negatively charged electrons outside the nucleus, an atom that has a hole from having lost an electron now has a net *positive* charge. The atom is called a positive *ion*. See Figure 2.3. It is also possible for a wandering electron to fall into a hole, thus returning the atom to a neutral state. In those cases, we say that a *recombination,* or *annihilation,* has occurred.

Current in Semiconductors

We have already noted that free electrons are available in a semiconductor to establish current flow through it. These electrons are called charge *carriers* because they carry negative charge from one location to another when they move. Holes are also charge carriers, in this case, positive charge carriers. Recall that an atom having a hole is positively charged. If a valence electron leaves one atom to occupy a hole in an adjacent atom, the atom it left becomes positively charged and the atom it joined becomes neutral. In effect, positive charge has moved from one atom to another. See Figure 2.4. The movement of holes through a crystal in this

Figure 2.2
Covalent bonding in a semiconductor
crystal.

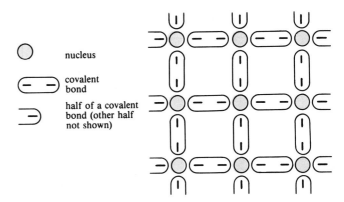

◯ nucleus

⊂− −⊃ covalent
bond

⊂− half of a covalent
bond (other half
not shown)

Figure 2.3
When a covalent bond is ruptured, a free electron is produced and the atom becomes a positively charged ion.

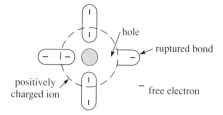

way is called hole current. Thus, there are two types of current in a semiconductor: electron current and hole current. Note that the movement of positive charge in one direction is equivalent to the movement of negative charge in the opposite direction, so the two components *add* to equal the total current in the material. Also note that a recombination (free electron falling into a hole) does not leave a hole behind, so there is no hole flow in that case.

There are no holes in a conductor, so electron current is the only type that can exist. This current results from the presence of an *electric field* through the material, created, for example, by an externally connected voltage source. The electric field drives electrons from the negative terminal of the voltage source through the material to the positive terminal. Current that exists due to the force of an electric field is called *drift* current. In a semiconductor, both electron current and hole current are created by electric field forces, so both electron drift and hole drift occur when a voltage source is connected across the ends of a semiconductor. A given electric field will cause electrons to flow in one direction and holes to flow in the opposite direction. If the same number of electrons and holes are subjected to the same electric field, the electrons will move faster than the holes (electrons are said to have greater *mobility*), so the component of drift current due to electrons will be greater than that due to holes.

Another type of current that can exist in a semiconductor is called *diffusion* current. Diffusion occurs when there is an imbalance in the number of carriers of a given type between two different regions of a semiconductor. Carriers tend to migrate from a region where there are many of their own type to a region where there are fewer, thus correcting the imbalance. If, for example, there are many more electrons at one end of a bar of semiconductor material than there are at the other end, electrons diffuse from the high-density end to the low-density end until their distribution is more or less uniform throughout. Unless the region containing the excess electrons is replenished with new electrons (as it often is in practical devices), diffusion current ceases when the imbalance has been corrected. Hole current can also be of the diffusion type.

Figure 2.4
Hole current. When the electron in position A is freed, a hole is left in its place. If the electron in position B moves into the hole at A, the hole, in effect, moves from A to B.

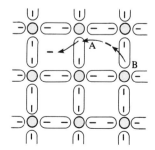

Figure 2.5
Structure of a silicon crystal containing a donor atom. The donor's nucleus is labeled D and the nuclei of the silicon atoms are labeled Si. Note the excess electron.

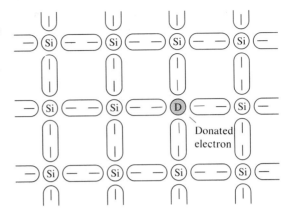

P and N Materials

Pure semiconductor material is said to be *intrinsic*. In practice, certain impurities are introduced into intrinsic material during the manufacturing process to give it new properties. The process of introducing impurities is called *doping,* and material that has been doped is called *extrinsic*. The purpose of doping is to create a semiconductor that has more (free) electrons than holes (N material) or more holes than electrons (P material). To create N material, the semiconductor is doped with impurity atoms that have five instead of four valence electrons. When each such atom joins the crystal structure, four of its valence electrons create the usual covalent bonds with other atoms, and the fifth electron is free. This type of impurity is called a *donor* because every such atom donates one free electron to the material. Figure 2.5 illustrates a donor atom in a crystal structure. To create P material, the semiconductor is doped with atoms that have three instead of four valence electrons. Each such atom forms covalent bonds with three neighbors only and thus creates one hole in the structure. These atoms are called *acceptors* because they can each accept one electron. Figure 2.6 illustrates this case.

It is important to note that both N and P materials are electrically *neutral*. Although every donor atom donates a free electron, the donor atom's nucleus brought with it just the right number of positively charged protons to neutralize

Figure 2.6
Structure of a silicon crystal containing an acceptor atom. The acceptor's nucleus is labeled A and the nuclei of the silicon atoms are labeled Si. Note the incomplete bond and resulting hole caused by the acceptor's presence.

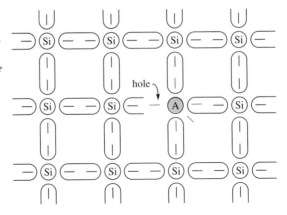

the charge carried by all its electrons. Similarly, each acceptor atom has the same number of protons as electrons and is also electrically neutral.

Although N material has more free electrons than holes, it does have *some* holes, and P material does have some free electrons. In N material, electrons are called the *majority* carriers and holes the *minority* carriers. In P material, holes are majority carriers and electrons are minority carriers. N material that has been very heavily doped, and which is therefore very conductive, is said to be N^+ and heavily doped P material is P^+.

2.2   The PN Junction Diode

PN Junctions

When a block of N material is constructed adjacent to a block of P material, the boundary between the two is called a PN junction. At the junction, holes diffuse from the P material into the N material and electrons diffuse from the N material into the P material. (Remember that diffusion occurs when there is an imbalance of carriers of a given type.) Every electron crossing the junction leaves behind a donor atom with a net positive charge and every hole leaves behind an acceptor atom with a net negative charge. See Figure 2.7. Consequently, after the diffusion there is a thin layer of positive ions on the N side of the junction and a thin layer of negative ions on the P side. There are no mobile charge carriers in this region, and it is called the *depletion* region because it is depleted of such charge. See Figure 2.8. The layers of opposite charge establish an electric field (like a voltage source) directed from the N material toward the P material. The direction of the field *opposes* the flow of further electron current from N to P and hole current from P to N. The voltage difference between the charged regions is therefore called a *barrier* voltage. The value of the barrier voltage depends on temperature and doping levels but is typically about 0.7 V for silicon and 0.3 V for germanium.

Suppose now that an external voltage source is connected across the P and N material as shown in Figure 2.9. Notice that the positive terminal of the source is connected to the P material and the negative terminal to the N material. The polarity of the external source thus *opposes* the barrier voltage at the PN junction

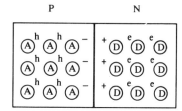

(a) Blocks of P and N materials at the instant they are joined; both blocks are initially neutral.

(b) The PN junction showing charged ions after hole and electron diffusion.

Figure 2.7
Formation of a PN junction. A = acceptor atom; h = associated hole; D = donor atom; e = associated electron. + = positively charged ion; − = negatively charged ion.

Figure 2.8
The electric field E across a PN junction inhibits diffusion current from the N to the P side. There are no mobile charge carriers in the depletion region (whose width is proportionally much smaller than that shown).

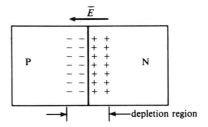

depletion region

and therefore enhances the flow of current through the material and across the junction. The PN junction is said to be *forward biased*. If the polarity of the external source is reversed (positive to N and negative to P) as shown in Figure 2.10, the barrier voltage is reinforced and very little current flows. In this case, the junction is said to be *reverse biased*. Notice that forward biasing the junction narrows the depletion region and reverse biasing widens it.

The Junction Diode

A *diode* is a PN junction and therefore has the property just described: It permits a generous flow of current in one direction (when forward biased) and permits virtually no current to flow in the opposite direction (when reverse biased). This property is responsible for many useful applications of diodes in electronic circuits. The P side of a diode is called the *anode* and the N side is called the *cathode*. Figure 2.11 illustrates forward and reverse biasing of a diode and shows the standard symbol for the device.

The *diode equation* permits us to compute the current through a diode as a function of the voltage V connected across it:

$$I = I_s \left(e^{V/\eta V_T} - 1 \right) \qquad (2.1)$$

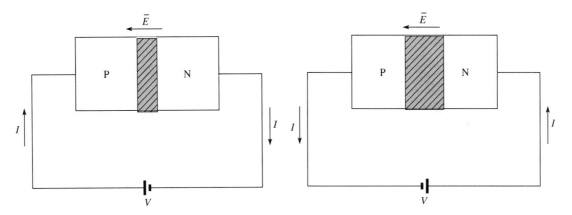

Figure 2.9
A voltage source V connected to forward bias a PN junction. The depletion region (shown shaded) is narrowed.

Figure 2.10
A voltage source V connected to reverse bias a PN junction. The depletion region (shown shaded) is widened. (Compare with Figure 2.9.)

Figure 2.11
Diode symbol and bias circuits.

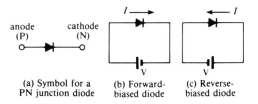

anode   cathode
(P)   (N)

(a) Symbol for a   (b) Forward-   (c) Reverse-
PN junction diode   biased diode   biased diode

where I = current through the diode, A

V = anode-to-cathode voltage across the diode, V (positive for forward bias, negative for reverse bias)

I_s = *saturation current*, A

η = *emission coefficient* (a function of V whose value also depends on the material: $1 \leqslant \eta \leqslant 2$)

V_T = *thermal voltage* $= \dfrac{kT}{q}$ V

k = Boltzmann's constant = 1.38×10^{-23} J/K

T = temperature in kelvin (K = 273 + °C)

q = electronic charge = 1.6×10^{-19} C

The current through a reverse-biased diode, called *reverse current*, is very small but not totally zero. The saturation current, I_s, in equation 2.1 is the current that flows through the reverse-biased diode (from cathode to anode) when voltage V is a few tenths of a volt negative. As we shall see in a forthcoming example, equation 2.1 produces a negative (reverse) current when voltage V is negative (reverse bias). When voltage V is a few tenths of a volt positive (forward bias), equation 2.1 will show that the positive (forward) current becomes quite large. The forward current becomes large when the forward voltage approaches about 0.7 V in silicon and about 0.3 V in germanium. Figure 2.12 illustrates this fact for a forward-biased silicon diode. Figure 2.13 shows diode current versus voltage under both forward-

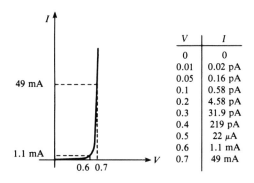

V	I
0	0
0.01	0.02 pA
0.05	0.16 pA
0.1	0.58 pA
0.2	4.58 pA
0.3	31.9 pA
0.4	219 pA
0.5	22 μA
0.6	1.1 mA
0.7	49 mA

Figure 2.12
Current versus voltage in a typical forward-biased silicon junction. $I_s = 0.1$ pA.

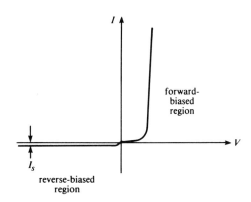

Figure 2.13
Current–voltage relations in a PN junction under forward and reverse bias. The negative current scale in the reverse-biased region is exaggerated.

and reverse-bias conditions. Example 2.7 at the end of the chapter demonstrates how SPICE can be used to compute and plot a diode characteristic curve.

Example 2.1

A silicon diode has saturation current 1 pA. Using the following values of η and assuming the temperature is 25°C ("room temperature"), find the current in the diode when

1. It is reverse biased by 0.1 V ($\eta = 2$).
2. It is reverse biased by 1.0 V ($\eta = 2$).
3. The anode is shorted to the cathode ($\eta = 2$).
4. It is forward biased by 0.5 V ($\eta = 1$).
5. It is forward biased by 0.7 V ($\eta = 1$).

Solution.   We first calculate the thermal voltage at $T = 273 + 25°C = 298$ kelvin:

$$V_T = \frac{kT}{q} = \frac{(1.38 \times 10^{-23})(298)}{1.6 \times 10^{-19}} = 0.0257 \text{ V}$$

1. Since the diode is reverse biased, we substitute $V = -0.1$ V in equation 2.1:

$$I = I_s (e^{V/\eta V_T} - 1) = (1 \text{ pA}) (e^{-0.1 \text{ V}/2(0.0257V)} - 1)$$
$$= (1 \text{ pA})(0.143 - 1) = -0.857 \text{ pA}$$

The negative result tells us that the current is reverse current, as expected. Note that the value of the exponential term (0.143) is relatively small, but not zero.

2. Substituting $V = -1$ V in (2.1), we find

$$I = 1 \text{ pA} (e^{-1 \text{ V}/2(0.0257V)} - 1) = (1 \text{ pA})(3.55 \times 10^{-9} - 1) = -1 \text{ pA}$$

Notice that the value of the exponential term is now for all practical purposes equal to 0 and $I = -I_s$. For all larger reverse voltages, I will remain at -1 pA.

3. Since the anode is shorted to the cathode, $V = 0$ and

$$I = (1 \text{ pA})(e^{0 \text{ V}/2(0.0257V)} - 1)$$
$$= (1 \text{ pA})(1 - 1) = 0 \text{ A}$$

As expected, the diode equation shows that the current through the diode is 0 when the voltage across it is 0.

4. Since the diode is forward biased, we substitute $V = 0.5$ V in (2.1) to obtain

$$I = (1 \text{ pA})(e^{0.5 \text{ V}/1(0.0257V)} - 1)$$
$$= (1 \text{ pA})(2.814 \times 10^8 - 1) \approx (10^{-12} \text{ A})(2.814 \times 10^8) = 0.2814 \text{ mA}$$

Notice that the value of the exponential term (1.035×10^8) in this case is so much larger than 1 that, for all practical purposes, $I = I_s e^{V/\eta V_T}$. For all forward-biasing voltages greater than a few tenths of a volt, this will be the case.

5. Substituting $V = 0.7$ V in (2.1) gives

$$I = (1 \text{ pA})(e^{0.7 \text{ V}/1(0.0257\text{V})} - 1)$$
$$= (1 \text{ pA})(6.75 \times 10^{11}) = 675 \text{ mA}$$

Notice how fast the current now increases for a very small increase in forward voltage: from 0.1035 mA to 675 mA when the forward voltage increases by just 0.1 V.

We can tell from the presence of temperature T in the diode equation that diode current depends on temperature. Furthermore, the value of I_s itself depends heavily on temperature. As a rule, I_s approximately doubles in value for every 10°C rise in temperature. Much of the art of electronic circuit design using semi-conductors is concerned with compensating for the effects of temperature changes on circuit performance.

If the reverse-biasing voltage across a diode is increased to a value called the *breakdown voltage,* the reverse current through the diode will no longer be limited to the small saturation value, I_s. When breakdown occurs, the diode conducts heavily in the reverse direction, limited only by whatever resistors or other components are in series with it. Breakdown does not necessarily result in permanent damage to a diode. If the reverse current is limited so that the power dissipation rating of the diode is not exceeded ($P = VI$), then no irreversible damage occurs. Figure 2.14 shows the increase in reverse current when the reverse voltage is near the breakdown value, V_{BR}. The value of V_{BR} depends on doping and other physical characteristics of a diode and may range in value from about 10 V to several hundred volts.

2.3   ANALYSIS OF DIODE CIRCUITS

R_{dc} and r_{ac}

We can tell from the I–V characteristic of the diode (Figure 2.13) that it is not a linear device, in the sense discussed in Chapter 1. Consequently, we can expect the dc resistance (R_{dc}, equation 1.6) and the ac resistance (r_{ac}, equation 1.7) to have different values at different points along the characteristic. This fact is illustrated in the next example.

Figure 2.14
A plot of the I–V relation for a diode, showing the sudden increase in reverse current near the reverse breakdown voltage.

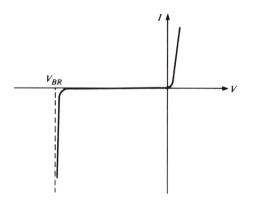

Example 2.2

Find R_{dc} and r_{ac} for the diode whose I–V characteristic is shown in Figure 2.15,

1. when the current through the diode is 4 mA;
2. when the voltage across the diode is 0.55 V.

Solution.

1. At the point on the characteristic where $I = 4$ mA, we see that the voltage V is approximately 0.59 V. Therefore,

$$R_{dc} = \frac{V}{I} = \frac{0.59 \text{ V}}{4 \text{ mA}} = 147.5 \ \Omega$$

The steepness of the curve in this region makes it difficult to obtain accurate values for ΔV and ΔI in the calculation of r_{ac}:

$$r_{ac} = \frac{\Delta V}{\Delta I} = \frac{V_2 - V_1}{I_2 - I_1} \approx \frac{0.61 \text{ V} - 0.58 \text{ V}}{6 \text{ mA} - 2 \text{ mA}} = \frac{0.03 \text{ V}}{4 \text{ mA}} = 7.5 \ \Omega$$

2. At the point where $V = 0.55$ V, $I \approx 0.8$ mA.

$$R_{dc} = \frac{0.55 \text{ V}}{0.8 \text{ mA}} = 687.5 \ \Omega$$

As shown in the figure,

$$r_{ac} = \frac{V_2 - V_1}{I_2 - I_1} = \frac{0.58 \text{ V} - 0.52 \text{ V}}{1.7 \text{ mA} - 0 \text{ mA}} = \frac{0.06 \text{ V}}{1.7 \text{ mA}} = 35.3 \ \Omega$$

Figure 2.15
(Example 2.2)

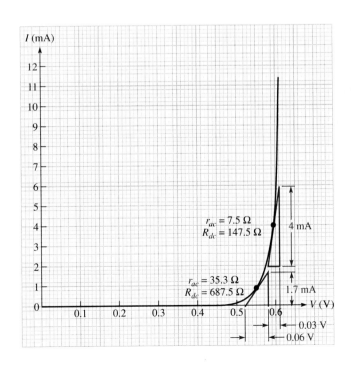

As this example demonstrates, the greater the current in the diode, the smaller its ac resistance. This property is apparent from the fact that the curve becomes steeper as current increases, meaning that the slope (conductance) increases and resistance decreases. Note the very significant decrease in ac resistance (from 35.3 Ω to 7.5 Ω) when the voltage increases by just 0.04 V. The characteristic becomes very nearly vertical when the forward-biasing voltage approaches 0.7 V.

Linearization of the *I–V* Characteristic

The region of the *I–V* characteristic where the curve turns upward and current begins to increase rapidly for small increases in voltage is called the *knee* of the characteristic. The knee in Figure 2.15 is in the vicinity of $V = 0.55$ V. The ac resistance of the diode at points above the knee is closely approximated by

$$r_{ac} \approx \frac{V_T}{I} \quad \text{ohms} \tag{2.2}$$

where V_T is the thermal voltage (equation 2.1) and I is the dc current through the diode. At room temperature, $V_T \approx 0.026$ V, so

$$r_{ac} \approx \frac{0.026 \text{ V}}{I} \quad \text{(room temperature)} \tag{2.3}$$

Using this equation for Example 2.2, we find the ac resistance when $I = 4$ mA to be

$$r_{ac} \approx \frac{0.026 \text{ V}}{4 \text{ mA}} = 6.5 \text{ } \Omega$$

This result is close to the approximation found graphically in the example (7.5 Ω).

When analyzing a circuit containing a diode that is biased above the knee, we can often assume, with little loss of accuracy, that the diode has a fixed dc voltage drop of 0.7 V for silicon and 0.3 V for germanium. In most practical circuits, we

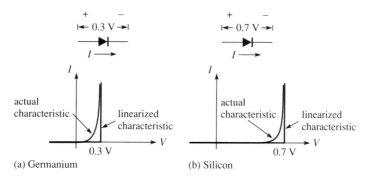

(a) Germanium   (b) Silicon

Figure 2.16
Approximating the I–V characteristic of a diode by linearizing it. The diode is assumed to conduct perfectly and have a fixed voltage drop when biased above the knee and to conduct zero current below the knee. Since each characteristic then consists of two pieces, it is said to be piecewise linear.

can also assume that there is zero current through a diode when it is biased below the knee. These assumptions amount to *linearizing* the I–V characteristic, i.e., approximating it by a vertical line that intersects the voltage axis at 0.3 V or 0.7 V. See Figure 2.16. Note that the linearized characteristic coincides with the horizontal axis until the voltage reaches either 0.3 V or 0.7 V. Since each approximation of the I–V characteristic consists of two line segments, or two "pieces," we say that each characteristic is *piecewise linear*. The next example demonstrates how linearization is applied to the analysis of a diode circuit containing both a dc and an ac source.

Example 2.3

Assuming the silicon diode in Figure 2.17(a) has a fixed dc voltage drop of 0.7 V, find the rms value of the ac voltage across the diode. Assume room temperature.

Solution.   Since there are two voltage sources, we can use the *superposition* principle to analyze the circuit, first with the dc source acting alone, and then with the ac source acting alone. (Superposition requires that the circuit consist of linear components, and since we are assuming that we are in the linear piece of the characteristic where it is vertical, we are justified in applying the principle.) Figure 2.17(b) shows the circuit when the ac source is replaced by a short circuit and the dc source acts alone. Applying Kirchhoff's voltage law around the circuit, we find the voltage drop V_R across the 2-kΩ resistor:

$$V_R = 5 \text{ V} - 0.7 \text{ V} = 4.3 \text{ V}$$

Figure 2.17
(Example 2.3)

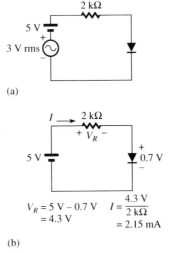

The dc current in the circuit is, therefore,

$$I = \frac{4.3 \text{ V}}{2 \text{ k}\Omega} = 2.15 \text{ mA}$$

Since 2.15 mA is the dc current in the diode, we can find its ac resistance from equation 2.3:

$$r_{ac} = \frac{0.026 \text{ V}}{2.15 \text{ mA}} = 12.1 \text{ }\Omega$$

Figure 2.17(c) shows the circuit with the ac source acting alone and with the diode replaced by its equivalent ac resistance. By the voltage-divider rule, the ac voltage across the diode is

$$V = \left(\frac{12.1 \text{ }\Omega}{2 \text{ k}\Omega + 12.1 \text{ }\Omega}\right)3 \text{ V rms} = 18 \text{ mV rms}$$

Example 2.8 at the end of the chapter shows how SPICE can be used to obtain a plot of the current through a diode in a circuit containing both a dc and an ac source.

Small- and Large-Signal Analysis

In Example 2.3, the voltage variation across the diode is quite small (18 mV rms, or about 51 mV peak-to-peak). This is an example of *small-signal analysis*. In small-signal analysis, we assume that current and voltage variations are small enough so that there is no appreciable change in the device characteristics, such as the slope of its I–V characteristic. In other words, variations occur over a portion of the characteristic that is small enough to be treated as linear.

In large-signal circuits, voltage and current variations occur over the entire range of the device's I–V characteristic. For example, if a silicon diode were in a circuit where the voltage across it periodically changed from, say, −1 V to +0.7 V, we would consider this to be large-signal operation. We will study examples of large-signal applications of diodes in Sections 2.4 and 2.5.

2.4   DIODE RECTIFIERS

As indicated in Section 2.3, a diode is said to operate under large-signal conditions when the current and voltage changes it undergoes extend over a substantial portion of its characteristic curve, including portions where there is a significant change in slope. In every practical large-signal application, the diode is operated both in the region where it is well forward biased (above the knee) and into the region where it is either reverse biased or biased near zero volts. We have seen that such large excursions will change the resistance of the diode from very small to very large values.

When the resistance of a diode changes from a very small to a very large value, it behaves very much like a *switch*. An ideal (perfect) switch has zero resistance when closed and infinite resistance when open. Similarly, an *ideal* diode for large-signal applications is one whose resistance changes between these same extremes. When analyzing such circuits, it is often helpful to think of

Figure 2.18

Linearized diode characteristic when it is assumed to be a perfect voltage-controlled switch with zero forward voltage drop.

the diode as a *voltage-controlled switch:* a forward-biasing voltage closes it, and a zero or reverse-biasing voltage opens it. Depending on the magnitudes of other voltages in the circuit, the 0.3- or 0.7-V drop across the diode when it is forward biased may or may not be negligible. If the voltage drop is negligibly small, the diode *I–V* characteristic can be linearized as shown in Figure 2.18. Here, the piecewise linear characteristic shows perfect conduction with zero voltage drop for all positive (forward-biasing) voltages and is an open circuit for all negative voltages. This is the characteristic of a perfect (voltage-controlled) switch.

One of the most common uses of a diode in large-signal operation is in a *rectifier* circuit. A rectifier is a device that permits current to flow through it in one direction only. It is easy to see how a diode performs this function when we think of it as a voltage-controlled switch. When the anode voltage is positive with respect to the cathode, i.e., when the diode is forward biased, the "switch is closed" and current flows through it from anode to cathode. If the anode becomes negative with respect to the cathode, the "switch is open" and no current flows. Of course, a *real* diode is not perfect, so there is in fact some very small reverse current that flows when it is reverse biased. Also, as we know, there is a nonzero voltage drop across the diode when it is forward biased (0.3 or 0.7 V), a drop that would not exist if it were a perfect switch.

Consider the rectifier circuit shown in Figure 2.19. We see in the figure that an ac voltage source is connected across a diode and a resistor R, the latter designed to limit current flow when the diode is forward biased. Notice that no dc source is present in the circuit. Therefore, during each positive half-cycle of the ac source voltage $e(t)$, the diode is forward biased and current flows through it in the direction shown. During each negative half-cycle of $e(t)$ the diode is reverse biased and no current flows. The waveforms of $e(t)$ and $i(t)$ are sketched in the figure. We see that $i(t)$ is a series of positive current pulses separated by intervals of zero current. Also sketched is the waveform of the voltage $v_R(t)$ that is developed across R as a result of the current pulses that flow through it. Note that the net effect of this circuit is the conversion of an ac voltage into a (pulsating) dc voltage, a fundamental step in the construction of a dc *power supply*.

If the diode in the circuit of Figure 2.19 is turned around, so that the anode is connected to the resistor and the cathode to the generator, then the diode will be

Figure 2.19

The diode used as a rectifier. Current flows only during the positive half-cycle of the input.

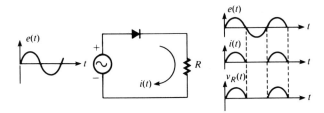

forward biased during the negative half-cycles of the sine wave. The current would then consist of a sequence of pulses representing current flow in a counter-clockwise, or negative, direction around the circuit.

Example 2.4

Assume that the silicon diode in the circuit of Figure 2.20 has a linearized characteristic like that shown in Figure 2.16(b). Find the peak values of the current $i(t)$ and the voltage $v_R(t)$ across the resistor when

1. $e(t) = 20 \sin \omega t$, and
2. $e(t) = 1.5 \sin \omega t$. In each case, sketch the waveforms for $e(t)$, $i(t)$, and $v_R(t)$.

Solution.

1. When $e(t) = 20 \sin \omega t$, the peak positive voltage generated is 20 V. At the instant $e(t) = 20$ V, the voltage across the resistor is 20 V $- 0.7$ V $= 19.3$ V, and the current is $i = 19.3$ V/(1.5 kΩ) $= 12.87$ mA. Figure 2.21 shows the resulting waveforms. Note that because of the characteristic assumed in Figure 2.16(b), the diode does not begin conducting until $e(t)$ reaches $+0.7$ V, and ceases conducting when $e(t)$ drops below 0.7 V. The time interval between the point where $e(t) = 0$ V and $e(t) = 0.7$ V is very short in comparison to the half-cycle of conduction time. From a practical standpoint, we could have assumed the characteristic in Figure 2.18, i.e., neglected the 0.7-V drop, and the resulting waveforms would have differed little from those shown.

Figure 2.20
(Example 2.4)

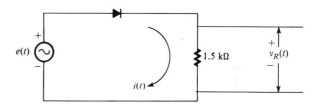

Figure 2.21
Diode current and voltage in the circuit of Figure 2.20. Note that the diode does not conduct until e(t) reaches 0.7 V, so short intervals of nonconduction occur during each positive half-cycle.

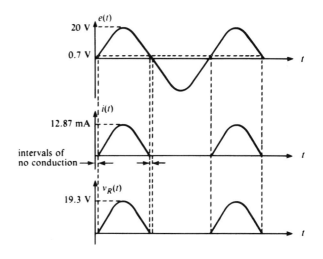

Figure 2.22
Diode current and voltage in the circuit of Figure 2.20 when the sine-wave peak is reduced to 1.5 V. Note that the intervals of nonconduction are much longer than those in Figure 2.21.

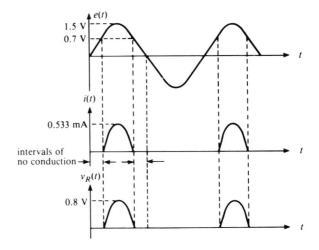

2. When $e(t) = 1.5 \sin \omega t$, the peak positive voltage generated is 1.5 V. At that instant, $v_R(t) = 1.5 \text{ V} - 0.7 \text{ V} = 0.8 \text{ V}$ and $i(t) = (0.8 \text{ V})/(1.5 \text{ k}\Omega) = 0.533 \text{ mA}$. The waveforms are shown in Figure 2.22. Note once again that the diode does not conduct until $e(t) = 0.7$ V. However, in this case, the time interval between $e(t) = 0$ V and $e(t) = 0.7$ V is a significant portion of the conducting cycle. Consequently, current flows in the circuit for significantly less time than one-half cycle of the ac waveform. In this case, it clearly would *not* be appropriate to use Figure 2.18 as an approximation for the characteristic curve of the diode.

2.5   ELEMENTARY DC POWER SUPPLIES

As already mentioned, an important application of diodes is in the construction of dc power supplies. It is instructive at this time to consider how diode rectification and waveform filtering, the first two operations performed by every power supply, are used to create an elementary dc power source. (If desired, this entire discussion can be deferred to a more detailed theoretical analysis in Chapter 13.)

The single diode in Figure 2.19 is called a *half-wave* rectifier because the waveforms it produces ($i(t)$ and $v_R(t)$) each represent half a sine wave. These half–sine waves are a form of pulsating dc and by themselves are of little practical use. (They can, however, be used for charging batteries, an application in which a steady dc current is not required.) Most practical electronic circuits require a dc voltage source that produces and maintains a *constant* voltage. For that reason, the pulsating half–sine waves must be converted to a steady dc level. This conversion is accomplished by *filtering* the waveforms. Filtering is a process in which selected frequency components of a complex waveform are *rejected* (filtered out) so that they do not appear in the output of the device (the filter) performing the filtering operation. The pulsating half–sine waves (like all periodic waveforms) can be regarded as waveforms that have both a dc component and ac components. Our purpose in filtering these waveforms for a dc power supply is to reject *all* the ac components.

Figure 2.23
Filter capacitor C effectively removes the ac components from the half-wave–rectified waveform.

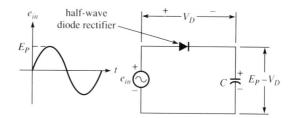

The simplest kind of filter that will perform the filtering task we have just described is a capacitor. Recall that a capacitor has reactance inversely proportional to frequency: $X_C = 1/2\pi fC$. Thus, if we connect a capacitor directly across the output of a half-wave rectifier, the ac components will "see" a low-impedance path to ground and will not therefore appear in the output. Figure 2.23 shows a filter capacitor, C, connected in this way. In this circuit the capacitor charges to the peak value of the rectified waveform, V_{PR}, so the output is the dc voltage V_{PR}. Note that $V_{PR} = E_P - V_D$, where E_P is the peak value of the sinusoidal input and V_D is the dc voltage drop across the diode (0.7 V for silicon).

In practice, a power supply must provide dc current to whatever load it is designed to serve, and this load current causes the capacitor to discharge and its voltage to drop. The capacitor discharges during the intervals of time between input pulses. Each time a new input pulse occurs, the capacitor recharges. Consequently, the capacitor voltage rises and falls in synchronism with the occurrence of the input pulses. These ideas are illustrated in Figure 2.24. The output waveform is said to have a *ripple voltage* superimposed on its dc level.

When the peak-to-peak value of the output ripple voltage, V_{PP}, is small compared to V_{dc} (a condition called *light loading*), we derive in Chapter 13 the follow-

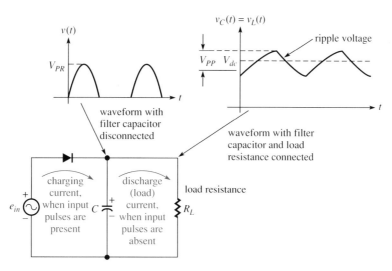

Figure 2.24
When load resistance R_L is connected across the filter capacitor, the capacitor charges and discharges, creating a load voltage that has a ripple voltage superimposed on a dc level.

ing equations for V_{dc} and V_{PP}:

$$V_{dc} = \frac{V_{PR}}{1 + \dfrac{1}{2f_r R_L C}} \tag{2.4}$$

$$V_{PP} = \frac{V_{dc}}{f_r R_L C} \tag{2.5}$$

where
V_{PR} = peak value of the rectified waveform $(E_P - V_D)$
f_r = frequency of the rectified waveform
R_L = load resistance
C = filter capacitance

Example 2.5

The sinusoidal input, e_{in}, in Figure 2.24 is 120 V rms and has frequency 60 Hz. The load resistance is 2 kΩ and the filter capacitance is 100 μF. Assuming light loading and neglecting the voltage drop across the diode:

1. Find the dc value of the load voltage.
2. Find the peak-to-peak value of the ripple voltage.

Solution

1. The peak value of the sinusoidal input voltage is $E_P = \sqrt{2}$ (120 V rms) = 169.7 V. Since the voltage drop across the diode can be neglected, $V_{PR} = E_P =$ 169.7 V. From equation 2.4,

$$V_{dc} = \frac{169.7 \text{ V}}{1 + \dfrac{1}{2(60 \text{ Hz})(2 \text{ k}\Omega)(100 \text{ } \mu\text{F})}} = 162.9 \text{ V}$$

2. From equation 2.5,

$$V_{PP} = \frac{162.9 \text{ V}}{(60 \text{ Hz})(2 \text{ k}\Omega)(100 \text{ } \mu\text{F})} = 13.57 \text{ V}$$

A *full-wave* rectifier effectively inverts the negative half-pulses of a sine wave to produce an output that is a sequence of positive half-pulses with no intervals between them. Figure 2.25 shows a widely used full-wave rectifier constructed

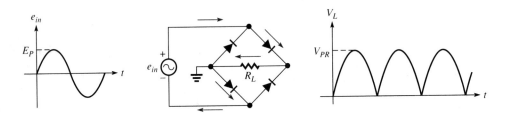

Figure 2.25
The full-wave bridge rectifier and output waveform. The arrows show the direction of current flow when e_{in} is positive.

Figure 2.26
Current flow in the full-wave bridge rectifier.

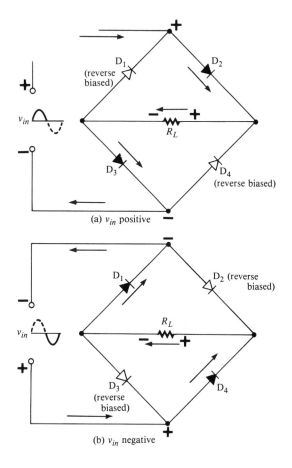

(a) v_{in} positive

(b) v_{in} negative

from four diodes and called a full-wave diode *bridge*. Also shown is the full-wave rectified output. Figure 2.26 demonstrates that current flows through R_L in the same direction when v_{in} is positive as it does when v_{in} is negative. In Figure 2.26 (a), v_{in} is positive, so D_2 and D_3 are forward biased, while D_1 and D_4 are reverse biased. Current therefore flows through R_L from right to left, as shown. In (b), v_{in} is negative, so D_1 and D_4 are forward biased, and D_2 and D_3 are reverse biased. Notice that current still flows through R_L from right to left. Therefore, the polarity of the voltage across R_L is always the same, confirming that full-wave rectification occurs. Note that the common side of the ac source *must* be isolated from the common side of the load voltage. Thus, the negative terminal of R_L in Figure 2.26(a) cannot be the same as the negative terminal of v_{in}. A transformer is often used to provide this isolation.

Since load current flows through two forward-biased diodes during each half-cycle of v_{in}, the peak rectified voltage across R_L is the peak input voltage reduced by 2(0.7 V) = 1.4 V:

$$V_{PR} = V_P - 1.4 \text{ V} \qquad\qquad (2.6)$$

As in the half-wave rectifier, the full-wave rectified waveform can be filtered by connecting a capacitor in parallel with load R_L. The advantage of the full-wave rectifier is that the capacitor does not discharge so far between input pulses

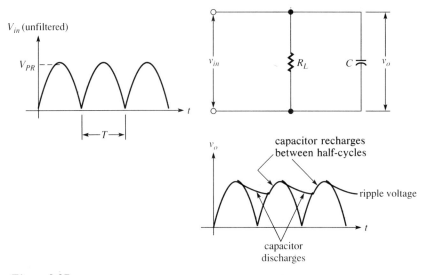

Figure 2.27
The ripple voltage in the filtered output of a full-wave rectifier is smaller than in the half-wave case because the capacitor recharges at shorter intervals. T = period of the full-wave rectified waveform (one half the period of the unrectified sine wave).

because a new charging pulse occurs every half-cycle instead of every full cycle. Consequently, the magnitude of the output ripple voltage is smaller. This fact is illustrated in Figure 2.27.

Equations 2.4 and 2.5 for V_{dc} and V_{PP} are valid for both half-wave and full-wave rectifiers. Note that f_r in those equations is the frequency of the *rectified* waveform, which in a full-wave rectifier is *twice* the frequency of the unrectified sine wave (see Figure 2.25). If the same input and component values used in Example 2.5 are used to compute V_{dc} and V_{PP} for a full-wave rectifier (f_r = 120 Hz), we find

$$V_{dc} = \frac{169.7 \text{ V}}{1 + \dfrac{1}{2(120 \text{ Hz})(2 \text{ k}\Omega)(100 \text{ }\mu\text{F})}} = 166.2 \text{ V}$$

and

$$V_{PP} = \frac{166.2 \text{ V}}{(120 \text{ Hz})(2 \text{ k}\Omega)(100 \text{ }\mu\text{F})} = 6.92 \text{ V}$$

Note that the peak-to-peak value of the ripple voltage is one-half that found for the half-wave rectifier. Example 2.9 at the end of the chapter shows how SPICE can be used to find the ripple voltage in the filtered output of a full-wave rectifier that is *not* lightly loaded.

Although the elementary power supplies we have described can be used in applications where the presence of some ripple voltage is acceptable, where the exact value of the output voltage is not critical, and where the load does not change appreciably, more sophisticated power supplies have more elaborate filters and special circuitry (voltage regulators) that maintain a constant output voltage under a variety of operating conditions. These refinements are discussed in detail in Chapter 13.

2.6   DIODE TYPES, RATINGS, AND SPECIFICATIONS

Discrete and Integrated Circuits

A *discrete* electronic device is a single component whose terminals are accessible for making external connections. For example, a discrete diode is a single diode mounted inside a *case* (enclosure) through which leads attached to the anode and cathode are brought out for making external electrical connections. Discrete circuits are constructed by interconnecting discrete components using external conductors such as wires or metallic paths on printed-circuit boards. By contrast, an *integrated circuit* may contain hundreds (or thousands) of devices embedded in a single semiconductor crystal (*chip*), and few if any of the individual devices are accessible for external connections. Integrated circuits are externally accessible at a limited number of terminals (*pins*) for making connections such as power sources, ground, and input and output signals. Individual components within the circuit are connected through internal conducting paths that are formed as part of the manufacturing process used to construct the circuit.

Diode Types and Ratings

Diodes are found in many integrated circuits and are commercially available in a wide range of discrete types designed for different kinds of service and a variety of applications. For example, *small-signal* diodes are small, general-purpose diodes with relatively low power dissipation ratings. *Rectifier* or *power* diodes are designed to carry larger currents and dissipate more power than small-signal diodes. They are used in power supply applications such as those discussed in Section 2.5.

Figure 2.28 illustrates some of the sizes and shapes that commercially available diodes may have. Each of those shown has a designation that identifies the standard case size it has (DO-4, C-15, etc.). Materials used for case construction include glass, plastic, and metal. Metal cases are used for large, rectifier-type

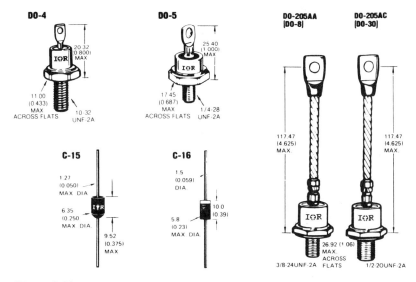

Figure 2.28
Discrete diode case styles (Courtesy of International Rectifier Corp.)

diodes to enhance the conduction of heat and improve their power dissipation capabilities.

There are two particularly important diode ratings that a designer using commercial, discrete diodes should know when selecting a diode for any application: the *maximum reverse voltage* (V_{RM}) and the *maximum forward current*. The maximum reverse voltage, also called the *peak inverse voltage* (PIV), is the maximum reverse-biasing voltage that the diode can withstand without breakdown. If the PIV is exceeded, the diode "breaks down" only in the sense that it readily conducts current in the reverse direction. As previously discussed, breakdown *may* result in permanent failure if the power dissipation rating of the device is exceeded. The maximum forward current is the maximum current that the diode can sustain when it is forward biased. Exceeding this rating will cause excessive heat to be generated in the diode and will lead to permanent failure. Manufacturers' ratings for the maximum forward current will specify whether the rating is for continuous, peak, average, or rms current, and they may provide different values for each. The symbols I_o and I_F are used to represent forward current.

Example 2.6

In the circuit of Figure 2.29, a rectifier diode is used to supply positive current pulses to the 100-Ω resistor load. The diode is available in the combinations of ratings listed in the table portion of the figure. Which is the least expensive diode that can be used for the application?

Solution   The applied voltage is 120 V rms. Therefore, when the diode is reverse biased by the *peak* negative value of the sine wave, it will be subjected to a maximum reverse-biasing voltage of (1.414)(120) = 169.7 V. The V_{RM} rating must be greater than 169.7 V.

The average value of the current is one-half the average value of a single sinusoidal pulse: $I_{AVG} = (\frac{1}{2})(0.637I_p)$ A, where I_p is the peak value of the pulse. (Note that the factor $\frac{1}{2}$ must be used because the pulse is present for only one-half of each full cycle.) The peak forward current in the example (neglecting the drop across the diode) is $I_p = (169.7 \text{ V})/(100 \ \Omega) = 1.697$ A. Therefore, the average forward current through the diode is $I_{AVG} = (\frac{1}{2})(0.637)(1.697) = 0.540$ A.

Figure 2.29
(Example 2.6)

V_{RM}	Max I_o (average)	Unit Cost
100 V	1.0 A	$0.50
150 V	2.0 A	1.50
200 V	1.0 A	2.00
200 V	2.0 A	3.00
500 V	2.0 A	3.50
500 V	5.0 A	5.00

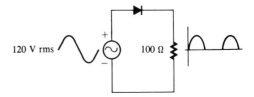

The least expensive diode having ratings adequate for the peak inverse voltage and average forward current values we calculated is the one costing $2.00.

Like many other manufactured electronic components, diodes are often identified by a standrd type number in accordance with JEDEC (Joint Electron Devices Engineering Council) specifications. JEDEC diode numbers have the prefix 1N, such as 1N4004, 1N2070A, etc. Not all manufacturers provide JEDEC numbers; many use their own commercial part numbers.

Diode Specifications

Figure 2.30 shows a typical specification sheet for a line of silicon rectifier diodes. The forward-current ratings are given as $I_{F(AV)}$ (average), and I_{FSM}, each in units of amperes. I_{FSM} is the maximum nonrepetitive forward current that the diode can sustain, that is, the maximum value of momentary or *surge* current it can conduct. Note that the I_{FSM} values are much larger than the $I_{F(AV)}$ values. The voltage ratings are specified by V_{RRM}, the maximum repetitive reverse voltage that each diode can sustain. Also note the large physical sizes and the metal cases of the stud-mounted rectifiers that are capable of conducting currents from 12 to 40 A.

2.7   SCHOTTKY DIODES

It is possible to create a junction having properties similar to those of a PN junction by bonding aluminum (Al) to suitably doped N-type silicon. The junction that results is called a *metal–semiconductor* (MS) junction. Like a PN junction, the MS junction presents a low resistance to current flow when it is forward biased (metal positive with respect to the N-type silicon) and a high resistance when reverse biased. A depletion region and a barrier potential are established in a metal–semiconductor junction by a mechanism similar to that described for PN junctions, except in this case carrier diffusion consists only of electrons diffusing from the semiconductor to the metal. The electrons accumulate at the metal surface and the depletion region exists only in the semiconductor side of the junction.

An important application of diodes is in *switching* circuits, where the bias on a diode is changed very rapidly from reverse to forward, or vice versa. In these applications it is necessary for a diode to *respond* very rapidly, that is, to change its nature very quickly from that of a high resistance to that of a low resistance, or vice versa. Metal–semiconductor junctions are able to respond more rapidly in these situations than are their PN counterparts because only majority carriers (electrons in the N-type silicon) are involved in the process. Diodes formed and used this way are called *Schottky barrier* diodes, or simply Schottky diodes. Figure 2.31 shows the special symbol used to represent a Schottky diode.

A metal–semiconductor junction with Schottky-diode properties can also be formed by bonding gold (Au) to P-type germanium. This device is called a *gold-bonded diode* and responds very rapidly in switching applications.

Ohmic Contacts

Aluminum is widely used in the fabrication of electronic devices to provide electrical contacts at semiconductor surfaces, where terminals can be attached or

General Purpose Stud Mounted Silicon Rectifiers

SILICON RECTIFIERS — GENERAL PURPOSE

Silicon rectifiers are available from less than 1 Amp to 3000 Amps, and voltages to 3000 volts. In addition to the standard industry packages, there are a number of other packages available as required. Two of these are shown below. These are ruggedly built devices with an excellent reputation for reliability and performance.

AXIAL LEAD SILICON RECTIFIERS — 750 mA TO 6 AMPS

$I_{F(AV)}$ (A) @ Max T_C (C)	750 mA (@ 25)	1 (@ 75)	1.5 (@ 40)	1.5 (@ 40)	2 (@ 100)	3 (@ 125)	4 (@ 120)	6 (@ 95)
I_{FSM} (A)	22	30	50	50	50	150	200	400
Notes	(1)(2)	(1)(2)	(1)(2)	(1)(2)	(2)	(2)	(2)	(2)
Case Style	DO-41	DO-41	DO-41	DO-41	DO-41	C-12	C-16	C-15
PART NUMBERS								
V_{RRM}								
50 Volts	—	1N4001	10D05	1N4816	20D05	30S1	40D1	60S05
100 Volts	—	1N4002	10D1	1N4817	20D1	30S2	40D2	60S1
200 Volts	1N2069 A	1N4003	10D2	1N4818	20D2	30S2	40D2	60S2
300 Volts	—	—	10D3	1N4819	—	30S3	—	—
400 Volts	1N2070 A	1N4004	10D4	1N4820	20D4	30S4	40D4	60S4
500 Volts	—	—	10D5	1N4821	—	30S5	—	60S5
600 Volts	1N2071 A	1N4005	10D6	1N4822	20D6	30S6	40D6	60S6
700 Volts	—	—	—	1N5052	—	—	—	—
800 Volts	—	1N4006	10D8	1N5053	20D8	30S8	40D8	60S8
1000 Volts	—	1N4007	10D10	1N5054	20D10	30S10	—	60S10

(1) Temperature given is ambient temperature.   (2) Also available on tape reel.

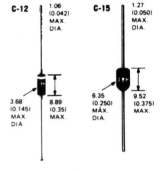

C-12   1.06 (0.042) MAX. DIA.   3.68 (0.145) MAX. DIA   8.89 (0.35) MAX.

C-15   1.27 (0.050) MAX. DIA.   6.35 (0.250) MAX. DIA.   9.52 (0.375) MAX.

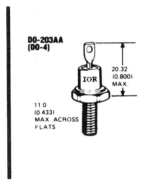

DO-203AA (DO-4)   20.32 (0.800) MAX.   11.0 (0.433) MAX. ACROSS FLATS

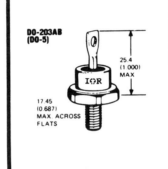

DO-203AB (DO-5)   25.4 (1.000) MAX   17.45 (0.687) MAX. ACROSS FLATS

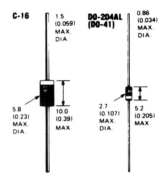

C-16   1.5 (0.059) MAX. DIA.   5.8 (0.23) MAX. DIA.   10.0 (0.39) MAX.

DO-204AL (DO-41)   0.86 (0.034) MAX. DIA.   2.7 (0.107) MAX. DIA.   5.2 (0.205) MAX.

STUD MOUNTED(4) SILICON RECTIFIERS — 12 TO 40 AMPS

$I_{F(AV)}$ (A) @ Max T_C (C)	12 (@ 150)	12 (@ 150)	12 (@ 150)	15 (@ 150)	16 (@ 150)	35 (@ 140)	40 (@ 140)	40 (@ 150)
I_{FSM} (A)	200	240	250	250	300	500	500	800
Notes	(5)	(5)	(5)	(5)	(5)	(5)	(5)	(5)
Case Style	DO-4	DO-4		DO-5	DO-4	DO-5	DO-5	DO-5
PART NUMBERS								
V_{RRM}								
50 Volts	12F5	1N1199A	12F5B	1N3208	16F5	1N1183	40HF5	1N1183A
100 Volts	12F10	1N1200A	12F10B	1N3209	16F10	1N1184	40HF10	1N1184A
150 Volts	—	1N1201A	—	—	—	1N1185	—	1N1185A
200 Volts	12F20	1N1202A†	12F20B	1N3210	16F20	1N1186	40HF20	1N1186A
300 Volts	—	1N1203	—	1N3211	—	1N1187	40HF30	1N1187A
400 Volts	12F40	1N1204A†	12F40B	1N3212	16F40	1N1188	40HF40	1N1188A
500 Volts	—	1N1205A	—	1N3213	—	1N1189	40HF50	1N1189A
600 Volts	12F60	1N1206A†	12F60B	1N3214	16F60	1N1190	40HF60	1N1190A
700 Volts	—	1N3670A	—	—	—	1N3765	40HF80	—
800 Volts	12F80	1N3671A	12F80B	—	16F80	1N3766	40HF100	—
900 Volts	—	1N3672A	—	—	—	1N3767	40HF120	—
1000 Volts	12F100	1N3673A	12F100B	—	16F100	1N3768	—	—

(4) Metric threads available on some stud packages.
(5) Cathode-to-stud. For anode-to-stud, add "R" to base number (example 12FLR, 40HFLR, 1N3889R).
† JAN and/or JAN-TX types available.

OTHER CONNECTIONS

In addition to the flex leads shown in the case style drawings, IR also has threaded stud, threaded hole and flag terminal top connections available for some case styles. Contact your local IR Distributor, IR Field Office or IR El Segundo offices for more information.

Figure 2.30
A typical rectifier data sheet (Courtesy of International Rectifier)

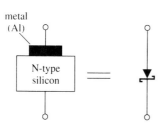

Figure 2.31
Construction and schematic symbol for a Schottky barrier diode.

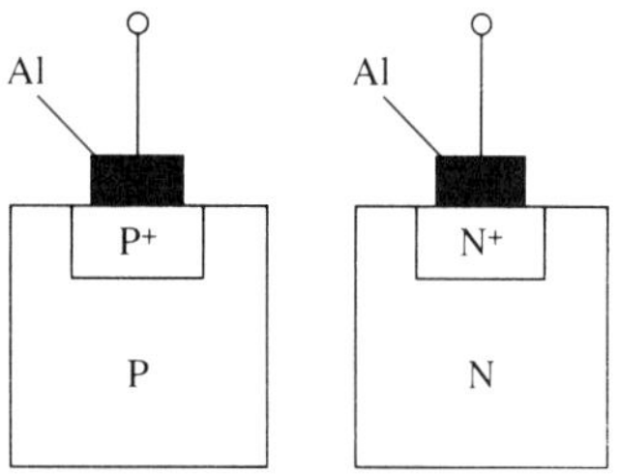

Figure 2.32
Ohmic contacts formed by using aluminum and N^+ or P^+ material.

where interconnections can be made to other devices. When aluminum is bonded to P-type silicon, no diode junction is formed The bond is simply called an *ohmic contact* because it exhibits a resistance that is independent of the voltage polarity across it.

To form a low-resistance ohmic contact, it is often the practice to join aluminum to a region of very heavily doped (highly conductive) semiconductor material. Usually the heavily doped P or N material is embedded in a region of more lightly doped material of the same type. Heavily doped N material is designated by N^+, and heavily doped P material by P^+. A bond between aluminum and N^+ silicon forms an ohmic contact rather than a Schottky diode. Ohmic contacts are illustrated in Figure 2.32.

Example 2.7

SPICE

Use SPICE to obtain a plot of diode current versus diode voltage for a forward-biasing voltage that ranges from 0.6 V through 0.7 V in 5-mV steps. The diode has saturation current 0.01 pA and emission coefficient 1.0.

Solution. Figure 2.33(a) shows a diode circuit that can be used by SPICE to perform a .DC analysis and generate the required plot. Note that VDUM is a dummy voltage source used as an ammeter to determine diode current. The polarity of VDUM is such that positive current values, I(VDUM), will be plotted. (If we plotted I(V1), current values would be negative.) Although the .MODEL statement specifies the saturation current, IS, and emission coefficient, N, the values used are the same as the default values so these could have been omitted from the statement. The .DC statement causes V1 to be stepped from 0.6 V through 0.7 V in 5-mV increments.

Figure 2.33(b) shows the resulting plot. (Portions of the complete printout produced by SPICE have been omitted to conserve space.) The plot has values of diode current, I(VDUM), scaled along the horizontal axis, and diode voltage, V1, along the vertical axis. By rotating the plot 90° counterclockwise, we see the conventional portrayal of current versus voltage, similar to that in Figure 2.12. Note that the forward current ranges from 0.1188 mA to 5.676 mA as the forward voltage ranges from 0.6 V to 0.7 V.

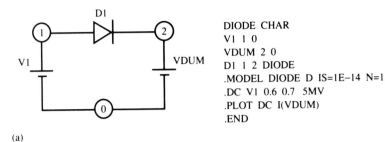

```
DIODE CHAR
V1 1 0
VDUM 2 0
D1 1 2 DIODE
.MODEL DIODE D IS=1E-14 N=1
.DC V1 0.6 0.7 5MV
.PLOT DC I(VDUM)
.END
```

(a)

```
DIODE CHAR
****        DC TRANSFER CURVES                    TEMPERATURE =    27.000 DEG C
***********************************************************************************
   V1            I(VDUM)
                     0.000D+00      2.000D-03      4.000D-03      6.000D-03  8.000D-03
              - - - - - - - - - - - - - - - - - - - - - - - - - - - - - - - - - - - -
 6.000D-01   1.188D-04  .*             .              .              .           .
 6.050D-01   1.441D-04  .*             .              .              .           .
 6.100D-01   1.749D-04  .*             .              .              .           .
 6.150D-01   2.122D-04  .*             .              .              .           .
 6.200D-01   2.574D-04  . *            .              .              .           .
 6.250D-01   3.123D-04  . *            .              .              .           .
 6.300D-01   3.789D-04  .   *          .              .              .           .
 6.350D-01   4.597D-04  .    *         .              .              .           .
 6.400D-01   5.578D-04  .     *        .              .              .           .
 6.450D-01   6.767D-04  .       *      .              .              .           .
 6.500D-01   8.211D-04  .         *    .              .              .           .
 6.550D-01   9.962D-04  .           *  .              .              .           .
 6.600D-01   1.209D-03  .             *.              .              .           .
 6.650D-01   1.466D-03  .             . *            .              .           .
 6.700D-01   1.779D-03  .             .   *          .              .           .
 6.750D-01   2.159D-03  .             .    *         .              .           .
 6.800D-01   2.619D-03  .             .       *      .              .           .
 6.850D-01   3.177D-03  .             .         *    .              .           .
 6.900D-01   3.855D-03  .             .            *.              .           .
 6.950D-01   4.677D-03  .             .              .    *         .           .
 7.000D-01   5.675D-03  .             .              .        *     .           .
              - - - - - - - - - - - - - - - - - - - - - - - - - - - - - - - - - - - -
```

(b)

Figure 2.33
(Example 2.7)

In PSpice, the dummy voltage source, VDUM, in Example 2.7 is not required, because the current in the diode can be specified directly as I(D1) (or i(d1)). Thus, the circuit can be described with just two nodes:

```
diode char
v1 1 0
d1 1 0 diode
.model diode d is=1e-14 n=1
.dc v1 0.6 0.7 5mv
.plot dc i(d1)
.end
```

This input file is written in lowercase letters for illustrative purposes, but PSpice, unlike the original Berkeley SPICE, does not distinguish between lowercase and uppercase letters, so either (or both) could have been used.

Example 2.8

SPICE

Use SPICE to obtain a plot of the diode current in Figure 2.34(a) versus time. Assume all diode parameters have their default values.

Solution. Figure 2.34(b) shows the circuit with a dummy voltage source inserted to obtain positive values of diode current. Notice that we require a *transient*

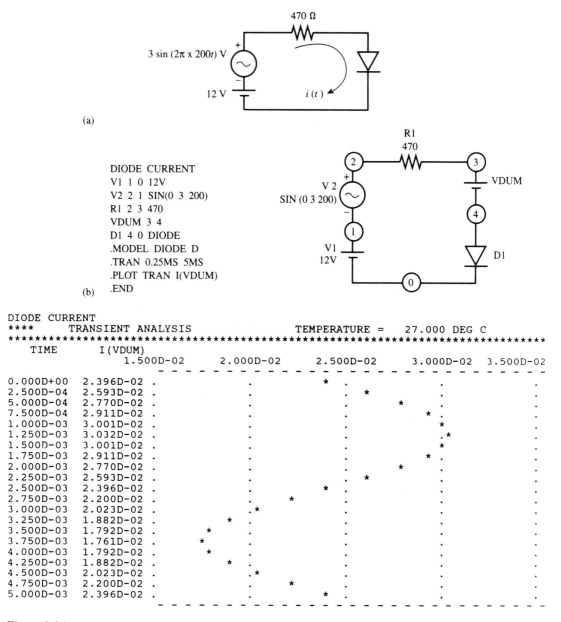

(a)

```
DIODE CURRENT
V1 1 0 12V
V2 2 1 SIN(0 3 200)
R1 2 3 470
VDUM 3 4
D1 4 0 DIODE
.MODEL DIODE D
.TRAN 0.25MS 5MS
.PLOT TRAN I(VDUM)
.END
```

(b)

```
DIODE CURRENT
****        TRANSIENT ANALYSIS                    TEMPERATURE =    27.000 DEG C
*****************************************************************************
    TIME        I(VDUM)
                    1.500D-02      2.000D-02      2.500D-02      3.000D-02  3.500D-02
                    - - - - - - - - - - - - - - - - - - - - - - - - - - - - - -
0.000D+00   2.396D-02  .                          *       .                .
2.500D-04   2.593D-02  .                          .    *                   .
5.000D-04   2.770D-02  .                          .           *            .
7.500D-04   2.911D-02  .                          .                *  .     .
1.000D-03   3.001D-02  .                          .                  *      .
1.250D-03   3.032D-02  .                          .                  .*     .
1.500D-03   3.001D-02  .                          .                  *      .
1.750D-03   2.911D-02  .                          .                *  .     .
2.000D-03   2.770D-02  .                          .           *            .
2.250D-03   2.593D-02  .                          .    *                   .
2.500D-03   2.396D-02  .                          *       .                .
2.750D-03   2.200D-02  .                  *       .                         .
3.000D-03   2.023D-02  .           .*             .                         .
3.250D-03   1.882D-02  .      *     .             .                         .
3.500D-03   1.792D-02  .    *       .             .                         .
3.750D-03   1.761D-02  .   *        .             .                         .
4.000D-03   1.792D-02  .    *       .             .                         .
4.250D-03   1.882D-02  .      *     .             .                         .
4.500D-03   2.023D-02  .           .*             .                         .
4.750D-03   2.200D-02  .                  *       .                         .
5.000D-03   2.396D-02  .                          *       .                .
                    - - - - - - - - - - - - - - - - - - - - - - - - - - - - - -
```

Figure 2.34
(Example 2.8)

(.TRAN) analysis to obtain a plot versus time and that the sinusoidal source is modeled by a SIN source. (Specifying an AC source and performing an .AC analysis would not produce a plot versus time.) Since the frequency of the source is 200 Hz, one period occupies 1/200 = 5 ms. Thus the control statement .TRAN 0.25MS 5MS will cause 5 ms/0.25 ms = 20 values of current to be plotted over one full period.

Shown next is the plot produced by a program run. (Portions of the complete printout produced by SPICE have been omitted to conserve space.) The peak value of the current is seen to be 30.32 mA, occurring at $t = 1.25$ ms. We can compare this result with that calculated using the method of Example 2.3: Since the dc current is (12 − 0.7 V)/470 Ω = 24 mA, the approximate value of r_{ac} is 0.026/(24 mA) = 1.08 Ω. Thus, the peak value of the current is

$$\frac{E - 0.7 \text{ V}}{R} + \frac{A}{R + r_{ac}} = \frac{11.3 \text{ V}}{470 \text{ }\Omega} + \frac{3 \text{ V}}{471 \text{ }\Omega} = 30.37 \text{ mA}$$

We see that there is very favorable agreement between the SPICE plot (30.32 mA) and the calculated value (30.37 mA).

Example 2.9

SPICE

A full-wave rectifier has $R_L = 120$ Ω. A 100-μF filter capacitor is connected in parallel with R_L. The input is a 60-Hz sine wave having peak value 30 V. The filter does *not* satisfy the lightly loaded criterion, so the equations for approximating the ripple voltage are not applicable. Use SPICE to determine the peak-to-peak ripple voltage.

Solution.   Figure 2.35(a) shows the SPICE circuit and input data file. Note that VIN is not connected to ground (node 0), since it must be applied across nodes 1 and 2. One-half the period of the 60-Hz input is 8.33 ms, so the .TRAN and .PLOT statements will produce a plot of the output extending over the first two half-cycles of input (0 to 17 ms). The results are shown in Figure 2.35(b). We see that the minimum and maximum values of the second pulse are 17.98 V and 28.40 V, respectively, so the peak-to-peak ripple voltage is 28.40 − 17.98 V = 10.42 V.

EXERCISES

Section 2.1   Semiconductor Theory

2.1   What two elements are the most widely used semiconductors? What property makes them useful for constructing electronic devices?

2.2   What is a covalent bond? What is the result of a covalent bond being ruptured?

2.3   Is an atom having a hole in a crystal structure charged positively or negatively? What is a recombination?

2.4   What is the difference between drift current and diffusion current?

2.5   What is the process used to create P and N materials? Name the impurity atoms used to create each type of material.

2.6   What are the majority carriers in N material? What are the minority carriers in P material? What does the notation N^+ mean?

Section 2.2   The PN Junction Diode

2.7   What is the voltage across a PN junction established by the diffusion of carriers

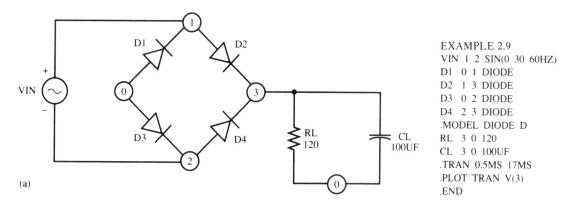

EXAMPLE 2.9
VIN 1 2 SIN(0 30 60HZ)
D1 0 1 DIODE
D2 1 3 DIODE
D3 0 2 DIODE
D4 2 3 DIODE
.MODEL DIODE D
RL 3 0 120
CL 3 0 100UF
.TRAN 0.5MS 17MS
.PLOT TRAN V(3)
.END

(a)

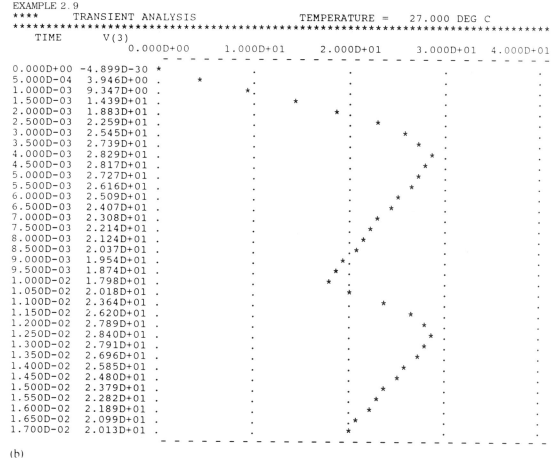

(b)

Figure 2.35
(Example 2.9)

across it called? What is the direction of the electric field due to that voltage?

2.8 Which terminals of a voltage source should be connected to which sides of a PN junction to reverse bias it?

2.9 A silicon diode has saturation current 2.4 pA and its temperature is 30°C.
  a. Find the current through the diode when it is reverse biased by 0.09 V. Assume $\eta = 2$.
  b. Find the current when it is reverse biased by 0.9 V. Assume $\eta = 2$.
  c. Find the current when it is forward biased by 0.55 V. Assume $\eta = 1$.
  d. Find the current when it is forward biased by 0.67 V. Assume $\eta = 1$.

2.10 A silicon diode has saturation current 1 pA and its temperature is 25°C. What value of forward-biasing voltage will cause a current of 10 mA to flow through it? Assume $\eta = 1$.

2.11 The saturation current in a certain diode at 25°C is 1.5 pA. What is its approximate value at 55°C?

2.12 A certain diode breaks down when its reverse voltage reaches 100 V. If it conducts

50 mA at breakdown, how much power does it consume?

Section 2.3 Analysis of Diode Circuits

2.13 For the diode whose I–V characteristic is shown in Figure 2.36.
  a. Find R_{dc} when $V = 0.55$ V.
  b. Find r_{ac} when $I = 4$ mA.

2.14 For the diode whose I–V characteristic is shown in Figure 2.36.
  a. Find R_{dc} when $I = 5$ mA.
  b. Find r_{ac} when $V = 0.5$ V.
  c. Find r_{ac} when $V = 0.3$ V.

2.15 Find the approximate ac resistance of a diode at room temperature when it is biased above its knee at $I = 2.5$ mA.

2.16 Find the approximate value of the ac resistance of the diode in Figure 2.37 at temperature 40°C. Assume the dc voltage drop across the diode is 0.65 V and that it is biased above its knee.

2.17 Find the rms value of the ac voltage across the diode in Figure 2.38 at room temperature. Assume the dc voltage drop across the diode is 0.7 V.

Figure 2.36
(Exercises 2.13 and 2.14)

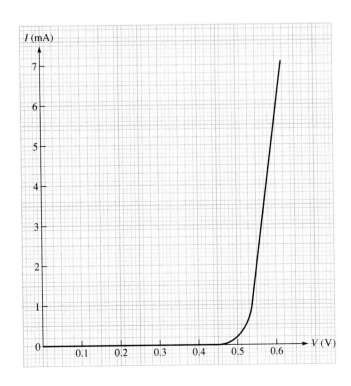

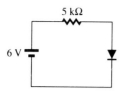

Figure 2.37
(Exercise 2.16)

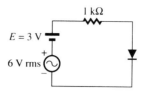

Figure 2.38
(Exercises 2.17 and 2.18)

2.18 To what value should E in Exercise 2.17 (Figure 2.38) be changed if it is desired to obtain an ac voltage of 40 mV rms across the diode?

Section 2.4 Diode Rectifiers

2.19 In the circuit shown in Figure 2.20, the 1.5-kΩ resistor is replaced by a 2.2-kΩ resistor. Assume that the silicon diode has a characteristic curve like that shown in Figure 2.16(b). If $e(t) = 2 \sin \omega t$, find the peak value of the current $i(t)$ and the voltage $v_R(t)$ across the resistor. Sketch the waveforms for $e(t)$, $i(t)$, and $v_R(t)$.

2.20 The silicon diode in Figure 2.39 has a characteristic curve like that shown in Figure 2.16(b). Find the peak values of the current $i(t)$ and the voltage $v_R(t)$ across the resistor. Sketch the waveforms for $e(t)$, $i(t)$, and $v_R(t)$.

2.21 The half-wave rectifier in Figure 2.24 has a 250-μF filter capacitor and a 1.5-kΩ load. The ac source is 120 V rms with frequency 60 Hz. The voltage drop across the silicon diode is 0.7 V. Assuming light loading, find
a. the dc value of the load voltage;
b. the peak-to-peak value of the ripple voltage.

2.22 The half-wave rectifier in Exercise 2.21 is replaced by a silicon full-wave rectifier, and a second 250-μF capacitor is connected in parallel with the filter capacitor. Assume light loading and do not neglect the voltage drop across the diodes. Find

a. the dc value of the load voltage;
b. the peak-to-peak value of the ripple voltage.
c. If the load resistance is decreased by a factor of 2, determine (without recalculating) the approximate factor by which the ripple voltage is changed.

Section 2.5 Diode Types, Ratings, and Specifications

2.23 In the circuit shown in Example 2.6 (Figure 2.29), suppose the load resistor R is changed to 47 Ω. What then is the least expensive of the diodes listed in the example that can be used in this application?

2.24 In the circuit shown in Example 2.6 (Figure 2.29), suppose the ac voltage is 100 V rms and the load resistor is changed to 68 Ω. What then is the least expensive of the diodes listed in the example that can be used in this application?

2.25 A rectifier diode is to be used in a power supply design where it must repeatedly withstand sine wave reverse voltages of 250 V rms and must conduct 0.6 A (average) of forward current. The forward surge current through the diode when the supply is first turned on will be 25 A. It is estimated that the diode case temperature (T_c) will be 30°C. Select a diode type number from Figure 2.30 that meets these requirements.

2.26 A rectifier diode is to be used in a large power supply where it must be capable of withstanding repeated reverse voltages of

Figure 2.39
(Exercise 2.20)

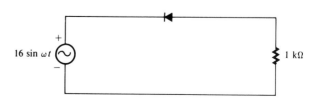

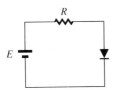

Figure 2.40
(Exercise 2.30)

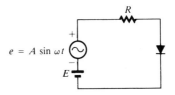

Figure 2.41
(Exercise 2.31)

450 peak volts. The forward current in the diode will average 13.5 A. Select a diode type number from Figure 2.30 that meets these requirements.

Section 2.6   Schottky Diodes

2.27   Name two specific materials that can be joined to form an MS junction. In which material does a depletion region form in such a junction?

2.28   What is the principal application for Schottky diodes? What aspect of the device's operating principle is responsible for the characteristic that makes it suitable for that application?

2.29   Give two examples of metal–semiconductor junctions that do not have diode properties.

DESIGN EXERCISES

2.30   The silicon diode in Figure 2.40 requires a minimum current of 5 mA to be above the knee of its $I–V$ characteristic. If the dc voltage source is 10 V, what is the largest value of resistance R that could be used in the circuit if operation must be at or above the knee?

2.31   In the circuit shown in Figure 2.41, $E = 9$ V, $R = 1$ kΩ, and $e = A \sin 1000t$ V. The circuit must be operated so that the peak value of the ac voltage drop across the diode is 10 mV. If the dc drop across the diode is fixed at 0.7 V, what should be the peak value (A) of the ac signal e?

2.32   In the rectifier circuit shown in Figure 2.19, $e(t) = A \sin 2\pi \times 100t$ V. Assume the voltage drop across the diode when it is conducting is 0.7 V. If conduction must begin during each positive half-cycle at an angle no greater than 5°, what is the minimum peak value A that the ac source must produce?

2.33   A full-wave rectifier with a capacitor filter supplies a dc value of 60 V to a 2-kΩ load. The ac voltage source driving the rectifier has frequency 60 Hz. What minimum value of filter capacitance must be used if the peak-to-peak ripple voltage cannot exceed 1 V?

SPICE EXERCISES

Note: In the exercises that follow, assume that all device parameters have their default values, unless otherwise specified.

2.34   Use SPICE to obtain a plot of the diode current in Figure 2.34 (a) versus time when the circuit is modified as follows: The dc voltage source is changed to 10 V, the ac voltage source is changed to 2 $\sin(2\pi \times 1000t)$ V, and the resistor is changed to 620 Ω. The plot should cover at least one full cycle of the ac signal. Compare the peak value of the current computed by SPICE with that predicted by calculation.

2.35   Use SPICE to obtain plots of the voltages, versus time, across the diode and across the resistor in the modified circuit described in Exercise 2.34.

2.36   Use SPICE to obtain a plot of voltage versus time across R in Figure 2.19 when $R = 1$ kΩ and
a. $e(t) = 2 \sin(2\pi \times 100t)$ V,
b. $e(t) = 40 \sin(2\pi \times 100t)$ V.
   Each plot should cover at least one full cycle of $e(t)$. Use the results to determine the peak value of the output in each case and comment on the validity of neglecting the drop across the diode in each case.

2.37   Use SPICE to obtain a plot of the diode current versus time in Figure 2.19 when $R = 500$ Ω and $e(t) = 1.5 \sin 6283t$ V. Use the results to determine the first *angle* (after 0°, at $t = 0$) at which significant conduction begins, assuming that any current less than 90 μA is essentially nonconduction.

Bipolar and Field-Effect Transistors

3.1 INTRODUCTION

The transistor is the workhorse of modern electronic circuits, both linear and digital. Its usefulness stems from its ability to *amplify*, that is, to increase the magnitude of voltage and current variations and/or to increase power levels in a circuit. Like diodes, transistors are available in discrete form for use in applications where heavy power dissipations occur, and are also widely found in integrated circuits. One integrated circuit may contain hundreds or even thousands of transistors.

Transistors can be classified as being either of the *bipolar* type or of the *field-effect* type. A bipolar transistor, known more precisely as a bipolar junction transistor (BJT), is so named because the current through it is due to the flow of both positively charged holes and negatively charged electrons (*bi*polar means having *two* polarities). In a field-effect transistor (FET), current flow is either due to holes alone or to electrons alone. Its name is derived from the fact that an externally applied voltage creates an electric field in the device, which controls the flow of charge through it.

The BJT was invented in 1947 by Bardeen, Brattain, and Shockley of Bell Laboratories. Because of its compactness, reliability, and particularly because of the fact that it does not eventually "burn out," it rapidly replaced vacuum tubes in all manner of electronic circuits. Field-effect transistors, first conceived by W. Shockley in 1952, came into widespread use in the early 1970s with advances in integrated-circuit technology that led to the development of microprocessors and semiconductor computer memories. The simplicity of their structure and their low power consumption made it possible to create large- and very large-scale (LSI and VLSI) integrated circuits that are used extensively today in digital applications. Discrete FETs are also available and are widely used in high-power applications.

3.2 BJT STRUCTURE AND BIASING

Structure

A bipolar junction transistor (BJT) is formed by alternating regions of P and N material that create *two* PN junctions. Figure 3.1 shows that there are two ways to

Figure 3.1
Two ways to alternate P and N materials to create two PN junctions.

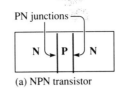

(a) NPN transistor

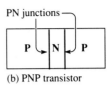

(b) PNP transistor

alternate P and N regions to form two junctions: NPN and PNP. Thus, there are two types of BJTs, NPN and PNP, and although both are found in discrete and integrated circuits, the NPN types predominate.

Figure 3.2 shows that the three regions of a transistor are called the *collector, base,* and *emitter* regions and that a terminal is attached to each region to make electrical connection to it. Notice in each type of transistor that the base is in the center, sandwiched between the collector and emitter regions. Accordingly, the junctions in each type of transistor are called the base–emitter junction and the collector–base junction. We should note that the physical structure of a practical BJT rarely consists of three laterally adjacent blocks as depicted in Figures 3.1 and 3.2 (more often, it consists of vertically stacked layers), but the figures accurately represent the alternating nature of the regions and make it easier to visualize current flow through them. The figures also accurately show the base region as being much thinner than the collector and emitter regions (even thinner, in reality, than shown here).

Biasing

To operate a BJT as a linear amplifier, it is necessary to bias the two PN junctions in a specific way. In both types of transistors, *the base–emitter junction must be forward biased and the collector–base junction must be reverse biased.* Recall that a PN junction is forward biased when the P side is made positive with respect to the N side, or, equivalently, when the N side is made negative with respect to the P side (*Positive to P, Negative to N*). In an NPN transistor, forward biasing the base–emitter junction means that the base (P) must be positive with respect to the emitter (N), and in a PNP transistor, the base (N) must be negative with respect to the emitter (P). Figure 3.3 shows how dc voltage sources are connected to forward bias the base–emitter junctions of both transistors. Also shown are the proper connections for reverse biasing the collector–base junctions (positive to N and negative to P). Note that the depletion region is wide at the reverse-biased junctions and narrow at the forward-biased junctions. Not shown in the figure are

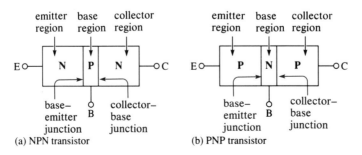

Figure 3.2
The collector, base, and emitter regions of NPN and PNP transistors.

Figure 3.3

Connecting voltage sources to forward-bias base–emitter junctions and reverse-bias collector–base junctions.

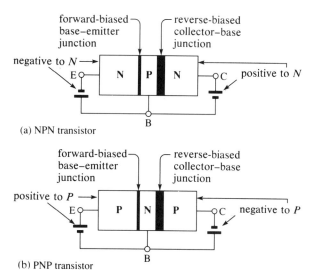

(a) NPN transistor

(b) PNP transistor

resistors that are connected in series with the voltage sources in practical circuits to limit the flow of current.

Current Flow in the BJT

In discussing the theory of current flow through a BJT, we will consider the NPN type only. The same theory can be applied to the PNP type by merely "reversing" all mention of carrier types (e.g., holes for electrons), polarities, and directions of (conventional) current flow. Figure 3.4 illustrates how current flows through a properly biased NPN transistor. The negative voltage at the emitter causes majority carriers (electrons) to diffuse into the emitter region. (The emitter is so named because it "emits" charge carriers.) The electrons diffuse into the P-type base region. Since the base region is very thin and is also lightly doped (meaning there is a relative scarcity of holes), very few electrons combine with holes in the base. Instead, the majority of the electrons are swept across the reverse-biased collector–base junction because the direction of the electric field at that junction promotes electron flow. (Recall that electrons in an electric field experience a force opposite to the direction of the field.) As discussed in Chapter 2, this type of current, caused by an electric field, is called drift current. At the same time, there is a flow of minority holes in the opposite direction, from collector to base. The electrons diffuse through the N-type collector region where they are "collected"

Figure 3.4

Current flow in a biased NPN transistor.

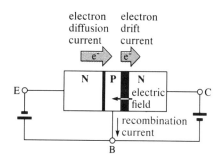

at the collector terminal to complete the circuit. The small number of electrons that *do* combine with holes in the base region constitute a small base current.

In Figure 3.4, note that electron flow is from emitter to collector. Therefore, *conventional* current flow in the NPN transistor is in the reverse direction: from collector to emitter. In the PNP transistor, the majority carriers are positively charged holes, which also flow from emitter to collector, so conventional current in the PNP device is from emitter to collector. Figure 3.5 shows the directions of all currents in both types of transistors. Also shown are the schematic symbols for both. Note that the arrowhead on the emitter in each schematic symbol points in the direction of conventional current flow, giving us a way to remember how to bias each type. The bias voltages across the junctions are designated V_{BE}—the voltage from base to emitter—and V_{CB}—the voltage from collector to base. In accordance with our rule for proper biasing (base–emitter junction forward biased and collector–base junction reverse biased), we see that the polarities of V_{BE} and V_{CB} must be as follows:

	E B C |↓/ **NPN**	E B C \\\|/ **PNP**
V_{BE}	positive	negative
V_{CB}	positive	negative

Recall that the voltage drop across a forward-biased PN junction is approximately 0.7 V in silicon and 0.3 V in germanium. Consequently, V_{BE} is approximately 0.7 V or 0.3 V in NPN transistors and −0.7 V or −0.3 V in PNP transistors.

In Figure 3.6, we represent each type of transistor as a single block so we can focus on the currents flowing in and out of each. Applying Kirchhoff's current law

Figure 3.5
Directions of current flow in NPN and PNP transistors and schematic symbols.

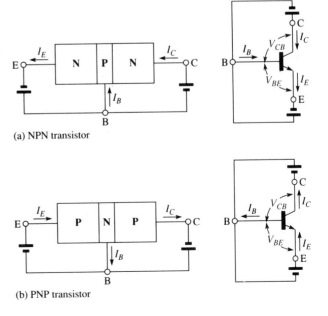

(a) NPN transistor

(b) PNP transistor

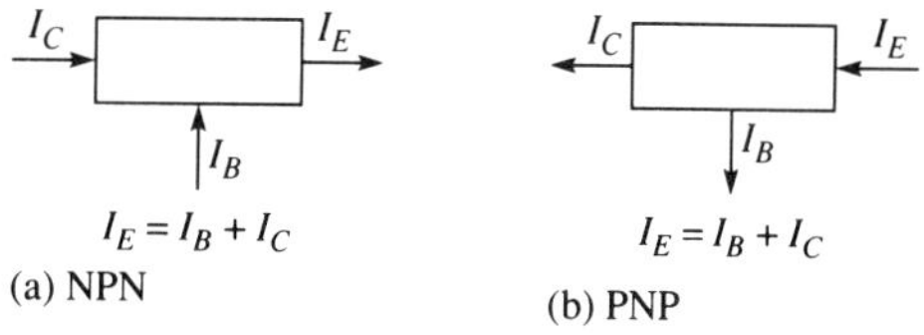

Figure 3.6
Kirchhoff's current law shows that $I_E = I_B + I_C$ for both types of transistors.

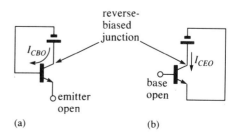

$I_E = I_B + I_C$

(a) NPN

$I_E = I_B + I_C$

(b) PNP

to each block (sum of currents entering equals sum of currents leaving), we see that the following equation always applies to both types of transistors:

$$I_E = I_C + I_B \qquad\qquad (3.1)$$

3.3   BJT CHARACTERISTICS

Reverse Currents

In our discussion of current flow in a BJT, we have overlooked the existence of the small *reverse current* that, as we know from our study of PN junctions, flows across a reverse-biased junction. The reverse current flowing from collector to base with the emitter open is designated I_{CBO}. See Figure 3.7(a). Note this standard way of identifying transistor currents using subscripts: the first two letters in the subscript represent the direction of current flow (from C to B) and the third letter (O) means that the remaining terminal (E) is open. Reverse current also flows from collector to emitter when the base is open. Thus, this current is designated I_{CEO}. See Figure 3.7(b). In modern transistors, particularly in integrated circuits, these reverse currents are so small that their effects can be largely neglected. In the discussion that follows, we omit consideration of reverse currents to preserve clarity and to maintain our focus on major concepts. We should note, however, that reverse currents *can* be troublesome in large, power-type transistors where heavy currents and high temperatures are encountered. A detailed discussion of the effects of reverse currents—and "leakage" currents, a similar phenomenon—can be found in other references. (See, for example, Bogart, *Electronic Devices and Circuits,* 3rd ed., Macmillan Publishing Company, 1993).

Alpha (α) and Beta (β)

Recall that the majority of the charge carriers crossing into the base region are swept into the collector region. Therefore, the emitter and collector currents are approximately equal: $I_E \approx I_C$, and the base current is very small. The ratio of collector current to emitter current in a transistor is called the alpha (α) of the

Figure 3.7
Reverse currents in an NPN transistor.

reverse-biased junction

I_{CBO}

emitter open

base open

I_{CEO}

(a)

(b)

transistor:

$$\alpha = \frac{I_C}{I_E} \qquad (3.2)$$

Since $I_E \approx I_C$, α is a number close to 1, but is always less than 1 since I_C is always less than I_E (as we can tell from equation 3.1). A typical value of α is 0.99.

Example 3.1

The emitter current in a BJT is 2 mA and the base current is 25 μA. Find the value of α for the transistor.

Solution.   From equation 3.1,

$$I_C = I_E - I_B = 2 \text{ mA} - 25 \text{ }\mu\text{A} = 1.975 \text{ mA}$$

From equation 3.2,

$$\alpha = \frac{I_C}{I_E} = \frac{1.975 \text{ mA}}{2 \text{ mA}} = 0.9875$$

Solving equation 3.2 for I_E gives $I_E = I_C/\alpha$. Substituting this into equation 3.1 gives

$$\frac{I_C}{\alpha} = I_C + I_B$$

$$I_C \left(\frac{1}{\alpha} - 1 \right) = I_B$$

$$\frac{I_C}{I_B} = \frac{1}{\left(\frac{1}{\alpha} - 1 \right)} = \frac{\alpha}{1 - \alpha} \qquad (3.3)$$

We see that the ratio of collector current to base current is $\alpha/(1 - \alpha)$, a ratio which is given the special name beta (β):

$$\beta = \frac{\alpha}{1 - \alpha} \qquad (3.4)$$

Since α is close to 1, $1 - \alpha$ is a small quantity, and β is therefore large. For example, if $\alpha = 0.995$, then

$$\beta = \frac{0.995}{1 - 0.995} = \frac{0.995}{0.005} = 190$$

Since $I_C = \beta I_B$, this result tells us that the collector current in a transistor having $\alpha = 0.995$ is 190 times greater than the base current. The value of β in typical transistors ranges from around 50 to 250 (corresponding to the much smaller range in α of 0.98 to 0.996). It is an end-of-chapter exercise to show that, given β, we can find α from:

$$\alpha = \frac{\beta}{\beta + 1} \qquad (3.5)$$

The relationship between I_C and I_B:

$$I_C = \beta I_B \qquad (3.6)$$

can give us some insight into how a transistor is capable of amplifying current variations. If we were to control the flow of base current and cause it to undergo some small variation (think of base current as an *input* to the transistor), then the collector current (*output*) would undergo a much larger variation, since it is β times greater than the base current. The next example demonstrates this point.

Example 3.2

A certain transistor has $\alpha = 0.992$. The ac base current supplied to the transistor is $i(t) = 20\sin(\omega t)$ μA. What is the peak value of the collector current?

Solution. From equation 3.4,

$$\beta = \frac{\alpha}{1 - \alpha} = \frac{0.992}{1 - 0.992} = 124$$

From equation 3.6,

$$I_C = \beta I_B = 124 I_B$$

The peak value of the sinusoidal base current is 20 μA. Thus, when the base current increases from 0 to 20 μA, the collector current increases from 124(0) = 0 to 124(20 μA) = 2.48 mA.

We see that the small current variation from 0 to 20 μA in the base produces the large current variation from 0 to 2.48 mA in the collector. This is the essence of current amplification.

Substituting $I_C = \beta I_B$ into equation 3.1 gives us another useful relationship:

$$I_E = I_C + I_B = \beta I_B + I_B$$
$$I_E = (\beta + 1)I_B \qquad (3.7)$$

Since β is usually much larger than 1, $\beta + 1 \approx \beta$ and we often use the approximation

$$I_E \approx \beta I_B = I_C \qquad (3.8)$$

For example, if $I_B = 10$ μA, and $\beta = 150$, then $I_C = \beta I_B = 150(10$ μA$) = 1.5$ mA, and

$$I_E = (\beta + 1)I_B = 151(10 \ \mu\text{A}) = 1.51 \text{ mA}$$

Thus,

$$I_E = 1.51 \text{ mA} \approx 1.5 \text{ mA} = I_C$$

This result confirms our previous discussion of the fact that the majority of the carriers emitted by the emitter cross the base region to the collector.

Collector Characteristics

Since a BJT is a *three-terminal* device, there are all together six electrical quantities that are of interest in our study of its behavior:

V_{BE} the base–emitter voltage

V_{CB} the collector–base voltage

V_{CE} the collector–emitter voltage

I_B the base current

I_C the collector current

I_E the emitter current

The value that any one of these quantities may have in a particular circuit affects the values of every other quantity to one extent or another. In other words, there are complex interactions between the voltages and currents in a transistor, interactions that we would like to be able to show *graphically* to gain insights into the nature of the device and how we can best use it in practical circuits. Although it is not possible to graph the variations and show the interactions of all six quantities in one plot, it *is* possible to show the relationships between any *three* quantities on a single graph. In this method of presentation, two of the variables are plotted along horizontal and vertical axes and the third is shown as a series of plots, one plot for each fixed value of the variable. We obtain what is called a *family* of characteristic curves.

Figure 3.8 shows a typical family of characteristic curves for an NPN transistor. Called the *collector characteristics* of the transistor, they are the most useful of the several sets of characteristic curves that can be constructed for a BJT, since they reveal properties most frequently needed in the design and analysis of transistor circuits. Note that collector current, I_C, is plotted along the vertical axis and collector-to-emitter voltage, V_{CE}, along the horizontal axis. Each plot corresponds to a fixed value of base current, I_B. We can specify the values of any two of these variables and use the characteristics to find the value of the third. This method of

Figure 3.8
A typical set of collector characteristics
for an NPN transistor.

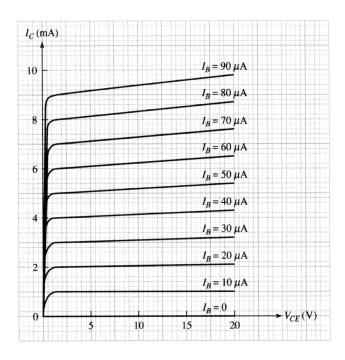

displaying device characteristics is widely used for showing interactions among three variables of many different types of three-terminal devices and this example should be studied carefully to gain facility with using and interpreting such presentations. (An instrument called a *transistor curve tracer*, similar to an oscilloscope, is commonly used in industry to generate and display families of characteristic curves.)

Example 3.3

Use the collector characteristics in Figure 3.8 to find

1. the value of β when $V_{CE} = 12.5$ V and $I_B = 40\ \mu$A;
2. the change in collector current when $V_{CE} = 5$ V and the base current changes from 10 μA to 30 μA.

Solution

1. Figure 3.9 shows the collector characteristics with a dot drawn at the point where the vertical (dashed) line representing $V_{CE} = 12.5$ V intersects the $I_B = 40\ \mu$A curve. At the point of intersection, we see that $I_C \approx 4.2$ mA. Therefore,

$$\beta = \frac{I_C}{I_B} = \frac{4.2\ \text{mA}}{40\ \mu\text{A}} = 105$$

2. As shown in Figure 3.9, a vertical line segment corresponding to $V_{CE} = 5$ V intersects the curve $I_B = 10\ \mu$A at a point where $I_C = 1$ mA and intersects the curve $I_B = 30\ \mu$A at a point where $I_C = 3$ mA. Thus,

$$\Delta I_C = 3\text{mA} - 1\ \text{mA} = 2\ \text{mA}$$

*Figure 3.9
(Example 3.3)*

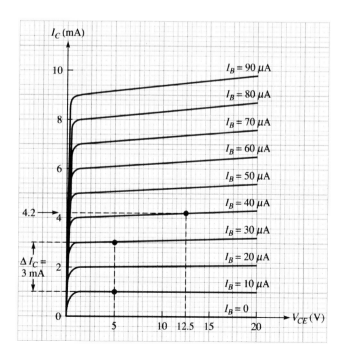

Examples 3.11 and 3.12 at the end of the chapter show how SPICE and PSpice can be used to plot a set of collector characteristics.

The β whose value was found in part (1) of Example 3.3 is called the *dc beta* (β_{dc}) of the transistor at the point where it is calculated. The *ac beta* of a transistor is defined to be the ratio of a *change* in collector current to the corresponding *change* in base current along a (vertical) line segment of constant V_{CE}:

$$\beta_{ac} = \frac{\Delta I_C}{\Delta I_B}\bigg|_{V_{CE} = \text{constant}} \tag{3.9}$$

In part (2) of the example, we found that $\Delta I_C = 2$ mA when $V_{CE} = 5$ V and $\Delta I_B = 30\ \mu\text{A} - 10\ \mu\text{A} = 20\ \mu\text{A}$. Therefore, in that example,

$$\beta_{ac} = \frac{2\ \text{mA}}{20\ \mu\text{A}}\bigg|_{V_{CE} = 5\ \text{V}} = 100$$

We see that $\beta_{dc} \approx \beta_{ac}$, a frequently used approximation. We should note that the values of β_{dc} and β_{ac} vary somewhat with the locations at which they are calculated. Verify, for example, that at the point in Figure 3.8 where $V_{CE} = 15$ V and $I_B = 90\ \mu\text{A}$, $\beta_{dc} \approx 107$. (see also Example 3.11.)

β_{ac} is sometimes called *small-signal beta* because it can be determined as the ratio of a small-signal ac collector current to a small-signal ac base current. In that context, we can regard ΔI_C and ΔI_B in equation 3.9 as the peak-to-peak values of each current.

Generally speaking, the greater the value of β, the better the transistor performs in many practical designs where it is used. Although it is useful to know the value of β, its variability makes it difficult to predict. Not only does it vary with changes in current, but also with changes in temperature and among transistors of the same type, due to small manufacturing variations. It is good practice in transistor circuit design to make circuit performance depend as little as possible on characteristics such as β that can vary over a wide range of values.

The BJT as a Linear Device

Figure 3.10 shows a set of collector characteristics with various *regions of operation* defined. By region of operation, we mean the totality of combinations of voltage and current that the transistor could experience in a particular type of application (linear, digital, etc.).

Recall from our study of the PN junction that a reverse-biased junction breaks down when the reverse voltage reaches a certain (breakdown) value. As can be seen in Figure 3.10, a transistor breaks down when the reverse-biasing voltage across the collector–base junction reaches a certain value. The sudden vertical rise in each characteristic curve indicates that there is a sudden increase in collector current when breakdown occurs, but, as in the diode, this breakdown is not necessarily destructive. Note that breakdown occurs at progressively smaller values of V_{CE} as base current increases. The region of the collector characteristics where breakdown occurs (to the right of the dashed line) is called the breakdown region.

Each characteristic curve in Figure 3.10 can be seen to rise rapidly when V_{CE} increases from zero to a few tenths of a volt. Each curve is then more or less

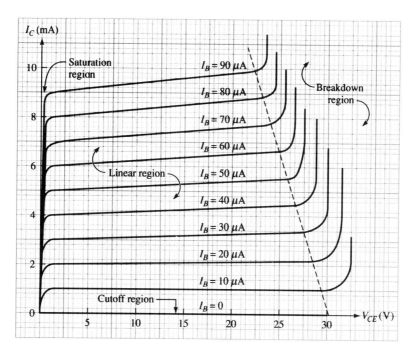

Figure 3.10
Collector characteristics showing regions of operation.

"flat" (horizontal). The region to the left of the curves before they flatten out is called the *saturation* region of the characteristics.

When there is zero collector current in a transistor, it is said to be *cut off*. When base current is zero (or negative, in an NPN transistor, i.e., flowing *out* of the base), the transistor is cut off. Accordingly, the region where this occurs is called the cutoff region.

The region of the collector characteristics where the curves are more or less flat, lying to the right of saturation, to the left of breakdown, and above cutoff, is called the *linear* region. This is the region in which the BJT is normally operated for linear applications, such as voltage amplification. (In digital applications, it is operated at just two values of V_{CE}: one in the saturation region (low voltage) and one in the cutoff region (high voltage)). The BJT is linear to the extent that *equal changes in base current create equal changes in collector current* no matter where those changes occur. (Think of base current as input and collector current as output.) The BJT whose collector characteristics are shown in Figure 3.10 very nearly satisfies this criterion (more so, in fact, than most practical transistors). For example, when $V_{CE} = 5$ V and I_B changes from 10 μA to 20 μA, we see that I_C changes from 1 mA to 2 mA, so $\Delta I_C = 1$ mA when $\Delta I_B = 10$ μA. When $V_{CE} = 14$ V and I_B changes from 70 μA to 80 μA, I_C changes from 7 mA to about 8.1 mA, so in this case $\Delta I_C = 1.1$ mA when $\Delta I_B = 10$ μA. The fact that the two values of ΔI_C are very nearly equal is an indication that the device is very nearly linear. We will discuss the consequences of nonlinearities in practical amplifiers in a later chapter. Note that the linearity criterion we have described here is equivalent to stating that the value of β_{ac} should be the same no matter where on the collector characteristics it is calculated.

The Quiescent Point

When we investigate voltage and current changes that occur in some region of transistor characteristics, it is convenient to regard the variation as occurring above and below some *fixed point*. For example, in the foregoing discussion where we investigated a change in I_B from 10 μA to 20 μA along the line $V_{CE} = 5$ V, we can say that the variation is centered on the fixed point at which $I_B = 15$ μA and $V_{CE} = 5$ V, i.e., that the variation is ± 5 μA from that point. Such a fixed point is known by various names, including *quiescent point, Q-point, bias point,* and *operating point.* It is the point representing the voltage and current in a device when the ac variation is zero. (Quiescent literally means quiet, calm, or nonchanging.) It is called the bias point because, in practical circuits, external resistors connected to the voltage sources that provide bias are selected to produce a specific voltage and current in a device when the ac input is zero. We will discuss such practical bias circuits in more detail in Chapter 5.

3.4   THE JUNCTION FIELD-EFFECT TRANSISTOR (JFET)

Introduction

The field-effect transistor (FET), like the bipolar junction transistor, is a three-terminal semiconductor device. However, the FET operates under principles completely different from those of the BJT. A field-effect transistor is called a *unipolar* device because the current through it results from the flow of only one of the two kinds of charge carriers: holes or electrons. The name *field effect* is derived from the fact that the current flow is controlled by an electric field set up in the device by an externally applied voltage.

There are two main types of FETs: the junction field-effect transistor (JFET) and the metal-oxide-semiconductor FET (MOSFET). Both types are fabricated as discrete components and as components of integrated circuits. The MOSFET is the most important component in modern digital integrated circuits, such as microprocessors and computer memories.

JFET Structure and Biasing

Figure 3.11 shows a diagram of the structure of a JFET and identifies the three terminals to which external electrical connections are made. As shown in the figure, a bar of N-type material has regions of P material embedded in each side. The two P regions are joined electrically and the common connection between

Figure 3.11
Structure of an N-channel JFET.

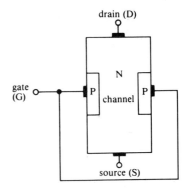

Figure 3.12
Reverse biasing the gate-to-source junctions causes the formation of depletion regions. V_{GS} is a small reverse-biasing voltage for the case illustrated.

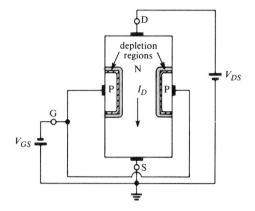

them is called the *gate* (G) terminal. A terminal at one end of the N-type bar is called the *drain* (D), and a terminal at the other end is called the *source* (S). The region of N material between the two opposing P regions is called the *channel*. The transistor shown in the figure is therefore called an *N-channel* JFET, the type that we will study initially, while a device constructed from a P-type bar with embedded N regions is called a *P-channel* JFET. As we develop the theory of the JFET, it may be helpful at first to think of the drain as corresponding to the collector of a BJT, the source as corresponding to the emitter, and the gate as corresponding to the base. As we shall see, the voltage applied to the gate controls the flow of current between drain and source, just as the signal applied to the base of a BJT controls the flow of current between collector and emitter.

When an external voltage is connected between the drain and the source of an N-channel JFET, so that the drain is positive with respect to the source, current is established by the flow of electrons through the N material from the source to the drain. (The source is so named because it is regarded as the origin of the electrons.) Thus, *conventional* current flows from drain to source and is limited by the resistance of the N material. In normal operation, an external voltage is applied between the gate and the source so that the PN junctions on each side of the channel are reverse biased. Thus, the gate (P) is made negative with respect to the source (N), as illustrated in Figure 3.12. Note in the figure that the reverse bias causes a pair of depletion regions to form in the channel.*

The width of the depletion regions in Figure 3.12 depends on the magnitude of the reverse-biasing voltage V_{GS}. The figure illustrates the case where V_{GS} is only a few tenths of a volt, so the depletion regions are relatively narrow. (V_{DS} is also assumed to be relatively small; we will investigate the effect of a large V_{DS} presently.) As V_{GS} is made more negative, the depletion regions expand and the width of the channel decreases. The reduction in channel width increases the resistance of the channel and thus decreases the flow of current I_D from drain to source.

To investigate the effect of increasing V_{DS} on the drain current I_D, let us suppose for the moment that the gate is shorted to the source ($V_{GS} = 0$). As V_{DS} is increased slightly above 0, we find that the current I_D increases in direct propor-

---

*In one variation of the JFET, the junctions are metal–semiconductor (Schottky) junctions, as discussed in Section 2.7, and the device is called a MESFET. Because of the superior switching capabilities of the Schottky junction, the MESFET finds use in digital and high-frequency applications.

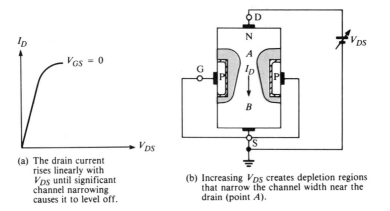

(a) The drain current rises linearly with V_{DS} until significant channel narrowing causes it to level off.

(b) Increasing V_{DS} creates depletion regions that narrow the channel width near the drain (point A).

Figure 3.13
Effects of increasing V_{DS} while the gate is shorted to the source ($V_{GS} = 0$).

tion to it, as shown in Figure 3.13(a). This is as we would expect, since increasing the voltage across the fixed-resistance channel simply causes an Ohm's law increase in the current through it. As we continue to increase V_{DS}, we find that noticeable depletion regions begin to form in the channel, as illustrated in Figure 3.13(b). Note that the depletion regions are broader near the drain end of the channel (in the vicinity of point A) than they are near the source end (point B). This is explained by the fact that current flowing through the channel creates a voltage drop along the length of the channel. Near the top of the channel, the channel voltage is very nearly equal to V_{DS}, so there is a large reverse-biasing voltage between the N channel and the P gate. As we proceed down the channel, less voltage is available because of the drop that accumulates through the resistive N material. Consequently, the reverse-biasing potential between channel and gate becomes smaller and the depletion regions become narrower as we approach the source. When V_{DS} is increased further, the depletion regions expand and the channel becomes very narrow in the vicinity of point A, causing the total resistance of the channel to increase. As a consequence, the rise in current is no longer directly proportional to V_{DS}. Instead, the current begins to level off, as shown by the curved portion of the plot in Figure 3.13(a).

Figure 3.14(a) shows what happens when V_{DS} is increased to a value large enough to cause the depletion regions to meet at a point in the channel near the drain end. This condition is called *pinch-off*. At the point where pinch-off occurs, the gate-to-channel junction is reverse biased by the value of V_{DS}, so (the negative of) this value is called the *pinch-off voltage*, V_p. The pinch-off voltage is an important JFET parameter, whose value depends on the doping and geometry of the device. V_p is always a negative quantity for an N-channel JFET and a positive quantity for a P-channel JFET. Figure 3.14(b) shows that the current reaches a maximum value at pinch-off and that it remains at that value as V_{DS} is increased beyond $|V_p|$. This current is called the *saturation* current and is designated I_{DSS}: the Drain-to-Source current with the gate Shorted.

Despite the implication of the name *pinch-off*, note again that current continues to flow through the device when V_{DS} exceeds $|V_p|$. The value of the current remains constant at I_{DSS} because of a kind of self-regulating or equilibrium process that controls the current when V_{DS} exceeds $|V_p|$: Suppose that an increase in V_{DS} did cause I_D to increase; then there would be in the channel an increased voltage

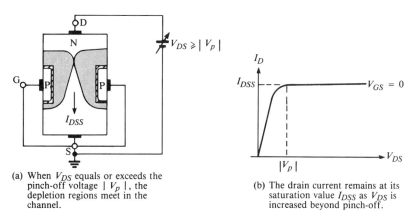

(a) When V_{DS} equals or exceeds the pinch-off voltage $| V_p |$, the depletion regions meet in the channel.

(b) The drain current remains at its saturation value I_{DSS} as V_{DS} is increased beyond pinch-off.

Figure 3.14
The N-channel JFET at pinch-off.

drop that would expand the depletion regions further and reduce the current to its original value. Of course, this change in current never actually occurs; I_D simply remains constant at I_{DSS}.

Drain Characteristics

A typical set of values for V_p and I_{DSS} are -4 V and 12 mA, respectively. Suppose we connect a JFET having those parameter values in the circuit shown in Figure 3.15(a). Note that the gate is no longer shorted to the source, but a voltage $V_{GS} = -1$ V is connected to reverse bias the gate-to-source junctions. The reverse bias causes the depletion regions to penetrate the channel farther along the entire length of the channel than they did when V_{GS} was 0. If we now begin to increase V_{DS} above 0, we find that the current I_D once more begins to increase linearly, as shown in Figure 3.15(b). Note that the slope of this line is not as steep as that of the $V_{GS} = 0$ line, because the total resistance of the narrower channel is greater than before. As we continue to increase V_{DS}, we find that the depletion regions

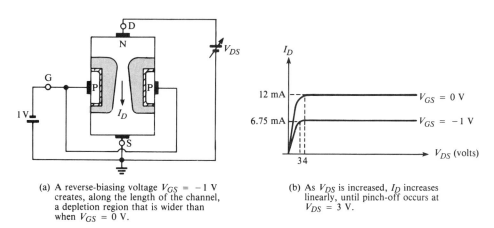

(a) A reverse-biasing voltage $V_{GS} = -1$ V creates, along the length of the channel, a depletion region that is wider than when $V_{GS} = 0$ V.

(b) As V_{DS} is increased, I_D increases linearly, until pinch-off occurs at $V_{DS} = 3$ V.

Figure 3.15
Effects of increasing V_{DS} when $V_{GS} = -1$ V.

again approach each other in the vicinity of the drain. This further narrowing of the channel increases its resistance and the current again begins to level off. Since there is already a 1-V reverse bias between the gate and the channel, the pinch-off condition, where the depletion regions meet, is now reached at $V_{DS} = 3$ V instead of 4 V ($V_{DS} = V_{GS} - V_p$). As shown in Figure 3.15, the current saturates at the lower value of 6.75 mA as V_{DS} is increased beyond 3 V.

If the procedure we have just described is repeated with V_{GS} set to -2 V instead of -1 V, we find that pinch-off is reached at $V_{DS} = 2$ V and that the current saturates at $I_D = 3$ mA. It is clear that increasing the reverse-biasing value of V_{GS} (making V_{GS} more negative) causes the pinch-off condition to occur at smaller values of V_{DS} and that smaller saturation currents result. Figure 3.16 shows the family of characteristic curves, the *drain characteristics*, obtained when the procedure is performed for $V_{GS} = 0, -1, -2, -3,$ and -4 V. The dashed line, which is parabolic, joins the points on each curve where pinch-off occurs. A value of V_{DS} on the parabola is called a *saturation voltage* $V_{DS(sat)}$. At any value of V_{GS}, the corresponding value of $V_{DS(sat)}$ is the difference between V_{GS} and V_p: $V_{DS(sat)} = V_{GS} - V_p$, as we have already described. The equation of the parabola is

$$I_D = I_{DSS}\left(\frac{V_{DS(sat)}}{V_p}\right)^2 \tag{3.10}$$

To illustrate, we have, in our example, $V_p = -4$ V and $I_{DSS} = 12$ mA; so at $V_{DS(sat)} = 3$ V we find

$$I_D = (12 \text{ mA})\left(\frac{3 \text{ V}}{-4 \text{ V}}\right)^2 = 6.75 \text{ mA}$$

which is the saturation current at the $V_{GS} = -1$ V line (see Figure 3.15(b)). Note in Figure 3.16 that the region to the right of the parabola is called the *pinch-off region*. This is the region in which the JFET is normally operated when used for

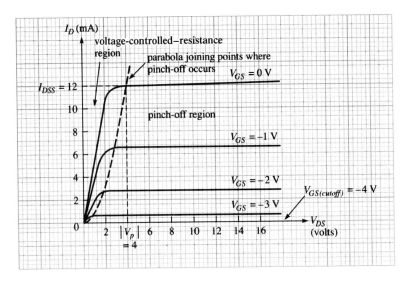

Figure 3.16
Drain characteristics of an N-channel JFET.

linear applications. It is also called the *active* region, or the *saturation* region. The region to the left of the parabola is called the *voltage-controlled–resistance* region, the *ohmic* region, or the *triode* region. In this region, the resistance between drain and source is controlled by V_{GS}, as we have previously discussed, and we can see that the lines become less steep (implying larger resistance) as V_{GS} becomes more negative. The device acts like a voltage-controlled resistor in this region, and there are some practical applications that exploit this characteristic.

The line drawn along the horizontal axis in Figure 3.16 shows that $I_D = 0$ when $V_{GS} = -4$ V, regardless of the value of V_{DS}. When V_{GS} reverse biases the gate-to-source junction by an amount equal to V_p, depletion regions meet along the entire length of the channel and the drain current is cut off. Since the value of V_{GS} at which the drain current is cut off is the same as V_p, the pinch-off voltage is also called the *gate-to-source cutoff voltage*. Thus, there are two ways to determine the value of V_p from a set of drain characteristics: It is the value of V_{DS} where I_D saturates when $V_{GS} = 0$; and it is the value of V_{GS} that causes all drain current to cease, i.e., $V_p = V_{GS(cutoff)}$.

One property of a field-effect transistor that makes it especially valuable as a voltage amplifier is the very high resistance at its gate. Since the path from gate to source is a reverse-biased PN junction, the only current that flows into the gate is the very small leakage current associated with a reverse-biased junction. Therefore, very little current is drawn from a signal source driving the gate, and the FET input looks like a very large resistance. A dc input resistance of several hundred megohms is not unusual. Although the gate of an N-channel JFET can be driven slightly positive, this action causes the input junction to be forward biased and radically decreases the gate-to-source resistance. In most practical applications, the sudden and dramatic decrease in resistance when the gate is made positive would not be tolerable to a signal source driving a FET.

Figure 3.17 shows the structure and drain characteristics of a typical P-channel JFET. Since the channel is P material, current is due to hole flow, rather than electron flow, between drain and source. The gate material is of course N type. Note that all voltage polarities are opposite those in the N-channel JFET. Figure 3.17(b) shows that positive values of V_{GS} control the amount of saturation current in the pinch-off region.

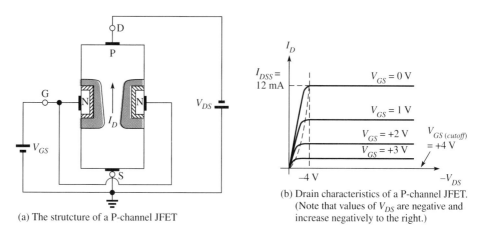

(a) The strutcture of a P-channel JFET

(b) Drain characteristics of a P-channel JFET. (Note that values of V_{DS} are negative and increase negatively to the right.)

Figure 3.17
Structure and characteristics of a P-channel JFET.

Figure 3.18
Schematic symbols for JFETs.

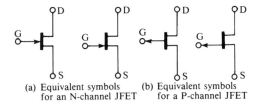

(a) Equivalent symbols
    for an N-channel JFET

(b) Equivalent symbols
    for a P-channel JFET

Figure 3.18 shows the schematic symbols used to represent N-channel and P-channel JFETs. Note that the arrowhead on the gate points into an N-channel JFET and outward for a P-channel device. The symbols showing the gate terminal off-center are used as a means of identifying the source: the source is the terminal drawn closest to the gate arrow. Some JFETs are manufactured so that the drain and source are interchangeable, and the symbols for these devices have the gate arrow drawn in the center.

Figure 3.19 shows the breakdown characteristics of an N-channel JFET. Breakdown occurs at large values of V_{DS} but note that the larger the magnitude of V_{GS}, the smaller the value of V_{DS} at which it occurs. This follows from the fact that the gate–channel junction is reverse biased by both the positive voltage on the N-type channel and the negative voltage on the P-type gate.

The Transfer Characteristic

The *transfer characteristic* of a JFET is a plot of output current versus input voltage, for a fixed value of output voltage. When the input to a JFET is the gate-to-source voltage and the output current is drain current, the transfer characteristic can be derived from the drain characteristics. It is only necessary to construct a vertical line on the drain characteristics (a line of constant V_{DS}) and to note the value of I_D at each intersection of the line with a line of constant V_{GS}. The values of I_D can then be plotted against the values of V_{GS} to construct the transfer characteristic. Figure 3.20 illustrates the process.

In Figure 3.20, the transfer characteristic is shown for $V_{DS} = 8$ V. As can be seen in the figure, this choice of V_{DS} means that all points are in the pinch-off region. For example, the point of intersection of the $V_{DS} = 8$ V line and the $V_{GS} = 0$ V line occurs at $I_D = I_{DSS} = 12$ mA. At $V_{DS} = 8$ V and $V_{GS} = -1$ V, we find $I_D = 6.75$ mA. Plotting these combinations of I_D and V_{GS} produces the parabolic transfer characteristic shown. The nonlinear shape of the transfer characteristic can be

Figure 3.19
Breakdown characteristics of an N-channel JFET.

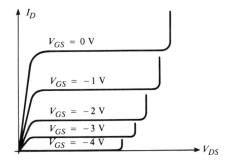

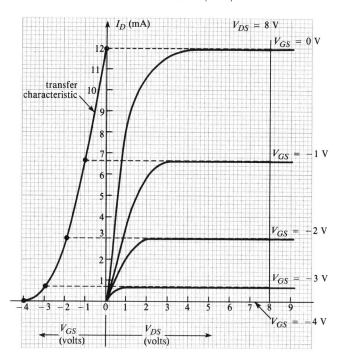

Figure 3.20
Construction of an N-channel transfer characteristic from the drain characteristics.

anticipated by observing that equal increments in the values of V_{GS} on the drain characteristics ($\Delta V_{GS} = 1$ V) do not produce equally spaced lines. (Recall from our discussion of BJT collector characteristics that this situation means that the device is not truly linear.) Note that the intercepts of the transfer characteristic are I_{DSS} on the I_D-axis and V_p on the V_{GS}-axis.

The equation for the transfer characteristic *in the pinch-off region* is, to a close approximation,

$$I_D = I_{DSS}\left(1 - \frac{V_{GS}}{V_p}\right)^2 \tag{3.11}$$

Note that equation 3.11 correctly predicts that $I_D = I_{DSS}$ when $V_{GS} = 0$ and that $I_D = 0$ when $V_{GS} = V_p$. The transfer characteristic is often called the *square-law* characteristic of a JFET and is used in some interesting applications to produce outputs that are nonlinear functions of inputs.

Example 3.13 at the end of the chapter shows how SPICE can be used to obtain a plot of the transfer characteristic of a JFET.

Example 3.4

An N-channel JFET has a pinch-off voltage of -4.5 V and $I_{DSS} = 9$ mA.

1. At what value of V_{GS} in the pinch-off region will I_D equal 3 mA?
2. What is the value of $V_{DS(sat)}$ when $I_D = 3$ mA?

Solution

1. We must solve equation 3.11 for V_{GS}:

$$\left(1 - \frac{V_{GS}}{V_p}\right)^2 = I_D/I_{DSS}$$

$$1 - V_{GS}/V_p = \sqrt{I_D/I_{DSS}}$$

$$V_{GS} = V_p(1 - \sqrt{I_D/I_{DSS}})$$

$$V_{GS} = -4.5[1 - \sqrt{(3 \text{ mA})/(9 \text{ mA})}] = -1.9 \text{ V}$$

2. Equation 3.10 relates I_D and $V_{DS(sat)}$. Solving for $V_{DS(sat)}$, we find

$$V_{DS(sat)} = \sqrt{(V_p)^2 I_D/I_{DSS}} = \sqrt{(4.5)^2(3 \text{ mA})/(9 \text{ mA})} = 2.6 \text{ V}$$

Note that we use the positive square root, since V_{DS} is positive for an N-channel JFET. For a P-channel JFET, we would use the negative root. The value of $V_{DS(sat)}$ could also have been determined from the fact that $V_{DS(sat)} = V_{GS} - V_p = -1.9 \text{ V} - (-4.5 \text{ V}) = 2.6 \text{ V}$.

Transconductance

The transfer characteristic in Figure 3.20 shows that we can regard the JFET as a *voltage-controlled* device: changing the gate-to-source voltage causes the drain current to change. In small-signal operation, we assume that the change in gate-to-source voltage occurs over a portion of the characteristic that is small enough to be treated as (approximately) linear. Recall that the quiescent point, or Q-point, is the point about which small-signal variations occur. As in the BJT, the Q-point is established by external resistors connected to be the voltage sources that provide bias.

Transconductance, g_m, is defined to be the ratio of output current to input voltage. Since current divided by voltage has the units of conductance, the units of g_m are siemens (S). The small-signal transconductance of a JFET is the ratio of a change in output current I_D to the change in input voltage V_{GS} that produced it:

$$g_m = \frac{\Delta I_D}{\Delta V_{GS}} \text{ siemens} \qquad (3.12)$$

Example 3.5

A JFET has small-signal transconductance equal to 8 mS. Find the peak-to-peak value of the drain current when the gate-to-source voltage varies according to $0.1\sin(\omega t)$ V.

Solution   The peak-to-peak variation of the gate-to-source voltage is $\Delta V_{GS} = 2(0.1 \text{ V}) = 0.2 \text{ V}$. From equation 3.12, the peak-to-peak value of the drain current is therefore

$$\Delta I_D = g_m \Delta V_{GS} = (8 \times 10^{-3} \text{ S})(0.2 \text{ V}) = 0.16 \text{ mA}$$

The value of g_m can be found graphically using a JFET transfer characteristic, since the latter is a plot of drain current versus gate-to-source voltage. Transconductance is thus the *slope* of the transfer characteristic at the quiescent point, as

Figure 3.21
Transconductance is the slope of the transfer characteristic. Note that the slope, and hence the value of g_m, increases with increasing I_D.

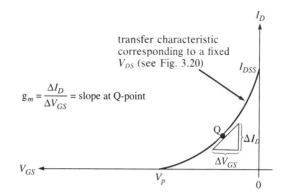

illustrated in Figure 3.21. As shown in the figure, g_m is calculated graphically by drawing a line tangent to the characteristic at the quiescent point and then measuring its slope. Note that transconductance increases with increasing values of I_D, that is, at Q-points lying further up the curve towards I_{DSS}, because the curve becomes steeper in that direction. Since the current in a JFET amplifier varies above and below the Q-point, and since the transconductance changes at every point along the characteristic, small-signal analysis is valid only if ΔI_D is small enough to make the change in g_m negligible.

It can be shown that under small-signal conditions the value of g_m is given by

$$g_m = \frac{2I_{DSS}}{|V_p|}\left(1 - \frac{V_{GS}}{V_p}\right) \text{ siemens} \tag{3.13}$$

where V_{GS} is the quiescent value of the gate-to-source voltage. Equation 3.13 can be used in conjunction with the equation for the transfer characteristic (Exercise 3.51) to express g_m in terms of I_D, the quiescent value of the drain current:

$$g_m = \frac{2I_{DSS}}{|V_p|}\sqrt{\frac{I_D}{I_{DSS}}} \text{ siemens} \tag{3.14}$$

Equation 3.14 clearly shows that the value of g_m increases with increasing I_D. It is a maximum when $I_D = I_{DSS}$, for which it is given the special symbol g_{mO}. From equation 3.14, with $I_D = I_{DSS}$, we have

$$g_{mO} = \frac{2I_{DSS}}{|V_p|} \text{ siemens} \tag{3.15}$$

Although a large value of g_m is desirable in most practical applications, we would never bias a JFET at $I_D = I_{DSS}$ for small-signal operation. Obviously, no signal variation above that Q-point could occur if that were the case.

Example 3.6

An N-channel JFET has the transfer characteristic shown in Figure 3.22. Determine, both (1) graphically and (2) analytically, the value of its transconductance when $I_D = 4$ mA. Also determine the value of g_{mO}.

Solution

1. As shown in Figure 3.22, a line is drawn tangent to the transfer characteristic at the point where $I_D = 4$ mA. Using values obtained from the figure, the slope of

Figure 3.22
(Example 3.6)

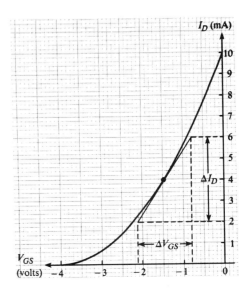

the tangent line is determined to be

$$g_m = \frac{\Delta I_D}{\Delta V_{GS}} = \frac{(6 \text{ mA}) - (2 \text{ mA})}{(2.1 \text{ V}) - (0.8 \text{ V})} = \frac{4 \times 10^{-3} \text{ A}}{1.3 \text{ V}} \approx 3.1 \times 10^{-3} \text{ S}$$

2. From Figure 3.22, it is evident that $V_p = -4$ V and $I_{DSS} = 10$ mA. Using equation 3.14 with $I_D = 4$ mA, we find

$$g_m = \frac{2(10 \times 10^{-3} \text{ A})}{4 \text{ V}} \sqrt{\frac{4 \times 10^{-3}}{10 \times 10^{-3}}} = 3.16 \times 10^{-3} \text{ S}$$

We see that there is good agreement between the graphical and algebraic solutions.

The value of g_{mO} is computed using equation 3.15:

$$g_{mO} = \frac{2(10 \times 10^{-3} \text{ A})}{4 \text{ V}} = 5 \times 10^{-3} \text{ S}$$

3.5   THE METAL-OXIDE-SEMICONDUCTOR FET (MOSFET)

The metal-oxide-semiconductor FET (MOSFET) is similar in many respects to its JFET counterpart, in that both have drain, gate, and source terminals, and both are devices whose channel conductivity is controlled by a gate-to-source voltage. The principal feature that distinguishes a MOSFET from a JFET is the fact that the gate terminal in a MOSFET is *insulated* from its channel region. For this reason, a MOSFET is often called an *insulated-gate* FET, or IGFET. There are two kinds of MOSFETs: the *depletion* type and the *enhancement* type, also referred to as depletion-*mode* and enhancement-*mode* MOSFETs. These names are derived from the two different ways that the conductivity of the channel can be altered by variations in V_{GS}, as we shall see.

Figure 3.23
Structure of an N-channel depletion-type MOSFET.

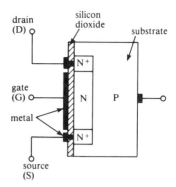

Depletion-Type MOSFETs

Figure 3.23 shows the structure of an N-channel, depletion-type MOSFET. A block of high-resistance, P-type silicon forms a *substrate,* in which are embedded two heavily doped N-type wells, or pockets, labeled N^+. A thin layer of silicon dioxide (SiO_2), which is an insulating material, is deposited along the surface. Metal contacts penetrate the silicon dioxide layer at the two N^+ wells and become the drain and source terminals. Between the two N^+ wells is a more lightly doped region of N material that forms the channel. Metal (aluminum) is deposited on the silicon dioxide opposite the channel and becomes the gate terminal. Note that the silicon dioxide insulates the gate from the channel. Going from gate to channel, we encounter *m*etal, *o*xide, and *s*emiconductor, in that sequence, which accounts for the name *MOSFET*. Notice that there is no PN junction formed between gate and channel, as there is in a JFET.

Figure 3.24 shows the normal mode of operation of a depletion-type, N-channel MOSFET. A voltage V_{DS} is connected between drain and source to make the drain positive with respect to the source. The substrate is usually connected to the source, as shown in the figure. When the gate is made negative with respect to the source by V_{GS}, the electric field it produces in the channel drives electrons away from a portion of the channel near the SiO_2 layer. This portion is thus *depleted* of carriers and the channel width is effectively narrowed. The narrower the channel, the greater its resistance and the smaller the current flow from drain to source. Thus the device behaves very much like an N-channel JFET, the principal difference being that the channel width is controlled by the action of the electric field rather than by the size of the depletion region of a PN

Figure 3.24
Operation of an N-channel, depletion-type MOSFET. The electric field produced by V_{GS} creates in the channel a region that is depleted of carriers.

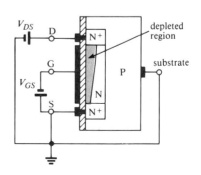

junction. Since there is no PN junction, the voltage V_{GS} can be made *positive* without any concern for the consequences of forward biasing a junction. In fact, making V_{GS} positive attracts more electrons *into* the channel and increases, or *enhances,* its conductivity. Thus, the gate voltage in a depletion-type MOSFET can be varied through both positive and negative voltages and the device can operate in both depletion and enhancement modes. For this reason, the depletion-type MOSFET is also called a *depletion-enhancement* type.

Although there is a PN junction between the N material and the P substrate, this junction is always reverse biased and very little substrate current flows. The substrate has little bearing on the operation of the device and we will hereafter

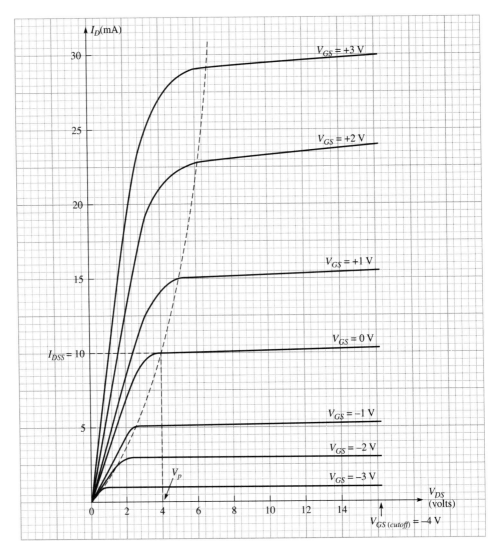

Figure 3.25
Drain characteristics of an N-channel depletion-type MOSFET, showing operation in the depletion and enhancement modes.

ignore its presence, other than to show its usual connection to the source. The resistance looking into the gate is extremely large, on the order of thousands of megohms, because there is no PN junction and no path for current to flow through the insulating layer separating the gate and the channel.

Because of the similarity of a depletion-type MOSFET to a JFET, we would expect it to have similar parameters and operating characteristics. This is indeed the case, as shown by the drain characteristics in Figure 3.25. (See page 66.) Note that current increases linearly with increasing V_{DS} until a pinch-off condition is reached. Beyond pinch-off, the drain current remains constant at a saturation value depending on V_{GS}. More negative values of V_{GS} cause pinch-off to be

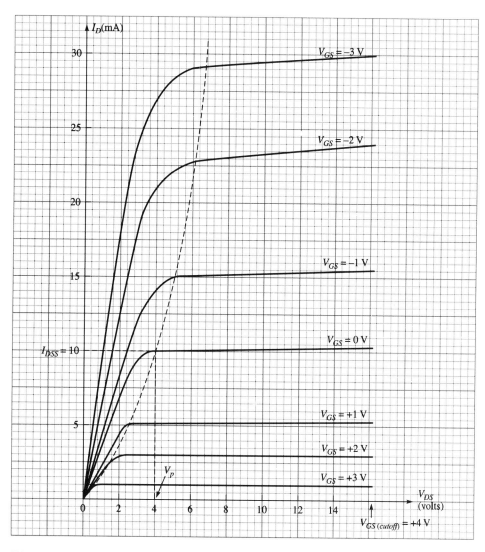

Figure 3.26
Drain characteristics of a P-channel depletion-type MOSFET, showing operation in the depletion and enhancement modes.

Figure 3.27
Schematic symbols for N- and P-channel
depletion-type MOSFETs.

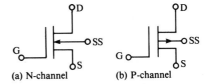

(a) N-channel   (b) P-channel

reached sooner and result in smaller values of saturation current. If $V_{GS} = 0$, the drain current saturates at I_{DSS} when $V_{DS} = -V_p$ volts. If V_{GS} is made sufficiently negative to deplete the entire channel, the drain current is completely cut off. The value of V_{GS} at which this occurs is the gate-to-source cutoff voltage, $V_{GS(cutoff)} = V_p$. Note that the characteristics in Figure 3.25 also show operation in the enhancement mode, where V_{GS} is positive. Figure 3.26 (see page 67) shows the drain

Figure 3.28
Transfer characteristics of N-channel
and P-channel depletion-type MOSFETs.

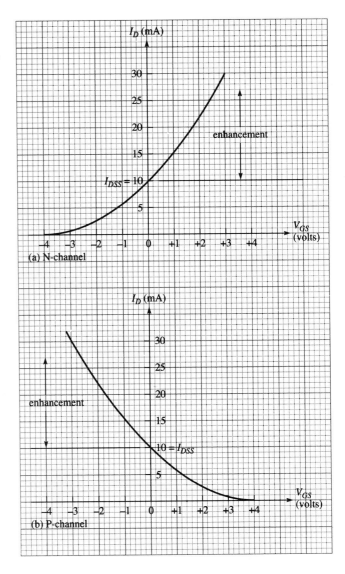

characteristics of a P-channel, depletion-type MOSFET. Notice that depletion occurs in this device when V_{GS} is positive and enhancement when V_{GS} is negative.

Figure 3.27 (see page 68) shows the schematic symbols for N- and P-channel depletion-type MOSFETs. The arrow on each symbol is drawn on the substrate terminal and its direction indicates whether the device is N-channel or P-channel. It points into the device for an N-channel and outward for a P-channel. Notice that the gate terminal is attached to a line that is separated from the rest of the symbol, to emphasize that the gate is insulated from the channel.

The square-law equation for the transfer characteristic of a depletion-type MOSFET is identical to that for a JFET:

$$I_D = I_{DSS}\left(1 - \frac{V_{GS}}{V_p}\right)^2 \tag{3.16}$$

This equation correctly predicts I_D when the depletion-type MOSFET is operated in the enhancement mode. Note, for example, that when V_{GS} is positive in an N-channel device, $(1 - V_{GS}/V_p) > 1$ (since V_p is negative), and therefore I_D is greater than I_{DSS}. Figure 3.28 (see page 68) shows transfer characteristics for N- and P-channel devices. Note in each case that I_D exceeds I_{DSS} in the enhancement mode.

Example 3.7

1. An N-channel depletion-type MOSFET has $I_{DSS} = 18$ mA and $V_p = -5$ V. Assuming that it is operated in the pinch-off region, find I_D when $V_{GS} = -3$ V, and again when $V_{GS} = +2.5$ V.
2. Repeat (1) if the MOSFET is P-channel and $V_p = +5$ V.

Solution

1. From equation 3.16, for $V_{GS} = -3$ V,

$$I_D = (18 \text{ mA})\left(1 - \frac{-3}{-5}\right)^2 = (18 \text{ mA})(0.4)^2 = 2.88 \text{ mA}$$

For $V_{GS} = +2.5$ V,

$$I_D = (18 \text{ mA})\left(1 - \frac{2.5}{-5}\right)^2 = (18 \text{ mA})(1.5)^2 = 40.5 \text{ mA}$$

2. From equation 3.16, for $V_{GS} = -3$ V,

$$I_D = (18 \text{ mA})\left(1 - \frac{-3}{5}\right)^2 = (18 \text{ mA})(1.6)^2 = 46.08 \text{ mA}$$

For $V_{GS} = +2.5$ V,

$$I_D = (18 \text{ mA})\left(1 - \frac{2.5}{5}\right)^2 = 4.5 \text{ mA}$$

Enhancement-Type MOSFETs

Recall that the channel of an N-channel depletion MOSFET is the region of N material between the drain and the source (see Figure 3.23). In the enhancement MOSFET, there is no N-type material between the drain and the source; instead,

Figure 3.29
*Enhancement-type MOSFET. The struc-
ture is similar to that of a depletion
MOSFET, but note the absence of N-
type material between drain and source.*

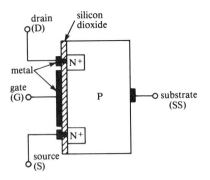

the P-type substrate extends all the way to the SiO₂ layer adjacent to the gate. This
structure is shown in Figure 3.29. Apart from the absence of the N-type channel,
the construction is the same as that of the depletion MOSFET.

Figure 3.30 shows the normal electrical connections between drain, gate, and
source. As in the depletion MOSFET, the substrate is usually connected to the
source. Notice that V_{GS} is connected so that *the gate is positive with respect to the
source*. The positive gate voltage attracts electrons from the substrate to the
region along the insulating layer opposite the gate. If the gate is made sufficiently
positive, enough electrons will be drawn into that region to convert it to N-type
material. Thus, an N-type channel will be formed between drain and source. The
P material is said to have been *inverted* to form an N-type channel. If the gate is
made still more positive, more electrons will be drawn into the region and the
channel will widen, making it more conductive. In other words, making V_{GS} more
positive *enhances* the conductivity of the channel and increases the flow of current from drain to source. Since electrons are induced into the channel to convert
it to N-type material, the MOSFET shown in Figures 3.29 and 3.30 is often called
an *induced N-channel* enhancement-type MOSFET. When this device is referred
to simply as an N-channel enhancement MOSFET, it is understood that the N
channel exists only when it is induced from the P substrate by a positive V_{GS}.

The induced N channel in Figure 3.30 does not become sufficiently conductive to allow drain current to flow until V_{GS} reaches a certain *threshold* voltage,
V_T. In modern silicon MOSFETs, the value of V_T is typically in the range from 1
to 3 V. Suppose that $V_T = 2$ V and that V_{GS} is set to some value greater than V_T,
say, 10 V. We will consider what happens when the drain-to-source voltage is
gradually increased above 0 V. As V_{DS} increases, the drain current increases

Figure 3.30
*The positive V_{GS} induces an N-type
channel in the substrate of an enhancement MOSFET.*

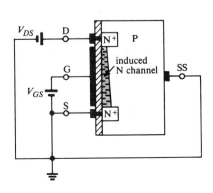

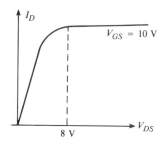

Figure 3.31

The drain current in an N-channel enhancement MOSFET increases with V_{DS} until $V_{DS} = V_{GS} - V_T (= 10 - 2 = 8$ V in this example).

because of normal Ohm's law action. The current rises linearly with V_{DS}, as shown in Figure 3.31. As V_{DS} continues to increase, we find that the channel becomes narrower at the drain end, as illustrated in Figure 3.30. This narrowing occurs because the *gate-to-drain* voltage becomes smaller when V_{DS} becomes larger, thus reducing the positive field at the drain end. For example, if $V_{GS} = 10$ V and $V_{DS} = 3$ V, then $V_{GD} = 10 - 3 = 7$ V. When V_{DS} is increased to 4 V, $V_{GD} = 10 - 4 = 6$ V. The positive gate-to-drain voltage decreases by the same amount that V_{DS} increases, so the electric field at the drain end is reduced and the channel is narrowed. As a consequence, the resistance of the channel begins to increase, and the drain current begins to level off. This leveling off can be seen in the curve of Figure 3.31. When V_{DS} reaches 8 V, then $V_{GD} = 10 - 8 = 2$ V $= V_T$. That is, the positive voltage at the drain end reaches the threshold voltage and the channel width at that end shrinks to zero. Further increases in V_{DS} do not change the shape of the channel and the drain current does not increase any further; i.e., I_D saturates. This action is quite similar to the saturation that occurs at pinch-off in a junction FET.

When the process we have just described is repeated with V_{GS} fixed at 12 V, we find that saturation occurs at $V_{DS} = 12 - 2 = 10$ V. Letting $V_{DS(sat)}$ represent the voltage at which saturation occurs, we have, in the general case,

$$V_{DS(sat)} = V_{GS} - V_T \qquad (3.17)$$

Figure 3.32 shows a set of drain characteristics resulting from repetitions of the process we have described, with V_{GS} set to different values of positive voltage. When V_{GS} is reduced to the threshold voltage $V_T = 2$ V, notice that I_D is reduced to 0 for all values of V_{DS}. The drain characteristics are similar to those of an N-channel JFET, except that all values of V_{GS} are positive in the case of the enhancement MOSFET. The enhancement MOSFET can be operated only in an enhancement mode, unlike the depletion MOSFET, which can be operated in both depletion and enhancement modes. The dashed, parabolic line shown on the characteristics in Figure 3.32 joins the saturation voltages, i.e., those satisfying equation 3.17. As in JFET characteristics, the region to the left of the parabola is called the *voltage-controlled–resistance* region where the drain-to-source resistance changes with V_{GS}. We will refer to the region to the right of the parabola as the *active* region. The device is normally operated in the active region for small-signal amplification.

Figure 3.33(a) shows the structure of a P-channel enhancement MOSFET and its electrical connections. Note that the substrate is N-type material and that a P-type channel is induced by a negative V_{GS}. The field produced by V_{GS} drives electrons away from the region near the insulating layer while attracting minority holes, and thus inverts the region to P material. Figure 3.33(b) shows a typical set of drain characteristics for the P-channel enhancement MOSFET. Note that all

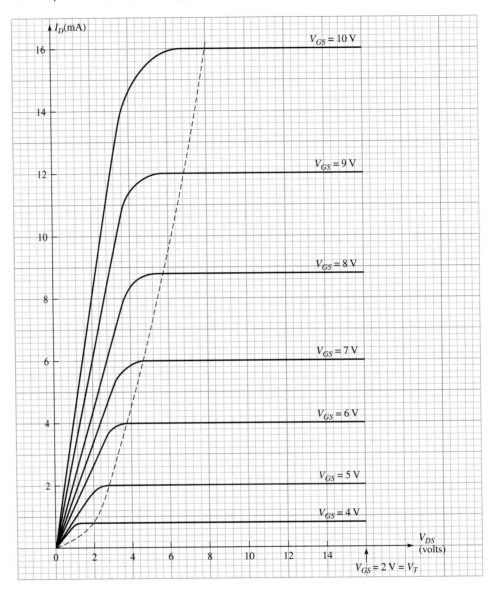

Figure 3.32
Drain characteristics of an induced N-channel enhancement MOSFET. Note that all values of V_{GS} are positive.

values of V_{GS} are negative and that the threshold voltage V_T is negative. N-channel and P-channel MOSFETs are often called *NMOS* and *PMOS* devices for short.

Figure 3.34 shows the schematic symbols used to represent N-channel and P-channel enhancement MOSFETs. As in previous FET symbols, the arrow is drawn pointing into the device for an N-channel FET and outward for a P-channel FET. To distinguish between the drain and source terminals, the gate terminal is drawn opposite the source terminal. The broken line symbolizes the fact that the channel is induced rather than being an inherent part of the structure.

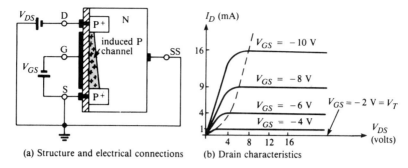

(a) Structure and electrical connections   (b) Drain characteristics

Figure 3.33
The induced P-channel enhancement MOSFET.

Enhancement MOSFET Transfer Characteristic

In the active region, the drain current and gate-to-source voltage are related by

$$I_D = 0.5\beta(V_{GS} - V_T)^2 \quad V_{GS} \geq V_T \tag{3.18}$$

where β is a constant whose value depends on the geometry of the device, among other factors. A typical value of β is 0.5×10^{-3} A/V^2. Figure 3.35 shows a plot of the transfer characteristic of an N-channel enhancement MOSFET for which $\beta = 0.5 \times 10^{-3}$ A/V^2 and $V_T = 2$ V.

The transconductance of the enhancement-type MOSFET can be found graphically using the transfer characteristic and the definition:

$$g_m = \left.\frac{\Delta I_D}{\Delta V_{GS}}\right|_{V_{DS} = \text{constant}} \text{siemens} \tag{3.19}$$

Figure 3.36 shows how g_m is computed as the slope of a line drawn tangent to the characteristic at the operating or Q-point. It is clear from the figure that the slope of the characteristic, and hence the value of g_m, changes as the Q-point is changed. Therefore small-signal analysis requires that the signal variation around the Q-point be confined to a limited range over which there is negligible change in g_m, i.e., to an essentially linear segment of the characteristic. Also, to ensure that the device is operated within its active region, the variation must be such that the following inequality is always satisfied:

$$|V_{DS}| > |V_{GS} - V_T| \tag{3.20}$$

It can be shown that under small-signal conditions the transconductance of an enhancement MOSFET can be determined from

$$g_m = \beta(V_{GS} - V_T) \text{ siemens} \tag{3.21}$$

Figure 3.34
Symbols for enhancement-type MOS-FETs.

(a) N-channel   (b) P-channel

Figure 3.35
Transfer characteristic for an enhancement NMOS FET. $\beta = 0.5 \times 10^{-3}$; $V_T = 2$ V.

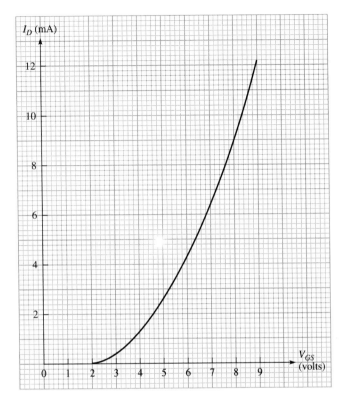

Figure 3.36
The transconductance of an enhancement MOSFET is the slope of a line drawn tangent to the transfer characteristic at the Q-point.

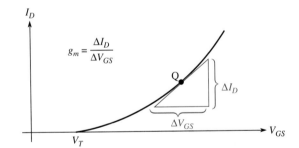

Example 3.8

An enhancement-type MOSFET has $\beta = 0.6 \times 10^{-3}$ A/V^2 and $V_T = 2$ V. At its quiescent point, $I_D = 6.5$ mA. Find the change in drain current when the gate-to-source voltage changes by ± 0.2 V from the Q-point.

Solution.   In order to find g_m using equation 3.21, we must first find the value of V_{GS} at the Q-point. Solving equation 3.18 for V_{GS}, we find

$$(V_{GS} - V_T)^2 = \frac{I_D}{0.5\,\beta}$$

$$V_{GS} - V_T = \sqrt{\frac{I_D}{0.5\,\beta}}$$

$$V_{GS} = \sqrt{\frac{I_D}{0.5\beta}} + V_T = \sqrt{\frac{6.5 \times 10^{-3}\ \text{A}}{0.5(0.6 \times 10^{-3}\ \text{A/V}^2)}} + 2\ \text{V} = 6.65\ \text{V}$$

Then, from equation 3.21,

$$g_m = \beta(V_{GS} - V_T) = (0.6 \times 10^{-3}\ \text{A/V}^2)(6.65\ \text{V} - 2\ \text{V}) = 2.79\ \text{mS}$$

Finally, since $\Delta V_{GS} = 2(0.2\ \text{V}) = 0.4\ \text{V}$, we find, from equation 3.19,

$$\Delta I_D = g_m \Delta V_{GS} = (2.79 \times 10^{-3}\ \text{S})(0.4\ \text{V}) = 1.12\ \text{mA}$$

3.6 INTEGRATED-CIRCUIT TECHNOLOGY

Applications and Classifications of Integrated Circuits

We have mentioned in several previous discussions that diodes and transistors are found in both *discrete-* and *integrated*-circuit form. Discrete components are packaged in individual enclosures and have leads that allow electrical connections to be made between component terminals, such as emitter, base, and collector. Discrete circuits are composed of discrete components, including resistors, capacitors, and/or inductors, interconnected by wires or through conducting paths on a printed circuit board. Figure 3.37(a) shows a typical discrete transistor circuit. In contrast, the components of an integrated circuit are all constructed on a single, tiny piece of semiconductor crystal, called a *chip,* which may contain hundreds of diodes, transistors, resistors, and/or capacitors. (Inductors are not constructed in integrated-circuit form.) The conducting paths that interconnect the components of an integrated circuit are contained entirely within the device, and the only leads that are brought out are those necessary for power supply connections, grounds, and circuit inputs and outputs. Figure 3.37(b) shows a typical integrated-circuit chip of the size that might well contain all of the discrete circuitry shown in Figure 3.37(a). An integrated circuit (IC) fabricated entirely on a single silicon chip is called a *monolithic* IC. A *hybrid* integrated circuit contains one or more monolithic circuits interconnected with external resistors and capacitors using *thin-film* or *thick-film* techniques.

One obvious advantage of integrated circuits over their discrete counterparts is the fact that very complex circuits can be constructed in very small packages, with attendant savings in wiring, assembly, cooling requirements, and material costs. Besides the economic benefits derived from miniaturization, integrated circuits are inherently more reliable than discrete circuits. Because the IC components are contained in a single rigid structure, an IC is not so susceptible as a discrete circuit to the kinds of mechanical failures that afflict the latter: ruptures and shorts in interconnecting paths, caused by shock and vibration; connector misalignments; solder-joint failures; and so forth. Furthermore, a good deal of back-up or redundant circuitry can be included in an IC at very little additional cost, thus ensuring satisfactory performance in the event that some components do fail. Back-up circuitry is feasible because circuit complexity is not so significant a cost factor as it is in discrete circuits, due to the nature of the manufacturing

(a)

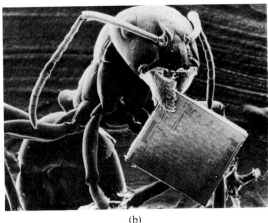
(b)

Figure 3.37
Discrete and integrated circuits. The ant holds in its mouth an integrated-circuit chip that could contain all the circuitry in the discrete circuit, to the left (Right: courtesy of Philips Science and Industry Division).

process. As we shall see in later discussions, the number of components that can be fabricated on a single chip is limited more by the need for maintaining isolation between components than by their cost. Although numerous steps are required to manufacture an IC, a large number of identical devices can be fabricated simultaneously, contributing to cost reduction.

Finally, integrated circuits are advantageous in high-frequency applications and in high-speed computer circuits because of the small distances that electrical signals travel between individual components. Long signal paths create delays and phase shifts that limit the frequency at which such circuits can operate.

Crystal Growth

The starting point in the manufacturing procedure for both discrete components and integrated circuits is the production of a single symmetric crystal, most often a silicon crystal. Silicon is obtained from certain chemical compounds, but is not initially in a form suitable for semiconductor devices. In its natural form, silicon is said to be *polycrystalline,* because it is composed of a large number of crystals having different orientations. To obtain a single crystal of uniform orientation, it is necessary to melt the polycrystalline silicon and then allow it to cool and solidify under certain closely controlled conditions. The location in the "melt" where cooling takes place and the rate at which cooling occurs are particularly critical. In one manufacturing process, a crucible of molten silicon is pulled slowly through a furnace and the crystal forms where the cooling melt emerges.

The development of a single crystal by controlled cooling of molten material is called *crystal growth*. A *seed* crystal is often used as a "starter" upon which the crystal is grown. In the process most often used for IC production, called the *Czochralski technique,* the seed crystal is brought into contact with the molten material and then slowly extracted while being rotated. The crystal formed using the Czochralski technique is in the shape of a cylinder up to 5 inches or more in diameter and several feet long. This cylindrical *ingot* is then sliced, using a diamond cutter, into thin *wafers,* about 0.5 mm thick. Figure 3.38(a) shows a typical

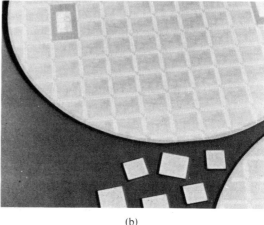

(a)   (b)

Figure 3.38
Left: An ingot of silicon crystal and wafers (courtesy of Monsanto Electronic Materials Company). Right: A wafer showing the individual chips (dies) cut from it (courtesy of American Microsystems, Inc., Pocatello, ID, 1985).

cylindrical ingot and a wafer obtained from it. The wafers are polished and each is used to produce a hundred or more identical IC chips. The finished chips, called *dies* or *dice*, are cut from the wafer (Figure 3.38(b)) and mounted in individual enclosures after the attachment of leads.

PN Device Fabrication—The Photolithographic Process

The method most widely used for integrated-circuit fabrication is now also used to produce discrete devices. The process to which we refer, called *batch production* using *photolithographic* methods, is a truly remarkable blend of ingenuity and precision, and is responsible for many dramatic achievements in the field of microminiature electronics. The manufacturing procedure is called *batch processing* because it permits the simultaneous production of many identical chips from a single wafer. This procedure has been made possible by technological advances in a number of diverse fields, including chemical process control, photochemistry, and computer-aided design and manufacturing (CAD/CAM).

The essence of the photolithographic process is the use of photographic methods to alter the characteristics of a special coating applied to the surface of a wafer. The coating is altered in an intricate pattern so that tiny regions of the crystal will become P- or N-type material when the wafer is later subjected to certain doping treatments. One such doping treatment is called *impurity diffusion,* whereby donor or acceptor impurities are allowed to diffuse through the surface pattern and enter the crystals at the desired locations. We will discuss some details of this important process and then return to the details of photolithography.

Recall that *carrier* diffusion is the migration of holes or electrons from a region where there is a surplus of one or the other to a region where there is a corresponding scarcity. The diffusion process used in batch production is a similar phenomenon, except that it is impurity *atoms* such as boron or phosphorus that are allowed to diffuse into semiconductor material to make it either P- or N-type. To enable atoms to diffuse into silicon, the material must be raised to a high temperature, on the order of 1000°C. At that temperature, silicon atoms leave the crystal structure, thus making it possible for migrating impurity atoms to occupy the vacancies. Boron atoms entering the crystal structure create holes and therefore produce P-type material, while phosphorus atoms supply excess electrons to create N material.

The first step in the photolithographic processing is to create a thin layer of silicon dioxide (SiO_2) on the surface of a silicon wafer. This is accomplished by placing the crystalline wafer in a furnace containing oxygen gas (O_2), which then reacts chemically with the Si to produce SiO_2. Impurity atoms cannot diffuse through the SiO_2 layer and into the Si crystal, so the SiO_2 is used to prevent the creation of P- and N-type regions where they are not desired. To create P and N regions where they are desired, "windows" must be made in the SiO_2 layer; that is, the SiO_2 must be removed at selected locations to permit impurity atoms to diffuse into the crystal. We say that windows are "opened" where the SiO_2 is removed.

The next step in the process is to deposit a coating of *photoresist* (PR) material on top of the SiO_2 layer. The photoresist is a photosensitive organic material that changes its composition (becomes polymerized) wherever it is exposed to ultraviolet light. The PR coating is exposed to ultraviolet light through a glass *mask* that *prevents* exposure (is opaque) at any location where a window is desired. Thus the mask creates a pattern of unexposed PR corresponding to regions where impurity diffusion will be allowed to occur. The unexposed PR material is removed and hydrofluoric acid (HF) is applied to the surface. The acid does not affect the regions covered by exposed PR but etches away the SiO_2 where it is not protected by the PR, thus creating the windows. Once the windows in the SiO_2 have been opened by the HF acid treatment, the exposed PR material is removed and the structure is ready for the impurity diffusion step. The wafer is placed in an oven and impurity atoms diffuse through the windows into the silicon crystal. The depth to which the diffused P or N layer penetrates the crystal, and its concentration (doping density), are determined by closely controlling the temperature of the process and the length of time it is allowed to occur. Because diffusion occurs only at those locations where windows have been opened, the process is often called *selective diffusion*. Figure 3.39 illustrates how the steps we have described so far could be used to create two PN junctions in an N-type crystal.

We should note that the photoresist described above is known as *negative* resist; positive photoresist produces patterns that are the exact opposite of those described.

The final steps in the photolithographic processing are those required to deposit metal contact surfaces where terminal leads can be attached and, in the case of integrated circuits, any metallic paths needed to interconnect devices. This phase of the procedure is called *metallization*. The metal most often used is aluminum (Al) because it adheres well to Si and SiO_2. Gold is sometimes used, but additional steps are then necessary to ensure adhesion. Vaporized aluminum is first deposited over the entire surface of a wafer that has undergone the diffusion processing described earlier. A layer of photoresist is then applied on top of the aluminum and exposed through a mask that defines the metallization pattern. As before, the unexposed PR material is removed and the aluminum is etched away from all regions where it is not covered by exposed PR.

The process we have described could be used to obtain hundreds of discrete PN diodes from a single wafer. The SiO_2 layer is generally left intact and the wafer is scribed, or cut, with a laser to separate the individual devices. The steps required for the fabrication of a more complex device, such as a transistor, are like

Figure 3.39
The creation of two PN junctions in an
N-type crystal using photolithography.
(The thickness of all layers is exagger-
ated for clarity.)

STEPS

CROSS SECTIONS

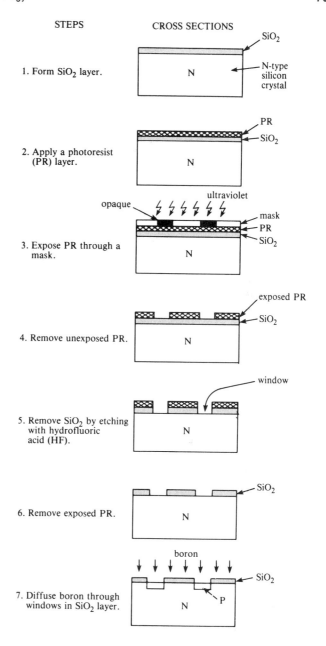

1. Form SiO$_2$ layer.

2. Apply a photoresist (PR) layer.

3. Expose PR through a mask.

4. Remove unexposed PR.

5. Remove SiO$_2$ by etching with hydrofluoric acid (HF).

6. Remove exposed PR.

7. Diffuse boron through windows in SiO$_2$ layer.

those for a single PN junction, but involve repeated oxidations and diffusions to create alternating layers of P and N regions. The crystal itself may be used as one of those layers (for example, as the N-type collector in an NPN transistor). A transistor whose base, emitter, and collector regions lie in parallel planes, one on top of the other, is said to be a *planar* transistor. Most integrated-circuit transistors are of this type. Figure 3.40 summarizes the steps involved in the fabrication of a planar NPN transistor. Note that three oxidation steps, two diffusion steps, and one metallization step are required. Four masks are required.

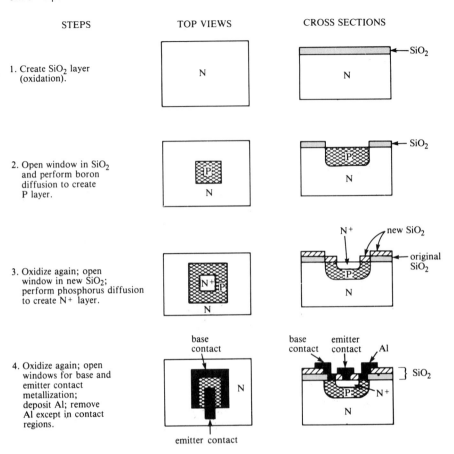

STEPS   TOP VIEWS   CROSS SECTIONS

1. Create SiO₂ layer (oxidation).

2. Open window in SiO₂ and perform boron diffusion to create P layer.

3. Oxidize again; open window in new SiO₂; perform phosphorus diffusion to create N+ layer.

4. Oxidize again; open windows for base and emitter contact metallization; deposit Al; remove Al except in contact regions.

Figure 3.40
Summary of steps involved in the fabrication of a planar NPN transistor.

All the steps we have described for fabricating PN devices are applicable to batch production of integrated-circuit chips, with a few variations and some additional steps we will describe presently. Of course, the diffusion and metallization masks necessary to produce a large number of complex devices must be very intricate and very precise. In fact, the preparation of these masks is one of the most expensive and time-consuming steps in the process. In many cases, the required precision can be obtained only through the use of computer-aided design techniques. Newer techniques employ a computer-controlled electron beam that "writes" a pattern directly on the wafer surface, thus eliminating the need for masking altogether. Much finer line widths can be achieved this way. Similar computer techniques are now also being employed to create very fine masks used in conventional photolithographic procedures.

As an alternative to impurity diffusion, a modern method called *ion implantation* is also being used to create P and N regions in a crystal. In this method, impurity ions are given high energies by an acceleration tube and literally driven through the surface of the wafer and embedded in the crystal. Since ion implantation is blocked by photoresist (or metal), the photolithographic methods we have already described can be used to control the pattern where P and N regions are formed. One advantage of ion implantation is that it can be performed at room temperature. The average depth to which the ions penetrate the crystal, called the *projected range,* can be closely controlled by adjusting the energy imparted to the ions in the accelerator tube. This degree of control is a second advantage of ion implantation: the doping level and depth of doping can be set quite accurately. Very thin regions of P or N material can be created very precisely, as is necessary in some IC designs. Ion implantation is used when these factors are critical to the operation of a circuit.

One disadvantage of ion implantation is that the high-energy ions rupture the crystal structure to a certain extent. The damage can be repaired by heating, a process called *annealing.*

Epitaxial and Buried Layers

In the fabrication of most monolithic integrated circuits, the processes we have described for producing tiny P and N regions in a silicon crystal are not applied directly to the wafer but rather to a thin crystal layer grown on the surface of the wafer. The original (wafer) crystal is called the *substrate* for the circuit, and the thin layer is called an *epitaxial* layer. Typically, the substrate is P-type silicon and the epitaxial layer is N-type silicon.

The substrate (wafer crystal) serves as the seed crystal for the growth of the epitaxial layer. The procedure for growing this new layer, called *epitaxy,* is a special case of *chemical vapor deposition* (CVD), in which gaseous chemicals are used to deposit a layer of solid material on a crystal. Besides an epitaxial layer, CVD can be used to deposit SiO_2 and polycrystalline silicon. Epitaxy is a special case because the solid deposited is a uniformly oriented crystal that aligns itself with the orientation of the crystal in the substrate. It occurs when the gas used is *silane* (SiH_4) and the temperature is greater than 1000°C.

In integrated circuits, the metallization pattern required to interconnect devices is deposited on the *surface* of the chip. For an NPN transistor, which is the most common type fabricated in ICs, the N-type epitaxial layer serves as the collector. To reduce the collector resistance in the path between the epitaxial collector region and the surface, a region of heavily doped (highly conductive) N^+ material is created *beneath* the epitaxial layer, and another from the epitaxial layer to the surface, where contact is made to the collector. Because of its location, the N^+ region lying beneath the epitaxial layer is called a *buried layer.* It is diffused into the substrate wafer before the epitaxial layer is created. In a later step, an N^+ region is diffused into the epitaxial layer from the wafer surface to provide the low-resistance collector-contacting path. Figure 3.41 summarizes the steps involved in the fabrication of a single NPN transistor in an epitaxial layer with a buried N^+ layer. The several oxidations and the masking steps required to perform all the diffusions are not shown explicitly in the figure. Diodes are fabricated in integrated circuits in exactly the same way as transistors. In fact, a diode is often formed by shorting the collector of a transistor to its base. Figure 3.42 illustrates this use.

Resistors

In a bipolar integrated circuit, each resistor is formed by diffusing a certain quantity of P- or N-type material into the epitaxial layer on the surface of an electrically isolated "island." The total resistance of each diffused resistor is determined by the geometry of the region in which it is formed (length, width, and depth) and the

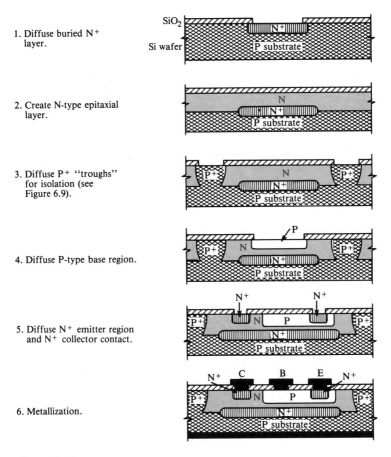

1. Diffuse buried N+ layer.

2. Create N-type epitaxial layer.

3. Diffuse P+ "troughs" for isolation (see Figure 6.9).

4. Diffuse P-type base region.

5. Diffuse N+ emitter region and N+ collector contact.

6. Metallization.

Figure 3.41
A summary of the steps required to fabricate a planar NPN transistor in an epitaxial layer. Note that the buried N+ layer is created before the epitaxial layer.

conductivity of the diffused material, i.e., the degree of doping. Once the resistors have been formed, they are interconnected with other components, as required, by metallization.

It is desirable to have all resistor diffusions occur at the same time that the diffusions for transistor base or emitter regions occur, thus avoiding the need for additional masking and diffusion steps. Therefore, the design of a resistor amounts to the specification of its *geometry* rather than the doping level of the diffused material, as the latter will be the same for all devices undergoing a diffusion step. Figure 3.43 shows a bar of diffused material forming a P-type

Figure 3.42
In integrated circuits, a diode is often fabricated by shorting the collector of a transistor to its base.

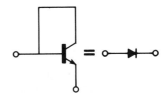

Figure 3.43
A resistor formed by diffusing P-type
material into an N-type epitaxial layer.
The dimensions shown are used in com-
puting the resistance.

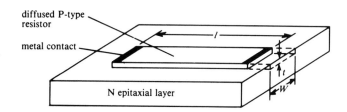

resistor in an N-type epitaxial layer and identifies the dimensions that determine its total resistance.

Recall that the resistance R between the ends of the bar shown in Figure 3.43 can be found from

$$R = \frac{\rho l}{A} \quad \text{ohms} \tag{3.22}$$

where ρ is the resistivity of the material in $\Omega \cdot$ m, and A is the cross-sectional area. As can be seen in the figure, $A = tW$, so

$$R = \frac{\rho l}{tW} \quad \text{ohms} \tag{3.23}$$

Consider now a bar of material having fixed thickness t and having $l = W$, i.e., a square section. For this bar, the quantities l and W in equation 3.23 cancel, and we obtain the special quantity R_s, called the *sheet resistance* (also called the *sheet resistivity*):

$$R_s = \frac{\rho}{t} \quad \text{ohms/square} \tag{3.24}$$

Note that R_s has the same value for *any* square section of a diffused layer having constant thickness and uniform resistivity. Note also that the units of ρ and t must be consistent in order for R_s to have the proper units. For example, if ρ is in ohm $\cdot$ meters, then t must be in meters. The thickness of an integrated-circuit layer is often expressed in *microns,* where 1 micron $= 1 \ \mu$m $= 10^{-6}$ m. It is the sheet resistance of a diffused layer that is of primary interest to an integrated-circuit designer. Its value, which can be measured using special techniques, is typically in the range from 2 to 10 ohms/square for emitter diffusions and 100 to 200 ohms/square for base diffusions.

In integrated-circuit technology, the property of a material called *conductivity* (σ) is used more often than resistivity. Conductivity is simply the reciprocal of resistivity:

$$\sigma = \frac{1}{\rho} \tag{3.25}$$

Since the units of ρ are $\Omega \cdot$ m, the units of σ are $1/(\Omega \cdot$ m$) =$ S/m (siemens per meter).

Example 3.9

A diffused P layer has a sheet resistance of 125 Ω/square. If the P material has a conductivity of 2000 S/m, find the thickness of the layer in microns.

Solution.   From equation 3.25, $\rho = 1/\sigma = 1/2000 = 5 \times 10^{-4} \ \Omega \cdot m$. From equation 3.24,

$$t = \frac{\rho}{R_s} = \frac{5 \times 10^{-4}}{125} = 4 \times 10^{-6} \ m = 4 \ \mu m$$

The ratio of the length of a diffused resistor to its width, l/W (see Figure 3.43), is called its *aspect ratio, a*. From equation 3.23, we have

$$R = \left(\frac{\rho}{t}\right)\left(\frac{l}{W}\right) = R_s a \tag{3.26}$$

Since the sheet resistance R_s is generally fixed by the requirements for base and emitter diffusion, it is the aspect ratio that designers control to obtain a resistance of desired value.

Example 3.10

It is desired to fabricate a 1.5-kΩ resistor using a diffused P layer having sheet resistance 200 Ω/square.

1. What aspect ratio should the resistor have?
2. What should be the total length of the diffused region if its width is 30 μm?

Solution

1. From equation 3.26, $a = R/R_s = 1500/200 = 7.5$.
2. From the definition of aspect ratio,

$$a = \frac{l}{W} = 7.5$$

$$l = 7.5W = 7.5(30 \ \mu m) = 225 \ \mu m = 0.225 \ mm$$

For a diffused resistor to have a large resistance, it is clear that it must have a large aspect ratio, l/W. This can be accomplished by using a zigzag or meandering pattern to increase its length, such as shown in Figure 3.44. In general, very large resistance values are difficult to obtain in integrated circuits and are avoided where possible. In some cases, large resistance values are obtained using ion implantation, which makes possible a very thin impurity layer (small value of t), and a corresponding large value for R_s. One advantage of this method is the saving of chip space that would otherwise be needed to obtain a large aspect ratio.

A resistor can be isolated from other IC components by connecting a suitable reverse-biasing potential around it. For example, the N material in Figure 3.44 could be connected to the highest positive voltage in the circuit. An alternative

Figure 3.44
The resistance of a diffused resistor can be increased by using a zigzag pattern that increases its aspect ratio.

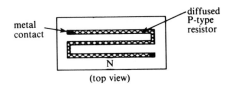

(top view)

Figure 3.45
A double-diffused integrated-circuit resistor.

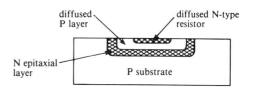

structure for an integrated-circuit resistor is the *double-diffused* resistor illustrated in Figure 3.45. To produce this structure, a layer of P material is diffused into the epitaxial layer at the same time that the base diffusion occurs, and the resistor is then formed as an N-type layer diffused into the P material when the emitter diffusion occurs.

It is difficult to obtain precise resistance values in integrated circuits. Tolerances are typically ±20%. On the other hand, two or more resistors of nearly equal value can be obtained with less difficulty. The production of components having equal or very similar characteristics is called *matching,* and IC resistors can be matched to within ±5% or better. Good matching is possible because resistors are created at the same time using the same diffusion material. In general, integrated-circuit designers avoid design procedures that require precise resistor values and take advantage of techniques that depend on matched values.

Capacitors

Recall that capacitance exists when two conducting regions are separated by an insulator, or dielectric. In the case of a reverse-biased PN junction, there is capacitance between the P and N sides because each side is conductive and the depletion region between them acts as a dielectric. Therefore, when a capacitor is required in an integrated-circuit design, it can be obtained by fabrication of a reverse-biased diode. (A forward-biased diode also has capacitance, but clearly does not block the flow of dc current like a capacitor.) Either the base–emitter or the base–collector junction of a transistor can be used, though the latter is preferred because of its higher reverse breakdown voltage. Capacitance values obtained this way are quite small, usually less than 20 pF. Tolerances are generally large, ±20%, but, as in the case of resistors, reasonably good matching can be achieved.

Recall the fundamental capacitance equation:

$$C = \frac{\varepsilon A}{d} \qquad (3.27)$$

where ε is the permittivity of the dielectric, A is the area of the parallel conducting surfaces, and d is the distance the surfaces are separated. The capacitance of a reverse-biased junction therefore depends directly on the area of the junction and inversely on the width of the depletion region. Since the width of the depletion region changes with the value of the reverse-biasing voltages (increases with increasing voltage), the capacitance also changes with bias voltage. This characteristic is undesirable for integrated-circuit applications but is exploited in some special applications where the diode serves as a voltage-controlled capacitor, called a *varicap diode,* or *varactor.*

Integrated-circuit capacitors can also be constructed using an area of metallization as one conducting surface, silicon dioxide as the dielectric, and an N+ diffused layer created during emitter diffusion as the other conducting surface. These capacitors have excellent characteristics, including low leakage and high

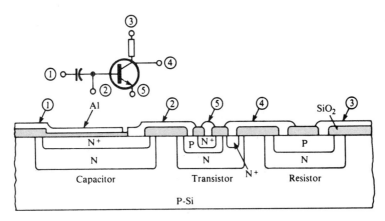

Figure 3.46
An integrated circuit containing interconnected transistor, resistor, and capacitor.

breakdown voltage. Capacitance values are again small, comparable to those of junction capacitors. Large capacitors are impractical in integrated circuits and are avoided in circuit design.

Figure 3.46 shows a cross-sectional view of an NPN transistor, a resistor, and a capacitor all constructed on a P substrate and interconnected to form a transistor amplifier. The capacitor is the SiO_2 dielectric type.

It is important to note that it is generally easier and more economical to construct transistors and diodes in integrated circuits than it is to construct resistors and capacitors. For this reason, designers use transistors and/or diodes in innovative ways to perform circuit functions that could be performed more easily in discrete circuits using resistor and capacitor networks.

Component Interconnections

We have already discussed how photolithographic methods are used to deposit an aluminum metallization pattern on the surface of a chip to interconnect components. As can be seen in Figure 3.46, the metallization makes contact with component terminals and crosses the SiO_2 layer where required to electrically join other component terminals. The metallization pattern is also used to create large metal areas called *pads* along the edges of the chip where external leads can be attached. These pads are electrically connected by the metallization to IC terminals that must be accessible from outside the package, such as power supply terminals, ground, input(s), and output(s). Figure 3.47 shows an enlarged top view of an IC chip in which the metallization pattern and the pads are clearly visible.

Interconnecting paths can also be formed using highly conductive polycrystalline silicon. This method is especially useful for joining components in complex structures involving several layers of circuitry. *Crossovers,* points where one conducting path crosses another without electrically contacting it, are avoided whenever possible by careful selection of the location of each component with respect to others (i.e., component layout, also called circuit *topology*). If a crossover is unavoidable, it is made at a resistor location, because the resistors are covered by SiO_2 and the conducting path can be placed on top of that protective coating. Figure 3.48 is a diagram illustrating this concept.

Figure 3.47
*Enlarged top view of an integrated-
circuit chip. The white areas are metallization. Note the pads along each side
(courtesy RCA Corporation).*

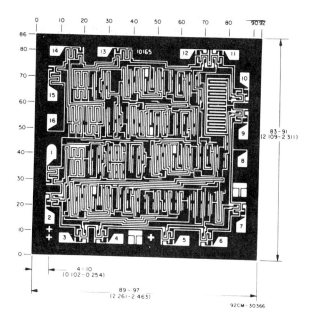

Packaging

Although some integrated circuits are available in cylindrical metal cases, such as
the TO-5 enclosures used to package discrete transistors, the majority of modern
ICs are packaged in the case type called a *dual in-line package* (DIP). These
rectangular cases have rigid metal tabs, called *pins*, that protrude from along each
side of the package (see Figure 3.49). Each pin is electrically connected to one of
the IC pads and therefore serves as a terminal to which external circuit connections can be made. The package can be conveniently inserted into a plugboard for
experimentation or into holes drilled in a printed circuit board. DIP packages are
available with several standard numbers of pins ranging from 4 to 64. Integrated
circuits are also packaged in "flatpak" form, where the pins protrude straight out
from the case instead of at right angles to it.

Figure 3.49 shows the standard DIP pin numbering system used by manufacturers to identify specific terminals and in schematic diagrams. Note that a groove
or dot on one end of the case identifies the top and that numbering proceeds down
the left side and up the right side.

Like discrete transistors and diodes, integrated circuits are given numbers for
identification purposes. However, the numbers do not have standard prefixes
such as 1N or 2N. A particular integrated circuit may be produced by several
different manufacturers and the identification number, which is usually printed on

Figure 3.48
*If conducting path AB must cross con-
ducting path CD without contacting it,
as shown in (a), it can be accomplished
as shown in (b), by routing path AB over
a resistor covered with SiO$_2$.*

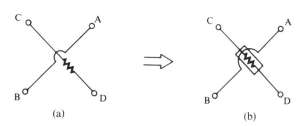

(a)   (b)

Figure 3.49
Integrated-circuit packages (courtesy RCA Corporation).

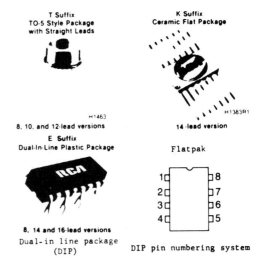

the case, will have certain additional letters or codes to identify the manufacturer. For example, the 741 monolithic IC operational amplifier is manufactured by RCA as a CA741 and by National Semiconductor as an LM741.

Example 3.11

SPICE

The *forward Early voltage* (named after J. M. Early) is a BJT parameter used by SPICE to determine how steeply the collector characteristics rise with increasing values of V_{CE}. It is the (absolute value of) the voltage at which the curves of constant I_B would intersect on the V_{CE}-axis if they were projected "backwards" (to the left), as shown in Figure 3.50. Note that the smaller the value of the forward Early voltage (*VAF*), the more steeply the characteristics rise. The default value of *VAF* in SPICE is infinite, for which value the curves are perfectly horizontal.

Use SPICE to obtain a set of collector characteristics for an NPN transistor having $VAF = 200$ V and an ideal maximum forward beta (*BF*) of 100. The characteristics should be plotted for V_{CE} ranging from 0 V to 50 V in 5-V steps and for I_B ranging from 0 to 40 μA in 10-μA steps. Use the results to determine the actual value of β at $V_{CE} = 25$ V and $I_B = 30$ μA.

Solution. Figure 3.51(a) shows the circuit and the SPICE input file. Note that a constant-current source is used to supply base current. The transistor .MODEL statement specifies that $\beta = 100$ and that the forward Early voltage VAF = 200 V. The .DC statement causes VCE to be stepped in 5-V increments from 0 through 50 V and I_B to be stepped in 10-μA increments from 0 through 40 μA.

The plot generated by SPICE is shown in Figure 3.51(b). This presentation is somewhat different than that shown in Figure 3.8, since the individual curves corresponding to different values of I_B must be displaced vertically by the printer. The curves have been filled in between the plotted points (asterisks) and labeled with the values of I_B to clarify the presentation. At $V_{CE} = 25$ V and $I_B = 30$ μA,

Figure 3.50
The forward Early voltage, VAF, is the intersection of the curves of the collector characteristics and determines how steeply the curves rise with increasing V_{CE}.

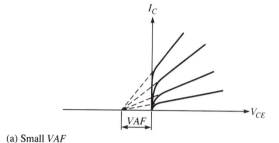

(a) Small *VAF*

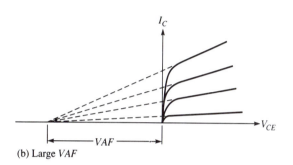

(b) Large *VAF*

we see that I(VDUM), which equals I_C, is 3.636 mA. Thus,

$$\beta = \frac{I_C}{I_B} = \frac{3.636 \text{ mA}}{30 \ \mu A} = 121.2$$

Note how the curves rise for increasing values of V_{CE} due to the effect of the forward Early voltage, meaning that β increases with increasing V_{CE} (and with increasing I_B).

Figure 3.51
(Example 3.11)

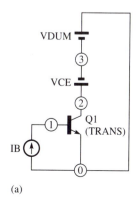

```
EXAMPLE 3.11
IB 0 1
VDUM 0 3
VCE 2 3
Q1 2 1 0 TRANS
.MODEL TRANS NPN BF=100 VAF=200
.DC VCE 0 50 5 IB 0 40U 10U
.PLOT DC I(VDUM)
.END
```

(a)

```
EXAMPLE 3.11
****      DC TRANSFER CURVES            TEMPERATURE =   27.000 DEG C
**********************************************************************************
     VCE          I(VDUM)
                   -2.000D-03    0.000D+00    2.000D-03    4.000D-03  6.000D-03
                   - - - - - - - - - - - - - - - - - - - - - - - - - - - - - -
  0.000D+00     7.026D-31  .
  5.000D+00     4.802D-10  .
  1.000D+01     1.024D-09  .
  1.500D+01     1.594D-09  .
  2.000D+01     2.189D-09  .
  2.500D+01     2.810D-09  .
  3.000D+01     3.456D-09  .
  3.500D+01     4.127D-09  .
  4.000D+01     4.824D-09  .
  4.500D+01     5.546D-09  .
  5.000D+01     6.293D-09  .
  0.000D+00    -9.901D-06  .
  5.000D+00     1.021D-03  .
  1.000D+01     1.046D-03  .
  1.500D+01     1.071D-03  .
  2.000D+01     1.096D-03  .
  2.500D+01     1.121D-03  .
  3.000D+01     1.146D-03  .
  3.500D+01     1.171D-03  .
  4.000D+01     1.196D-03  .
  4.500D+01     1.221D-03  .
  5.000D+01     1.246D-03  .
  0.000D+00    -1.980D-05  .
  5.000D+00     2.042D-03  .
  1.000D+01     2.092D-03  .
  1.500D+01     2.142D-03  .
  2.000D+01     2.192D-03  .
  2.500D+01     2.242D-03  .
  3.000D+01     2.292D-03  .
  3.500D+01     2.342D-03  .
  4.000D+01     2.392D-03  .
  4.500D+01     2.442D-03  .
  5.000D+01     2.492D-03  .
  0.000D+00    -2.970D-05  .
  5.000D+00     3.063D-03  .
  1.000D+01     3.138D-03  .
  1.500D+01     3.213D-03  .
  2.000D+01     3.288D-03  .
  2.500D+01     3.363D-03  .
  3.000D+01     3.438D-03  .
  3.500D+01     3.513D-03  .
  4.000D+01     3.588D-03  .
  4.500D+01     3.663D-03  .
  5.000D+01     3.738D-03  .
  0.000D+00    -3.960D-05  .
  5.000D+00     4.084D-03  .
  1.000D+01     4.184D-03  .
  1.500D+01     4.284D-03  .
  2.000D+01     4.384D-03  .
  2.500D+01     4.484D-03  .
  3.000D+01     4.584D-03  .
  3.500D+01     4.684D-03  .
  4.000D+01     4.784D-03  .
  4.500D+01     4.884D-03  .
  5.000D+01     4.984D-03  .
```

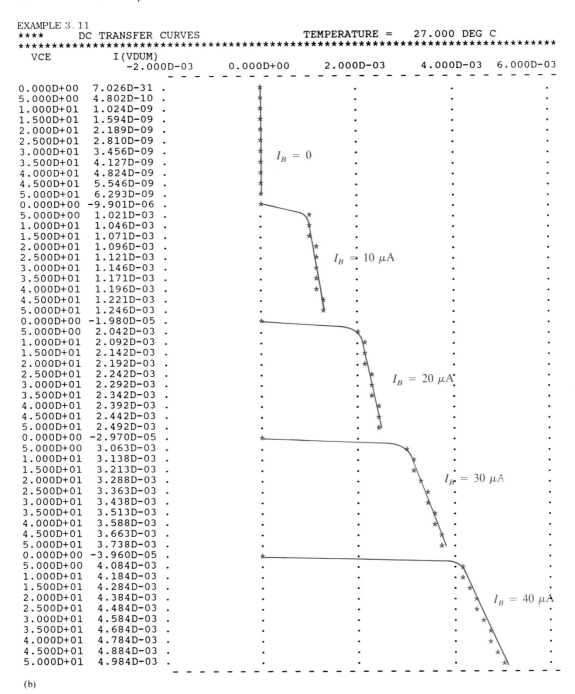

$I_B = 0$

$I_B = 10\ \mu A$

$I_B = 20\ \mu A$

$I_B = 30\ \mu A$

$I_B = 40\ \mu A$

(b)

Figure 3.51
(Continued)

Example 3.12

PSPICE	Use PSpice to obtain a set of collector characteristics for the 2N2222A transistor over the same range of values of V_{CE} and I_B as in Example 3.11.

Solution. The 2N2222A transistor is in the *library* of the evaluation (student) version of PSpice and can be referred to directly in the input data file, as shown below. The value of *VAF* used by PSpice for the 2N2222A is 74.03 V. In PSpice it is not necessary to insert a dummy voltage source to obtain values of collector current, since IC(Q1) identifies collector current in transistor Q1. Otherwise, the input data file is the same as that for Example 3.11:

```
EXAMPLE 3.12
IB 0 1
VCE 2 0
Q1 2 1 0 Q2N2222A
.LIB EVAL.LIB
.DC VCE 0 50 5 IB 0 40U 10U
.PLOT DC IC (Q1)
.END
```

Execution of the program produces a set of characteristic curves similar to those in Figure 3.51. However, since the value of *VAF* is smaller than in Example 3.11 (74.03 V versus 200 V), the curves rise more steeply at higher values of I_B.

Example 3.13

SPICE	Use SPICE to obtain a plot of the transfer characteristic of an N-channel JFET having $I_{DSS} = 10$ mA and $V_p = -2$ V. The characteristic should be plotted for $V_{DS} = 10$ V.

Solution. Figure 3.52(a) shows a SPICE circuit that can be used to obtain the desired characteristic. The value of BETA in the .MODEL statement is found from

$$\beta = \frac{I_{DSS}}{V_p^2} = \frac{10 \times 10^{-3} \text{ A}}{(-2 \text{ V})^2} = 2.5 \times 10^{-3} \text{ A/V}^2$$

Note that we step VGS from 0 to -2 V in 0.1-V increments. The voltage source labeled VGS has its positive terminal connected to the gate of the FET, but the stepped voltages should all be negative. (Alternatively, we could reverse the polarity of VGS and step it through positive voltages.) The voltage source labelled VIDS is a dummy source used to obtain positive values of drain current. The .PLOT statement produces a plot of (VIDS) (i.e., I_{DS}) versus V_{GS}, which constitutes a transfer characteristic. (In PSpice, the dummy voltage source is not required, and we can request a plot of ID(JX).)

Figure 3.52(b) shows the plot produced by SPICE. (Rotate it 180° to obtain the orientation shown in Figure 3.20.) Note that $I_D = 10$ mA $= I_{DSS}$ when $V_{GS} = 0$

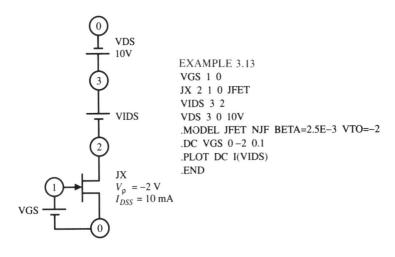

EXAMPLE 3.13
VGS 1 0
JX 2 1 0 JFET
VIDS 3 2
VDS 3 0 10V
.MODEL JFET NJF BETA=2.5E–3 VTO=–2
.DC VGS 0 –2 0.1
.PLOT DC I(VIDS)
.END

(a)

```
EXAMPLE 3.13
****         DC  TRANSFER  CURVES                    TEMPERATURE =     27.000 DEG C
*******************************************************************************************
   VGS              I(VIDS)
                    0.000D+00      5.000D-03       1.000D-02      1.500D-02   2.000D-02
                - - - - - - - - - - - - - - - - - - - - - - - - - - - - - - - - - -
0.000D+00     1.000D-02  .                               *                .           .
1.000D-01     9.025D-03  .                          *    .                .           .
2.000D-01     8.100D-03  .                      *        .                .           .
3.000D-01     7.225D-03  .                   *           .                .           .
4.000D-01     6.400D-03  .                *              .                .           .
5.000D-01     5.625D-03  .             *                 .                .           .
6.000D-01     4.900D-03  .          *                    .                .           .
7.000D-01     4.225D-03  .       *   .                   .                .           .
8.000D-01     3.600D-03  .     *     .                   .                .           .
9.000D-01     3.025D-03  .   *       .                   .                .           .
1.000D+00     2.500D-03  .  *        .                   .                .           .
1.100D+00     2.025D-03  . *         .                   .                .           .
1.200D+00     1.600D-03  .  *        .                   .                .           .
1.300D+00     1.225D-03  . *         .                   .                .           .
1.400D+00     9.000D-04  . *         .                   .                .           .
1.500D+00     6.250D-04  . *         .                   .                .           .
1.600D+00     4.000D-04  .*          .                   .                .           .
1.700D+00     2.250D-04  .*          .                   .                .           .
1.800D+00     1.000D-04  *           .                   .                .           .
1.900D+00     2.500D-05  *           .                   .                .           .
2.000D+00     1.201D-11  *           .                   .                .           .
                - - - - - - - - - - - - - - - - - - - - - - - - - - - - - - - - - -
```

(b)

Figure 3.52
(Example 3.13)

and that $I_D \approx 0$ when $V_{GS} = V_p$. Although the values of V_{GS} printed down the left margin should all be negative, SPICE prints only the absolute values of the independent variable associated with a .PLOT statement, which is V_{GS} in this example.

EXERCISES

Section 3.2 BJT Structure and Biasing

3.1   What type of charge carrier (hole or electron) is the majority carrier in
        a. the emitter region of an NPN transistor?
        b. the collector region of a PNP transistor?

3.2   For linear applications,
        a. which junction in an NPN transistor should be forward biased?
        b. which junction in a PNP transistor should be reverse biased?

3.3   For linear applications, what should be the polarity of the voltage
        a. from base to emitter in an NPN transistor?
        b. from collector to base in a PNP transistor?
        c. from collector to base in an NPN transistor?

3.4   What type of current (drift or diffusion) occurs
        a. in the collector region of a BJT?
        b. in the emitter region of a BJT?
        c. across the collector–base junction of a BJT?

3.5   Draw the schematic symbols for an NPN transistor and a PNP transistor. Show how external dc voltage sources should be connected to each to provide proper biasing. Draw arrows showing the direction of conventional current flow between the collector and emitter.

3.6   If the base voltage on an NPN transistor is +5 V with respect to ground and the emitter voltage is +6 V with respect to ground, is the base–emitter junction properly biased for linear applications? Explain.

3.7   Assuming each transistor is properly biased for linear applications, what is the approximate value of V_{BE} in
        a. a silicon NPN transistor?
        b. a germanium PNP transistor?

3.8   If the junction voltages in each of the following transistors have the values listed, which transistors are properly biased for linear applications?
        a. Si NPN; $V_{BE} = -0.7$ V, $V_{CB} = 10$ V.
        b. Ge PNP; $V_{BE} = -0.3$ V, $V_{CB} = -6$ V.
        c. Ge NPN; $V_{BE} = 0.3$ V, $V_{CB} = 9$ V.
        d. Si PNP; $V_{BE} = -0.7$ V, $V_{CB} = 12$ V.

3.9   An NPN transistor has $I_C = 8$ mA and $I_E = 8.01$ mA.
        a. What is the value of I_B?
        b. Draw a schematic symbol for the transistor and draw conventional current arrows labeled with the values of I_C, I_E, and I_B.

3.10   A PNP transistor has $I_E = 15$ mA and $I_B = 100$ μA.
        a. What is the value of I_C?
        b. Draw a schematic symbol for the transistor and draw conventional current arrows labeled with the values of I_C, I_E, and I_B.

Section 3.3 BJT Characteristics

3.11   Define fully the quantity represented by the term I_{CBO}.

3.12   Write the symbol used to represent the collector-to-emitter current in a BJT when the base is open. What type of current is it?

3.13   The collector current in a BJT is 7.6 mA and the base current is 40 μA. Find the value of α for the transistor.

3.14   If 97.8% of the holes entering the base of a PNP transistor cross into the collector, what is the value of α for the transistor?

3.15   The emitter current in a transistor is 25 mA and the collector current is 24.75 mA. What is the value of β for the transistor?

3.16   If 1% of the electrons entering the base of an NPN transistor recombine with holes in the base, what is the value of β for the transistor?

3.17   Using equation 3.4, derive equation 3.5: $\alpha = \beta/(\beta + 1)$.

3.18   The β of a certain type of transistor may range from 100 to 225. What is the corresponding range in the value of α?

3.19   If a transistor has $\beta = 100$, what is the percent error in the value of I_E if it is assumed that $I_E = I_C$?

$$\left(\% \text{ error} = \frac{\text{true } I_E - \text{assumed } I_E}{\text{true } I_E} \times 100\%\right)$$

        Hint: Use equation 3.7.

3.20   Using equation 3.7, derive the equation $I_E = I_B/(1 - \alpha)$.

3.21   For the transistor whose collector characteristics are shown in Figure 3.53, find
        a. the value of I_C when $V_{CE} = 10$ V and $I_B = 50$ μA;
        b. the value of β when $V_{CE} = 30$ V and $I_B = 150$ μA.

3.22   For the transistor whose collector characteristics are shown in Figure 3.53, find the

Figure 3.53
(Exercises 3.21–3.24)

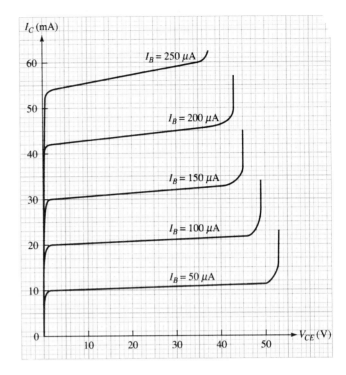

change in I_C when I_B changes from 50 μA to 100 μA while V_{CE} is held constant at 10 V.

3.23 For the transistor whose collector characteristics are shown in Figure 3.53, find
  a. the value of β_{dc} at V_{CE} = 20 V and I_B = 250 μA;
  b. the value of β_{ac} at V_{CE} = 30 V and I_B = 150 μA;
  c. the approximate voltage V_{CE} at which the transistor *begins* to break down when I_B = 150 μA.

3.24 For the transistor whose collector characteristics are shown in Figure 3.53, find
  a. the approximate value of β_{dc} when V_{CE} = 25 V and I_B = 75 μA; (It will be necessary to *interpolate*.)
  b. the approximate value of β_{ac} at V_{CE} = 20 V and I_B = 225 μA.
  c. the value of V_{CE} at which the transistor is fully into breakdown when I_B = 150 μA.

3.25 Name the four operating regions that can be defined on a set of collector characteristics and briefly describe the properties of the BJT in each.

3.26 State the criterion for linearity of a BJT in terms of its value of β_{ac}.

3.27 When the base current in a certain BJT changes from 10 μA to 40μA, the collector current changes from 1.5 mA to 6 mA. What are the quiescent values of I_B and I_C?

3.28 Define and give two other names for quiescent point.

Section 3.4 The Junction Field-Effect Transistor (JFET)

3.29 In terms of charge carriers, what is the principal difference between a BJT and an FET? In terms of the principle of operation, what is the significance of the word that the letter F stands for in an FET?

3.30 What are the two types of JFETs? What are the three terminals of a JFET named?

3.31 What type of charge carrier flows in an N-channel JFET? Name the terminals from and to which the carriers flow.

3.32 Between which two regions of a JFET is there a PN junction? How is the junction biased in normal operation?

3.33 For normal operation, what should be the polarity (positive or negative) of V_{GS} in
  a. an N-channel JFET
  b. a P-channel JFET?

3.34 What is the effect on the resistance of the channel of a JFET when the depletion regions expand in size? What causes the depletion regions to expand in size?

3.35 Write the symbol used to represent the current between drain and source when the gate is connected directly to the source.

3.36 What is the polarity of V_p in
a. an N-channel JFET?
b. a P-channel JFET?

3.37 The magnitude of the pinch-off voltage in a certain JFET is 3.5 V and the saturation current is 10 mA. Find the drain current when $V_{DS(sat)} = 2.5$ V.

3.38 At what value of $V_{DS(sat)}$ will the drain current of the JFET in Exercise 3.37 equal 4 mA?

3.39 For the JFET whose drain characteristics are shown in Figure 3.54, find approximate values for
a. the drain current when $V_{DS} = 10$ V and $V_{GS} = -2$ V;
b. the values of V_p and I_{DSS}.

3.40 For the JFET whose drain characteristics are shown in Figure 3.54, find approximate values for
a. the drain current when $V_{DS} = 15$ V and $V_{GS} = -1$ V;
b. the channel resistance when $V_{GS} = 0$ V and $V_{DS} = 1$ V.

3.41 Write the equation for the transfer characteristic of the JFET whose drain characteristics are shown in Figure 3.54. (Your equations should contain the numerical values of all constants used in the general equation of a transfer characteristic.)

3.42 Use the drain characteristics in Figure 3.54 to construct a transfer characteristic for the JFET when $V_{DS} = 5$ V.

3.43 An N-channel JFET has $I_{DSS} = 13$ mA and $V_p = -4.5$ V.
a. Find the drain current when $V_{GS} = -2.5$ V.

Figure 3.54
(Exercises 3.39 to 3.42)

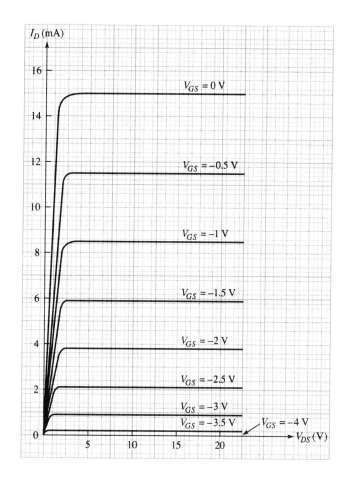

b. At what value of V_{GS} does the drain current equal
  i. 13 mA?
  ii. 6.5 mA?
  iii. 0 mA?

3.44 A P-channel JFET has $I_{DSS} = 12.5$ mA and $V_p = 5$ V.
  a. Find the drain current when $V_{GS} = 3$ V.
  b. At what value of V_{GS} does the drain current equal
    i. 12.5 mA?
    ii. 10 mA?
    iii. 0 mA?

3.45 Using the JFET transfer characteristic shown in Figure 3.55,
  a. find the values of I_{DSS} and V_p;
  b. find the approximate value of the drain current when the gate-to-source voltage is −2 V;
  c. write the equation for the transfer characteristic and use it to solve (b).

3.46 Using the JFET transfer characteristic shown in Figure 3.55,
  a. find the values of I_{DSS} and V_p;
  b. find the approximate value of the gate-to-source voltage for which the drain current equals 4 mA;
  c. write the equation for the transfer characteristic and use it to solve (b).

3.47 Refer to the transfer characteristic in Figure 3.56.
  a. What type of JFET has this characteristic?
  b. Find the values of I_{DSS} and V_p.
  c. Find the approximate value of V_{GS} for which $I_D = 4$ mA.
  d. Write the equation for the transfer characteristic and use it to solve (c).

3.48 Refer to the transfer characteristic in Figure 3.56.
  a. Find the approximate value of I_D when $V_{GS} = 0.5V_p$.
  b. Write the equation for the transfer characteristic and use it to solve (a).

3.49 Use the transfer characteristic in Figure 3.55.
  a. Graphically determine the approximate value of g_m when $V_{GS} = -1$ V.
  b. Find the value of g_m analytically when $V_{GS} = -1$ V.
  c. Approximately what change in I_D will occur when the quiesecent value of V_{GS} is −1 V and the total change in V_{GS} is 0.3 V?

3.50 Use the transfer characteristic shown in Figure 3.56.
  a. Graphically determine the approximate value of g_m when $I_D = 4$ mA.

Figure 3.55
(Exercises 3.45, 3.46, and 3.49)

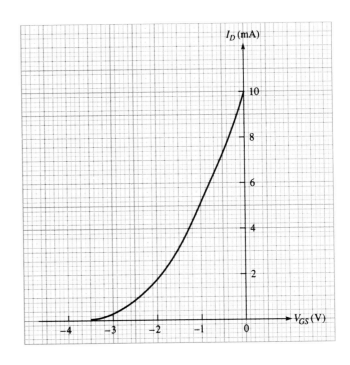

Figure 3.56
(Exercises 3.47, 3.48, and 3.50)

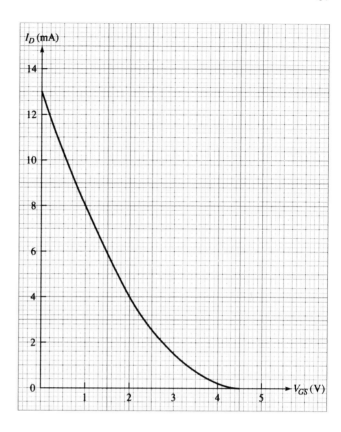

b. Find the value of g_m analytically when $I_D = 4$ mA.
c. Find the value of g_{mo} analytically.
d. Approximately what change in I_D will occur when the quiescent drain current is 4 mA and the total change in V_{GS} is 0.4 V?

3.51 Use equations 3.11 and 3.13 to derive equation 3.14:

$$g_m = \frac{2\,I_{DSS}}{|V_p|}\sqrt{\frac{I_D}{I_{DSS}}}$$

3.52 What is the value of the transconductance of a JFET when
a. $V_{GS} = 0$ V?
b. $V_{GS} = V_p$?

Section 3.5 The Metal-Oxide-Semiconductor FET (MOSFET)

3.53 In terms of its construction, what is the principle difference between a MOSFET and a JFET? What is another name for a MOSFET?

3.54 Briefly describe the meaning and origin of the prefix *MOS* in the term *MOSFET*.

3.55 Briefly explain how and why a depletion-type MOSFET can be operated in an enhancement mode.

3.56 What should be the polarity of the gate-to-source voltage in order to operate
a. an N-channel depletion-type MOSFET in the enhancement mode?
b. a P-channel depletion-type MOSFET in the depletion mode?

3.57 An N-channel depletion-type MOSFET has $I_{DSS} = 15$ mA and $V_p = 3.75$ V. Assuming it is operated in the pinch-off region, find I_D when
a. $V_{GS} = -1$ V;
b. $V_{GS} = 0$ V;
c. $V_{GS} = +1$ V.

3.58 A P-channel depletion-type MOSFET has $V_p = 3$ V. What value of V_{GS} will make the drain current equal twice the value of I_{DSS}?

3.59 In terms of its construction, what is the principle difference between an N-channel depletion-type MOSFET and an N-channel enhancement-type MOSFET?

3.60 For normal operation, what should be the polarity of the gate-to-source voltage of
a. an enhancement-type N-channel MOSFET?
b. an enhancement-type P-channel MOSFET?

3.61 Using the drain characteristics in Figure 3.32, find an approximate value for I_D when $V_{DS} = 10$ V and $V_{GS} = 9$ V.

3.62 Using the drain characteristics in Figure 3.32, determine the approximate value of $V_{DS(sat)}$ when $V_{GS} = 10$ V
a. graphically;
b. analytically.

3.63 Using the transfer characteristic in Figure 3.35, find the approximate value of the transconductance when $V_{GS} = 7$ V
a. graphically;
b. analytically.

3.64 A MOSFET having $V_T = -2.5$ V and $\beta = 0.55 \times 10^{-3}$ A/V^2 is biased at a Q-point where $I_D = 8$ mA. What change in V_{GS} is required to create a variation of ± 1 mA in I_D?

Section 3.6 Integrated-Circuit Technology

3.65 What are two major advantages of integrated circuits in comparison to discrete circuits? Can you think of an application in which a discrete circuit would be more appropriate than an integrated circuit? Explain.

3.66 What is the difference between polycrystalline silicon and the silicon used in integrated-circuit fabrication? Briefly describe one process used to convert one type to the other.

3.67 Integrated-circuit manufacturing involves the production of *ingots*, *wafers*, and *chips*. How are these related to each other? What is another name for a chip?

3.68 How is impurity diffusion different from carrier diffusion? What environmental requirement is necessary to achieve impurity diffusion?

3.69 What material is used to shield a silicon crystal from impurity diffusion? How is this material applied to the surface of a wafer?

3.70 Briefly define and/or describe the nature of each of the following terms and discuss its relation to batch processing using photolithographic methods:
a. window
b. photoresist
c. mask
d. hydrofluoric acid
e. metallization

3.71 Name a principal material used in each of the following processes:
a. oxidation
b. etching
c. P-layer diffusion
d. N-layer diffusion
e. metallization

3.72 Name and describe a way of creating impurity layers in a crystal without using impurity diffusion. What are the principal advantages of this method in comparison to impurity diffusion? What is one disadvantage?

3.73 What is an epitaxial layer? Describe one way in which it can be created.
What is a buried layer and what is its purpose? When and how is it formed?

3.74 The conductivity of the P material used to construct the diffused resistor shown in Figure 3.57 is 1250 S/m.
a. Find the sheet resistance of the diffused layer.
b. Find the resistance of the resistor.

3.75 The diffused layer used to construct the resistor shown in Figure 3.58 has sheet resistance 50 Ω/square. What is the resistance of the resistor?

3.76 A diffused resistor is to be fabricated using P material having resistivity 4×10^{-4} $\Omega \cdot$ m. The thickness of the layer is 5 μm. If the resistor has width 0.025 mm and is to have total resistance 800 Ω, what should its total length be? What is its aspect ratio?

3.77 What technique is used to create an impurity layer having a very large sheet resistance? What capability of this technique makes high sheet resistance possible?

3.78 Describe two methods used to fabricate capacitors in monolithic integrated circuits.

3.79 Find the capacitance of a capacitor consisting of two conducting surfaces separated by a layer of silicon dioxide 1 μm thick. The surfaces have dimensions 80 μm $\times$ 80 μm and the permittivity (dielectric constant) of silicon dioxide is 6.6×10^{-11} F/m.

3.80 Draw a schematic diagram of the circuit corresponding to that shown in the cross-sectional view in Figure 3.59. (No buried layers are shown.)

Figure 3.57
(Exercise 3.74)

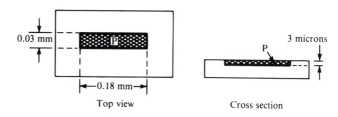

Top view   Cross section

Figure 3.58
(Exercise 3.75)

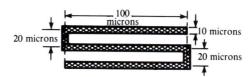

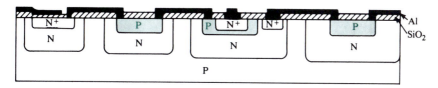

Figure 3.59
(Exercise 3.80)

SPICE EXERCISES

3.81 Use SPICE (or PSpice) to obtain a set of collector characteristics for a PNP transistor having an ideal maximum forward beta of 110 and forward Early voltage 150 V. Choose appropriate increments in stepping V_{CE} from 0 to −50 V and I_B from 0 to 40 μA. Use the results to determine the actual value of β_{dc} at $I_B = 20$ μA and $V_{CE} = -20$ V.

3.82 Use SPICE (or PSpice) to obtain a transfer characteristic for a P-channel JFET having $I_{DSS} = 18$ mA and $V_p = 3$ V. Use $V_{DD} = 18$ V. Use the results to determine the value of I_D when $V_{GS} = 1$ V and compare with the theoretical value predicted by equation 3.11.

3.83 Use SPICE (or PSpice) to obtain a set of drain characteristics for an induced N-channel (enhancement) MOSFET having $\beta = 0.5 \times 10^{-3}$ A/V^2 and $V_T = 2$ V. Choose appropriate increments in stepping V_{DS} from 0 to 18 V and V_{GS} from 2 to 10 V. Use the results to determine I_D at $V_{GS} = 5$ V and $V_{DS} = 9$ V and compare with the theoretical value predicted by equation 3.18.

Amplifier Fundamentals

4.1 INTRODUCTION

Amplifier Inputs and Outputs

An amplifier must supply, as output, the voltage, current, and power required by its *load* to perform some useful function. Examples of loads include loudspeakers, electric motors, antennas, instruments, data displays, and the input circuitry of other amplifiers. The input to an amplifier is a *signal source* that, by itself, is incapable of "driving" such a load. An example of a signal source is a microphone, which typically generates a few millivolts and which cannot therefore drive a speaker that may require several volts and consume from 5 or 10 to several hundred watts of power.

Implicit in the concept of amplification is the notion that input signals undergo *variations* and that the output should faithfully reproduce those variations at higher voltage and/or current levels. (*Fidelity* is the term used to describe the extent to which output variations resemble input variations, as in a *high-fidelity* audio system.) Thus, we study amplifiers from the standpoint of their ability to reproduce *ac* signals. Rarely, if ever, is an amplifier used simply to increase a fixed dc level. Even the input to a so-called dc amplifier used, for example, to drive a dc motor, will undergo some change from time to time.

Most electrical signals requiring amplification are complex waveforms having many different *frequency components*. For example, the signals representing speech and music generated by a microphone have frequency components in the range from 20 Hz to 20 kHz—the *audio-frequency range*. Such complex waveforms can be treated as the sum of a large number of *sine waves,* each sine wave having a different frequency, amplitude, and phase angle. (Theoretically, it is possible to reconstruct *any* waveform by summing an appropriate number of sine waves.) In our study of amplifiers, we will often consider signals to be *single*-frequency sine waves, rather than complex waveforms. However, since we are dealing with linear amplifiers, we can invoke the principle of superposition: The output of the amplifier is the sum of the outputs obtained by amplifying each single-frequency component of the input separately. Figure 4.1 illustrates this idea

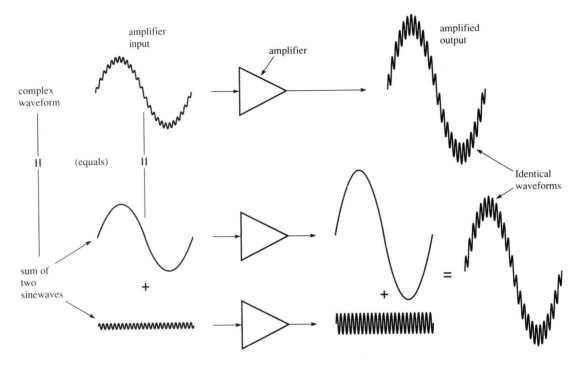

Figure 4.1
By the superposition principle, the output of an amplifier whose input is a complex waveform is the sum of the outputs obtained by amplifying each frequency component separately.

for an input signal that contains just two frequency components. Although we will not actually perform such summations, it is reassuring to know that we *could* do so, and to know that our ability to predict the behavior of an amplifier at a single frequency is sufficient to be able to predict it for a complex waveform containing many frequencies.

The Purpose of Bias

The input and output of many amplifiers, particularly single-transistor amplifiers, can have one polarity only. In other words, the input and output voltages must always be positive or must always be negative. In order to accommodate this requirement, the ac variations of the input and output are superimposed on dc levels, called *bias voltages*. Thus, for example, an ac variation of ±1 V could appear as a variation below and above a bias level of +5 V, so the total voltage, ranging from +4 V to +6 V, would always remain positive. Recall that a quiescent point, or Q-point, specifies the voltage about which output variations occur, and we see now why it is also called a bias point. Figure 4.2 shows an example of input and output waveforms having ac variations superimposed on dc bias voltages. By superimposed, we mean *added*, so we can represent an ac voltage superimposed on a dc level as a mathematical sum:

$$v(t) = V_{dc} + V_p\sin\omega t \qquad (4.1)$$

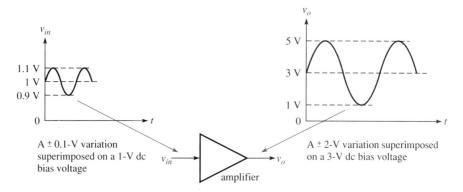

Figure 4.2
Examples of input and output waveforms having ac variations superimposed on dc (bias) voltages.

where V_{dc} is the bias voltage (quiescent voltage) and V_p is the peak value of the sinusoidal ac component having frequency $\omega = 2\pi f$ rad/s. Note that the voltage $v(t)$ varies between a minimum value of $V_{dc} - V_p$ and a maximum value of $V_{dc} + V_p$. V_{dc} is also called the *dc component* of the waveform and $V_p \sin \omega t$ the ac component.

Example 4.1

The ac component of the output voltage of an amplifier is $6\sin 1000t$ V.

1. If the output voltage of the amplifier cannot be less than $+2$ V, what is the minimum value that the output bias voltage can have?
2. If the bias voltage is made equal to the value found in (1), what is the maximum value of the output voltage? What is its average value?
3. What is the effective (rms) value of the ac component of the output?
4. What is the period of the ac component of the output?

Solution

1. The minimum value of the output voltage is $V_{dc} - V_p$. Since $V_p = 6$ V, we have

$$2 \text{ V} = V_{dc} - 6 \text{ V}$$
$$V_{dc} = 8 \text{ V}$$

2. The maximum value of the output voltage is $V_{dc} + V_p = 8 \text{ V} + 6 \text{ V} = 14 \text{ V}$. Thus, the output varies between 2 V and 14 V, a ± 6 V variation around 8 V dc.
   Recall that the average value of a waveform, also called the dc value, equals the value that a dc instrument would indicate when used to measure the waveform. Thus, in our case, the average value is the dc component, 8 V.
3. Since the peak value of the ac component is 6 V, the effective value is

$$\frac{\sqrt{2}}{2} (6 \text{ V}) = 4.24 \text{ V rms}$$

4.

$$\omega = 2\pi f = 1000 \text{ rad/s}$$

$$f = \frac{1000 \text{ rad/s}}{2\pi} = 159.15 \text{ Hz}$$

$$T = \frac{1}{f} = \frac{1}{159.15 \text{ Hz}} = 6.28 \text{ ms}$$

DC and AC Equivalent Circuits

When analyzing an amplifier in which a signal having both a dc and an ac component present, the equivalent circuits we use in the analysis may differ considerably, depending on whether we are solving for dc voltages and currents or for ac voltages and currents. Recall again that the principle of superposition in a linear circuit allows us to find the voltage or current anywhere in a circuit due to a single voltage source acting alone, i.e., with all other sources set to zero. Thus, if we wish to find only the ac voltage across a component, we can set all dc voltage sources to zero (replace them by short circuits). In amplifier circuits, it is almost always true that one side of every dc voltage source is connected to ground, so we are saying, in effect, that *dc voltage sources are ac short circuits to ground.* Similarly, ac voltage sources are dc short circuits. DC current sources are open circuits to ac, and ac current sources are open circuits to dc. The next example demonstrates that these facts lead to quite different dc and ac equivalent circuits. In the context of amplifiers, an ac equivalent circuit is often called a *small-signal equivalent circuit.*

Example 4.2

The circuit in Figure 4.3(a) is designed to produce an ac voltage having a dc component across the 6-kΩ resistor.

1. Find the ac voltage across the 6-kΩ resistor.
2. Find the dc voltage across the 6-kΩ resistor.
3. Find the total voltage (dc + ac) across the 6-kΩ resistor.

Solution

1. As far as the ac source is concerned, the 14-V dc source appears as a short circuit to ground. Thus, for the ac analysis, the 3-kΩ and 6-kΩ resistors are in parallel, as shown in Figure 4.3(b). By the voltage-divider rule, we see that the ac component of the voltage across the 6-kΩ resistor is

$$\left(\frac{2 \text{ k}\Omega}{2 \text{ k}\Omega + 12 \text{ k}\Omega}\right)(2.8\sin\omega t) = \left(\frac{2 \text{ k}\Omega}{14 \text{ k}\Omega}\right)(2.8\sin\omega t) = 0.4\sin\omega t \text{ V}$$

2. As far as the dc source is concerned, the ac source appears as a short circuit to ground. Thus, for the dc analysis, the 6-kΩ and 12-kΩ resistors are in parallel, as shown in Figure 4.3(c). By the voltage-divider rule, we see that the dc component of the voltage across the 6-kΩ resistor is

$$\left(\frac{4 \text{ k}\Omega}{4 \text{ k}\Omega + 3 \text{ k}\Omega}\right)14 \text{ V dc} = \left(\frac{4 \text{ k}\Omega}{7 \text{ k}\Omega}\right)14 \text{ V dc} = 8 \text{ V dc}$$

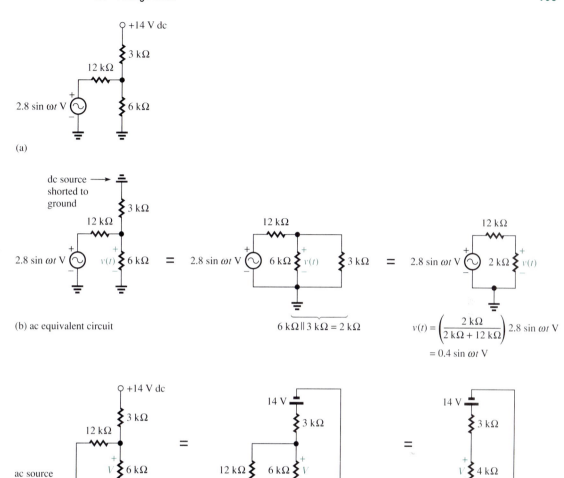

Figure 4.3
(Example 4.2)

3. By the superposition principle, the total voltage across the 6-kΩ resistor is the sum of the dc and ac components:

$$V(t) = 8 + 0.4 \sin \omega t \text{ V}$$

4.2 VOLTAGE GAIN

The voltage gain of an amplifier, A_v, is the ratio of the ac component of its output voltage to the ac component of its input voltage. The *magnitude* of the voltage gain $|A_v|$ (which excludes any consideration of the phase relationship between input and output) is the ratio of the magnitude of the ac output voltage to the magnitude of the ac input voltage. Voltage magnitudes can be expressed in terms of effective values, V(rms), peak values, V(pk), or peak-to-peak values, V(pk-pk):

$$|A_v| = \frac{V_o(\text{rms})}{V_{in}(\text{rms})} = \frac{V_o(\text{pk})}{V_{in}(\text{pk})} = \frac{V_o(\text{pk-pk})}{V_{in}(\text{pk-pk})} \qquad (4.2)$$

where the subscripts o and in refer to output and input values, respectively. It must be emphasized that all quantities in equation 4.2 are values of *ac components* only; be certain to exclude dc bias values in all computations and measurements. Note that gain is a dimensionless quantity, since the units of numerator and denominator are the same.

Example 4.3

The input signal to an amplifier is $v_{in}(t) = 0.7 + 0.5\sin 10^6 t$ V and the output voltage is $v_o(t) = 6 + 3\sin 10^6 t$ V.

1. Find the magnitude of the voltage gain.
2. Assuming the amplifier is linear, what is the peak-to-peak value of the output voltage when the input is changed to $v_{in}(t) = 0.8 + 0.6\sin 10^6 t$?
3. What are the input and output bias voltages (of the original, unchanged signals)?

Solution

1. Using equation 4.2, with $V_{in}(\text{pk}) = 0.05$ V and $V_o(\text{pk}) = 3$ V, we find

$$|A_v| = \frac{3 \text{ V}}{0.05 \text{ V}} = 60$$

2. Recall that one of the properties of linearity is that the ratio of output to input (gain) does not depend on the value of input for which it is calculated. Consequently, the value of $|A_v|$ remains at 60 when the input changes to its new value. Since the new value of $V_{in}(\text{pk-pk})$ is $2(0.06 \text{ V}) = 0.12$ V, we have, from equation 4.2, $V_o(\text{pk-pk}) = |A_v|V_{in}(\text{pk-pk}) = 60(0.12 \text{ V}) = 7.2$ V.
3. Comparing $v_{in}(t)$ and $v_o(t)$ with equation 4.1: $v(t) = V_{dc} + V_p\sin\omega t$, we see that the input and output bias voltages are 0.7 V and 6 V, respectively.

Input and Output Resistance

The input resistance of an amplifier is the total equivalent resistance that the amplifier presents to a signal source when the source is connected to the input of the amplifier. Thus, input resistance, r_{in}, is input voltage divided by input current, which tells us one way we could measure its value: Connect a signal source to the input, measure the current i_{in} that flows into the amplifier, and divide voltage by current:

$$r_{in} = \frac{v_{in}}{i_{in}} \text{ ohms} \qquad (4.3)$$

Note that we use lower-case letters in our definition, with the usual implication that all quantites have ac (small-signal) values, including r_{in}. The dc input resistance, R_{in}, is dc input voltage divided by dc input current, and it, in general, has a different value from r_{in}:

$$R_{in} = \frac{V_{dc}}{I_{dc}} \text{ ohms} \qquad (4.4)$$

Example 4.4

The network in Figure 4.4 represents the equivalent circuit at the input of a certain amplifier. Assuming the capacitive reactance, $|X_C|$, of C is negligibly small at the frequency of operation, find

1. r_{in};
2. R_{in}.

Figure 4.4
(Example 4.4) Capacitor C bypasses the 1-kΩ resistor.

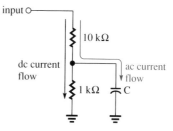

Solution

1. Since $|X_C|$ is negligibly small, the parallel combination of the 1-kΩ resistor and C essentially equals 0 ohms *for ac signals*. (C is said to be a *bypass* capacitor for the 1-kΩ resistor; the ac signal is bypassed to ground through the capacitor.) Therefore, the total ac resistance at the input terminal is r_{in} = 10 kΩ + 0 Ω = 10 kΩ.
2. Since the capacitor is an *open circuit* to dc, R_{in} = 10 kΩ + 1 kΩ = 11 kΩ.

The output resistance of an amplifier is the equivalent internal resistance that is effectively in series with the output of the amplifier. Like input resistance, output resistance can have both a dc and an ac value. We will be primarily concerned with the ac value, r_o. Output resistance is the Thevenin equivalent resistance of the amplifier with respect to the output terminals. (Recall that this is the resistance measured at the output terminals when all internal sources are set to

Figure 4.5
Representation of input and output circuits of a voltage amplifier.

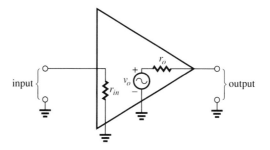

zero.) We will discuss a practical method for measuring r_o presently. It is convenient to think of the output side of an amplifier in terms of the Thevenin equivalent circuit at its output terminals, since that circuit is simply a voltage source having v_o (output voltage) in series with r_o. Figure 4.5 shows how we can represent an amplifier having input and output resistance. (See page 107).

Example 4.5

An amplifier has $|A_v| = 50$. When the effective value of the ac input is 40 mV rms and the output of the amplifier is *short-circuited,* the ac current in the short circuit is found to be 20 mA rms. What is the output resistance of the amplifier?

Solution

$$V_o(\text{rms}) = |A_v|V_{in}(\text{rms}) = 50(40 \text{ mV rms}) = 2 \text{ V rms}$$

Since the output of the amplifier is short-circuited, the only resistance limiting current flow is r_o (see Figure 4.5). Therefore,

$$r_o = \frac{2 \text{ V rms}}{20 \text{ mA rms}} = 100 \ \Omega$$

Despite the implication of the word *gain,* it is possible for the output voltage of an amplifier to have a smaller magnitude than the input voltage. In such a case we say that the amplifier *attenuates* (reduces) the signal voltage, and as a consequence, the value of $|A_v|$ is less than 1. For example, if $|A_v| = 0.5$, we know that the magnitude of the output voltage is one-half the magnitude of the input voltage. The "gain" of a resistive voltage divider is always less than (or at most equal to) 1. Referring to Figure 4.6, we see by the voltage-divider rule that

$$V_o = \left(\frac{R_2}{R_1 + R_2}\right) V_{in} \tag{4.4}$$

$$\frac{V_o}{V_{in}} = \frac{R_2}{R_1 + R_2} \leq 1 \tag{4.5}$$

Source and Load Resistance

Every signal source has its own internal resistance, which we will call *source resistance* and designate r_S. When analyzing amplifier circuits, we can represent a signal source as a voltage source, v_S, in series with its resistance, r_S. (v_S and r_S can be thought of as the Thevenin equivalent values of the source.) Since there is a vast variety of electronic devices that could (and do) serve as signal sources for amplifiers, practical values of source resistance may range from a few tenths of an

Figure 4.6
A resistive voltage divider has "gain"
$R_2/(R_1 + R_2)$ *which is always* ≤ 1.

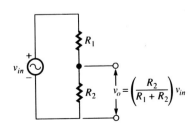

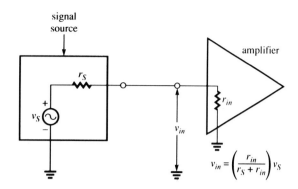

Figure 4.7
Signal-source resistance r_S and input resistance r_{in} form a resistive voltage divider, making $v_{in} \leq v_S$.

ohm to several megohms. One amplifier may serve as the signal source for another, in which case the source resistance is the output resistance of the amplifier serving as the source. A typical value for the source resistance of a laboratory-type function generator is 50 Ω.

Figure 4.7 shows a signal source connected to the input of an amplifier having input resistance r_{in}. Note that r_S and r_{in} are in *series*. Consequently, they form a voltage divider, and the voltage v_{in} at the amplifier input is *smaller* than the value of v_S. By voltage-divider action,

$$v_{in} = \left(\frac{r_{in}}{r_s + r_{in}}\right) v_S$$

$$\frac{v_{in}}{v_S} = \frac{r_{in}}{r_S + r_{in}}$$

(4.6)

The load resistance, R_L, connected to the output of an amplifier is the total equivalent resistance presented by the load to the amplifier. R_L could be, for example, the input resistance of another amplifier. Figure 4.8 shows a load resistance connected to the output of an amplifier having output resistance r_o. Note that R_L and r_o are in series and that the load voltage, v_L, is thus smaller than the output voltage of the amplifier:

$$v_L = \left(\frac{R_L}{r_o + R_L}\right) v_o$$

$$\frac{v_L}{v_o} = \frac{R_L}{r_o + R_L}$$

(4.7)

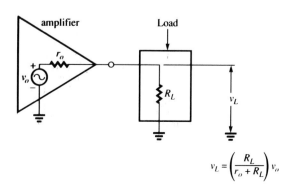

Figure 4.8
Output resistance r_o and load resistance R_L form a resistive voltage divider, making $v_L \leq v_o$.

Figure 4.9
Measuring the output resistance r_o of an amplifier. $r_o = R_L$ when the measured voltage equals 0.5 v_o.

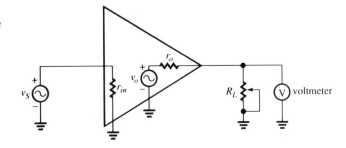

With the help of equation 4.7, we can deduce a method for measuring the output resistance of an amplifier, a method that is commonly used for that purpose. With an input signal applied to the amplifier, we first measure the output voltage when there is no load connected to the amplifier (output open-circuited). The measured voltage is v_o. We then connect an adjustable-resistance load to the output and measure the voltage across it. (Refer to Figure 4.9.) The resistance of the load is adjusted until the voltage across it equals $0.5v_o$ (one-half the open-circuit output voltage). The load is then removed and its resistance measured. The measured value of load resistance equals r_o, since, by equation 4.7, $v_L = 0.5v_o$ when $R_L = r_o$:

$$v_L = \left(\frac{R_L}{R_L + r_o}\right) v_o = \left(\frac{R_L}{R_L + R_L}\right) v_o = 0.5v_o$$

As a matter of practice, we should note that this method is not recommended for an amplifier that has a very small output resistance, since the heavy loading (large load current) that may be necessary to reduce the load voltage to $0.5v_o$ could damage the amplifier.

Like other resistances in an amplifier circuit, load resistance can have both a dc and an ac value. AC load resistance is the total equivalent resistance presented to an ac signal and is designated r_L. Calculation of its value is illustrated in the next example.

Example 4.6

The network in Figure 4.10 represents the equivalent circuit of the load connected to the output of an amplifier. Assuming the capacitive reactance, $|X_C|$, of the capacitor is negligibly small at the frequency of operation, find

1. the dc load resistance, R_L;
2. the ac load resistance, r_L.

Figure 4.10
(Example 4.6)

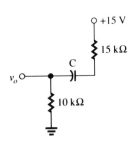

Solution
1. Since the capacitor is an open circuit to dc, the 15-kΩ resistor is isolated from the dc component of the output voltage and $R_L = 10$ kΩ.
2. Recall that a dc voltage source is an ac short circuit. Therefore, as far as the ac component of the signal is concerned, the 15-kΩ resistor is shorted to ground. Since $|X_C|$ is negligibly small, the 15-kΩ resistor appears in parallel with the 10-kΩ resistor, and

$$r_L = \frac{(10 \text{ k}\Omega)(15 \text{ k}\Omega)}{10 \text{ k}\Omega + 15 \text{ k}\Omega} = 6 \text{ k}\Omega$$

We have seen that there is a reduction in voltage between the signal source and input of an amplifier, and again between the output and load of the amplifier, due to the voltage divisions that occur at input and output. Consequently, the voltage gain *between source and load* (which is of primary interest, from a practical standpoint) is smaller than the voltage gain of the amplifier alone. Referring to Figure 4.11, we see that

$$v_L = v_o \left(\frac{R_L}{R_L + r_o}\right)$$

and, since $v_o = v_{in} A_v$,

$$v_L = v_{in} A_v \left(\frac{R_L}{R_L + r_o}\right)$$

Substituting $v_S \left(\dfrac{r_{in}}{r_S + r_{in}}\right)$ for v_{in} gives

$$v_L = v_S \left(\frac{r_{in}}{r_S + r_{in}}\right) A_v \left(\frac{R_L}{r_L + r_o}\right)$$

or,

$$\frac{v_L}{v_S} = \left(\frac{r_{in}}{r_S + r_{in}}\right) A_v \left(\frac{R_L}{r_L + r_o}\right)$$

Figure 4.11
The voltage gain between source and load, v_L/v_S, is A_v reduced by the voltage-division ratios at input and output.

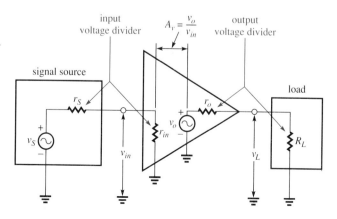

In terms of voltage magnitudes,

$$\frac{V_L}{V_S} = \underbrace{\frac{r_{in}}{r_S + r_{in}}}_{\substack{\text{voltage-division} \\ \text{ratio at input}}} |A_v| \underbrace{\frac{R_L}{r_o + R_L}}_{\substack{\text{voltage-division} \\ \text{ratio at output}}} \tag{4.8}$$

where V_L and V_S are the magnitudes of the load and source voltages (effective, peak, or peak-to-peak values). We will henceforth use the notation V with an appropriate subscript to represent the magnitude of the ac component of a particular voltage, without necessarily specifying it to be an effective, peak, or peak-to-peak value.

Example 4.7

The amplifier shown in Figure 4.12 has $|A_v| = 120$. Find

1. V_{in}/V_S;
2. V_L/V_o;
3. V_L/V_S;
4. V_L when $V_S = 25$ mV pk-pk.

Figure 4.12
(Example 4.7)

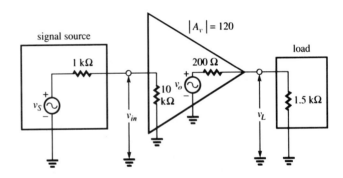

Solution

1. Voltage division at the input gives

$$V_{in} = \left(\frac{10 \text{ k}\Omega}{10 \text{ k}\Omega + 1 \text{ k}\Omega}\right) V_S = 0.909 \ V_S$$

$$\frac{V_{in}}{V_S} = 0.909$$

2. Voltage division at the output gives

$$V_L = \left(\frac{1.5 \text{ k}\Omega}{1.5 \text{ k}\Omega + 200 \ \Omega}\right) V_o = 0.882 \ V_o$$

$$\frac{V_L}{V_o} = 0.882$$

3. V_L/V_S is the amplifier gain, $|A_v|$, reduced by the two voltage-division ratios:

$$\frac{V_L}{V_S} = 0.909 \, |A_v| \, 0.882 = 0.803 \, |A_v| = 0.803(120) = 96.3$$

4. $V_L = 96.3 V_S = 96.3(25 \text{ mV pk-pk}) = 2.41 \text{ V pk-pk}$

To obtain the maximum possible voltage amplification between source and load, the voltage-division ratios at the input and output should have values as close as possible to 1, their maximum possible values. As far as input voltage division is concerned, we can deduce from equation 4.6, $v_{in}/v_S = r_{in}/(r_S + r_{in})$, that r_{in} should not have as large a value as possible, since

$$r_{in} \gg r_S \Rightarrow r_S + r_{in} \approx r_{in}$$

$$\Rightarrow \frac{r_{in}}{r_S + r_{in}} \approx \frac{r_{in}}{r_{in}} = 1$$

To illustrate, suppose $r_S = 1 \text{ k}\Omega$ and $r_{in} = 1 \text{ M}\Omega$. Then $r_{in}/(r_S + r_{in}) = 1 \text{ M}\Omega/1.001$ $\text{M}\Omega = 0.999 \approx 1$. On the other hand, if $r_{in} = r_S = 1 \text{ k}\Omega$, then $r_{in}/(r_S + r_{in}) = 0.5$, and we have already lost 50% of the overall amplification possible. Of course, the voltage-division ratio also equals 1 if $r_S = 0$, but in practice, we may not be able to control or specify values of source resistance. We conclude that a desirable property of a voltage amplifier is that it have as *large* an input resistance as possible (ideally infinite).

As far as output voltage division is concerned, we can deduce from equation 4.7, $v_L/v_o = R_L/(r_o + R_L)$, that r_o should have a small a value as possible, since

$$r_o \ll R_L \Rightarrow r_o + R_L \approx R_L$$

$$\Rightarrow \frac{R_L}{r_o + R_L} \approx \frac{R_L}{R_L} = 1$$

To illustrate, suppose $r_o = 1 \text{ }\Omega$ and $R_L = 100 \text{ }\Omega$. Then, $R_L/(r_o + R_L) = 100 \text{ }\Omega/101 \text{ }\Omega = 0.99 \approx 1$. Of course the output voltage-division ratio is also close to 1 if R_L is very large, but we may not be able to control that circumstance. We conclude that another desirable property of a voltage amplifier is that it have as *small* an output resistance as possible (ideally zero).

Example 4.8

An amplifier having input resistance 20 kΩ is driven from a fixed signal source that generates a 50-mV rms signal and that has source resistance 5 kΩ. The voltage gain of the amplifier is $|A_v| = 90$. In a certain application, the loads driven by the amplifier may vary from 50 Ω to 4 kΩ. What is the maximum permissible output resistance of the amplifier if it must deliver at least 3 V rms to every load?

Solution. When there is no load connected to the amplifier (output open-circuited), the output voltage of the amplifier is determined by voltage division at the input only:

$$V_o = \frac{r_{in}}{r_S + r_{in}} |A_v| \, V_S$$

$$= \left(\frac{20 \text{ k}\Omega}{5 \text{ k}\Omega + 20 \text{ k}\Omega}\right) 90(50 \text{ mV rms}) = 3.6 \text{ V rms}$$

When load R_L is connected to the output, voltage division at the output gives

$$V_L = \left(\frac{R_L}{R_L + r_o}\right) V_o = \left(\frac{R_L}{R_L + r_o}\right) 3.6 \text{ V rms}$$

The load voltage V_L will have its smallest value when R_L is smallest, namely, 50 Ω. For that case, we require that

$$V_L = \left(\frac{50 \text{ Ω}}{50 \text{ Ω} + r_o}\right) 3.6 \text{ V rms} = 3 \text{ V rms}$$

Solving for r_o gives $r_o = 10$ Ω.

Phase Inversion

In our discussion of voltage gain we have thus far disregarded phase relations and considered magnitude, $|A_v|$, only. In a more general formulation, voltage gain is treated as a *phasor* quantity that specifies both the magnitude of the gain and its phase. The phase of the voltage gain equals the phase angle of the output voltage relative to the input voltage. A negative phase means that the output lags the input and a positive phase means that the output leads. Until we study the effects of capacitance on amplifier performance (Chapter 6, Frequency Response), we will not generally be concerned with phase relations. However, there is one phase relationship that is so commonly encountered that it deserves mention here. When the output voltage of an amplifier is shifted in phase by 180° from the input voltage, the amplifier is said to *invert* the signal, or to perform *phase inversion*. The amplifier is called an inverting amplifier. The voltage gain of an inverting amplifier is expressed in phasor form as $|A_v| \underline{/180°}$ (which is equivalent to $|A_v| \underline{/-180°}$), or, more simply, as $-|A_v|$. For example, an amplifier having voltage gain -10 amplifies the magnitude of an input signal by 10 and shifts its phase by 180°. Figure 4.13 shows an example of input and output waveforms of an inverting amplifier. Notice that, except for magnitude, the output is the mirror image of the

Figure 4.13
An example of input and output wave-forms in an inverting amplifier having voltage gain $-|A_v|$*. The output is 180° out of phase with the input.*

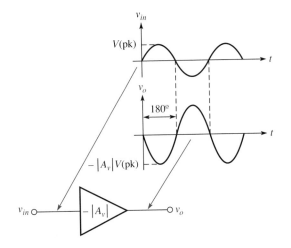

input: It reaches its negative peak when the input is at its positive peak and vice versa. The name *inverter* clearly stems from the fact that the output appears to be an upside-down version of the input.

4.3 CURRENT GAIN

When studying current relations in an amplifier, it is convenient to represent the output circuit of the amplifier as an equivalent current source (the Norton equivalent circuit). Recall that the equivalent current source has $i_o = v_o/r_o$, as shown in Figure 4.14(a). Figure 4.14(b) shows the amplifier with its output in current-source form. The current gain A_i of an amplifier is defined to be the ratio of the ac output current to the ac input current:

$$A_i = \frac{i_o}{i_{in}} \tag{4.9}$$

The magnitude of the current gain is the ratio of the magnitude (effective, peak, or peak-to-peak value) of the ac output current to the magnitude of the ac input current:

$$|A_i| = \frac{I_o(\text{rms})}{I_{in}(\text{rms})} = \frac{I_o(\text{pk})}{I_{in}(\text{pk})} = \frac{I_o(\text{pk-pk})}{I_{in}(\text{pk-pk})} \tag{4.10}$$

From Figure 4.14(b), we see that $v_{in} = i_{in}r_{in}$ and $v_o = i_o r_o$. Therefore,

$$i_{in} = \frac{v_{in}}{r_{in}} \quad \text{and} \quad i_o = \frac{v_o}{r_o}$$

Figure 4.14
Representation of the output of an amplifier as an equivalent current source.

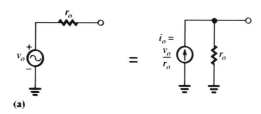

(a)

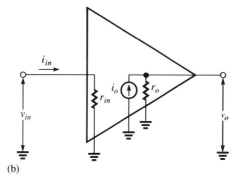

(b)

Note carefully that v_o is the *open-circuit* output voltage of the amplifier. Substituting for i_{in} and i_o in equation 4.9, we find

$$A_i = \frac{i_o}{i_{in}} = \frac{v_o/r_o}{v_{in}/r_{in}} = A_v \frac{r_{in}}{r_o}$$

or,

$$|A_i| = |A_v| \frac{r_{in}}{r_o} \tag{4.11}$$

Example 4.9

When the input current to an amplifier is $i_{in}(t) = 100 + 50\sin\omega t$ μA, the ac component of the output voltage is 15 V pk. If $r_{in} = 10$ kΩ and $r_o = 1.5$ kΩ, find

1. $|A_i|$;
2. $|A_v|$.

Solution

1.

$$I_o(\text{pk}) = \frac{V_o(\text{pk})}{r_o} = \frac{15\ \text{V}}{1.5\ \text{k}\Omega} = 10\ \text{mA pk}$$

Therefore,

$$|A_i| = \frac{I_o(\text{pk})}{I_{in}(\text{pk})} = \frac{10\ \text{mA}}{50\ \mu A} = 200$$

2. From equation 4.11,

$$|A_v| = \frac{|A_i|}{r_{in}/r_o} = \frac{200}{10\ \text{k}\Omega/1.5\ \text{k}\Omega} = 30$$

As in the case of an amplifier's output circuit, it is convenient to represent the signal source as an equivalent current source when studying current gain: $i_S = v_S/r_S$. Figure 4.15 shows the amplifier with current sources for both the output circuit and the signal source. Also shown are the input resistance of the amplifier and

Figure 4.15
Amplifier circuit with signal source and amplifier output represented by current sources.

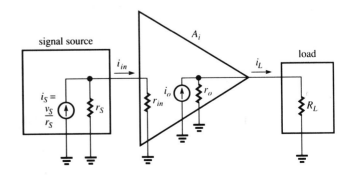

load resistance R_L. We see that r_S and r_{in} are in parallel and therefore form a *current divider*. Consequently, not all of the current produced by the source reaches the input to the amplifier. By the current-divider rule, the current into the amplifier, i_{in}, is

$$i_{in} = \left(\frac{r_S}{r_S + r_{in}}\right) i_S \tag{4.12}$$

Since r_o and R_L also form a current divider, not all of the current at the output of the amplifier reaches the load. By current division,

$$i_L = \left(\frac{r_o}{R_L + r_o}\right) i_o \tag{4.13}$$

We see that the overall current gain between source and load, i_L/i_S, equals A_i reduced by the two current-division ratios:

$$\frac{i_L}{i_S} = \left(\frac{r_S}{r_S + r_{in}}\right) A_i \left(\frac{r_o}{R_L + r_o}\right)$$

or,

$$\frac{I_L}{I_S} = \left(\frac{r_S}{r_S + r_{in}}\right) |A_i| \left(\frac{r_o}{R_L + r_o}\right) \tag{4.14}$$

Examination of the input current-division ratio, $r_S/(r_S + r_{in})$, reveals that r_{in} should be as *small* as possible to make the ratio close to 1:

$$r_{in} \ll r_S \Rightarrow r_S + r_{in} \approx r_{in}$$

$$\Rightarrow \frac{r_S}{r_S + r_{in}} \approx \frac{r_S}{r_S} = 1$$

Similarly, the output current-division ratio, $r_o/(R_L + r_o)$, shows that r_o should be as *large* as possible:

$$r_o \gg R_L \Rightarrow R_L + r_o \approx r_o$$

$$\Rightarrow \frac{r_o}{R_L + r_o} \approx \frac{r_o}{r_o} = 1$$

Summarizing, a *current* amplifier should have a small input resistance (ideally zero) and a large output resistance (ideally infinite), characteristics which are exactly opposite those of a voltage amplifier.

Example 4.10

The voltage gain of the amplifier in Figure 4.16(a) is $|A_v| = 40$. Find the overall current gain, I_L/I_S, and the load current I_L.

Solution

From equation 4.11, the current gain of the amplifier is

$$|A_i| = |A_v|\frac{r_{in}}{r_o} = 40 \left(\frac{4\text{ k}\Omega}{500\text{ }\Omega}\right) = 320$$

Figure 4.16
(Example 4.10)

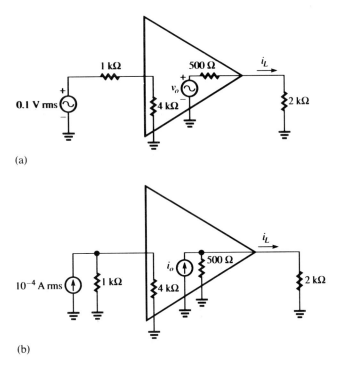

(a)

(b)

The equivalent source current is

$$I_S = \frac{V_S}{r_S} = \frac{0.1 \text{ V rms}}{1 \text{ k}\Omega} = 0.1 \text{ mA rms}$$

Note carefully that I_S is *not* the current that flows from voltage source v_S in Figure 4.16(a); I_S is by definition equal to V_S/r_S, which is not the same as input current I_{in} flowing into the amplifier. Figure 4.16(b) shows the amplifier with equivalent current sources for the signal source and amplifier output. By equation 4.14,

$$\frac{I_L}{I_S} = \left(\frac{1 \text{ k}\Omega}{1 \text{ k}\Omega + 4 \text{ k}\Omega}\right) 320 \left(\frac{500 \text{ }\Omega}{500 \text{ }\Omega + 2 \text{ k}\Omega}\right) = 12.8$$

Therefore,

$$I_L = 12.8I_S = 12.8(0.1 \text{ mA rms}) = 1.28 \text{ mA rms}$$

Note that the amplifier's voltage gain in this example is reduced from 40 to V_L/V_S = (4/5)40(20/25) = 25.6, while the current gain is reduced much more severely, from 320 to 12.8, due to resistance ratios that favor voltage gain but are very unfavorable for current gain.

The load current in this example can also be obtained using the original voltage sources shown in Figure 4.16(a). We see that V_{in} = [4 kΩ/(4 kΩ + 1 kΩ)] (0.1 V rms) = 0.08 V rms. Therefore, V_o = 40(0.08 V rms) = 3.2 V rms, and I_L = 3.2 V rms/(2 kΩ + 500 Ω) = 1.28 mA rms. Note that I_L is *not* the same as I_o.

4.4   POWER GAIN

Recall that average power P can be computed using any of the relationships:

$$P = V(\text{rms})I(\text{rms}) = \frac{V(\text{pk}) \, I(\text{pk})}{2}$$

$$= I^2 \, (\text{rms}) \, R = \frac{I^2(\text{pk}) \, R}{2} \qquad (4.15)$$

$$= \frac{V^2(\text{rms})}{R} = \frac{V^2(\text{pk})}{2 \, R} \text{ watts}$$

where V and I refer to the magnitudes of the voltage across and current through resistance R. (The equations involving rms values are valid for any waveform and those involving peak values are valid for sinusoidal waveforms only.)

The power delivered to the input of an amplifier, P_{in}, can be found using any of the equations in (4.15) with $V = V_{in}$, $I = I_{in}$, and $R = r_{in}$. Similarly, output power, P_o, can be found using $V = V_o$, $I = I_o$, and $R = r_o$. The power gain of an amplifier is

$$A_p = \frac{P_o}{P_{in}} \qquad (4.16)$$

Using $P = V(\text{rms})I(\text{rms})$, we can write (4.16) as

$$A_p = \frac{V_o(\text{rms})I_o(\text{rms})}{V_{in}(\text{rms})I_{in}(\text{rms})} = |A_v| \, |A_i| \qquad (4.17)$$

Since $|A_i| = |A_v| \, (r_{in}/r_o)$, we can also write

$$A_p = |A_v| \, |A_i| = |A_v| \, |A_v| \left(\frac{r_{in}}{r_o}\right)$$

$$= |A_v|^2 \left(\frac{r_{in}}{r_o}\right) \qquad (4.18)$$

Also,

$$A_p = |A_v| \, |A_i| = \frac{|A_i|}{(r_{in}/r_o)} |A_i| = |A_i|^2 \left(\frac{r_o}{r_{in}}\right) \qquad (4.19)$$

Using equations 4.15, numerous other equivalent expressions can be derived for calculating power gain, some of which are left as end-of-chapter exercises. Note that power gain has magnitude only; that is, there is no phase angle associated with it.

Example 4.11

The amplifier shown in Figure 4.17 has $|A_v| = 50$.

1. Find $|A_i|$.
2. Find A_p using equation 4.17 and verify that equations 4.18 and 4.19 give the same result.

Figure 4.17
(Example 4.9)

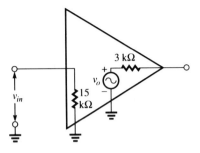

Solution

1. From equation 4.11,

$$|A_i| = |A_v| \left(\frac{r_{in}}{r_o}\right) = 50 \left(\frac{15 \text{ k}\Omega}{3 \text{ k}\Omega}\right) = 250$$

2. From equation 4.17,

$$A_p = |A_v| \, |A_i| = (50)(250) = 12,500$$

From equation 4.18,

$$A_p = |A_v|^2 \left(\frac{r_{in}}{r_o}\right) = (50)^2 \left(\frac{15 \text{ k}\Omega}{3 \text{ k}\Omega}\right) = 12,500$$

From equation 4.19,

$$A_p = |A_i|^2 \left(\frac{r_o}{r_{in}}\right) = (250)^2 \left(\frac{3 \text{ k}\Omega}{15 \text{ k}\Omega}\right) = 12,500$$

We have seen that voltage and current division at the input and output of an amplifier reduces the transfer of voltage and current from source to load. Therefore, we would expect that power transfer between source and load is similarly affected. Such is indeed the case: Power gain between source and load may be less than the total power gain available in the amplifier. The total power delivered by the source to the input of the amplifier is

$$P_{in} = I_{in}^2(\text{rms}) \, r_{in} = \frac{V_{in}^2(\text{rms})}{r_{in}} = V_{in}(\text{rms})I_{in}(\text{rms}) \qquad (4.20)$$

The total power delivered to the load is

$$P_L = I_L^2(\text{rms})R_L = \frac{V_L^2(\text{rms})}{R_L} = V_L(\text{rms})I_L(\text{rms}) \qquad (4.21)$$

Using equations 4.20 and 4.21, we can calculate the power gain between input and load, P_L/P_{in}.

Maximum Power Transfer

Recall from the *maximum power transfer theorem* that maximum power is transferred from a source to a load when the load resistance equals the source resistance. Therefore, maximum power is transferred from a signal source to the input of an amplifier when $r_{in} = r_S$, and maximum power is transferred from the output of the amplifier to the load when $R_L = r_o$. When $r_{in} = r_S$, the signal source is said to be *matched* to the input of the amplifier, and when $R_L = r_o$, the output is said to be matched to the load.

We must be careful how we interpret the matching criterion for maximum power transfer. If we were given total latitude in our choice of *all* resistance values, then we would choose to make the signal source ideal (a voltage source with $r_S = 0$ or a current source with $r_S = \infty$), for in that case all of the power available from the source would be transferred to the amplifier, regardless of the value of r_{in}. Similarly, we would make the output of the amplifier ideal in order to transfer all power to the load, regardless of the value of R_L. In practice, we achieve maximum power transfer under the constraint that r_S and r_o have specific values that we are unable to change. Under that constraint, the best we can achieve is to make r_{in} equal to r_S and R_L equal to r_o.

Example 4.12

The amplifier shown in Figure 4.18 has $A_p = 2 \times 10^5$. Find the power P_L delivered to the load when

1. $r_{in} = 2$ kΩ and $R_L = 3$ kΩ;
2. $r_{in} = 500$ Ω and $R_L = 600$ Ω.

Figure 4.18
(Example 4.12)

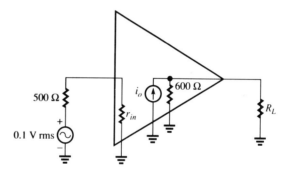

Solution

1.

$$I_{in} = \frac{V_S}{r_S + r_{in}} = \frac{0.1 \text{ V rms}}{500 \ \Omega + 2 \ \text{k}\Omega} = 40 \ \mu\text{A rms}$$

$$P_{in} = I_{in}^2 r_{in} = (40 \ \mu\text{A rms})^2 \ 2 \ \text{k}\Omega = 3.2 \ \mu\text{W}$$

$$P_o = A_p P_{in} = (2 \times 10^5) \ 3.2 \ \mu\text{W} = 0.64 \text{ W}$$

$$P_o = I_o^2(\text{rms}) \ r_o \Rightarrow I_o = \sqrt{\frac{0.64 \text{ W}}{600 \ \Omega}} = 32.66 \text{ mA rms}$$

$$I_L = \left(\frac{r_o}{r_o + R_L}\right) I_o = \left(\frac{600 \ \Omega}{600 \ \Omega + 3 \ \text{k}\Omega}\right) 32.66 \text{ mA rms}$$

$$= 5.44 \text{ mA rms}$$

$$P_L = I_L^2 R_L = (5.44 \text{ mA rms})^2 \ 3 \ \text{k}\Omega = 88.89 \text{ mW}$$

2. Notice that making $r_{in} = 500\ \Omega$ matches the input of the amplifier to the signal source, and making $R_L = 600\ \Omega$ matches the load to the output of the amplifier. In that case,

$$I_{in} = \frac{V_S}{r_S + r_{in}} = \frac{0.1\ \text{V rms}}{500\ \Omega + 500\ \Omega} = 100\ \mu\text{A rms}$$

$$P_{in} = I_{in}^2 r_{in} = (100\ \mu\text{A rms})^2\ 500\ \Omega = 5\ \mu\text{W}$$

$$P_o = A_p P_{in} = (2 \times 10^5)\ 5\ \mu\text{W} = 1\ \text{W}$$

$$I_o = \sqrt{\frac{P_o}{R_L}} = \sqrt{\frac{1\ \text{W}}{600\ \Omega}} = 40.82\ \text{mA rms}$$

$$I_L = \left(\frac{r_o}{r_o + R_L}\right)I_o = \left(\frac{600\ \Omega}{600\ \Omega + 600\ \Omega}\right)40.82\ \text{mA rms}$$

$$= 20.41\ \text{mA rms}$$

$$P_L = I_L^2 R_L = (20.41\ \text{mA rms})^2\ 600\ \Omega = 0.25\ \text{W}$$

We see that matching the amplifier to source and load increases the power delivered to the load in this example by a factor of 0.25 W/88.89 mW ≈ 2.8.

<hr>

4.5   METHODS OF COUPLING

In the context of electronic devices, *coupling* means connecting the output of one device to the input of another. In many practical amplifiers, special circuitry is required to facilitate coupling because the output of one device may not be directly compatible with the input of another. A common example of such an incompatibility occurs when the dc level, or bias voltage, at the output of one device is different from the dc level at the input to the other device. Some means must be used to couple the ac component of the voltage while isolating the dc voltages, i.e., while preventing dc current from flowing from one device to the other. Signal sources and loads must often be isolated from dc. A speaker, for example, presents a very small resistance to dc and cannot be connected directly to the output of an amplifier having a dc component because it would effectively short the dc component to ground.

Capacitor Coupling

As we know, a capacitor blocks the flow of dc current (when it becomes fully charged, after a brief transient), so capacitors are widely used for coupling, particularly in discrete circuits. While blocking the flow of dc current, a capacitor readily permits the flow of ac current (provided its capacitance is sufficiently large, about which we will have more to say later). Although capacitor coupling is not used within integrated-circuit amplifiers, it *is* used in some applications to couple signal sources to the inputs and loads to the outputs of integrated circuits. Figure 4.19 illustrates capacitor coupling and shows how dc levels are blocked from a signal source and from a load. Coupling capacitors are sometimes called blocking capacitors when they are used for that purpose.

Figure 4.20 shows a coupling capacitor used to isolate the dc bias voltage at the output of one amplifier from the dc bias voltage at the input to another. Since there is a dc voltage difference across the capacitor, it must have a *dc–working-*

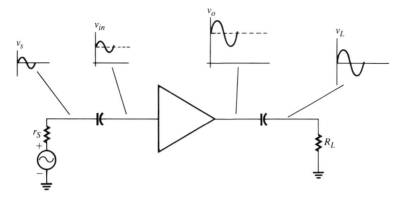

Figure 4.19
Coupling capacitors are used to block the flow of dc current between the amplifier and the signal source and between the amplifier and load.

voltage (DCWV) rating that is at least equal to that voltage difference. In practice, the DCWV rating should be greater than the voltage difference to provide a margin of safety:

$$\text{DCWV} > |V_1 - V_2| \tag{4.22}$$

where $|V_1 - V_2|$ is the magnitude of the dc voltage difference across the capacitor. Coupling capacitors are often large, electrolytic types, so care must be used to ensure that the positive side of the capacitor is connected to the input or output terminal having the *more positive* voltage.

In order to "pass" the ac components of a signal while blocking the dc component, a coupling capacitor must have negligibly small reactance at the lowest frequency component of the signal. Recall that the magnitude of capacitive reactance is

$$|X_C| = \frac{1}{\omega C} = \frac{1}{2\pi f C} \text{ ohms} \tag{4.23}$$

At the lowest signal frequency, the reactance of an input coupling capacitor should be much smaller than the sum of the output (or source) resistance to which it is connected on one side and the input (or load) resistance to which it is connected on the other side. If the total resistance value is small, and if it is necessary to operate at low frequencies, a coupling capacitor may have to have an impractically large capacitance to be effective. We will study the effects of coupling capacitors on the performance of amplifiers in more detail in Chapter 6.

Figure 4.20
The dc voltage across the coupling capacitor is $V_1 - V_2$ and its DCWV rating must be at least that large. If the capacitor is electrolytic (polarized), its positive side must be connected to the more positive dc voltage (V_1 in the figure).

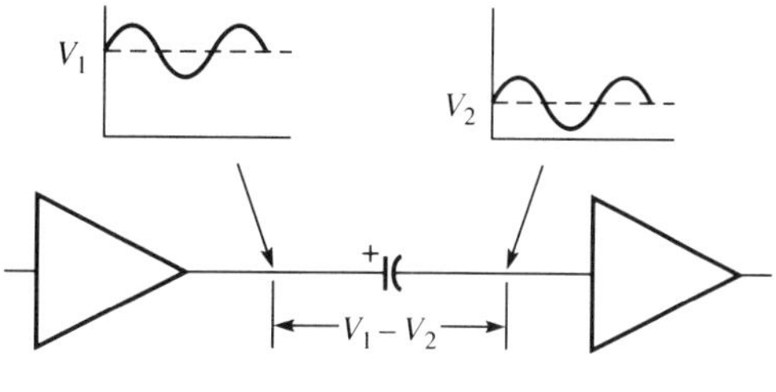

Example 4.13

A coupling capacitor is to be used to couple the output of an amplifier whose output resistance is 1 kΩ to the input of another amplifier whose input resistance is 14 kΩ. The bias voltage at the output of the first amplifier is 9 V dc and the bias voltage at the input to the second is 1 V dc.

1. If the amplifiers are used for audio-frequency signals, what should be the capacitance of the coupling capacitor?
2. What should be the DCWV rating of the capacitor?

Solution

1. Recall that the audio-frequency range is considered to be 20 Hz–20 kHz, so the coupling capacitor must have small reactance at the lowest frequency in that range: 20 Hz. The sum of the input and output resistances is 1 kΩ + 14 kΩ = 15 kΩ. To make $|X_C| \ll$ 15 kΩ at 20 Hz, let us choose $|X_C|$ = 150 Ω (one one-hundredth of 15 kΩ). Then,

$$|X_C| = 150 \ \Omega = \frac{1}{2\pi(20 \text{ Hz})C}$$

$$\Rightarrow C = \frac{1}{2\pi(20 \text{ Hz})(150 \ \Omega)} = 53.1 \ \mu\text{F}$$

In practice, we would probably choose C = 50 μF, which would make $|X_C|$ equal to only 159 Ω at 20 Hz, an acceptable value. This example demonstrates the practical disadvantage of capacitor coupling at low frequencies: large (bulky, expensive) capacitors may be required. In the present example, amplifier redesign may be a viable alternative, or, we might settle for a 5-μF capacitor, making $|X_C|$ = 1.5 kΩ at 20 Hz.

2. The dc working voltage of the capacitor should be at least

$$|V_1 - V_2| = 9 \text{ V} - 1 \text{ V} = 8 \text{ V dc}$$

We would probably choose DCWV = 10 V dc. The physical size of a capacitor increases dramatically with increases in its DCWV, so this is another important practical consideration in the design and specification of coupling requirements.

Transformer Coupling

Another method for coupling an ac signal from one point in a circuit to another is through the use of a transformer. Recall that a transformer responds to ac signals only; that is, the current in the primary winding must be *changing* in order for corresponding changes to be induced in the secondary winding. Thus, any dc component in the primary winding has no effect on the secondary winding. In short, a transformer blocks the flow of dc current while passing ac signals, and in that respect accomplishes the same task as a coupling capacitor. In a typical application, the primary winding could be connected to the output side of an amplifier and the secondary side connected to the load, so the dc bias level in the amplifier's output would not appear in the load.

The advantages of transformer coupling include low dc power dissipation and the capability for designing a turns ratio that results in maximum power transfer. The principal disadvantages are the bulk and cost of the transformer itself and the fact that transformers are not generally capable of passing a wide range of ac frequencies. Transformers are frequently used in applications where the range of frequencies is limited (*narrowband* systems) such as radio-frequency (rf) amplifiers.

As previously noted, an advantage of using a transformer for coupling is that the turns ratio can be designed to achieve *impedance matching* and therefore to maximize power transfer. Recall that the turns ratio of a transformer is defined by

$$\text{turns ratio} = \frac{N_p}{N_s} = \frac{e_p}{e_s} \tag{4.24}$$

where N_p, N_s = number of turns in the primary and secondary windings, respectively

e_p, e_s = primary and secondary voltages

Figure 4.21 shows a transformer whose primary is driven by a signal source v_1 having resistance r_1 and whose secondary has a load resistance r_L connected across it. Assuming an ideal transformer (zero winding resistance and no power loss), it is easy to show that the resistance r_{in} looking into the primary winding is

$$r_{in} = \left(\frac{N_p}{N_s}\right)^2 r_L \tag{4.25}$$

To achieve maximum power transfer from the source to the load in Figure 4.21, it is necessary (by the maximum power transfer theorem) that the resistance r_{in} seen by the signal source be equal to the source resistance r_1. Thus, from equation 4.25, we require

$$r_1 = r_{in} = \left(\frac{N_p}{N_s}\right)^2 r_L \tag{4.26}$$

Solving (4.26) for $N_p N_s$, we find the turns ratio necessary to achieve maximum power transfer:

$$\frac{N_p}{N_s} = \sqrt{\frac{r_1}{r_L}} \tag{4.27}$$

In practical amplifiers, the signal source driving the primary winding often has a large output resistance, and the load resistance may be a much smaller value. In that case, the ratio r_1/r_L is greater than 1 and therefore, by equation 4.27, the turns ratio N_p/N_s is greater than 1. As a consequence, the coupling transformer is typically a *step-down* transformer, meaning the secondary voltage is less than the primary voltage.

Commercially available transformers are often specified in terms of their impedance ratios rather than their turns ratios. In these cases, the turns ratio can be

Figure 4.21
A transformer whose primary winding is driven by the signal source v_1 and whose secondary winding has load r_L connected across it.

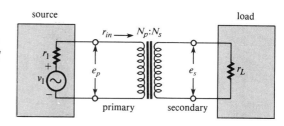

determined using equation 4.27, i.e., by taking the square root of the specified impedance ratio. For example, if the specifications on a transformer state that it has primary impedance 16 kΩ and secondary impedance 16 Ω (typical values for matching a transistor output stage to a speaker), then

$$\frac{N_p}{N_s} = \sqrt{\frac{Z_p}{Z_s}} = \sqrt{\frac{16 \times 10^3}{16}} = 31.62$$

Example 4.14

Figure 4.22 shows a transformer used to match a 100-kΩ source to a 250-Ω load.

1. What should be the turns ratio of the transformer?
2. With the matching transformer in place, what is the rms voltage across the load?
3. What is the maximum power that can be delivered to the load?

Figure 4.22
(Example 4.14)

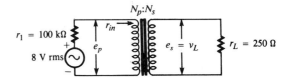

Solution

1. From equation 4.27,

$$\frac{N_p}{N_s} = \sqrt{\frac{r_1}{r_L}} = \sqrt{\frac{100 \times 10^3}{250}} = 20$$

Thus the primary winding should have 20 times as many turns as the secondary winding.

2. From equation 4.25,

$$r_{in} = (N_p/N_s)^2 r_L = (20)^2(250) = 100 \text{ k}\Omega$$

We see that the resistance looking into the primary winding is 100 kΩ, as it should be when load and source are matched by the transformer. Since the source resistance and r_{in} form a voltage divider across the 8-V-rms signal source, the voltage across the primary winding is *one-half* the signal voltage:

$$e_p = \left(\frac{r_{in}}{r_1 + r_{in}}\right)e_s = \left[\frac{100 \text{ k}\Omega}{(100 \text{ k}\Omega) + (100 \text{ k}\Omega)}\right](8 \text{ V rms}) = 4 \text{ V rms}$$

From equation 4.24,

$$e_s = \left(\frac{1}{N_p/N_s}\right)e_p = \left(\frac{1}{20}\right)4 \text{ V rms} = 0.2 \text{ V rms}$$

We see that the transformer steps the primary voltage down by a factor of 20. (The current is stepped up by a factor of 20.)

3. Maximum power transfer occurs when the source is matched to the load, and with the matching transformer in place, we have

$$p_{max} = P_L = \frac{e_s^2}{r_L} = \frac{v_L^2}{r_L} = \frac{(0.2 \text{ V})^2}{250} = 0.16 \text{ mW}$$

(When the source and load are matched, the same amount of power is dissipated in the source as is delivered to the load; verify that fact in this example.)

Direct-Coupling

As the name implies, direct-coupling is the coupling method in which amplifiers, sources, and/or loads are connected directly together. For example, two amplifiers are directly coupled when the output of one is connected directly to the input of the other. In such a case, it is clear that the dc level at both points is exactly the same, and any change in dc level at one point creates an identical change in dc level at the other. For that reason, the amplifiers are said to be dc, which can mean either direct-coupled or direct current.

Direct-coupling is the method used in integrated circuits, since coupling capacitors and transformers are totally impractical in such small space (a single $1\text{-}\mu\text{F}$ capacitor has a greater volume than an entire integrated-circuit chip!). It is apparent that dc amplifier circuitry must be designed so that the output bias level of one amplifier is suitable to serve as the input bias level of the one to which it is coupled. Most differential and operational amplifiers, which we will study in considerable detail in later chapters, are direct-coupled.

Example 4.15

PSpice

Use PSpice to determine the voltage gain v_L/v_S and current gain i_L/i_S of an amplifier having $|A_v| = 150$, $r_{in} = 1 \text{ k}\Omega$, and $r_o = 500 \text{ }\Omega$. The amplifier is driven from a signal source that has $r_S = 200 \text{ }\Omega$ and the amplifier load is $R_L = 2 \text{ k}\Omega$. Assume the source voltage has magnitude 0.1 V.

Solution.   Figure 4.23 shows the circuit and input data file. Note that the amplifier is modeled by a *voltage-controlled voltage source* named EAMP. The statement

```
EAMP 3 0 2 0 150
```

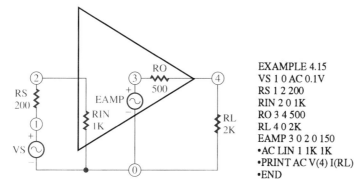

Figure 4.23
(Example 4.15)

EXAMPLE 4.15
VS 1 0 AC 0.1V
RS 1 2 200
RIN 2 0 1K
RO 3 4 500
RL 4 0 2K
EAMP 3 0 2 0 150
•AC LIN 1 1K 1K
•PRINT AC V(4) I(RL)
•END

means that the amplifier's output voltage (v_o) is developed across nodes 3 and 0, its input (controlling) voltage is across nodes 2 and 0 (i.e., across r_{in}), and that the voltage gain is 150. We arbitrarily perform an ac analysis at 1 kHz. Execution of the program gives V(4) = 10 V and I(RL) = 5 mA. Thus, v_L = 10 V and i_L = 5 mA. Therefore,

$$\frac{v_L}{v_S} = \frac{10\text{ V}}{0.1\text{ V}} = 100$$

and, since $i_S = v_S/r_S = 0.1\text{ V}/200\ \Omega = 0.5$ mA,

$$\frac{i_L}{i_S} = \frac{5\text{ mA}}{0.5\text{ mA}} = 10$$

Example 4.16

SPICE

Assuming that the transformer in Figure 4.22 is ideal and that the secondary winding has inductance 50 mH, use SPICE to find the primary and secondary voltages and currents when the input voltage and the turns ratio are the same as in Example 4.14. Assume that the frequency of the input is 50 kHz.

Solution.   The turns ratio is related to the primary and secondary inductance, L_p and L_s, by

$$\frac{N_p}{N_s} = \sqrt{\frac{L_p}{L_s}}, \quad \text{or} \quad L_p = \left(\frac{N_p}{N_s}\right)^2 L_s$$

Therefore, since N_p/N_s = 20 and L_s = 50 mH,

$$L_p = (20)^2(50 \times 10^{-3}\text{ H}) = 20\text{ H}$$

Figure 4.24(a) shows the circuit in its SPICE format. Note that we cannot isolate the secondary circuit from the primary because SPICE requires that there be a dc path to ground from every node. Thus, the secondary winding must have one side connected to node 0. The statement K1 LP LS 1 makes the transformer ideal, since it specifies the coupling coefficient, k, to be 1. (In PSpice, k must be < 1, but can be set to 0.999, for example.) As is required to simulate an iron-core transformer in SPICE, the primary reactance at 50 kHz ($2\pi f L_p$ = 6.28 MΩ) is much greater than the 100-kΩ source impedance, and the secondary reactance ($2\pi f L_s$ = 1.57 kΩ) is much greater than the 250-Ω load impedance. Dummy voltage sources VIP and VIS have been inserted to measure the primary and secondary currents. The results of the analysis are shown in Figure 4.24(b). We see that v_p = V(3) = 4 V and v_s = V(5) = 0.2 V, in agreement with Example 4.14. The primary current is i_p = I(VIP) = 40 μA and the secondary current is i_s = I(VIS) = 0.8 mA. Note that $N_p/N_s = i_s/i_p$ = 20.

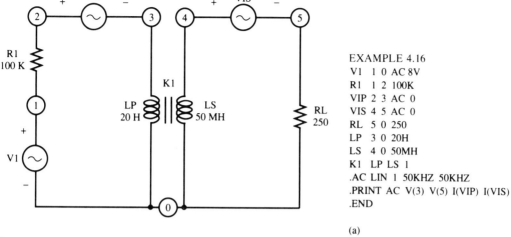

EXAMPLE 4.16
V1 1 0 AC 8V
R1 1 2 100K
VIP 2 3 AC 0
VIS 4 5 AC 0
RL 5 0 250
LP 3 0 20H
LS 4 0 50MH
K1 LP LS 1
.AC LIN 1 50KHZ 50KHZ
.PRINT AC V(3) V(5) I(VIP) I(VIS)
.END

(a)

```
EXAMPLE 4.16
****       AC ANALYSIS                              TEMPERATURE =    27.000 DEG C
*********************************************************************************
    FREQ        V(3)         V(5)          I(VIP)        I(VIS)
  5.000E+04    4.000E+00    2.000E-01    4.000E-05    8.000E-04
```

(b)

Figure 4.24
(Example 4.16)

EXERCISES

Section 4.1   Introduction

4.1   The voltage waveform at the output of an amplifier is $v(t) = 10 + 5\sin 10^4 t$ V.
   a. What is the dc component of the voltage?
   b. What is the peak-to-peak value of the ac component?
   c. Find the minimum and maximum output voltages.
   d. What is the frequency of the waveform, in hertz?
   e. What is the average value of the waveform?
   f. What is the average value of the ac component?
   g. Sketch the waveform versus time. Be sure to label significant values on both axes.

4.2   The voltage at the input to an amplifier has a dc bias level of 1.5 V and a sinusoidal ac component with frequency 2.5 kHz. The effective value of the ac component is 141.4 mV rms.
   a. Write the mathematical expression for the input voltage.
   b. What are the minimum and maximum values of the input voltage?
   c. What is the peak-to-peak value of the ac component?
   d. Sketch the waveform versus time. Be sure to label significant values on both axes.

4.3   Figure 4.25 shows the waveform of a current having dc and ac components. Write the mathematical expression for the current as a function of time.

4.4   Write the mathematical expression for the voltage waveform in Figure 4.26 as a function of time.

4.5   The ac component of the output voltage of an amplifier is $4\sin(2\pi \times 100t)$ V.
   a. If the output voltage cannot be less than 1 V, what is the minimum value that the output bias voltage can have?
   b. If the bias voltage is made equal to the value found in (a), what is the maximum value of the output voltage?
   c. What is the period of the ac component of the voltage?

4.6   The bias voltage at the output of an amplifier is −6 V. If the output voltage cannot be

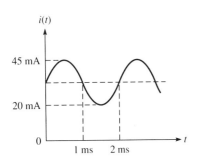

Figure 4.25
(Exercise 4.3)

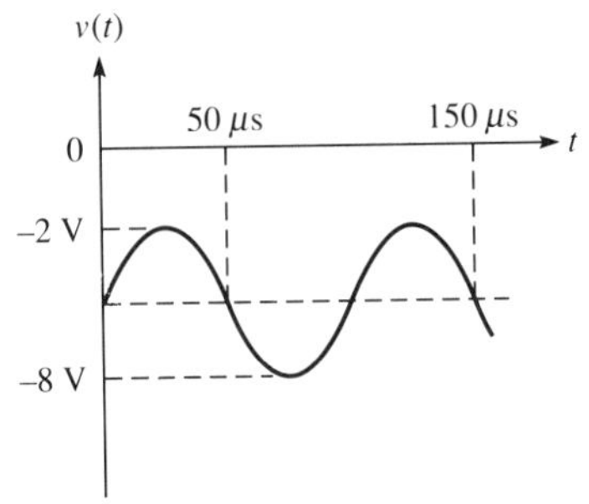

Figure 4.26
(Exercise 4.4)

greater (more positive) than −1 V,
  a. What is the maximum peak-to-peak value that the output voltage can have?
  b. What is the minimum value of the output voltage when it has a peak-to-peak value of 3 V?

4.7   For the network shown in Figure 4.27, find
  a. the dc component of the voltage v across R_1;
  b. the ac component of the voltage v across R_1;
  c. the total voltage v (dc + ac) across R_1.

4.8   For the network shown in Figure 4.28, find
  a. the dc component of the current i in the 10-kΩ resistor;
  b. the ac component of the current i in the 10-kΩ resistor;
  c. the total current (dc + ac) i in the 10-kΩ resistor.

4.9   Write a mathematical expression for the total current $i(t)$ (including dc and ac components) in the 200-Ω resistor in Figure 4.29.

4.10   Write a mathematical expression for the total voltage $v(t)$ (including dc and ac components) across the 100-kΩ resistor in Figure 4.30.

Section 4.2   Voltage Gain

4.11   The input voltage to an amplifier is $1 + 0.01\sin\omega t$ V and the output voltage is $5 + 1.5\sin\omega t$ V.
  a. What is the voltage gain $|A_v|$ of the amplifier?
  b. Assuming the amplifier is linear, what is the effective value of the ac component of the output voltage when the input is changed to $1.5 + 0.015\sin\omega t$?

4.12   When the effective value of the ac component of the input to an amplifier is 35 mV rms, the peak-to-peak value of the output voltage is 8.41 V pk-pk.
  a. Find the voltage gain $|A_v|$ of the amplifier.
  b. Assuming the amplifier is linear, find the effective value of the ac component of the output voltage when the peak-to-peak value of the input voltage is 50 mV pk-pk.
  c. Assuming the amplifier is linear, what peak-to-peak input voltage is required to obtain a 10-V pk-pk output voltage?

4.13   The network shown in Figure 4.31 represents the equivalent circuit at the input of a

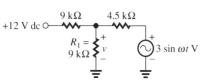

Figure 4.27
(Exercise 4.7)

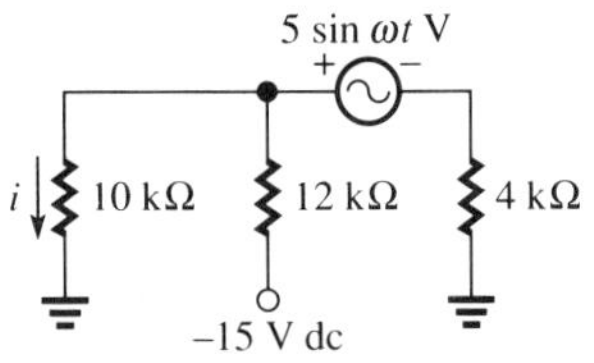

Figure 4.28
(Exercise 4.8)

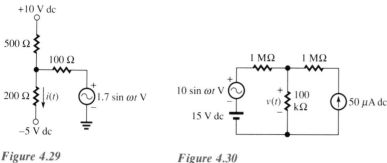

Figure 4.29
(Exercise 4.9)

Figure 4.30
(Exercise 4.10)

certain amplifier. Assuming the reactance of the capacitor is negligibly small at the frequency of operation, find

a. r_{in};
b. R_{in}.

4.14 When the input voltage to a certain amplifier is $v_{in}(t) = 0.4 + 0.2\sin\omega t$ V, the input current is $i_{in}(t) = 800 + 20\sin\omega t$ μA. Find

a. r_{in};
b. R_{in}.

4.15 A certain amplifier has $r_{in} = 1.5$ kΩ and $R_{in} = 5$ kΩ. Write the mathematical expression for the input current, $i_{in}(t)$, when the input voltage is $v_{in}(t) = 5 + 0.6\sin\omega t$ V.

4.16 Figure 4.32 shows measurements that were made in an experiment designed to determine the value of the input resistance, r_{in}, of the amplifier. Find r_{in}.

4.17 The ac input current to an amplifier having output resistance 2.5 kΩ is 24 μA rms. The voltage gain of the amplifier is $|A_v| = 75$. When a short circuit is connected from the output to ground, the current in the short circuit is 22.5 mA rms. Find the output resistance of the amplifier.

4.18 To determine the output resistance of an amplifier by measuring the ac current that flows in a short circuit connected between its output and ground, it is necessary to short the ac component of the output without affecting the dc component. Describe how this could be achieved.

4.19 A signal source has $r_S = 1.2$ kΩ and generates 1 V pk-pk when its output is open. What is the input voltage to an amplifier having $r_{in} = 5$ kΩ when the source is connected to the input of the amplifier?

4.20 A signal source is connected to the input of an amplifier having input resistance 15 kΩ. Find the maximum permissible source resistance if the input voltage to the amplifier must be at least 80% of the source voltage.

4.21 What is the smallest load resistance that can be driven by an amplifier whose output resistance is 1 kΩ if the load voltage must be at least 75% of the amplifier's (unloaded) output voltage?

4.22 An amplifier having $r_{in} = 13$ kΩ and $r_o = 600$ Ω drives a 2-kΩ load. The signal source has $v_S = 25$ mV pk-pk and $r_S = 500$ Ω.

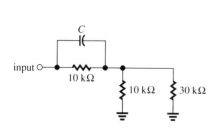

Figure 4.31
(Exercise 4.13)

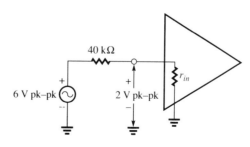

Figure 4.32
(Exercise 4.16)

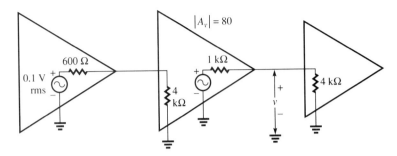

Figure 4.33
(Exercise 4.23)

What must be the voltage gain $|A_v|$ of the amplifier if the load voltage must be 2.89 V pk-pk?

4.23 Find the effective value of the ac voltage v at the output of the middle amplifier in Figure 4.33.

4.24 Find the effective value of the ac voltage v at the output of the amplifier in Figure 4.34. The capacitor has negligibly small reactance at the frequency of operation.

4.25 Find the peak-to-peak value of the ac voltage v at the output of the amplifier in Figure 4.35. The capacitor has negligibly small reactance at the frequency of operation.

4.26 The voltage divider consisting of R_1 and R_2 in Figure 4.36 is to be designed so that the peak value of the ac load voltage is $v_L = 2.5$ V pk. If $R_1 = 500$ Ω, what should be the value of R_2? (Hint: Find the Thevenin equivalent circuit of the signal source.)

4.27 The ac component of the input to an amplifier having voltage gain -50 is $v_{in}(t) = 0.06\sin(2\pi \times 2 \times 10^5 t)$ V.

a. Write a mathematical expression for the ac component of the output voltage.
b. Sketch the waveforms of the ac components of the input and output voltages versus time.

4.28 Figure 4.37 shows the waveforms of the ac components of the input and output voltages of an amplifier.

a. Write the mathematical expressions for the input and output waveforms.
b. Find the voltage gain of the amplifier.

Section 4.3 Current Gain

4.29 The input current to an amplifier is $i_{in}(t) = 50 + 10\sin\omega t$ μA. The ac component of the open-circuit output voltage has peak value 2.5 V pk. The output resistance of the amplifier is $r_o = 1$ kΩ.

a. Find and draw the equivalent current source at the amplifier's output.
b. Find the current gain $|A_i|$.

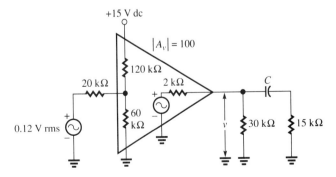

Figure 4.34
(Exercise 4.24)

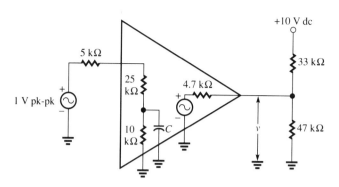

Figure 4.35
(Exercise 4.25)

4.30 The peak-to-peak values of the ac input and output voltages of an amplifier are V_{in} = 0.4 V pk-pk and V_o = 16 V pk-pk. The input and output resistances are r_{in} = 1.5 kΩ and r_o = 500 Ω.
a. Find the voltage gain $|A_v|$.
b. Find the current gain $|A_i|$.

4.31 The input resistance of a certain amplifier is three times as great as its output resistance. If the current gain of the amplifier is $|A_i|$ = 180, what is the voltage gain, $|A_v|$?

4.32 An amplifier has output resistance r_o = 500 Ω and voltage gain $|A_v|$ = 25. What input resistance should it have if the ac component of its output current must be 150 times greater than the ac component of its input current?

4.33 Find the current gain I_L/I_S between source and load in Figure 4.38.

4.34 Find the current gain I_L/I_S between source and load in Figure 4.39 (Recall that I_S is the value of an equivalent current source.)

4.35 The amplifier in Figure 4.15 has $|A_i|$ = 40, r_{in} = 1 kΩ, and r_o = 6 kΩ. The magnitude of the ac source current is 0.1 mA rms and r_S = 10 kΩ. What is the maximum load R_L that can be connected to the output of the amplifier if the ac current delivered to the load must have magnitude at least 2 mA rms?

4.36 Find the peak-to-peak value of the ac load current in R_L in Figure 4.40.

Section 4.4   Power Gain

4.37 An amplifier has $|A_i|$ = 45, r_{in} = 15 kΩ, and r_o = 5 kΩ. Find the power gain A_p.

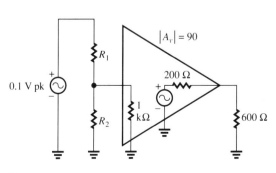

Figure 4.36
(Exercise 4.26)

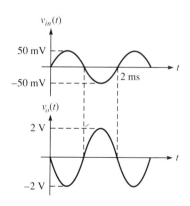

Figure 4.37
(Exercise 4.28)

Figure 4.38
(Exercise 4.33)

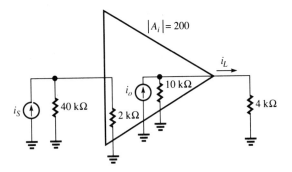

Figure 4.39
(Exercise 4.34)

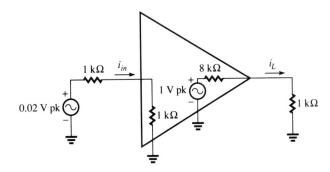

Figure 4.40
(Exercise 4.36)

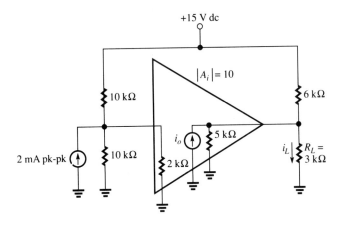

Figure 4.41
(Exercise 4.41)

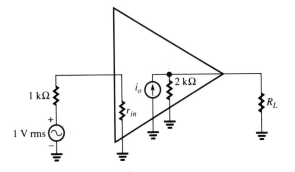

Figure 4.42
(Exercise 4.42)

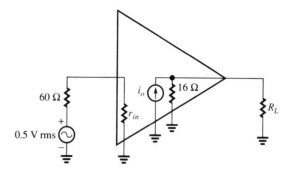

4.38 An amplifier has $|A_v| = 30$, $r_{in} = 12$ kΩ, and $r_o = 1$ kΩ. If the power delivered to the input of the amplifier is 0.5 mW, what is the output power?

4.39 Derive the following equation for power gain:

$$A_p = \frac{(V_o/I_{in})^2}{r_o r_{in}}$$

4.40 Derive the following equation for power gain:

$$A_p = g_m^2 r_o r_{in}$$

(Hint: Recall that transconductance is the ratio of output current to input voltage.)

4.41 The amplifier shown in Figure 4.41 has $A_p = 8 \times 10^5$. Find the power delivered to the load when
   a. $r_{in} = 500$ Ω and $R_L = 10$ kΩ.
   b. $r_{in} = 1$ kΩ and $R_L = 2$ kΩ.

4.42 The amplifier shown in Figure 4.42 has $A_p = 4 \times 10^5$. Find the power delivered to the load when
   a. $r_{in} = 1$ kΩ and $R_L = 50$ Ω.
   b. the input and output of the amplifier are matched to the source and load, respectively.

Section 4.5 Methods of Coupling

4.43 With respect to the coupling circuit shown in Figure 4.43.

   a. What minimum DCWV rating should the capacitor have?
   b. If the minimum signal frequency in the circuit is 500 Hz, what should be the minimum capacitance of the capacitor?

4.44 Coupling capacitor C_1 in Figure 4.44 has a DCWV rating of 20 V. The frequency range of signals in the circuit is 1 kHz–10 kHz.
   a. What is the maximum positive value that voltage V should have?
   b. What is the minimum value that R_L should have?

4.45 A signal source having resistance 90 kΩ drives the primary winding of a transformer whose turns ratio is 6:1. What value of load resistance across the secondary would result in maximum power delivered to the load?

4.46 A 200-Ω load resistor is connected across the secondary of a transformer whose turns ratio is 86:20. If maximum power is being transferred to the load, what is the signal-source impedance on the primary side of the transformer?

4.47 In the transformer circuit shown in Figure 4.45, find the power dissipated in r_1 and the power delivered to r_L. Are source and load matched by the transformer?

4.48 **a.** Find the turns ratio in Exercise 4.47 that will result in maximum power delivered to the load.
   b. Using the turns ratio found in (a), find

Figure 4.43
(Exercise 4.43)

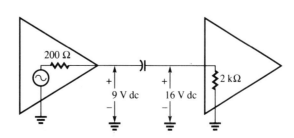

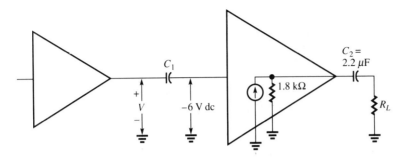

Figure 4.44
(Exercise 4.44)

the power dissipated in r_1 and the power delivered to r_L.

4.49 **a.** What should be the turns ratio of the transformer in Figure 4.46 in order to deliver maximum power to r_L?
   b. Using the turns ratio found in (a), find the power dissipated in r_1 and the power delivered to r_L.

DESIGN EXERCISES

4.50 An amplifier must deliver 10 V pk-pk to a 500-Ω load. The amplifier has input resistance 2 kΩ and output resistance 100 Ω. It is driven from a signal source that has resistance 200 Ω and that produces 0.2 V pk-pk (open circuit). What voltage gain $|A_v|$ should the amplifier have?

4.51 Figure 4.47 shows the input circuit of a certain amplifier. The input signal has frequency components in the range from 500 Hz to 15 kHz. The magnitude of the current in bypass capacitor C must be at *least* 0.9 $|i_{in}|$ at *all* frequencies of the input. What value of capacitance should be used? Assume the capacitor must be selected

from a line of standard-valued capacitors having capacitances 0.1 μF, 0.5 μF, 1 μF, 5 μF, and 10 μF.

4.52 An amplifier is driven from a constant-current source having resistance 2 kΩ. The amplifier must deliver 0.5 A pk to a 50-Ω load when the current source generates 5 mA pk. If the amplifier has input resistance 1 kΩ and output resistance 100 Ω, what value of current gain $|A_i|$ should it have?

4.53 An amplifier must deliver 6 V rms ac to a 100-Ω load when its input is $i_{in} = 1.5\sin(1000t)$ mA. The amplifier has input resistance 800 Ω. What power gain P_L/P_{in} must the amplifier provide?

SPICE EXERCISES

4.54 Use SPICE (or PSpice) to find the power gain P_L/P_{in} of an amplifier having $|A_v| = 80$. The amplifier has input resistance 2 kΩ and output resistance 250 Ω. It drives a load resistance of 1 kΩ. The signal source driving the amplifier has negligibly small resistance.

4.55 The current-source equivalent circuit of the output of an amplifier has $I_o = 0.2$ A pk and

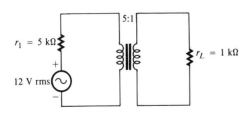

Figure 4.45
(Exercises 4.47 and 4.48)

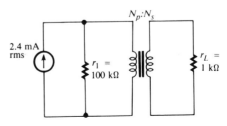

Figure 4.46
(Exercise 4.49)

Figure 4.47
(Exercise 4.51)

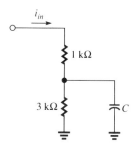

$r_o = 50 \ \Omega$. The amplifier has $r_{in} = 1.5 \ \text{k}\Omega$ and drives a 600-Ω load. The signal source driving the amplifier has $v_S = 0.1$ V pk and $r_S = 2.5 \ \text{k}\Omega$. Using SPICE (or PSpice) to model the amplifier as a *voltage-controlled current source*, find the voltage gain v_L/v_S and current gain i_L/i_S. (Hint: To model the amplifier, it will be necessary to calculate its *transconductance* parameter: I_o/V_{in}, where V_{in} is the voltage across r_{in}.)

4.56 Use SPICE (or PSpice) to show that less power is delivered to the load in Example 4.14 when the turns ratio of the transformer is smaller and when it is greater than the turns ratio necessary for matching.

4.57 Use SPICE (or PSpice) to verify the design in Exercise 4.50.

4.58 Use SPICE (or PSpice) to verify the design in Exercise 4.51. (Let the input be a single-frequency sine wave.)

Transistor Amplifiers

5

5.1 THE COMMON TERMINAL OF AN AMPLIFIER

The input and output voltages of an amplifier typically have the same reference (ground, or *common*) point. In other words, the input and output voltages are both measured with respect to the same point in the amplifier circuit. This fact is illustrated in Figure 5.1, where the amplifier is shown simply as a block and the input and output voltages are both referenced to a point identified by the conventional ground symbol. Since the reference point in the circuit is electrically connected to both the input and output, that point is said to be common to both input and output.

The Common Terminal in Single-Transistor Amplifiers

As we know, a bipolar junction transistor has three terminals: emitter, base, and collector. A single-transistor amplifier can be constructed using any one of these three terminals as the common point. The three amplifier configurations that result are shown in Figure 5.2. Only small-signal (ac) input and output voltages are shown in the figure: The dc circuitry necessary to provide proper bias is omitted for the moment, but it must of course be included in a practical amplifier and will be discussed presently. The three possible amplifier configurations are called *common emitter* (CE), *common base* (CB), and *common collector* (CC), according to which terminal is used as the common point. (The terminology *grounded* emitter, etc., is also sometimes used, but the term *common* is preferred.) Note that small-signal input and output voltages are identified by lower-case letters whose subscripts denote the two amplifier terminals across which each voltage appears. For example, the output voltage of the CE configuration is v_{ce}, the voltage between collector and emitter. The second subscript of any input or output voltage will always refer to the terminal which is common in the configuration used (e in CE, b in CB, and c in CC). Although the figure shows only NPN transistors, each configuration has a PNP counterpart whose small-signal designations are identical to those shown. For example, v_{be} is the small-signal input to both an NPN and a PNP CE amplifier. As we shall see, the only difference between PNP and NPN amplifiers is the polarity of the dc voltage sources used to bias them. The amplifying capabilities of the three circuits and their small-signal

Figure 5.1
*Input and output voltages are referenced
to the same point in an amplifier circuit,
so that point is said to be* common *to
both input and output.*

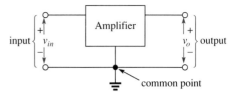

characteristics (such as input and output impedance) differ according to the configuration used. The common-emitter configuration is the most useful and most widely used of the three, so we choose it to begin our analysis of single-transistor amplifiers.

5.2   COMMON-EMITTER BIAS CIRCUITS

Figure 5.3 shows the simplest way to provide bias for NPN and PNP transistors designed to operate as common-emitter amplifiers. Recall that the base–emitter junction must be forward biased and the collector–base junction reverse biased. In each circuit shown, a single dc *supply* voltage, designated V_{CC}, is connected to provide the proper polarity of the voltage across each junction (see Figures 3.3 and 3.5). In each case, the base–emitter junction is clearly forward biased by the supply voltage and the collector–base junction is reverse biased, provided the dc voltage drop across R_B ($I_B R_B$) is greater than the dc voltage drop across R_C ($I_C R_C$)—the usual situation. Since the base–emitter junctions are forward biased,

Figure 5.2
*The three transistor amplifier configura-
tions.*

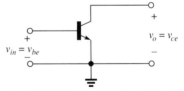

(a) Common-emitter (CE) configuration

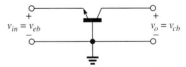

(b) Common-base (CB) configuration

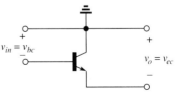

(c) Common-collector (CC) configuration

Figure 5.3
An elementary bias circuit using a single power supply for a common-emitter amplifier.

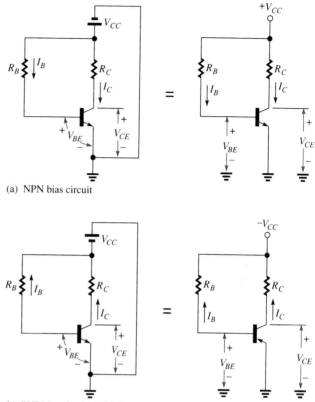

(a) NPN bias circuit

(b) PNP bias circuit. With the reference polarities shown, V_{BE} and V_{CE} are negative.

the values of V_{BE} can be assumed to be either 0.7 V or 0.3 V (Si or Ge) with the following polarities:

	Si	Ge
NPN	0.7 V	0.3 V
PNP	−0.7 V	−0.3 V

Note that V_{CE} is negative in the PNP amplifier.

The base current in each transistor is, from Ohm's law, the difference in voltage across R_B divided by R_B:

$$I_B = \frac{V_{CC} - V_{BE}}{R_B} \tag{5.1}$$

From equation 3.6, the quiescent (bias) value of the collector current is then

$$I_C = \beta I_B \tag{5.2}$$

The quiescent value of V_{CE} is found by writing Kirchhoff's voltage law around the loop containing V_{CC} and R_C:

$$V_{CE} + I_C R_C = V_{CC}$$
$$V_{CE} = V_{CC} - I_C R_C \qquad\qquad (5.3)$$

Example 5.1

For the circuit shown in Figure 5.4,

1. Find the quiescent values of I_C and V_{CE}.
2. Verify that the collector–base junction is reverse biased.

Figure 5.4
(Example 5.1)

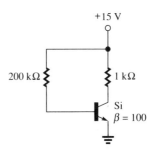

Solution

1. Since the transistor is a silicon NPN type, $V_{BE} = 0.7$ V. Then, from equation 5.1,

$$I_B = \frac{V_{CC} - V_{BE}}{R_B} = \frac{15 \text{ V} - 0.7 \text{ V}}{200 \text{ k}\Omega} = 71.5 \ \mu\text{A}$$

The quiescent collector current is

$$I_C = \beta I_B = 100(71.5 \ \mu\text{A}) = 7.15 \text{ mA}$$

From equation 5.3, the quiescent collector voltage is

$$V_{CE} = V_{CC} - I_C R_C = 15\text{V} - (7.15 \text{ mA})(1 \text{ k}\Omega) = 7.85 \text{ V}$$

2. The collector-to-base voltage, V_{CB}, is

$$V_{CB} = V_C - V_B$$

where V_C is the collector-to-ground voltage (the same as V_{CE}) and V_B is the base-to-ground voltage, 0.7 V. Thus,

$$V_{CB} = V_C - V_B = 7.85 \text{ V} - 0.7 \text{ V} = 7.15 \text{ V}$$

Since V_{CB} is positive, the N-type collector is positive with respect to the P-type base, and the collector–base junction is therefore reverse biased.

When equations 5.1 through 5.3 are used to find the quiescent point of a PNP amplifier, current values will be negative when negative voltages (V_{CC} and V_{BE}) are used in the equations. Since we assume base current flowing into the base is positive and collector current flowing from collector to emitter is positive, the significance of negative currents is simply that (conventional) current flows in the opposite directions in a PNP transistor (out of the base and from emitter to collector). For example, if the transistor in Example 5.1 had been PNP, then, with $V_{CC} = -15$ V and $V_{BE} = -0.7$ V, we would find

$$I_B = \frac{-15 \text{ V} - (-0.7 \text{ V})}{200 \text{ k}\Omega} = -71.5 \ \mu\text{A}$$

$$I_C = \beta I_B = 100(-71.5 \ \mu\text{A}) = -7.15 \text{ mA}$$

$$V_{CE} = V_{CC} - I_C R_C = -15 \text{ V} - (-7.15 \text{ mA})(1 \text{ k}\Omega) = -7.85 \text{ V}$$

The Load Line

When we think of V_{CE} and I_C as *variables,* equation 5.3 (repeated below) shows how those variables must be related in a *particular* bias circuit, i.e., for particular (constant) values of V_{CC} and R_C:

$$\underset{\text{variable}}{V_{CE}} = \underset{\text{constant}}{V_{CC}} - \underset{\text{variable constant}}{I_C R_C}$$

$$(\text{analogous to} \quad \underset{}{x} = \underset{}{c_1} - \underset{}{y c_2})$$

As in any such equation, picking any value of I_C determines (by computing from the equation) what the value of V_{CE} must be. In short, equation 5.3 can be used to determine *every* combination of values of I_C and V_{CE} that could possibly exist in a particular bias circuit. To illustrate, V_{CE} and I_C in the bias circuit of Figure 5.4 (Example 5.1) are related by

$$V_{CE} = 15 \text{ V} - I_C(10^3 \ \Omega)$$

Thus, for example, if we wanted to know the value of V_{CE} when I_C equaled, say, 0.5 mA, we could readily solve this equation to find $V_{CE} = 10$ V. Equation 5.3 is in the form of the equation of a straight line plotted on a set of $I_C - V_{CE}$-axes. It is called the *load line* for the circuit. To plot a load line (or any straight line) on a set of axes, it is only necessary to find the *intercepts* of the line: the points where the line intersects each axis. Setting $I_C = 0$ in equation 5.3, we find the intercept on the V_{CE}-axis:

$$V_{CE} = V_{CC} - (0)R_C = V_{CC}$$

Setting $V_{CE} = 0$ and solving for I_C, we find the intercept on the I_C-axis:

$$0 = V_{CC} - I_C R_C$$

$$I_C = \frac{V_{CC}}{R_C}$$

Figure 5.5 shows the plot of a load line on a set of $I_C - V_{CE} =$ axes. The quiescent point Q is of course one point on the plot of a load line.

Recall that the collector characteristics of a BJT are plots of I_C versus V_{CE} for different values of I_B (Figure 3.8). Since the load line is an equation relating I_C to V_{CE}, we can plot it on the same axes as the collector characteristics. The collector characteristics show the relationship between I_C and V_{CE} for a particular transistor, and the load line shows how I_C is related to V_{CE} for a particular bias circuit.

Figure 5.5
The load line is a plot of the straight line
whose equation is $V_{CE} = V_{CC} - I_C R_C$.
Every point on the load line represents
one possible combination of current I_C
and voltage V_{CE} that can exist in a par-
ticular bias circuit.

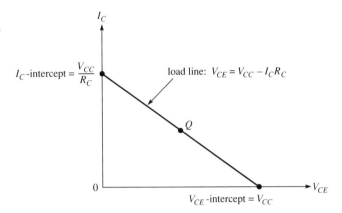

Thus, plotting a load line on a set of collector characteristics allows us to determine graphically the behavior of a particular transistor in a particular bias circuit. The next example illustrates graphical analysis of a CE amplifier.

Example 5.2

The silicon transistor in the CE amplifier shown in Figure 5.6(a) has the collector characteristics shown in Figure 5.6(b).

1. Construct the load line on the collector characteristics and use it to find the quiescent point.
2. Compare the results of (1) with values found by computation (using the method of Example 5.1).

Solution

1. The equation of the load line is

$$V_{CE} = 15 \text{ V} - I_C(1.5 \text{ k}\Omega)$$

The intercepts of the load line are

$$(I_C = 0) \quad V_{CE} = 15 \text{ V}$$

$$(V_{CE} = 0) \quad I_C = \frac{V_{CC}}{R_C} = \frac{15 \text{ V}}{1.5 \text{ k}\Omega} = 10 \text{ mA}$$

Drawing a straight line through the intercepts produces the load line shown in Figure 5.6(b). The quiescent base current is

$$I_B = \frac{15 \text{ V} - 0.7 \text{ V}}{286 \text{ k}\Omega} = 50 \text{ } \mu\text{A}$$

The quiescent point is therefore at the intersection of the load line with the $I_B = 50 \text{ } \mu\text{A}$ curve on the characteristics, as shown by the point labeled Q in the figure. The coordinates of the Q-point, which are the quiescent values of I_C and V_{CE}, are seen to be $I_C \approx 5.1 \text{ mA}$ and $V_{CE} \approx 7.3 \text{ V}$.

2. At the Q-point,

$$\beta_{\text{dc}} = \frac{I_C}{I_B} = \frac{5.1 \text{ mA}}{50 \text{ } \mu\text{A}} = 102$$

Figure 5.6
(Example 5.2)

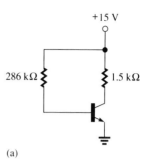

(a)

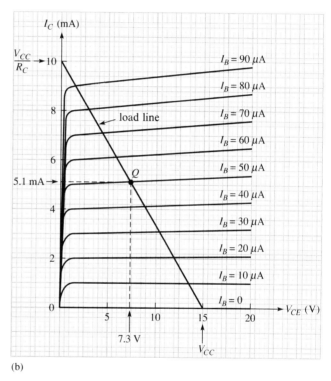

(b)

From equation 5.2, the quiescent value of I_C is

$$I_C = \beta I_B = 102(50 \ \mu A) = 5.1 \text{ mA}$$

From equation 5.3, the quiescent value of V_{CE} is

$$V_{CE} = V_{CC} - I_C R_C = 15 \text{ V} - (5.1 \text{ mA})(1.5 \text{ k}\Omega) = 7.35 \text{ V}$$

We see that the computed values compare favorably with those found graphically.

When the CE configuration is operated as a small-signal amplifier, small changes in the input voltage cause corresponding changes in the input (base) current. Changes in base current cause changes in output current and output voltage: I_C and V_{CE}. For any value of I_B, the values of I_C and V_{CE} can be found at the intersection of the load line and the curve corresponding to the value of I_B. Suppose, for example, that I_B in Example 5.2 increases to 60 μA. Locate the $I_B = 60$ μA curve in Figure 5.6(b) and verify that at its intersection with the load line, $I_C \approx 6.1$ mA and $V_{CE} \approx 5.8$ V. Similarly, if I_B decreases to 40 μA, then the intersection of $I_B = 40$ μA with the load line gives $I_C \approx 4.1$ mA and $V_{CE} \approx 8.8$ V. Thus, we can visualize operation of the small-signal amplifier as a point that moves up and down the load line, tracing out all possible values of I_C and V_{CE} as I_B changes between its minimum and maximum values. This variation is illustrated in Figure 5.7. We see that as base current changes (sinusoidally) between 40 μA and 60 μA, the values of I_C and V_{CE} change sinusoidally between their minimum and maximum values. The peak value of the sinusoidal output voltage $v_o(t)$ is one-half its peak-to-peak value: ½(8.8 V − 5.8 V) = 1.5 V, and its dc component is the quiescent voltage 7.3 V. Thus,

$$v_o(t) = 7.3 + 1.5\sin\omega t \text{ V}$$

Note that the current gain of the amplifier is

$$A_i = \frac{\Delta I_C}{\Delta I_B} = \beta_{ac} = \frac{6.1 \text{ mA} - 4.1 \text{ mA}}{60 \text{ } \mu\text{A} - 40 \text{ } \mu\text{A}} = 100$$

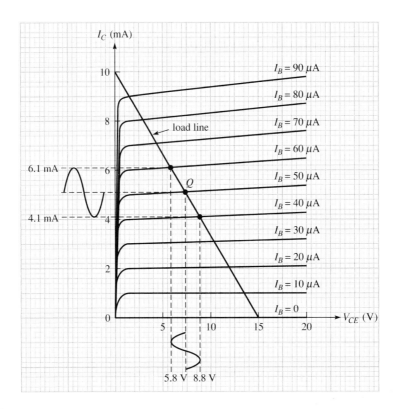

Figure 5.7
As the input current (I_B) varies between 40 μA and 60 μA, the output current (I_C) varies between 4.1 mA and 6.1 mA, and the output voltage (V_{CE}) varies between 5.8 V and 8.8 V. Visualize a point moving up and down the load line to trace out these values.

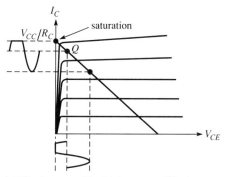

(a) Clipping caused by driving an amplifier into saturation. The Q-point is set too high on the load line.

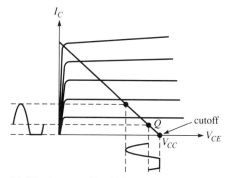

(b) Clipping caused by driving an amplifier to cutoff. The Q-point is set too low on the load line.

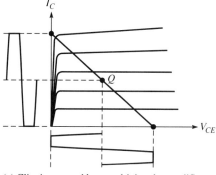

(c) Clipping caused by overdriving the amplifier.

Figure 5.8
Clipping is the flattening of peaks due to an amplifier being driven into saturation and/or cutoff.

The total possible variation of output current and voltage in a transistor amplifier is limited at the top end of the load line by the onset of *saturation* (where $V_{CE} \approx 0$ and $I_C \approx I_C(\text{sat}) = V_{CC}/R_C$) and at the bottom end by the onset of *cutoff* (where $I_C \approx 0$ and $V_{CE} \approx V_{CC}$. If the Q-point is set too close to either one of these extremes, the total possible output variation (called the output *swing*) will be restricted. If the input signal is too large, the amplifier may be driven into saturation or cutoff, or both, and the output waveform will have flattened peaks, a type of distortion called *clipping*. Figure 5.8 illustrates these facts.

An Improved CE Bias Circuit

The practical difficulty with the bias circuit shown in Figure 5.3 is that the quiescent point depends heavily on the value of β. This dependency is apparent from equation 5.2: $I_C = \beta I_B$. In Chapter 3, we discussed the fact that β varies with temperature, collector current, and among transistors of the same type. This variability makes it difficult to predict or maintain a specific quiescent point in the bias circuit of Figure 5.3. For example, if the bias circuit is designed for a transistor having $\beta = 50$ and a new transistor having $\beta = 150$ is substituted for it, the quiescent current will increase by 300%, an increase that might well drive the transistor into saturation.

Figure 5.9(a) shows a bias circuit that, when properly designed, greatly reduces the dependency of the quiescent point on the value of β. Proper design requires that the value of the dc input resistance, R_{in}, looking directly into the base of the transistor be much greater than the parallel equivalent resistance of R_1 and R_2, $R_1 \| R_2$. It can be shown that

$$R_{in} = (\beta + 1)R_E \approx \beta R_E \qquad (5.4)$$

so we require that

$$\beta R_E \gg R_1 \| R_2 \qquad (5.5)$$

By making $R_1 \| R_2$ small, we can ensure that this inequality holds for a range of values of β. Figure 5.9(b) shows the voltage source that is the (Thevenin) equivalent of the bias network consisting of resistors R_1 and R_2 and the supply voltage, V_{CC}. When condition 5.5 is met, R_{in} does not *load* the voltage source; that is, the base voltage V_B is not substantially affected by the exact value of R_{in}, so long as R_{in} is large. Consequently, the base voltage and quiescent point do not depend heavily on the value of β. As shown in Figure 5.9(b), the Thevenin equivalent voltage is $V_{CC}R_2/(R_1 + R_2)$. When condition 5.5 is met, the voltage source is approximately ideal (has a small internal resistance compared to the "load" (R_{in})

Figure 5.9
An improved bias circuit.

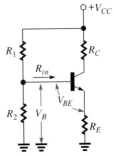

(a) Schematic diagram

Thevenin equivalent circuit

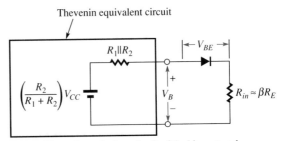

(b) The Thevenin equivalent circuit of the bias network at the base of the transistor. When $R_{in} \gg R_1 \| R_2$,

$$V_B \approx \left(\frac{R_2}{R_1 + R_2} \right) V_{CC}.$$

connected across it), and in that case, neglecting V_{BE},

$$V_B \approx \left(\frac{R_2}{R_1 + R_2}\right) V_{CC} \qquad (5.6)$$

It is clear that V_B is simply the voltage across R_2 when R_1 and R_2 are treated as an unloaded voltage divider across V_{CC}. This method of providing bias is often called the voltage-divider method. Rules of thumb that are often used in the design of a voltage-divider bias circuit are:

1. The emitter-to-ground voltage should be approximately $0.1V_{CC}$.
2. The current in R_2 should be approximately 10 times the base current, a condition that can be met by making $R_2 \approx \beta(\min)R_E/10$, where $\beta(\min)$ is the smallest value of β the transistor may have.

To analyze a properly designed voltage-divider bias circuit, we first calculate V_B using equation 5.6. Then, the voltage V_E between emitter and ground is

$$V_E = V_B - V_{BE} \qquad (5.7)$$

where $V_{BE} = 0.3$ V (Ge) or 0.7 V (Si). Refer to Figure 5.10, which shows the sequence of calculations. The emitter current I_E is then the emitter voltage divided by R_E:

$$I_E = \frac{V_E}{R_E} \qquad (5.8)$$

Since $I_C = \alpha I_E \approx I_E$, the quiescent collector current is

$$I_C \approx I_E$$

Writing Kirchhoff's voltage law from V_{CC} through R_C and R_E to ground, we find

$$V_{CC} = I_C R_C + V_{CE} + I_E R_E$$

The quiescent value of V_{CE} is then

$$\begin{aligned} V_{CE} &= V_{CC} - I_C R_C - I_E R_E \\ &\approx V_{CC} - I_C(R_C + R_E) \end{aligned} \qquad (5.9)$$

Figure 5.10
Calculating the quiescent values of I_C and V_{CE} in a voltage-divider bias circuit. The circled numbers show the sequence of calculations.

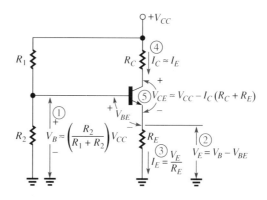

Example 5.3

The silicon transistor in Figure 5.11 has $\beta = 150$. Find the quiescent values of I_C and V_{CE}.

Figure 5.11
(Example 5.3)

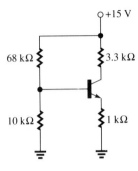

Solution.   We first verify that the circuit is properly designed:

$$R_{in} \approx \beta R_E = 150(1 \text{ k}\Omega) = 150 \text{ k}\Omega$$

$$R_1 \| R_2 = \frac{(68 \text{ k}\Omega)(10 \text{ k}\Omega)}{68 \text{ k}\Omega + 10 \text{ k}\Omega} = 8.72 \text{ k}\Omega$$

We see that condition 5.5 ($\beta R_E \gg R_1 \| R_2$) is satisfied, so we may proceed with the analysis illustrated in Figure 5.10:

1. $V_B = \left(\dfrac{R_2}{R_1 + R_2}\right) V_{CC} = \left(\dfrac{10 \text{ k}\Omega}{10 \text{ k}\Omega + 68 \text{ k}\Omega}\right) 15 \text{ V} = 1.92 \text{ V}$

2. $V_E = V_B - V_{BE} = 1.92 \text{ V} - 0.7 \text{ V} = 1.22 \text{ V}$

3. $I_E = \dfrac{V_E}{R_E} = \dfrac{1.22 \text{ V}}{1 \text{ k}\Omega} = 1.22 \text{ mA}$

4. $I_C \approx I_E = 1.22 \text{ mA}$

5. $V_{CE} \approx V_{CC} - I_C(R_C + R_E) = 15 \text{ V} - (1.22 \text{ mA})(1 \text{ k}\Omega + 3.3 \text{ k}\Omega) = 9.84 \text{ V}$

If the value of β is known, we can find the quiescent point in Figure 5.9(a) more accurately by taking into account both V_{BE} and the loading effect of $R_{in} \approx \beta R_E$ on the voltage divider. Whether condition 5.5 is met or not, we can show (Exercise 5.16) that

$$I_C = \beta \left[\frac{\left(\dfrac{R_2}{R_1 + R_2}\right) V_{CC} - V_{BE}}{R_1 \| R_2 + \beta R_E} \right] \qquad \text{(5.10)}$$

where $V_{BE} \approx 0.7 \text{ V}$ for silicon and 0.3 V for germanium.

Current Mirrors

The current mirror is a bias method used frequently in integrated circuits. Its name is derived from the fact that it is capable of *reflecting* (duplicating) the

Figure 5.12
The current mirror used to provide bias in integrated circuits. $I_C \approx I$, so I is reflected to Q_2.

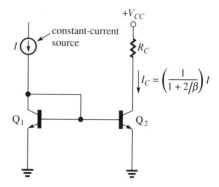

current flowing in a constant-current source to one or more amplifiers. Figure 5.12 shows the basic circuit. *I* is a constant-current source and Q_1 is a *diode-connected transistor,* a transistor that behaves like a diode because its collector is connected directly to its base. Q_2 is the transistor amplifier we wish to bias. When the transistors are closely *matched* (have the same values of α, β, V_{BE}, etc.), as they usually are in an integrated circuit, it can be shown* that bias current I_{C2} in Q_2 is

$$I_{C2} = \left(\frac{1}{1 + 2/\beta}\right) I \qquad (5.11)$$

If β is relatively large, then $1 + 2/\beta \approx 1$ and $I_{C2} \approx I$. Thus, the current-source current *I* is reflected to the transistor. More than one transistor can be biased by the same mirror, as will be demonstrated in the next example. If the mirror is used to provide bias to *n* transistors, the current I_C in each is

$$I_C = \left(\frac{1}{1 + (n + 1)/\beta}\right) I \qquad (5.12)$$

Example 5.4

Find the quiescent values of I_C and V_{CE} in each of Q_1, Q_2, and Q_3 in Figure 5.13,

1. assuming the current in the current source is exactly reflected to each transistor;
2. assuming each transistor has $\beta = 160$ and using equation 5.12.

Solution

1. $I_{C1} = I_{C2} = I_{C3} = 1.5$ mA
   $V_{CE1} = 10$ V $- (1.5$ mA$)(2.5$ k$\Omega) = 6.25$ V
   $V_{CE2} = 10$ V $- (1.5$ mA$)(3$ k$\Omega) = 5.5$ V
   $V_{CE3} = 10$ V $- (1.5$ mA$)(4.2$ k$\Omega) = 3.7$ V
2. From equation 5.12, with $n = 3$,

$$I_{C1} = I_{C2} = I_{C3} = \left[\frac{1}{1 + \left(\frac{3 + 1}{160}\right)}\right] (1.5 \text{ mA}) = 1.46 \text{ mA}$$

*See, for example, Theodore F. Bogart. Jr., *Electronic Devices and Circuits,* 3rd ed. (New York: Macmillan Publishing Co., 1993), p. 234.

Figure 5.13
(Example 5.4)

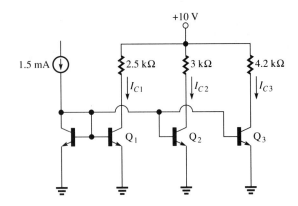

$$V_{CE1} = 10 \text{ V} - (1.46 \text{ mA})(2.5 \text{ k}\Omega) = 6.35 \text{ V}$$
$$V_{CE2} = 10 \text{ V} - (1.46 \text{ mA})(3 \text{ k}\Omega) = 5.62 \text{ V}$$
$$V_{CE3} = 10 \text{ V} - (1.46 \text{ mA})(4.2 \text{ k}\Omega) = 3.87 \text{ V}$$

5.3   THE CE SMALL-SIGNAL AMPLIFIER MODEL

Small-Signal Transistor Parameters

Recall that small-signal operation of an electronic device means the device is used in such a way that current and voltage variations are limited to a small (essentially linear) range of its characteristic curve(s). Small-signal *parameters* are ac properties such as ac resistance and gain that are derived from small-signal operation of the device. We have already studied (Chapter 3) one small-signal parameter of a BJT: β.

Another important small-signal parameter of a BJT is its *emitter resistance*, r_e. By definition, small-signal emitter resistance is small-signal base-to-emitter voltage divided by small-signal emitter current:

$$r_e = \frac{v_{be}}{i_e} \text{ ohms} \tag{5.13}$$

Recall that the base–emitter junction of a BJT is a forward-biased PN junction. In Chapter 2 we studied the ac resistance of a forward-biased PN junction in the context of a diode and learned (equation 2.3) that

$$r_{ac} = \frac{V_T}{I} \approx \frac{0.026 \text{ V}}{I} \text{ (at room temperature)}$$

where I is the dc current flowing through the junction. Figure 5.14(a) shows the base–emitter junction of a BJT represented as a forward-biased diode. Since the dc current through the junction is the emitter current I_E, we have

$$r_e \approx \frac{0.026 \text{ V}}{I_E} \text{ (at room temperature)} \tag{5.14}$$

Figure 5.14(b) shows the junction replaced by its small-signal resistance, r_e.

Figure 5.14
The small-signal emitter resistance r_e.

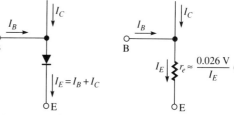

(a) Representation of the base–emitter junction of a BJT as a forward-biased diode

(b) The base–emitter junction replaced by its small-signal resistance r_e

Figure 5.15
Input circuit of a BJT in the CE configuration.

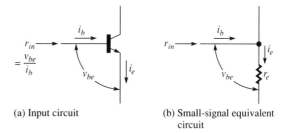

(a) Input circuit

(b) Small-signal equivalent circuit

We wish next to determine the small-signal resistance looking into the base of the BJT, since that is the input terminal for the common-emitter configuration. See Figure 5.15(a). Figure 5.15(b) shows the small-signal circuit with the base–emitter junction replaced by its small-signal equivalent, r_e. Now,

$$r_{in} = \frac{v_{be}}{i_b} = \frac{v_{be}}{i_e/(\beta + 1)} = (\beta + 1)\frac{v_{be}}{i_e} = (\beta + 1)r_e \qquad (5.15)$$

Equation 5.15 expresses the important fact that the input resistance of a CE amplifier is r_e magnified by the factor $\beta + 1$. Although r_e itself is typically small (usually less than 50 Ω), the quantity $r_{in} = (\beta + 1)r_e$ is much larger, a desirable characteristic for a voltage amplifier. The quantity $(\beta + 1)r_e$ is often designated r_π. Since $\beta + 1 \approx \beta$,

$$r_{in} = r_\pi = (\beta + 1)r_e \approx \beta r_e \qquad (5.16)$$

Any external resistance connected in series with the emitter is also magnified by the factor $\beta + 1$. For example, in the bias circuit of Figure 5.9(a),

$$r_{in} = (\beta + 1)(r_e + R_E) \qquad (5.17)$$

Example 5.5

Find the small-signal input resistance of the CE amplifiers in Figures 5.6 and 5.9 (Examples 5.2 and 5.3). Assume $\beta = 100$ in both cases.

Solution.   In Example 5.2, we found that $I_C = 5$ mA $\approx I_E$. Therefore,

$$r_e = \frac{0.026 \text{ V}}{I_E} = \frac{0.026 \text{ V}}{5 \text{ mA}} = 5.2 \text{ }\Omega$$

$$r_{in} = (\beta + 1)r_e = 101(5.2 \text{ }\Omega) = 525 \text{ }\Omega$$

In Example 5.3, $I_E = 1.22$ mA. Therefore,

$$r_e = \frac{0.026 \text{ V}}{1.22 \text{ mA}} = 21.3 \text{ }\Omega$$

and

$$r_{in} = (\beta + 1)(r_e + R_E) = 101(21.3 \text{ }\Omega + 1 \text{ k}\Omega) = 103 \text{ k}\Omega$$

Another parameter that is sometimes useful in analyzing BJT amplifiers is small-signal *collector resistance*, r_c:

$$r_c = \frac{v_{cb}}{i_c} \tag{5.18}$$

Collector resistance is generally quite large, so large, in fact, that in practical work it is often assumed to be infinite.

A Small-Signal Transistor Model

A transistor *model* is an equivalent circuit that has properties similar or identical to a transistor. Many models have been developed, some that are very complex but very accurate representations of a transistor, and others that are simpler, easier to analyze, and sufficiently accurate for many practical purposes. Figure 5.16 shows one of the simpler models for a transistor in the common-emitter configuration. Notice that it contains a *current-controlled current source* whose value is βi_b. The source produces a small-signal current whose value depends on the small-signal base current i_b. For example, if $\beta = 100$ then, when $i_b = 50$ μA, the current source produces $\beta i_b = 100(50 \text{ }\mu\text{A}) = 5$ mA, and if i_b changes to 75 μA, the current source changes to $100(75 \text{ }\mu\text{A}) = 7.5$ mA. We see that the current source in the model reflects the fact that $i_c = \beta i_b$ in the transistor. The input resistance looking into the base is clearly $r_{in} = (\beta + 1)r_e$, as shown by equation 5.15, and the output resistance is r_c/β.

Figure 5.16 is not a model for a *complete* small-signal amplifier because it does not show, among other things, the external resistors that must be connected to provide bias. Figure 5.17(a) shows a schematic diagram of a small-signal amplifier using the bias method of Figure 5.3 and with a signal source connected to the input. For the moment, we omit any source resistance (r_s) that may be present.

Figure 5.16
Small-signal model for a BJT in the common-emitter configuration.

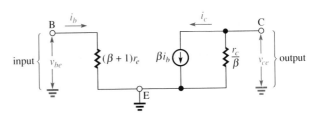

Figure 5.17
*Small-signal CE amplifier and its equiva-
lent circuit.*

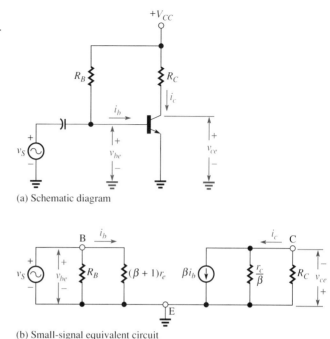

(a) Schematic diagram

(b) Small-signal equivalent circuit

Notice that the signal source is capacitor-coupled to the amplifier, as discussed in Chapter 4. For purposes of small-signal analysis, we assume that the signal frequency is high enough to make the capacitive reactance of the coupling capacitor negligibly small. Figure 5.17(b) shows the small-signal equivalent circuit for the amplifier, where the transistor in part (a) of the figure has been replaced by its model (Figure 5.16). Note these important points:

1. Since the capacitor is assumed to have negligibly small reactance, it does not appear in the model, having been replaced by a short circuit.
2. The dc power source, V_{CC}, does not appear in the model. Recall that a dc voltage source is a short circuit to ac signals. Thus, as far as small signals are concerned, R_C is connected between the collector and ground. For the same reason, R_B is connected between the base and ground.

We can use the model of Figure 5.17(b) to determine the voltage gain of the amplifier. Let us first make this important observation: The polarity of v_{ce} is opposite that of v_{be}. Therefore, the amplifier *inverts* the signal; that is, there is a 180° phase shift between input and output. This also follows from the equation of the load line:

$$V_{CE} = V_{CC} - I_C R_C$$

An increase in V_{BE} causes an increase in I_B, which causes an increase in I_C ($I_C = \beta I_B$), and from the equation for the load line we see that an increase in I_C causes V_{CE} to decrease. Recapitulating, an increase in input voltage (V_{BE}) causes a decrease in output voltage (V_{CE}), so the amplifier inverts. Returning to Figure 5.17(b), we see that

$$v_{in} = v_S = v_{be} = (\beta + 1)r_e i_b \tag{5.19}$$

Also,

$$v_o = v_{ce} = R_C\|(r_c/\beta)i_c \tag{5.20}$$

where $R_C\|(r_c/\beta)$ is the parallel equivalent resistance of R_C and r_c/β. Thus,

$$
\begin{aligned}
A_v &= \frac{v_o}{v_{in}} = \frac{v_{ce}}{v_{be}} = \frac{-R_C\|(r_c/\beta)\overbrace{\beta i_c}}{(\beta + 1)r_e i_b} = \frac{-R_C\|(r_c/\beta)\overbrace{\beta i_b}}{(\beta + 1)r_e i_b} \\
&= \frac{-R_C\|(r_c/\beta)\beta}{(\beta + 1)r_e}
\end{aligned}
\tag{5.21}
$$

where the minus sign denotes phase inversion. We can invoke two reasonable approximations to simplify equation 5.21. First, R_C is generally much smaller in practice than r_c/β, so

$$R_C\|(r_c/\beta) \approx R_C \tag{5.22}$$

Also,

$$\frac{\beta}{\beta + 1} \approx 1 \tag{5.23}$$

Thus,

$$A_v \approx \frac{-R_C\beta}{(\beta + 1)r_e} \approx -\frac{R_C}{r_e} \tag{5.24}$$

Notice that the current gain of a transistor in the CE configuration is

$$A_i = \frac{i_c}{i_b} = \beta \tag{5.25}$$

Example 5.6

For the amplifier in Figure 5.17(a), the silicon transistor has $\beta = 75$, $r_c = 15$ MΩ, $V_{CC} = 9$ V, $R_B = 220$ kΩ, and $R_C = 1$ kΩ.

1. Draw the small-signal equivalent circuit of the amplifier.
2. Verify the validity of approximations 5.22 and 5.23 and find the approximate voltage gain.
3. What is the peak-to-peak output voltage when $v_S = 0.01\sin\omega t$ V?

Solution

1. We must first find the value of r_e, which means we must find the dc emitter current, I_E, in order to use equation 5.14.

$$I_B = \frac{V_{CC} - V_{BE}}{R_B} = \frac{9 \text{ V} - 0.7 \text{ V}}{220 \text{ k}\Omega} = 37.73 \text{ }\mu\text{A}$$

$$I_E \approx I_C = \beta I_B = 75(37.73 \text{ }\mu\text{A}) = 2.83 \text{ mA}$$

Then, from equation 5.14,

$$r_e \approx \frac{0.026 \text{ V}}{I_E} = \frac{0.026 \text{ V}}{2.83 \text{ mA}} = 9.2 \text{ }\Omega$$

Figure 5.18
(Example 5.6)

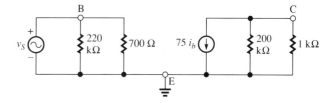

The input resistance at the base is

$$r_{in} = (\beta + 1)r_e = 76(9.2 \ \Omega) = 700 \ \Omega$$

Also,

$$r_c/\beta = \frac{15 \ M\Omega}{75} = 200 \ k\Omega$$

The small-signal equivalent circuit is shown in Figure 5.18.
2. $R_C\|(r_c/\beta) = 1 \ k\Omega\|200 \ k\Omega = 995 \ \Omega \approx 1 \ k\Omega = R_C$

$$\frac{\beta}{\beta + 1} = \frac{75}{76} = 0.987 \approx 1$$

Since the approximations are valid, we use equation 5.24 to find the approximate voltage gain:

$$A_v \approx -\frac{R_C}{r_e} = \frac{-1000 \ \Omega}{9.2 \ \Omega} = -109$$

3. The peak-to-peak value of the input voltage is 2(0.01 V) = 0.02 V. Therefore, the peak-to-peak value of the output voltage is

$$|A_v|(0.02 \ V) = 109(0.02 \ V) = 2.18 \ V \ \text{pk-pk}$$

Source and Load Resistance

Our small-signal analysis of the CE amplifier up to this point has not considered the effect of signal-source resistance and load resistance on the overall voltage gain of the amplifier (from source to load). Figure 5.19(a) shows the circuit with those components connected. Note that load resistance R_L is capacitor-coupled to the output of the amplifier at the collector terminal. To distinguish between the external components and the transistor amplifier itself, we now refer to the latter as the amplifier *stage*. The terms r_{in}(stage) and r_o(stage) refer to the input and output resistance of the amplifier stage, as shown in the figure. Figure 5.19(b) shows the small-signal equivalent circuit when the transistor is replaced by its small-signal model. Note that we omit the output coupling capacitor and that R_L is in parallel with R_C. We readily see that

$$r_{in}(\text{stage}) = R_B\|(\beta + 1)r_e \tag{5.26}$$

and

$$r_o(\text{stage}) = R_C\|(r_c/\beta) \approx R_C \tag{5.27}$$

Recall from Chapter 4 that voltage division between the source and input resistance and between the load and output resistance of the amplifier reduces the overall voltage gain v_L/v_S. Substituting r_{in}(stage) for r_{in} and r_o(stage) for r_o in

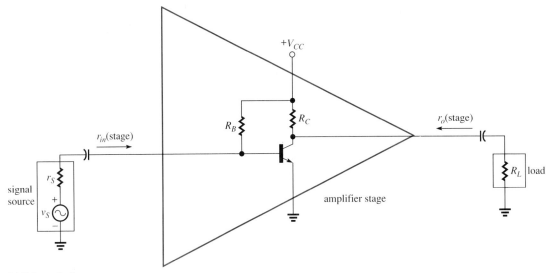

(a) Schematic diagram

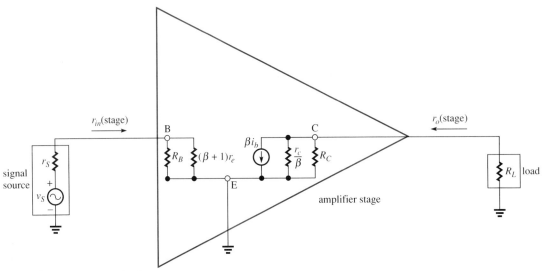

(b) Small-signal equivalent circuit

Figure 5.19
The small-signal CE amplifier with load and source resistance.

equation 4.8, we have

$$\frac{v_L}{v_S} = \frac{r_{in}(\text{stage})}{r_S + r_{in}(\text{stage})} A_v \frac{R_L}{r_o(\text{stage}) + R_L} \tag{5.28}$$

Substituting equations 5.24, 5.26, and 5.27 into 5.28, we find

$$\frac{v_L}{v_S} = \frac{R_B \| (\beta + 1)r_e}{r_s + R_B \| (\beta + 1)r_e} \left(\frac{-R_C}{r_e}\right) \frac{R_L}{R_C + R_L} \tag{5.29}$$

Note that the two rightmost terms in equation 5.29 can be written

$$-\frac{1}{r_e}\left(\frac{R_C R_L}{R_C + R_L}\right) = -\frac{R_C\|R_L}{r_e} = -\frac{r_L}{r_e} \tag{5.30}$$

where $r_L = R_C\|R_L$ is called the *ac load resistance*, since R_C and R_L are indeed in parallel in the ac (small-signal) equivalent circuit. In terms of r_L, (5.29) becomes

$$\frac{v_L}{v_S} = \frac{-R_B\|(\beta + 1)r_e}{r_S + R_B\|(\beta + 1)r_e}\left(\frac{r_L}{r_e}\right) \tag{5.31}$$

Example 5.7

Find the voltage gain of the amplifier in Example 5.6 when the signal source has resistance 400 Ω and the load resistance is 4 kΩ.

Solution.   In Example 5.6, we found that $(\beta + 1)r_e = 700\ \Omega$ and $A_v = -109$.

$$r_{in}(\text{stage}) = R_B\|(\beta + 1)r_e = 220\ \text{k}\Omega\|700\ \Omega \approx 700\ \Omega$$

From equation 5.28,

$$\frac{v_L}{v_S} = \frac{700\ \Omega}{400\ \Omega + 700\ \Omega}(-109)\frac{4\ \text{k}\Omega}{1\ \text{k}\Omega + 4\ \text{k}\Omega} = -55.5$$

Note that $r_L = R_C\|R_L = 1\ \text{k}\Omega\|4\ \text{k}\Omega = 800\ \Omega$, so, with $r_e = 9.2\ \Omega$, we can also compute

$$\frac{v_L}{v_S} = \frac{700\ \Omega}{400\ \Omega + 700\ \Omega}\left(\frac{-r_L}{r_e}\right) = \frac{700}{1100}\left(\frac{-800}{9.2\ \Omega}\right) = -55.3$$

which, except for small roundoff error, is the same result.

Figure 5.20 is a summary of the bias and small-signal equations for the CE amplifier.

Figure 5.21(a) shows a CE amplifier using the improved voltage-divider bias method. Note that capacitor C_E is connected in parallel with R_E. Called an *emitter-bypass capacitor,* its purpose is to present a low-impedance path between the emitter and ground and thus effectively connect the emitter directly to ground, as far as ac signals are concerned. As we shall presently see, omitting C_E reduces the voltage gain of the amplifier but increases its input resistance. C_E has no effect on bias calculations since it is an open circuit to dc. Figure 5.21(b) shows the small-signal equivalent circuit of the amplifier with the transistor replaced by its small-signal model. Note that the voltage-divider resistors R_1 and R_2 are in parallel as far as ac is concerned, since each has one side connected to ac ground (V_{CC} is ac ground to R_1). If we let $R_B = R_1\|R_2$, we obtain a small-signal equivalent circuit that is *identical* to that studied earlier (Figure 5.19(b)). Consequently, the equations for $r_{in}(\text{stage})$ (5.26), $r_o(\text{stage})$ (5.27), and voltage gain (5.29) for that circuit are the same for this one, with $R_1\|R_2$ substituted for R_B.

If the emitter-bypass capacitor is omitted, then the total resistance seen by ac signals in the emitter circuit is $r_e + R_E$, instead of just r_e. In that case, the ac input resistance looking into the base of the transistor is, from equation 5.17,

$$r_{in} = (\beta + 1)(R_E + r_e) \tag{5.32}$$

Figure 5.20
Summary of bias and small-signal equations for the CE amplifier.

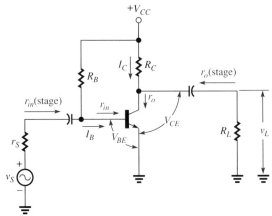

BIAS EQUATIONS

$V_{BE} = 0.7$ V (Si) 0.3 V (Ge)

$I_B = \dfrac{V_{CC} - V_{BE}}{R_B}$

$I_C = \beta I_B$

$V_{CE} = V_{CC} - I_C R_C$

SMALL-SIGNAL EQUATIONS

$r_e = \dfrac{0.026 \text{ V}}{I_E}$ \quad (room temperature)

$r_{in} = r_\pi = (\beta + 1)r_e \approx \beta r_e$

$r_{in}(\text{stage}) = R_B \| r_{in}$

$r_o = r_c / \beta$

$r_o(\text{stage}) = r_o \| R_C \approx R_C$

$A_v = \dfrac{v_{ce}}{v_{be}} = -\dfrac{r_o(\text{stage})}{r_e} \approx -\dfrac{R_C}{r_e}$

$\dfrac{v_L}{v_S} = \dfrac{r_S}{r_S + r_{in}(\text{stage})} A_v \dfrac{R_L}{r_o(\text{stage}) + R_L}$

$\approx \dfrac{-r_S}{r_S + r_{in}(\text{stage})} \left(\dfrac{r_L}{r_e}\right)$, where $r_L = R_C \| R_L$

$A_i = \dfrac{i_c}{i_b} = \beta$

Also,

$$r_{in}(\text{stage}) = R_B \| r_{in} \tag{5.33}$$

where $R_B = R_1 \| R_2$.

The voltage gain of the amplifier is found by substituting $r_e + R_E$ for r_e in equation 5.29:

$$\frac{v_L}{v_S} = \frac{-R_B \|(\beta + 1)(r_e + R_E)}{r_S + R_B \|(\beta + 1)(r_e + R_E)} \left[\frac{R_C}{r_e + R_E}\right] \frac{R_L}{R_C + R_L} \tag{5.34}$$

where $R_B = R_1 \| R_2$.

Figure 5.21
Small-signal CE amplifier using voltage-divider bias.

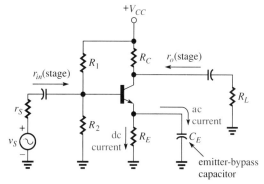

(a) Schematic diagram

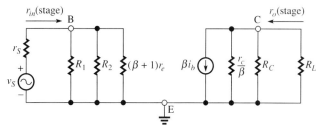

(b) Small-signal equivalent circuit. Compare with Figure 5.19 and note
that we can let R_B in Figure 5.19 equal $R_1 \| R_2$.

Example 5.8

Find the voltage gain v_L/v_S in Figure 5.22

1. with C_E connected as shown;
2. with C_E removed.

Solution

1. We first find the quiescent value of I_E.

$$\beta R_E = 100(470 \ \Omega) = 47 \ \text{k}\Omega$$
$$R_1 \| R_2 = 100 \ \text{k}\Omega \| 10 \ \text{k}\Omega = 9.09 \ \text{k}\Omega$$

Figure 5.22
(Example 5.8)

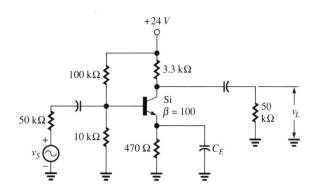

We do not have $\beta R_E \gg R_1 \| R_2$, so we must use equation 5.10 to find I_C:

$$I_C = \beta \left[\frac{\left(\dfrac{R_2}{R_1 + R_2}\right) V_{CC} - V_{BE}}{R_1 \| R_2 + \beta R_E} \right] = 100 \left[\frac{\left(\dfrac{10\ \text{k}\Omega}{10\ \text{k}\Omega + 100\ \text{k}\Omega}\right) 24\ \text{V} - 0.7\ \text{V}}{9.09\ \text{k}\Omega + 47\ \text{k}\Omega} \right]$$

$$= 2.64\ \text{mA} \approx I_E$$

Then,

$$r_e = \frac{0.026\ \text{V}}{2.64\ \text{mA}} = 7.8\ \Omega$$

The ac load resistance is

$$r_L = R_C \| R_L = 3.3\ \text{k}\Omega \| 50\ \text{k}\Omega = 3.1\ \text{k}\Omega$$

Also,

$$R_B = R_1 \| R_2 = 10\ \text{k}\Omega \| 100\ \text{k}\Omega = 9.1\ \text{k}\Omega$$
$$r_{in}(\text{stage}) = R_B \| (\beta + 1) r_e = 9.1\ \text{k}\Omega \| (101)(7.8\ \Omega)$$
$$= 9.1\ \text{k}\Omega \| 788\ \Omega = 725\ \Omega$$

Therefore, the voltage gain v_L/v_S is

$$\frac{v_L}{v_S} = \frac{-r_{in}(\text{stage})}{r_S + r_{in}(\text{stage})} \left[\frac{r_L}{r_e}\right] = -\left(\frac{725\ \Omega}{50\ \text{k}\Omega + 725\ \Omega}\right)\left(\frac{3.1\ \text{k}\Omega}{7.8\ \Omega}\right)$$
$$= -(0.0143)(397) = -5.68$$

Note that the voltage gain is diminished significantly (by the factor 0.0143) due to the fact that the source resistance is large compared to the input resistance.

2. Omitting C_E has no effect on r_e and r_L. However,

$$r_{in}(\text{stage}) = R_B \| (\beta + 1)(r_e + R_E) = 9.1\ \text{k}\Omega \| (101)(7.8\ \Omega + 470\ \Omega)$$
$$= 9.1\ \text{k}\Omega \| 48.3\ \text{k}\Omega = 7.66\ \text{k}\Omega$$

and

$$\frac{v_L}{v_S} = \frac{-r_{in}(\text{stage})}{r_S + r_{in}(\text{stage})} \left[\frac{R_C}{r_e + R_E}\right]\left(\frac{R_L}{R_C + R_L}\right)$$
$$= \frac{-7.66\ \text{k}\Omega}{50\ \text{k}\Omega + 7.66\ \text{k}\Omega} \left(\frac{3.3\ \text{k}\Omega}{7.8\ \Omega + 470\ \Omega}\right)\left(\frac{50\ \text{k}\Omega}{3.3\ \text{k}\Omega + 50\ \text{k}\Omega}\right)$$
$$= -(0.133)(6.48) = -0.86$$

Although the gain reduction due to voltage division at the input is not so severe as in part (1) (due to increased input resistance), omission of C_E has caused the gain factor to be reduced from 397 to 6.48, resulting in an overall gain that is much smaller than in (1). This reduction is called emitter *degeneration*.

Figure 5.23 is a summary of the bias and small-signal equations for the CE amplifier using the voltage-divider bias method.

5.4   THE COMMON-COLLECTOR AMPLIFIER

Bias Circuit

Figure 5.24 shows bias circuitry for the common-collector amplifier. Notice that the collector is connected directly to V_{CC}, which, as usual, is ac ground. The

Figure 5.23
Summary of bias and small-signal equations for the CE amplifier with voltage-divider bias.

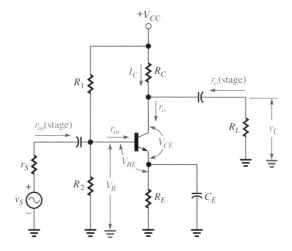

BIAS EQUATIONS

$V_{BE} = 0.7 \text{ V (Si)}\ \ 0.3 \text{ V (Ge)}$

$$I_C = \beta \left[\frac{R_2 V_{CC}/(R_1+R_2) - V_{BE}}{R_1 \| R_2 + \beta R_E} \right]$$

$$V_B \approx \left(\frac{R_2}{R_1+R_2} \right) V_{CC} \quad (\beta R_E >> R_1 \| R_2)$$

$V_E = V_B - V_{BE}$

$I_E = V_E/R_E \approx I_C$

$V_{CE} = V_{CC} - I_C (R_C + R_E)$

SMALL-SIGNAL EQUATIONS

$$r_e = \frac{0.026 \text{ V}}{I_E} \quad \text{(room temperature)}$$

$r_{in} = r_\pi = (\beta + 1)r_e \approx \beta r_e \quad (R_E \text{ bypassed by } C_E)$

$r_{in}(\text{stage}) = R_1 \| R_2 \| r_{in} = (\beta + 1)(r_e + R_E) \quad (R_E \text{ not bypassed})$

$r_o = r_c/\beta$

$r_o(\text{stage}) = r_o \| R_C \approx R_C$

$$A_v = \frac{v_{ce}}{v_{be}} = -\frac{r_o(\text{stage})}{r_e} \approx -\frac{R_C}{r_e} \quad (R_E \text{ bypassed by } C_E)$$

$$= -\frac{r_o(\text{stage})}{r_e + R_E} \approx -\frac{R_C}{r_e + R_E} \quad (R_E \text{ not bypassed})$$

$$\frac{v_L}{v_S} = \frac{r_S}{r_S + r_{in}(\text{stage})} A_v \frac{R_L}{r_o(\text{stage}) + R_L}$$

$$= \frac{-r_S}{r_S + r_{in}(\text{stage})} \left(\frac{r_L}{r_e} \right), \text{ where } r_L = R_C \| R_L \quad (R_E \text{ bypassed by } C_E)$$

$$A_i = \frac{i_c}{i_b} = \beta$$

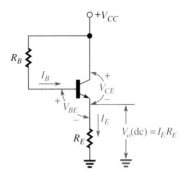

Figure 5.24
Common-collector bias circuit.

collector is therefore common to small-signal input (v_{bc}) and output (v_{ec}) voltages. Writing Kirchhoff's voltage law from V_{CC} through R_B, V_{BE}, and R_E, we obtain

$$V_{CC} = I_B R_B + V_{BE} + I_E R_E \qquad (5.34)$$

Substituting $I_E = (\beta + 1)I_B$ and solving for I_B, we obtain

$$I_B = \frac{V_{CC} - V_{BE}}{R_B + (\beta + 1)R_E} \qquad (5.35)$$

The quiescent value of the output current, I_E, is then

$$I_E = (\beta + 1)I_B \qquad (5.36)$$

The ac output voltage is between emitter and ground (which is the same as emitter-to-collector since the collector is at ac ground), so the dc, or bias, voltage for the output is the dc voltage across I_E:

$$V_o(\text{dc}) = I_E R_E = V_E \qquad (5.37)$$

Note that

$$V_{CC} = V_{CE} + I_E R_E$$

so

$$V_o(\text{dc}) = I_E R_E = V_{CC} - V_{CE} \qquad (5.38)$$

Small-Signal Circuit

Figure 5.25(a) shows the CC amplifier with signal source and load connected. Figure 5.25(b) shows the small-signal equivalent circuit. By writing Kirchhoff's voltage law from the base through the emitter to ground, it can be shown (Exercise 5.27) that

$$v_{in} = v_{bc} = (\beta + 1)i_b(r_e + r_L) \qquad (5.39)$$

where $r_L = R_E \| R_L$ is the ac load resistance. Since $v_o = v_{ec} = i_e r_L = (\beta + 1)i_b r_L$, we have

$$A_v = \frac{v_o}{v_{in}} = \frac{(\beta + 1)i_b r_L}{(\beta + 1)i_b(r_e + r_L)} = \frac{r_L}{r_e + r_L} \qquad (5.40)$$

Figure 5.25
The common-collector small-signal amplifier.

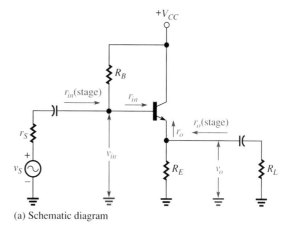

(a) Schematic diagram

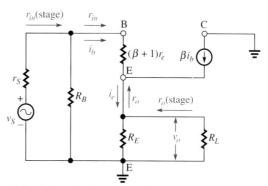

(b) Small-signal equivalent circuit

Note that A_v in the CC configuration is positive (noninverting) and has a value that is *always less than 1*. It is often true that $r_e \ll r_L$, in which case (5.40) becomes

$$A_v \approx 1 \qquad (r_e \ll r_L) \tag{5.41}$$

When $A_v \approx 1$, the output voltage is approximately equal to the input voltage and in phase with it, and we say that the output *follows* the input. The CC configuration is often called an *emitter follower* because the output (emitter voltage) follows (equals) the input. Although the voltage gain is less than 1, the current gain is large:

$$A_i = \frac{i_e}{i_b} = \frac{(\beta + 1)i_b}{i_b} = \beta + 1 \tag{5.42}$$

One useful property of an emitter follower is that it has a very large input resistance:

$$r_{in} = (\beta + 1)(r_e + r_L) \tag{5.43}$$

From Figure 5.25(b), we see that

$$r_{in}(\text{stage}) = R_B \| r_{in} = R_B \| (\beta + 1)(r_e + r_L) \tag{5.44}$$

Taking into account the voltage division that takes place at the input, we find the overall voltage gain v_L/v_S to be

$$\frac{v_L}{v_S} = \frac{r_{in}(\text{stage})}{r_S + r_{in}(\text{stage})} A_v = \left(\frac{R_B\|r_{in}}{r_S + R_B\|r_{in}}\right) \frac{r_L}{r_L + r_e} \tag{5.45}$$

The output resistance of the emitter follower is the generally small value

$$r_o = r_e + \frac{R_B\|r_S}{\beta + 1} \tag{5.46}$$

$$r_o(\text{stage}) = R_E\|r_o \tag{5.47}$$

It is often the case that $R_B\|r_S/(\beta + 1) \ll r_e$ (note that $R_B\|r_S = 0$ when $r_S = 0$), from which it follows that

$$r_o \approx r_e$$

and

$$r_o(\text{stage}) = R_E\|r_o \approx R_E\|r_e \approx r_e \tag{5.48}$$

The large input resistance and small output resistance of the emitter follower make it useful as a *buffer* amplifier between a source having large source resistance and a low-resistance load. The next example demonstrates such an application.

Example 5.9

An amplifier having an output resistance of 1 kΩ is to drive a 50-Ω load, as shown in Figure 5.26(a). The amplifier has $A_v = 140$.

1. Find the voltage gain v_L/v_S with the 50-Ω load connected directly to the output of the amplifier.
2. Find v_L/v_S when the emitter follower is inserted between the amplifier and the load, as shown in Figure 5.26(b).

Solution

1. The voltage gain with the 50-Ω load connected directly to the amplifier is severely reduced by the voltage division that occurs at the output:

$$\frac{v_L}{v_S} = \left(\frac{R_L}{r_S + R_L}\right) A_v = \left(\frac{50 \text{ }\Omega}{1 \text{ k}\Omega + 50 \text{ }\Omega}\right) 140 = 6.67$$

2. To find the input resistance of the emitter follower, we must find r_e, which means that we must first find the dc bias current I_E:

$$I_B = \frac{V_{CC} - V_{BE}}{R_B + (\beta + 1)R_E} = \frac{10 \text{ V} - 0.7 \text{ V}}{470 \text{ k}\Omega + 101(3.3 \text{ k}\Omega)} = 11.58 \text{ }\mu\text{A}$$

$$I_E = (\beta + 1)I_B = 101(11.58 \text{ }\mu\text{A}) = 1.17 \text{ mA}$$

Therefore,

$$r_e = \frac{0.026 \text{ V}}{I_E} = \frac{0.026 \text{ V}}{1.17 \text{ mA}} = 22.2 \text{ }\Omega$$

The ac load resistance for the emitter follower is

$$r_L = R_E\|R_L = 3.3 \text{ k}\Omega\|50 \text{ }\Omega \approx 50 \text{ }\Omega$$

Figure 5.26
(*Example 5.9*)

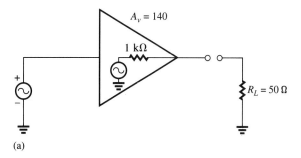

(a)

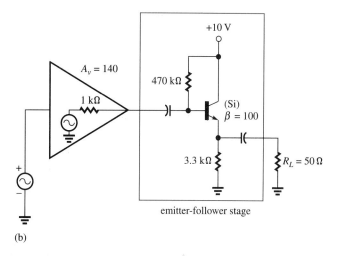

emitter-follower stage

(b)

From equation 5.44,

$$r_{in}(\text{stage}) = 470 \text{ k}\Omega \| (101)(22.2 \ \Omega + 50 \ \Omega)$$
$$= 470 \text{ k}\Omega \| 7292 \ \Omega = 7.18 \text{ k}\Omega$$

As far as the emitter follower is concerned, the signal-source resistance r_S is the output resistance (1 kΩ) of the amplifier driving it. Therefore, from equation 5.45, the overall voltage gain of the emitter follower is

$$\left(\frac{7.18 \text{ k}\Omega}{1 \text{ k}\Omega + 7.18 \text{ k}\Omega}\right)\left(\frac{50 \ \Omega}{50 \ \Omega + 22.2 \ \Omega}\right) = 0.607$$

Therefore, the voltage gain between the signal source and load is

$$\frac{v_L}{v_S} = (0.607)(140) = 85$$

We see that insertion of the emitter follower improved the voltage gain from 6.67 to 85, a 1174% increase. End-of-chapter exercise 5.30 shows that the overall power gain is also greatly increased when the emitter follower is used.

Figure 5.27 is a summary of the bias and small-signal equations for the common-collector amplifier.

Figure 5.27
Summary of bias and small-signal equations for the CC amplifier.

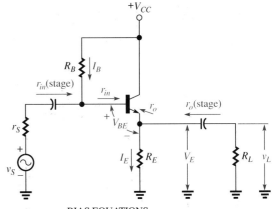

BIAS EQUATIONS

$V_{BE} = 0.7 \text{ V (Si)}\ \ 0.3 \text{ V (Ge)}$

$I_B \approx \dfrac{V_{CC} - V_{BE}}{R_B + (\beta + 1)R_E}$

$I_E = (\beta + 1)I_B$

$V_E = I_E R_E = V_{CC} - V_{CE}$

SMALL-SIGNAL EQUATIONS

$r_e = \dfrac{0.026 \text{ V}}{I_E}$ (room temperature)

$r_{in} = (\beta + 1)(r_e + r_L)$ where $r_L = R_E \| R_L$

$r_{in}(\text{stage}) = R_B \| r_{in}$

$r_o = r_e + \dfrac{R_B \| r_S}{\beta + 1} \approx r_e$

$r_o(\text{stage}) = R_E \| r_o \approx r_e$

$A_v = \dfrac{r_L}{r_e + r_L}$

$\quad \approx 1 \quad (r_e << r_L)$

$\dfrac{v_L}{v_S} = \left(\dfrac{R_B \| r_{in}}{r_S + R_B \| r_{in}}\right)\dfrac{r_L}{r_L + r_e}$

5.5   THE COMMON-BASE AMPLIFIER

Bias Circuit

Figure 5.28 shows a voltage-divider bias circuit for a common-base amplifier using a single power supply. The procedure used to find the bias point is identical to that used for the CE amplifier employing voltage-divider bias:

$$I_C = \beta \left[\frac{\left(\dfrac{R_2}{R_1 + R_2}\right)V_{CC} - V_{BE}}{R_1 \| R_2 + \beta R_E} \right] \tag{5.49}$$

Figure 5.28
Voltage-divider bias circuit for a CB amplifier.

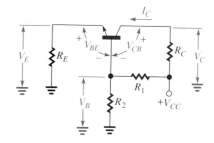

If $\beta R_E \gg R_1 \| R_2$,

$$V_B \approx \left(\frac{R_2}{R_1 + R_2}\right) V_{CC}, \quad V_E = V_B - V_{BE}, \quad \text{and} \quad I_C \approx I_E = \frac{V_E}{R_E} \qquad (5.50)$$

$$V_C = V_{CC} - I_C R_C \qquad (5.51)$$

$$V_{CB} = V_{CC} - I_C R_C - I_E R_E - V_{BE} \approx V_{CC} - I_C (R_C + R_E) - V_{BE} \qquad (5.52)$$

Small-Signal Amplifier

Figure 5.29(a) shows the CB amplifier with source and load connected. Notice that bias resistor R_2 is *bypassed* by capacitor C_B to keep the base of the transistor at ac ground. Figure 5.29(b) shows the small-signal equivalent circuit. Note that neither R_1 nor R_2 appear in this circuit since R_2 is bypassed to ground and R_1 is connected between R_2 and V_{CC} (ac ground). The input resistance looking into the emitter is the generally small value r_e, which is found in the usual way:

$$r_{in} = r_e \approx \frac{0.026 \text{ V}}{I_E} \qquad (5.53)$$

Figure 5.29
The small-signal CB amplifier.

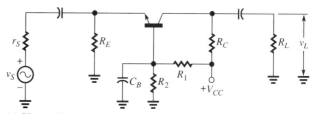

(a) CB amplifier with source and load connected

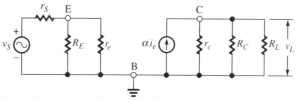

(b) Small-signal equivalent circuit of (a)

Thus,

$$r_{in}(\text{stage}) = R_E \| r_e \approx r_e \qquad (5.54)$$

The output resistance is the very large collector resistance r_c:

$$r_o = r_c \qquad (5.55)$$

and

$$r_o(\text{stage}) = R_C \| r_c \approx R_C \qquad (5.56)$$

The voltage gain A_v is

$$A_v = \frac{v_{cb}}{v_{eb}} = \frac{\alpha i_e R_C}{i_e R_E} \approx \frac{R_C}{r_e} \qquad (5.57)$$

and the overall voltage gain is

$$\frac{v_L}{v_S} = \frac{r_{in}(\text{stage})}{r_S + r_{in}(\text{stage})} A_v \frac{R_L}{r_o(\text{stage}) + R_L}$$

$$\approx \left(\frac{r_e}{r_S + r_e}\right)\left(\frac{R_C}{r_e}\right)\left(\frac{R_L}{R_C + R_L}\right) = \frac{r_e}{r_S + r_e}\left(\frac{r_L}{r_e}\right) = \frac{r_L}{r_S + r_e} \qquad (5.58)$$

where r_L is the ac load resistance, $R_C \| R_L$.

The current gain is

$$A_i = \frac{i_c}{i_e} = \alpha \qquad (5.59)$$

We see that the CB amplifier is noninverting (the voltage gain is positive), has small input resistance, large output resistance, and current gain less than 1. These properties limit its usefulness as either a voltage or a current amplifier. However, it does find use in some high-frequency applications where a moderately large, noninverting voltage gain results in an improved frequency response. (The inverting gain of the CE amplifier creates *Miller-effect capacitance* which limits the bandwidth of the amplifier, a phenomenon that will be discussed in Chapter 6.)

Figure 5.30 is a summary of the bias and small-signal equations for the CB amplifier.

Table 5.1 compares, in a very general way, the important small-signal characteristics of the three BJT configurations.

Table 5.1

	Common Base	Common Emitter	Common Collector
Voltage Gain	Large (noninverting)	Large (inverting)	≈ 1 (noninverting)
Current Gain	≈ 1	Large	Large
Power Gain	Moderate	Large	Small
Input Resistance	Small	Moderate	Large
Output Resistance	Large	Moderate	Small

Figure 5.30
Summary of bias and small-signal equations for the CB amplifier.

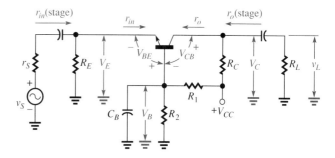

BIAS EQUATIONS

$V_{BE} = 0.7 \text{ V (Si)} \quad 0.3 \text{ V (Ge)}$

$$I_C = \beta \left[\frac{\left(\dfrac{R_2}{R_1 + R_2} \right) V_{CC} - V_{BE}}{R_1 \| R_2 + \beta R_E} \right]$$

$$V_B \approx \left(\frac{R_2}{R_1 + R_2} \right) V_{CC} \quad (\beta R_E >> R_1 \| R_2)$$

$V_E = V_B - V_{BE}$

$I_C \approx I_E = V_E / R_E$

$V_C = V_{CC} - I_C R_C$

$V_{CB} = V_{CC} - I_C R_C - I_E R_E - V_B \approx V_{CC} - I_C (R_C + R_E) - V_B$

SMALL-SIGNAL EQUATIONS

$r_{in} = r_e$

$r_{in}(\text{stage}) = R_E \| r_e \approx r_e$

$r_o = r_c$

$r_o(\text{stage}) = R_C \| r_c \approx R_C$

$$A_v = \frac{R_C}{r_e}$$

$$\frac{v_L}{v_S} = \frac{r_L}{r_S + r_e} \quad \text{where } r_L = R_C \| R_L$$

Example 5.10

DESIGN

Design a silicon, single-transistor amplifier to deliver approximately 1 V pk-pk to a capacitor-coupled 1-kΩ load when the input is 20 mV pk-pk. The signal source has negligible resistance but cannot supply a current greater than 20 μA pk-pk. The β of the transistor used may range from 100 to 200. Use standard-valued 5% resistors in the design. A 24-V dc power supply is available.

Solution.   The required voltage gain is

$$\frac{1 \text{ V pk-pk}}{20 \text{ mV pk-pk}} = 50$$

The input resistance of the amplifier must be at least

$$\frac{20 \text{ mV pk-pk}}{20 \ \mu\text{A pk-pk}} = 1 \text{ k}\Omega$$

It is clear that a CC configuration cannot be used, since the voltage gain must be greater than 1. The small input resistance of the CB configuration ($\approx r_e$) also eliminates it from consideration. Therefore, we elect to design the amplifier in a CE configuration. We first determine the value of r_e that is necessary to achieve a 1-kΩ input resistance with the smallest possible β:

$$\beta_{min} r_e \geqslant 1 \text{ k}\Omega$$

$$r_e \geqslant \frac{1 \text{ k}\Omega}{100} = 10 \ \Omega$$

Realizing that bias resistors in the base circuit will reduce the value of r_{in} (stage), we choose a value of r_e slightly larger than 10 Ω, say 12 Ω. Then, r_{in} (min) = (100)(12 Ω) = 1.2 kΩ.

To determine the value of R_C required we calculate (from equation 5.30):

$$A_v \approx 50 = \frac{r_L}{r_e} \qquad \text{where } r_L = R_C \| R_L \text{ and } R_L = 1 \text{ k}\Omega$$

$$\Rightarrow 50 = \frac{\dfrac{(R_C)(1 \text{ k}\Omega)}{R_C + 1 \text{ k}\Omega}}{12 \ \Omega}$$

Solving for R_C gives 1.5 kΩ.

The quiescent value of I_E is found from

$$r_e = \frac{0.026 \text{ V}}{I_E} \Rightarrow I_E = \frac{0.026 \text{ V}}{12 \ \Omega} = 2.17 \text{ mA}$$

Since the range of values of β is relatively large, we need a well-stabilized bias circuit and therefore choose a large value for R_E, say 2 kΩ. Thus, $\beta_{min} R_E =$ 100(2 kΩ) = 200 kΩ. We will bypass R_E with a capacitor so that it does not reduce the voltage gain.

We will use a voltage-divider bias circuit that satisfies $R_1 \| R_2 \ll \beta_{min} R_E = 200$ kΩ. Let $R_2 = 15$ kΩ. (Note that the rule-of-thumb value for R_2 would be $\beta(min)R_E/$ 10 = 20 kΩ.) Now,

$$V_E = I_E R_E = (2.17 \text{ mA})(2 \text{ k}\Omega) = 4.34 \text{ V}$$

and

$$V_B = V_E + 0.7 \text{ V} = 4.34 \text{ V} + 0.7 \text{ V} = 5.04 \text{ V}$$

We can now solve equation 5.6 to find the value of bias resistor R_1:

$$\left(\frac{15 \text{ k}\Omega}{R_1 + 15 \text{ k}\Omega}\right) 24 \text{ V} = 5.04 \text{ V}$$

Solving for R_1 gives 56.4 kΩ. Choosing the closest standard-value 5% resistor, we use $R_1 = 56$ kΩ. The completed design is shown in Figure 5.31.

To verify that the design satisfies the performance requirements for the resis-

Figure 5.31
(Example 5.10)

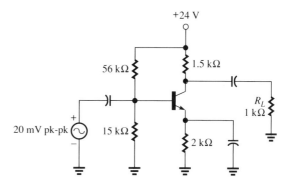

tor values chosen and for the range of possible β values, we first use equation 5.10 to find the actual value of I_C:

$\beta = 100$

$$I_C = 100 \left[\frac{\left(\dfrac{15 \text{ k}\Omega}{15 \text{ k}\Omega + 56 \text{ k}\Omega} \right) 24 \text{ V} - 0.7 \text{ V}}{15 \text{ k}\Omega \| 56 \text{ k}\Omega + 100(2 \text{ k}\Omega)} \right] = 2.06 \text{ mA}$$

$$r_e = \frac{0.026 \text{ V}}{2.06 \text{ mA}} = 12.6 \text{ }\Omega$$

$$A_v = \frac{-r_L}{r_e} = \frac{-1.5 \text{ k}\Omega \| 1 \text{ k}\Omega}{12.6 \text{ }\Omega} = -47.6$$

$$r_{in}(\text{stage}) = R_1 \| R_2 \| \beta r_e = 56 \text{k}\Omega \| 15 \text{ k}\Omega \| 100(12.6 \text{ }\Omega) = 1.14 \text{ k}\Omega$$

$\beta = 200$

$$I_C = 200 \left[\frac{\left(\dfrac{15 \text{ k}\Omega}{15 \text{ k}\Omega + 56 \text{ k}\Omega} \right) 24 \text{ V} - 0.7 \text{ V}}{15 \text{ k}\Omega \| 56 \text{ k}\Omega + 200(2 \text{ k}\Omega)} \right] = 2.12 \text{ mA}$$

$$r_e = \frac{0.026 \text{ V}}{2.12 \text{ mA}} = 12.3 \text{ }\Omega$$

$$A_v = \frac{-1.5 \text{ k}\Omega \| 1 \text{ k}\Omega}{12.3 \text{ }\Omega} = -48.8$$

$$r_{in}(\text{stage}) = 56 \text{ k}\Omega \| 15 \text{ k}\Omega \| 200(12.3 \text{ }\Omega) = 2.04 \text{ k}\Omega$$

Example 5.11

PSpice

Use PSpice to determine the quiescent point, voltage gain v_L/v_S, and the current gain i_L/i_S of the amplifier shown in Figure 5.32(a). The transistor has $\beta = 100$; let all other parameters default. Assume that the signal frequency is 10 kHz. (At this frequency, the reactance of the 10-μF coupling capacitors is negligibly small and will have very little affect on the voltage gain.)

Solution. Figure 5.32(b) shows the amplifier redrawn in a PSpice format with node numbers assigned. Also shown is the input data file. Note that it is not necessary to perform a dc analysis to obtain quiescent values, since PSpice auto-

Figure 5.32
(Example 5.11)

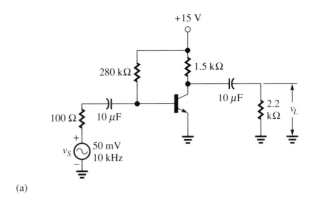

(a)

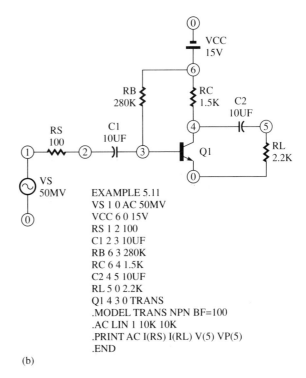

EXAMPLE 5.11
VS 1 0 AC 50MV
VCC 6 0 15V
RS 1 2 100
C1 2 3 10UF
RB 6 3 280K
RC 6 4 1.5K
C2 4 5 10UF
RL 5 0 2.2K
Q1 4 3 0 TRANS
.MODEL TRANS NPN BF=100
.AC LIN 1 10K 10K
.PRINT AC I(RS) I(RL) V(5) VP(5)
.END

(b)

matically performs a dc analysis and provides dc voltages at the nodes when an ac analysis is requested. Execution of this program gives the following dc voltages:

```
NODE  VOLTAGE  NODE  VOLTAGE  NODE  VOLTAGE  NODE  VOLTAGE
(1)   0.0000   (2)   0.0000   (3)   .8612    (4)   7.4015
(5)   0.0000   (6)   15.0000
```

Thus, the quiescent point is

$$V_{CE} = V(4) = 7.4015 \text{ V}$$

$$I_C = \frac{V(6) - V(4)}{R_C} = \frac{15 \text{ V} - 7.4015 \text{ V}}{1.5 \text{ k}\Omega} = 5.066 \text{ mA}$$

The results of the ac analysis are

```
I(RS)      I(RL)      V(5)       VP(5)
8.201E-5  3.319E-3  7.301E+0  -1.798E+2
```

Notice that the phase angle of the output voltage, VP(5) = −179.8°, is very nearly 180°, confirming that the CE amplifier inverts. Thus,

$$\frac{v_L}{v_S} = \frac{-V(5)}{50 \text{ mV}} = \frac{-7.301 \text{ V}}{50 \text{ mV}} = -146.02$$

$$\frac{i_L}{i_S} = \frac{I(RL)}{I(RS)} = \frac{3.319 \text{ mA}}{82.01 \text{ } \mu\text{A}} = 40.47$$

5.6 COMMON-SOURCE JFET AMPLIFIER BIAS

Fixed Bias

Like a bipolar transistor, a JFET used as an ac amplifier must be biased in order to create a dc output voltage around which ac variations can occur. When a JFET is connected in the *common-source* configuration, the input voltage is V_{GS} and the output voltage is V_{DS}. Therefore, the bias circuit must set dc (quiescent) values for the drain-to-source voltage V_{DS} and drain current I_D. Figure 5.33 shows one method that can be used to bias N-channel and P-channel JFETs.

Notice in Figure 5.33 that a dc supply voltage V_{DD} is connected to supply drain current to the JFET through resistor R_D, and that another dc voltage is used to set the gate-to-source voltage V_{GS}. This biasing method is called *fixed bias* because the gate-to-source voltage is fixed by the constant voltage applied across those terminals. Writing Kirchhoff's voltage law around the output loops in Figure 5.33, we find

$$V_{DS} = V_{DD} - I_D R_D \qquad \text{(N-channel)}$$
$$V_{DS} = -V_{DD} + I_D R_D \qquad \text{(P-channel)}$$

(5.60)

When using these equations, always substitute a positive value for V_{DD} to ensure that the correct sign is obtained for V_{DS}. V_{DS} should always turn out to be a positive quantity in an N-channel JFET and a negative quantity in a P-channel JFET. For example, in an N-channel device where V_{DD} is +15 V from drain to ground, if I_D = 10 mA, and R_D = 1 kΩ, we have V_{DS} = 15 − (10 mA)(1 kΩ) = +5

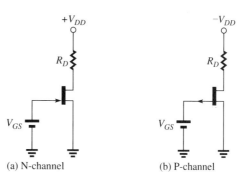

Figure 5.33
Fixed-bias circuits for N- and P-channel JFETs.

(a) N-channel   (b) P-channel

V. For a P-channel device where V_{DD} is -15 V from drain to ground, $V_{DS} = -15 + (10 \text{ mA})(1 \text{ k}\Omega) = -5$ V. Equations 5.60 are the equations of the dc load lines for N- and P-channel JFETs and each can be plotted on a set of drain characteristics to determine a Q-point. This technique is the same as the one we used to determine the Q-point in a BJT bias circuit. The load line intersects the V_{DS}-axis at V_{DD} and the I_D-axis at V_{DD}/R_D.

Example 5.12

The JFET in the circuit of Figure 5.34 has the drain characteristics shown in Figure 5.35. Find the quiescent values of I_D and V_{DS} when (1) $V_{GS} = -1.5$ V, and (2) $V_{GS} = -0.5$ V.

Figure 5.34
(Example 5.12)

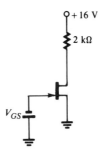

Solution

1. The load line intersects the V_{DS}-axis at $V_{DD} = +16$ V and the I_D-axis at $I_D = (16 \text{ V})/(2 \text{ k}\Omega) = 8$ mA. It is plotted on Figure 5.35.
   At the intersection of the load line with $V_{GS} = -1.5$ V (labeled Q$_1$), we find the quiescent values $I_D \approx 3.9$ mA and $V_{DS} \approx 8.4$ V.
2. The load line is, of course, the same as in part (1). Changing V_{GS} to -0.5 V moves the Q-point to the point labeled Q$_2$ in Figure 5.35. Here we see that $I_D \approx 6.7$ mA and $V_{DS} \approx 2.6$ V.

Part 2 of the preceding example illustrates an important result. Note that changing V_{GS} to -0.5 V in the bias circuit of Figure 5.34 caused the Q-point to move out of the pinch-off region and into the voltage-controlled–resistance region. As we have already mentioned, the Q-point must be located in the pinch-off region for normal amplifier operation. *To ensure that the Q-point is in the pinch-off region, the quiescent value of* $|V_{DS}|$ *must be greater than* $|V_p| - |V_{GS}|$. The pinch-off voltage for the device whose characteristics are given in Figure 5.35 can be seen to be approximately -4 V. Since $|V_{GS}| = 0.5$ V and the quiescent value of V_{DS} at Q$_2$ is only 2.6 V, we do not satisfy the requirement $|V_{DS}| > |V_p| - |V_{GS}|$. Q$_2$ is therefore in the variable-resistance region.

Of course, the quiescent value of I_D can also be determined using the transfer characteristic of a JFET. Since the transfer characteristic is a plot of I_D versus V_{GS}, it is only necessary to locate the V_{GS} coordinate and read the corresponding value of I_D directly. The value of V_{DS} can then be determined using equation 5.60. While graphical techniques for locating the bias point are instructive and provide insights to the way in which the circuit variables affect each other, the quiescent

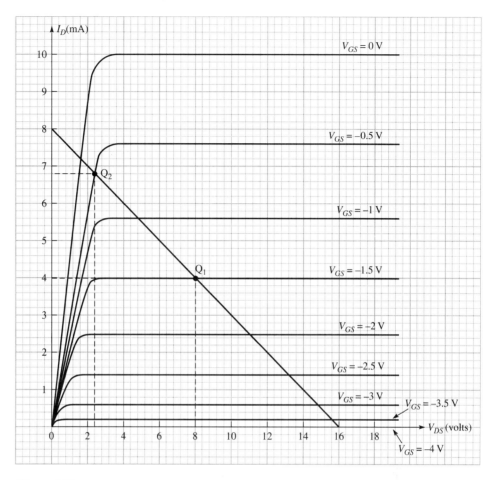

Figure 5.35
(Example 5.12)

values of I_D and V_{DS} can be calculated using a straightforward computation, if the values of V_p and I_{DSS} are known. The next example illustrates that the square-law characteristic is used in this computation.

Example 5.13

Given that the JFET in Figure 5.34 has $I_{DSS} = 10$ mA and $V_p = -4$ V, compute the quiescent values of I_D and V_{DS} when $V_{GS} = -1.5$ V. Assume that it is biased in the pinch-off region.

Solution. From equation 3.11,

$$I_D = I_{DSS}(1 - V_{GS}/V_p)^2 = (10 \text{ mA}) \left(1 - \frac{-1.5}{-4}\right)^2 = 3.9 \text{ mA}$$

From equation 5.60, $V_{DS} = V_{DD} - I_D R_D = 16 \text{ V} - (3.9 \text{ mA})(2 \text{ k}\Omega) = 8.2 \text{ V}$. These results are in close agreement with those obtained graphically in Example 5.11.

Note that it was necessary to assume that the JFET is biased in the pinch-off region, to justify the use of equation 3.11. If the computation had produced a value of V_{DS} less than $|V_p| - |V_{GS}| = 2.5$ V, we would have had to conclude that the device is not biased in pinch-off and would then have had to use another means to find the Q-point.

Self-Bias

The values of I_{DSS} and V_p are likely to vary widely among JFETs of a given type. A variation of 50% is not unusual. When the fixed-bias circuit is used to set a Q-point, a change in the parameter values of the JFET for which the circuit was designed (caused, for example, by substitution of another JFET) can result in an intolerable shift in quiescent values. Suppose, for example, that a JFET having parameters $I_{DSS} = 13$ mA and $V_p = -4.3$ V is substituted into the bias circuit of Example 5.12 (Figure 5.34), with V_{GS} once again set to -1.5 V. Then

$$I_D = (13 \text{ mA}) \left(1 - \frac{-1.5}{-4.3}\right)^2 = 5.51 \text{ mA}$$

$$V_{DS} = 16 - (5.51 \text{ mA})(2 \text{ k}\Omega) = 4.98 \text{ V}$$

These results show that I_D increases 41.3% over the value obtained in Example 5.11 and that V_{DS} decreases 68.7%. Note also that the value of V_{DS} (4.98 V) is now perilously near the pinch-off voltage (4.3 V). We conclude that the fixed-bias circuit does not provide good Q-point stability against changes in JFET parameters.

Figure 5.36 shows a bias circuit that provides improved stability and requires only a single supply voltage. This bias method is called *self-bias*, because the voltage drop across R_S due to the flow of quiescent current determines the quiescent value of V_{GS}. We can understand this fact by realizing that the current I_D in resistor R_S creates the voltage $V_S = I_D R_S$ at the source terminal, with respect to ground. For the N-channel JFET, this means that the source is positive with respect to the gate, since the gate is grounded. In other words, the gate is negative (by $I_D R_S$ volts) with respect to the source, as required for biasing an N-channel JFET: $V_{GS} = -I_D R_S$. For the P-channel device, the gate is positive by $I_D R_S$ volts, with respect to the source: $V_{GS} = I_D R_S$.

The equations

$$V_{GS} = -I_D R_S \qquad \text{(N-channel)} \tag{5.61}$$

$$V_{GS} = I_D R_S \qquad \text{(P-channel)} \tag{5.62}$$

Figure 5.36
Self-bias circuits.

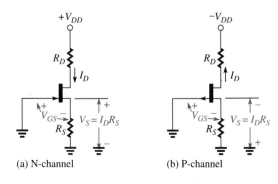

(a) N-channel   (b) P-channel

describe straight lines when plotted on V_{GS}–I_D-axes. (Verify these equations by writing Kirchhoff's voltage law around each gate-to-source loop in Figure 5.36.) Each line is called the *bias line* for its respective type. The quiescent value of I_D in the self-bias circuit can be determined graphically by plotting the bias line on the same set of axes with the transfer characteristic. The intersection of the two locates the Q-point. In effect, we solve the bias-line equation and the square-law equation simultaneously by finding the point where their graphs intersect. The quiescent value of V_{DS} can be found by summing voltages (writing Kirchhoff's voltage law) around the output loops in Figure 5.36:

$$V_{DS} = V_{DD} - I_D(R_D + R_S) \qquad \text{(N-channel)}$$
$$V_{DS} = -V_{DD} + I_D(R_D + R_S) \qquad \text{(P-channel)}$$

(5.63)

The next example illustrates the graphical procedure.

Example 5.14

The transfer characteristic of the JFET in Figure 5.37 is given in Figure 5.38. Determine the quiescent values of I_D and V_{DS} graphically.

Figure 5.37
(Example 5.14)

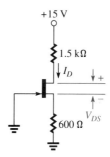

Solution. Since $R_S = 600 \ \Omega$, the equation of the bias line is $V_{GS} = -600I_D$. It is clear that the bias line always passes through the origin ($I_D = 0$ when $V_{GS} = 0$), so (0,0) is one point on the line. To determine another point on the line, choose a convenient value of V_{GS} and solve for I_D. In this example, if we let $V_{GS} = -3$, then

$$I_D = \frac{-V_{GS}}{600 \ \Omega} = \frac{-(-3 \ \text{V})}{600 \ \Omega} = 5 \ \text{mA}$$

Thus, $(-3 \ \text{V}, 5 \ \text{mA})$ is another point on the bias line. We can then draw a straight line between the two points $(0,0)$ and $(-3 \ \text{V}, 5 \ \text{mA})$ and note where that line intersects the transfer characteristic. The line is plotted on the transfer characteristic shown in Figure 5.38. We note that it intersects the characteristic at $I_D \approx 3$ mA, which is the quiescent drain current. The corresponding value of V_{GS} is seen to be approximately -1.8 V. The quiescent value of V_{DS} is found from equation 5.63:

$$V_{DS} = 15 \ \text{V} - (3 \ \text{mA})[(1.5 \ \text{k}\Omega) + (0.6 \ \text{k}\Omega)] = 8.7 \ \text{V}$$

Figure 5.38
(Example 5.14)

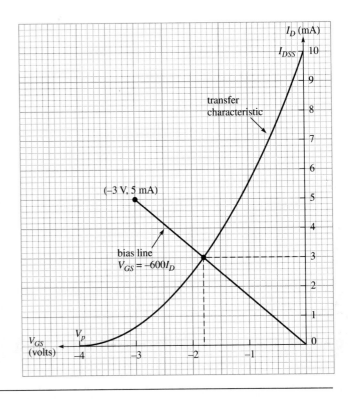

General Algebraic Solution—Self-Bias

The quiescent values of I_D and V_{GS} in the self-bias circuit can also be computed algebraically by solving the bias-line equation and the square-law equation simultaneously. To perform the computation, we must know the values of I_{DSS} and V_p. As in the fixed-bias case, the results are valid only if the Q-point is in the pinch-off region, i.e., if $|V_{DS}| > |V_p| - |V_{GS}|$. We must therefore assume that to be the case, but discard the results if the computation reveals the quiescent value of $|V_{DS}|$ to be less than $|V_p| - |V_{GS}|$. Equations 5.64 give the general form of the algebraic solution for the quiescent values of I_D, V_{DS}, and V_{GS} in the self-bias circuit. Since absolute values are used in the computations, the equations are valid for both P-channel and N-channel devices.

General algebraic solution for the bias point of self-biased JFET circuits

$$I_D = \frac{-B - \sqrt{B^2 - 4AC}}{2A}$$

where $A = R_S^2$

$$B = -\left(2|V_p|R_S + \frac{V_p^2}{I_{DSS}}\right)$$

$$C = V_p^2$$

$$|V_{DS}| = |V_{DD}| - I_D(R_D + R_S) \qquad \text{See note 1.}$$
$$|V_{GS}| = I_D R_S \qquad \text{See note 2.}$$

Note 1. V_{DS} is positive for an N-channel JFET and negative for a P-channel JFET.

Note 2. V_{GS} is negative for an N-channel JFET and positive for a P-channel JFET.   **(5.64)**

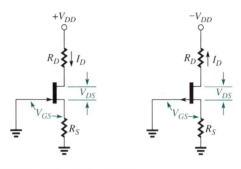

Example 5.15

Use equations 5.64 to find the bias point that was determined graphically in Example 5.14.

Solution. As shown in Figure 5.37, $R_S = 600\ \Omega$ and $R_D = 1.5\ \mathrm{k\Omega}$. Also, the transfer characteristic in Figure 5.38 shows that $I_{DSS} = 10\ \mathrm{mA}$ and $V_p = -4\ \mathrm{V}$. Thus, with reference to equations 5.64, we find

$$A = R_S^2 = (600)^2 = 3.6 \times 10^5$$

$$B = -\left(2|V_p|R_S + \frac{V_p^2}{I_{DSS}}\right) = -\left[2(4)(600) + \frac{(-4)^2}{10 \times 10^{-3}}\right] = -6.4 \times 10^3$$

$$C = V_p^2 = (-4)^2 = 16$$

$$I_D = \frac{-B - \sqrt{B^2 - 4AC}}{2A}$$

$$= \frac{6.4 \times 10^3 - \sqrt{40.96 \times 10^6 - 4(3.6 \times 10^5)(16)}}{2(3.6 \times 10^5)} = 3.0\ \mathrm{mA}$$

$$|V_{DS}| = |V_{DD}| - I_D(R_D + R_S) = 15\ \mathrm{V} - 3\ \mathrm{mA}(1.5\ \mathrm{k\Omega} + 600\ \Omega) = 8.7\ \mathrm{V}$$

$$|V_{GS}| = I_D R_S = (3\ \mathrm{mA})(600\ \Omega) = 1.8\ \mathrm{V}$$

Since the JFET is N-channel, $V_{GS} = -1.8\ \mathrm{V}$. These results agree well with those found in Example 5.13. Since $|V_{DS}| = 8.7\ \mathrm{V} > |V_p| - |V_{GS}| = 4\ \mathrm{V} - 1.8\ \mathrm{V} = 2.2\ \mathrm{V}$, we know the bias point is in the pinch-off region and the results are valid.

To demonstrate that the self-bias method provides better stability than the fixed-bias method, let us compare the shift in the quiescent value of I_D that occurs using each method, when the JFET parameters of the previous example are changed to $I_{DSS} = 12\ \mathrm{mA}$ and $V_p = -4.5\ \mathrm{V}$. In each case, we will assume that the initial bias point (using a JFET with $I_{DSS} = 10\ \mathrm{mA}$ and $V_p = -4\ \mathrm{V}$) is set so that

$I_D = 3$ mA and that a JFET having the new parameters is then substituted in the circuit. We have already seen that $I_D = 3$ mA when $V_{GS} = -1.8$ V, so let us suppose that a fixed-bias circuit has V_{GS} set to -1.8 V. When I_{DSS} changes to 12 mA and V_p to -4.5 V, with V_{GS} fixed at -1.8 V, we find that the new value of I_D in the fixed-bias circuit is

$$I_D = I_{DSS}\left(1 - \frac{V_{GS}}{V_p}\right)^2 = (12 \times 10^{-3} \text{ A})\left(1 - \frac{1.8}{4.5}\right)^2 = 4.32 \text{ mA}$$

This change in I_D from 3 mA to 4.32 mA represents a 44% increase.

Suppose now that the JFET parameters in the self-bias circuit change by the same amount: $I_{DSS} = 12$ mA and $V_p = -4.5$ V. Using equations 5.64, we find $I_D = 3.46$ mA. In this case, the increase in I_D is 15.3%, less than half that of the fixed-bias design.

Figure 5.39 shows the transfer characteristic of the JFET having $I_{DSS} = 10$ mA and $V_p = -4$ V, and the transfer characteristic corresponding to the JFET with $I_{DSS} = 12$ mA and $V_p = -4.5$ V. The self-bias line $V_{GS} = -600I_D$ is shown intersecting both characteristics. Note that it intersects these characteristics at the values of I_D previously calculated for the self-bias circuit: 3 mA and 3.46 mA. Also shown is the vertical line corresponding to $V_{GS} = -1.8$ V—i.e., the line corresponding to the fixed-bias condition. This line intersects the characteristics at the two values previously calculated for the fixed-bias circuit: 3 mA and 4.32 mA. It is apparent in this figure why the self-bias method produces a smaller change in I_D than the fixed-bias method when a change in JFET parameters results in a differ-

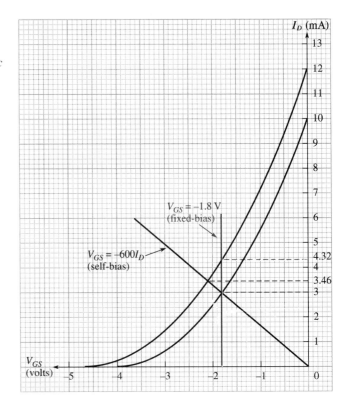

Figure 5.39
The quiescent value of I_D is initially 3 mA for both the fixed- and self-bias circuits. When the transfer characteristic changes, there is a smaller change in I_D for the self-bias circuit than for the fixed-bias circuit.

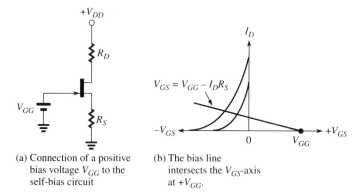

(a) Connection of a positive bias voltage V_{GG} to the self-bias circuit

(b) The bias line intersects the V_{GS}-axis at $+V_{GG}$.

Figure 5.40
A positive gate voltage V_{GG} reduces the slope of the bias line and improves bias stability.

ent transfer characteristic: The smaller the slope of the bias line, the smaller the change in I_D.

Voltage-Divider Bias

Our interpretation of Figure 5.39 reveals that good bias stability against changes in JFET parameters can be achieved by making the slope of the bias line as small as possible. The slope of this line becomes smaller as R_S is made larger, but large values of R_S can result in unacceptably small values of I_D. One way to obtain a bias line having a small slope (large value of R_S) and still maintain a respectable amount of drain current is to connect a *positive* voltage V_{GG} to the gate (of an N-channel JFET) in the self-bias circuit.

This arrangement is shown in Figure 5.40(a). The effect of V_{GG} is to shift the intercept of the bias line on the horizontal axis to V_{GG}, as shown in Figure 5.40(b). The equation of this bias line is

$$V_{GS} = V_{GG} - I_D R_S \qquad (5.65)$$

In practice, the positive gate voltage is obtained from a voltage divider connected across the gate from the positive supply voltage V_{DD}. For a P-channel JFET, the gate voltage is made negative and is obtained from the negative supply $-V_{DD}$. These connections are shown in Figure 5.41. Since the input resistance at the gate is very large, it is not necessary to consider its loading effect on the voltage divider. Thus, the gate-to-ground voltage V_G for the N-channel JFET is determined from

$$V_G = \left(\frac{R_2}{R_1 + R_2}\right) V_{DD} \qquad (5.66)$$

(For a P-channel device, $V_G = -R_2 V_{DD}/(R_1 + R_2)$.) The bias-line equations for N- and P-channel devices are then

$$V_{GS} = V_G - I_D R_S \qquad \text{(N-channel)} \qquad (5.67)$$

$$V_{GS} = V_G - I_D R_S \qquad \text{(P-channel)} \qquad (5.68)$$

Figure 5.41
Biasing the gate using a voltage divider rather than a separate V_{GG}.

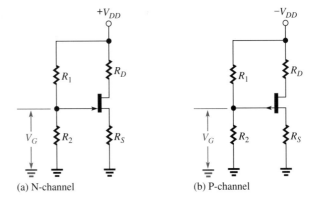

(a) N-channel   (b) P-channel

Note that V_G is a positive number in equation 5.67 and a negative number in equation 5.68.

Example 5.16

The circuit in Figure 5.41(a) has $V_{DD} = 10$ V, $R_1 = 3.3$ MΩ, $R_2 = 680$ kΩ, $R_S = 500$ Ω, and $R_D = 1$ kΩ. If $V_{DS} = 3$ V, find I_D and V_{GS}.

Solution. Since $V_{DS} = V_{DD} - I_D(R_D + R_S)$, we have

$$3 \text{ V} = 10 \text{ V} - I_D(1 \text{ k}\Omega + 500 \text{ }\Omega)$$

$$I_D = \frac{7 \text{ V}}{1.5 \text{ k}\Omega} = 4.67 \text{ mA}$$

From equation 5.66,

$$V_G = \left(\frac{R_2}{R_1 + R_2}\right) V_{DD} = \left(\frac{680 \text{ k}\Omega}{3.3 \text{ M}\Omega + 680 \text{ k}\Omega}\right) 10 \text{ V} = 1.71 \text{ V}$$

From equation 5.67,

$$V_{GS} = V_G - I_D R_S = 1.71 \text{ V} - (4.67 \text{ mA})(500 \text{ }\Omega) = -0.625 \text{ V}$$

General Algebraic Solution—Voltage-Divider Bias

The general form of the algebraic solution for the bias point in a voltage-divider bias circuit can be found by solving the square-law equation (3.11) simultaneously with the bias-line equation (5.67 or 5.68). The results, shown in equations 5.69, are valid for both P-channel and N-channel devices, since absolute values are used in the computations. The computed values must be checked to verify that the solution is in the pinch-off region: $|V_{DS}| > |V_p| - |V_{GS}|$. Note that equations 5.69 reduce to equations 5.64 for self-bias when 0 is substituted for V_G.

> **General algebraic solution for the bias point of**
> **JFET circuits using voltage-divider bias**
>
> $$I_D = \frac{-B - \sqrt{B^2 - 4AC}}{2A}$$

where $A = R_S^2$

$$B = -\left[2(|V_p| + |V_G|)R_S + \frac{V_p^2}{I_{DSS}}\right]$$

$$C = (|V_p| + |V_G|)^2$$

$$|V_G| = \frac{R_2}{R_1 + R_2}|V_{DD}|$$

$$|V_{DS}| = |V_{DD}| - I_D(R_D + R_S) \qquad \text{See note 1.}$$

$$|V_{GS}| = |V_G| - I_D R_S \qquad \qquad \text{See note 2.} \qquad \textbf{(5.69)}$$

Note 1. V_{DS} is positive for an N-channel JFET and negative for a P-channel JFET.

Note 2. V_{GS} is negative for an N-channel JFET and positive for a P-channel JFET.

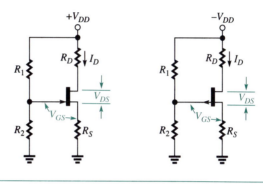

Example 5.17

The P-channel JFET in Figure 5.42 has the transfer characteristic shown in Figure 5.43. Determine the quiescent value of I_D (1) graphically and (2) algebraically.

Solution

1. To find the equation of the bias line, we must first find the gate-to-ground voltage V_G:

$$V_G = \left(\frac{47 \times 10^3}{188 \times 10^3 + 47 \times 10^3}\right)(-20 \text{ V}) = -4 \text{ V}$$

Figure 5.42
(Example 5.17)

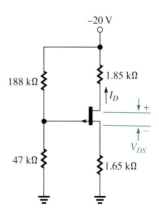

Figure 5.43
(Example 5.17)

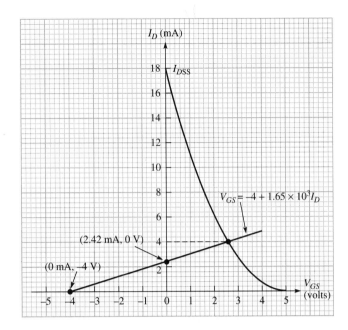

Then, from equation 5.68, the bias line is

$$V_{GS} = -4 + 1.65 \times 10^3 I_D \qquad (5.70)$$

This line intersects the V_{GS}-axis at -4 V. A second point that is easy to find on the line is its intercept on the I_D-axis. Let $V_{GS} = 0$, and solve for I_D:

$$I_D = \frac{4}{1.65 \times 10^3} = 2.42 \text{ mA}$$

The bias line joining the points (0 mA, -4 V) and (2.42 mA, 0 V) is shown drawn on the transfer characteristic in Figure 5.43. It can be seen that the bias line intersects the characteristic at the quiescent value $I_D \approx 4$ mA.

2. From Figure 5.42, $R_D = 1.85$ kΩ, $R_S = 1.65$ kΩ, and $V_{DD} = 20$ V. From the transfer characteristic in Figure 5.43, we see that $V_p = 5$ V and $I_{DSS} = 18$ mA. We have already found (in part 1) that $V_G = -4$ V. With reference to equations 5.69,

$$A = R_S^2 = (1.65 \times 10^3)^2 = 2.7225 \times 10^6$$

$$B = -\left[2(|V_p| + |V_G|)R_S + \frac{V_p^2}{I_{DSS}}\right]$$

$$= -[2(5 + 4)(1.65 \times 10^3) + 5^2/18 \times 10^{-3}] = -31.09 \times 10^3$$

$$C = (|V_p| + |V_G|)^2 = (5 + 4)^2 = 81$$

Substituting these values into the equation for I_D, we find $I_D = 4.02$ mA. Then

$$|V_{DS}| = |V_{DD}| - I_D(R_D + R_S)$$
$$= 20 \text{ V} - 4.02 \text{ mA} (1.85 \text{ k}\Omega + 1.65 \text{ k}\Omega) = 5.93 \text{ V}$$

Since the JFET is P-channel, we know $V_{DS} = -5.93$ V.

$$|V_{GS}| = |V_G| - I_D R_S = 4 \text{ V} - (4.02 \text{ mA})(1.65 \text{ k}\Omega) = 2.63 \text{ V}$$

Since 5.93 V > 5 V $-$ 2.63 V = 2.37 V, we know the solution is valid.

Example 5.23 at the end of this chapter shows how to design a voltage-divider bias circuit to compensate for a specified change in the transfer characteristic of a JFET.

5.7 SMALL-SIGNAL COMMON-SOURCE JFET AMPLIFIERS

Recall from Chapter 3 that the *transconductance* of a JFET is defined to be the ratio of small-signal drain current to small-signal drain-to-source voltage:

$$g_m = \frac{i_d}{v_{gs}} \text{ siemens} \tag{5.71}$$

We will incorporate this parameter, along with small-signal output resistance, into a model that can be used to determine important characteristics, such as voltage gain, of the common-source amplifier. Small-signal output resistance is defined by

$$r_o = \frac{v_{ds}}{i_d} \text{ ohms} \tag{5.72}$$

This parameter is also called the *drain resistance*, r_d, or r_{ds}. Its value can be determined graphically from a set of drain characteristics (Figure 3.16), by finding the ratio

$$r_o = \frac{\Delta V_{DS}}{\Delta I_D}\bigg|_{V_{GS} = \text{constant}} \tag{5.73}$$

However, the lines of constant V_{GS} are so nearly horizontal in the pinch-off region that it is difficult to obtain accurate values. In any case, the value is generally so large that it has little effect on the computation of voltage gain in practical circuits. Values of r_d range from about 50 kΩ to several hundred kΩ in the pinch-off region.

Figure 5.44 shows the small-signal equivalent circuit of the common-source JFET, incorporating the parameters we have discussed. The current source having value $g_m v_{gs}$ is a voltage-controlled current source, since the current it produces depends on the (input) voltage v_{gs}. It is apparent in the figure that $i_d = g_m v_{gs}$, in agreement with the definition of g_m:

$$g_m = \frac{i_d}{v_{gs}} \Rightarrow i_d = g_m v_{gs}$$

Notice that there is an open circuit shown between the gate and source terminals. Since the gate-to-source junction is reverse biased in normal operation, the extremely large resistance between those terminals can be assumed to be infinite in most practical situations.

Figure 5.45 shows a common-source JFET amplifier with fixed bias V_{GG}. R_G is a large resistance connected in series with V_{GG} to prevent the dc source from

Figure 5.44
Small-signal equivalent circuit of a common-source JFET.

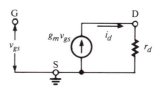

Figure 5.45
A common-source amplifier with fixed bias.

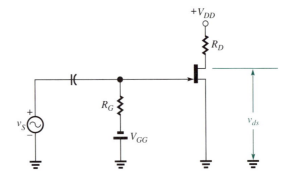

shorting the ac signal to ground. (Recall that a dc source is an ac short circuit.) The input resistance of the JFET is so large that there is negligible dc voltage division at the gate; that is, the major part of V_{GG} appears from gate-to-source instead of across R_G, so $V_{GS} \approx V_{GG}$. From another viewpoint, the dc gate current is so small that there is negligible drop across R_G. The input coupling capacitor serves the same purpose it does in a BJT amplifier, namely, to provide dc isolation between the signal source and the FET. For the moment, we ignore any signal-source resistance (r_S) and assume that the output is open $(R_L = \infty)$.

The total gate-to-source voltage is the sum of the small-signal source voltage, v_S, and the bias voltage V_{GG}. For example, if $V_{GG} = -2$ V and v_S is a sine wave having peak value 0.1 V, then $v_{gs} = -2 + 0.1 \sin \omega t$ volts. Thus, v_{gs} varies between the extreme values $-2 - 0.1 = -2.1$ V and $-2 + 0.1 = -1.9$ V. When v_{gs} goes more positive (toward -1.9 V), the drain current increases. This increase in i_d causes the output voltage v_{ds} to *decrease*, since

$$V_{DS} = V_{DD} - I_D R_D \tag{5.74}$$

We conclude that an increase in the input signal voltage causes a decrease in the output voltage and that the output is therefore 180° out of phase with the input.

Figure 5.46 shows the equivalent circuit of the common-source amplifier in Figure 5.45, with the JFET replaced by its small-signal equivalent. The coupling capacitor is assumed to have negligible impedance at the signal frequency we are considering (for now), so it is replaced by a short circuit. As usual, all dc sources are treated as ac short circuits to ground. Notice that we have attached a minus sign to the controlled current source, so its value now reads $-g_m v_{gs}$. The minus sign denotes the phase inversion between input and output, as we have described.

It is clear from Figure 5.46 that

$$v_{ds} = i_d(r_d \parallel R_D) \tag{5.75}$$

Since $i_d = -g_m v_{gs}$, we have

$$v_{ds} = -g_m v_{gs}(r_d \parallel R_D) \tag{5.76}$$

Figure 5.46
The small-signal equivalent circuit of the common-source amplifier shown in Figure 5.45.

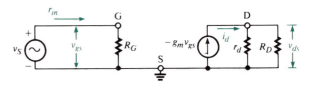

There is no signal-source resistance, so $v_S = v_{gs}$, and the voltage gain is therefore

$$A_v = \frac{v_{ds}}{v_S} = \frac{v_{ds}}{v_{gs}} = -g_m(r_d \| R_D) \tag{5.77}$$

In most practical amplifier circuits, the value of r_d is much greater than that of R_D, so $r_d \| R_D \approx R_D$ and a good approximation for the voltage gain is

$$\frac{v_{ds}}{v_{gs}} \approx -g_m R_D \tag{5.78}$$

It is important to remember that a JFET amplifier is operated in its pinch-off region. Therefore, the bias point and the voltage variations around the bias point must always satisfy $|V_{DS}| \geq |V_p| - |V_{GS}|$.

So long as the signal source does not drive the gate positive with respect to the source, the reverse-biased gate-to-source resistance is very large and the input resistance to the amplifier is essentially R_G. The equivalent circuit in Figure 5.46 is based on this assumption, and it can be seen that $r_{in} = R_G$.

Example 5.18

The JFET in the amplifier circuit shown in Figure 5.47 has $I_{DSS} = 12$ mA, $V_p = -4$ V, and $r_d = 100$ kΩ.

1. Find the quiescent values of I_D and V_{DS}.
2. Find g_m.
3. Draw the ac equivalent circuit.
4. Find the voltage gain.

Solution

1. I_D is found using the equation of the transfer characteristic:

$$I_D = I_{DSS} \left(1 - \frac{V_{GS}}{V_p}\right)^2 = (12 \text{ mA}) \left(1 - \frac{-2 \text{ V}}{-4 \text{ V}}\right)^2 = 3 \text{ mA}$$

From equation 5.74, $V_{DS} = 15 - (3 \text{ mA})(2.2 \text{ k}\Omega) = 8.4$ V. Since $V_{DS} = 8.4$ V and $|V_p| - |V_{GS}| = 4 - 2 = 2$ V, we have $|V_{DS}| > |V_p| - |V_{GS}|$, confirming that the JFET is biased in its pinch-off region.

2. From equation 3.14,

$$g_m = \frac{2(12 \text{ mA})}{4 \text{ V}} \sqrt{\frac{3 \text{ mA}}{12 \text{ mA}}} = 3 \times 10^{-3} \text{ S}$$

Figure 5.47
(Example 5.18)

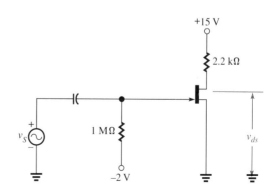

Figure 5.48
(Example 5.18)

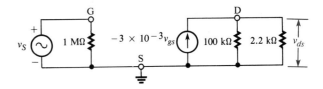

3. The ac equivalent circuit is shown in Figure 5.48.
4. Since $r_d \parallel R_D = (100 \text{ k}\Omega) \parallel (2.2 \text{ k}\Omega) \approx 2.2 \text{ k}\Omega$, we have, from equation 5.78,

$$\frac{v_{ds}}{v_S} = \frac{v_{ds}}{v_{gs}} \approx (-3 \times 10^{-3} \text{ S})(2.2 \times 10^3 \ \Omega) = -6.6$$

As the preceding example illustrates, the voltage gain obtainable from a JFET amplifier is generally smaller than that which can be obtained from its BJT counterpart. The principal advantage of the JFET amplifier is its very large input resistance.

Bias-Stabilized Small-Signal Amplifiers

The JFET amplifier we have been studying employs the fixed-bias method for setting the Q-point. Recall that this method makes the Q-point sensitive to parameter changes and is therefore undesirable in any application where JFET parameters may vary. Figure 5.49 shows a common-source JFET amplifier using the improved bias method that incorporates self-bias and a voltage divider across the gate. Note the *source-bypass capacitor* connected in parallel with R_S. This capacitor serves the same purpose as the emitter-bypass capacitor in a BJT common-emitter amplifier, namely, to eliminate the ac degeneration that would otherwise occur due to part of the output signal being dropped across the resistor. As far as the ac signal (v_{ds}) is concerned, the source is grounded, and there is no loss across R_S. Of course, the *dc* voltage V_{DS} is unaffected by the capacitor, so R_S continues to serve its role in providing self-bias for the JFET.

Figure 5.50 shows the ac equivalent circuit of the bias-stabilized amplifier in Figure 5.49. Notice that R_1 and R_2 appear in parallel to ac signals, so $r_{in} = R_1 \parallel R_2$. These resistors are made quite large to maintain a large input resistance to the amplifier.

Figure 5.49
A common-source amplifier with improved bias stabilization.

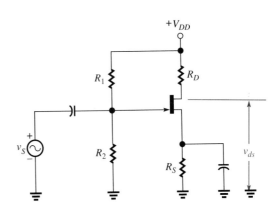

Figure 5.50
The ac equivalent circuit of the bias-stabilized common-source amplifier.

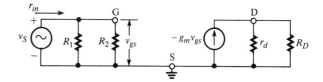

The voltage gain of the bias-stabilized amplifier is derived in exactly the same way and has exactly the same value as the fixed-bias amplifier, assuming no signal-source resistance: $A_v = -g_m(r_d \parallel R_D) \approx -g_m R_D$.

When the JFET amplifier is biased using only the self-biasing resistor (no voltage divider), the voltage gain is again computed using $A_v = -g_m(r_d \parallel R_D)$. A resistor R_G is connected between gate and ground to provide dc continuity for the gate circuit. This resistor can be made very large to maintain a large input resistance to the amplifier. The reverse current through the reverse-biased gate–source junction is very small, so the voltage drop across R_G is negligible. However, if an extremely large resistance is used with a JFET having excessive leakage current, the drop should be taken into account. For example, if $R_G = 10$ MΩ and $I_{DS(reverse)} = 0.1$ μA, the gate voltage will be increased by $(10$ m$\Omega)(0.1$ μA$) = 1$ V with respect to ground. Figure 5.51 shows the self-biased JFET amplifier and its ac equivalent circuit.

To avoid confusion, we will hereafter refer to resistance associated with the signal source as *signal-source resistance*, which we will designate by r_S, and we will continue to refer to resistance in series with the source terminal of the JFET as simply *source resistance*, R_S. When signal-source resistance is present, there is the usual voltage division between r_S and the input resistance of the amplifier, r_{in}. When a load resistor R_L is capacitor-coupled to the output, there is also a voltage division between the amplifier output resistance $r_d \parallel R_D$ and R_L. Thus,

$$\frac{v_L}{v_S} = \left(\frac{r_{in}}{r_S + r_{in}}\right) A_v \left(\frac{R_L}{R_L + r_d \parallel R_D}\right) \qquad (5.79)$$

Figure 5.52 shows the common-source amplifier in each of the three bias arrangements with load and signal-source resistances included. Also shown are the ac equivalent circuits of each. Note in each case that r_S is in series with the amplifier input and R_L is in parallel with the amplifier output.

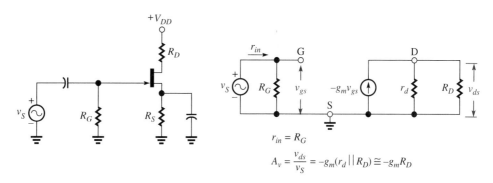

Figure 5.51
The self-biased JFET amplifier and its ac equivalent circuit.

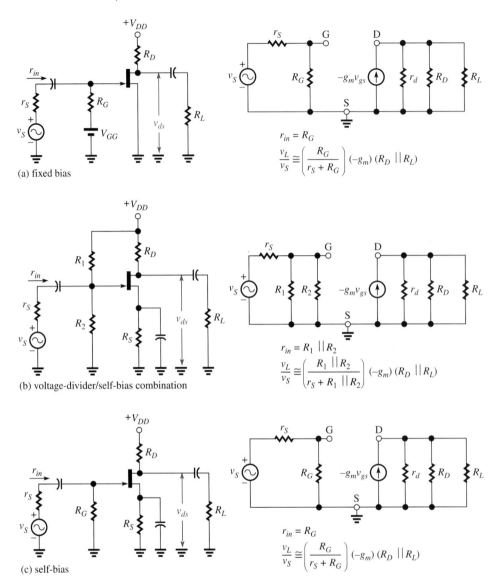

Figure 5.52
Common-source amplifiers with load and signal-source resistances included and their ac equivalent circuits.

For the fixed-bias amplifier (Figure 5.52(a)), equation 5.79 becomes

$$\frac{v_L}{v_S} = \left(\frac{R_G}{r_S + R_G}\right)\left[-g_m(r_d \,\|\, R_D)\right]\left(\frac{R_L}{R_L + r_d \,\|\, R_D}\right) \tag{5.80}$$

$$\approx \left(\frac{R_G}{r_S + R_G}\right)(-g_m R_D)\left(\frac{R_L}{R_L + R_D}\right)$$

$$= \left(\frac{R_G}{r_S + R_G}\right)(-g_m)(R_D \,\|\, R_L) \tag{5.81}$$

For the voltage-divider/self-biased amplifier (Figure 5.52(b)), equation 5.79 becomes

$$\frac{v_L}{v_S} = \left(\frac{R_1 \| R_2}{r_S + R_1 \| R_2}\right) [-g_m(r_d \| R_D)] \left(\frac{R_L}{R_L + r_d \| R_D}\right) \tag{5.82}$$

$$\approx \left(\frac{R_1 \| R_2}{r_S + R_1 \| R_2}\right) (-g_m R_D) \left(\frac{R_L}{R_L + R_D}\right)$$

$$= \left(\frac{R_1 \| R_2}{r_S + R_1 \| R_2}\right) (-g_m)(R_D \| R_L) \tag{5.83}$$

For the self-biased amplifier (Figure 5.52(c)), equation 5.79 becomes

$$\frac{v_L}{v_S} = \left(\frac{R_G}{r_S + R_G}\right) (-g_m)(r_d \| R_D) \left(\frac{R_L}{R_L + r_d \| R_D}\right) \tag{5.84}$$

$$\approx \left(\frac{R_G}{r_S + R_G}\right) (-g_m)(R_D \| R_L) \tag{5.85}$$

Of course, all of the gain relations derived above are exactly the same for P-channel JFET amplifiers.

Example 5.19

The JFET shown in Figure 5.53 has transconductance 4000 μS at its bias point. Its drain resistance is 100 kΩ. Assuming small-signal conditions, find the overall voltage gain, v_L/v_S.

Figure 5.53
(Example 5.19)

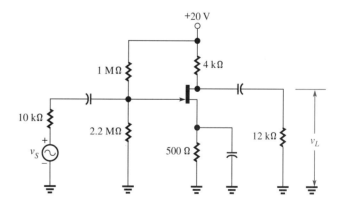

Solution

$$R_1 \| R_2 = (1 \text{ M}\Omega) \| (2.2 \text{ M}\Omega) = 687.5 \text{ k}\Omega$$
$$r_d \| R_D = (100 \text{ k}\Omega) \| (4 \text{ k}\Omega) = 3.846 \text{ k}\Omega$$

Using equation 5.82, we find

$$\frac{v_L}{v_S} = \left(\frac{687.5 \text{ k}\Omega}{10 \text{ k}\Omega + 687.5 \text{ k}\Omega}\right) [(-4 \text{ mS})(3.846 \text{ k}\Omega)] \left(\frac{12 \text{ k}\Omega}{12 \text{ k}\Omega + 3.846 \text{ k}\Omega}\right) = -11.48$$

Assuming that the approximations that lead to equation 5.83 are valid, we obtain from that equation

$$\frac{v_L}{v_S} = \left(\frac{687.5 \text{ k}\Omega}{10 \text{ k}\Omega + 687.5 \text{ k}\Omega}\right)(-4 \text{ mS})(4 \text{ k}\Omega \parallel 12 \text{ k}\Omega) = -11.83$$

The results found using the exact and approximate equations are quite close, certainly close enough for all practical work.

5.8   THE COMMON-DRAIN JFET AMPLIFIER

Figure 5.54 shows a JFET connected as a common-drain amplifier. Note that the drain terminal is connected directly to the supply voltage V_{DD}, so the drain is at ac ground. Since the input and output signals are taken with respect to ground, the drain is common to both, which accounts for the name of the configuration. Since the ac output signal is between the source terminal and ground, it is the same signal measured from source to drain as measured across R_S. Note that the JFET is biased using the combined self-bias/voltage-divider method, though either of the other two bias methods we studied for the common-source configuration could also be used. As can be seen from the figure,

$$v_{in} = v_{gs} + v_L \tag{5.86}$$

Therefore, the gate-to-source voltage is

$$v_{gs} = v_{in} - v_L \tag{5.87}$$

Figure 5.55 shows the ac equivalent circuit of the common-drain amplifier. Although this circuit appears at first glance to be the same as the common-source configuration (Figure 5.52(b)), note that v_{in} is *not* v_{gs}. Rather, v_{in} is related to v_{gs} by equations 5.86 and 5.87. Neglecting the signal-source resistance r_S for the moment, we derive the voltage gain v_L/v_{in} as follows:

$$v_L = g_m v_{gs}(r_d \parallel R_S \parallel R_L) \tag{5.88}$$

Figure 5.54
A common-drain JFET amplifier.

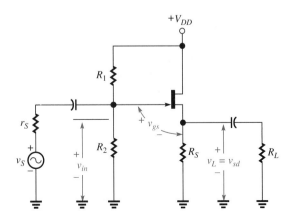

Figure 5.55
The small-signal equivalent circuit for the common-drain amplifier.

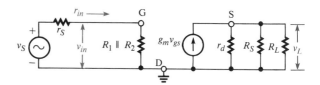

Substituting $v_{gs} = v_{in} - v_L$ (equation 5.87) into (5.88), we obtain

$$v_L = g_m(v_{in} - v_L)(r_d \parallel R_S \parallel R_L)$$
$$= g_m v_{in}(r_d \parallel R_S \parallel R_L) - g_m v_L(r_d \parallel R_S \parallel R_L)$$
$$v_L + g_m v_L(r_d \parallel R_S \parallel R_L) = g_m v_{in}(r_d \parallel R_S \parallel R_L)$$

Solving for v_L/v_{in}, we obtain

$$\frac{v_L}{v_{in}} = \frac{g_m(r_d \parallel R_S \parallel R_L)}{1 + g_m(r_d \parallel R_S \parallel R_L)} \tag{5.89}$$

It is clear from equation 5.89 that the voltage gain of the common-drain configuration is always less than 1. Note that there is no phase inversion between input and output. When $g_m(r_d \parallel R_S \parallel R_L) \gg 1$ (often the case), equation 5.89 shows that

$$\frac{v_L}{v_{in}} \approx 1 \tag{5.90}$$

In other words, the load voltage is approximately the same as the input voltage, in magnitude and phase, and we say that the output *follows* the input. For this reason, the common-drain amplifier is called a *source follower*.

The input resistance of the source follower can be seen from Figure 5.54 to be the same as it is for a common-source amplifier:

$$r_{in} = R_1 \parallel R_2 \tag{5.91}$$

Similarly, r_{in} is the same as it is for a common-source amplifier when the other bias arrangements are used. When the voltage division between signal-source resistance and r_{in} is taken into account, we find the overall voltage gain to be

$$\frac{v_L}{v_S} = \left(\frac{r_{in}}{r_S + r_{in}}\right)\left[\frac{g_m(r_d \parallel R_S \parallel R_L)}{1 + g_m(r_d \parallel R_S \parallel R_L)}\right] \tag{5.92}$$

The output resistance of the source follower (r_o(stage), looking from R_L toward the source terminal) is R_S in parallel with the resistance "looking into" the JFET at the source terminal. The resistance looking into the JFET at the source terminal can be found using equation 5.89. Since we are looking to the left of R_S and R_L, the parallel combination of those resistances is not relevant in the equation. Substituting r_d for $r_d \parallel R_S \parallel R_L$, equation 5.89 becomes

$$A_v = \frac{v_o}{v_{in}} = \frac{g_m r_d}{1 + g_m r_d}$$

Now,

$$r_o = \frac{v_o}{i_o} = \frac{A_v v_{in}}{i_o} = \frac{A_v}{g_m} = \frac{r_d}{1 + g_m r_d}$$

Thus,

$$r_o(\text{stage}) = R_S \left\| \left(\frac{r_d}{1 + g_m r_d} \right) \right. \tag{5.93}$$

It is almost always true that $g_m r_d \gg 1$, so (5.93) reduces to

$$r_o(\text{stage}) \approx R_S \left\| \left(\frac{1}{g_m} \right) \right. \tag{5.94}$$

Like the BJT emitter follower, the source follower is used primarily as a buffer amplifier because of its large input resistance and small output resistance.

Example 5.20

The JFET in Figure 5.56 has $g_m = 5 \times 10^{-3}$ S and $r_d = 100$ kΩ. Find

    the input resistance;
    the voltage gain; and
    the output resistance of the amplifier stage.

*Figure 5.56
(Example 5.20)*

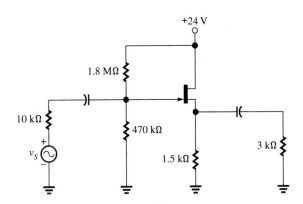

Solution

1. $r_{in} = R_1 \| R_2 = (1.8 \text{ M}\Omega) \| (470 \text{ k}\Omega) = 372.7$ kΩ.
2. Since $r_d \| R_S \| R_L = (100 \text{ k}\Omega) \| (1.5 \text{ k}\Omega) \| (3 \text{ k}\Omega) = 990$ Ω, we have, from equation 5.92,

$$\frac{v_L}{v_S} = \left(\frac{372.7 \text{ k}\Omega}{10 \text{ k}\Omega + 372.7 \text{ k}\Omega} \right) \left[\frac{(5 \times 10^{-3} \text{ S})(990 \text{ }\Omega)}{1 + (5 \times 10^{-3} \text{ S})(990 \text{ }\Omega)} \right] = 0.81$$

3. From equation 5.93,

$$r_o(\text{stage}) = (1.5 \text{ k}\Omega) \left\| \left(\frac{100 \text{ k}\Omega}{1 + (5 \times 10^{-3} \text{ S})(100 \text{ k}\Omega)} \right) \right.$$

$$= (1.5 \text{ k}\Omega) \| (199.6 \text{ }\Omega) = 176.2 \text{ }\Omega$$

Note that the approximation for r_o (equation 5.94) gives a nearly equal result:

$$r_o(\text{stage}) \approx R_S \| (1/g_m) = (1.5 \text{ k}\Omega) \| 200 = 176.5 \text{ }\Omega$$

5.9   MOSFET AMPLIFIER BIAS CIRCUITS

Although enhancement MOSFETs are most widely used in digital integrated circuits (and require no bias circuitry in those applications), they can, and occasionally do, find applications in small-signal amplifiers. Figure 5.57 shows one way to bias an enhancement NMOS for such an application. This circuit appears to be identical to the bias circuit we used for an N-channel JFET (Figure 5.41), but it is quite different in principle. The resistor R_S does not provide self-bias as it does in the JFET circuit. Self-bias is not possible with enhancement devices; it can occur only in depletion devices. In Figure 5.57, the resistor R_S is used to provide feedback for bias stabilization, in the same way that the emitter resistor does in a BJT bias circuit. The larger the value of R_S, the less sensitive the bias point is to changes in MOSFET parameters caused by temperature changes or by device replacement. Recall that R_S in a JFET self-bias circuit also provides this beneficial effect: We saw (Figure 5.39) that the greater the value of R_S, the less steep the bias line.

Figure 5.58 shows the voltage drops in the enhancement MOSFET bias circuit. R_1 and R_2 form a voltage divider that determines the gate-to-ground voltage V_G:

$$V_G = \left(\frac{R_2}{R_1 + R_2}\right) V_{DD} \tag{5.95}$$

The voltage divider is not loaded by the very large input resistance of the MOSFET, so the values of R_1 and R_2 are usually made very large to keep the ac input resistance of the stage large. Writing Kirchhoff's voltage law around the gate-to-source loop, we find

$$V_{GS} = V_G - I_D R_S \quad \text{(NMOS)} \tag{5.96}$$

For a PMOS device, V_G and V_{GS} are negative, so equation 5.96 would be written

$$V_{GS} = V_G + I_D R_S \quad \text{(PMOS)} \tag{5.97}$$

(Note that I_D is considered positive in both equations.) Writing Kirchhoff's voltage law around the drain-to-source loop, we find

$$V_{DS} = V_{DD} - I_D(R_D + R_S) \quad \text{(NMOS)} \tag{5.98}$$

Again regarding I_D as positive in both the NMOS and PMOS devices, the counterpart of equation 5.98 for a PMOS device is

$$V_{DS} = -|V_{DD}| + I_D(R_D + R_S) \quad \text{(PMOS)} \tag{5.99}$$

Figure 5.57
A bias circuit for an enhancement MOS-FET.

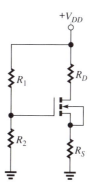

Figure 5.58
Voltage drops in the enhancement
NMOS bias circuit.

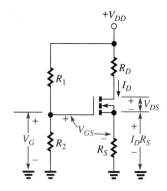

V_{DS} is negative in a PMOS circuit; note that the absolute value of V_{DD} must be used in equation 5.99 to obtain the correct sign for V_{DS}.

Equation 5.96 can be rewritten in the form

$$I_D = -(1/R_S)V_{GS} + V_G/R_S \qquad (5.100)$$

Equation 5.100 is seen to be the equation of a straight line on I_D–V_{GS}-axes. It intercepts the I_D-axis at V_G/R_S and the V_{GS}-axis at V_G. The line can be plotted on the same set of axes as the transfer characteristics of the device, and the point of intersection locates the bias values of I_D and V_{GS}.

General Algebraic Solution

We can obtain general algebraic expressions for the bias points in PMOS and NMOS circuits by solving equation 3.18 simultaneously with equation 5.96 or 5.97 for I_D. The results are shown as equations 5.101 and are valid for both NMOS and PMOS devices.

General algebraic solution for the bias point of NMOS and PMOS circuits

$$|V_G| = \frac{R_2}{R_1 + R_2}|V_{DD}|$$

$$I_D = \frac{-B - \sqrt{B^2 - 4AC}}{2A}$$

where $A = R_S^2$

$$B = -2\left[(|V_G| - |V_T|)R_S + \frac{1}{\beta}\right]$$

$$C = (|V_G| - |V_T|)^2$$

$|V_{DS}| = |V_{DD}| - I_D(R_D + R_S)$ See note 1.

$|V_{GS}| = |V_G| - I_D R_S$ See note 2.

1. V_{DS} is positive for an NMOS FET and negative for a PMOS FET.
2. V_{GS} is positive for an NMOS FET and negative for a PMOS FET.

(5.101)

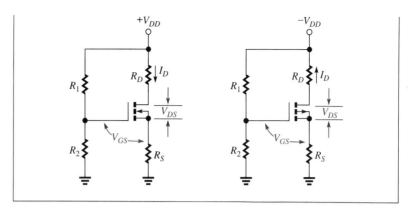

Example 5.21

The transfer characteristic of the NMOS FET in Figure 5.59 is given in Figure 5.60 ($\beta = 0.5 \times 10^{-3}$ and $V_T = 2$ V). Determine values of V_{GS}, I_D, and V_{DS} at the bias point (1) graphically and (2) algebraically.

Figure 5.59
(Example 5.21)

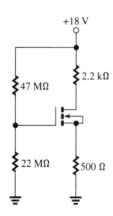

Solution

1. From equation 5.95,

$$V_G = \left(\frac{22 \times 10^6}{47 \times 10^6 + 22 \times 10^6}\right) 18 \text{ V} = 5.74 \text{ V}$$

Substituting in equation 5.100, we have

$$I_D = -2 \times 10^{-3} V_{GS} + 11.48 \times 10^{-3}$$

This equation intersects the I_D-axis at 11.48 mA and the V_{GS}-axis at $V_G = 5.74$ V. It is shown plotted with the transfer characteristic in Figure 5.60. The two plots intersect at the quiescent point, where the values of I_D and V_{GS} are approximately $I_D = 2.0$ mA and $V_{GS} = 4.6$ V. The corresponding quiescent value of V_{DS} is found from equation 5.98:

$$V_{DS} = 18 \text{ V} - (2.0 \text{ mA})[(2.2 \text{ k}\Omega) + (0.5 \text{ k}\Omega)] = 12.6 \text{ V}$$

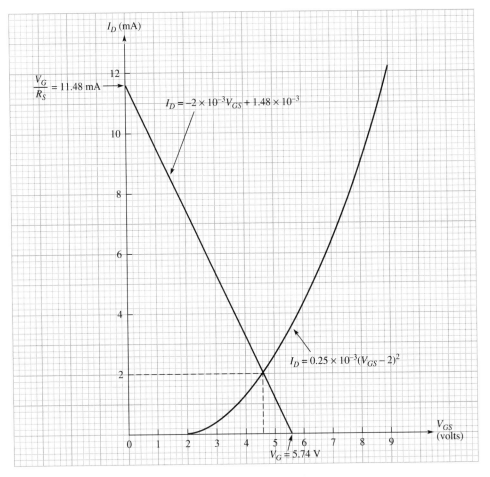

Figure 5.60
(Example 5.21)

In order for this analysis to be valid, the Q-point must be in the active region; that is, we must have $V_{DS} > V_{GS} - V_T$. In our example, we have $V_{DS} = 12.6$ V and $V_{GS} - V_T = 2.6$ V, so we know the results are valid. The validity criterion can be expressed for both NMOS and PMOS FETs as $|V_{DS}| > |V_{GS} - V_T|$.

2. We have already found $V_G = 5.74$ V. Using $R_S = 500$ Ω, $R_D = 2.2$ kΩ, $V_{DD} = 18$ V, $V_T = 2$ V, and $\beta = 0.5 \times 10^{-3}$, we have, with reference to equations 5.101,

$$A = (500)^2 = 2.5 \times 10^5$$
$$B = -2[(5.74 - 2)500 + 1/(0.5 \times 10^{-3})] = -7.74 \times 10^3$$
$$C = (5.74 - 2)^2 = 13.9876$$

Substituting these values into the equation for I_D, we find $I_D = 1.927$ mA. Then, $V_{DS} = 18$ V $- (1.927$ mA$)(2.2$ kΩ $+ 500$ Ω$) = 12.8$ V and $V_{GS} = 5.74$ V $- (1.927$ mA$)(500$ Ω$) = 4.78$ V. These results agree reasonably well with those obtained graphically in part 1.

Figure 5.61
Use of a feedback resistor R_G to bias an enhancement MOSFET.

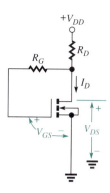

Feedback Bias

Figure 5.61 shows another way to bias an NMOS FET. The resistor R_G, which is usually very large, is connected between drain and gate and carries no current because of the very large (essentially infinite) resistance at the gate. Since there is no voltage drop across R_G, $V_{GS} = V_{DS}$. We can therefore be sure that $V_{DS} > V_{GS} - V_T$, which ensures that the device is biased in the active region. R_G provides negative feedback to stabilize the bias point. For example, if I_D increases for any reason, then there is a greater drop across R_D and V_{DS} decreases. But, since $V_{GS} = V_{DS}$, this means that V_{GS} also decreases. The decrease in V_{GS} causes a decrease in I_D, thus counteracting the original increase in I_D.

From Figure 5.61, it is clear that

$$V_{DS} = V_{DD} - I_D R_D \qquad (5.102)$$

Since $V_{GS} = V_{DS}$, the equation of the transfer characteristic (equation 3.18) can be written as

$$I_D = 0.5\beta(V_{DS} - V_T)^2 \qquad (5.103)$$

General Algebraic Solution—Feedback Bias

By solving equations 5.102 and 5.103 simultaneously for I_D, we can obtain the general algebraic solution for the bias point in a MOSFET circuit employing feedback bias. (The derivation is Exercise 5.61.) Equations 5.104 show the results, valid for both NMOS and PMOS circuits.

**General algebraic solution for the bias point
of PMOS and NMOS circuits with feedback bias**

$$I_D = \frac{-B - \sqrt{B^2 - 4AC}}{2A}$$

where $A = R_D^2$

$$B = -2\left[(|V_{DD}| - |V_T|)R_D + \frac{1}{\beta}\right]$$

$$C = (|V_{DD}| - |V_T|)^2$$

$|V_{DS}| = |V_{GS}| = |V_{DD}| - I_D R_D$ (positive for NMOS, negative for PMOS)   **(5.104)**

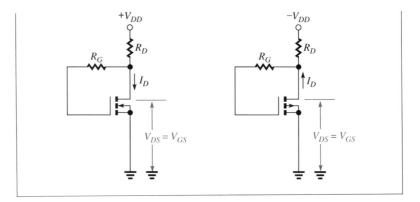

5.10   SMALL-SIGNAL MOSFET AMPLIFIERS

The most convenient small-signal model for a MOSFET, like that of a JFET, incorporates the transconductance of the device. Recall that the characteristics of a depletion-type MOSFET are quite similar to those of a JFET, the only difference being that the depletion MOSFET can be operated in both the depletion and enhancement modes. As we might therefore expect, the small-signal model for the depletion MOSFET is identical to the JFET model, and it is used in the same way to analyze a depletion-type MOSFET amplifier. The value of the transconductance can be found graphically and algebraically in the same way it is found for a JFET.

Figure 5.62 shows a common-source enhancement MOSFET amplifier and its

Figure 5.62
A common-source NMOS amplifier and its equivalent circuit.

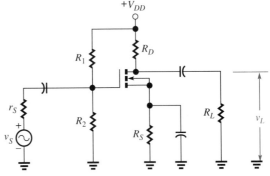

(a) An N-channel enhancement MOSFET (NMOS) amplifier
with voltage-divider bias

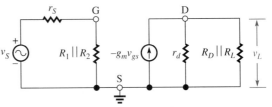

(b) The small-signal equivalent circuit of (a)

small-signal equivalent circuit. The MOSFET is biased using the voltage-divider method discussed earlier (Figure 5.57). Notice that the small-signal equivalent circuit of the NMOS amplifier is identical to that of the JFET amplifier (Figure 5.52(b)). Consequently, all gain and impedance relations are the same as those derived for the JFET amplifier:

$$r_{in} = R_1 \parallel R_2 \qquad\qquad (5.105)$$

$$\frac{v_L}{v_S} = \left(\frac{R_1 \parallel R_2}{r_S + R_1 \parallel R_2}\right)(-g_m)(r_d \parallel R_D \parallel R_L) \qquad\qquad (5.106)$$

Example 5.22

The MOSFET shown in Figure 5.63 has the following parameters: $V_T = 2$ V, $\beta = 0.5 \times 10^{-3}$, $r_d = 75$ kΩ. It is biased at $I_D = 1.93$ mA.

1. Verify that the MOSFET is biased in its active region.
2. Find the input resistance.
3. Draw the small-signal equivalent circuit and find the voltage gain v_L/v_S.

Figure 5.63
(Example 5.22)

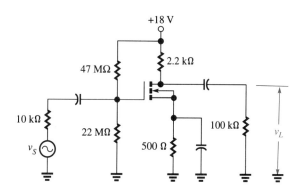

Solution

1. $V_{DS} = V_{DD} - I_D(R_D + R_S) = 18$ V $- (1.93$ mA$)(2.2$ kΩ $+ 500$ Ω$) = 12.78$ V

$$V_G = \left(\frac{22\ \text{MΩ}}{47\ \text{MΩ} + 22\ \text{MΩ}}\right)18\ \text{V} = 5.74\ \text{V}$$

Using equation 5.96 to find V_{GS}, we have

$$V_{GS} = 5.74\ \text{V} - (1.93\ \text{mA})(500\ \Omega) = 4.78\ \text{V}$$

$$|V_{GS} - V_T| = |4.78\ \text{V} - 2\ \text{V}| = 2.78\ \text{V}$$

Therefore, condition 3.20 is satisfied:

$$12.78\ \text{V} = |V_{DS}| > |V_{GS} - V_T| = 2.78\ \text{V}$$

and we conclude that the MOSFET is biased in its active region.

2. $r_{in} = R_1 \parallel R_2 = (47\ \text{MΩ}) \parallel (22\ \text{MΩ}) = 15\ \text{MΩ}.$

Figure 5.64
(Example 5.22) The small-signal equivalent circuit of Figure 5.63.

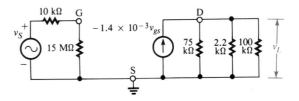

3. From equation 3.21,

$$g_m = 0.5 \times 10^{-3}(4.79 \text{ V} - 2 \text{ V}) = 1.4 \times 10^{-3} \text{ S}$$

The small-signal equivalent circuit is shown in Figure 5.64. From equation 5.106,

$$
\begin{aligned}
\frac{v_L}{v_S} &= \left(\frac{15 \times 10^6}{10 \times 10^3 + 15 \times 10^6}\right) \\
&\quad \times (-1.4 \times 10^{-3})[(75 \times 10^3) \,\|\, (2.2 \times 10^3) \,\|\, (100 \times 10^3)] \\
&= (0.999)(-1.4 \times 10^{-3})(2.09 \times 10^3) = -2.92
\end{aligned}
$$

Example 5.23

DESIGN

An N-channel JFET is to be biased at $V_{DS} = 6$ V using a 15-V supply. The nominal FET characteristic has $V_p = -3.5$ V and $I_{DSS} = 13.5$ mA. The quiescent drain current should not vary more than ± 0.5 mA from a nominal value of 6 mA when the JFET characteristic changes from $V_p = -3$ V to $V_p = -4$ V, with a corresponding change in I_{DSS} from 12 mA to 15 mA. Find values for R_D, R_S, R_1, and R_2 in a voltage-divider bias design.

Find the actual range of I_D and V_{DS} over the range of the JFET characteristic when the standard-valued 5% resistors closest to the calculated values are used in the design. Assume the resistors have their nominal values.

Solution.   Recall that the steepness of the slope of the bias line (Figure 5.40) determines how much the quiescent point changes when the JFET characteristic is changed. In the voltage-divider bias circuit, the steepness of the slope depends on the value of V_G: the gate voltage, which is the intercept of the bias line on the horizontal axis of the transfer characteristic. Figure 5.65 shows how to calculate V_G for an allowed range of bias points when the JFET characteristic changes over a specified range:

$$|V_G| = \frac{I_{D1}(|V_{GS2}| - |V_{GS1}|)}{I_{D2} - I_{D1}} - |V_{GS1}| \tag{5.107}$$

where (V_{GS1}, I_{D1}) and (V_{GS2}, I_{D2}) are the quiescent points over the range of change of the transfer characteristic. In our example, $I_{D1} = 5.5$ mA and $I_{D2} = 6.5$ mA (6 mA $\pm$ 0.5 mA). To use equation 5.107, we must find the corresponding values of V_{GS1} and V_{GS2}. Solving the square-law equation (3.11) for V_{GS} (Example 3.4) gives

$$
\begin{aligned}
V_{GS} &= V_P (1 - \sqrt{I_D/I_{DSS}}) \\
V_{GS1} &= (-3 \text{ V})(1 - \sqrt{5.5 \text{ mA}/12 \text{ mA}}) = -0.97 \text{ V} \\
V_{GS2} &= (-4 \text{ V})(1 - \sqrt{6.5 \text{ mA}/15 \text{ mA}}) = -1.37 \text{ V}
\end{aligned}
$$

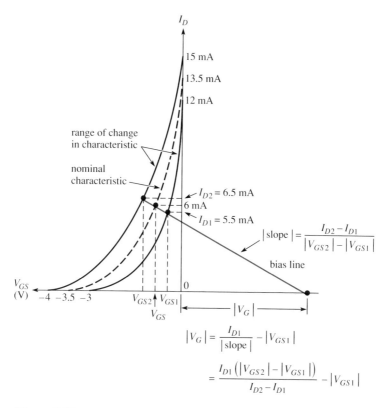

Figure 5.65
Computing the value of $|V_G|$ for a voltage-divider bias circuit when a specified range of bias points, (V_{GS2}, I_{D2}) to (V_{GS1}, I_{D1}), is allowed (Example 5.23).

Then, from equation 5.107,

$$|V_G| = \frac{5.5\ \text{mA}(1.37\ \text{V} - 0.97\ \text{V})}{6.5\ \text{mA} - 5.5\ \text{mA}} - 0.97\ \text{V} = 1.23\ \text{V}$$

The nominal value of V_{GS} is

$$V_{GS} = (-3.5\ \text{V})(1 - \sqrt{6\ \text{mA}/13.5\ \text{mA}} = -1.17\ \text{V}$$

We can now solve the bias-line equation (5.96) for R_S:

$$V_{GS} = V_G - I_D R_S$$
$$-1.17\ \text{V} = 1.23\ \text{V} - (6\ \text{mA})R_S$$
$$R_S = 400\ \Omega$$

Solving the load-line equation (5.98) for R_D gives

$$R_D = \frac{V_{DD} - V_{DS} - I_D R_S}{I_D} = \frac{15\ \text{V} - 6\ \text{V} - (6\ \text{mA})(400\ \Omega)}{6\ \text{mA}} = 1.1\ \text{k}\Omega$$

Choose $R_2 = 330$ kΩ. Solving equation 5.95 for R_1, we find

$$R_1 = \frac{R_2(|V_{DD}| - |V_G|)}{|V_G|} = \frac{(330 \text{ k}\Omega)(15 \text{ V} - 1.23 \text{ V})}{1.23 \text{ V}} = 3.69 \text{ M}\Omega$$

The standard-valued 5% resistors closest to the calculated values are $R_S = 390 \ \Omega$, $R_D = 1.1$ kΩ, $R_1 = 3.6$ MΩ, and $R_2 = 330$ kΩ. Using these values in equations 5.69, we find that I_D ranges from 5.65 mA to 6.65 mA and V_{DS} ranges from 5.09 V to 6.58 V over the range of the JFET characteristic.

Example 5.24

PSpice

Use PSpice to find the quiescent point and voltage gain v_L/v_S of the MOSFET amplifier shown in Figure 5.66(a). The MOSFET has threshold voltage $V_T = 2$ V and $\beta = 0.5 \times 10^{-3}$. Assume a signal frequency of 10 kHz. (At this frequency, the reactance of the 10-μF coupling capacitors is negligibly small and has very little effect on the voltage gain.)

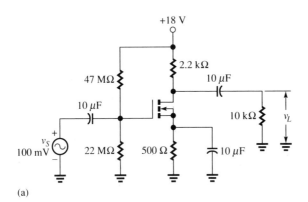

(a)

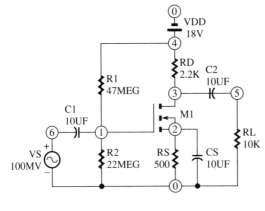

(b)

```
EXAMPLE 5.23
M1 3 1 2 2 MOSFET
.MODEL MOSFET NMOS
        VTO=2 KP=0.5E-3
R1 4 1 47MEG
R2 1 0 22MEG
RS 2 0 500
RD 3 4 2.2K
RL 5 0 10K
C1 6 1 10UF
C2 3 5 10UF
CS 2 0 10UF
VDD 4 0 18V
VS 6 0 AC 100MV
.AC LIN 1 10K 10K
.PRINT AC V(5)
.END
```

Figure 5.66
(Example 5.24)

Solution. The circuit is redrawn in a PSpice format with node numbers assigned as shown in Figure 5.66(b). Note that V_T is entered as VTO = 2 and β as KP = 0.5E-3. The quiescent point is found from the dc analysis that PSpice performs automatically when an ac analysis is requested:

NODE	VOLTAGE	NODE	VOLTAGE	NODE	VOLTAGE	NODE	VOLTAGE
(1)	5.7391	(2)	0.9632	(3)	13.7620	(4)	18.0000
(5)	0.0000	(6)	0.0000				

Thus,

$$V_{DS} = V(3) - V(2) = 13.762 \text{ V} - 0.9632 \text{ V} = 12.7988 \text{ V}$$

and

$$I_D = \frac{V_{DD} - V_D}{R_D} = \frac{18 \text{ V} - V(3)}{2.2 \text{ k}\Omega} = \frac{18 \text{ V} - 13.762 \text{ V}}{2.2 \text{ k}\Omega} = 1.926 \text{ mA}$$

The ac analysis shows that V(5) = 0.2503 V. Thus,

$$\frac{v_L}{v_S} = \frac{0.2503 \text{ V}}{100 \text{ mV}} = 2.503$$

(Although we did not find the phase angle of the output, we know that the common-source configuration inverts the signal, so the voltage gain is actually −2.503.)

EXERCISES

Section 5.1 The Common Terminal of an Amplifier

5.1 Draw schematic diagrams (the small-signal (ac) circuits only) showing PNP transistors in common-emitter, common-base, and common-collector configurations. Label input and output voltages.

5.2 When the second subscript of a voltage designation is omitted, it is understood that the voltage is measured from the terminal represented by the first subscript to the common terminal. For example, v_b is the same as v_{bc} in a common-collector amplifier. A certain amplifier has input voltage v_e and output voltage v_c. What configuration does the amplifier have (CE, CB, or CC)?

Section 5.2 Common-Emitter Bias Circuits

5.3 **a.** Find the quiescent values of I_C and V_{CE} in the amplifier shown in Figure 5.67.
b. Verify that the collector–base junction is reverse biased.

5.4 Repeat Exercise 5.3 for the circuit shown in Figure 5.68.

Figure 5.67
(Exercises 5.3, 5.6, 5.7, 5.19, and 5.21)

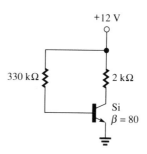

Figure 5.68
(Exercises 5.4 and 5.5)

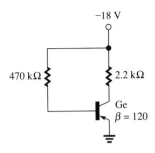

5.5   To what value should R_B in Exercise 5.4 be changed in order to make $V_{CE} = -11.13$ V?

5.6   To what value should R_C in Exercise 5.3 be changed in order to make $V_{CE} = 7.89$ V?

5.7   a. Write the equation for the load line of the circuit in Figure 5.67 (Exercise 5.3) and sketch it on a set of $I_C - V_{CE}$-axes.

    b. What is the value of I_C when $V_{CE} = 4$ V?

Figure 5.69
(Exercises 5.9 and 5.10)

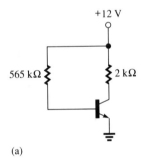

(a)

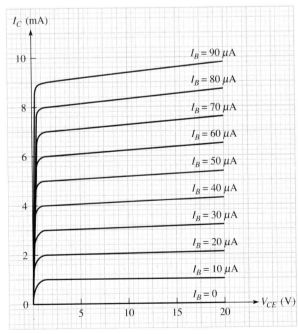

(b)

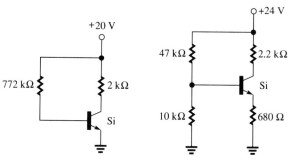

Figure 5.70
(Exercise 5.10)

Figure 5.71
(Exercise 5.13)

5.8   The general form for the equation of a
       straight line in the x-y plane is $y = mx + b$,
       where m is the slope of the line and b is its
       intercept on the y-axis.
       a. Express the load-line equation (equation
          5.3) in the general form of a straight line
          (with I_C analogous to y).
       b. What is the slope of the load-line equa-
          tion? What are the units of the slope?
5.9   The silicon transistor in Figure 5.69(a) has
       the collector characteristics shown in Fig-
       ure 5.69(b).
       a. Plot the load line on the characteristics
          and use it to determine the quiescent
          point.
       b. Find the quiescent point by computa-
          tion.
5.10   Repeat Exercise 5.10 for the circuit in Fig-
       ure 5.70. (The collector characteristics are
       the same as those in Exercise 5.9.)
5.11   Write the mathematical expression for the

       total output (collector) current whose sinus-
       oidal variation is shown in Figure 5.7.
5.12   If the Q-point in Figure 5.7 were located at
       the intersection of the load line and the
       $I_B = 20 \ \mu A$ curve in Figure 5.7, what would
       be the maximum peak-to-peak output cur-
       rent and output voltage that would be possi-
       ble without clipping?
5.13   Find the quiescent values of I_C and V_{CE} in
       Figure 5.71. (Assume the resistance looking
       into the base is much greater than the paral-
       lel equivalent resistance of the voltage-
       divider resistors.)
5.14   Repeat Exercise 5.13 for the circuit shown
       in Figure 5.72. (Note that $V_E \approx V_B + 0.3$ V.)
5.15   Assuming the transistor in Exercise 5.13
       has $\beta = 150$, find more accurate values for
       I_C and V_{CE} by taking into account the load-
       ing of the voltage divider.
5.16   Use the Thevenin equivalent circuit in Fig-

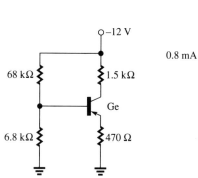

Figure 5.72
(Exercise 5.14)

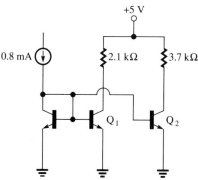

Figure 5.73
(Exercise 5.17)

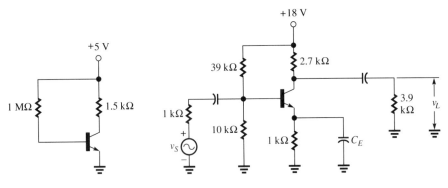

Figure 5.74
(Exercise 5.22)

Figure 5.75
(Exercise 5.23)

ure 5.9(b) to derive equation 5.10. (Hint: Write Kirchhoff's voltage law in terms of I_B and solve for I_B. Then, $I_C = \beta I_B$.)

5.17 Find the quiescent values of I_C and V_{CE} in Q_1 and Q_2 in Figure 5.73:
  a. assuming an ideal current mirror;
  b. taking into account that the β of each transistor is 100.

5.18 A current mirror is to be used to reflect a 1-mA current source to several transistors in an integrated circuit. The transistors all have $\beta = 100$. If the actual current in each transistor cannot be less than 0.9 mA, what is the maximum number of transistors to which current can be reflected?

Section 5.3 The CE Small-Signal Amplifier Model

5.19 Find the small-signal input resistance looking into the base of the transistor in Figure 5.67 (Exercise 5.3).

5.20 Find the small-signal input resistance looking into the base of the transistor in Figure 5.71 (Exercise 5.13). Assume the transistor has $\beta = 150$.

5.21 **a.** Draw the small-signal equivalent circuit of the amplifier in Figure 5.67 (Exercise 5.3). The transistor has $r_c = 8$ MΩ. Label all component values.
  b. Find an approximate value for the small-signal voltage gain, A_v.

5.22 The silicon transistor in Figure 5.74 has $\beta = 200$ and $r_c = 12$ MΩ.
  a. Draw the small-signal equivalent circuit. Label all component values.
  b. Find an approximate value for the voltage gain, A_v.
  c. Find the current gain, A_i.

5.23 The silicon transistor in Figure 5.75 has $\beta = 150$ and $r_c = 18$ MΩ.
  a. Draw the small-signal equivalent circuit. Label all component values.
  b. Find an approximate value for the volt-

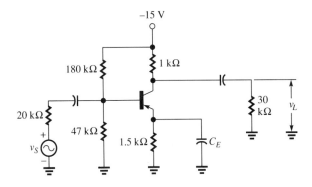

Figure 5.76
(Exercise 5.24)

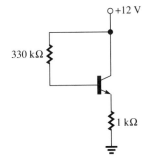

Figure 5.77
(Exercise 5.25)

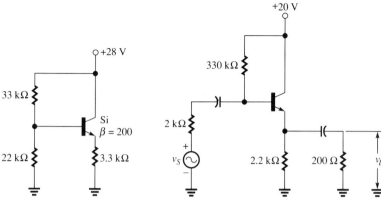

Figure 5.78
(Exercise 5.26)

Figure 5.79
(Exercise 5.29)

age gain, v_L/v_S, with C_E connected as shown.

c. Repeat (b) with C_E removed.

5.24 Repeat Exercise 5.23 for the amplifier shown in Figure 5.76. The silicon transistor has $\beta = 80$ and $r_c = 14$ MΩ.

Section 5.4 The Common-Collector Amplifier

5.25 Find the quiescent values of I_E and V_E in Figure 5.77. Also find V_{CE}.

5.26 Repeat Exercise 5.25 for the circuit in Figure 5.78. (Hint: V_B and V_E are found in the same way as they are in a CE voltage-divider bias circuit.)

5.27 Derive equation 5.39.

5.28 If $R_E \gg R_L$ in a CC amplifier, what is the approximate value of the voltage gain A_v when $R_L = r_e$? (This situation occurs in a *differential* amplifier.)

5.29 a. Find the voltage gain v_L/v_S of the amplifier in Figure 5.79.
b. Find r_o(stage).

5.30 Find P_L/P_{in} in each of parts (a) and (b) of Figure 5.26 (Example 5.9). The input to the

first amplifier is $v_{in} = 10$ mV rms and it has input resistance 1 kΩ.

Section 5.5 The Common-Base Amplifier

5.31 For the amplifier shown in Figure 5.80, find
a. the quiescent values of I_C and V_{CB};
b. r_{in}(stage);
c. v_L/v_S.

5.32 Repeat Exercise 5.31 for the amplifier shown in Figure 5.81.

Section 5.6 Common-Source JFET Amplifier Bias

5.33 The JFET in the circuit of Figure 5.82 has the drain characteristics shown in Figure 5.35. Find the quiescent values of I_D and V_{DS} when (a) $V_{GS} = -2$ V, and (b) $V_{GS} = 0$ V. Which, if either, of the Q-points is in the pinch-off region?

5.34 Using Figure 5.35, determine the value of V_{GS} in Exercise 5.33 that would be required to obtain $V_{DS} = 8$ V.

5.35 The JFET shown in Figure 5.83 has $I_{DSS} = $

Figure 5.80
(Exercise 5.31)

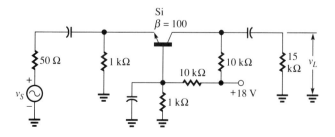

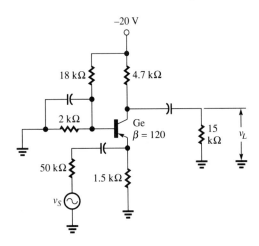

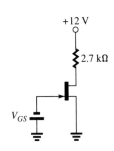

Figure 5.81
(Exercise 5.32)

Figure 5.82
(Exercise 5.33)

14 mA and $V_p = -5$ V. Algebraically determine the quiescent values of I_D and V_{DS} for (a) $V_{GS} = -3.6$ V, (b) $V_{GS} = -3$ V, and (c) $V_{GS} = -1.5$ V. In each case, check the validity of your results by verifying that the quiescent point is in the pinch-off region. Identify any cases that do not meet that criterion and for which the results are therefore not valid.

5.36   Repeat Exercise 5.35 when R_D is changed to 1 kΩ. Does this modification affect the validity of the results in any of the three cases (a), (b), or (c)? Explain.

5.37   Figure 5.85 shows the transfer characteristic of the JFET in the circuit of Figure 5.84. Graphically determine the quiescent values of I_D and V_{GS}. Compute the quiescent value of V_{DS} based on your results, and verify their validity.

5.38   Algebraically determine the quiescent values of I_D, V_{GS}, and V_{DS} in the circuit of Exercise 5.37. (Refer to Figure 5.85 to obtain values for I_{DSS} and V_p.)

5.39   Figure 5.87 shows the transfer characteristic for the JFET in the circuit of Figure 5.86.
   a. Graphically determine the quiescent values of I_D and V_{GS}. Compute the quiescent value of V_{DS} based on your results, and verify their validity.
   b. Determine the quiescent values of I_D, V_{GS}, and V_{DS} algebraically.

5.40   Figure 5.39 shows the possible range of the transfer characteristics for each of the JFETs in Figure 5.88. Find the change in I_D that could be expected in each bias circuit if the JFET characteristics changed over their possible range.

5.41   The JFET in the circuit of Figure 5.89 has

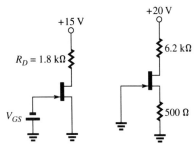

Figure 5.83
(Exercise 5.35)

Figure 5.84
(Exercise 5.37)

Figure 5.85
(Exercise 5.37)

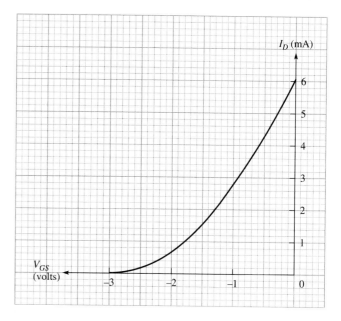

Figure 5.86
(Exercise 5.39)

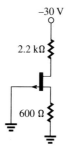

Figure 5.87
(Exercise 5.39)

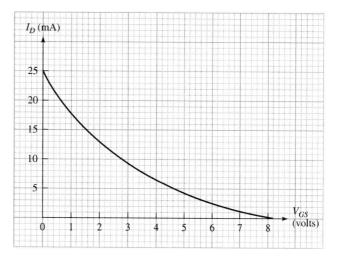

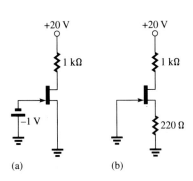

Figure 5.88
(Exercise 5.40)

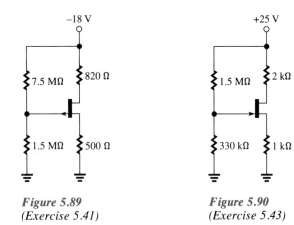

Figure 5.89
(Exercise 5.41)

Figure 5.90
(Exercise 5.43)

the transfer characteristic shown in Figure 5.43. Graphically determine the quiescent value of I_D and use it to determine the quiescent value of V_{DS}. Verify the validity of your results.

5.42 Algebraically determine the quiescent values of I_D and V_{DS} in Exercise 5.41. (Refer to Figure 5.43 to determine I_{DSS} and V_p.)

5.43 Figure 5.91 shows the transfer characteristic of the JFET in the circuit of Figure 5.90.
  a. Graphically determine the quiescent value of I_D. Verify the validity of your result.

b. Algebraically determine the quiescent values of I_D and V_{DS}. Verify the validity of your results.

5.44 It is desired to modify the circuit of Exercise 5.43 (Figure 5.90) so that the bias line intersects the transfer characteristic at $V_{GS} = -2.8$ V and the V_{GS}-axis at +3 V. Assuming that the JFET has the same transfer characteristic (Figure 5.91), answer the following:
  a. If R_1 is to retain its original value (1.5 MΩ), what should be the new value of R_2?

Figure 5.91
(Exercise 5.43)

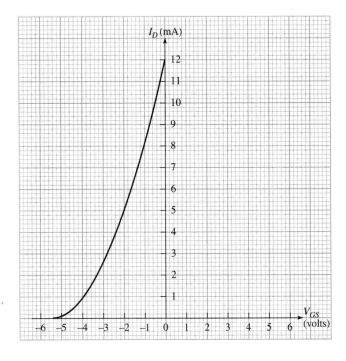

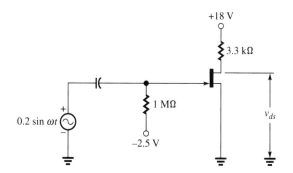

Figure 5.92
(Exercise 5.46)

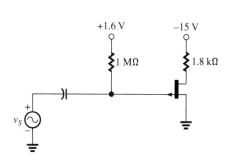

Figure 5.93
(Exercise 5.47)

b. What should be the new value of R_S?
c. What is the new value of V_{DS}?
d. Is the JFET biased in its pinch-off region?

Section 5.7   Small-Signal Common-Source JFET Amplifiers

5.45 The following measurements were taken from a curve tracer display of the output characteristics of a P-channel JFET along the line $V_{GS} = +2.5$ V: $I_D = 4.2$ mA at $V_{DS} = -4.5$ V; $I_D = 4.3$ mA at $V_{DS} = -12$ V. What is the drain resistance at $V_{GS} = 2.5$ V?

5.46 The JFET in the amplifier shown in Figure 5.92 has $I_{DSS} = 16$ mA, $V_p = -4.5$ V, and $r_d = 80$ kΩ.
a. Find the quiescent values of I_D and V_{DS}.
b. Find g_m.
c. Draw the ac equivalent circuit.

d. What is the peak value of the ac component of v_{ds}?

5.47 The JFET in the amplifier shown in Figure 5.93 has $I_{DSS} = 14$ mA, $V_p = 3$ V, and $r_d = 120$ kΩ.
a. Find the quiescent values of I_D and V_{DS}.
b. Find g_m.
c. Draw the ac equivalent circuit.
d. Find the voltage gain.

5.48 The JFET in the amplifier shown in Figure 5.94 has $g_m = 4.2 \times 10^{-3}$ S and $r_d = 98$ kΩ.
a. Find the input resistance of the amplifier.
b. Find the voltage gain.

5.49 The JFET in the amplifier shown in Figure 5.95 has $r_d = 50$ kΩ, $V_p = 5$ V, and $I_{DSS} = 15$ mA. The dc source-to-ground voltage is -5.15 V.
a. Find the voltage gain.
b. Find the input resistance of the amplifier.

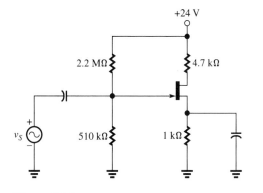

Figure 5.94
(Exercise 5.48)

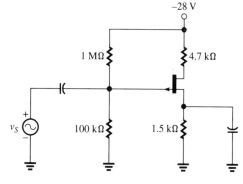

Figure 5.95
(Exercise 5.49)

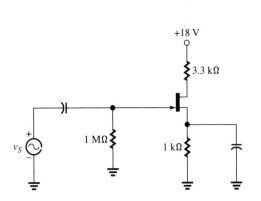

Figure 5.96
(Exercise 5.50)

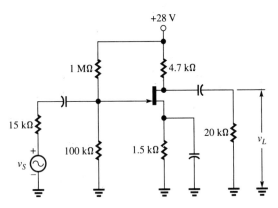

Figure 5.97
(Exercise 5.51)

5.50 The JFET in the amplifier shown in Figure 5.96 has $r_d = 75$ kΩ, $V_p = -3.6$ V, and $I_{DSS} = 9$ mA. The dc voltage drop across the 3.3-kΩ resistor is 6.38 V. Find the voltage gain of the amplifier.

5.51 The JFET in the amplifier shown in Figure 5.97 has $r_d = 100$ kΩ and $g_m = 2871$ μS. Find the voltage gain v_L/v_S.

5.52 The JFET in the amplifier shown in Figure 5.98 has $r_d = 60$ kΩ, $V_p = 3.9$ V, and $I_{DSS} = 10$ mA. The quiescent value of V_{DS} is -9.24 V. Find the voltage gain v_L/v_S.

Section 5.8 The Common-Drain JFET Amplifier

5.53 The JFET in the amplifier shown in Figure 5.99 has $g_m = 0.004$ S and $r_d = 90$ kΩ.
a. Find the voltage gain v_L/v_S.
b. Find the input resistance.

5.54 The JFET in the amplifier shown in Figure 5.100 has $g_m = 5200$ μS and $r_d = 80$ kΩ.
a. Find the voltage gain v_L/v_S.
b. Find the output resistance.

Section 5.9 MOSFET Amplifier Bias Circuits

5.55 In the bias circuit of Figure 5.57, $R_1 = 2.2$ MΩ, $R_2 = 1$ MΩ, $V_{DD} = 28$ V, $R_D = 2.7$ kΩ, and $R_S = 600$ Ω. If $V_{GS} = 5.5$ V, find (a) I_D and (b) V_{DS}.

5.56 In the bias circuit of Figure 5.57, $R_1 = 470$ kΩ, $V_{DD} = 20$ V, $R_D = 1.5$ kΩ, $R_S = 220$ Ω, $I_D = 6$ mA, and $V_{GS} = 6$ V. Find (a) V_{DS} and (b) R_2.

5.57 The MOSFET shown in Figure 5.101 has the transfer characteristic shown in Figure 3.35. Graphically determine the quiescent values of I_D and V_{GS}. Find the quiescent value of V_{DS}.

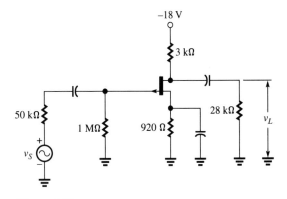

Figure 5.98
(Exercise 5.52)

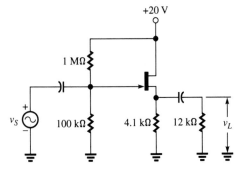

Figure 5.99
(Exercise 5.53)

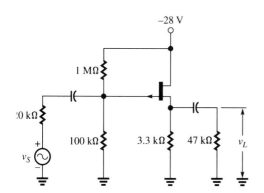

Figure 5.100
(Exercise 5.54)

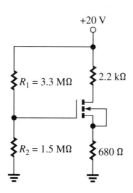

Figure 5.101
(Exercise 5.57)

5.58 Algebraically determine I_D, V_{GS}, and V_{DS} in Exercise 5.57. Verify that the MOSFET is operating in its active region.

5.59 It is desired to modify the circuit of Exercise 5.57 so that the load line intersects the transfer characteristic (Figure 3.35) at $V_{GS} = 5$ V and the V_{GS}-axis at 7 V.
  a. If R_2 is to retain its original value (1.5 MΩ), what should be the new value of R_1?
  b. What should be the new value of R_S?
  c. What is the new value of V_{DS}?
  d. Is the new bias point in the active region?

5.60 The MOSFET in Figure 5.102 has $\beta = 0.62 \times 10^{-3}$ and $V_T = -2.4$ V. Algebraically determine the quiescent values of I_D, V_{GS}, and V_{DS}. Verify the validity of your results.

5.61 Derive equations 5.104.

5.62 The MOSFET in Figure 5.103 has $\beta = 0.5 \times$

10^{-3} and $V_T = 2$ V. Find the quiescent values of V_{DS}, V_{GS}, and I_D. Is the device in the active region?

5.63 How would the solution to Exercise 5.62 be affected if R_G were reduced from 22 MΩ to 10 MΩ?

Section 5.10   Small-Signal MOSFET Amplifiers

5.64 The MOSFET shown in Figure 5.104 has $r_d = 100$ kΩ and a transconductance of 3×10^{-3} S.
  a. Draw the ac equivalent circuit.
  b. Find the voltage gain v_L/v_S.

5.65 The MOSFET shown in Figure 5.105 has $V_T = 2.4$ V, $\beta = 0.62 \times 10^{-3}$, and $r_d = 120$ kΩ. It is biased at $I_D = 5$ mA.
  a. Find the input resistance.
  b. Find the voltage gain v_L/v_S.

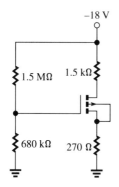

Figure 5.102
(Exercise 5.60)

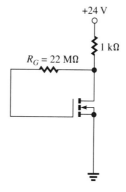

Figure 5.103
(Exercise 5.62)

Figure 5.104
(Exercise 5.64)

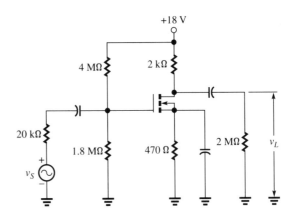

DESIGN EXERCISES

5.66 a. Design a single base-resistor bias circuit for an NPN silicon transistor in a common-emitter configuration. The nominal β of the transistor is 80 and the supply voltage is +24 V. The bias point is to be $I_C = 5$ mA, and $V_{CE} = +10$ V. Use standard-valued resistors with 10% tolerance and draw a schematic diagram of your design.

b. Calculate the actual bias point assuming the 10% resistors have their nominal values.

c. Calculate the range of possible values that the bias point could have if the value of β changed over the range from 50 to 100. Assume the resistors have their nominal values.

5.67 a. Design a single base-resistor bias circuit for an NPN silicon transistor having a nominal β of 100, to be used in a common-emitter configuration. The bias

point is to be $I_C = 1$ mA, and $V_{CE} = 5$ V. The supply voltage is 15 V. Use standard-valued 5% resistors and draw a schematic diagram of your design.

b. Calculate the possible range of values of the bias point taking into consideration *both* the resistor tolerances and a possible variation in β from 30 to 150. Interpret and comment on your results.

5.68 Design a voltage-divider bias circuit for an NPN transistor in a CE configuration that sets the quiescent values of V_{CE} and I_C to 7 V and 1.2 mA. The supply voltage is 18 V. Assume that the silicon transistor has a value of β that may range from 80 to 200. Use standard 5% resistor values closest to your design calculations and analyze the final circuit to find the actual Q-point.

5.69 Design a voltage-divider bias circuit for an NPN transistor in a CE configuration that sets the quiescent values of V_{CE} and I_C to 10 V and 2 mA. The supply voltage is 30 V. Assume that the smallest value of β that the

Figure 5.105
(Exercise 5.65)

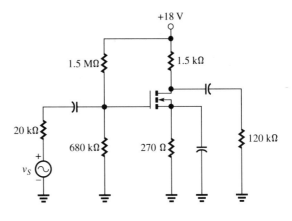

silicon transistor used in the circuit might have is 100. Use standard 10% resistor values closest to your design calculations and analyze the final circuit to find the actual Q-point. Assuming that $r_S = 0$ and $R_L = \infty$, calculate the ac voltage gain of the circuit, with the emitter resistor bypassed by a capacitor.

5.70 Design a silicon, single-transistor amplifier that will deliver approximately 2.5 V pk-pk to a capacitor-coupled 10-kΩ load when the signal source generates 25 mV pk-pk. The signal source has resistance 1 kΩ and the transistor has a β of at least 100. A +15 V dc power supply is available. Use standard-valued 5% resistors and calculate the actual load voltage delivered by your design.

5.71 a. Design a bias circuit for an NPN silicon transistor in a common-collector configuration. The nominal β for the transistor is 100, and the supply voltage is 30 V. The bias point is to be $I_C = 10$ mA, and $V_{CE} = 12$ V. Use standard-valued resistors having 5% tolerance and draw a schematic diagram of your design.

b. Calculate the *minimum* value that V_{CE} could have if *both* the resistor tolerances and a variation in β from 60 to 120 are taken into account. (*Hint:* Use equations 5.27 to derive the expression

$$V_{CE} = V_{CC} - \frac{V_{CC} - V_{BE}}{\dfrac{R_B}{(\beta + 1)R_E} + 1}$$

5.72 Design a silicon, single-transistor buffer amplifier that is capable of delivering at least 0.02 A pk-pk to a capacitor-coupled 50-Ω load when a signal source having resistance 3 kΩ produces 2 V pk-pk. A transistor having a β of at least 200 will be used. A 20-V dc power supply is available. Use standard-valued 5% resistors and calculate the actual current delivered by your design.

5.73 A P-channel JFET is to be biased at the nominal values $I_D = 3$ mA and $V_{DS} = -5$ V. The quiescent value of I_D should not vary outside the range from 2.8 mA to 3.5 mA when the JFET characteristic changes over a range in which $I_{DSS} = 10$ mA and $V_p = 1.5$ V to $I_{DSS} = 15$ mA and $V_p = 3$ V. (The nominal characteristic has $I_{DSS} = 12$ mA and $V_p = 2$ V.)

a. Design the circuit, using an 18-V supply.

b. Find the actual range of I_D and V_{DS} over the range of the JFET characteristic when standard valued 5% resistors clos-

est to the calculated values are used. Assume the resistors have their nominal values.

c. Determine if the bias point remains in the pinch-off region over the range of variation when the resistors in (b) are used.

SPICE EXERCISES

5.74 Use SPICE to determine the quiescent point, voltage gain v_L/v_S, and current gain i_L/i_S of the common-emitter amplifier shown in Figure 5.19(a). The transistor has a beta of 200. Component values are: $r_S = 1$ kΩ, $R_B = 470$ kΩ, $R_C = 1$ kΩ, and $R_L = 10$ kΩ. Both coupling capacitors are 10 μF, and $V_{CC} = 18$ V. The analysis should be performed at 10 kHz.

5.75 Use SPICE to determine (approximately) the maximum temperature that the transistor in Figure 5.21(a) can be operated without creating visible distortion (clipping) in the load voltage across R_L when the input is a sine wave having peak value 1.8 V and frequency 10 kHz. Component values in the circuit are: $r_S = 100$ Ω, $R_1 = 24$ kΩ, $R_2 = 4.7$ kΩ, $R_C = 1$ kΩ, $R_E = 220$ Ω, and $R_L = 10$ kΩ. Both coupling capacitors are 10 μF. $V_{CC} = 24$ V. The "ideal maximum forward beta" of the transistor is 100, the "forward Early voltage" is 130 V, and the "forward and reverse beta temperature exponent" is 1.5.

5.76 To investigate the effect of β on the current mirror shown in Figure 5.13 (Example 5.4), use SPICE to determine V_{CE} in Q_1, Q_2, and Q_3 for the case where the β of every transistor is 100 and again for the case where every β is 200. In each case, compare the results obtained from SPICE with the results shown in the example for the ideal current mirror.

5.77 Use SPICE to determine the quiescent point, voltage gain v_L/v_S, and current gain i_L/i_S of the common-collector amplifier shown in Figure 5.25(a). The transistor has a beta of 150. Component values are: $r_S = 1$ kΩ, $R_B = 180$ kΩ, $R_E = 1$ kΩ, and $R_L = 300$ Ω. The input coupling capacitor is 10 μF, and the output coupling capacitor is 100 μF. $V_{CC} = 9$ V.

5.78 Use SPICE to determine r_{in}(stage) and r_{in} of the common-collector amplifier in Exercise 5.79. (*Hint:* Find ac voltages and currents at appropriate points in the circuit.)

5.79 Use SPICE to solve Exercise 5.38.

5.80 Use SPICE to solve Exercise 5.42.

5.81 Use SPICE to find the voltage gain, v_L/v_S, and input resistance, r_{in}(stage), of the amplifier shown in Figure 5.98. The JFET has $V_p = 3.9$ V and $I_{DSS} = 10$ mA. The input and output coupling capacitors are each 1 μF and the source-bypass capacitor is 10 μF. The input signal v_S has frequency 10 kHz.

5.82 The JFET in the amplifier shown in Figure 5.99 has $V_p = 2.5$ V and $I_{DSS} = 12$ mA. Use SPICE to find the voltage gain v_L/v_S and the input resistance, r_{in}(stage). The input cou-pling capacitor is 1 μF and the output coupling capacitor is 10 μF. The input signal v_S has frequency 10 kHz.

5.83 **a.** Use SPICE to verify the voltage gain obtained in Example 5.21 for the MOSFET amplifier. Obtain a plot of the load voltage over one full cycle when the input, v_S, is a 20-kHz sine-wave voltage with peak value 0.5 V. The input and output coupling capacitors are 1 μF and the source-bypass capacitor is 10 μF.

b. Obtain the plot described in (a) when the source bypass capacitor is removed, and comment on the results.

Frequency Response

6.1 DEFINITIONS AND BASIC CONCEPTS

The *frequency response* of an electronic device or system is the variation it causes, if any, in the level of its output signal when the frequency of the signal is changed. In other words, it is the manner in which the device *responds* to changes in signal frequency. Variation in the level (amplitude, or rms value) of the output signal is usually accompanied by a variation in the *phase angle* of the output relative to the input, so the term *frequency response* also refers to phase shift as a function of frequency. (Phase shift versus frequency is sometimes called *phase response*.) Figure 6.1 shows an amplifier whose frequency response causes small output amplitudes at both low and high frequencies. Notice that the input signal amplitude is the same at each frequency, but the output signal amplitude changes with frequency. Thus, the *gain* of the amplifier is a function of frequency. In this example the gain is small at the low frequency and small at the high frequency.

The frequency response of an amplifier is usually presented in the form of a graph that shows output amplitude (or, more often, voltage gain) plotted versus frequency. Phase-angle variation is sometimes plotted on the same graph. Figure 6.2 shows a typical plot of the voltage gain of an ac amplifier versus frequency. Notice that the gain is 0 at dc (zero frequency), then rises as frequency increases, levels off for further increases in frequency, and then begins to drop again at high frequencies.

The frequency range over which the gain is more or less constant ("flat") is called the *midband range,* and the gain in that range is designated A_m. As shown in Figure 6.2, the low frequency at which the gain equals $(\sqrt{2}/2) A_m \approx 0.707A_m$ is called the *lower cutoff frequency* and is designated f_1. The high frequency at which the gain once again drops to $0.707A_m$ is called the *upper cutoff frequency* and is designated f_2. The *bandwidth* of the amplifier is defined to be the difference between the upper and lower cutoff frequencies:

$$\text{bandwidth} = \text{BW} = f_2 - f_1 \qquad \textbf{(6.1)}$$

The points on the graph in Figure 6.2 where the gain is $0.707A_m$ are often called *half-power points,* and the cutoff frequencies are sometimes called *half-*

Figure 6.1
An amplifier whose frequency response
is such that the output signal amplitude
is small at both low and high frequencies

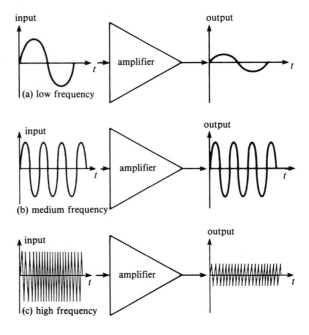

(a) low frequency

(b) medium frequency

(c) high frequency

power frequencies, because the output power of the amplifier at cutoff is one-half of its output power in the midband range. To demonstrate this fact, suppose that an rms output voltage V is developed across R ohms in the midband range. Then the output power in midband is

$$P_{(midband)} = \frac{V^2}{R} \text{ watts} \qquad (6.2)$$

At each of the cutoff frequencies, the output voltage is $(\sqrt{2}/2)V$, so the power is

$$P_{(at\ cutoff)} = \frac{\left(\frac{\sqrt{2}}{2}V\right)^2}{R} = \frac{0.5\ V^2}{R} = 0.5\ P_{(midband)}$$

Figure 6.2
A typical amplifier's frequency response,
showing gain versus frequency

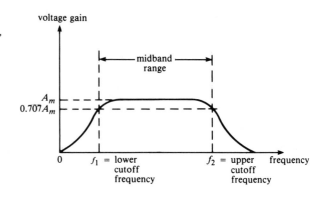

Example 6.1

An audio amplifier has a lower cutoff frequency of 20 Hz and an upper cutoff frequency of 20 kHz. (This is the frequency range of sound waves—the audio frequency range.) The amplifier delivers 20 W to a 12-Ω load at 1 kHz.

1. What is the bandwidth of the amplifier?
2. What is the rms load voltage at 20 kHz?
3. What is the rms load voltage at 2 kHz?

Solution

1. BW $= f_2 - f_1 = 20 \times 10^3$ Hz $- 20$ Hz $= 19,980$ Hz
2. Since 1 kHz is in the midband range, the midband power is 20 W. At the 20-kHz cutoff frequency, the power is $\frac{1}{2}(20) = 10$ W, so

$$\frac{V^2}{12} = 10$$
$$V^2 = 120$$
$$V = \sqrt{120} = 10.95 \text{ V rms}$$

3. Assuming that the load voltage is exactly the same throughout the midband range (not always the case in practice), the output at 2 kHz will be the same as that at 1 kHz, and we can use equation 6.2 with $P = 20$ W to solve for V. Alternatively, the midband voltage equals the voltage at cutoff *divided* by 0.707:

$$V_{(midband)} = \frac{10.95 \text{ V rms}}{0.707} = 15.49 \text{ V rms}$$

Amplitude and Phase Distortion

The signal passed through an ac amplifier is usually a complex waveform containing many different frequency components rather than a single-frequency ("pure") sine wave. For example, audio-frequency signals such as speech and music are combinations of many different sine waves occurring simultaneously with different amplitudes and different frequencies, in the range from 20 Hz to 20 kHz. As another example, any *periodic* waveform, such as a square wave or a triangular wave, can be shown to be the sum of a large number of sine waves whose amplitudes and frequencies can be determined mathematically. As discussed in Chapter 1, it is enough to know how a linear amplifier treats any single sine wave to know how it treats sums of sine waves (by the superposition principle).

In order for an output waveform to be an amplified version of the input, *an amplifier must amplify every frequency component in the signal by the same amount.* For example, if an input signal is the sum of a 0.5-V-rms, 100-Hz sine wave and a 0.2-V-rms, 1-kHz sine wave, then an amplifier having gain 10 must amplify each frequency component by 10, so that the output consists of a 5-V-rms, 100-Hz sine wave and a 2-V-rms, 1-kHz sine wave. If the frequency response of an amplifier is such that the gain at one frequency is different than it is at another frequency, the output will be *distorted,* in the sense that it will not have the same shape as the input waveform. This alteration in waveshape is called *amplitude*

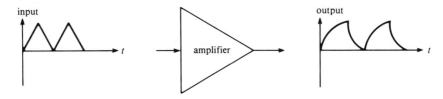

Figure 6.3
The output waveform is a distorted version of the input. The amplifier has a frequency response that is inadequate for the frequency components of the input waveform

distortion. Figure 6.3 shows the distortion that results when a triangular waveform is passed through an amplifier having an inadequate frequency response. In this example, the high-frequency components in the waveform fall beyond the upper cutoff frequency of the amplifier, so they are not amplified by the same amount as low-frequency components.

It can be seen that knowledge of the frequency response of an amplifier is important in determining whether it will distort a signal having known frequency components. The bandwidth must cover the entire range of frequency components in the signal if undistorted amplification is to be achieved. In general, "jagged" waveforms and signals having abrupt changes in amplitude, such as square waves and pulses, contain very broad ranges of frequencies and require wide-bandwidth amplifiers.

An amplifier will also distort a signal if it causes components having different frequencies to be shifted by different *times*. For example, if an amplifier shifts one component by 1 ms, it must shift every component by 1 ms. This means that the *phase* shift at each frequency must be proportional to frequency. Distortion caused by failure to shift phase in this way is called phase distortion. In most amplifiers, phase distortion occurs at the same frequencies where amplitude distortion occurs, because phase shifts are not proportional to frequency outside the midband range.

Amplitude and phase distortion should be contrasted with *nonlinear* distortion. Nonlinear distortion results when an amplifier's gain depends on signal *amplitude* rather than frequency, that is, when a plot of output voltage versus input voltage is not a straight line. See, for example, Figure 1.4 (Example 1.2), where we observed that the value of $A_v = \Delta V_o / \Delta V_{in}$ depends upon where along the curve it is calculated. The effect of nonlinear distortion is to create frequency components in the output that were not present in the input signal. These new components are integer multiples of the frequency components in the input and are called *harmonic* frequencies. For example, if the input signal were a pure 1-kHz sine wave, the output would be said to contain third and fifth harmonics if it contained 3-kHz and 5-kHz components in addition to the 1-kHz *fundamental*. Such distortion is often called *harmonic distortion.* Clipping, discussed in Chapter 5, is another example of harmonic distortion.

6.2   DECIBELS AND LOGARITHMIC PLOTS

Decibels

Frequency-response data are often presented in *decibel* form. Recall that decibels (dB) are the units used to compare two power levels in accordance with the definition

$$dB = 10 \log_{10} \frac{P_2}{P_1} \qquad (6.3)$$

The two power levels, P_1 and P_2, are often the input and output power of a system, respectively, in which case equation 6.3 defines the power gain of the system in decibels. If $P_2 > P_1$, then equation 6.3 gives a *positive* number, and if $P_2 < P_1$, the result is *negative*, signifying a reduction in power. If $P_2 = P_1$, the result is 0 dB, since $\log_{10}(1) = 0$.

Let R_1 be the resistance across which the power P_1 is developed and R_2 be the resistance across which P_2 is developed. Then, since $P = V^2/R$, we have, from equation 6.3,

$$dB = 10 \log_{10} \frac{(V_2^2/R_2)}{(V_1^2/R_1)} \qquad (6.4)$$

where V_2 is the rms voltage across R_2 and V_1 is the rms voltage across R_1. If the resistance values are the *same* at the two points where the power comparison is made ($R_1 = R_2 = R$), then equation 6.4 becomes

$$dB = 10 \log_{10} \left(\frac{V_2^2/R}{V_1^2/R} \right) = 10 \log_{10} \left(\frac{V_2}{V_1} \right)^2 = 20 \log_{10} \left(\frac{V_2}{V_1} \right) \qquad (6.5)$$

Equation 6.5 gives power gain (or loss) in terms of the voltage levels at two points in a circuit, but it must be remembered that the equation is valid for power comparison only if the resistances at the two points are equal. *The same equation is used to compare* voltage *levels regardless of the resistance values at the two points.* In other words, it is common practice to compute voltage gain as

$$dB(\text{voltage gain}) = 20 \log_{10} \left(\frac{V_2}{V_1} \right) \qquad (6.6)$$

If the resistances R_1 and R_2 are equal, then the power gain in dB equals the voltage gain in dB.

Example 6.2

The amplifier shown in Figure 6.4 has input resistance 1500 Ω and drives a 100-Ω load. If the input current is 0.632 mA rms and the load voltage is 30 V rms, find
1. the power gain in dB; and
2. the voltage gain in dB.

Figure 6.4
(Example 6.2)

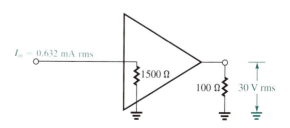

Solution

1. $P_{in} = I_{in}^2 r_{in} = (0.632 \times 10^{-3} \text{ A rms})^2 (1500 \ \Omega) = 0.6 \text{ mW}$

   $P_L = \dfrac{V_L^2}{R_L} = \dfrac{(30 \text{ V rms})^2}{100 \ \Omega} = 9 \text{ W}$

   From equation 6.3,

   $$\text{power gain} = 10 \log_{10} \left(\frac{9 \text{ W}}{0.6 \times 10^{-3} \text{ W}} \right) = 10 \log_{10} (15 \times 10^3) = 41.76 \text{ dB}$$

2. Since $V_{in} = I_{in} r_{in} = (0.632 \times 10^{-3} \text{ A})(1500 \ \Omega) = 0.948 \text{ V rms}$, we have from equation 6.6,

   $$\text{voltage gain} = 20 \log_{10} \left(\frac{30}{0.948} \right) = 20 \log_{10} (31.645) = 30 \text{ dB}$$

It is helpful to remember that a two-to-one change in voltage corresponds to approximately 6 dB, and a ten-to-one change corresponds to 20 dB. The sign ($\pm$) depends on whether the change represents an increase or a decrease in voltage. Suppose, for example, that $V_1 = 8$ V rms. If this voltage is doubled ($V_2 = 16$ V rms), then

$$20 \log_{10} \left(\frac{16 \text{ V}}{8 \text{ V}} \right) = 20 \log_{10}(2) \approx 6 \text{ dB}$$

If V_1 is halved ($V_2 = 4$ V rms), then

$$20 \log_{10} \left(\frac{4 \text{ V}}{8 \text{ V}} \right) = 20 \log_{10}(0.5) \approx -6 \text{ dB}$$

If V_1 is increased by a factor of 10 ($V_2 = 80$ V rms), then

$$20 \log_{10} \left(\frac{80 \text{ V}}{8 \text{ V}} \right) = 20 \log_{10}(10) = 20 \text{ dB}$$

If V_1 is reduced by a factor of 10 ($V_2 = 0.8$ V rms), then

$$20 \log_{10} \left(\frac{0.8 \text{ V}}{8 \text{ V}} \right) = 20 \log_{10}(0.1) = -20 \text{ dB}$$

Every time a voltage is doubled, an additional 6 dB is *added* to the voltage gain, and every time it is increased by a factor of 10, an additional 20 dB is added to the voltage gain. For example, a gain of $100 = 10 \times 10$ corresponds to 40 dB and a gain of 4 corresponds to $2 \times 2 = 12$ dB. As another example, a gain of $400 = 2 \times 2 \times 10 \times 10$ corresponds to $(6 + 6 + 20 + 20)$ dB $= 52$ dB. Similarly, a reduction in voltage by a factor of $0.05 = (1/2)(1/10)$ corresponds to $-6 - 20 = -26$ dB. Common logarithms (base 10) can be computed on most scientific-type calculators, and the reader should become familar with the calculator's use for that purpose and for computing inverse logarithms. For reference and comparison purposes, Table 6.1 shows the decibel values corresponding to some frequently encountered ratios between 0.001 and 1000.

Table 6.1

(V_2/V_1)	dB = $20 \log_{10}(V_2/V_1)$	(V_2/V_1)	dB = $20 \log_{10}(V_2/V_1)$
0.001	−60	2	6
0.002	−54	4	12
0.005	−46	8	18
0.008	−42	10	20
0.01	−40	20	26
0.02	−34	40	32
0.05	−26	80	38
0.08	−22	100	40
0.1	−20	200	46
0.2	−14	400	52
0.5	−6	800	58
0.8	−2	1000	60
1.0	0		

Example 6.3

The input voltage to an amplifier is 4 mV rms. At point 1 in the amplifier, the voltage gain with respect to the input is −4.2 dB and at point 2 the voltage gain with respect to point 1 is 18.5 dB. Find

1. the voltage at point 1;
2. the voltage at point 2; and
3. the voltage gain in dB at point 2, with respect to the input.

Solution.   Let V_i = the input voltage (4×10^{-3} V rms), V_1 = the rms voltage at point 1, and V_2 = the rms voltage at point 2. Figure 6.5 shows diagrams of the decrease and increase in gain as we progress through the amplifier.

1. Between the input and point 1 we have

$$20 \log_{10} \left(\frac{V_1}{4 \times 10^{-3}} \right) = -4.2$$

$$\log_{10} \left(\frac{V_1}{4 \times 10^{-3}} \right) = -0.21$$

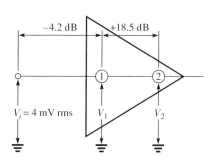
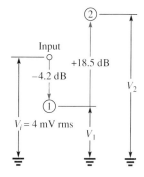

Figure 6.5
(Example 6.3)

Taking the inverse log (antilog) of both sides,

$$\frac{V_1}{4 \times 10^{-3}} = 0.617$$

$$V_1 = 2.46 \text{ mV rms}$$

(On most scientific calculators, the inverse log of -0.21 can be computed directly by entering a sequence such as -0.21, *inverse*, *log*; it can also be found on a calculator having the y^x function by computing $10^{-0.21}$.)

2. Between point 1 and point 2 we have

$$20 \log_{10} \left(\frac{V_2}{2.466 \times 10^{-3}}\right) = 18.5$$

$$\log_{10} \left(\frac{V_2}{2.466 \times 10^{-3}}\right) = 0.925$$

$$V_2 = 2.466 \times 10^{-3} \text{ antilog } (0.925) = 20.75 \text{ mV rms}$$

3. Between the input and point 2,

$$\text{voltage gain} = 20 \log_{10} \left(\frac{V_2}{V_i}\right) = 20 \log_{10} \left(\frac{20.75 \times 10^{-3}}{4 \times 10^{-3}}\right) = 14.3 \text{ dB}$$

Notice that this result is the same as -4.2 dB $+$ 18.5 dB; the overall gain in dB is the *sum* of the intermediate dB gains.

It must be remembered that decibels are derived from a *ratio* and therefore represent a comparison of one voltage or power level to another. It is correct to speak of voltage or power *gain* in terms of decibels, but it is meaningless to speak of output *level* in dB, unless the reference level is specified. Popular publications and the broadcast media frequently abuse the term *decibel* because no reference level is reported. Do not be confused by this practice.

It is common practice in some technical fields to use one standard reference level for all decibel computations. For example, the power level 1 mW is used extensively as a reference. When the reference is 1 mW, the decibel unit is written dBm:

$$\text{dBm} = 10 \log_{10} \left(\frac{P}{10^{-3} \text{ W}}\right) \tag{6.7}$$

Note that 0 dBm corresponds to a power level of 1 mW. Another standard reference is 1 W:

$$\text{dBW} = 10 \log_{10} \left(\frac{P}{1 \text{ W}}\right) = 10 \log_{10} P \tag{6.8}$$

When the voltage reference is 1 V, voltage gain in decibels is written dBV:

$$\text{dBV} = 20 \log_{10} \left(\frac{V}{1 \text{ V}}\right) = 20 \log_{10} V \tag{6.9}$$

The *neper* is a logarithmic unit based on the natural log (ln) of a ratio:

$$A_p \text{ (nepers)} = \frac{1}{2} \ln \left(\frac{P_2}{P_1}\right)$$

$$A_v \text{ (nepers)} = \ln \left(\frac{V_2}{V_1}\right)$$

(6.10)

Semilog and Log-Log Plots

It is a convenient and widely followed practice to plot the logarithm of frequency-response data rather than actual data values. If the logarithm of frequency is plotted along the horizontal axis, a wide frequency range can be displayed on a convenient size of paper without losing resolution at the low-frequency end. For example, if it were necessary to scale frequencies directly on average-sized graph paper over the range from 1 Hz to 10 kHz, each small division might represent 100 Hz. It would then be impossible to plot points in the range from 1 Hz to 10 Hz, where the lower cutoff frequency might well occur. When the horizontal scale represents logarithms of frequency values, the low-frequency end is expanded and the high-frequency end is compressed.

One way in which a logarithmic frequency scale can be obtained is to compute the logarithm of each frequency and then label conventional graph paper with those logarithmic values. An easier way is to use specially designed *log paper*, on which coordinate lines are logarithmically spaced. It is then necessary only to label each line directly with an actual frequency value. If only one axis of the graph paper has a logarithmically spaced scale and the other has conventional linear spacing, the paper is said to be *semilog* graph paper. If both the horizontal and vertical axes are logarithmic, it is called *log-log* paper.

Any ten-to-one range of values is called a *decade*. For example, each of the frequency ranges 1 Hz to 10 Hz, 10 kHz to 100 kHz, 500 Hz to 5 kHz, and 0.02 Hz to 0.2 Hz is a decade. Figure 6.6 shows a sample of log-log graph paper on which two full decades can be plotted along each axis. This sample is called *2-cycle–by–2-cycle*, or simply *2 × 2*, log-log paper. Log-log graph paper is available with different numbers of decades along each axis, including 4 × 2, 5 × 3, 3 × 3, and so forth. Notice that each decade along each axis occupies the same amount of space. Several horizontal and vertical decades are identified on the figure. Log-log graph paper is usually printed with identical scale values along each decade. The user must relabel the divisions in accordance with the actual decade values that are appropriate for the data to be plotted. Suppose, for example, that the gain of an amplifier varies from 2 to 60 over the frequency range from 150 Hz to 80 kHz. Then the frequency axis must cover the three decades 100 Hz to 1 kHz, 1 kHz to 10 kHz, and 10 kHz to 100 kHz, and the gain axis must cover the two decades 1 to 10 and 10 to 100. 3 × 2 graph paper would be required. In Figure 6.6, the axes are arbitrarily labeled with decades 0.1–1 and 1–10 (vertical), and 10–100 and 100–1000 (horizontal).

An *octave* is any two-to-one range of values, such as 5–10, 80–160, and 1000–2000. Notice in Figure 6.6 that every octave occupies the same length. Notice also that the value *zero* does not appear on either axis of the figure. Zero can never appear on a logarithmic scale, no matter how many decades are represented, because log(0) = $-\infty$.

Semilog graph paper is used to plot gain in dB versus the logarithm of frequency. When the frequency axis is logarithmic and the vertical axis is linear with

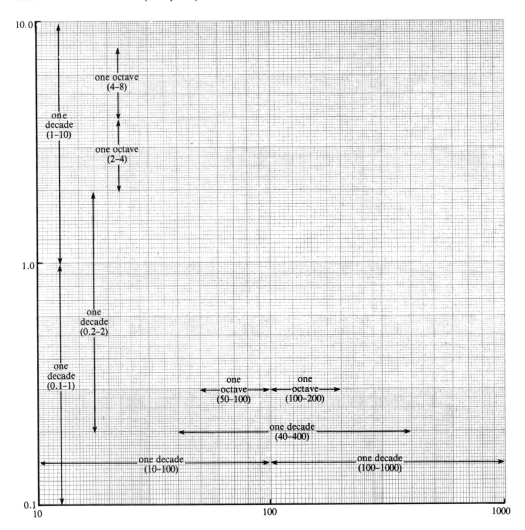

Figure 6.6
Two-cycle-by-2-cycle (2 × 2) log-log graph paper, showing some typical octaves and decades along each axis

its divisions labeled in decibels, the graph paper is essentially the same as log-log paper. Thus, a plot of an amplifier's frequency response will have the same general shape when constructed on either type of graph paper. Semilog graph paper is also used to plot phase shift on a linear scale versus the logarithm of frequency. Graphs of frequency response plotted against the logarithm of frequency are called Bode (pronounced bō-dē) plots.

The cutoff frequency is the frequency at which the gain on a frequency-response plot is 3 dB less than the midband gain. At cutoff, the gain is said to be "3 dB down," and the cutoff frequencies are often called *3-dB frequencies.* The value is 3 dB because the output voltage is $\sqrt{2}/2$ times its value at midband,

and

$$20 \log_{10}\left[\frac{(\sqrt{2}/2)V_m}{V_m}\right] = 20 \log_{10}(\sqrt{2}/2) \approx -3 \text{ dB} \qquad (6.11)$$

where V_m = the midband voltage.

Example 6.4

Figure 6.7 shows the voltage gain of an amplifier plotted versus frequency on log-log paper. Find approximate values for

1. the midband gain, in dB;
2. the gain in dB at the cutoff frequencies;
3. the bandwidth;
4. the gain in dB at a frequency 1 decade below the lower cutoff frequency;
5. the gain in dB at a frequency 1 octave above the upper cutoff frequency; and
6. the frequencies at which the gain is down 15 dB from its midband value.

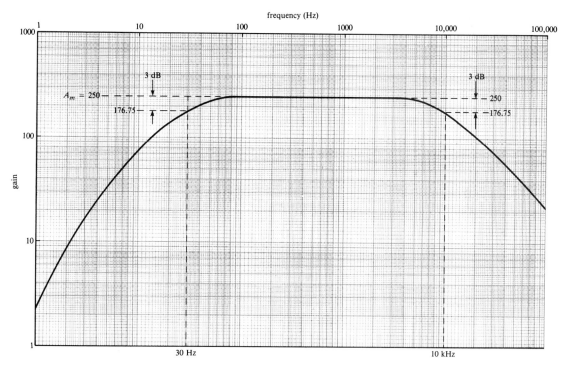

Figure 6.7
(Example 6.4)

Solution

1. As shown in Figure 6.7, the magnitude of the midband gain is approximately 250. Therefore, $A_m = 20 \log_{10}250 = 47.96$ dB.

2. Since the gain at each cutoff frequency is 3 dB less than the gain at midband, A_v(at cutoff) = 47.96 dB − 3 dB = 44.96 dB.

3. The magnitude of the gain at each cutoff frequency is $(\sqrt{2}/2)250 = 176.78$. As shown in Figure 6.7, this value of gain is reached at $f_1 = 30$ Hz and at $f_2 = 10$ kHz. Therefore, BW $= f_2 − f_1 = (10$ kHz$) − (30$ Hz$) = 9970$ Hz.

4. The frequency 1 decade below f_1 is $(0.1)f_1 = 3$ Hz. From Figure 6.7, the magnitude of the gain at this frequency is approximately 16, so $A_v = 20 \log_{10} 16 = 24.08$ dB.

5. The frequency 1 octave above f_2 is $2f_2 = 20$ kHz. From Figure 6.7, the magnitude of the gain at 20 kHz is approximately 105, so $A_v = 20 \log_{10} 105 = 40.42$ dB.

6. The gain 15 dB below midband is $(47.96$ dB$) − (15$ dB$) = 32.96$ dB.

$$20 \log_{10} A_v = 32.96$$
$$\log_{10} A_v = 1.648$$
$$A_v = \text{antilog}(1.648) = 44.46$$

From Figure 6.7, the frequencies at which $A_v = 44.46$ are approximately 6.5 Hz and 50 kHz.

One-*nth* Decade and Octave Intervals

In many practical investigations, including computer-generated frequency-response data, it is necessary to specify logarithmic frequency intervals within one decade or within one octave. For example, we may want ten logarithmically spaced frequencies within one decade. These frequencies are said to be at *one-tenth decade* intervals. Similarly, three logarithmically spaced frequencies in one octave are said to be at *one-third octave* intervals. The frequencies in one-*nth* decade intervals beginning at frequency f_1 are

$$10^x, \; 10^{x+\frac{1}{n}}, \; 10^{x+\frac{2}{n}}, \; 10^{x+\frac{3}{n}}, \; \ldots$$

where $x = \log_{10} f_1$.

The frequencies in one-*nth* octave intervals beginning at frequency f_1 are

$$2^x, \; 2^{x+\frac{1}{n}}, \; 2^{x+\frac{2}{n}}, \; 2^{x+\frac{3}{n}}, \; \ldots$$

where $x = \log_2 f_1$.

6.3   SERIES CAPACITANCE AND LOW-FREQUENCY RESPONSE

The lower cutoff frequency of an amplifier is affected by capacitance connected in *series* with the signal flow path. The most important example of series-connected capacitance is the amplifier's input and output coupling capacitors. At low frequencies, the reactance of these capacitors becomes very large, so a significant portion of the ac signal is dropped across them. As frequency approaches 0 (dc), the capacitive reactance approaches infinity (open circuit), so the coupling capacitors perform their intended role of blocking all dc current flow. In previous discussions, we have assumed that the signal frequency was high enough that the capacitive reactance of all coupling capacitors was negligibly small, but we will now

Figure 6.8
The input resistance of the amplifier and the coupling capacitor form an RC network that reduces the amplifier's input signal at low frequencies.

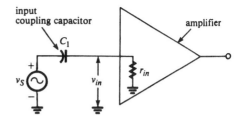

consider how large reactances at low frequencies affect the overall voltage gain.

Figure 6.8 shows the capacitor–resistor combination formed by the coupling capacitor and the input resistance at the input side of an amplifier. We omit consideration of any signal-source resistance for the moment. Notice that r_{in} and the capacitive reactance of C_1 form a voltage divider across the amplifier input. The amplifier input voltage, v_{in}, is found from the voltage-divider rule:

$$v_{in} = \left(\frac{r_{in}}{r_{in} - jX_{C_1}}\right)v_S \qquad (6.12)$$

where

$$X_{C_1} = \frac{1}{\omega C_1} \text{ ohms}$$

$$\omega = 2\pi f \text{ radians/second}$$

From equation 6.12, we can determine the *magnitude* (amplitude) of v_{in}, which we will designate $|v_{in}|$, as a function of ω:

$$|v_{in}| = \frac{r_{in}}{\sqrt{r_{in}^2 + X_{C_1}^2}}|v_S| = \frac{r_{in}}{\sqrt{r_{in}^2 + \left(\frac{1}{\omega C_1}\right)^2}}|v_S| = \frac{\omega r_{in} C_1}{\sqrt{1 + (\omega r_{in} C_1)^2}}|v_S| \qquad (6.13)$$

Equation 6.13 shows that $|v_{in}| = 0$ when $\omega = 0$ (dc) and that $|v_{in}|$ approaches $|v_S|$ in value as ω becomes very large. At the frequency $\omega = 1/(r_{in}C_1)$ rad/s, we have, from equation 6.13,

$$|v_{in}| = \frac{\left(\frac{1}{r_{in}C_1}\right)(r_{in}C_1)}{\sqrt{1 + \left[\left(\frac{1}{r_{in}C_1}\right)(r_{in}C_1)\right]^2}}|v_S| = \frac{1}{\sqrt{1+1}}|v_S| = \frac{1}{\sqrt{2}}|v_S| \approx 0.707\,|v_S|$$

This result shows that the amplifier input voltage falls to 0.707 times the source voltage when the frequency is reduced to $1/(r_{in}C_1)$ rad/s. Therefore, if there are no other frequency-sensitive components affecting the signal level, the overall gain from source to output is 0.707 times its midband value, meaning that $1/(r_{in}C_1)$ rad/s is the lower cutoff frequency. It is an exercise at the end of this chapter to show that the lower cutoff frequency is the frequency at which the capacitive reactance X_{C_1} equals the resistance r_{in}.

If source resistance r_S is present, then equation 6.13 becomes

$$|v_{in}| = \frac{\omega r_{in} C_1}{\sqrt{1 + [\omega(r_{in} + r_S)C_1]^2}}|v_S| \qquad (6.14)$$

and the lower cutoff frequency is

$$\omega_1 = \frac{1}{(r_{in} + r_S)C_1} \text{ rad/s} \qquad (6.15)$$

or

$$f_1 = \frac{1}{2\pi(r_{in} + r_S)C_1} \text{ Hz} \qquad (6.16)$$

The cutoff frequency defined by equations 6.15 and 6.16 is the frequency at which the ratio $|v_{in}|/|v_S|$ is 0.707 times its *midband* value, namely, 0.707 times $r_{in}/(r_S + r_{in})$. The ratio $|v_{in}|/|v_S|$ can be written in terms of the cutoff frequency f_1 and the signal frequency f as follows:

$$\frac{|v_{in}|}{|v_S|} = K\left(\frac{f}{\sqrt{f_1^2 + f^2}}\right) = K\left(\frac{1}{\sqrt{1 + (f_1/f)^2}}\right) \qquad (6.17)$$

where

$$K = \frac{r_{in}}{r_S + r_{in}}$$

Let us now use v_S as a phase angle reference ($\underline{/v_S} = 0°$) and compute the phase of v_{in} as a function of frequency. From equation 6.12,

$$\underline{/v_{in}} = -\arctan\left(\frac{-X_{C_1}}{r_{in}}\right) + \underline{/v_S} = \arctan\left(\frac{X_{C_1}}{r_{in}}\right) = \arctan\left(\frac{1}{\omega r_{in} C_1}\right) \qquad (6.18)$$

Again, if source resistance r_S is present, we find

$$\underline{/v_{in}} = \arctan\left[\frac{1}{\omega(r_{in} + r_S)C_1}\right] \qquad (6.19)$$

At $\omega = 1/(r_{in} + r_S)C_1$, equation 6.19 becomes

$$\underline{/v_{in}} = \arctan(1) = 45°$$

Thus, v_{in} leads v_S by 45° at cutoff. The phase shift in terms of f_1 and the signal frequency f is

$$\underline{/v_{in}} = \arctan(f_1/f) \qquad (6.20)$$

Note that $\underline{/v_{in}}$ approaches 90° as f approaches 0.

Example 6.5

The amplifier shown in Figure 6.9 has midband gain $|v_L|/|v_S|$ equal to 90. Find

1. the lower cutoff frequency;
2. the voltage gain $|v_L|/|v_S|$, in dB, at the cutoff frequency; and
3. the voltage gain $|v_L|/|v_{in}|$.

Solution

1. From equation 6.16,

$$f_1 = \frac{1}{2\pi(150 \text{ Ω} + 750 \text{ Ω})(0.5 \times 10^{-6} \text{ F})} = 353.67 \text{ Hz}$$

Figure 6.9
(Example 6.5)

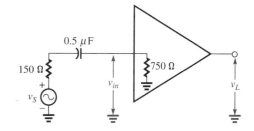

2. At midband, the gain in dB is $20 \log_{10}(90) = 39.08$ dB. At cutoff, the gain is 3 dB less than its midband value: 39.08 dB $- 3$ dB $= 36.08$ dB.

3. At midband, the reactance of the coupling capacitor is negligible, so the overall voltage gain $|v_L|/|v_S|$ is

$$\frac{|v_L|}{|v_S|} = \frac{|v_{in}|}{|v_S|}\frac{|v_L|}{|v_{in}|} = \left(\frac{r_{in}}{r_S + r_{in}}\right)\frac{|v_L|}{|v_{in}|}$$

Thus,
$$90 = \left(\frac{750\ \Omega}{150\ \Omega + 750\ \Omega}\right)\frac{|v_L|}{|v_{in}|}$$

$$\frac{|v_L|}{|v_{in}|} = \frac{90}{\left(\dfrac{750}{900}\right)} = 108$$

Figure 6.10 shows *normalized* plots of the gain of an RC network connected so that the capacitor is in series with the signal flow and the output is taken across the resistor. This is the configuration at the input of the capacitor-coupled amplifier, as shown in Figure 6.8. The gain of the network approaches 1 (0 dB) at high frequencies, which corresponds to the condition $|v_{in}|/|v_S| = 1$, when the capacitive reactance is negligibly small. Of course, the *overall* gain of an amplifier having a capacitor-coupled input may be greater than 1 at frequencies above cutoff, but the shape of the frequency response is the same as that shown in Figure 6.10. The only difference is that the gain above cutoff is A_m—the amplifier's midband gain—rather than 1 (0 dB).

Note that the plots in Figure 6.10 extend 2 decades below and 1 decade above the cutoff frequency, which is labeled f_1. Figure 6.10(b) shows that the gain is "down" approximately 7 dB at a frequency 1 octave below cutoff ($0.5f_1$) and is down 20 dB 1 decade below cutoff (at $0.1f_1$). Figure 6.10(c) shows that the phase shift is $63.4°$ one octave below f_1 and $87.1°$ one decade below f_1. The gain is down 0.04 dB and the phase shift is $5.7°$ at a frequency 1 decade above cutoff ($10f_1$). The gain plots also show the straight line *asymptotes* that the gain approaches at frequencies below cutoff. An asymptote is often used to approximate the gain response. Note that it "breaks" downward at f_1, where its deviation from the actual response curve is the greatest (3 dB). The frequency at which an asymptote breaks (f_1 in this example) is called a *break* frequency. *The asymptote has a slope of 6 dB/octave, or 20 dB/decade.*

When using these plots, remember that they are valid for only a *single* resistor–capacitor (RC) combination. We will study the effects of multiple RC combinations in the signal flow path in a later discussion. The plots in Figure 6.10 apply to any RC network in which the capacitor is in series with the signal path and the

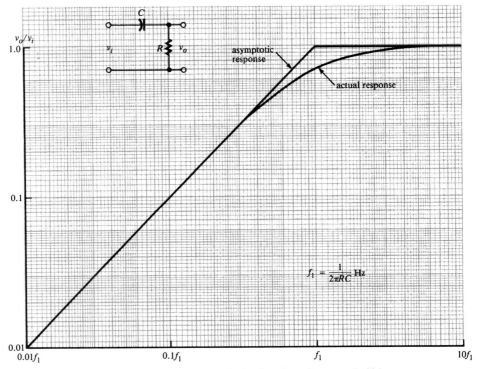

(a) Normalized frequency-response plot showing gain vs. frequency of a high-pass RC network. f_1 = lower cutoff frequency. Both gain and frequency are plotted on logarithmic scales.

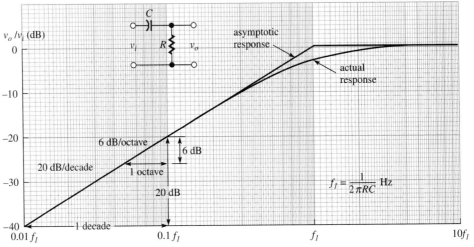

(b) Gain vs. frequency for the high-pass network, plotted on semilog graph paper. Note that the vertical axis is linear and is scaled in dB.

Figure 6.10
Normalized gain and phase plots for a high-pass RC network

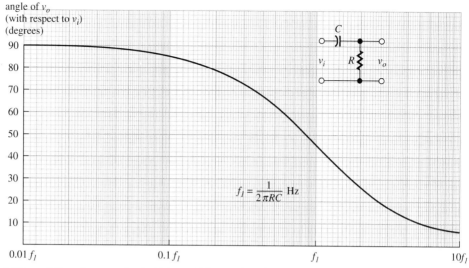

(c) Phase shift vs. frequency for the high-pass RC network. v_o leads v_i.

Figure 6.10
(Continued)

output is taken across the resistor. Such a network is called a *high-pass filter,* because, as the plots show, the gain is constant at high frequencies (above cutoff) and "falls off," at a constant rate, below cutoff. To find the overall gain of an RC network in which part of the resistance (r_S) is in the signal source, the gain determined from the plot must be multiplied by the factor $K = r_{in}/(r_S + r_{in})$.

Example 6.6

The amplifier shown in Figure 6.11 has voltage gain $|v_L|/|v_{in}| = 120$. Calculate

1. the lower cutoff frequency due to the input coupling capacitor;
2. the gain $|v_L|/|v_S|$ 1 octave below cutoff; and
3. the phase shift 1 decade above cutoff.

Use Figure 6.10 to find approximate values for

4. the asymptotic gain $|v_L|/|v_S|$ 1 octave below cutoff;
5. the gain $|v_L|/|v_S|$, in dB, at $f = 10.85$ Hz; and
6. the frequency at which the phase shift is 20°.

Figure 6.11
(Example 6.6)

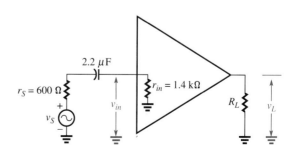

Solution

1. From equation 6.16,

$$f_1 = \frac{1}{2\pi(1400\ \Omega + 600\ \Omega)(2.2 \times 10^{-6}\ \text{F})} = 36.17\ \text{Hz}$$

2. The frequency 1 octave below cutoff is $f_1/2 = 18.09$ Hz. From equation 6.14,

$$\frac{|v_{in}|}{|v_S|} = \frac{2\pi(18.09)(1400)(2.2 \times 10^{-6})}{\sqrt{1 + [2\pi(18.09)(1400 + 600)(2.2 \times 10^{-6})]^2}} = 0.313$$

Therefore, $|v_L|/|v_S|$ at 18.09 Hz is

$$\frac{|v_L|}{|v_S|} = \frac{|v_{in}|}{|v_S|}\frac{|v_L|}{|v_{in}|} = 0.313(120) = 37.6$$

Note that this result can be computed more directly by using (6.17) and recognizing that, when f is 1 octave below f_1, the quantity f_1/f equals 2:

$$\frac{|v_{in}|}{|v_S|} = \left(\frac{r_{in}}{r_S + r_{in}}\right)\frac{1}{\sqrt{1 + (f_1/f)^2}} = \left(\frac{1400\ \Omega}{2000\ \Omega}\right)\frac{1}{\sqrt{1 + 2^2}} = 0.313$$

3. At 1 decade above cutoff, $f = 10f_1$, or $f_1/f = 0.1$. Assuming that the amplifier does not cause any phase shift beyond that due to the coupling capacitor, the phase shift is, from equation 6.20, arctan(0.1) = 5.71°.

4. One octave below cutoff, $f = 0.5f_1$. From Figure 6.10(a), the gain at the intersection of the asymptote and the $0.5f_1$ coordinate line is 0.5. Therefore, the overall asymptotic gain is

$$\frac{|v_L|}{|v_S|} = 0.5\left(\frac{r_{in}}{r_S + r_{in}}\right)120 = 0.5\left(\frac{1400\ \Omega}{2000\ \Omega}\right)120 = 42$$

(Compare with the actual gain of 37.6, computed in part 2.)

5. At $f = 10.85$ Hz, $f/f_1 = 10.85/36.17 = 0.3$, so $f = 0.3f_1$. From Figure 6.10(b), the gain curve intersects the 0.3 coordinate line at approximately -11 dB. The midband gain $|v_L|/|v_S|$ in decibels is

$$20 \log_{10}\left[\left(\frac{1400\ \Omega}{600\ \Omega + 1400\ \Omega}\right)(120)\right] = 38.49\ \text{dB}$$

Therefore, the gain at 10.85 Hz is 38.49 dB $-$ 11 dB = 27.49 dB.

6. From Figure 6.10(c), the 20° phase coordinate intersects the curve at approximately $2.8f_1$. Therefore, the phase shift is 20° at (2.8)(36.17 Hz) = 101.28 Hz.

The smaller the desired lower cutoff frequency of an amplifier, the larger the input coupling capacitor must be. If the input resistance of the amplifier is small, the required capacitance may be impractically large. For example, if an audio amplifier having input resistance 100 Ω is to have a lower cutoff frequency of 15 Hz, then, assuming that $r_S = 0$, the coupling capacitor must have value

$$C_1 = \frac{1}{2\pi(100\ \Omega)(15\ \text{Hz})} = 106\ \mu\text{F}$$

Figure 6.12

The output and load resistance of the amplifier and the output coupling capacitor form an RC network that reduces the signal reaching the load at low frequencies.

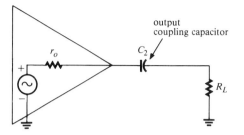

On the other hand, if the input resistance is 10 kΩ, the required capacitance is 1.06 μF, a much more reasonable value.

Figure 6.12 shows the resistance–capacitance combination formed by the output resistance, output coupling capacitor, and load resistance of an amplifier. The effect of the output coupling capacitor on the frequency response is the same as that of the input coupling capacitor: At low frequencies, there is a significant voltage drop across the capacitive reactance, so less signal is delivered to the load. The lower cutoff frequency due to the output coupling capacitor (C_2) is found in the same way in which we found f_1 due to C_1. It is the frequency at which the capacitive reactance X_{C_2} equals the resistance $r_o + R_L$. To distinguish between the two lower frequencies, we will hereafter write $f_1(C_1)$ and $f_1(C_2)$:

$$f_1(C_1) = \frac{1}{2\pi(r_{in} + r_S)C_1} \tag{6.21}$$

$$f_1(C_2) = \frac{1}{2\pi(r_o + R_L)C_2} \tag{6.22}$$

If the lower cutoff frequency were determined by C_2 alone, then the frequency response in the vicinity of cutoff would have the same appearance as the normalized plots shown in Figure 6.10. However, when both input and output coupling capacitors are present, the low-frequency response is significantly different because both capacitors contribute *simultaneously* to a reduction in gain. Let us first consider the case where the frequencies $f_1(C_1)$ and $f_1(C_2)$ are considerably different, at least a decade apart. Then, to a good approximation, the overall frequency response will have a lower cutoff frequency, f_1, equal to the *larger* of $f_1(C_1)$ and $f_1(C_2)$. Figure 6.13 shows the frequency response for the case $f_1(C_1) = 10$ Hz and $f_1(C_2) = 100$ Hz. The midband gain is assumed to be 20 dB. It is clear that $f_1 = 100$ Hz, because the gain at that frequency is $20 - 3 = 17$ dB. The gain at 10 Hz is much lower (approximately -3 dB) because at frequencies below 100 Hz *both* C_1 and C_2 cause gain reduction.

Note that in Figure 6.13(a) the gain asymptote has slope 20 dB/decade (6 dB/octave) between 10 Hz and 100 Hz, but breaks downward at 10 Hz with a slope of 40 dB/decade (12 dB/octave). There are, therefore, two break frequencies in this example, i.e., two frequencies where the asymptote changes slope: 10 Hz and 100 Hz. The phase shift is approximately 51° at $f_1 = 100$ Hz and 129° at 10 Hz and approaches 180° as frequency approaches 0. The response characteristics shown in Figure 6.13 are valid for two high-pass RC networks connected in series, provided the networks are *isolated* from each other, in the sense that one does not load the other. In our illustration, the two networks are assumed to be isolated by an amplifier having gain 20 dB.

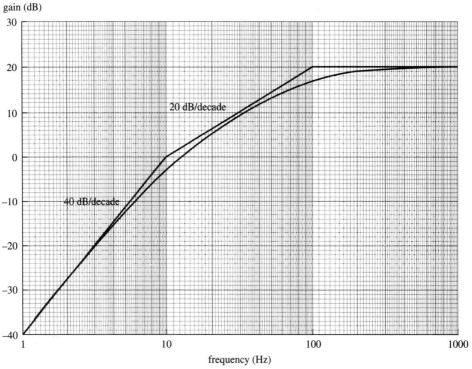

(a) Gain vs. frequency for an amplifier having $f_1(C_1) = 10$ Hz and $f_1(C_2) = 100$ Hz.
The lower cutoff frequency in this case is the larger of the two, $f_1 = 100$ Hz.
Note that there are two "break" frequencies. $A_m = 20$ dB.

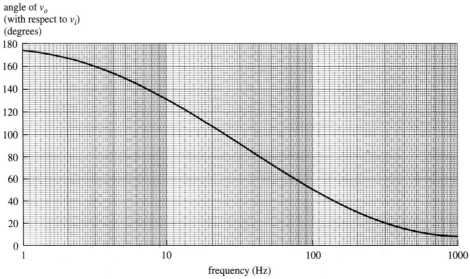

(b) Phase shift vs. frequency for the amplifier whose gain plot is shown in Figure
6.13(a). Note that the phase angle approaches 180° as frequency approaches 0.

Figure 6.13
Gain and phase versus frequency for an amplifier having two break frequencies

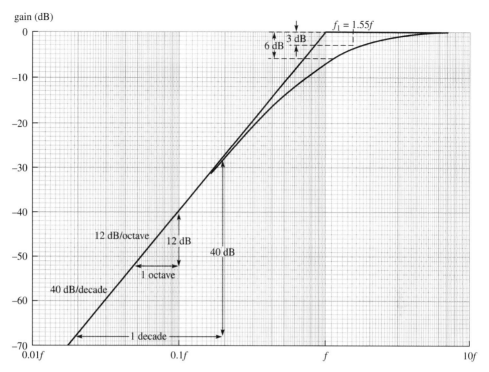

(a) Gain vs. frequency for two isolated, high-pass RC networks having the same cutoff frequency, f_1.
  Note that $f_1 = 1.55f$. For this plot, the midband (high-frequency) gain is 0 dB.

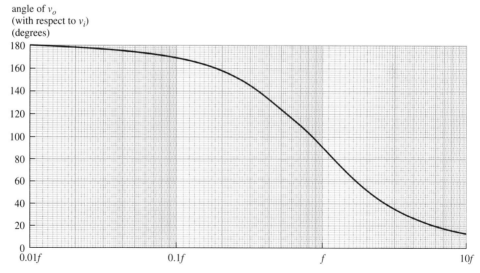

(b) Phase shift vs. frequency for the two high-pass RC networks whose gain is shown
  in Figure 6.14(a). Note that the output leads the input by 90° at frequency f.

Figure 6.14
Normalized gain and phase response for the case $f_1(C_1) = f_1(C_2) = f$

If the frequencies $f_1(C_1)$ and $f_1(C_2)$ are closer than one decade to each other, then the overall lower cutoff frequency is somewhat higher than the larger of the two. In such cases, it is usually adequate to assume that f_1 equals the larger of the two. The exact value of f_1 can be found using the following rather cumbersome equation (derived in Appendix B):

$$f_1 = \sqrt{\frac{2ab}{-(a+b) + \sqrt{(a+b)^2 + 4ab}}} \tag{6.23}$$

where
$$a = f_1^2(C_1)$$
$$b = f_1^2(C_2)$$

If $f_1(C_1) = f_1(C_2)$, equation 6.23 can be used to show that the overall lower cutoff frequency is 1.55 times the value of either. For example, if $f_1(C_1) = f_1(C_2) = 100$ Hz, then $f_1 = 155$ Hz. Figure 6.14, on page 241, shows the normalized gain and phase response for the special case $f_1(C_1) = f_1(C_2)$. Note that the asymptote breaks downward at $f = f_1(C_1) = f_1(C_2)$ and has a slope of 40 dB/decade, or 12 dB/octave. The actual response is 6 dB below the asymptote at that frequency. The cutoff frequency is $f_1 = 1.55f$, where the gain is down 3 dB. Note that the break frequency (f) is not the same as the cutoff frequency (f_1) in this case. The phase shift is 90° at f and approaches 180° as frequency approaches 0. Once again, the plots shown in Figure 6.14 are valid only when the two high-pass RC networks whose response they represent are isolated from each other. In our case, we assume that the isolation is provided by an amplifier (having unity gain, since the gain in Figure 6.14 approaches 0 dB at high frequencies).

Example 6.7

The amplifier shown in Figure 6.15 has midband gain $|v_L|/|v_S| = 140$. Find

1. the approximate lower cutoff frequency;
2. the gain $|v_o|/|v_{in}|$;
3. the lower cutoff frequency when C_2 is changed to 50 μF;
4. the approximate gain $|v_L|/|v_S|$, in dB, at 2.9 Hz with $C_2 = 50$ μF; and
5. the value that C_2 would have to be in order to obtain a lower cutoff frequency of approximately 100 Hz.

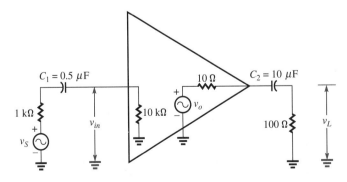

Figure 6.15
(Example 6.7)

Solution

1. From equation 6.21,

$$f_1(C_1) = \frac{1}{2\pi(1 \times 10^3 + 10 \times 10^3)(0.5 \times 10^{-6})} = 28.94 \text{ Hz}$$

From equation 6.22,

$$f_1(C_2) = \frac{1}{2\pi(10 + 100)(10 \times 10^{-6})} = 144.7 \text{ Hz}$$

Therefore, $f_1 \approx 144.7$ Hz. (The actual value, computed from equation 6.23, is 151 Hz.)

2.

$$\frac{|v_L|}{|v_S|} = \frac{|v_{in}|}{|v_S|} \frac{|v_o|}{|v_{in}|} \frac{|v_L|}{|v_o|}$$

$$140 = \left[\frac{10 \text{ k}\Omega}{(10 \text{ k}\Omega) + (1 \text{ k}\Omega)} \right] \frac{|v_o|}{|v_{in}|} \left[\frac{100 \text{ }\Omega}{(10 \text{ }\Omega) + (100 \text{ }\Omega)} \right]$$

$$\frac{|v_o|}{|v_{in}|} = \frac{140}{\left(\dfrac{1 \times 10^4}{1.1 \times 10^4} \right) \left(\dfrac{1 \times 10^2}{1.1 \times 10^2} \right)} = 169.4$$

3.

$$f_1(C_2) = \frac{1}{2\pi(10 + 100)(50 \times 10^{-6})} = 28.94 \text{ Hz}$$

Therefore, $f_1(C_1) = f_1(C_2) = 28.94$ Hz, so $f_1 = 1.55(28.94 \text{ Hz}) = 44.86$ Hz.

4. 2.9 Hz is approximately 1 decade below the break frequency of 28.94 Hz. Since the gain falls at the rate of 40 dB/decade below the break frequency, at 2.9 Hz it will be 40 dB less than A_m:

$$A_m = 20 \log_{10} 140 = 42.9 \text{ dB}$$

$$A_v(\text{at } 2.9 \text{ Hz}) = 42.9 \text{ dB} - 40 \text{ dB} = 2.9 \text{ dB}$$

5. Using the approximation $f_1 \approx f_1(C_2) = 100$ Hz,

$$100 \text{ Hz} = \frac{1}{2\pi(10 \text{ }\Omega + 100 \text{ }\Omega)C_2}$$

$$C_2 = \frac{1}{2\pi(10 + 100)(100)} = 14.5 \text{ }\mu\text{F}$$

6.4   SHUNT CAPACITANCE AND HIGH-FREQUENCY RESPONSE

Capacitance that provides an ac path between an amplifier's signal flow path and ground is said to *shunt* the signal. The most common form of shunt capacitance is that which exists between the terminals of an electronic device due to its structural characteristics. Recall, for example, that a PN junction has capacitance between its terminals because the depletion region forms a dielectric separating the two conductive regions of P and N material. Capacitance between device terminals is called *interelectrode* capacitance. Shunt capacitance is also created by wiring, terminal connections, solder joints, and any other circuit structure where conducting regions are close to each other. This type of capacitance is called *stray* capacitance.

Figure 6.16
Shunt capacitance C_A in parallel with the input of an amplifier affects its high-frequency response.

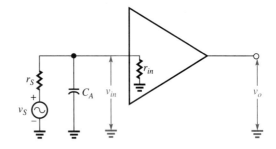

Shunt capacitance affects the high-frequency performance of an amplifier because at high frequencies the small capacitive reactance diverts the signal to ground, or to some other point besides the load. (We should note that a semiconductor device has inherent frequency limitations that may impose more severe restrictions on high-frequency operation than the effect of shunt capacitance alone.) Figure 6.16 shows shunt capacitance C_A between the input side of an amplifier and ground (in parallel with r_{in}). Notice that we can now neglect series coupling capacitance, because we are considering only high-frequency operation.

In the midband frequency range, the effect of C_A in Figure 6.16 can be neglected because the frequency is not high enough to make the capacitive reactance small. In other words, the midband frequency range is that range of frequencies that are high enough to neglect coupling capacitance and low enough to neglect shunt capacitance. It is apparent in the figure that, in the midband frequency range,

$$\frac{v_{in}}{v_S} = \frac{r_{in}}{r_S + r_{in}} \tag{6.24}$$

When the frequency is high enough to consider the effect of C_A, we must replace r_{in} in (6.24) by the parallel combination of r_{in} and $-jX_{C_A}$:

$$\frac{v_{in}}{v_S} = \frac{-jX_{C_A} \parallel r_{in}}{r_S + (-jX_{C_A} \parallel r_{in})} \tag{6.25}$$

At very high frequencies, where X_{C_A} becomes very small, the parallel combination of X_{C_A} and r_{in} becomes very small, and the net effect is the same as if the input impedance of the amplifier were made small. As we know, the consequence of that result is that the overall gain is reduced due to the voltage division between r_S and the input impedance. Our goal now is to find the frequency at which that reduction in gain equals 0.707 times the midband value given by equation 6.24. From equation 6.25,

$$\frac{v_{in}}{v_S} = \frac{\left(\dfrac{\dfrac{-jr_{in}}{\omega C_A}}{r_{in} - j/\omega C_A}\right)}{\left(r_S + \dfrac{\dfrac{-jr_{in}}{\omega C_A}}{r_{in} - j/\omega C_A}\right)} = \frac{\dfrac{-jr_{in}}{\omega C_A}}{r_S r_{in} - \dfrac{j(r_S + r_{in})}{\omega C_A}} \tag{6.26}$$

$$\frac{|v_{in}|}{|v_S|} = \frac{\dfrac{r_{in}}{\omega C_A}}{\sqrt{r_S^2 r_{in}^2 + \left(\dfrac{r_S + r_{in}}{\omega C_A}\right)^2}} = \frac{r_{in}}{\sqrt{(\omega C_A r_S r_{in})^2 + (r_S + r_{in})^2}}$$

$$= \frac{\dfrac{r_{in}}{r_S + r_{in}}}{\sqrt{[\omega C_A(r_S \parallel r_{in})]^2 + 1}} \tag{6.27}$$

The upper cutoff frequency, ω_2, is the frequency at which the expression in (6.27) equals $(\sqrt{2}/2)[r_{in}/(r_S + r_{in})]$:

$$\frac{\dfrac{r_{in}}{r_S + r_{in}}}{\sqrt{[\omega_2 C_A(r_S \parallel r_{in})]^2 + 1}} = \frac{\sqrt{2}}{2}\left(\frac{r_{in}}{r_S + r_{in}}\right)$$

$$\frac{1}{[\omega_2 C_A(r_S \parallel r_{in})]^2 + 1} = \left(\frac{\sqrt{2}}{2}\right)^2 = 0.5$$

$$\omega_2 = \frac{1}{(r_S \parallel r_{in})C_A} \text{ rad/s}$$

or

$$f_2 = \frac{1}{2\pi(r_S \parallel r_{in})C_A} \text{ Hz} \tag{6.28}$$

Equation 6.28 shows that the upper cutoff frequency due to C_A is inversely proportional to the parallel combination of r_S and r_{in}. Thus, to achieve a large bandwidth it is necessary to make $r_S \parallel r_{in}$ as small as possible, preferably by making r_S small. Theoretically, if $r_S = 0$, then $r_S \parallel r_{in} = 0$, and $f_2 = \infty$. In practice, it is not possible to have $r_S = 0$, but a very small value of r_S can increase the upper cutoff frequency significantly (not always a desirable practice, as we shall see in later discussions).

An RC network in which the resistor is in series with the signal flow path and the output is taken across the shunt capacitor is called a *low-pass filter*. The upper cutoff frequency f_2 is determined by $f_2 = 1/(2\pi RC)$ Hz. Comparing with equation 6.28 we see that the input of an amplifier having shunt capacitance C_A is the same as a low-pass filter having $R = r_S \parallel r_{in}$ and $C = C_A$. Normalized gain and phase plots for the low-pass RC network are shown in Figure 6.17. Note that the gain plot is the mirror image of that of the high-pass RC network shown in Figure 6.9. The asymptote shows that the gain falls off at the rate of 6 dB/octave, or 20 dB/decade, at frequencies above f_2. The phase shift approaches $-90°$ at frequencies above f_2. Note that negative phase means that the output voltage *lags* the input voltage, in contrast to the high-pass RC network, where the output leads the input. The gain and phase equations as functions of the ratio f/f_2 are

$$|A| = \frac{1}{\sqrt{1 + (f/f_2)^2}} \tag{6.29}$$

and

$$\underline{/A} = -\arctan(f/f_2) \tag{6.30}$$

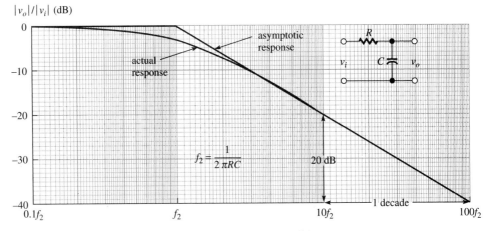

(a) Gain vs. frequency for the low-pass RC network. f_2 = upper cutoff frequency.

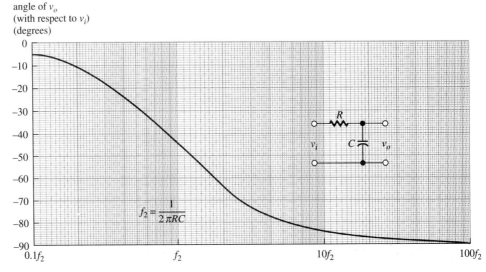

(b) Phase shift vs. frequency for the low-pass RC network. Note that v_o lags v_i.

Figure 6.17
Gain and phase versus frequency for a low-pass RC network

where $f_2 = 1/(2\pi RC)$, $R = r_S \parallel r_{in}$, and $C = C_A$, for an amplifier having input shunt capacitance C_A.

Example 6.8

The amplifier shown in Figure 6.18 has midband gain $|v_L|/|v_S| = 40$ dB. Calculate

1. the upper cutoff frequency;
2. the gain $|v_L|/|v_S|$, in dB, at $f = 20$ MHz; and
3. the phase shift at $f = f_2$.

Figure 6.18
(Example 6.8)

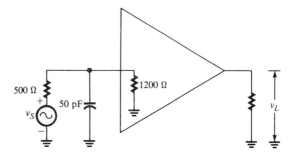

Use Figure 6.17 to find

4. the frequency at which $|v_L|/|v_S|$ is 35 dB; and

5. the frequency at which the output voltage lags the input voltage by 30°.

Solution

1. Since $r_S \parallel r_{in} = 500 \ \Omega \parallel 1200 \ \Omega = 353 \ \Omega$, we have, from equation 6.28,

$$f_2 = \frac{1}{2\pi(353 \ \Omega)(50 \times 10^{-12} \ \text{F})} = 9.02 \ \text{MHz}$$

2. From equation 6.29, the gain of the equivalent RC network at the amplifier input when $f = 20$ MHz is

$$|A| = \frac{1}{\sqrt{1 + \left(\dfrac{20 \times 10^6 \ \text{Hz}}{9.02 \times 10^6 \ \text{Hz}}\right)^2}} = 0.441$$

$$20 \log_{10}(0.441) = -7.11 \ \text{dB}$$

Therefore, the overall gain $|v_L|/|v_S|$ at $f = 20$ MHz is (40 dB) − (7.11 dB) = 32.89 dB.

3. From equation 6.30, when $f = f_2$, $\underline{/A} = -\arctan(f_2/f_2) = -\arctan(1) = -45°$. We conclude that *the output of a low-pass RC network lags the input by 45° at the cutoff frequency.* (See Figure 6.17(b).)

4. An overall gain of 35 dB corresponds to a 5-dB drop in gain from the midband value of 40 dB. From Figure 6.17(a), the gain of the equivalent RC network at the amplifier input is down 5 dB at approximately $f = 1.5f_2 = 1.5(9.02 \ \text{MHz}) = 13.53$ MHz.

5. From Figure 6.17(b), the frequency at which the phase shift is −30° is approximately $f = 0.57f_2 = 0.57(9.02 \ \text{MHz}) = 5.14$ MHz.

Figure 6.19 shows shunt capacitance C_B connected across the output of an amplifier. The output resistance r_o of the amplifier, the load resistance R_L, and the capacitance C_B form a low-pass RC network that has the same effect on the high-frequency response as the low-pass network at the input: the gain falls off because the impedance to ground decreases with increasing frequency. The upper cutoff frequency due to C_B is derived in the same way as that due to C_A. Hereafter, we will distinguish between the two frequencies by using the notation $f_2(C_A)$ and

Figure 6.19
An amplifier having shunt capacitance
C_B across its output

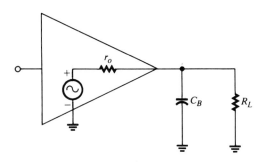

$f_2(C_B)$:

$$f_2(C_A) = \frac{1}{2\pi(r_S \parallel r_{in})C_A} \text{ hertz} \tag{6.31}$$

$$f_2(C_B) = \frac{1}{2\pi(r_o \parallel R_L)C_B} \text{ hertz} \tag{6.32}$$

When there is shunt capacitance at both the input and the output of an amplifier, the frequency response is different than it would be if only one were present. If $f_2(C_A)$ and $f_2(C_B)$ are not close in value, then the actual upper cutoff frequency, f_2, is approximately equal to the *smaller* of $f_2(C_A)$ and $f_2(C_B)$. The gain is asymptotic to a line that falls off at 20 dB/decade (6 dB/octave) between the smaller of $f_2(C_A)$ and $f_2(C_B)$ and is asymptotic to a line that falls off at 40 dB/decade (12 dB/octave) at frequencies above the larger of the two. See Figure 6.20(a). At the higher frequencies, the gain falls off at twice the rate it would for a single low-pass network, because both C_A and C_B contribute *simultaneously* to gain reduction. The total phase shift approaches $-180°$ at high frequencies. The exact value of the upper cutoff frequency in terms of $f_2(C_A)$ and $f_2(C_B)$ can be found (from Appendix B):

$$f_2 = \sqrt{\frac{-(a + b) + \sqrt{(a + b)^2 + 4ab}}{2ab}} \tag{6.33}$$

where
$$a = 1/f_2^2(C_A)$$
$$b = 1/f_2^2(C_B)$$

If $f_2(C_A) = f_2(C_B) = f$, equation 6.33 can be used to show that the cutoff frequency is $f_2 = 0.645f$. In that case, there is but one break frequency, $f = f_2(C_A) = f_2(C_B)$, and the single asymptote has slope 40 dB/decade, or 12 dB/octave. See Figure 6.20(b). When the two frequencies are equal, the total phase shift is $-90°$ at that frequency.

Thevenin Equivalent Circuits at Input and Output

We note that the same *form* of equation is used to compute both the lower and upper cutoff frequencies of an amplifier: $f = 1/2\pi RC$. In the case of f_1, C is either input or output coupling capacitance and in the case of f_2, C is input or output shunt capacitance. In *both* cases, R is the Thevenin equivalent resistance seen by the capacitance. This observation provides an easy way to remember the equa-

Figure 6.20
High-frequency gain and phase response when shunt capacitance is present at both the input and the output of an amplifier. $f_2(C_A)$ = break frequency due to input shunt capacitance; $f_2(C_B)$ = break frequency due to output shunt capacitance.

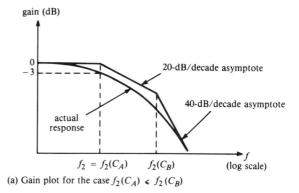

(a) Gain plot for the case $f_2(C_A) < f_2(C_B)$

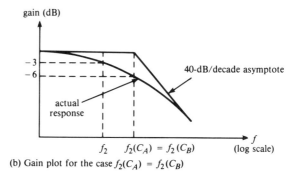

(b) Gain plot for the case $f_2(C_A) = f_2(C_B)$

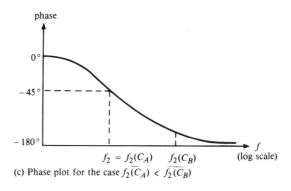

(c) Phase plot for the case $f_2(\overline{C_A}) < f_2(\overline{C_B})$

tions for calculating f_1 and f_2:

$$f = \frac{1}{2\pi r_{TH} C} \text{ hertz} \qquad (6.34)$$

where r_{TH} is the Thevenin equivalent resistance with respect to the capacitor terminals at input or output. Figure 6.21 illustrates this point. Recall that the Thevenin equivalent resistance is found by open-circuiting the capacitor terminals and computing the total equivalent resistance looking into those terminals when

Figure 6.21
Examples of the use of equation 6.34
(f = 1/2πr_{TH}C) to find lower and upper
cutoff frequencies

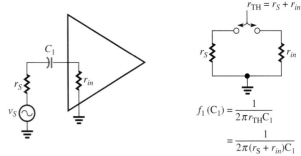

$$f_1(C_1) = \frac{1}{2\pi r_{TH}C_1}$$

$$= \frac{1}{2\pi(r_S + r_{in})C_1}$$

(a) Using equation 6.34 to find the lower cutoff
frequency due to input coupling capacitance

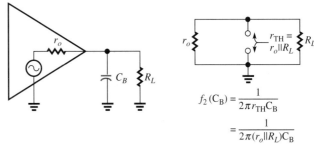

$$f_2(C_B) = \frac{1}{2\pi r_{TH}C_B}$$

$$= \frac{1}{2\pi(r_o\|R_L)C_B}$$

(b) Using equation 6.34 to find the upper cutoff
frequency due to output shunt capacitance

all voltage sources are replaced by short circuits. Figure 6.21(a) shows how equation 6.34 is applied to calculate the lower cutoff frequency due to input coupling
capacitance, C_1, and Figure 6.21(b) shows how the equation is applied to calculate
the upper cutoff frequency due to output shunt capacitance, C_B. As an exercise,
verify that the equations for $f_1(C_2)$ and $f_2(C_A)$ can also be found using the
Thevenin equivalent resistance with respect to the capacitor terminals.

Miller-Effect Capacitance

Figure 6.22 shows an amplifier having capacitance C_C connected between its input
and output terminals. The most common example of such capacitance is interelectrode capacitance, as, for example, between the base and the collector of a com-

Figure 6.22
Capacitance C_C connected between the
input and the output of an amplifier
affects its high-frequency response due
to the Miller effect.

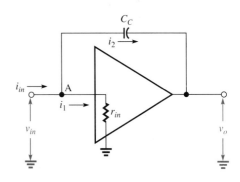

mon-emitter amplifier or between the gate and the drain of a common-source amplifier. This capacitance forms a *feedback* path for ac signals and it can have a significant influence on the high-frequency response of an amplifier.

Writing Kirchhoff's current law at the node labeled A in Figure 6.22, we have

$$i_{in} = i_1 + i_2 \tag{6.35}$$

The current i_2 that flows in the capacitor is the difference in voltage across it divided by the capacitive reactance:

$$i_2 = \frac{v_{in} - v_o}{-jX_{C_C}} \tag{6.36}$$

It is clear that the current i_1 flowing into the amplifier is

$$i_1 = \frac{v_{in}}{r_{in}} \tag{6.37}$$

Substituting (6.36) and (6.37) into (6.38),

$$i_{in} = \frac{v_{in}}{r_{in}} + \frac{v_{in} - v_o}{-jX_{C_C}} \tag{6.38}$$

Let the amplifier gain be $A_v = v_o/v_{in}$. Substituting $v_o = A_v v_{in}$ in (6.36) gives

$$i_{in} = \frac{v_{in}}{r_{in}} + \frac{v_{in} - A_v v_{in}}{-jX_{C_C}} = v_{in}\left(\frac{1}{r_{in}} + \frac{1 - A_v}{-jX_{C_C}}\right) = v_{in}\left[\frac{1}{r_{in}} + j\omega C_C(1 - A_v)\right]$$

Then

$$i_{in} = v_{in}(y_1 + y_2) \tag{6.39}$$

where $y_1 = 1/r_{in}$ and $y_2 = \omega C_C(1 - A_v)$. Equation 6.39 shows that the input admittance consists of the conductance component $1/r_{in}$ and the capacitive susceptance component $\omega C_C(1 - A_v)$. Note that *this input admittance is exactly the same as it would be if a capacitance having value $C_C(1 - A_v)$ were connected between input and ground,* instead of capacitance C_C connected between input and output. In other words, as far as the input signal is concerned, capacitance connected between input and output has the same effect as that capacitance "magnified" by the factor $(1 - A_v)$ and connected so that it shunts the input. This magnification of feedback capacitance, reflected to the input, is called the *Miller effect,* and the magnified value $C_C(1 - A_v)$ is called the *Miller capacitance, C_M.* Miller capacitance is relevant only for an *inverting* amplifier, so A_v is a negative number and the magnification factor $(1 - A_v)$ equals one *plus* the magnitude of A_v.

By a derivation similar to the foregoing, it can be shown that capacitance in the feedback path is also reflected to the output side of an amplifier. In this case, the effective shunt capacitance at the output is $(1 - 1/A_v)C_C$. Once again, A_v is negative, so the magnitude of the reflected capacitance is $(1 + 1/|A_v|)C_C$. Since the increase in capacitance is inversely proportional to gain, the effect is much less significant than that of the capacitance reflected to the input.

The computation of the value of the Miller capacitance, C_M, is complicated by the fact that the gain A_v itself depends on C_M. At high frequencies, the Miller capacitance reduces the gain, just as any other shunt capacitance does, and the gain reduction in turn reduces the Miller capacitance. As a first approximation, the midband gain can be used to compute C_M: $C_M \approx C_C(1 - A_m)$. This computation will always be conservative, in the sense that it will predict an upper cutoff frequency that is less than the actual value of f_2.

The total shunt capacitance at the input is the sum of the Miller capacitance and any other input-to-ground capacitance that may be present. Also, the total shunt capacitance at the output is the sum of the reflected capacitance and any other output-to-ground capacitance present. The effect of Miller capacitance is illustrated in the next example.

Example 6.9

The inverting amplifier shown in Figure 6.23 has midband gain $|v_L|/|v_{in}| = -200$. Find its upper cutoff frequency.

Figure 6.23
(Example 6.9)

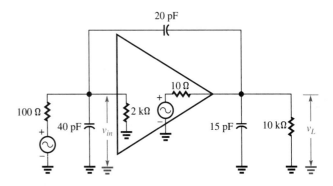

Solution.   We will use the midband gain $|v_L|/|v_{in}|$ to determine the Miller capacitance. Notice that this is the gain *between the points where the 20-pF capacitance is connected*. The Miller capacitance is, therefore, $C_M \approx (20 \text{ pF})[1 - (-200)] = 4020$ pF. The total capacitance C_A shunting the input is then $C_A = (40 \text{ pF}) + (4020 \text{ pF}) = 4060$ pF. The cutoff frequency due to C_A is

$$f_2(C_A) = \frac{1}{2\pi[(2 \times 10^3 \ \Omega) \parallel 100 \ \Omega]4060 \times 10^{-12} \text{ F}} = 411.6 \text{ kHz}$$

The total capacitance shunting the output is 15 pF + (1 + 1/200)20 pF $\approx$ 35 pF. The cutoff frequency due to the output capacitance is

$$f_2(C_B) = \frac{1}{2\pi[(10 \times 10^3 \ \Omega) \parallel 10 \ \Omega]35 \times 10^{-12} \text{ F}} = 455 \text{ MHz}$$

The upper cutoff frequency is the smaller of the two: $f_2 = 411.6$ kHz.

6.5   TRANSIENT RESPONSE

The *transient response* of an electronic amplifier or system is the output waveform that results when the input is a pulse or a sudden change in level. Since the transient response is a waveform, it is presented as a plot of voltage versus time,

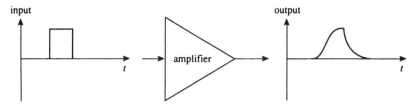

Figure 6.24
The output waveform is the transient response of the amplifier to a pulse-type input.

in contrast to frequency response, which is plotted versus frequency. Figure 6.24 shows a typical transient response.

The transient response of an amplifier is completely dependent on its frequency response, and vice versa. In other words, if two amplifiers have identical frequency responses, they will have identical transient responses, and vice versa. A pulse can be regarded as consisting of an infinite number of frequency components, so the transient waveform represents the amplifier's ability (or inability) to amplify all frequency components equally and to phase-shift all components equally. Theoretically, if an amplifier had infinite bandwidth, its transient response would be an exact duplicate of the input pulse. It is not possible for an amplifier to have infinite bandwidth, so the transient response is always a distorted version of the input pulse. It is necessary for an amplifier to have a wide bandwidth (to be a *wideband,* or *broadband,* amplifier) in order for it to amplify pulse or square-wave signals with a minimum of distortion. *Square-wave testing* is sometimes used to check the frequency response of an amplifier, as shown in Figure 6.25. The figure shows typical waveforms that result when a square wave is applied to an amplifier whose frequency response causes attenuation of either low- or high-frequency components. "Low" or "high" frequency in any given case means low or high in relation to the square-wave frequency, f_s.

As a concrete example of how frequency response is related to transient response, consider the low-pass RC network shown in Figure 6.26. The figure shows the output transient when the input is an abrupt change in level (called a

Figure 6.25
Typical outputs resulting from square-wave testing of an amplifier. f_s = square-wave frequency. In (a) and (b), only low frequencies are attenuated, and in (c) and (d), only high frequencies are attenuated. In both cases, the attenuation outside cutoff is 20 dB/decade.

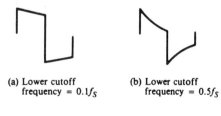

(a) Lower cutoff
frequency = $0.1f_s$

(b) Lower cutoff
frequency = $0.5f_s$

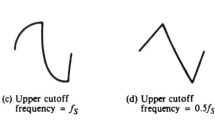

(c) Upper cutoff
frequency = f_s

(d) Upper cutoff
frequency = $0.5f_s$

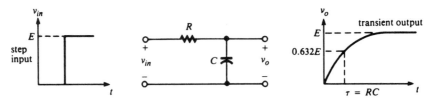

(a) Transient response to a step input

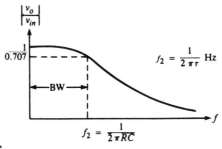

(b) Frequency response

Figure 6.26
Relationship between transient response and frequency response. Note that the band-width is inversely proportional to the time constant of the transient.

step input), such as might occur when a dc voltage is switched into the network. Recall that the time constant, $\tau = RC$ seconds, is the time required for the transient output to reach 63.2% of its final value. Also recall that the upper cutoff frequency of the network is $f_2 = 1/(2\pi RC)$ Hz. Thus, $f_2 = 1/(2\pi\tau)$ Hz. Since the network passes all frequencies below f_2, down to dc, its bandwidth equals $f_2 - 0 = f_2$. Summarizing,

$$\text{BW} = f_2 = \frac{1}{2\pi\tau} \text{ Hz} \tag{6.40}$$

Equation 6.40 shows that the cutoff frequency and the bandwidth, which are frequency-response characteristics, are *inversely* proportional to the time constant, which is a characteristic of the transient response. It is generally true for all electronic devices that the time required for the transient response to rise to a certain level is inversely proportional to bandwidth.

Recall that rise time, t_r, is the time required for a waveform to change from 10% of its final value to 90% of its final value. A widely used approximation that relates the rise time of the transient response of an amplifier to its bandwidth is

$$t_r \approx \frac{0.35}{\text{BW}} \text{ seconds} \tag{6.41}$$

where BW is the bandwidth, in hertz. Equation 6.41 is used when the lower cutoff frequency is 0 (dc) or very small, so that the bandwidth is essentially the same as the upper cutoff frequency. The relationship is exact if the high-frequency response beyond cutoff is the same as that of the single RC low-pass network (Figure 6.26).

Example 6.10

The specifications for a certain oscilloscope state that the rise time of the vertical amplifier is 8.75 ns. What is the approximate bandwidth of the amplifier?

Solution.   From equation 6.41,

$$\text{BW} \approx \frac{0.35}{t_r} = \frac{0.35}{8.75 \times 10^{-9}} = 40 \text{ MHz}$$

6.6   FREQUENCY RESPONSE OF MULTISTAGE AMPLIFIERS

Gain Relations in Multistage Amplifiers

In many applications, a single amplifier cannot furnish all the gain that is required to drive a particular kind of load. For example, a speaker represents a "heavy" load in an audio amplifier system, and several amplifier *stages* may be required to "boost" a signal originating at a microphone or magnetic tape head to a level sufficient to provide a large amount of power to the speaker. We hear of *preamplifiers, power amplifiers,* and *output amplifiers,* all of which constitute stages of amplification in such a system. Actually, each of these components may itself consist of a number of individual transistor amplifier stages. Amplifiers that create voltage, current, and/or power gain through the use of two or more stages are called *multistage* amplifiers.

When the output of one amplifier stage is connected to the input of another, the amplifier stages are said to be in *cascade*. Figure 6.27 shows two stages connected in cascade. To illustrate how the overall voltage gain of the combination is computed, let us assume that the input to the first stage is 10 mV rms and that the voltage gain of each stage is $A_1 = A_2 = 20$, as shown in the figure. The output of the first stage is $A_1 v_{i1} = 20 \, (10 \text{ mV rms}) = 200 \text{ mV rms}$. Thus, the input to the second stage is 200 mV rms. The output of the second stage is, therefore, $A_2 v_{i2} = 20(200 \text{ mV rms}) = 4 \text{ V rms}$. Therefore, the overall voltage gain is

$$A_v = \frac{v_{o2}}{v_{i1}} = \frac{4 \text{ V rms}}{10 \text{ mV rms}} = 400$$

Notice that $A_v = A_1 A_2 = (20)(20) = 400$.

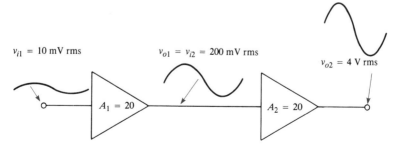

$v_{i1} = 10 \text{ mV rms}$

$v_{o1} = v_{i2} = 200 \text{ mV rms}$

$v_{o2} = 4 \text{ V rms}$

$A_1 = 20$

$A_2 = 20$

Figure 6.27
Two amplifier stages connected in cascade

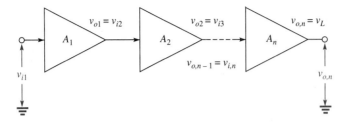

Figure 6.28
n amplifier stages connected in cascade. The output voltage of each stage is the input voltage to the next stage.

Figure 6.28 shows an arbitrary number (n) of stages connected in cascade. Note that the output of each stage is the input to the succeeding one ($v_{o1} = v_{i2}$, $v_{o2} = v_{i3}$, etc.). We assume that each stage gain $A_1, A_2 . . . , A_n$ is the value of the voltage gain between input and output of a stage *with all other stages connected* (more about that important assumption later). By a simple extension of the two-amplifier case just discussed, it is easy to see that the overall gain $v_{o,n}/v_{i1}$ is the *product* of the individual stage gains (not the sum!):

$$\frac{v_{o,n}}{v_{i1}} = A_n A_{n-1} \cdot \cdot \cdot A_2 A_1 \qquad (6.42)$$

In general, any one or more of the stage gains can be negative, signifying, as usual, that the stage causes a 180° phase inversion. It follows from equation 6.42 that the cascaded amplifiers will cause the output of the last stage ($v_{o,n}$) to be out of phase with the input to the first stage (v_{i1}) if there is an *odd* number of inverting stages, and will cause $v_{o,n}$ to be in phase with v_{i1} if there is an even (or zero) number of inversions.

To find the overall voltage gain of the cascaded system in decibels, we ignore the algebraic signs of each stage gain and compute

$$20 \log_{10} \left(\frac{v_{o,n}}{v_{i1}}\right) = 20 \log_{10}(A_n A_{n-1} \cdot \cdot \cdot A_2 A_1)$$

$$= 20 \log_{10} A_n + 20 \log_{10} A_{n-1} + \cdot \cdot \cdot + 20 \log_{10} A_2 + 20 \log_{10} A_1$$
$$= A_n \text{ (dB)} + A_{n-1} \text{ (dB)} + \cdot \cdot \cdot + A_2 \text{ (dB)} + A_1 \text{ (dB)} \qquad (6.43)$$

Equation 6.43 shows that the overall voltage gain in dB is the *sum* of the individual stage gains expressed in decibels. Equations similar to (6.42) and (6.43) for overall current gain and overall power gain in terms of individual stage gains are easily derived.

Equation 6.43 does not include the effect of source or load resistance on the overall voltage gain. Source resistance r_S causes the usual voltage division to take place at the input to the first stage, and load resistance r_L causes voltage division to occur between r_L and the output resistance of the last stage. Under those circumstances, the overall voltage gain between load and signal source becomes

$$\frac{v_L}{v_S} = \left(\frac{r_{i1}}{r_S + r_{i1}}\right) A_n A_{n-1} \cdot \cdot \cdot A_2 A_1 \left(\frac{r_L}{r_{o,n} + r_L}\right) \qquad (6.44)$$

where r_{i1} = input resistance to first stage and $r_{o,n}$ = output resistance of last stage.

Example 6.11

Figure 6.29 shows a three-stage amplifier and the ac rms voltages at several points in the amplifier. Note that v_1 is the input voltage delivered by a signal source having zero resistance and that v_3 is the output voltage with no load connected.

1. Find the voltage gain of each stage and the overall voltage gain v_3/v_1.
2. Repeat (1) in terms of decibels.
3. Find the overall voltage gain v_L/v_S when the multistage amplifier is driven by a signal source having resistance 2000 Ω and the load is 25 Ω. Stage 1 has input resistance 1 kΩ and stage 3 has output resistance 50 Ω.
4. What would be the gain v_L/v_S in dB for the conditions of (3) if the voltage gain of the second stage were reduced by 6 dB?
5. What is the power gain in decibels under the conditions of (3) (measured between the input to the first stage and the load)?
6. What is the overall current gain i_L/i_1 under the conditions of (3)?

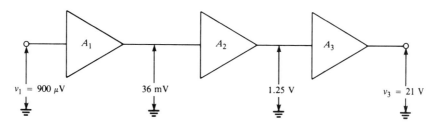

Figure 6.29
(Example 6.11)

Solution

1. $A_1 = (36 \text{ mV})/(900 \ \mu\text{V}) = 40$
    $A_2 = (1.25 \text{ V})/(36 \text{ mV}) = 34.722$
    $A_3 = (21 \text{ V})/(1.25 \text{ V}) = 16.8$
    $v_3/v_1 = A_1A_2A_3 = (40)(34.722)(16.8) = 23{,}333$

    Note that the product of the voltage gains equals the overall voltage gain, which, in this example, can also be calculated directly: $v_3/v_1 = (21 \text{ V})/(900 \ \mu\text{V}) = 23{,}333$.

2. $A_1 \text{ (dB)} = 20 \log_{10} 40 = 32.04 \text{ dB}$
    $A_2 \text{ (dB)} = 20 \log_{10}(34.722) = 30.81 \text{ dB}$
    $A_3 \text{ (dB)} = 20 \log_{10}(16.8) = 24.51 \text{ dB}$
    $v_3/v_1 \text{ (dB)} = A_1 \text{ (dB)} + A_2 \text{ (dB)} + A_3 \text{ (dB)} = 87.36 \text{ dB}$

    Again, note that $20 \log_{10}(v_3/v_1) = 20 \log_{10}[21/(900 \times 10^{-6})] = 20 \log_{10} (23{,}333) = 87.36 \text{ dB}$.

3. From equation 6.44,

$$\frac{v_L}{v_S} = \left(\frac{1000 \ \Omega}{2000 \ \Omega + 1000 \ \Omega}\right) (23{,}333) \left(\frac{25 \ \Omega}{50 \ \Omega + 25 \ \Omega}\right) = 2592.5$$

4. The (original) voltage gain with load and source connected is $20 \log_{10}(2592.5) = 68.27$ dB. A 6-dB reduction in the gain of stage 2 therefore results in an overall gain of $(68.27$ dB$) - (6$ dB$) = 62.27$ dB.

5. When a signal-source resistance of 2000 Ω is inserted in series with the input, v_1 becomes

$$v_1 = \left(\frac{1000\ \Omega}{2000\ \Omega + 1000\ \Omega}\right)(900\ \mu V) = 300\ \mu V$$

The input power is, therefore,

$$P_i = \frac{v_1^2}{r_{i1}} = \frac{(300 \times 10^{-6}\ V)^2}{1000\ \Omega} = 90\ pW$$

The voltage across the 25-Ω load is then

$$v_L = v_1(A_1 A_2 A_3)\left(\frac{R_L}{r_{o3} + R_L}\right)$$

$$= (300\ \mu V)(40)(34.722)(16.8)\left(\frac{25\ \Omega}{50\ \Omega + 25\ \Omega}\right) = 2.33\ V$$

The output power developed across the load resistance is, therefore,

$$P_o = \frac{v_L^2}{R_L} = \frac{(2.33\ V)^2}{25\ \Omega} = 0.217\ W$$

Finally,

$$A_p\ (dB) = 10 \log_{10}\left(\frac{P_o}{P_i}\right) = 10 \log_{10}\left(\frac{0.217\ W}{90 \times 10^{-12}\ W}\right) = 93.82\ dB$$

6. Recall that $A_p = A_v A_i$. Using the results calculated in (5), the power gain between the input to the first stage and the load is

$$A_p = \frac{P_o}{P_i} = \frac{0.217\ W}{90 \times 10^{-12}\ W} = 2.41 \times 10^9$$

The voltage gain between the input to the first stage and the load is

$$A_v = \frac{v_L}{v_1} = \frac{2.33\ V}{300\ \mu V} = 7766$$

Therefore,

$$A_i = \frac{A_p}{A_v} = \frac{2.41 \times 10^9}{7766} = 3.1 \times 10^5$$

It is important to remember that the gain equations we have derived are based on the *in-circuit* values of $A_1, A_2, \ldots$, that is, on the stage gains that result when all other stages are connected. Thus, we have assumed that each value of stage gain takes into account the loading the stage causes on the previous stage and the loading presented to it by the next stage (except we assumed that A_1 did not include loading by r_S and A_n did not include loading by r_L). If we know the *open-circuit* (unloaded) voltage gain of each stage and its input and output resistances, we can calculate the overall gain by taking into account the loading effects of each stage on another. Theoretically, the load presented to a given stage may depend on *all* of the succeeding stages lying to its right, since the input resistance of any one stage depends on its output load resistance, which in turn is the input resistance to the next stage, and so forth. In practice, we can usually ignore this cumulative loading effect of stages beyond the one immediately connected to a given stage, or assume that the input resistance that represents the load of one stage to a preceding one is given for the condition that all succeeding stages are connected.

To illustrate the ideas we have just discussed, Figure 6.30 shows a three-stage amplifier for which the individual *open-circuit* voltage gains A_{o1}, A_{o2}, and A_{o3} are assumed to be known, as well as the input and output resistances of each stage. From the voltage division that occurs at each node in the system, it is apparent that the following relations hold:

$$v_1 = \left(\frac{r_{i1}}{r_S + r_{i1}}\right) v_S$$

$$v_2 = A_{o1} v_1 \left(\frac{r_{i2}}{r_{o1} + r_{i2}}\right)$$

$$v_3 = A_{o2} v_2 \left(\frac{r_{i3}}{r_{o2} + r_{i3}}\right)$$

$$v_L = A_{o3} v_3 \left(\frac{r_L}{r_{o3} + r_L}\right)$$

Combining these relations leads to

$$\frac{v_L}{v_S} = \left(\frac{r_{i1}}{r_S + r_{i1}}\right) A_{o1} \left(\frac{r_{i2}}{r_{o1} + r_{i2}}\right) A_{o2} \left(\frac{r_{i3}}{r_{o2} + r_{i3}}\right) A_{o3} \left(\frac{r_L}{r_{o3} + r_L}\right) \qquad \textbf{(6.45)}$$

As might be expected, equation 6.45 shows that the overall voltage gain of the multistage amplifier is the product of the open-circuit stage gains multiplied by the voltage-division ratios that account for the loading of each stage. Notice that a *single* voltage-division ratio accounts for the loading between any pair of stages.

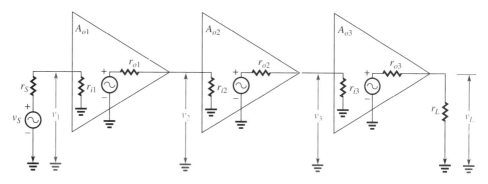

Figure 6.30
A three-stage amplifier. A_{o1}, A_{o2}, and A_{o3} are the open-circuit (unloaded) voltage gains of the respective stages.

In other words, it is *not* correct to compute loading effects twice: once by regarding an input resistance as the load on a previous stage and again by regarding the output resistance of that previous stage as the source resistance for the next stage.

Example 6.12

The open-circuit (unloaded) voltage gains of three amplifier stages and their input and output resistances are shown in Table 6.2. If the three stages are cascaded and the first is driven by a 10-mV-rms signal source having resistance 12 kΩ, what is the voltage across a 12-Ω load connected at the output of the third stage?

Table 6.2
(Example 6.12)

Amplifier Stage	Unloaded Voltage Gain (dB)	Input Resistance (kΩ)	Output Resistance (kΩ)
1	24	10	4.7
2	20	20	1.5
3	12	1.5	0.02

Solution

$$20 \log_{10} A_{o1} = 24$$
$$\log_{10} A_{o1} = 1.2$$
$$A_{o1} = \text{antilog } (1.2) = 15.85$$

Similarly,

$$A_{o2} = \text{antilog } (1) = 10$$
$$A_{o3} = \text{antilog } (0.6) = 3.98$$

Then, from equation 6.45,

$$\frac{v_L}{v_S} = \left(\frac{10 \text{ k}\Omega}{12 \text{ k}\Omega + 10 \text{ k}\Omega}\right) 15.85 \left(\frac{20 \text{ k}\Omega}{4.7 \text{ k}\Omega + 20 \text{ k}\Omega}\right) 10 \left(\frac{1.5 \text{ k}\Omega}{1.5 \text{ k}\Omega + 1.5 \text{ k}\Omega}\right)$$
$$\times (3.98) \left(\frac{12 \text{ }\Omega}{20 \text{ }\Omega + 12 \text{ }\Omega}\right) = 43.53$$

Therefore, $v_L = 43.53 v_S = 43.53 \ (10 \text{ mV rms}) = 0.4353$ V rms.

Frequency Response

It is possible for every stage in a multistage amplifier to have a different lower cutoff frequency. If these lower break frequencies are not close in value, the overall lower cutoff frequency is approximately equal to the largest of the individual cutoff frequencies. Similarly, the overall upper cutoff frequency is approximately equal to the smallest of the individual upper cutoff frequencies, provided they are not close in value.

In practice, a multistage amplifier may have some lower break frequencies that are equal, or close in value, and others that are not. The same is true for upper break frequencies. In these situations, computation of the actual lower and upper

Table 6.3
The lower and upper cutoff frequencies of a multistage amplifier consisting of n stages, each having lower cutoff frequency f_1 and upper cutoff frequency f_2

Number of Stages n	$f_{1(overall)}$	$f_{2(overall)}$
1	f_1	f_2
2	$1.55f_1$	$0.64f_2$
3	$1.96f_1$	$0.51f_2$
4	$2.30f_1$	$0.43f_2$
5	$2.59f_1$	$0.39f_2$

cutoff frequencies of a multistage amplifier is a very complex problem. From a practical standpoint, the cutoff frequencies are best determined experimentally or by use of a computer program that computes the overall frequency response.

For the special cases where all stages of a multistage amplifier have identical lower cutoff frequencies or identical upper cutoff frequencies, the overall cutoff frequencies can be calculated readily:

$$f_{1(overall)} = \frac{f_1}{\sqrt{2^{1/n} - 1}} \tag{6.46}$$

$$f_{2(overall)} = f_2\sqrt{2^{1/n} - 1} \tag{6.47}$$

where $f_{1(overall)}$ = overall lower cutoff frequency of the multistage amplifier
$f_{2(overall)}$ = overall upper cutoff frequency of the multistage amplifier
n = number of stages having identical lower cutoff frequencies and/or identical upper cutoff frequencies
f_1 = lower cutoff frequency of each stage
f_2 = upper cutoff frequency of each stage

Table 6.3 shows values of $f_{1(overall)}$ and $f_{2(overall)}$ in terms of f_1 and f_2, for n ranging from 1 to 5. Note that when $n = 2$, $f_{1(overall)} = 1.55f_1$ and $f_{2(overall)} = 0.64f_2$, in agreement with our previous discussion for the case of two identical break frequencies in a single stage. The table confirms what should be intuitively clear: the greater the number of identical stages, the larger the lower cutoff frequency and the smaller the upper cutoff frequency. In other words, cascading stages with identical frequency-response characteristics reduces the overall bandwidth of a multistage amplifier. When n stages having identical frequency response are cascaded, the overall frequency response falls off along asymptotes having slopes $20n$ dB/decade ($6n$ dB/octave) at frequencies outside the midband range. The break frequencies are in all cases equal to the cutoff frequencies of a single stage.

Example 6.13

A multistage audio amplifier is to be constructed using four identical stages. What should be the lower and upper cutoff frequencies of each stage if the overall lower and upper cutoff frequencies are to be 20 Hz and 20 kHz, respectively?

Solution. From Table 6.3, with $n = 4$, $2.3f_1 = 20$, and $0.43f_2 = 20 \times 10^3$. Therefore, $f_1 = 20/2.3 = 8.7$ Hz and $f_2 = 20 \times 10^3/0.43 = 46.5$ kHz. This example shows that each stage must have a bandwidth of approximately 46 kHz to achieve an overall bandwidth of approximately 20 kHz.

Example 6.14

Sketch the asymptotic frequency response in the low-frequency range of the two-stage amplifier in Figure 6.31 the way it would appear when v_L/v_S is plotted on log-log axes. The midband value of v_L/v_S is 500.

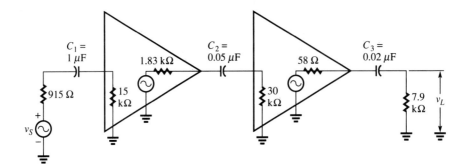

Figure 6.31
(Example 6.14)

Solution. The lower break frequency due to coupling capacitor C_1 at the input to the first stage is

$$f_1(C_1) = \frac{1}{2\pi(r_S + r_{in1})C_1} = \frac{1}{2\pi(915\ \Omega + 16\ k\Omega)1\ \mu F} = 10\ Hz$$

Similarly, the lower break frequency due to C_2 is

$$f_1(C_2) = \frac{1}{2\pi(1.83\ k\Omega + 30\ k\Omega)(0.05\ \mu F)} = 100\ Hz$$

The lower break frequency due to output coupling capacitor C_3 is

$$f_1(C_3) = \frac{1}{2\pi(58\ \Omega + 7.9\ k\Omega)(0.02\ \mu F)} = 1\ kHz$$

Since each break frequency is separated from the next by 1 decade, the overall lower cutoff frequency is approximately equal to the largest of the three: $f_1(C_3) = f_1 = 1\ kHz$.

Figure 6.32 shows a sketch of the frequency response on log-log axes. Note that the gain falls off at 20 dB/decade between $f_1(C_3)$ and $f_1(C_2)$, at 40 dB/decade between $f_1(C_2)$ and $f_1(C_1)$, and at 60 dB/decade at frequencies below $f_1(C_1)$. Since $f_1(C_2)$ is 1 decade below $f_1(C_3)$, the gain at $f_1(C_2)$ is (asymptotically) one-tenth the gain at $f_1(C_3)$, i.e., 20 dB down from the midband value:

$$\frac{v_L}{v_S}\ (\text{at }100\ Hz) = \frac{1}{10}\ (500) = 50$$

Since the gain falls at the rate of 40 dB/decade between 100 Hz and 10 Hz, it falls a total of 40 dB in that decade. In other words, the gain at 10 Hz is one one-

Figure 6.32
(Example 6.14)

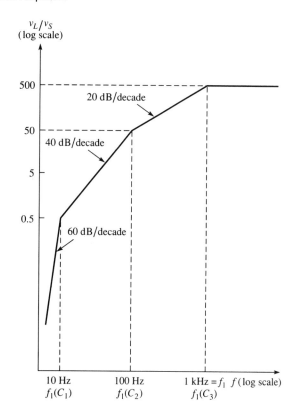

hundredth of its value at 100 Hz:

$$\frac{v_L}{v_S} \text{ (at 10 Hz)} = \frac{1}{100} (50) = 0.5$$

6.7   FREQUENCY RESPONSE OF BJT AMPLIFIERS

Low-Frequency Response of BJT Amplifiers

We have learned that the lower cutoff frequency of an amplifier is approximately equal to the larger of $f_1(C_1)$ and $f_1(C_2)$, where

$$f_1(C_1) = \frac{1}{2\pi(r_S + r_{in})C_1} \text{ hertz} \qquad (6.48)$$

and

$$f_1(C_2) = \frac{1}{2\pi(r_o + R_L)C_2} \text{ hertz} \qquad (6.49)$$

In a BJT amplifier, the term r_{in} appearing in equation 6.48 is r_{in}(stage), the resistance seen by the source when looking into the amplifier. Its value depends on the transistor configuration, the bias resistors, and the values of the transistor parameters. Similarly, r_o in equation 6.49 is r_o(stage) and its value depends on particular amplifier characteristics. The next example illustrates the application of equations 6.48 and 6.49 to determine the lower cutoff frequency of a fixed-bias common-emitter amplifier.

Example 6.15

Find the lower cutoff frequency of the amplifier shown in Figure 6.33. Assume that the resistance looking into the base of the transistor is 1500 Ω and that the transistor output resistance at the collector, r_o, is 100 kΩ.

Figure 6.33
(Example 6.15)

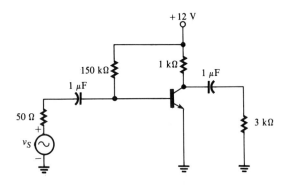

Solution.   Since r_{in}(stage) = (150 kΩ) $\|$ (1.5 kΩ) $\approx$ 1.5 kΩ, we have, from equation 6.48,

$$f_1(C_1) = \frac{1}{2\pi(50 + 1500)(1 \times 10^{-6})} = 102.7 \text{ Hz}$$

Since r_o(stage) = (100 kΩ) $\|$ (1 kΩ) $\approx$ 1 kΩ, we have, from equation 6.49,

$$f_1(C_2) = \frac{1}{2\pi(1 \times 10^3 + 3 \times 10^3)(1 \times 10^{-6})} = 39.8 \text{ Hz}$$

Therefore, $f_1 \approx$ 103 Hz.

Figure 6.34 shows a common-emitter amplifier having an emitter bypass capacitor C_E designed to eliminate signal degeneration, as discussed in Chapter 5. Recall that ac signal degeneration occurs in the absence of the bypass capacitor because of the ac voltage drop across R_E. (See Example 5.8.) At sufficiently high frequencies, the reactance of the capacitor is negligible, so the emitter is effectively at ac ground and there is no signal loss across R_E. At low frequencies, the reactance of C_E can become significant, to the extent that the voltage gain is reduced. Thus, we can define a lower cutoff frequency due to C_E, $f_1(C_E)$, that affects the amplifier's low-frequency response in the same way as the coupling capacitors, C_1 and C_2. If the coupling capacitors are large enough to have negligible effect, then the gain will fall to $\sqrt{2}/2$ times its midband value at the frequency where the reactance of C_E equals the resistance R_e looking into node A in Figure 6.34. R_e is the same as the output resistance of an emitter follower and was

Figure 6.34
The lower cutoff frequency due to the bypass capacitor, $f_l(C_E)$, is the frequency where the reactance of C_E equals the resistance R_e.

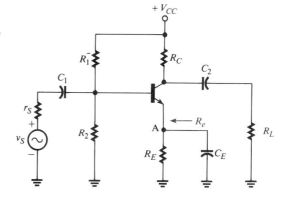

previously expressed as r_o(stage) (see equation 5.47):

$$R_e = R_E \left\| \left(\frac{r_S \| R_B}{\beta + 1} + r_e \right) \right. \tag{6.50}$$

where $R_B = R_1 \| R_2$. Thus, $f_l(C_E)$ is the frequency at which $\dfrac{1}{[2\pi f_l(C_E)]C_E} = R_e$, or

$$f_l(C_E) = \frac{1}{2\pi R_e C_E} \tag{6.51}$$

We see that there are three possible break frequencies in the amplifier of Figure 6.34: $f_l(C_1)$, $f_l(C_2)$, and $f_l(C_E)$. If these are all distinct, and reasonably well removed from each other, the lower cutoff frequency will be the largest of the three. At frequencies below the smallest of the three, the gain will fall off at a rate of 60 dB/decade (18 dB/octave), because all three capacitors will be contributing simultaneously to gain reduction.

Example 6.16

Find the lower cutoff frequency of the amplifier shown in Figure 6.35. Assume that the transistor has $\beta = 100$, $r_e = 12\ \Omega$, and $r_o = 50\ \text{k}\Omega$.

Figure 6.35
(Example 6.16)

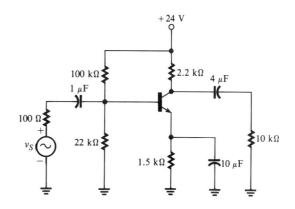

Solution

$$r_{in}(\text{stage}) = R_1 \parallel R_2 \parallel \beta r_e = (100 \text{ k}\Omega) \parallel (22 \text{ k}\Omega) \parallel (100)(12 \ \Omega) = 1.1 \text{ k}\Omega$$

$$f_1(C_1) = \frac{1}{2\pi(1.1 \times 10^3 + 100)(1 \times 10^{-6})} = 133 \text{ Hz}$$

$$r_o(\text{stage}) = (50 \text{ k}\Omega) \parallel (2.2 \text{ k}\Omega) = 2.1 \text{ k}\Omega$$

$$f_1(C_2) = \frac{1}{2\pi(2.1 \times 10^3 + 10 \times 10^3)(4 \times 10^{-6})} = 3.3 \text{ Hz}$$

From equation 6.50,

$$R_e = (1.5 \text{ k}\Omega) \parallel \left[\frac{(100 \ \Omega) \parallel (100 \text{ k}\Omega) \parallel (22 \text{ k}\Omega)}{101} + (12 \ \Omega) \right]$$

$$\approx (1.5 \text{ k}\Omega) \parallel (13 \ \Omega) \approx 13 \ \Omega$$

From equation 6.51,

$$f_1(C_E) = \frac{1}{2\pi(13)(10 \times 10^{-6})} = 1.22 \text{ kHz}$$

The lower cutoff frequency is the largest of the three computed frequencies, so $f_1 = f_1(C_E) = 1.22$ kHz.

This example demonstrates that the emitter-bypass capacitor is generally the most troublesome, from the standpoint of achieving a low cutoff frequency, because of the small value of R_e. To obtain a value of $f_1(C_E)$ equal to the lowest audio frequency, 20 Hz, would require the impractically large value of $C_E = 2450$ μF in this example.

Example 6.17

SPICE

Use SPICE to obtain a plot of the frequency response of the amplifier in Example 6.16 over the frequency range from 1 Hz through 10 kHz.

Solution.   Figure 6.36 shows the circuit of Figure 6.35 redrawn in the SPICE format as well as the input data file. Since the β of the transistor is 100, we can allow BETA in the .MODEL statement to have its default value (100). All other parameter values are allowed to default, since none have significant bearing on the low-frequency response. Note that the .AC statement specifies ten frequencies per decade for each of the four decades from 1 Hz through 10 kHz.

The log-log plot of the frequency response produced by SPICE is shown in Figure 6.37(a), on page 268. Since we are allowing the ac source VIN to have its default value of 1 V, the output voltages are unrealistically large. Recall that SPICE does not consider practical voltage limitations and clipping when performing an ac analysis. The results show that the midband gain is approximately 132.7 (at 10 kHz, about one decade above the lower cutoff frequency; the actual gain may be slightly greater at higher frequencies). Thus, the gain will be approximately 0.707(132.7) = 93.8 at the lower cutoff frequency. The frequency at which the output is nearest 93.8 is 1.259 kHz (92.09). A more accurate estimate of the lower cutoff frequency can be made by restricting the range and increasing the number of frequencies at which the output is calculated in the vicinity of cutoff.

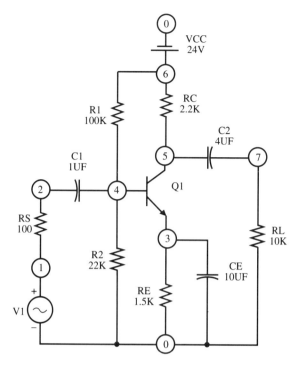

EXAMPLE 6.17
```
V1    1 0 AC
RS    1 2 100
C1    2 4 1UF
R1    6 4 100K
R2    4 0 22K
VCC   6 0 24V
RC    5 6 2.2K
Q1    5 4 3 TRAN
.MODEL TRAN NPN
RE    3 0 1.5K
CE    3 0 10UF
C2    5 7 4UF
RL    7 0 10K
.AC DEC 10 1HZ 10KHZ
.PLOT AC V(7)
.END
```

Figure 6.36
(Example 6.17)

Figure 6.37(b), on page 269, shows the results of a .PRINT statement when the .AC statement is changed to LIN 20 1KHZ 1.5KHZ, producing output at 20 linearly spaced frequencies between 1 kHz and 1.5 kHz. We see that the lower cutoff frequency is between 1.289 KHZ and 1.316 kHz. These results agree well with the value calculated in Example 6.16.

The principal consideration in the design and analysis of amplifiers in the low-frequency region is the (Thevenin equivalent) resistance in series with each capacitor used in the circuit. We have seen that a break frequency occurs whenever the frequency becomes low enough to make the capacitive reactance equal to the equivalent resistance at the point of connection. Therefore, in any BJT amplifier, the capacitor that is most critical in determining the lower cutoff frequency is the one that "sees" the smallest resistance. In the case of the common-emitter amplifier, C_E is that capacitor. In a common-base amplifier, the input resistance $(r_{in} \approx r_e)$ is quite small, so the input coupling capacitor, C_1, is the most crucial. In the common-collector amplifier, the output resistance is quite small, so the output coupling capacitor, C_2, is crucial. Of course, a large source resistance in the case of a CB amplifier, or a large load resistance in the case of a CC amplifier, will mitigate these circumstances, since each is in series with the affected capacitor.

Figure 6.38 shows common-base and common-collector amplifiers and gives the equations for the break frequencies due to each coupling capacitor. These are derived in a straightforward manner by solving for the frequency at which the

```
EXAMPLE 6.17
****      AC ANALYSIS                        TEMPERATURE =    27.000 DEG C
**********************************************************************************
    FREQ        V(7)
                    1.000D-02      1.000D+00      1.000D+02      1.000D+04  1.000D+06
                    - - - - - - - - - - - - - - - - - - - - - - - - - - - - - -
1.000D+00   3.501D-02 .     *            .              .              .          .
1.259D+00   5.409D-02 .      *           .              .              .          .
1.585D+00   8.251D-02 .       *          .              .              .          .
1.995D+00   1.237D-01 .         *        .              .              .          .
2.512D+00   1.814D-01 .        *         .              .              .          .
3.162D+00   2.591D-01 .          *       .              .              .          .
3.981D+00   3.595D-01 .          *       .              .              .          .
5.012D+00   4.848D-01 .           *      .              .              .          .
6.310D+00   6.375D-01 .            *.    .              .              .          .
7.943D+00   8.210D-01 .            *.    .              .              .          .
1.000D+01   1.042D+00 .             *    .              .              .          .
1.259D+01   1.310D+00 .             .*   .              .              .          .
1.585D+01   1.640D+00 .             . *  .              .              .          .
1.995D+01   2.052D+00 .             . *  .              .              .          .
2.512D+01   2.568D+00 .             .  * .              .              .          .
3.162D+01   3.218D+00 .             .   *.              .              .          .
3.981D+01   4.037D+00 .             .   *.              .              .          .
5.012D+01   5.070D+00 .             .    *              .              .          .
6.310D+01   6.371D+00 .             .    .*             .              .          .
7.943D+01   8.007D+00 .             .    .*             .              .          .
1.000D+02   1.006D+01 .             .    . *            .              .          .
1.259D+02   1.264D+01 .             .    .  *           .              .          .
1.585D+02   1.587D+01 .             .    .  *           .              .          .
1.995D+02   1.990D+01 .             .    .   *          .              .          .
2.512D+02   2.489D+01 .             .    .    *         .              .          .
3.162D+02   3.102D+01 .             .    .    *         .              .          .
3.981D+02   3.845D+01 .             .    .     *        .              .          .
5.012D+02   4.727D+01 .             .    .      * .     .              .          .
6.310D+02   5.745D+01 .             .    .      * .     .              .          .
7.943D+02   6.872D+01 .             .    .       *.     .              .          .
1.000D+03   8.053D+01 .             .    .       *.     .              .          .
1.259D+03   9.209D+01 .             .    .        *     .              .          .
1.585D+03   1.026D+02 .             .    .        *     .              .          .
1.995D+03   1.114D+02 .             .    .        *     .              .          .
2.512D+03   1.183D+02 .             .    .        .*    .              .          .
3.162D+03   1.234D+02 .             .    .        .*    .              .          .
3.981D+03   1.270D+02 .             .    .        .*    .              .          .
5.012D+03   1.294D+02 .             .    .        .*    .              .          .
6.310D+03   1.310D+02 .             .    .        .*    .              .          .
7.943D+03   1.320D+02 .             .    .        .*    .              .          .
1.000D+04   1.327D+02 .             .    .        .*    .              .          .
                    - - - - - - - - - - - - - - - - - - - - - - - - - - - - - -
```

(a)

Figure 6.37
(Example 6.17) *Figure continued on page 269.*

capacitive reactance of each capacitor equals the Thevenin equivalent resistance in series with it.

Example 6.18

A certain transistor has $\beta = 100$, $r_C = 100$ kΩ, and $r_e = 25$ Ω. The transistor is used in each of the circuits shown in Figure 6.38. Find the approximate lower cutoff frequency in each circuit, given the following component values:

1. (Common-base circuit)

$$r_S = 100\ \Omega \qquad R_C = 1\ \text{k}\Omega \qquad C_1 = 2.2\ \mu\text{F}$$
$$R_E = 10\ \text{k}\Omega \qquad R_L = 15\ \text{k}\Omega \qquad C_2 = 1\ \mu\text{F}$$

```
EXAMPLE 6.17
****          AC ANALYSIS                            TEMPERATURE =    27.000 DEG C
****************************************************************************************
      FREQ          V(7)
    1.000E+03      8.053E+01
    1.026E+03      8.186E+01
    1.053E+03      8.315E+01
    1.079E+03      8.441E+01
    1.105E+03      8.563E+01
    1.132E+03      8.682E+01
    1.158E+03      8.797E+01
    1.184E+03      8.909E+01
    1.211E+03      9.018E+01
    1.237E+03      9.123E+01
    1.263E+03      9.226E+01
    1.289E+03      9.325E+01
    1.316E+03      9.422E+01
    1.342E+03      9.515E+01
    1.368E+03      9.606E+01
    1.395E+03      9.694E+01
    1.421E+03      9.780E+01
    1.447E+03      9.863E+01
    1.474E+03      9.944E+01
    1.500E+03      1.002E+02
```

(b)

Figure 6.37
(Continued)

2. (Common-collector circuit)

$$r_S = 100 \ \Omega \qquad R_E = 1 \ \text{k}\Omega \qquad C_1 = 1 \ \mu\text{F}$$
$$R_1 = 33 \ \text{k}\Omega \qquad R_L = 50 \ \Omega \qquad C_2 = 10 \ \mu\text{F}$$
$$R_2 = 10 \ \text{k}\Omega$$

Solution

1. $f_1(C_1) = \dfrac{1}{2\pi(100 + 25 \parallel 10 \times 10^3)(2.2 \times 10^{-6})} = 579 \ \text{Hz}$

   $f_1(C_2) = \dfrac{1}{2\pi(1 \times 10^3 \parallel 100 \times 10^3 + 15 \times 10^3)(1 \times 10^{-6})} = 9.9 \ \text{Hz}$

   Therefore, $f_1 = f_1(C_1) = 579 \ \text{Hz}$.

2. $r_{in}(\text{stage}) = (33 \ \text{k}\Omega) \parallel (10 \ \text{k}\Omega) \parallel (100[(25 \ \Omega) + (1 \ \text{k}\Omega) \parallel (50 \ \text{k}\Omega)]$

   $\qquad\qquad = (7.67 \ \text{k}\Omega) \parallel (7.26 \ \text{k}\Omega) = 3.73 \ \text{k}\Omega$

   $f_1(C_1) = \dfrac{1}{2\pi(100 + 3.73 \times 10^3)(1 \times 10^{-6})} = 41.6 \ \text{Hz}$

   $R_e = (1 \ \text{k}\Omega) \parallel \left[\dfrac{(100 \ \Omega) \parallel (33 \ \text{k}\Omega) \parallel (10 \ \text{k}\Omega)}{100} + (25 \ \Omega) \right] \approx 25 \ \Omega$

   $f_1(C_2) = \dfrac{1}{2\pi(25 + 50)(10 \times 10^{-6})} = 212.2 \ \text{Hz}$

   Therefore, $f_1 \approx f_1(C_2) = 212.2 \ \text{Hz}$.

Design Considerations

Capacitors in practical discrete circuits are generally bulky and costly, or else they are unreliable if quality is sacrificed for cost. Therefore, a principal objective

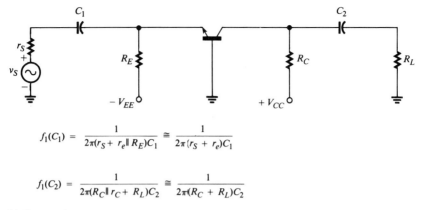

$$f_1(C_1) = \frac{1}{2\pi(r_S + r_e \| R_E)C_1} \cong \frac{1}{2\pi(r_S + r_e)C_1}$$

$$f_1(C_2) = \frac{1}{2\pi(R_C \| r_C + R_L)C_2} \cong \frac{1}{2\pi(R_C + R_L)C_2}$$

(a) Common base

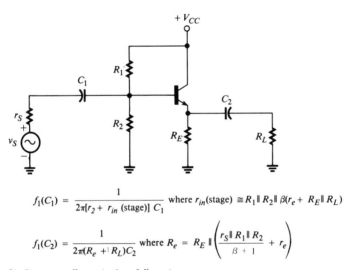

$$f_1(C_1) = \frac{1}{2\pi[r_2 + r_{in}\ (\text{stage})]\ C_1} \quad \text{where } r_{in}(\text{stage}) \cong R_1 \| R_2 \| \beta(r_e + R_E \| R_L)$$

$$f_1(C_2) = \frac{1}{2\pi(R_e + \cdot R_L)C_2} \quad \text{where } R_e = R_E \| \left(\frac{r_S \| R_1 \| R_2}{\beta + 1} + r_e \right)$$

(b) Common collector (emitter follower)

Figure 6.38
CB and CC amplifiers and the equations for $f_1(C_1)$ and $f_1(C_2)$ for each

in the design of discrete BJT amplifiers is selection of the smallest capacitors possible, consistent with the low-frequency response desired. Equations 6.51 through 6.53 can be used to find minimum capacitor values for CE, CB, and CC amplifiers in terms of break frequencies and circuit parameters. As noted earlier, each configuration has one capacitor whose value is the most critical: The one connected to the point where the equivalent resistance is smallest. This capacitor should be selected first to ensure that the lower cutoff frequency is at least as low as the break frequency it produces. The other capacitor(s) can then be selected to set break frequencies a decade or so below cutoff. (With two identical break frequencies occurring 1 decade below the third, the gain is actually reduced by a factor of 0.7, rather than 0.707, at the third frequency.)

Capacitor values necessary to achieve specified break frequencies in a common emitter (CE) amplifier

$$C_E{}^* = \frac{1}{2\pi R_e f_1(C_E)}$$

where $R_e = R_E \left\| \left(\frac{r_S \| R_1 \| R_2}{\beta + 1} + r_e \right) \right.$.

$$C_1 = \frac{1}{2\pi [r_{in}(\text{stage}) + r_S] f_1(C_1)}$$

where $r_{in}(\text{stage}) = R_1 \| R_2 \| \beta r_e$.

$$C_2 = \frac{1}{2\pi [r_o(\text{stage}) + R_L] f_1(C_2)}$$

where $r_o(\text{stage}) = r_o \| R_C$.

(6.51)

* Select first.

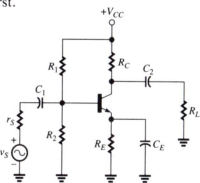

Capacitor values necessary to achieve specified break frequencies in a common base (CB) amplifier

$$C_1{}^* = \frac{1}{2\pi (r_S + r_e \| R_E) f_1(C_1)}$$

$$C_2 = \frac{1}{2\pi (R_C \| r_C + R_L) f_1(C_2)}$$

* Select first.

(6.52)

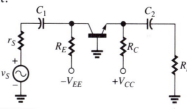

Capacitor values necessary to achieve specified break frequencies in a common collector (CC) amplifier

$$C_2{}^* = \frac{1}{2\pi(R_e + R_L)f_1(C_2)}$$

where $R_e = R_E \left\| \left(\dfrac{r_S \| R_1 \| R_2}{\beta + 1} + r_e\right)\right.$.

$$C_1 = \frac{1}{2\pi[r_{in}(\text{stage}) + r_S]f_1(C_1)}$$

where $r_{in}(\text{stage}) = R_1 \| R_2 \| \beta(r_e + R_E \| R_L)$.

* Select first.

(6.53)

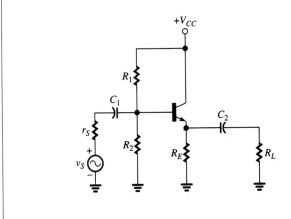

Example 6.19

A CE amplifier having $R_1 = 330$ kΩ, $R_2 = 47$ kΩ, $R_C = 3.3$ kΩ, and $R_E = 1.8$ kΩ is to have a lower cutoff frequency of 50 Hz. The amplifier drives a 10-kΩ load and is driven from a signal source whose resistance is 600 Ω. The transistor has $\beta = 90$ and $r_o = 100$ kΩ and is biased at $I_E = 0.5$ mA. Find coupling and bypass capacitors necessary to meet the low-frequency response requirement. Assume that capacitors must be selected from a line having standard values of 1 μF, 1.5 μF, 2.2 μF, 3.3 μF, 4.7 μF, 10 μF, 50 μF, 75 μF, and 100 μF.

Solution.   To obtain a value for C_E, we must first find the value of r_e:

$$r_e \approx \frac{0.026}{I_E} = \frac{0.026}{0.5 \text{ mA}} = 52 \ \Omega$$

Then, from equations 6.51,

$$R_e = 1.8 \text{ k}\Omega \left\| \left[\frac{(600 \ \Omega) \| (330 \text{ k}\Omega) \| (47 \text{ k}\Omega)}{91} + 52 \ \Omega \right] = 58.6 \ \Omega \right.$$

and

$$C_E = \frac{1}{2\pi(58.6\ \Omega)(50\ \text{Hz})} = 54.36\ \mu\text{F}$$

To ensure that the lower cutoff frequency is no greater than 50 Hz, we choose the standard-value capacitor with the next *higher* capacitance, rather than the one closest to 54.36 μF. Thus, we choose $C_E = 75\ \mu$F.

To find C_1 and C_2, we let $f_1(C_1) = f_1(C_2) = 50$ Hz/10 = 5 Hz and use equations 6.51 to calculate

$$r_{in}(\text{stage}) = 330\ \text{k}\Omega\ \|\ 47\ \text{k}\Omega\ \|\ (90)(52\ \Omega) = 4.2\ \text{k}\Omega$$

$$C_1 = \frac{1}{2\pi(4.2\ \text{k}\Omega\ +\ 600\ \Omega)5\ \text{Hz}} = 6.63\ \mu\text{F}$$

$$r_o(\text{stage}) = 100\ \text{k}\Omega\ \|\ 3.3\ \text{k}\Omega = 3.19\ \text{k}\Omega$$

$$C_2 = \frac{1}{2\pi(3.19\ \text{k}\Omega\ +\ 10\ \text{k}\Omega)5\ \text{Hz}} = 2.41\ \mu\text{F}$$

Again choosing standard capacitors with the next highest values, we select $C_1 = 10\ \mu$F and $C_2 = 3.3\ \mu$F.

It is a SPICE programming exercise at the end of this chapter to find the actual reduction in gain at 50 Hz when these standard-value capacitors are used. Our conservative choices result in an actual lower cutoff frequency of about 40 Hz.

High-Frequency Response of BJT Amplifiers

Figure 6.39 shows a common-emitter amplifier having interelectrode capacitance designated C_{be}, C_{bc}, and C_{ce}. Since we are now considering high-frequency performance, the emitter bypass capacitor effectively shorts the emitter terminal to ground, so C_{be} and C_{ce} are input-to-ground and output-to-ground capacities, respectively. We can apply the general equations developed earlier to determine the upper cutoff frequency due to the interelectrode capacitances:

$$f_2(C_A) = \frac{1}{2\pi(r_s\ \|\ r_{in})C_A} \tag{6.54}$$

where $C_A = C_{be} + C_M = C_{be} + C_{bc}(1 - A_v)$ and $r_{in} = r_{in}(\text{stage}) = R_1\ \|\ R_2\ \|\ (\beta r_e)$.

Figure 6.39
A common-emitter amplifier showing the interelectrode capacitances of the transistor

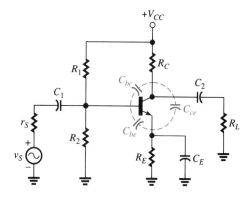

Note that the Miller-effect capacitance, C_M, is due to C_{bc} because the latter is connected between the input (base) and output (collector) of the common-emitter configuration. The cutoff frequency due to the output shunt capacitance is

$$f_2(C_B) = \frac{1}{2\pi(r_o \parallel R_L)C_B} \qquad (6.55)$$

where $C_B = C_{ce} + C_{bc}(1 - 1/A_v)$ and $r_o = r_o(\text{stage}) = r_C \parallel R_C$. Any wiring or stray capacitance that shunts the input or output must be added to C_A or C_B.

The high-frequency performance of a transistor is affected by certain internal characteristics that may or may not limit the upper cutoff to a frequency less than that predicted by equations 6.54 or 6.55. The β of a BJT is frequency dependent and its value begins to drop at the rate of 20 dB/decade at frequencies beyond a certain frequency called the *beta cutoff frequency,* f_β. By definition, f_β is the frequency at which the β of a transistor is $\sqrt{2}/2$ times its low-frequency value. Figure 6.40 shows a plot of β versus frequency as it would appear on log-log paper. The low- (or mid-) frequency value of β is designated β_m. If f_β is significantly lower than $f_2(C_A)$ and $f_2(C_B)$, then the upper cutoff frequency of an amplifier in the common-emitter configuration will be determined by f_β.

If the frequency is increased beyond f_β, β continues to decrease until it eventually reaches a value of 1. As shown in Figure 6.40, the frequency at which $\beta = 1$ is designated f_T. Note that the asymptote's slope of -20 dB/decade is the same as a slope of -1 on log-log scales, since 20 dB is a ten-to-one change along the vertical axis and a decade is a ten-to-one change along the horizontal axis. Therefore, in Figure 6.40, the difference between the logarithms of β_m and 1 must be the same as the difference between the logarithms of f_T and f_β:

$$\log \beta_m - \log(1) = \log f_T - \log f_\beta$$
$$\log(\beta_m/1) = \log(f_T/f_\beta)$$

Taking the antilog of both sides leads to the important result

$$f_\beta = \frac{f_T}{\beta_m} \qquad (6.56)$$

Since β decreases with frequency, we would expect α to do the same. This is, in fact, the case, and the frequency at which α falls to $\sqrt{2}/2$ times its low-frequency value is called the *alpha cutoff frequency,* f_α. Because f_α is generally much larger than f_β, the common-base amplifier is not as frequency limited by parameter cutoff as is the common-emitter amplifier. A commonly used approximation is

$$f_\alpha \approx f_T \qquad (6.57)$$

Figure 6.40
The β of a transistor becomes smaller at high frequencies. At frequencies above f_β, its value falls off at the rate of 20 dB/ decade, and reaches value 1 at the frequency designated f_T.

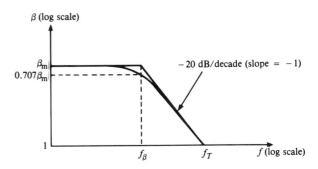

In specification sheets, f_T is often called the *gain-bandwidth product*. We will discuss this interpretation of a unity-gain frequency in Chapter 8. In terms of internal junction capacitances, f_T can be found from

$$f_T = \frac{1}{2\pi r_e (C_\pi + C_\mu)}$$

where C_π is the base-to-emitter junction capacitance and C_μ is the base-to-collector junction capacitance. Recall that junction capacitance depends on the width of the depletion region, which in turn depends on the bias voltage. Since r_e also depends on bias conditions, f_T is clearly a function of bias and is usually so indicated on specification sheets.

Example 6.20

The transistor in Figure 6.41 has a low-frequency β of 120, $r_e = 20\ \Omega$, $r_o = 100\ \text{k}\Omega$, and $f_T = 100$ MHz. The interelectrode capacitances are $C_{be} = 40$ pF, $C_{bc} = 1.5$ pF, and $C_{ce} = 5$ pF. There is wiring capacitance equal to 4 pF across the input and 8 pF across the output. Find the approximate upper cutoff frequency.

Solution.   To determine the Miller-effect capacitance, we must find the midband voltage gain of the transistor. Note that the *overall* voltage gain v_L/v_S is *not* used in this computation, because the gain reduction due to voltage division at the input does not affect the Miller capacitance. (It would, if the feedback capacitance were connected all the way back to v_S.)

$$A_v \approx \frac{-r_L}{r_e} = \frac{-(4\ \text{k}\Omega)\ \|\ (20\ \text{k}\Omega)}{20\ \Omega} = -166.67$$

We use the midband gain as a conservative approximation in determining C_M:

$$C_M \approx C_{bc}(1 - A_v) = (1.5\ \text{pF})[1 - (-166.67)] = 251.5\ \text{pF}$$

The total capacitance shunting the input is, therefore,

$$C_A = C_{be} + C_M + C_{wiring} = (40\ \text{pF}) + (251.5\ \text{pF}) + (4\ \text{pF}) = 295.5\ \text{pF}$$

Then, since

$$r_{in}(\text{stage}) = R_1\ \|\ R_2\ \|\ \beta r_e = (100\ \text{k}\Omega)\ \|\ (22\ \text{k}\Omega)\ \|\ (2.4\ \text{k}\Omega) = 2.1\ \text{k}\Omega$$

Figure 6.41
(Example 6.20)

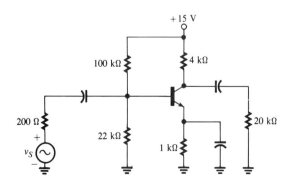

we have, from equation 6.54,

$$f_2(C_A) = \frac{1}{2\pi(200 \parallel 2.1 \times 10^3)295.5 \times 10^{-12}} = 2.95 \text{ MHz}$$

Since $r_o(\text{stage}) = r_o \parallel R_C = (100 \text{ k}\Omega) \parallel (4 \text{ k}\Omega) = 3.8 \text{ k}\Omega$

and
$$\begin{aligned} C_B &= C_{ce} + C_{bc}(1 - 1/A_v) + C_{wiring} \\ &= (5 \text{ pF}) + (1.5 \text{ pF})(1 + 1/167) + (8 \text{ pF}) = 14.5 \text{ pF} \end{aligned}$$

we have, from equation 6.55,

$$f_2(C_B) = \frac{1}{2\pi(3.8 \times 10^3 \parallel 20 \times 10^3)(14.5 \times 10^{-12})} = 3.44 \text{ MHz}$$

From equation 6.56,

$$f_\beta = \frac{f_T}{\beta_m} = \frac{100 \text{ MHz}}{120} = 833 \text{ kHz}$$

The upper cutoff frequency is the smallest of the three computed frequencies, in this case, f_β: $f_2 = f_\beta = 833$ kHz.

Since the common-base amplifier is noninverting, the shunt capacitance at its input does not have a Miller-effect component. Therefore, the CB amplifier is not as severely limited as its CE counterpart in terms of the upper cutoff frequency $f_2(C_A)$, and CB amplifiers are often used in high-frequency applications. We have also noted that f_α is generally much higher than f_β. The common-collector amplifier (emitter follower) is also noninverting and unaffected by Miller capacitance. It, too, is generally superior to the CE configuration for high-frequency applications.

6.8   FREQUENCY RESPONSE OF FET AMPLIFIERS

Low-Frequency Response of FET Amplifiers

Figure 6.42 shows a common-source JFET amplifier biased using the combination of self-bias and a voltage divider, as studied in Chapter 5. For our purposes now, the FET could also be a MOSFET. The capacitors that affect the low-frequency response are the input coupling capacitor, C_1, the output coupling capacitor, C_2, and the source bypass capacitor, C_S. Applying equations 6.48 and 6.49 to this amplifier configuration, we find

$$f_1(C_1) = \frac{1}{2\pi(r_S + r_{in})C_1} \qquad (6.58)$$

where $r_{in} = r_{in}(\text{stage}) \approx R_1 \parallel R_2$, and

$$f_1(C_2) = \frac{1}{2\pi(r_o + R_L)C_2} \qquad (6.59)$$

where $r_o = r_o(\text{stage}) = r_d \parallel R_D$. The bypass capacitor C_S affects low-frequency response because at low frequencies its reactance is no longer small enough to eliminate degeneration. The cutoff frequency due to C_S is the frequency at which

Figure 6.42
The low-frequency response of an FET amplifier is affected by capacitors C_1, C_2, and C_S.

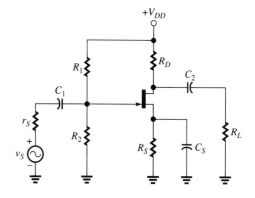

the reactance of C_S equals the resistance looking into the junction of R_S and C_S at the source terminal, namely, $R_S \parallel (1/g_m)$. Thus,

$$f_1(C_S) = \frac{1}{2\pi[R_S \parallel (1/g_m)]C_S} \tag{6.60}$$

Since the bias resistors R_1 and R_2 are usually very large in an FET amplifier, the input coupling capacitor C_1 can be smaller than its counterpart in a BJT CE amplifier, and C_1 does not generally determine the low-frequency cutoff. The small value of $R_S \parallel (1/g_m)$ makes the source bypass capacitor the most troublesome, in that a large amount of capacitance is required to achieve a small value of $f_1(C_S)$.

Example 6.21

The FET shown in Figure 6.43 has $g_m = 3.4$ mS and $r_d = 100$ kΩ. Find the approximate lower cutoff frequency.

Solution

$$r_{in}(\text{stage}) \approx (1.5 \text{ M}\Omega) \parallel (330 \text{ k}\Omega) = 270 \text{ k}\Omega$$

$$f_1(C_1) = \frac{1}{2\pi(20 \times 10^3 + 270 \times 10^3)(0.02 \times 10^{-6})} = 27.4 \text{ Hz}$$

$$r_o(\text{stage}) = (100 \text{ k}\Omega) \parallel (2 \text{ k}\Omega) = 1.96 \text{ k}\Omega$$

Figure 6.43
(Example 6.21)

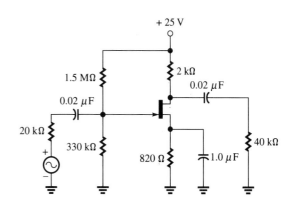

$$f_1(C_1) = \frac{1}{2\pi(1.96 \times 10^3 + 40 \times 10^3)(0.02 \times 10^{-6})} = 189.6 \text{ Hz}$$

$$R_S \parallel (1/g_m) = 820 \parallel 294.1 = 216.5 \ \Omega$$

$$f_1(C_S) = \frac{1}{2\pi(216.5)(1 \times 10^{-6})} = 735.1 \text{ Hz}$$

The actual lower cutoff frequency is approximately equal to the largest of $f_1(C_1)$, $f_1(C_2)$, and $f_1(C_S)$, namely, $f_1 \approx f_1(C_S) = 735.1$ Hz.

Figure 6.44 shows JFET and MOSFET amplifiers in the common-drain (source-follower) configuration and gives the equations for $f_1(C_1)$ and $f_1(C_2)$. These equations are obtained in a straightforward manner using the same low-frequency–cutoff theory we have discussed for general amplifiers.

For design purposes, each of the equations for $f_1(C_1)$, $f_1(C_2)$, or $f_1(C_S)$ in each of the three configurations is easily solved for C_1, C_2, or C_S, giving a set of capacitor equations similar to equations 6.51 through 6.53.

High-Frequency Response of FET Amplifiers

Like the BJT common-emitter amplifier, the FET common-source amplifier is affected by Miller capacitance that often determines the upper cutoff frequency. Figure 6.45 shows a common-source amplifier and the interelectrode capacitance that affects its high-frequency performance. Since R_S is completely bypassed at high frequencies, C_{gs} and C_{ds} are effectively shunting the input and output, respectively, to ground. The equations for $f_2(C_A)$ and $f_2(C_B)$ are obtained directly from the general high-frequency–cutoff theory we have studied:

$$f_2(C_A) = \frac{1}{2\pi(r_S \parallel r_{in})C_A} \tag{6.61}$$

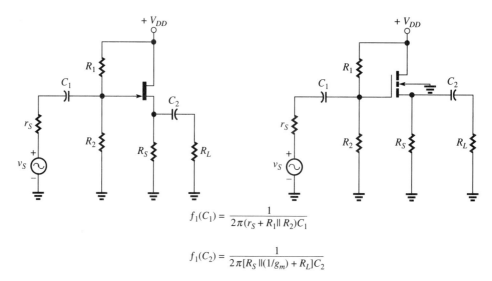

$$f_1(C_1) = \frac{1}{2\pi(r_S + R_1 \parallel R_2)C_1}$$

$$f_1(C_2) = \frac{1}{2\pi[R_S \parallel(1/g_m) + R_L]C_2}$$

Figure 6.44
Low-frequency–cutoff equations for the common-drain amplifier

Figure 6.45
Common-source amplifier with interelec-
trode capacitance that affects high-
frequency response

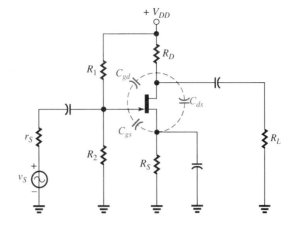

where $C_A = C_{gs} + C_M = C_{gs} + C_{gd}(1 - A_v)$ and $r_{in} = r_{in}(\text{stage}) \approx R_1 \parallel R_2$, and

$$f_2(C_B) = \frac{1}{2\pi(r_o \parallel R_L)C_B} \qquad (6.62)$$

where $C_B = C_{ds} + C_{gd}(1 - 1/A_v)$ and $r_o = r_o(\text{stage}) = r_d \parallel R_D$.

Example 6.22

The JFET in Figure 6.45 has $g_m = 3.2$ mS, $r_d = 100$ kΩ, $C_{gs} = 4$ pF, $C_{ds} = 0.5$ pF, and $C_{gd} = 1.2$ pF. The wiring capacitance shunting the input is 2.5 pF and that shunting the output is 4 pF. The component values in the circuit are

$$
\begin{aligned}
r_S &= 20 \text{ k}\Omega & R_D &= 2 \text{ k}\Omega \\
R_1 &= 1.5 \text{ M}\Omega & R_S &= 820 \text{ }\Omega \\
R_2 &= 330 \text{ k}\Omega & R_L &= 40 \text{ k}\Omega
\end{aligned}
$$

Find the approximate upper cutoff frequency.

Solution

$$A_v = -g_m(r_d \parallel R_D \parallel R_L) = -3.2 \times 10^{-3}[(100 \text{ k}\Omega) \parallel (2 \text{ k}\Omega) \parallel (40 \text{ k}\Omega)] = -6$$
$$C_M = (1.2 \text{ pF})[1 - (-6)] = 8.4 \text{ pF}$$
$$C_A = C_{gs} + C_M + C_{(wiring)} = (4 \text{ pF}) + (8.4 \text{ pF}) + (2.5 \text{ pF}) = 14.9 \text{ pF}$$
$$r_{in}(\text{stage}) = (1.5 \text{ M}\Omega) \parallel (330 \text{ k}\Omega) = 270 \text{ k}\Omega$$
$$f_2(C_A) = \frac{1}{2\pi(20 \times 10^3 \parallel 270 \times 10^3)(14.9 \times 10^{-12})} = 573.6 \text{ kHz}$$
$$
\begin{aligned}
C_B &= C_{ds} + C_{gd}(1 - 1/A_v) + C_{(wiring)} \\
&= (0.5 \text{ pF}) + (1.2 \text{ pF})(1 + 1/6) + (4 \text{ pF}) = 5.9 \text{ pF}
\end{aligned}
$$
$$r_o(\text{stage}) = (100 \text{ k}\Omega) \parallel (2 \text{ k}\Omega) = 1.96 \text{ k}\Omega$$
$$f_2(C_B) = \frac{1}{2\pi(1.96 \times 10^3 \parallel 40 \times 10^3)(5.9 \times 10^{-12})} = 14.43 \text{ MHz}$$

Therefore, $f_2 = f_2(C_A) = 573.6$ kHz.

Example 6.23

PSPICE

Use PSpice and Probe to determine the lower and upper cutoff frequencies of the MOSFET amplifier shown in Figure 6.46(a). The capacitances C_{gs}, C_{gd}, and C_{ds} shown in the figure represent stray capacitance and should be included in the input circuit file. The MOSFET transistor has VTO = 2 and KP = 0.5E-3. Let other parameters default.

Solution.   The PSpice circuit and input circuit file are shown in Figure 6.46(b). A plot of the voltage across the load resistance, V(7), as displayed by Probe, is shown in Figure 6.46(c). We see that the maximum load voltage is 2.5 V. Therefore, the lower and upper cutoff frequencies occur where the load voltage is 0.707(2.5 V) = 1.767 V. As shown in the Probe display, cursor C1 is moved to a point on the plot where the voltage equals 1.7672 V, and we see that the lower cutoff frequency is 1.4342 kHz. C2 is moved to 1.7667 V at the high-frequency end of the plot, and we see that the upper cutoff frequency is 8.0475 MHz.

Figure 6.46
(Example 6.23)

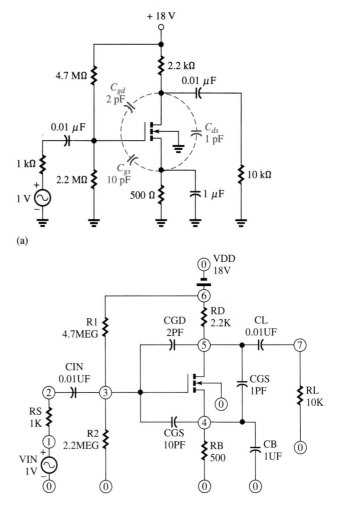

(a)

Figure 6.46
(Continued)

EXAMPLE 6.23
VIN 1 0 AC 1V
VDD 6 0 18V
RS 1 2 1K
CIN 2 3 0.01UF
R1 6 3 4.7MEG
R2 3 0 2.2MEG
RD 5 6 2.2K
RB 4 0 500
CGD 3 5 2PF
CGS 3 4 10PF
CDS 5 4 1PF
CB 4 0 1UF
CL 5 7 0.01UF
RL 7 0 10K
M1 5 3 4 0 MOSFET
.MODEL MOSFET NMOS VTO=2 KP=0.5E-3
.AC DEC 10 1K 10MEG
.END

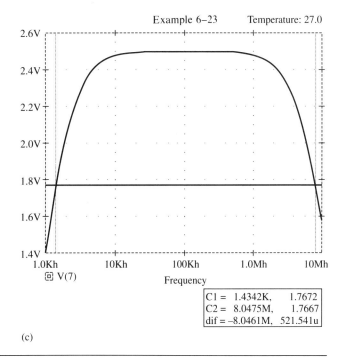

(c)

EXERCISES

Note: All references to rms values in the exercises that follow refer to the *ac* components of amplifier inputs and outputs.

Section 6.1 Definitions and Basic Concepts

6.1 The lower cutoff frequency of a certain amplifier is 120 kHz and its upper cutoff frequency is 1 MHz. The peak-to-peak output voltage in the midband frequency range is 2.4 V p–p and the output power at 120 kHz is 0.4 W. Find

a. the bandwidth,
b. the rms output voltage at 1 MHz,
c. the output power in the midband frequency range, and
d. the output power at 1 MHz.

6.2 The voltage at the input of an amplifier is 15 mV rms. The amplifier delivers 0.02 A rms to a 12-Ω load at 1 kHz. The input resistance of the amplifier is 1400 Ω. Its lower cutoff frequency is 50 Hz and its bandwidth is 9.95 kHz. Find
  a. the upper cutoff frequency, f_2;
  b. the output power at 50 Hz;
  c. the midband power gain; and
  d. the power gain at f_2.

6.3 The input signal to a certain amplifier has a 1 kHz component with amplitude 0.2 V rms and a 4-kHz component with amplitude 0.05 V rms. In the amplifier's output, the 1-kHz component has amplitude 0.5 V rms and is delayed with respect to the 1-kHz input component by 0.25 ms. If the output is to be an undistorted version of the input, what should be the amplitude and delay of the 4-kHz output component?

Section 6.2 Decibels and Logarithmic Plots

6.4 Find the power gain P_2/P_1 in dB corresponding to each of the following:
  a. $P_1 = 4$ mW, $P_2 = 1$ W;
  b. $P_1 = 0.2$ W, $P_2 = 80$ mW;
  c. $P_1 = 5000$ μW, $P_2 = 5$ mW; and
  d. $P_1 = 0.4 P_2$.

6.5 A certain amplifier has a power gain of 42 dB. The output power is 16 W and the input resistance is 1 kΩ. What is the rms input voltage?

6.6 Find the voltage gain v_2/v_1 in dB corresponding to each of the following:
  a. $v_1 = 120$ μV rms, $v_2 = 40$ mV rms;
  b. $v_1 = 1.8$ V rms, $v_2 = 18$ V peak-to-peak;
  c. $v_1 = 0.707$ V rms, $v_2 = 1$ V rms;
  d. $v_1 = 5 \times 10^3 v_2$; and
  e. $v_1 = v_2$.

6.7 The amplifier shown in Figure 6.47 has $v_S = 30$ mV rms and $v_L = 1$ V rms. Find

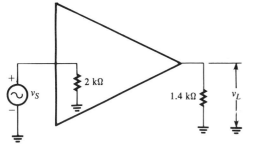

Figure 6.47
(Exercise 6.7)

  a. the power gain in decibels; and
  b. the voltage gain in decibels.

6.8 Repeat Exercise 6.7 if the load resistor is 2 kΩ.

6.9 The voltage gain of an amplifier is 18 dB. If the output voltage is 6.8 V rms, and the input and load resistances both equal 800 Ω, find
  a. the input voltage; and
  b. the power gain in decibels.

6.10 The output voltage of an amplifier is 12.5 dB above its level at point "A" in the amplifier. The amplifier input voltage is 6.4 dB below the voltage at point "A." If the output voltage is 2.4 V rms, find
  a. the voltage at point A;
  b. the input voltage; and
  c. the amplifier voltage gain, in decibels.

6.11 The input voltage of a certain amplifier is 0.36 V rms. Without performing any computations, estimate the output voltage that would correspond to each of the following gains: (a) 20 dB, (b) 46 dB, (c) −6 dB, and (d) 60 dB.

6.12 An amplifier has output power 28 dBm and output voltage 34 dBV. Find
  a. its output power in watts; and
  b. its output voltage in V rms.

6.13 In frequency response measurements made on an amplifier, the gain was found to vary from 0.06 to 75 over a frequency range from 7 Hz to 3.3 kHz. What kind of log-log graph paper would be required to plot this data (how many cycles on each axis)?

6.14 The following frequency response measurements were made on an amplifier:

Frequency	Voltage Gain	Phase Shift
2 Hz	4.0	84.3°
4 Hz	7.8	78.6°
8 Hz	14.9	68.2°
15 Hz	24.0	53.1°
30 Hz	33.3	33.7°
40 Hz	35.8	14.0°
100 Hz	39.2	6.0°
200 Hz	39.8	3.4°
500 Hz	40.0	−3.4°
1 kHz	39.2	−11.3°
2 kHz	37.1	−21.8°
4 kHz	31.2	−38.6°
8 kHz	21.2	−58.0°
10 kHz	17.9	−63.4°

Plot the gain data on log-log graph paper and the phase data on semilog graph paper. Use your plots to estimate the following:

a. the lower and upper cutoff frequencies;

b. the (positive) phase shift when the gain is 20;

c. the gain when the phase shift is −30°; and

d. the gain 2 octaves below the lower cutoff frequency.

6.15 The following frequency response measurements were made on an amplifier:

Frequency	Voltage Gain	Phase Shift
150 Hz	23.2 dB	89.6°
200 Hz	25.4 dB	68.2°
400 Hz	29.9 dB	51.3°
600 Hz	31.7 dB	39.8°
800 Hz	32.6 dB	29.0°
1 kHz	33.0 dB	22.8°
2 kHz	33.7 dB	6.4°
5 kHz	34.0 dB	−12.7°
8 kHz	32.9 dB	−28.1°
10 kHz	32.4 dB	−33.7°
12 kHz	31.9 dB	−38.7°
18 kHz	30.1 dB	−50.2°
20 kHz	29.6 dB	−53.1°
40 kHz	24.9 dB	−69.4°
100 kHz	17.4 dB	−81.5°
200 kHz	11.5 dB	−85.7°
500 kHz	3.5 dB	−88.3°
1 MHz	−2.5 dB	−89.1°

Plot the gain and phase data on semilog graph paper and use your plots to estimate the following:

a. the lower and upper cutoff frequencies;

b. the magnitude of the output voltage when the phase shift is −60° (the input voltage is 20 mV rms);

c. the (positive) phase shift when the input voltage is 15 mV rms and the output voltage is 0.336 V rms; and

d. the magnitude of the voltage gain 1 decade above the upper cutoff frequency.

Section 6.3 Series Capacitance and Low-Frequency Response

6.16 Show that the lower cutoff frequency $f_1(C_1)$ in Figure 6.8 is the frequency at which the capacitive reactance X_{C_1} equals the resistance r_{in}.

6.17 The amplifier shown in Figure 6.48 has midband voltage gain $|v_L|/|v_S|$ equal to 80. Calculate

a. the lower cutoff frequency;

b. the gain $|v_L|/|v_S|$ at 300 Hz;

c. the phase shift 1 decade below cutoff; and

d. the frequency at which the gain $|v_L|/|v_S|$ is 10 dB down from its midband value.

6.18. Repeat Exercise 6.17 using the normalized plots in Figure 6.10 for (b), (c), and (d).

6.19. The amplifier shown in Figure 6.49 has voltage gain $|v_L|/|v_{in}|$ equal to 200. Calculate

a. the lower cutoff frequency;

b. the frequency at which v_L leads v_S by 75°; and

c. the voltage gain $|v_L|/|v_S|$ 2 octaves above cutoff.

6.20 The amplifier shown in Figure 6.50 has voltage gain $|v_L|/|v_{in}|$ equal to 180. What should be the value of the coupling capacitor C in order that the gain $|v_L|/|v_S|$ be no less than 100 at 20 Hz?

6.21 The amplifier shown in Figure 6.51 has midband gain $|v_L|/|v_S|$ equal to 160. Find the (approximate) values of

a. the lower cutoff frequency;

b. the gain $|v_o|/|v_{in}|$; and

c. a value for the input coupling capacitor that would cause the asymptotic gain plot to have a single break frequency.

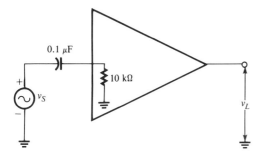

Figure 6.48
(Exercise 6.17)

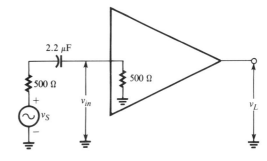

Figure 6.49
(Exercise 6.19)

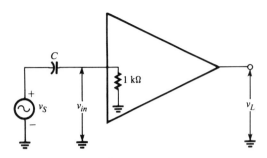

Figure 6.50
(Exercise 6.20)

6.22 The amplifier shown in Figure 6.52 has a very large (essentially infinite) input impedance and a very small (essentially 0) output impedance. What should be the gain $|v_o|/|v_{in}|$ if it is desired to have $|v_L| = 1$ V rms when $|v_S| = 20$ mV rms at 100 Hz?

6.23 The amplifier shown in Figure 6.53 has midband gain $|v_L|/|v_S| = 100$. Calculate
a. the approximate lower cutoff frequency;
b. the approximate gain $|v_L|/|v_S|$, in decibels, at $f = 120$ Hz; and
c. the approximate gain $|v_L|/|v_S|$, in decibels, at $f = 77.6$ Hz.
   Use Figure 6.14 to find
d. the phase shift 1 octave below the cutoff frequency; and
e. the frequency at which v_L leads v_S by 160°.

6.24 Find values for the input and output coupling capacitors in Exercise 6.23 that will result in a single break frequency and a lower cutoff frequency at 40 Hz.

Section 6.4 Shunt Capacitance and High-Frequency Response

6.25 The amplifier shown in Figure 6.54 has midband gain $|v_L|/|v_S| = 150$. Calculate

a. the upper cutoff frequency;
b. $|v_L|/|v_{in}|$ at midband;
c. $|v_L|/|v_S|$ at 10 MHz; and
d. the phase angle of the output with respect to v_S, at 10 MHz.

6.26 For the amplifier of Exercise 6.25, use Figure 6.17 to find approximate values of
a. $|v_L|/|v_S|$ 1 octave above cutoff;
b. the frequency at which $|v_L|/|v_S| = 75$; and
c. the frequency at which v_L lags v_S by 60°.

6.27 What is the maximum permissible value of the source resistance for the amplifier of Exercise 6.25 if the upper cutoff frequency must be at least 1 MHz?

6.28 Design a low-pass RC network, consisting of a single resistor and a single capacitor, whose output voltage is 0.5 V rms when the input is a 1-V-rms signal having frequency 120 kHz. Draw the schematic diagram, label input and output terminals, and show the component values of your design.

6.29 The amplifier shown in Figure 6.55 has midband gain $|v_L|/|v_S| = 84$.
a. Find the approximate upper cutoff frequency.
b. Find the midband value of $|v_o|/|v_{in}|$.
c. Sketch the asymptotic Bode plot of $|v_L|/|v_S|$ as it would appear if plotted on log-log graph paper. (It is not necessary to use log-log graph paper for your sketch.) Label important frequency and gain coordinates on your sketch.

6.30 a. What is the approximate upper cutoff frequency of the amplifier shown in Figure 6.56?
b. What value of R_L would result in the upper cutoff frequency being equal to 64.5% of that calculated in (a)?

6.31 The amplifier shown in Figure 6.57 has midband gain $|v_o|/|v_{in}|$ equal to -160. Find the approximate upper cutoff frequency.

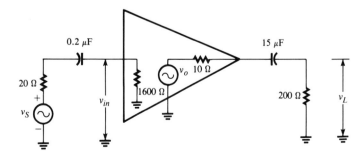

Figure 6.51
(Exercise 6.21)

Figure 6.52
(Exercise 6.22)

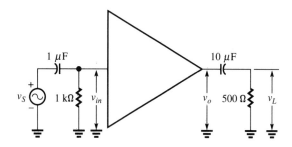

Figure 6.53
(Exercise 6.23)

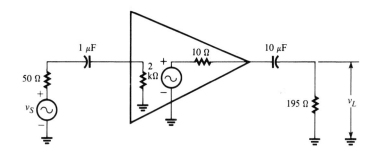

Figure 6.54
(Exercise 6.25)

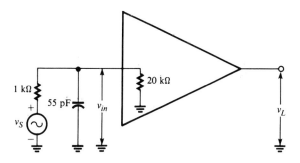

Figure 6.55
(Exercise 6.29)

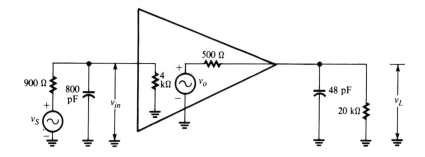

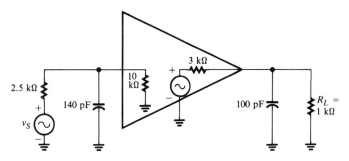

Figure 6.56
(Exercise 6.30)

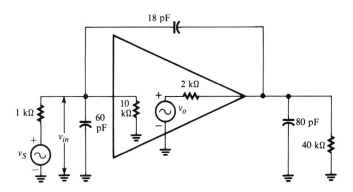

Figure 6.57
(Exercise 6.31)

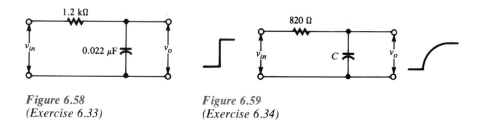

Figure 6.58   **Figure 6.59**
(Exercise 6.33)   **(Exercise 6.34)**

6.32   For what value of amplifier gain $|v_o|/|v_{in}|$ in
Exercise 6.31 would the frequency re-
sponse have a single break frequency?

Section 6.5   Transient Response

6.33   a. What is the bandwidth of the low-pass
network shown in Figure 6.58?
   b. What is the rise time at the output when
the input is a step voltage?
   c. How would the bandwidth be affected
by an increase in capacitance? In resistance?

d. How would the rise time be affected by
an increase in capacitance? In resistance?

6.34   The rise time of the output of the network
shown in Figure 6.59 is 2.7 μs. What is the
value of C?

Section 6.6   Frequency Response of
Multistage Amplifiers

6.35   The in-circuit voltage gains of the stages in
a multistage amplifier are shown in Figure
6.60. Find

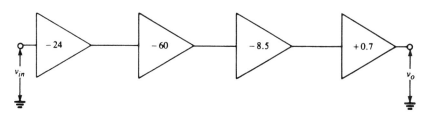

Figure 6.60
(Exercises 6.35 and 6.36)

a. the overall voltage gain, v_o/v_{in}; and
b. the voltage gain that would be necessary in a fifth stage, which, if added to the cascade, would make the overall voltage gain 100 dB.

6.36 The in-circuit voltage gains of the stages in a multistage amplifier are shown in Figure 6.60 (The gain of the first stage does not include loading by the signal source and that of the fourth stage does not include loading by a load resistor.) The input resistance to the first stage is 20 kΩ and the output resistance of the fourth stage is 20 Ω. The amplifier is driven by a signal source having resistance 25 kΩ and a 12-Ω load is connected to the output of the fourth stage. If the source voltage is $v_s = 5$ mV rms, find
a. the load voltage, v_L;
b. the power gain in dB, between the input to the first stage and the load.

6.37 It is desired to construct a three-stage amplifier whose overall voltage gain is 53.98 dB. The in-circuit voltage gains of the first two stages are to be equal, and the voltage gain of the third stage is to be one-half that of each of the first two. What should be the voltage gain of each stage?

6.38 The open-circuit (unloaded) voltage gains of the stages in the multistage amplifier shown in Figure 6.61 are $A_{o1} = -42$, $A_{o2} = -26$, and $A_{o3} = 1.8$. Find the overall voltage gain v_L/v_S.

6.39 A multistage amplifier is to be constructed using four identical stages, each of which has lower cutoff frequency 15 Hz and upper cutoff frequency 30 kHz.
a. What will be the lower and upper cutoff frequencies of the multistage amplifier?
b. If the midband, in-circuit voltage gain of each stage is 8.2, what will be the asymptotic voltage gain of the multistage amplifier at 7.5 Hz? At 300 kHz?

6.40 A multistage amplifier consists of three identical stages, each of which has bandwidth 250 kHz. The bandwidth of the multistage amplifier is 40 kHz. What are the lower and upper cutoff frequencies of each stage?

6.41 A multistage amplifier is to be constructed using six identical stages, each of which has lower cutoff frequency 50 kHz and upper cutoff frequency 1 MHz. What will be the bandwidth of the multistage amplifier?

6.42 The in-circuit lower and upper cutoff fre-

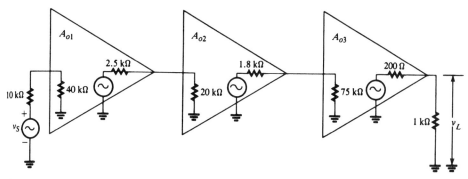

Figure 6.61
(Exercise 6.38)

quencies of each stage of a three-stage amplifier are as follows:

	Stage 1	Stage 2	Stage 3
Lower	75 Hz	5 Hz	1 kHz
Upper	25 kHz	125 kHz	1 MHz

a. Sketch the overall asymptotic frequency response of the amplifier as it would appear if plotted on log-log axes. Label break frequencies and the rates of attenuation in dB/decade.

b. What are the approximate lower and upper cutoff frequencies of the amplifier? What is its bandwidth?

c. Should this amplifier be used as a high-fidelity audio amplifier? Explain.

Section 6.7 Frequency Response of BJT Amplifiers

6.43 The transistor shown in Figure 6.62 has $\beta = 90$, $r_e = 15\ \Omega$, and $r_o = 125\ k\Omega$. Find the approximate lower cutoff frequency.

6.44 The transistor shown in Figure 6.63 has $r_e = 10\ \Omega$, $\beta = 100$, and $r_o = 100\ k\Omega$. Find the approximate lower cutoff frequency.

6.45 The transistor shown in Figure 6.64 has $r_C = 500\ k\Omega$ and $r_e = 18\ \Omega$. Find the approximate lower cutoff frequency.

6.46 The transistor shown in Figure 6.65 has $r_e = 20\ \Omega$ and $\beta = 150$. Find the approximate lower cutoff frequency.

6.47 The transistor shown in Figure 6.66 has $r_e = 25\ \Omega$, $r_o = 50\ k\Omega$, $f_\beta = 0.8$ MHz, and a midband β equal to 100. Find the approximate upper cutoff frequency.

Figure 6.62
(Exercise 6.43)

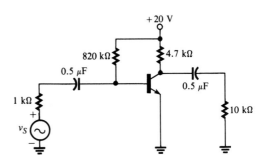

Figure 6.63
(Exercise 6.44)

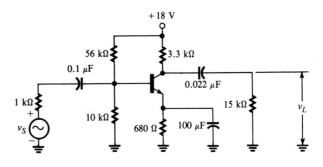

Figure 6.64
(Exercise 6.45)

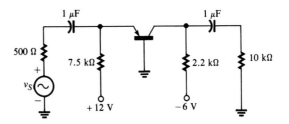

Figure 6.65
(Exercise 6.46)

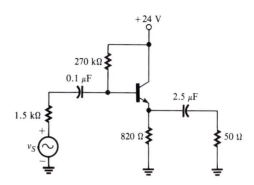

Figure 6.66
(Exercise 6.47)

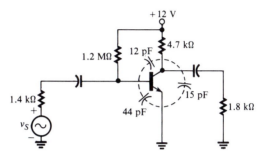

Figure 6.67
(Exercise 6.50)

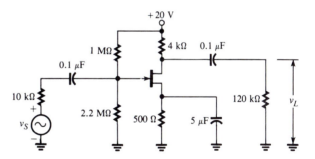

6.48 Repeat Exercise 6.47 using the following parameters: $r_e = 25\ \Omega$, $r_o = 50\ k\Omega$, $f_T = 80$ MHz, $\beta_{(midband)} = 100$, $C_{bc} = 2$ pF, $C_{be} = 22$ pF, $C_{ce} = 15$ pF.

6.49 How much input-to-ground wiring capacitance could be tolerated in the amplifier of Exercise 6.47 without the upper cutoff frequency falling below 100 kHz?

Section 6.8 Frequency Response of FET Amplifiers

6.50 The JFET shown in Figure 6.67 has $g_m = 4 \times 10^{-3}$ S and $r_d = 100\ k\Omega$.
  a. Find the approximate lower cutoff frequency.

  b. Sketch asymptotic gain versus frequency as it would appear if plotted on log-log graph paper. Label the break frequencies on your sketch. It is not necessary to show actual gain values, but indicate the rate of gain reduction corresponding to each asymptote.

6.51 The MOSFET shown in Figure 6.68 has $g_m = 3 \times 10^{-3}$ S and $r_d = 75\ k\Omega$. Find the approximate lower cutoff frequency.

6.52 What is the minimum value of R_L in Exercise 6.51 that could be used without the lower cutoff frequency being greater than 10 Hz?

6.53 The MOSFET shown in Figure 6.69 has $g_m = 4000\ \mu$S and $r_d = 60\ k\Omega$. Find the approximate upper cutoff frequency.

Figure 6.68
(Exercise 6.51)

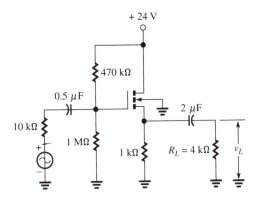

Figure 6.69
(Exercise 6.53)

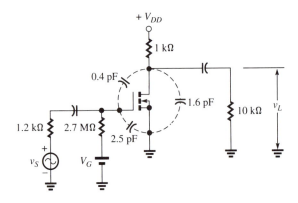

6.54   What is the maximum input wiring capacitance (to ground) that could be tolerated in the amplifier of Exercise 6.53 if the upper cutoff frequency must be at least 1 MHz?

DESIGN EXERCISES

6.55   Following the design procedure given in Section 6.7 (Example 6.19), redesign the amplifier in Exercise 6.44 (i.e., find new capacitor values) so that the lower cutoff frequency is approximately 250 Hz. Assume that capacitors must be selected from a line having standard values 0.1 μF, 0.2 μF, 0.4 μF, 1 μF, 2 μF, 4 μF, 10 μF, 20 μF, 40 μF, and 100 μF.

6.56   The common-base amplifier in Figure 6.38(a) has $r_S = 100 \ \Omega$, $R_E = 1 \ \text{k}\Omega$, $R_C = 4.7 \ \text{k}\Omega$, $R_L = 10 \ \text{k}\Omega$, $V_{EE} = -2$ V, and $V_{CC} = 15$ V. Select capacitor values so that the lower cutoff frequency is approximately 250 Hz. Assume that capacitors must be selected from a line having standard values 0.1 μF,

0.5 μF, 1 μF, 5 μF, 7.5 μF, 10 μF, 15 μF, 50 μF, and 100 μF.

6.57   The emitter follower in Figure 6.38(b) has $r_S = 600 \ \Omega$, $R_1 = 220 \ \text{k}\Omega$, $R_2 = 47 \ \text{k}\Omega$, $R_E = 1 \ \text{k}\Omega$, $R_L = 500 \ \Omega$, and $V_{CC} = 15$ V. The β of the transistor is 100.
a.   Find capacitor values so that the lower cutoff frequency is approximately 50 Hz. Assume that capacitors must be selected from a line having standard values 1 μF, 1.5 μF, 2.2 μF, 3.3 μF, 4.7 μF, 10 μF, 15 μF, 22 μF, and 33 μF.
b.   Find the approximate lower cutoff frequency when the standard-value capacitors are used.
c.   Find the approximate lower cutoff frequency when the standard-value capacitors are used and the β of the transistor changes to 200.

SPICE EXERCISES

6.58   The amplifier in Figure 6.34 has $r_S = 1$ kΩ, $R_1 = 100$ kΩ, $R_2 = 22$ kΩ, $R_C = 3$ kΩ, $R_E = $

2 kΩ, R_L = 20 kΩ, C_1 = 0.22 μF, C_2 = 0.22 μF, C_E = 15 μF, and V_{CC} = 24 V. The β of the transistor is 120. Use SPICE to find the lower cutoff frequency and to obtain a plot of the low-frequency response.

6.59   Use SPICE to determine the actual factor by which the gain of the amplifier in Example 6.19 is reduced at 50 Hz. The supply voltage is V_{CC} = 15 V. Also find the actual lower cutoff frequency.

6.60   Use SPICE to solve Exercise 6.45.

6.61   The amplifier in Figure 6.45 has r_S = 10 kΩ, R_1 = 1.5 MΩ, R_2 = 330 kΩ, R_D = 2 kΩ, R_S = 820 Ω, and R_L = 20 kΩ. The input and output coupling capacitors are each 1 μF and the source bypass capacitor is 10 μF. The JFET has interelectrode capacitance C_{gd} = 1.5 pF, C_{gs} = 6 pF, and C_{ds} = 5 pF. The JFET parameters are I_{DSS} = 10 mA and V_p = −2 V. Use SPICE to obtain a plot of the high-frequency response of the amplifier and to determine its upper cutoff frequency. (Use discrete capacitors to represent interelectrode capacitance in the SPICE model.)

Introduction to Differential and Operational Amplifiers

7

7.1   DIFFERENTIAL AMPLIFIERS

Differential amplifiers are widely used in linear integrated circuits. They are a fundamental component of every *operational amplifier,* which, as we shall learn, is an extremely versatile device with a broad range of practical applications. We will discuss the circuit theory of differential amplifiers briefly, in preparation for a more comprehensive investigation of the capabilities (and limitations) of operational amplifiers.

Difference Voltages

A differential amplifier is also called a *difference* amplifier because it amplifies the difference between two signal voltages. Let us refine the notion of a *difference voltage* by reviewing some simple examples. We have already encountered difference voltages in our study of transistor amplifiers. Recall, for example, that the collector-to-emitter voltage of a BJT is the difference between the collector-to-ground voltage and the emitter-to-ground voltage:

$$V_{CE} = V_C - V_E \qquad (7.1)$$

The basic idea here is that a difference voltage is the mathematical difference between two other voltages, each of whose values is given with respect to ground. Suppose the voltage at point A in a circuit is 12 V with respect to ground and the voltage at point B is 3 V with respect to ground. The notation V_{AB} for the difference voltage means the voltage that would be measured if the positive side of a voltmeter were connected to point A and the negative side to point B; in this case, $V_{AB} = V_A - V_B = 12 - 3 = 9$ V. If the voltmeter connections were reversed, we would measure $V_{BA} = V_B - V_A = 3 - 12 = -9$ V. Thus, $V_{BA} = -V_{AB}$.

To help get used to thinking in terms of difference voltages, consider the system shown in Figure 7.1, where two identical amplifiers are driven by two different signal voltages. Although a differential amplifier does not behave in exactly the same way as this amplifier arrangement, the concepts of input and output difference voltages are similar. The two signal input voltages, v_1 and v_2, are shown as sine waves, one greater in amplitude than the other. For illustrative

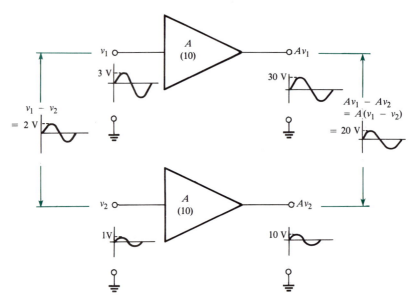

Figure 7.1
The amplification of difference voltages

purposes, their peak values are 3 V and 1 V, respectively. If the voltage gain of each amplifier is A, then the amplifier outputs are Av_1 and Av_2. The input difference voltage, $v_{12} = v_1 - v_2$, is a sine wave and the output difference voltage, $Av_1 - Av_2 = A(v_1 - v_2)$, is seen to be an amplified version of the input difference voltage. In our illustration, the gain A is 10 and the input difference voltage is (3 V pk) − (1 V pk) = 2 V pk. The output difference voltage is $A(v_1 - v_2) = 10(3\text{ V} - 1\text{ V}) = 10(2\text{ V}) = 20$ V pk.

7.2   BJT AND FET DIFFERENTIAL AMPLIFIERS

The BJT Differential Amplifier

The principal feature that distinguishes a differential amplifier from the configuration shown in Figure 7.1 is that a signal applied to one input of a differential amplifier induces a voltage with respect to ground on the amplifier's other output. This fact will become clear in our study of the voltage and current relations in the amplifier.

Figure 7.2 shows the basic BJT version of a differential amplifier. Two transistors are joined at their emitter terminals, where a constant-current source is connected to supply bias current to each. Note that each transistor is basically in a common-emitter configuration, with an input supplied to its base and an output taken from its collector. The two base terminals are the two signal inputs to the differential amplifier, v_{i1} and v_{i2}, and the two collectors are the two outputs, v_{o1} and v_{o2}, of the differential amplifier. Thus, the differential input voltage is $v_{i1} - v_{i2}$ and the differential output voltage is $v_{o1} - v_{o2}$.

Figure 7.3 shows the schematic symbol for the differential amplifier. Since there are two inputs and two outputs, the amplifier is said to have a *double-ended* (or double-sided) input and a double-ended output.

Figure 7.2
The basic BJT differential amplifier. The two transistors can be regarded as CE amplifiers having a common connection at their emitters. The base terminals are the inputs to the differential amplifier and the collectors are the outputs.

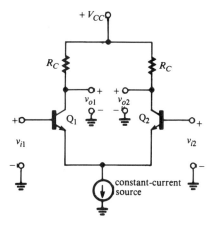

Figure 7.3
Schematic symbol for the differential amplifier

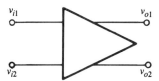

We will postpone, temporarily, our analysis of the dc bias levels in the amplifier and focus on its behavior as a small-signal amplifier. Toward that end, we will determine the output voltage at each collector due to each input voltage acting *alone,* that is, with the opposite input grounded, and then apply the superposition principle to determine the outputs due to both inputs acting simultaneously. Figure 7.4 shows the amplifier with input 2 grounded ($v_{i2} = 0$) and a small signal applied to input 1. The ideal current source presents an infinite impedance (open circuit) to an ac signal, so we need not consider its presence in our small-signal analysis. We also assume the ideal situation of perfectly matched transistors, so Q_1 and Q_2 have identical values of β, r_e, etc. Since Q_1 is essentially a common-emitter amplifier, the voltage at its collector (v_{o1}) is an amplified and *inverted*

Figure 7.4
The small-signal voltages in a differential amplifier when one input is grounded. Note that v_{e1} is in phase with v_{i1}, and that v_{o1} is out of phase with v_{i1}.

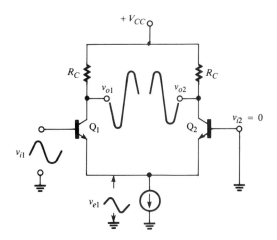

version of its input, v_{i1}. Note that there is also an ac voltage v_{e1} developed at the emitter of Q_1. This voltage is in phase with v_{i1} and exists because of emitter-follower action across the base–emitter junction of Q_1.

Now, the voltage v_{e1} is developed across the emitter resistance r_e looking into the emitter of Q_2 (in parallel with the infinite resistance of the current source). Therefore, as far as the emitter-follower action of Q_1 is concerned, the load resistance seen by Q_1 is r_e. Since the emitter resistance of Q_1 is itself r_e, it follows from equation 5.40 that the emitter-follower gain is 1/2:

$$A_v = \frac{r_L}{r_L + r_e} = \frac{r_e}{r_e + r_e} = 0.5$$

Therefore, v_{e1} is in phase with, and one-half the magnitude of, v_{i1}. Now, it is clear that v_{e1} is the emitter-to-ground voltage of *both* transistors. *When v_{e1} goes positive, the base-to-emitter voltage of Q_2 goes negative by the same amount.* In other words, $v_{be2} = v_{b2} - v_{e1} = 0 - v_{e1}$. (Since the base of Q_2 is grounded, its base-to-emitter voltage is the same as the negative of its emitter-to-ground voltage.) We see that even though the base of Q_2 is grounded, there exists an ac base-to-emitter voltage on Q_2 that is out of phase with v_{e1} and therefore out of phase with v_{i1}. Consequently, there is an ac output voltage v_{o2} produced at the collector of Q_2 and it is out of phase with v_{o1}.

Since both transistors are identical, they have equal gain and the output v_{o2} has the same magnitude as v_{o1}. To verify this last assertion, and to help solidify all the important ideas we have presented so far, let us study the specific example illustrated in Figure 7.5. We assume that v_{i1} (which is the base-to-ground voltage of Q_1) is a 100-mV-pk sine wave, and that each transistor has voltage gain -100, where, as usual, the minus sign denotes phase inversion. By "transistor voltage gain," we mean the collector voltage divided by the *base-to-emitter* voltage.

Since the emitter-follower gain of Q_1 is 0.5, v_e is a 0.5(100 mV) = 50-mV-pk sine wave. The peak value of v_{be1} is therefore $v_{b1} - v_{e1} = (100 \text{ mV}) - (50 \text{ mV}) = 50$ mV. When v_{be1} is at this 50-mV peak, v_{o1} is $-100(50 \text{ mV}) = -5$ V, that is, an inverted 5-V-pk sine wave. At this same point in time, where v_e is at its 50-mV

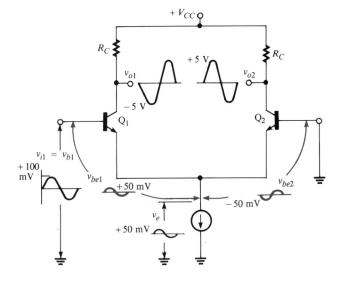

Figure 7.5
Each transistor has identical voltage gain -100, and the outputs at the collectors are -100 times their respective base-to-emitter voltages.

peak, the base-to-emitter voltage of Q_2 is $0 - (50 \text{ mV}) = -50 \text{ mV}$ pk. Therefore, v_{o2} is $(-100)(-50 \text{ mV}) = +5 \text{ V}$ pk; that is, v_{o2} is a 5-V-pk sine wave in phase with v_{i1} and out of phase with v_{o1}.

Note in Figure 7.5 that the input difference voltage is $v_{i1} - v_{i2} = (100 \text{ mV}) - 0 = 100 \text{ mV}$ pk, and the output difference voltage is 10 V pk, since v_{o1} and v_{o2} are out of phase. Therefore, the magnitude of the *difference voltage gain* $(v_{o1} - v_{o2})/(v_{i1} - v_{i2})$ is 100. So, while the voltage gain v_o/v_i for each side is only 50, the difference voltage gain is the same as the gain v_c/v_{be} of each transistor. We will refine and generalize this idea later, when we finish our small-signal analysis using superposition.

In many applications, the two inputs of a differential amplifier are driven by signals that are equal in magnitude and out of phase: $v_{i2} = -v_{i1}$. Continuing our analysis of the amplifier, let us now ground input 1 ($v_{i1} = 0$) and assume that there is a signal applied to input 2 equal to and out of phase with the v_{i1} signal we previously assumed. Since the transistors are identical and the circuit is completely symmetrical, the outputs have exactly the same relationships to the inputs as they had before: v_{o2} is out of phase with v_{i2} and v_{o1} is in phase with v_{i2}. These relationships are illustrated in Figure 7.6.

When we compare Figure 7.6 with Figure 7.4, we note that the v_{o1} outputs are identical, as are the v_{o2} outputs. In other words, driving the two inputs with equal but out-of-phase signals reinforces, or duplicates, the signals at the two outputs. By superposition, each output is the sum of the voltages resulting from each input acting alone, so the outputs are exactly twice the level they would be if only one input signal were present. These ideas are summarized in Figure 7.7.

In many applications, the output of a differential amplifier is taken from just one of the transistor collectors, v_{o1}, for example. In this case the input is a difference voltage and the output is a voltage with respect to ground. This use of the amplifier is called *single-ended output* operation and the voltage gain in that mode is

$$A_{v(single\text{-}ended\ output)} = \frac{v_{o1}}{v_{i1} - v_{i2}} \tag{7.2}$$

Figure 7.6
The differential amplifier with v_{i1} grounded and a signal input v_{i2}. Compare with Figure 7.4. Note that v_{i2} here is the opposite phase from v_{i1} in Figure 7.4 and that v_{o1} and v_{o2} are the same as in Figure 7.4.

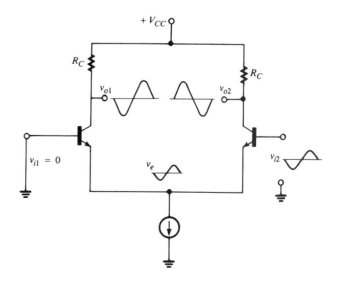

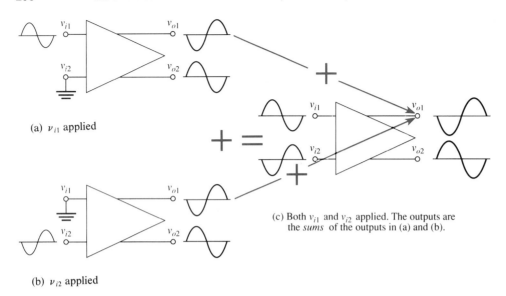

(a) v_{i1} applied

(b) v_{i2} applied

(c) Both v_{i1} and v_{i2} applied. The outputs are the *sums* of the outputs in (a) and (b).

Figure 7.7
By the superposition principle, the output v_{o1} when both inputs are applied is the sum of the v_{o1} outputs due to each signal acting alone. Likewise for v_{o2}.

The next example demonstrates that the single-ended output gain is one-half the difference voltage gain. To distinguish between these terms, we will hereafter refer to $(v_{o1} - v_{o2})/(v_{i1} - v_{i2})$ as the *double-ended voltage gain*.

Example 7.1

The magnitude of the voltage gain (v_c/v_{be}) for each transistor in Figure 7.2 is 100. If v_{i1} and v_{i2} are out-of-phase, 100-mV-pk signals applied simultaneously to the inputs, find

1. the peak values of v_{o1} and v_{o2};
2. the magnitude of the double-ended voltage gain $(v_{o1} - v_{o2})/(v_{i1} - v_{i2})$; and
3. the magnitude of the single-ended output gain $v_{o1}/(v_{i1} - v_{i2})$.

Solution

1. As demonstrated in Figure 7.5, the peak value of each output is 5 V when one input is driven and the other is grounded. Since the outputs are doubled when the inputs are equal and out of phase, each output is 10 V pk.

2. Since $v_{i1} = -v_{i2}$, the input difference voltage is $v_{i1} - v_{i2} = 2v_{i1} = 200$ mV pk. Similarly, $v_{o1} = -v_{o2}$, so the output difference voltage is $v_{o1} - v_{o2} = 2v_{o1} = 20$ V pk. Therefore, the magnitude of $(v_{o1} - v_{o2})/(v_{i1} - v_{i2})$ is $(20 \text{ V})/(200 \text{ mV}) = 100$.

3. The magnitude of the single-ended output gain is

$$|A_{v(single\text{-}ended\ output)}| = \left|\frac{v_{o1}}{v_{i1} - v_{i2}}\right| = \frac{10 \text{ V}}{200 \text{ mV}} = 50$$

Since v_{o1} is out of phase with $(v_{i1} - v_{i2})$, the correct specification for the single-ended output gain is -50. If the single-ended output is taken from the *other* side (v_{o2}), which is out of phase with v_{o1}, then the gain $v_{o2}/(v_{i1} - v_{i2})$ is $+50$.

Note once again that the double-ended voltage gain is the same as the voltage gain v_c/v_{be} for each transistor. *Note also that the single-ended output gain is one-half the double-ended gain.* Since the output difference voltage $v_{o1} - v_{o2}$ is out of phase with the input difference voltage $v_{i1} - v_{i2}$, the correct specification for the double-ended voltage gain is -100.

It should now be clear that if the two inputs are driven by equal *in-phase* signals, the output at each collector will be exactly 0, and the output difference voltage will be 0. Of course, in this case, the input difference voltage is also 0. These ideas are illustrated in Figure 7.8.

We can now derive general expressions for the double-ended and single-ended output voltage gains in terms of the circuit parameters. Figure 7.9 shows one side of the differential amplifier with the other side replaced by its emitter resistance, r_e. Recall that this is the resistance in series with the emitter of Q_1 when the input to Q_2 is grounded. We are again assuming that the current source has infinite resistance.

Neglecting the output resistance r_o at the collector of Q_1, we can use the familiar approximation (equation 5.24) for the voltage gain of the transistor:

$$\frac{v_{o1}}{v_{be1}} \approx \frac{-R_C}{r_e} \tag{7.3}$$

where r_e is the emitter resistance of Q_1. It is clear from Figure 7.9 that the voltage gain v_{o1}/v_{i1} is

$$\frac{v_{o1}}{v_{i1}} \approx \frac{-R_C}{2r_e} \tag{7.4}$$

where the quantity $2r_e$ is in the denominator because we assume that the emitter resistances of Q_1 and Q_2 are equal. Equations 7.3 and 7.4 confirm our previous conclusion that the transistor voltage gain is twice the value of the gain v_{o1}/v_{i1}. Also, we have already shown that the double-ended (difference) voltage gain equals the transistor gain and that the single-ended output gain is one-half that value. Therefore, we conclude that

$$\frac{v_{o1} - v_{o2}}{v_{i1} - v_{i2}} \approx \frac{-R_C}{r_e} \tag{7.5}$$

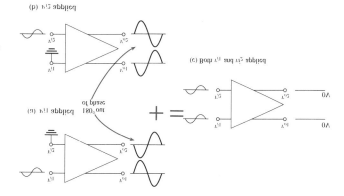

Figure 7.8
The outputs of the differential amplifier are 0 when the two inputs are equal and in phase.

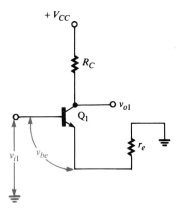

Figure 7.9
When the input to Q_2 is grounded, there is resistance r_e in series with the emitter of Q_1.

and

$$\frac{v_{o1}}{v_{i1} - v_{i2}} \approx \frac{-R_C}{2r_e} \qquad (7.6)$$

We should note that these gain relations are valid irrespective of the magnitudes and phase relations of the two inputs v_{i1} and v_{i2}. We have considered only the two special cases where v_{i1} and v_{i2} are equal and in phase and where they are equal and out of phase, but equations 7.5 and 7.6 hold under any circumstances. Note also that v_{o1} and v_{o2} will always have the same amplitude and be out of phase with each other. Thus,

$$\frac{v_{o2}}{v_{i1} - v_{i2}} \approx \frac{R_C}{2r_e} \qquad (7.7)$$

The small-signal *differential input resistance* is defined to be the input difference voltage divided by the total input current. Imagine a signal source connected *across* the input terminals, so the same current that flows out of the source into one input of the amplifier flows out of the other input and returns to the source. The signal-source voltage, which is the input difference voltage, divided by the signal-source current, is the differential input resistance. Since the total small-signal resistance in the path from one input through both emitters to the other input is $2r_e$, the differential input resistance is

$$r_{id} = 2(\beta + 1)r_e \qquad (7.8)$$

Figure 7.10 shows the dc voltages and currents in the ideal differential amplifier. Since the transistors are identical, the source current I divides equally between them, and the emitter current in each is, therefore,

$$I_E = I_{E1} = I_{E2} = I/2 \qquad (7.9)$$

The dc output voltage at the collector of each transistor is

$$V_{o1} = V_{CC} - I_{C1}R_C$$

$$V_{o2} = V_{CC} - I_{C2}R_C$$

Since $I_C \approx I_E = I/2$ in each transistor, we have

$$V_{o1} = V_{o2} \approx V_{CC} - (I/2)R_C \qquad (7.10)$$

Figure 7.10
*DC voltages and currents in an ideal
differential amplifier*

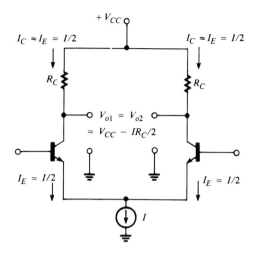

To determine the ac emitter resistance of each transistor, we can use equation 7.9
and the familiar approximation $r_e \approx 0.026/I_E$ to obtain

$$r_{e1} = r_{e2} = r_e \approx \frac{0.026}{I_E} = \frac{0.026}{I/2} \tag{7.11}$$

Example 7.2

For the ideal differential amplifier shown in Figure 7.11, find

1. the dc output voltages V_{o1} and V_{o2};
2. the single-ended output gain $v_{o1}/(v_{i1} - v_{i2})$; and
3. the double-ended gain $(v_{o1} - v_{o2})/(v_{i1} - v_{i2})$.

Solution

1. The emitter current in each transistor is $I_E = I/2 = (2 \text{ mA})/2 = 1 \text{ mA} \approx I_C$.
   Therefore, $V_{o1} = V_{o2} = V_{CC} - I_C R_C = 15 \text{ V} - (1 \text{ mA})(6 \text{ k}\Omega) = 9 \text{ V}$.

Figure 7.11
(Example 7.2)

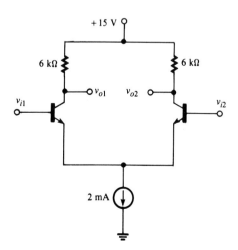

2. The emitter resistance of each transistor is

$$r_e \approx \frac{0.026}{I_E} = \frac{0.026}{1\ \text{mA}} = 26\ \Omega$$

Therefore, from equation 7.6,

$$\frac{v_{o1}}{v_{i1} - v_{i2}} \approx \frac{-R_C}{2r_e} = \frac{-6\ \text{k}\Omega}{52\ \Omega} = -115.4$$

3. From equation 7.5,

$$\frac{v_{o1} - v_{o2}}{v_{i1} - v_{i2}} = \frac{-R_C}{r_e} = \frac{-6\ \text{k}\Omega}{26\ \Omega} = -230.8$$

The FET Differential Amplifier

Many differential amplifiers are constructed using field-effect transistors because of the large impedance they present to input signals. This property is exceptionally important in many applications, including operational amplifiers, instrument amplifiers, and charge amplifiers. A large voltage gain is also important in these applications, and although the FET does not produce much gain, an FET differential amplifier is often the first stage in a multistage amplifier whose overall gain is large. Because FETs are easily fabricated in integrated-circuit form, FET differential amplifiers are commonly found in linear integrated circuits.

Figure 7.12 shows a JFET differential amplifier, and it can be seen that it is basically the same configuration as its BJT counterpart. The two JFETs operate as common-source amplifiers with their source terminals joined. A constant-current source provides bias current.

The derivations of the gain equations for the JFET amplifier are completely parallel to those for the BJT version. A source-to-ground voltage is developed at the common source connection by source-follower action. With one input grounded, the output resistance and load resistance of the source follower are both equal to $1/g_m$ (assuming matched devices), so the source-follower gain is 0.5. Therefore, as shown in Figure 7.13, one-half the input voltage is developed across $1/g_m$, and the current is

$$i_d = \frac{v_i/2}{1/g_m} = \frac{v_i g_m}{2}$$

Figure 7.12
The JFET differential amplifier

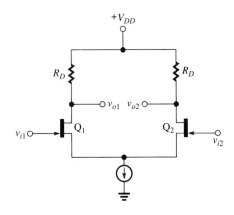

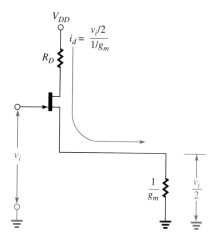

Figure 7.13
Source-follower action with gain 0.5
results in one-half the input voltage
developed across $(1/g_m)$ ohms.

The output voltage is then

$$v_o = -i_d R_D = \frac{-v_i g_m R_D}{2} \tag{7.12}$$

from which we find the voltage gain

$$\frac{v_o}{v_i} = \frac{-g_m R_D}{2} \tag{7.13}$$

Applying the superposition principle in the same way we did for the BJT version, we readily find that

$$\frac{v_{o1}}{v_{i1} - v_{i2}} = \frac{-g_m R_D}{2} \tag{7.14}$$

and

$$\frac{v_{o1} - v_{o2}}{v_{i1} - v_{i2}} = -g_m R_D \tag{7.15}$$

Like the gain equations for the BJT differential amplifier, equations 7.14 and 7.15 show that the double-ended (difference) voltage gain is the same as the gain of one transistor, and the single-ended output gain is one-half that value.

Example 7.3

The matched transistors in Figure 7.14 have $I_{DSS} = 12$ mA and $V_p = -2.5$ V. Find

1. the dc output voltages v_{o1} and v_{o2};
2. the single-ended output gain $v_{o1}/(v_{i1} - v_{i2})$; and
3. the double-ended gain $(v_{o1} - v_{o2})/(v_{i1} - v_{i2})$.

Solution

1. The dc current in each JFET is $I_D = (1/2)(6$ mA$) = 3$ mA. Therefore, $V_{o1} = V_{o2} = V_{DD} - I_D R_D = 15$ V $- (3$ mA$)(3$ k$\Omega) = 6$ V.

Figure 7.14
(Example 7.3)

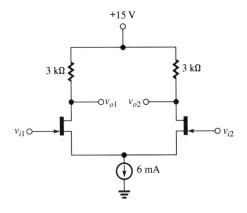

2.
$$g_m = \frac{2I_{DSS}}{|V_p|} \sqrt{\frac{I_D}{I_{DSS}}} = \frac{2(12 \text{ mA})}{2.5 \text{ V}} \sqrt{\frac{3 \text{ mA}}{12 \text{ mA}}} = 4.8 \text{ mS}$$

From equation 7.14,

$$\frac{v_{o1}}{v_{i1} - v_{i2}} = \frac{-g_m R_D}{2} = \frac{-(4.8 \text{ mS})(3 \text{ k}\Omega)}{2} = -7.2$$

3. From equation 7.15,

$$\frac{v_{o1} - v_{o2}}{v_{i1} - v_{i2}} = -g_m R_D = -(4.8 \text{ mS})(3 \text{ k}\Omega) = -14.4$$

7.3   COMMON-MODE PARAMETERS

One attractive feature of a differential amplifier is its ability to reject signals that are *common* to both inputs. Since the outputs are amplified versions of the difference between the inputs, any voltage component that appears identically in both signal inputs will be "differenced out," that is, will have zero level in the outputs. (We have already seen that the outputs are exactly 0 when both inputs are identical, in-phase signals.) Any dc or ac voltage that appears simultaneously in both signal inputs is called a *common-mode* signal. The ability of an amplifier to suppress, or zero-out, common-mode signals is called *common-mode rejection*. An example of a common-mode signal whose rejection is desirable is electrical *noise* induced in both signal lines, a frequent occurrence when the lines are routed together over long paths. Another example is a dc level common to both inputs, or common dc fluctuations caused by power-supply variations.

In the ideal differential amplifier, any common-mode signal will be completely cancelled out and therefore have no effect on the output signals. In practical amplifiers, mismatched components and certain other nonideal conditions result in imperfect cancellation of common-mode signals. Figure 7.15 shows a differential amplifier in which a common-mode signal v_{cm} is applied to both inputs. Ideally, the output voltages should be 0, but in fact some small component of v_{cm} may appear. The differential *common-mode gain*, A_{cm}, is defined to be the ratio of the output difference voltage caused by the common-mode signal to the common-

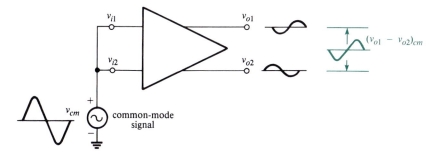

Figure 7.15
If the differential amplifier were ideal, both outputs would be 0 when the inputs have the same (common-mode) signal. In reality, there is a small common-mode output, as shown.

mode signal itself:

$$A_{cm} = \frac{(v_{o1} - v_{o2})_{cm}}{v_{cm}} \qquad (7.16)$$

We can also define a single-ended common-mode gain as the ratio of $(v_{o1})_{cm}$ or $(v_{o2})_{cm}$ to v_{cm}. Obviously the ideal amplifier has common-mode gain equal to 0.

A widely used specification and figure of merit for a differential amplifier is its *common-mode rejection ratio* (CMRR), defined to be the ratio of the magnitude of its differential (difference-mode) gain A_d to the magnitude of its common-mode gain:

$$\text{CMRR} = \frac{|A_d|}{|A_{cm}|} \qquad (7.17)$$

The value of the CMRR is often given in decibels:

$$\text{CMRR} = 20 \log_{10} \left| \frac{A_d}{A_{cm}} \right| \qquad (7.18)$$

Example 7.4

When the inputs to a certain differential amplifier are $v_{i1} = 0.1 \sin \omega t$ and $v_{i2} = -0.1 \sin \omega t$, it is found that the outputs are $v_{o1} = -5 \sin \omega t$ and $v_{o2} = 5 \sin \omega t$. When both inputs are $2 \sin \omega t$, the outputs are $v_{o1} = -0.05 \sin \omega t$ and $v_{o2} = 0.05 \sin \omega t$. Find the CMRR in dB.

Solution. We will use the peak values of the various signals for our gain computations, but note carefully how the minus signs are used to preserve phase relations. The difference-mode gain is

$$A_d = \frac{v_{o1} - v_{o2}}{v_{i1} - v_{i2}} = \frac{-5 - 5}{0.1 - (-0.1)} = \frac{-10}{0.2} = -50$$

The common-mode gain is

$$A_{cm} = \frac{(v_{o1} - v_{o2})_{cm}}{v_{cm}} = \frac{-0.05 - 0.05}{2} = \frac{-0.1}{2} = -0.05$$

The common-mode rejection ratio is

$$\text{CMRR} = \frac{|A_d|}{|A_{cm}|} = \frac{50}{0.05} = 1000$$

Expressing this result in dB, we have CMRR = $20 \log_{10}(1000)$ = 60 dB.

7.4   INTRODUCTION TO OPERATIONAL AMPLIFIERS

An operational amplifier is basically a differential amplifier modified by the addition of circuitry that improves its performance and gives it certain special features. The most important characteristics of an operational amplifier are as follows:

1. It is a dc (direct-coupled, direct-current) amplifier.
2. It should have a very large voltage gain—ideally, infinite.
3. It should have a very large input impedance—ideally, infinite.
4. It should have a very small output impedance—ideally zero.
5. The output should be exactly zero V when the inputs are zero V.
6. The output must be capable of both positive and negative voltage swings.
7. It should have a very large CMRR.
8. It is operated with a single-ended output and differential input (although one input is often grounded, as we shall presently see).
9. It should meet whatever special requirements are demanded by a particular application; these include parameters such as noise level, frequency response, and slew rate, which we will discuss in Chapter 8.

The name *operational* amplifier is derived from amplifier applications that the preceding characteristics make possible: the performance of precise mathematical operations on input signals, including voltage summation, subtraction, and integration. Characteristics 2 and 3 are particularly important for those kinds of operations, and in Chapter 9 we shall explore how these and other features contribute to many other useful applications of operational amplifiers. Our present interest is in the circuit methods used to expand a differential amplifier into an operational amplifier.

The input stage of every operational amplifier is a differential amplifier. To achieve a large input impedance, the differential stage may be constructed with field-effect transistors, or it may employ certain additional circuitry, such as emitter followers, to increase the impedance seen at each input. Components in the input stage should be very closely matched to achieve the best possible balance in the differential operation. This is important to ensure that the output of the operational amplifier is a precise representation of the input difference voltage, that the output is exactly zero when the inputs are zero, and that the CMRR is large. Ideally, component characteristics should be independent of temperature. Any changes that do occur should track one another; that is, device parameters should change in the same way and at the same rate under temperature variations. Matching and tracking of characteristics is, of course, best accomplished in integrated-circuit construction, and virtually all modern operational amplifiers are integrated circuits. Many designs include temperature-compensation circuitry to minimize the effects of temperature.

Voltage gain is achieved through the use of multistage amplifiers, at least one of which is usually another differential stage. At some point in the multistage amplification, the output becomes single-ended. We have seen that single-ended gain is one-half the double-ended value, so this conversion results in an undesirable loss of voltage gain. Some designs incorporate circuitry that eliminates this loss in a clever fashion, but we will not take time to detail the somewhat complex theory involved.

To permit the output voltage to swing through both positive and negative values, both a positive and a negative supply voltage are required. These are usually equal-valued, opposite-polarity supplies, a typical example being ±15 V. To obtain zero output voltage when the inputs are zero, the amplifier must incorporate *level-shifting circuitry* that eliminates any nonzero bias voltage that would otherwise appear in the output. Of course, this cannot be accomplished with a coupling capacitor, because we require dc response. Level shifting is usually accomplished near or at the output stage.

The Ideal Operational Amplifier

To facilitate the understanding of basic operational amplifier theory, it is convenient to assume initially that we are dealing with an *ideal* operational amplifier. For the theory that we will develop in the balance of this chapter, we define an ideal operational amplifier to be one that has the following attributes:

1. It has infinite gain.
2. It has infinite input impedance.
3. It has zero output impedance.

Although no real amplifier* can satisfy any of these requirements, we will see that most modern amplifiers have such large gains and input impedances, and such small output impedances, that a negligibly small error results from assuming ideal characteristics. A detailed study of the ideal amplifier will therefore be beneficial in terms of understanding how practical amplifiers are used as well as in building some important theoretical concepts that have broad implications in many areas of electronics.

Figure 7.16 shows the standard symbol for an operational amplifier. Note that the two inputs are labeled "+" and "−" and the input signals are correspondingly designated v_i^+ and v_i^-. In relation to our previous discussion of differential amplifiers, these inputs correspond to v_{i1} and v_{i2}, respectively, when the single-ended output is v_{o2} (see Figure 7.2). In other words, if the inputs are out-of-phase signals, the amplifier output will be in phase with v_i^+ and out of phase with v_i^-. For this reason, the + input is called the *noninverting* input and the − input is called the *inverting input*. In many applications, one of the amplifier inputs is grounded, so v_o is in phase with the input if the signal is connected to the noninverting terminal, and v_o is out of phase with the input if the signal is connected to the inverting input. These ideas are summarized in the table accompanying Figure 7.16.

At this point, a legitimate question that may have already occurred to the reader is this: If the gain is infinite, how can the output be anything other than a severely clipped waveform? Theoretically, if the amplifier has infinite gain, an

---

*We will hereafter use the word *amplifier* with the understanding that operational amplifier is meant. Some authors use the term *op-amp*.

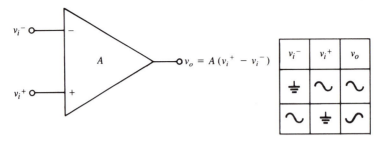

Figure 7.16
Operational amplifier symbol, showing inverting (−) and noninverting (+) inputs

infinitesimal input voltage must result in an infinitely large output voltage. The answer, of course, is that the gain is not truly infinite, just very large. Nevertheless, it *is* true that a very small input voltage will cause the amplifier output to be driven all the way to its extreme positive or negative voltage limit. The practical answer is that an operational amplifier is seldom used in such a way that the full gain is applied to an input. Instead, external resistors are connected to and around the amplifier in such a way that the signal undergoes vastly smaller amplification. The resistors cause gain reduction through signal *feedback,* which we will soon study in considerable detail.

7.5   THE INVERTING AMPLIFIER

Consider the configuration shown in Figure 7.17. In this very useful application of an operational amplifier, the noninverting input is grounded, v_{in} is connected through R_1 to the inverting input, and feedback resistor R_f is connected between the output and v_i^-. Let A denote the voltage gain of the amplifier: $v_o = A(v_i^+ - v_i^-)$. Since $v_i^+ = 0$, we have

$$v_o = -Av_i^- \tag{7.19}$$

(Note that $v_{in} \neq v_i^-$.) We wish to investigate the relation between v_o and v_{in} when the magnitude of A is infinite.

Figure 7.18 shows the voltages and currents that result when signal v_{in} is connected. From Ohm's law, the current i_1 is simply the difference in voltage

Figure 7.17
An operational-amplifier application in which signal v_{in} is connected through R_1. Resistor R_f provides feedback. $v_o/v_i^- = -A$.

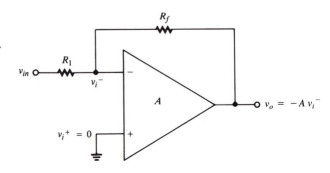

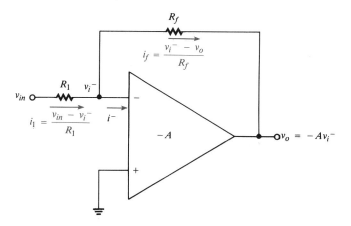

Figure 7.18
Voltages and currents resulting from the application of the signal voltage v_{in}

across R_1, divided by R_1:

$$i_1 = (v_{in} - v_i^-)/R_1 \tag{7.20}$$

Similarly, the current i_f is the difference in voltage across R_f, divided by R_f:

$$i_f = (v_i^- - v_o)/R_f \tag{7.21}$$

Writing Kirchhoff's current law at the inverting input, we have

$$i_1 = i_f + i^- \tag{7.22}$$

where i^- is the current entering the amplifier at its inverting input. However, the ideal amplifier has infinite input impedance, which means i^- must be 0. So (7.22) is simply

$$i_1 = i_f \tag{7.23}$$

Substituting (7.20) and (7.21) into (7.23) gives

$$\frac{v_{in} - v_i^-}{R_1} = \frac{v_i^- - v_o}{R_f}$$

or

$$\frac{v_{in}}{R_1} - \frac{v_i^-}{R_1} = \frac{v_i^-}{R_f} - \frac{v_o}{R_f} \tag{7.24}$$

From equation 7.19,

$$v_i^- = -\frac{v_o}{A} \tag{7.25}$$

If we now invoke the assumption that $A = \infty$, we see that $-v_o/A = 0$ and, therefore,

$$v_i^- = 0 \qquad \text{(ideal amp, with } A = \infty) \tag{7.26}$$

Substituting $v_i^- = 0$ into (7.24) gives

$$\frac{v_{in}}{R_1} = \frac{-v_o}{R_f}$$

or

$$\frac{v_o}{v_{in}} = \frac{-R_f}{R_1} \qquad (7.27)$$

We see that the gain is negative, signifying that the configuration is an *inverting* amplifier. Equation 7.27 also reveals the exceptionally useful fact that the magnitude of v_o/v_{in} *depends only on the ratio of the resistor values* and not on the amplifier itself. Provided the amplifier gain and impedance remain quite large, variations in amplifier characteristics (due, for example, to temperature changes or manufacturing tolerance) do not affect v_o/v_{in}. For example, if $R_1 = 10$ kΩ and $R_f = 100$ kΩ, we can be certain that $v_o = -[(100 \text{ k}\Omega)/(10 \text{ k}\Omega)]v_{in} = -10\ v_{in}$, i.e., that the gain is as close to -10 as the resistor precision permits. The gain v_o/v_{in} is called the *closed-loop gain* of the amplifier, while A is called the *open-loop gain*. In this application, we see that an extremely large open-loop gain, perhaps 10^6, is responsible for giving us the very predictable, though much smaller, closed-loop gain equal to 10. This is the essence of most operational-amplifier applications: trade the very large gain that is available for less spectacular but more precise and predictable characteristics.

In our derivation, we used the infinite-gain assumption to obtain $v_i^- = 0$ (equation 7.26). In real amplifiers, having very large, but finite, values of A, v_i^- is a very small voltage, near 0. For that reason, the input terminal where the feedback resistor is connected in the inverting configuration is said to be at *virtual ground*. For *analysis* purposes, we often assume that $v_i^- = 0$, but we cannot actually ground that point. Since v_i^- is at virtual ground, the impedance seen by the signal source generating v_{in} is R_1 ohms.

Example 7.5

Assuming that the operational amplifier in Figure 7.19 is ideal, find

1. the rms value of v_o when v_{in} is 1.5 V rms;
2. the rms value of the current in the 25-kΩ resistor when v_{in} is 1.5 V rms; and
3. the output voltage when $v_{in} = -0.6$ V dc.

Figure 7.19
(Example 7.5)

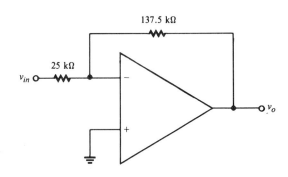

137.5 kΩ

25 kΩ

v_{in}

v_o

Solution

1. From equation 7.27,

$$\frac{v_o}{v_{in}} = \frac{-R_f}{R_1} = \frac{-137.5 \text{ k}\Omega}{25 \text{ k}\Omega} = -5.5$$

Thus, $|v_o| = 5.5|v_{in}| = 5.5(1.5 \text{ V rms}) = 8.25 \text{ V rms}$.

2. Since $v_i^- \approx 0$ (virtual ground), the current in the 25-kΩ resistor is

$$i = \frac{v_{in}}{R_1} = \frac{1.5 \text{ V rms}}{25 \text{ k}\Omega} = 60 \; \mu\text{A rms}$$

3. $v_o = (-5.5)v_{in} = (-5.5)(-0.6 \text{ V}) = 3.3 \text{ V dc}$. Notice that the output is a positive dc voltage when the input is a negative dc voltage, and vice versa.

7.6   THE NONINVERTING AMPLIFIER

Figure 7.20 shows another useful application of an operational amplifier, called the *noninverting* configuration. Notice that the input signal v_{in} is connected directly to the noninverting input and that resistor R_1 is connected from the inverting input to ground. Under the ideal assumption of infinite input impedance, no current flows into the inverting input, so $i_1 = i_f$. Thus,

$$\frac{v_i^-}{R_1} = \frac{v_o - v_i^-}{R_f} \tag{7.28}$$

Now, as shown in the figure,

$$v_o = A(v_i^+ - v_i^-) \tag{7.29}$$

Solving (7.29) for v_i^- gives

$$v_i^- = v_i^+ - v_o/A \tag{7.30}$$

Letting $A = \infty$, the term v_o/A goes to 0, and we have

$$v_i^- = v_i^+ \tag{7.31}$$

Figure 7.20
The operational amplifier in a noninverting configuration

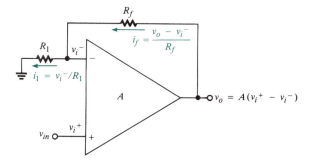

Substituting v_i^+ for v_i^- in (7.28) gives

$$\frac{v_i^+}{R_1} = \frac{v_o - v_i^+}{R_f} \tag{7.32}$$

Solving for v_o/v_i^+ and recognizing that $v_i^+ = v_{in}$ lead to

$$\frac{v_o}{v_{in}} = 1 + \frac{R_f}{R_1} = \frac{R_1 + R_f}{R_1} \tag{7.33}$$

We saw (equation 7.26) that when an operational amplifier is connected in an inverting configuration, with $v_i^+ = 0$, the assumption $A = \infty$ gives $v_i^- = 0$ (virtual ground), i.e., $v_i^- = v_i^+$. Also, in the noninverting configuration, the same assumption gives the same result: $v_i^- = v_i^+$ (equation 7.31). Thus, we reach the important general conclusion that feedback in conjunction with a very large voltage gain *forces* the voltages at the inverting and noninverting inputs to be approximately equal.

Equation 7.33 shows that the closed-loop gain of the noninverting amplifier, like that of the inverting amplifier, depends only on the values of external resistors. A further advantage of the noninverting amplifier is that the input impedance seen by v_{in} is infinite, or at least extremely large in a real amplifier. The inverting and noninverting amplifiers are used in voltage *scaling* applications, where it is desired to multiply a voltage precisely by a fixed constant, or scale factor. The multiplying constant in the inverting amplifier is R_f/R_1 (which may be less than 1), and it is $1 + R_f/R_1$ (which is always greater than 1) in the noninverting amplifier. A wide range of constants can be realized with convenient choices of R_f and R_1 when the gain ratio is R_f/R_1, which is not so much the case when the gain ratio is $1 + R_f/R_1$. For that reason, the inverting amplifier is more often used in precision scaling applications.

The reader may wonder why it would be desirable or necessary to use an amplifier to multiply a voltage by a number less than 1, since this can also be accomplished using a simple voltage divider. The answer is that the amplifier provides power gain to drive a load. Also, the ideal amplifier has zero output impedance, so the output voltage is not affected by changes in load impedance. Figure 7.21 shows a special case of the noninverting amplifier, used in applications where power gain and impedance isolation are of primary concern. Notice that $R_f = 0$ and $R_1 = \infty$, so, by equation 7.33, the closed-loop gain is $v_o/v_{in} = 1 + R_f/R_1 = 1$. This configuration is called a *voltage follower* because v_o has the same magnitude and phase as v_{in}. Like a BJT emitter follower, it has large input im-

Figure 7.21
The voltage follower

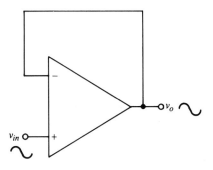

pedance and small output impedance, and is used as a buffer amplifier between a high-impedance source and a low-impedance load.

Example 7.6

DESIGN

In a certain application, a signal source having 60 kΩ of source impedance produces a 1-V-rms signal. This signal must be amplified to 2.5 V rms and drive a 1-kΩ load. Assuming that the phase of the load voltage is of no concern, design an operational-amplifier circuit for the application.

Solution. Since phase is of no concern and the required voltage gain is greater than 1, we can use either an inverting or noninverting amplifier. Suppose we decide to use the inverting configuration and arbitrarily choose $R_f = 250$ kΩ. Then,

$$\frac{R_f}{R_1} = 2.5 \Rightarrow R_1 = \frac{R_f}{2.5} = \frac{250 \text{ k}\Omega}{2.5} = 100 \text{ k}\Omega$$

Note, however, that the signal source sees an impedance equal to $R_1 = 100$ kΩ in the inverting configuration, so the usual voltage division takes place and the input to the amplifier is actually

$$v_{in} = \left(\frac{R_1}{R_1 + r_S}\right) (1 \text{ V rms}) = \left[\frac{100 \text{ k}\Omega}{(100 \text{ k}\Omega) + (60 \text{ k}\Omega)}\right] (1 \text{ V rms}) = 0.625 \text{ V rms}$$

Therefore, the magnitude of the amplifier output is

$$v_o = \frac{R_f}{R_1} (0.625 \text{ V rms}) = \frac{250 \text{ k}\Omega}{100 \text{ k}\Omega} (0.625 \text{ V rms}) = 1.5625 \text{ V rms}$$

Clearly the large source impedance is responsible for a reduction in gain, and it is necessary to redesign the amplifier circuit to compensate for this loss. (Do this, as an exercise.)

In view of the fact that the source impedance may not be known precisely or may change if a replacement source is used, a far better solution is to design a noninverting amplifier. Since the input impedance of this design is extremely large, the choice of values for R_f and R_1 will not depend on the source impedance. Letting $R_f = 150$ kΩ, we have

$$1 + \frac{R_f}{R_1} = 2.5$$

$$\frac{R_f}{R_1} = 1.5$$

$$R_1 = \frac{R_f}{1.5} = \frac{150 \text{ k}\Omega}{1.5} = 100 \text{ k}\Omega$$

The completed design is shown in Figure 7.22. Since we can assume that the amplifier has zero output impedance, we do not need to be concerned with voltage division between the amplifier output and the 1-kΩ load.

Figure 7.22
(Example 7.6)

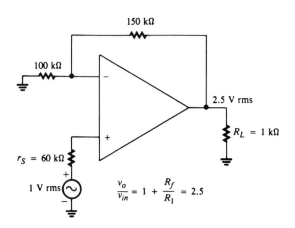

$$\frac{v_o}{v_{in}} = 1 + \frac{R_f}{R_1} = 2.5$$

EXERCISES

Section 7.1 Differential Amplifiers

7.1 For each of the following pairs of voltages, write the mathematical expression for the difference voltage $(v_1 - v_2)$ and sketch it versus time.

   a. $v_1 = 4 \sin \omega t$, $v_2 = 2$
   b. $v_1 = 10 \sin \omega t$, $v_2 = 6 \sin \omega t$
   c. $v_1 = 0.02 \sin \omega t$, $v_2 = -0.02 \sin \omega t$
   d. $v_1 = 4 + 0.01 \sin \omega t$, $v_2 = 4 - 0.01 \sin \omega t$

Section 7.2 BJT and FET Differential Amplifiers

7.2 The voltage gain of each transistor in the ideal differential amplifier shown in Figure

7.4 is $v_o/v_{be} = -160$. If v_{i1} is a 40-mV-peak sine wave and $v_{i2} = 0$, find
   a. the peak value of v_{e1};
   b. the peak value of v_{o1};
   c. the peak value of v_{o2}; and
   d. the voltage gain v_{o1}/v_{i1}.

7.3 A 40-mV-peak sine wave that is out of phase with v_{i1} is applied to v_{i2} in the amplifier in Exercise 7.2. Find
   a. the single-ended voltage gains $v_{o1}/(v_{i1} - v_{i2})$ and $v_{o2}/(v_{i1} - v_{i2})$; and
   b. the double-ended voltage gain $(v_{o1} - v_{o2})/(v_{i1} - v_{i2})$.

7.4 Repeat Exercise 7.3 if the signal applied to v_{i2} is a 40-mV-peak sine wave that is *in phase* with v_{i1}. (Think carefully.)

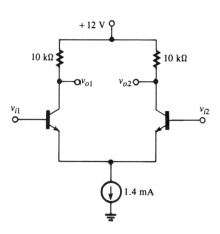

Figure 7.23
(Exercise 7.7)

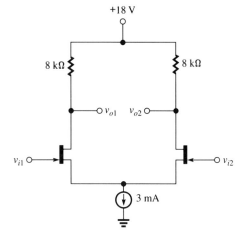

Figure 7.24
(Exercise 7.9)

7.5 The ideal BJT differential amplifier shown in Figure 7.2 is biased so that 0.75 mA flows in each emitter. If $R_C = 9.2$ kΩ, find
   a. the single-ended voltage gains $v_{o1}/(v_{i1} - v_{i2})$ and $v_{o2}/(v_{i1} - v_{i2})$; and
   b. the double-ended voltage gain $(v_{o1} - v_{o2})/(v_{i1} - v_{i2})$.

7.6 If the transistors in the differential amplifier of Exercise 7.5 each have $\beta = 120$, find the differential input resistance of the amplifier.

7.7 The β of each transistor in the ideal differential amplifier shown in Figure 7.23 is 100. Find
   a. the dc output voltages V_{o1} and V_{o2};
   b. the double-ended voltage gain; and
   c. the differential input resistance.

7.8 The current source in Exercise 7.7 is changed to 1 mA. If v_{i1} is a 16-mV-peak sine wave and $v_{i2} = 0$, find the peak values of v_{o1} and v_{o2}.

7.9 The FETs in the ideal differential amplifier shown in Figure 7.24 have $I_{DSS} = 10$ mA and $V_p = -2$ V. Find
   a. the dc output voltages V_{o1} and V_{o2};
   b. the single-ended output gain $v_{o1}/(v_{i1} - v_{i2})$; and
   c. the output difference voltage when the input difference voltage is 50 mV rms.

7.10 The inputs to the differential amplifier in Exercise 7.9 are $v_{i1} = 65$ mV rms and $v_{i2} = 10$ mV rms. v_{i1} and v_{i2} are in phase. Find the rms values of v_{o1} and v_{o2}.

Section 7.3   Common-Mode Parameters

7.11 A differential amplifier has CMRR = 68 dB and a differential mode gain of 175. Find the rms value of the output difference voltage when the common-mode signal is 1.5 mV rms.

7.12 The noise signal common to both inputs of a differential amplifier is 2.4 mV rms. When an input difference voltage of 0.1 V rms is applied to the amplifier, each output must have 4 V rms of signal level. Assuming that the amplifier produces no noise and that the noise component in the output difference voltage must be no greater than 500 μV rms, what is the minimum CMRR, in dB, that the amplifier should have?

Section 7.5   The Inverting Amplifier

7.13 Find the output of the ideal operational amplifier shown in Figure 7.25 for each of the following input signals:
   a. $v_{in} = 120$ mV dc
   b. $v_{in} = 0.5 \sin \omega t$ V
   c. $v_{in} = -2.5$ V dc
   d. $v_{in} = 4 - \sin \omega t$ V
   e. $v_{in} = 0.8 \sin (\omega t + 75°)$ V

7.14 Assume that the feedback resistance in Exercise 7.13 is doubled and the input resistance is halved. Find the output for each of the following input signals:
   a. $v_{in} = -60.5$ mV dc
   b. $v_{in} = 500 \sin \omega t$ μV
   c. $v_{in} = -0.16 + \sin \omega t$ V
   d. $v_{in} = -0.2 \sin (\omega t - 30°)$ V

7.15 Find the current in the feedback resistor for each part of Exercise 7.14.

7.16 The amplifier in Exercise 7.13 is driven by a signal source whose output resistance is 40 kΩ. The source voltage is 2.2 V rms. What is the rms value of the amplifier's output voltage?

7.17 Design an inverting operational-amplifier circuit that will provide an output of 10 V rms when the input is a 1-V-rms signal origi-

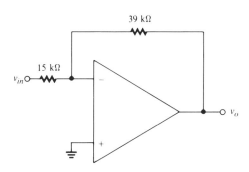

Figure 7.25
(Exercise 7.13)

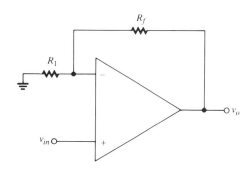

Figure 7.26
(Exercise 7.18)

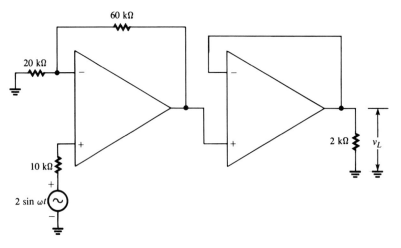

Figure 7.27
(Exercise 7.20)

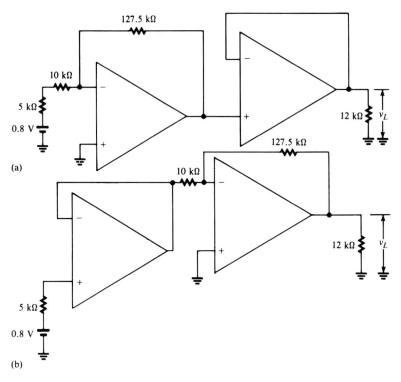

(a)

(b)

Figure 7.28
(Exercise 7.21)

nating at a source having 10 kΩ source resistance.

Section 7.6   The Noninverting Amplifier

7.18   The input to the ideal operational amplifier
shown in Figure 7.26 is 0.5 V rms. Find the
rms value of the output for each of the following combinations of resistor values:
  a. $R_1 = R_f = 10$ kΩ
  b. $R_1 = 20$ kΩ, $R_f = 100$ kΩ
  c. $R_1 = 100$ kΩ, $R_f = 20$ kΩ
  d. $R_f = 10R_1$

7.19   Repeat Exercise 7.18 for each of the following resistor combinations:
  a. $R_1 = 125$ kΩ, $R_f = 1$ MΩ
  b. $R_1 = 220$ kΩ, $R_f = 47$ kΩ
  c. $R_1/R_f = 0.1$
  d. $R_1/R_f = 10$

7.20   Assuming ideal operational amplifiers, find
the load voltage v_L in Figure 7.27.

7.21   Assuming ideal operational amplifiers, find
the load voltage v_L in each part of Figure
7.28.

SPICE EXERCISES

7.22   The JFET differential amplifier in Figure
7.12 has $V_{DD} = 24$ V, $R_D = 3$ kΩ, and a
6-mA constant current source supplying
bias current. To investigate the effects of
parameter variability on the balance of the
amplifier, use SPICE to find the dc output
voltages, V_{o1} and V_{o2}, and the currents in
each source, I_{s1} and I_{s2}, when
  a. Q_1 and Q_2 are perfectly matched, each
having $V_p = -2$ V and $I_{DSS} = 12$ mA;
  b. Q_1 has $V_p = -1.8$ V and Q_2 has $V_p =
-2.2$ V, while $I_{DSS} = 12$ mA for both;
  c. the conditions of (b) exist, except 100-Ω
resistors are inserted in series with the
source of each JFET.

For each case, use the results of the SPICE
simulations to find the difference in the dc
output voltages and the difference in the
source currents. In particular, compare
cases (b) and (c) and comment.

8

Practical Operational Amplifiers—Characteristics and Limitations

8.1   FEEDBACK THEORY

In Chapter 7 we discussed some useful applications of operational amplifiers that were assumed to have infinite open-loop gains. Of course, no practical amplifier has infinite gain. We wish now to investigate the performance of practical inverting and noninverting amplifier configurations using amplifiers with finite open-loop gains. Toward that end, we must undertake a study of the theory of feedback. In the process, we will discover some important consequences of its use. Feedback theory is widely used to study the behavior of electronic components as well as complex systems in many different technical fields, so it is important to develop an appreciation and understanding of its underlying principles.

Feedback in the Noninverting Amplifier

We begin our study of feedback principles with an analysis of the noninverting amplifier. Figure 8.1 shows that configuration along with an equivalent block diagram on which we can identify the signal and feedback paths. The block labeled A represents the amplifier and its open-loop gain, and the block labeled β is the feedback path. The quantity β is called the *feedback ratio* and represents the portion of the output voltage that is fed back to the input. For example, if $\beta = 0.5$, then a voltage equal to one-half the output level is fed back to the input. Notice the special symbol where the input and feedback paths come together. This symbol represents the differential action at the amplifier's input. It is usually called a *summing* junction, although in our case it is performing a differencing operation, as indicated by the $+$ and $-$ symbols. The output of the junction, which is the input to the amplifier, is seen to be $v_e = v_{in} - v_f$. v_e is often called the *error* voltage. Note that it corresponds to $v_{in} - v_i^-$ in the noninverting amplifier, and under ideal conditions is equal to 0 (equation 7.31). The feedback voltage $v_f = \beta v_o$ corresponds to v_i^- in the amplifier circuit. Since the feedback voltage subtracts from the input voltage, the amplifier is said to have *negative feedback*.

With reference to Figure 8.1(b), we see that

$$v_o = A(v_{in} - v_f) \qquad \textbf{(8.1)}$$

Figure 8.1
Block-diagram representation of the noninverting amplifier. Identify corresponding voltages in the two diagrams.

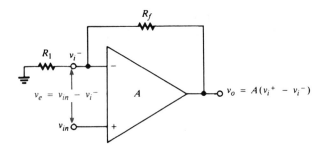

(a) Noninverting amplifier

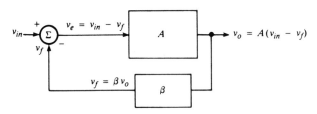

(b) Block-diagram representation of (a)

and

$$v_f = \beta v_o \qquad (8.2)$$

Substituting (8.2) into (8.1) gives $v_o = A(v_{in} - \beta v_o) = Av_{in} - A\beta v_o$, or $v_o(1 + A\beta) = Av_{in}$. Thus,

$$\frac{v_o}{v_{in}} = \frac{A}{1 + A\beta} = \frac{1/\beta}{1 + 1/A\beta} \qquad (8.3)$$

Equation 8.3 is a very important and very useful result. It expresses the closed-loop gain v_o/v_{in} as a function of the open-loop gain A and the feedback ratio β. We can now apply this result to the noninverting amplifier in Figure 8.1(a). Notice that R_f and R_1 form a voltage divider across the output of the amplifier, so

$$v_i^- = \left(\frac{R_1}{R_1 + R_f}\right) v_o \qquad (8.4)$$

Since v_i^- is the voltage fed back from the output, and $v_f = \beta v_o$, we conclude that

$$\beta = \frac{R_1}{R_1 + R_f} \qquad \text{(noninverting amplifier)} \qquad (8.5)$$

Substituting into (8.3), we find

$$\frac{v_o}{v_{in}} = \frac{(R_1 + R_f)/R_1}{1 + 1/A\beta} = \frac{(R_1 + R_f)/R_1}{1 + (R_1 + R_f)/AR_1} \qquad (8.6)$$

Equation 8.6 gives us the means for investigating how significant the value of open-loop gain A is in the determination of the closed-loop gain v_o/v_{in}. First, note that when $A = \infty$, (8.6) reduces to $v_o/v_{in} = (R_1 + R_f)/R_1$, which is exactly the same result we obtained in Section 7.6 for the ideal, noninverting amplifier (equation 7.33). Notice also that

$$\frac{v_o}{v_{in}} = \frac{1}{\beta} \qquad \text{(ideal noninverting amplifer, } A = \infty) \tag{8.7}$$

Equation 8.7 can also be obtained by letting $A = \infty$ in equation 8.3.
    The next example shows how finite values of A affect the value of v_o/v_{in}.

Example 8.1

Find the closed-loop gain of the amplifier in Figure 8.2 when (1) $A = \infty$, (2) $A = 10^6$, and (3) $A = 10^3$.

Figure 8.2
(Example 8.1)

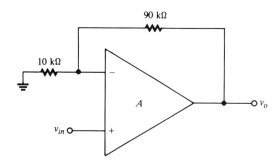

90 kΩ

10 kΩ

A

v_{in}

v_o

Solution

1. The feedback ratio is

$$\beta = \frac{R_1}{R_1 + R_f} = \frac{10 \text{ k}\Omega}{(10 \text{ k}\Omega) + (90 \text{ k}\Omega)} = 0.1$$

Therefore, the closed-loop gain when $A = \infty$ is $v_o/v_{in} = 1/\beta = 1/0.1 = 10$.

2. Using equation 8.3, the closed-loop gain when $A = 10^6$ is

$$\frac{v_o}{v_{in}} = \frac{A}{1 + A\beta} = \frac{10^6}{1 + 10^6(0.1)} = 9.99990$$

We see that v_o/v_{in} is for all practical purposes the same value when $A = 10^6$ as it is when $A = \infty$.

3. When $A = 10^3$,

$$\frac{v_o}{v_{in}} = \frac{10^3}{1 + 10^3(0.1)} = 9.90099$$

We see that reducing A to 1000 creates a discrepancy of about 1% with respect to the value of v_o/v_{in} when $A = \infty$.

Equation 8.3 shows that the closed-loop gain of a real amplifier also departs from that of an ideal amplifier when the value of β becomes very small. Small values of β correspond to large closed-loop gains.

Example 8.2

An operational amplifier has open-loop gain $A = 10,000$. Compare its closed-loop gain with that of an ideal amplifier when (1) $\beta = 0.1$, and (2) $\beta = 0.001$.

Solution

1. $\beta = 0.1$. For $A = \infty$, $v_o/v_{in} = 1/\beta = 10$. For $A = 10^4$,

$$\frac{v_o}{v_{in}} = \frac{A}{1 + A\beta} = \frac{10^4}{1 + 10^4(0.1)} = 9.99$$

2. $\beta = 0.001$. For $A = \infty$, $v_o/v_{in} = 1/\beta = 1000$. For $A = 10^4$,

$$\frac{v_o}{v_{in}} = \frac{A}{1 + A\beta} = \frac{10^4}{1 + 10^4(10^{-3})} = 909.09$$

We see that when $\beta = 0.001$, v_o/v_{in} departs more from the ideal case than it does when $\beta = 0.1$.

The last two examples have shown that the closed-loop gain departs from the ideal value of $1/\beta$ when A is small or when β is small. We can deduce that fact from another examination of equation 8.3:

$$\frac{v_o}{v_{in}} = \frac{1/\beta}{1 + 1/A\beta}$$

Clearly both A and β should be large if we want v_o/v_{in} to equal $1/\beta$. The product $A\beta$ is called the *loop gain* and is very useful in predicting the behavior of a feedback system. The name *loop gain* is derived from its definition as the product of the gains in the feedback model as one travels around the loop from amplifier input, through the amplifier, and through the feedback path (with the summing junction open).

Negative feedback improves the performance of an amplifier in several ways. In the case of the noninverting amplifier, it can be shown that the input resistance seen by the signal source (looking directly into the + terminal) is

$$r_{in} = (1 + A\beta)r_{id} \approx A\beta r_{id} \tag{8.8}$$

where r_{id} is the differential input resistance of the amplifier. This equation shows that the input resistance is r_{id} multiplied by the factor $1 + A\beta$, which is usually much greater than 1 and is approximately equal to the loop gain $A\beta$. For example, if $r_{id} = 20$ kΩ, $A = 10^5$, and $\beta = 0.01$, then $r_{in} \approx (20 \times 10^3\ \Omega)(10^5)(0.01) = 20$ MΩ, a very respectable value. In the case of the voltage follower, $\beta = 1$, and r_{id} is multiplied by the full value of A, which accounts for the extremely large input resistance it can provide in buffer applications.

The closed-loop output resistance of the noninverting amplifier is also improved by negative feedback:

$$r_o(\text{stage}) = \frac{r_o}{1 + A\beta} \approx \frac{r_o}{A\beta} \tag{8.9}$$

where r_o is the open-loop output resistance of the amplifier. Equation 8.9 shows that the output resistance is decreased by the same factor by which the input resistance is increased. A typical value for r_o is 75 Ω, so with $A = 10^5$ and $\beta =$

0.01, we have $r_o(\text{stage}) \approx 75\ \Omega/10^3 = 0.075\ \Omega$, which is very close to the ideal value of 0.

Feedback in the Inverting Amplifier

To investigate the effect of open-loop gain A and feedback ratio β on the closed-loop gain of the inverting amplifier, let us recall equations 7.24 and 7.25 from Section 7.5:

$$\frac{v_{in}}{R_1} - \frac{v_i^-}{R_1} = \frac{v_i^-}{R_f} - \frac{v_o}{R_f} \tag{8.10}$$

$$v_i^- = -v_o/A \tag{8.11}$$

Substituting (8.11) into (8.10) gives

$$\frac{v_{in}}{R_1} + \frac{v_o}{AR_1} = \frac{-v_o}{AR_f} - \frac{v_o}{R_f} \tag{8.12}$$

Exercise 8.6 at the end of this chapter is included to show that (8.12) can be solved for v_o/v_{in} with the result

$$\frac{v_o}{v_{in}} = \frac{-R_f/R_1}{1 + (R_1 + R_f)/AR_1} \tag{8.13}$$

Once again we see that the closed-loop gain reduces to the ideal amplifier value, $-R_f/R_1$, when $A = \infty$. Notice that the denominator of (8.13) is the same as that of (8.6), the equation for the closed-loop gain of the noninverting amplifier. Furthermore, the quantity $R_1/(R_1 + R_f)$ is also the feedback ratio β for the inverting amplifier. This fact is illustrated in Figure 8.3, which shows the feedback paths of both configurations when their signal inputs are grounded. Think of the amplifier output as a source that generates the feedback signal. By the superposition principle, we can analyze the contribution of the feedback source by grounding all

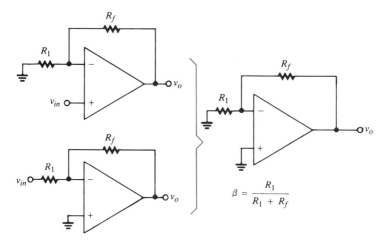

Figure 8.3
When v_{in} is grounded in both the inverting and noninverting amplifiers, it can be seen that the feedback paths are identical.

other signal sources. When this is done, as shown in Figure 8.3, we see that the feedback voltage in both configurations is developed across the R_1–R_f voltage divider, and $\beta = R_1/(R_1 + R_f)$ in both cases. In view of this fact, we can write (8.13) as

$$\frac{v_o}{v_{in}} = \frac{-R_f/R_1}{1 + 1/A\beta} \tag{8.14}$$

Toward developing a feedback model for the inverting amplifier, consider the block diagram shown in Figure 8.4. This diagram is quite similar to Figure 8.1(b) for the noninverting amplifier, except that we now denote the open loop gain by $-A$. Note also that the summing junction now *adds* its two inputs. Since v_o is inverted, so is the feedback voltage, and adding a negative voltage to v is the same as subtracting a positive one from it. In other words, we still have a negative-feedback situation. Notice that we use v to represent an arbitrary input voltage, rather than v_{in}, because we will have to make some adjustments in this model before it can truly represent the inverting amplifier.

As shown in Figure 8.4,

$$v_o = -A(v + \beta v_o) \tag{8.15}$$

Solving for v_o/v, we find

$$\frac{v_o}{v} = \frac{-A}{1 + A\beta} = \frac{-1/\beta}{1 + 1/A\beta} \tag{8.16}$$

Comparing (8.16) with the equation we have already developed for the inverting amplifier (equation 8.14), we see that they differ slightly. We must therefore adjust the model so that it produces the same result as (8.14). Equation 8.16 for the model can be written

$$\frac{v_o}{v} = \frac{\dfrac{-(R_1 + R_f)}{R_1}}{1 + 1/A\beta} \tag{8.17}$$

If the right side of (8.17) is multiplied by the factor $R_f/(R_1 + R_f)$, we obtain

$$\frac{v_o}{v} = \left[\frac{\dfrac{-(R_1 + R_f)}{R_1}}{1 + 1/A\beta}\right] \frac{R_f}{(R_1 + R_f)} = \frac{-R_f/R_1}{1 + 1/A\beta} \tag{8.18}$$

Equation 8.18 shows that multiplication of the model equation by the constant $R_f/(R_1 + R_f)$ gives us exactly the same result (equation 8.14 with $v_{in} = v$) that we

Figure 8.4
First step in the development of a feed-back model for the inverting amplifer

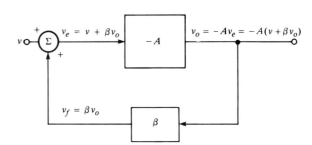

Figure 8.5
The complete feedback model for the inverting amplifier

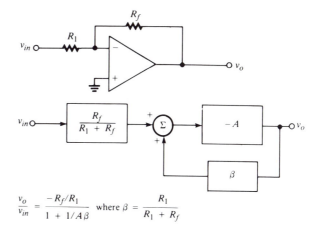

$$\frac{v_o}{v_{in}} = \frac{-R_f/R_1}{1 + 1/A\beta} \quad \text{where } \beta = \frac{R_1}{R_1 + R_f}$$

obtain for the inverting amplifier. Therefore, we modify the block-diagram model in Figure 8.4 by adding a block that multiplies the input by $R_f/(R_1 + R_f)$. The complete feedback model is shown in Figure 8.5.

As can be seen in Figure 8.5, the loop gain for the inverting amplifier is $A\beta$, the same as that for the noninverting amplifier. From equation 8.14, it is apparent that the greater the loop gain, the closer the closed-loop gain is to its value in the ideal inverting amplifier, $-R_f/R_1$.

It can be shown that the input resistance seen by the signal source driving the inverting amplifier is

$$r_{in} = R_1 + \frac{R_f}{1 + A} \approx R_1 \tag{8.19}$$

This equation confirms that the input resistance is R_1 for the ideal inverting amplifier, where $A = \infty$. It also shows that the input resistance decreases with increasing values of A.

As with the noninverting amplifier, the output resistance of the inverting amplifier is decreased by negative feedback. In fact, the relationship between output resistance and loop gain is the same for both:

$$r_o(\text{stage}) = \frac{r_o}{1 + A\beta} \approx \frac{r_o}{A\beta} \tag{8.20}$$

Example 8.3

The amplifier shown in Figure 8.6 has open-loop gain equal to -2500 and open-loop output resistance $100 \ \Omega$. Find

1. the magnitude of the loop gain;
2. the closed-loop gain;
3. the input resistance seen by v_{in}; and
4. the closed-loop output resistance.

Solution

1. $$\beta = R_1/(R_1 + R_f) = (1.5 \text{ k}\Omega)/[(1.5 \text{ k}\Omega) + (150 \text{ k}\Omega)] = 9.90099 \times 10^{-3}$$
   $$\text{loop gain} = A\beta = (2.5 \times 10^3)(9.90099 \times 10^{-3}) = 24.75$$

Figure 8.6
(Example 8.3)

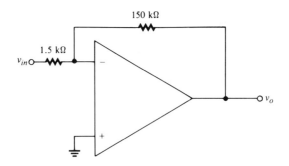

2. From equation 8.14,

$$\frac{v_o}{v_{in}} = \frac{-R_f/R_1}{1 + 1/A\beta} = \frac{-(150 \text{ k}\Omega)/(1.5 \text{ k}\Omega)}{1 + 1/24.75} = -96.12$$

Note that this value is about 4% less than $-R_f/R_1 = -100$.

3. From equation 8.19,

$$r_{in} = R_1 + \frac{R_f}{1 + A} = (1.5 \text{ k}\Omega) + \frac{150 \ \Omega}{1 + 2500} = 1560 \ \Omega$$

4. From equation 8.20,

$$r_o(\text{stage}) = \frac{r_o}{1 + A\beta} = \frac{100}{1 + 24.75} = 3.88 \ \Omega$$

Example 8.17 at the end of the chapter shows how SPICE can be used to verify the results of this example.

We should note once again that the same relationship between actual and ideal closed-loop gain applies to inverting and noninverting amplifiers. This relationship is

$$\text{actual } \frac{v_o}{v_{in}} = \frac{(\text{ideal closed-loop gain})}{1 + 1/A\beta} \qquad (8.21)$$

where (ideal closed-loop gain) is the closed-loop gain v_o/v_{in} that would result if the amplifier were ideal. We saw this relationship in equations 8.6 and 8.14, repeated here:

$$\frac{v_o}{v_{in}} = \frac{(R_1 + R_f)/R_1}{1 + 1/A\beta} \qquad \text{(noninverting amplifier)}$$

$$\frac{v_o}{v_{in}} = \frac{-R_f/R_1}{1 + 1/A\beta} \qquad \text{(inverting amplifier)}$$

In both cases, the numerator is the closed-loop gain that would result if the amplifier were ideal. Also in both cases, the greater the value of the loop gain $A\beta$, the closer the actual closed-loop gain is to the ideal closed-loop gain.

Although we have demonstrated this relationship only for inverting and non-inverting amplifiers, it is a fact that equation 8.21 applies to a wide variety of amplifier configurations, many of which we shall be examining in future discussions.

8.2   HARMONIC DISTORTION AND FEEDBACK

Harmonic Distortion

Recall that any periodic waveform, sinusoidal or otherwise, can be represented as the sum of an infinite number of sine waves having different amplitudes, frequencies, and phase relations. The mathematical technique called *Fourier analysis* is concerned with finding the exact amplitudes, frequencies, and phase angles of the sine-wave components that reproduce a given waveform when they are added together. Of course, if the waveform is itself a pure sine wave, then all other frequency components have zero amplitudes.

Apart from a possible dc component (the average value), the lowest frequency component in the infinite sum is a sine wave having the same frequency as the periodic waveform itself. This component is called the *fundamental* and usually has an amplitude greater than that of any other frequency component. The other frequency components are called *harmonics,* and each has a frequency that is an *integer multiple* of the fundamental frequency. For example, the second harmonic of a 3-kHz waveform has frequency 6 kHz, the third harmonic has frequency 9 kHz, and so forth. Not all harmonics need be present. For example, a square wave contains only odd harmonics (third, fifth, seventh, etc.).

As we have indicated in previous discussions, nonlinear amplifier characteristics are responsible for harmonic distortion of an output signal. This distortion is in fact the creation of harmonic frequencies that would not otherwise be present in the output when the input is a pure sine wave. The extent to which a particular harmonic component distorts a signal is specified by the ratio of its amplitude to the amplitude of the fundamental component, expressed as a percentage:

$$\% \; n\text{th harmonic distortion} = \%D_n = \frac{A_n}{A_1} \times 100\% \qquad (8.22)$$

where A_n is the amplitude of the nth harmonic component and A_1 is the amplitude of the fundamental. For example, if a 1-V-peak, 10-kHz sine wave is distorted by the addition of a 0.1-V-peak, 20-kHz sine wave and a 0.05-V-peak, 30-kHz sine wave, then it has

$$\%D_2 = \frac{0.1}{1} \times 100\% = 10\% \text{ second harmonic distortion}$$

and

$$\%D_3 = \frac{0.05}{1} \times 100\% = 5\% \text{ third harmonic distortion}$$

The *total harmonic distortion* (THD) is the square root of the sum of the squares of all the individual harmonic distortions:

$$\%\text{THD} = \sqrt{D_2^2 + D_3^2 + \cdots} \times 100\% \qquad (8.23)$$

Special instruments, called *distortion analyzers,* are available for measuring the total harmonic distortion in a waveform.

Example 8.4

The principal harmonics in a certain 10-V-peak, 15-kHz signal having a 10-Vpk fundamental are the second and fourth. All other harmonics are negligibly small.

If the THD is 12% and the amplitude of the second harmonic is 0.5 V, what is the amplitude of the 60-kHz harmonic?

Solution

$$THD = 0.12 = \sqrt{D_2^2 + D_4^2}$$
$$0.0144 = D_2^2 + D_4^2$$
$$D_2 = 0.5/10 = 0.05 \quad (5\%)$$
$$0.0144 = (0.05)^2 + D_4^2$$
$$D_4 = \sqrt{0.0144 - 0.0025} = 0.1091 \quad (10.91\%)$$
$$A_4 = D_4 A_1 = (0.1091)(10) = 1.091 \text{ V}$$

Using Negative Feedback to Reduce Distortion

One of the important benefits of negative feedback is that it reduces distortion caused by amplifier nonlinearities. Having defined a quantitative measure of distortion, we can now undertake a quantitative investigation of the degree to which feedback affects harmonic distortion in the output of an amplifier. Let us begin by recognizing that a nonlinear amplifier is essentially an amplifier whose gain changes with signal level. Figure 8.7(a) shows the transfer characteristic of an ideal, distortionless amplifier, in which the gain, i.e., the slope of the characteristic, $\Delta V_o/\Delta V_{in} = 50$, is constant. Figure 8.7(b) shows a nonlinear transfer characteristic in which the slope, and hence the gain, increases with increasing signal level. We see that the gain at $V_{in} = 0.2$ V is 50 while the gain at 0.4 V is 100. There is a 100% increase in gain over that range, so serious output distortion is to be expected.

In Section 8.1, we showed that the closed-loop gain of an amplifier having open-loop gain A and feedback ratio β can be found from

$$\frac{v_o}{v_{in}} = \frac{A}{1 + A\beta} \tag{8.24}$$

Let us suppose we introduce negative feedback into the amplifier with the nonlinear characteristic shown in Figure 8.7(b). Assume that $\beta = 0.05$. Then the gain at $v_{in} = 0.2$ V becomes

$$\frac{v_o}{v_{in}} = \frac{50}{1 + 50(0.05)} = 14.29$$

and the gain at $v_{in} = 0.4$ V becomes

$$\frac{v_o}{v_{in}} = \frac{100}{1 + 100(0.05)} = 16.67$$

Notice that the *change* in gain (16.7%) is now much less than it was without feedback (100%). The effect of feedback has been to "linearize" the transfer characteristics somewhat, as shown in Figure 8.7(c). We conclude that less-severe distortion will result. Of course, the penalty we pay for this improved performance is an overall reduction in gain.

To compute the reduction in harmonic distortion caused by negative feedback, consider the models shown in Figure 8.8. Figure 8.8(a) shows how amplifier distortion (without feedback) can be represented as a distortionless amplifier hav-

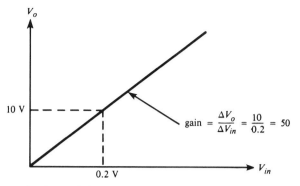

(a) Transfer characteristic of an ideal, distortionless amplifier having gain 50

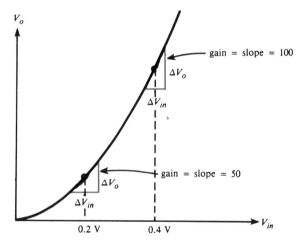

(b) Nonlinear transfer characteristic

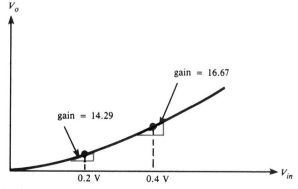

(c) Negative feedback ($\beta = 0.05$) reduces the gain but linearizes the characteristic

Figure 8.6
Transfer characteristics and the effect of negative feedback

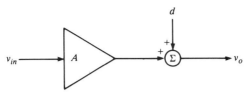

(a) Amplifier distortion represented by summing a distortion component with the amplifier output

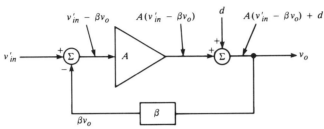

(b) Model for an amplifier with distortion component d and negative feedback ratio β

Figure 8.8
Amplifier distortion models

ing a distortion component d added to its output. For example, d could represent the level of a second-harmonic sine wave added to an otherwise undistorted fundamental. The output of the amplifier is

$$v_o = Av_{in} + d \tag{8.25}$$

so the harmonic distortion is

$$D = \frac{d}{Av_{in}} \tag{8.26}$$

Figure 8.8(b) shows the amplifier when negative feedback is connected. We realize that the negative feedback will reduce the closed-loop gain, so the output level will be reduced if v_{in} remains the same as in (a). But, for comparison purposes, we want the output levels in both cases to be the same, since the amount of distortion depends on that level. Accordingly, we assume that v_{in} in (b) is increased to v'_{in}, as necessary to make the distortion component d equal to its value in (a). As shown in the figure,

$$v_o = A(v'_{in} - \beta v_o) + d \tag{8.27}$$

Solving for v_o, we find

$$v_o = \frac{Av'_{in}}{1 + A\beta} + \frac{d}{1 + A\beta} \tag{8.28}$$

When we adjusted v_{in} to v'_{in}, we did it so that the amplifier outputs in (a) and (b) of the figure would be the same:

$$Av_{in} = \frac{A}{1 + A\beta} v'_{in} \tag{8.29}$$

or

$$v'_{in} = (1 + A\beta)v_{in} \qquad (8.30)$$

Substituting (8.30) into (8.28), we find

$$v_o = Av_{in} + \frac{d}{1 + A\beta} \qquad (8.31)$$

The harmonic distortion is now

$$D = \frac{d/(1 + A\beta)}{Av_{in}} = \frac{d}{(1 + A\beta)Av_{in}} \qquad (8.32)$$

Comparing with equation 8.26, we see that the distortion has been reduced by the factor $1/(1 + A\beta)$.

It is important to remember that it was necessary to increase the input level in our analysis in order to achieve the improved performance. From a practical standpoint, that means that a relatively distortion-free *preamplifier* may be necessary to compensate for the loss in gain caused by negative feedback around the output amplifier.

Example 8.5

An amplifier has a voltage gain of 120 and generates 20% harmonic distortion with no feedback.

1. How much negative feedback should be used if it is desired to reduce the distortion to 2%?
2. How much preamplifier gain will have to be provided in cascade with the output amplifier to maintain an overall gain equal to that without feedback?

Solution

1. From equation 8.32,

$$D = \frac{d}{(1 + A\beta)Av_{in}}$$

where $d/Av_{in} = 0.2$ is the distortion without feedback. Then,

$$0.02 = \frac{0.2}{1 + A\beta} = \frac{0.2}{1 + 120\beta}$$

Solving for β gives $\beta = 0.075$. Thus, the negative feedback voltage must be 7.5% of the output voltage.

2. The closed-loop gain with 7.5% feedback is

$$\frac{v_o}{v_{in}} = \frac{A}{1 + A\beta} = \frac{120}{1 + 120(0.075)} = 12$$

To maintain an overall gain of 120, the preamplifier gain, A_p, must be such that $12A_p = 120$, or $A_p = 10$.

8.3   FREQUENCY RESPONSE

Stability

When the word *stability* is used in connection with a high-gain amplifier, it usually means the property of behaving like an amplifier rather than like an *oscillator*. An oscillator is a device that spontaneously generates an ac signal becuase of *positive* feedback. We will study oscillator theory in a later chapter, but for now it is sufficient to know that oscillations are easily induced in high-gain, wide-bandwidth amplifiers, due to positive feedback that occurs through reactive elements. As we know, an operational amplifier has very high gain, so precautions must be taken in its design to ensure that it does not oscillate, i.e., to ensure that it remains stable. Large gains at high frequencies tend to make an amplifier unstable because those properties enable positive feedback through stray capacitance.

To ensure stable operation, most operational amplifiers have internal *compensation* circuitry that causes the open-loop gain to diminish with increasing frequency. This reduction in gain is called ''rolling-off'' the amplifier. Sometimes it is necessary to connect external roll-off networks to reduce high-frequency gain even more rapidly. Because the dc and low-frequency open-loop gain of an operational amplifier is so great, the gain roll-off must begin at relatively low frequencies. As a consequence, the open-loop bandwidth of an operational amplifier is generally rather small.

In most operational amplifiers, the gain over the usable frequency range rolls off at the rate of −20 dB/decade, or −6 dB/octave. Recall from Chapter 6 that this rate of gain reduction is the same as that of a single low-pass RC network. Any device whose gain falls off like that of a single RC network is said to have a *single-pole* frequency response, a name derived from advanced mathematical theory. Beyond a certain very high frequency, the frequency response of an operational amplifier exhibits further break frequencies, meaning that the gain falls off at greater rates, but for most practical analysis purposes we can treat the amplifier as if it had a single-pole response.

The Gain-Bandwidth Product

Figure 8.9 shows a typical frequency response characteristic for the open-loop gain of an operational amplifier having a single-pole frequency response, plotted

Figure 8.9

Frequency response of the open-loop gain of an operational amplifier; f_c = cutoff frequency, f_t = unit-gain frequency

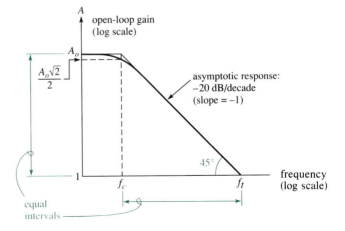

on log-log scales. We use f_c to denote the cutoff frequency, which, as usual, is the frequency at which the gain A falls to $\sqrt{2}/2$ times its low-frequency, or dc, value (A_0). Frequency f_t is the *unity-gain frequency,* the frequency at which the gain equals 1. Note that the slope of -20 dB/decade is the same as a slope of -1 on log-log scales, since 20 dB is a ten-to-one change along the vertical axis and a decade is a ten-to-one change along the horizontal axis. Therefore, along the asymptote shown in Figure 8.9, the difference between the logarithms of A_0 and 1 must be the same as the difference between the logarithms of f_t and f_c:

$$\log A_0 - \log 1 = \log f_t - \log f_c$$

or,

$$\log A_0 - 0 = \log f_t/f_c$$

Taking the antilog of both sides gives

$$A_0 = f_t/f_c$$

or,

$$f_t = A_0 f_c \qquad (8.33)$$

Thus, the frequency at which the amplifier gain falls to 1 is the product of the cutoff frequency and the low-frequency gain A_0.

Since the amplifier is dc (lower cutoff frequency = 0), the bandwidth equals f_c. The term $A_0 f_c$ in equation 8.33 is called the *gain-bandwidth* product. In specifications, a value may be quoted either for the gain-bandwidth product, or for its equivalent, the unity-gain frequency.

The significance of the gain-bandwidth product is that it makes it possible for us to compute the bandwidth of an amplifier when it is operated in one of the more useful closed-loop configurations. Obviously, a knowledge of the upper-frequency limitation of a certain amplifier configuration is vital information when designing it for a particular application. The relationship between closed-loop bandwidth (BW_{CL}) and the gain-bandwidth product is closely approximated by

$$BW_{CL} = f_t \beta = A_0 f_c \beta \qquad (8.34)$$

where β is the feedback ratio.

Example 8.6

Each of the amplifiers shown in Figure 8.10 has an open-loop, gain-bandwidth product equal to 1×10^6. Find the cutoff frequencies in the closed-loop configurations shown.

Figure 8.10
(Example 8.6)

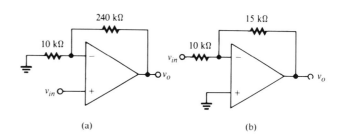

(a) (b)

Solution

1. In Figure 8.10(a), $\beta = R_1/(R_1 + R_f) = (10 \text{ k}\Omega)/[(10 \text{ k}\Omega) + (240 \text{ k}\Omega)] = 0.04$. From equation 8.34, $\text{BW}_{\text{CL}} = f_t\beta = (10^6)(0.04) = 40 \text{ kHz}$.

   Since the amplifier is dc, the closed-loop cutoff frequency is the same as the closed-loop bandwidth, 40 kHz.

2. In Figure 8.10, $\beta = R_1/(R_1 + R_f) = (10 \text{ k}\Omega)/[(10 \text{ k}\Omega) + (15 \text{ k}\Omega)] = 0.4$. Then $\text{BW}_{\text{CL}} = 10^6(0.4) = 400 \text{ kHz}$.

It is worthwhile noting that in the case of the *noninverting* amplifier, the fact that the ideal closed-loop gain is $1/\beta$ makes equation 8.34 equivalent to

$$\text{BW}_{\text{CL}} = f_t/(\text{ideal closed-loop gain}) \tag{8.35}$$

or (ideal closed-loop gain) × (closed-loop bandwidth) = gain-bandwidth product. To illustrate the validity of this expression, refer to part (1) of Example 8.6. Here, the ideal closed-loop gain is $(R_1 + R_f)/R_1 = (250 \text{ k}\Omega)/(10 \text{ k}\Omega) = 25$, so 25 × (closed-loop bandwidth) $= 10^6$, which yields

$$\text{closed-loop bandwidth} = \text{BW}_{\text{CL}} = 10^6/25 = 40 \text{ kHz} \qquad \text{(correct)}$$

Equation 8.35 is *not* valid for the inverting amplifier. In part (2) of Example 8.6, we have

$$\text{ideal closed-loop gain} = R_f/R_1 = (15 \text{ k}\Omega)/(10 \text{ k}\Omega) = 1.5$$

If we now apply equation 8.35, we obtain

$$\text{BW}_{\text{CL}} = 10^6/1.5 = 666.6 \text{ kHz} \qquad \text{(incorrect)}$$

Although some authors interpret the gain-bandwidth product to be the product of closed-loop gain and closed-loop bandwidth regardless of configuration, we have seen that this interpretation yields a bandwidth for the inverting amplifier that is larger than its actual value. At large values of closed-loop gain, the bandwidths of the inverting and noninverting amplifiers are comparable, but at low gains the noninverting amplifier has a larger bandwidth. For example, when the closed-loop gain is 1, the bandwidth of the noninverting amplifier is twice that of the inverting amplifier.

Figure 8.11 shows a typical set of frequency response plots for a noninverting

Figure 8.11
A typical set of frequency response plots for a noninverting amplifier

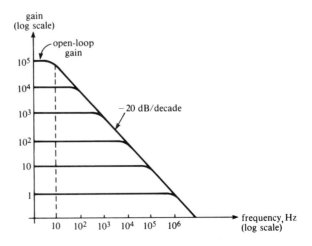

amplifier, as the gain ranges from its open-loop value of 10^5 to a closed-loop value of 1. This figure clearly shows how the bandwidth decreases as the closed-looped gain increases. Notice that the bandwidth at maximum (open-loop) gain is only 10 Hz.

Example 8.7

With reference to the amplifier whose frequency response is shown in Figure 8.11, find

1. the unity-gain frequency;
2. the gain-bandwidth product;
3. the bandwidth when the feedback ratio is 0.02; and
4. the closed-loop gain at 0.4 MHz when the feedback ratio is 0.04.

Solution

1. In Figure 8.11, it is apparent that the open-loop gain equals 1 when the frequency is 1 MHz. Thus, $f_t = 1$ MHz.
2. Gain-bandwidth product $= A_0 f_c = f_t = 10^6$.
3. From equation 8.34, $BW_{CL} = f_t\beta = 10^6(0.02) = 20$ kHz.
4. $BW_{CL} = f_t\beta = 10^6(0.04) = 40$ kHz. Thus, the closed-loop cutoff frequency is 40 kHz. Since the amplifier is noninverting, the closed-loop gain is $1/\beta = 25$. Since 0.4 MHz is 1 decade above the cutoff frequency, the gain is down 20 dB from 25, or down by a factor of $1/10$: $0.1(25) = 2.5$.

User-Compensated Amplifiers

As noted earlier, many commercially available amplifiers have internal compensation circuitry to make the frequency response roll off at 6 dB/octave (20 dB/decade) over the entire frequency range from f_c to f_t (Figure 8.9). Some amplifiers do not have such circuitry and must be compensated by connecting external roll-off networks. These networks, typically RC circuits, are selected by the user to ensure that the frequency response is 6 dB/octave at the closed-loop gain at which the amplifier is to be operated. Manufacturers' specifications usually include equations for determining the values of the components of external roll-off networks, based on the closed-loop gain desired.

Figure 8.12(a) shows a typical frequency response of an uncompensated amplifier. Compensation is particularly critical when the amplifier is to be operated with a small closed-loop gain, since the bandwidth is then very large and the amplifier's roll-off rate may be 12 or 18 dB/octave, rates that jeopardize stability. Figure 8.12(a) also shows an example of a response that has been compensated to roll off at 6 dB/octave when a particular value of closed-loop gain, A_{CL}, is desired. Note that the roll-off rate would be 12 dB/octave if the uncompensated amplifier were used at that value of closed-loop gain. It is clear that compensation reduces the bandwidth of the amplifier. However, it is generally true that user compensation results in a wider bandwidth than can be achieved with an internally compensated amplifier that rolls off at 6 dB/octave over its entire range. Figure 8.12(b)

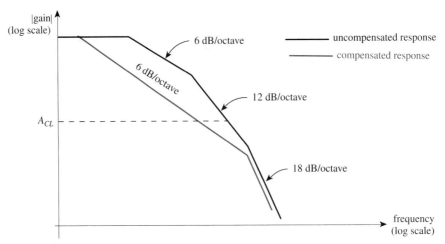

(a) Typical uncompensated and compensated frequency responses. The compensation ensures that the roll-off rate is 6 dB/octave at a desired closed-loop gain, A_{CL}.

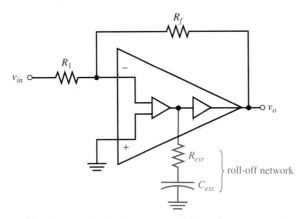

(b) Typical external roll-off network, called *lag phase compensation*, and used to create a 6 dB/octave roll-off

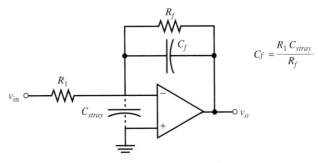

(c) Use of feedback capacitance C_f to compensate for shunt capacitance at the input (lead compensation)

Figure 8.12
User-compensated amplifiers

shows a typical RC network used for external compensation. Note that this network is usually connected to an internal stage of the amplifier at an external terminal that may be identified as "roll-off," "phase," or "frequency compensation."

The external compensation shown in Figure 8.12(b) is called *lag phase* compensation. Figure 8.12(c) shows an example of *lead* compensation, used to offset the effects of input and stray capacitance. The feedback capacitor C_f is selected so that the break frequency due to the combination of R_f and the input shunt capacitance equals the break frequency due to R_f and C_f:

$$\frac{1}{2\pi R_1 C_{stray}} = \frac{1}{2\pi R_f C_f}$$

$$C_f = \frac{R_1 C_{stray}}{R_f} \tag{8.36}$$

8.4 SLEW RATE

We have discussed the fact that internal compensation circuitry used to ensure amplifier stability also affects the frequency response and places a limit on the maximum operating frequency. The capacitor(s) in this compensation circuitry limit amplifier performance in still another way. When the amplifier is driven by a step or pulse-type signal, the capacitance must charge and discharge rapidly in order for the output to "keep up with," or track, the input. Since the voltage across a capacitor cannot be changed instantaneously, there is an inherent limit on the *rate* at which the output voltage can change. The maximum possible rate at which an amplifier's output voltage can change, in volts per second, is called its *slew rate*.

It is not possible for *any* waveform, input or output, to change from one level to another in *zero* time. An instantaneous change corresponds to an *infinite rate of change*, which is not realizable in any physical system. Therefore, in our investigation of performance limitations imposed by an amplifier's slew rate, we need concern ourselves only with inputs that undergo a total change in voltage, ΔV, over some nonzero time interval, Δt. For simplicity, we will assume that the change is linear with respect to time; that is, it is a *ramp*-type waveform, as illustrated in Figure 8.13. The rate of change of this kind of waveform is the change in voltage divided by the length of time that it takes for the change to occur:

$$\text{rate of change} = \frac{V_2 - V_1}{t_2 - t_1} = \frac{\Delta V}{\Delta t} \text{ volts/second} \tag{8.37}$$

Figure 8.13
The rate of change of a linear, or ramp, signal is the change in voltage divided by the change in time.

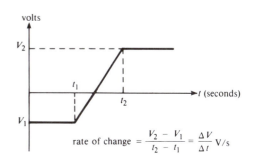

Since the value specified for the slew rate of an amplifier is the maximum rate at which its output can change, we cannot drive the amplifier with any kind of input waveform that would require the output to exceed that rate. For example, if the slew rate is 10^6 V/s (a typical value), we could not drive an amplifier having unity gain with a signal that changes from -5 V to $+5$ V in 0.1 μs, because that would require the output to change at the rate $\Delta V/\Delta t = (10 \text{ V})/(10^{-7} \text{ s}) = 10^8$ V/s. Similarly, we could not drive an amplifier having a gain of 10 with an input that changes from 0 V to 1 V in 1 μs because that would require the output to change from 0 V to 10 V in 1 μs, giving $\Delta V/\Delta t = 10/10^{-6} = 10^7$ V/s. When we say we "could not" drive the amplifier with these inputs, we simply mean that we could not do so and still expect the output to be a faithful replica of the input.

In specifications, the slew rate is often quoted in the units volts per microsecond. Of course, 1 V/μs is the same as 10^6 V/s: $(1 \text{ V})/(10^{-6} \text{ s}) = 10^6$ V/s.

Example 8.8

The operational amplifier in Figure 8.14 has a slew rate specification of 0.5 V/μs. If the input is the ramp waveform shown, what is the maximum closed-loop gain that the amplifier can have without exceeding its slew rate?

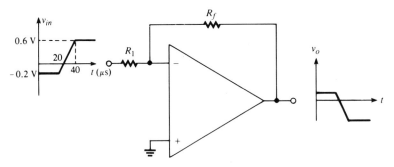

Figure 8.14
(Example 8.8)

Solution.   The rate of change of the input is

$$\frac{\Delta V}{\Delta t} = \frac{V_2 - V_1}{t_2 - t_1} = \frac{0.6 \text{ V} - (-0.2)\text{V}}{(40 - 20) \times 10^{-6} \text{ s}} = 4 \times 10^4 \text{ V/s}$$

Since the slew rate is 0.5 V/μs $= 5 \times 10^5$ V/s, the maximum permissible gain is

$$\frac{5 \times 10^5 \text{ V}}{4 \times 10^4 \text{ V}} = 12.5$$

Notice that the amplifier is connected in an inverting configuration, so the output changes from positive to negative. The inversion is of no consequence as far as slew rate is concerned. With a gain of -12.5, the output will change from

$(-12.5)(-0.2 \text{ V}) = +2.5 \text{ V}$ to $(-12.5)(0.6 \text{ V}) = -7.5 \text{ V}$ in 20 μs, giving

$$\frac{\Delta V}{\Delta t} = \frac{10 \text{ V}}{20 \ \mu\text{s}} = 0.5 \text{ V}/\mu\text{s}$$

the specified slew rate.

Slew rate is a performance specification used primarily in applications where the waveforms are large-signal pulses or steps that cause the output to swing through a substantial part of its total range ($\pm V_{CC}$ volts). *However,* the slew rate imposes a limitation on output rate of change regardless of the nature of the signal waveform. In particular, if the signal is sinusoidal, or a complex waveform containing many different frequencies, we must be certain that no large-amplitude, high-frequency component will require the output to exceed the slew rate. High-frequency signals change (continuously) at rapid rates, and if their amplitudes are so large that the slew rate specification is exceeded, distortion will result. It is especially important to realize that a frequency component may be within the bandwidth of the amplifier, as determined in Section 8.3, but may have such a large amplitude that it must be excluded because of slew rate limitations. The converse is also true: A high-frequency signal that does not exceed the slew rate may have to be excluded because it is outside the amplifier bandwidth. In other words, the maximum frequency at which an amplifier can be operated depends on both the bandwidth and the slew rate, the latter being a function of amplitude as well as frequency. In a later discussion, we will summarize the criteria for determining the operating frequency range of an amplifier based on both slew rate and bandwidth limitations.

When the output of an amplifier is the sine-wave voltage $v_o(t) = K \sin \omega t$, it can be shown using calculus (differentiating with respect to t) that the signal has a maximum rate of change given by

$$\text{rate of change (max)} = K\omega \text{ volts/second} \qquad \textbf{(8.38)}$$

where K is the peak amplitude of the sine wave, in volts, and ω is the angular frequency, in radians/second. (We use K to represent amplitude, rather than the conventional A, to avoid confusion with the symbol for gain.) Equation 8.38 clearly shows that the rate of change is proportional to both the amplitude and the frequency of the signal. If S is the specified slew rate of an amplifier, then we must have

$$K\omega \leq S \quad \text{or} \quad K(2\pi f) \leq S \qquad \textbf{(8.39)}$$

This inequality allows us to solve for the maximum frequency, $f_S(\text{max})$, that the slew-rate limitation permits at the output of an amplifier:

$$f_S(\text{max}) = \frac{S}{2\pi K} \text{ hertz} \quad \text{or} \quad \omega_S(\text{max}) = \frac{S}{K} \text{ radius/second} \qquad \textbf{(8.40)}$$

We emphasize again that $f_S(\text{max})$ is the frequency limit imposed by the slew rate *alone,* i.e., disregarding bandwidth limitations. Also, equation 8.40 applies to sinusoidal signals only. When dealing with complex waveforms containing many different frequency components, the slew rate should be at *least* as great as that necessary to satisfy equation 8.40 for the highest frequency component. Depending on phase relations, maximum rates of change may actually be additive.

Example 8.9

The operational amplifier in Figure 8.15 has a slew rate of 0.5 V/μs. The amplifier must be capable of amplifying the following input signals: $v_1 = 0.01 \sin(10^6 t)$, $v_2 = 0.05 \sin(350 \times 10^3 t)$, $v_3 = 0.1 \sin(200 \times 10^3 t)$, and $v_4 = 0.2 \sin(50 \times 10^3 t)$.

1. Determine whether the output will be distorted due to slew-rate limitations on any input.
2. If so, find a remedy (other than changing the input signals).

Figure 8.15
(Example 8.9)

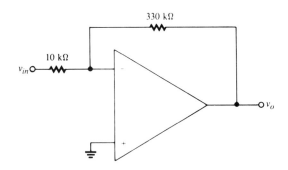

Solution

1. We must check each frequency to verify that $\omega \leq \omega_S(\text{max}) = S/K$ rad/s. Note that K is the peak amplitude at the *output* of the amplifier, so each input amplitude must be multiplied by the closed-loop gain before the check is performed. Assuming that the closed-loop gain is ideal, we have $v_o/v_{in} = -R_f/R_1 = -(330 \text{ k}\Omega)/(10 \text{ k}\Omega) = -33$. Thus, the upper limit on the ω of each signal component will be $S/(33 K_i)$, where K_i is the peak amplitude of the input.

$$v_1: \quad \omega_S(\text{max}) = S/(33K_i) = 0.5 \times 10^6/(33)(0.01) = 1.515 \times 10^6 \text{ rad/s}$$

Since the actual angular frequency ω of v_1 is 10^6 rad/s and $10^6 < 1.515 \times 10^6$, v_1 will not cause distortion, i.e.,

$$\omega = 10^6 < 1.515 \times 10^6 \quad \text{(ok)}$$
$$v_2: \quad \omega_S(\text{max}) = S/(33K_i) = 0.5 \times 10^6/(33)(0.05) = 303.03 \times 10^3 \text{ rad/s}$$
$$\omega = 350 \times 10^3 > 303.03 \times 10^3 \quad \text{(not ok)}$$
$$v_3: \quad \omega_S(\text{max}) = S/(33K_i) = 0.5 \times 10^6/(33)(0.1) = 151.5 \times 10^3 \text{ rad/s}$$
$$\omega = 200 \times 10^3 > 151.5 \times 10^3 \quad \text{(not ok)}$$
$$v_4: \quad \omega_S(\text{max}) = S/(33K_i) = 0.5 \times 10^6/(33)(0.2) = 75.75 \times 10^3 \text{ rad/s}$$
$$\omega = 50 \times 10^3 < 75.75 \times 10^3 \quad \text{(ok)}$$

We see that v_2 and v_3 would both cause the slew-rate specification of the amplifier to be exceeded. Consequently, the output will be distorted.

2. Since we cannot change the input signal amplitudes or frequencies, there are only two remedies: (1) find an amplifier with a greater slew rate, or (2) reduce the closed-loop gain of the present amplifier. We will investigate both remedies.

   a. The slew rate of a new amplifier must satisfy *both* $S/(33)(0.05) \geq 350 \times 10^3$ (for v_2) and $S/(33)(0.1) \geq 200 \times 10^3$ (for v_3). These inequalities are equivalent to

$$S \geq 0.5775 \times 10^6 \ \text{V/s} \quad \text{and} \quad S \geq 0.66 \times 10^6 \ \text{V/s}$$

Therefore, we must use an amplifier with a slew rate of at least 0.66×10^6 V/s, or 0.66 V/μs.

b. If we use the present amplifier, we must reduce its closed-loop gain G so that it satisfies *both* $0.5 \times 10^6/0.05G \geq 350 \times 10^3$ (for v_2) and $0.5 \times 10^6/0.1G \geq 200 \times 10^3$ (for v_3). These inequalities are equivalent to

$$G \leq 28.57 \quad \text{and} \quad G \leq 25$$

Therefore, the maximum closed-loop gain is 25. This limit can be achieved by changing the 330-kΩ resistor in Figure 8.15 to 250 kΩ.

To ensure that an operational-amplifier circuit will not distort a signal component having frequency f, we require that *both* of the following conditions be satisfied:

$$f \leq \text{BW}_{\text{CL}} \tag{8.41}$$

$$f \leq S/2\pi K \tag{8.42}$$

If the signal is a complex waveform containing multiple frequency components, the highest frequency in the signal should satisfy both conditions. Note that both conditions depend on closed-loop gain: Large gains reduce BW_{CL} and increase the value of K, so the greater the closed-loop gain, the more severe the restrictions.

Example 8.10

The operational amplifier in Figure 8.16 has a unity-gain frequency of 1 MHz and a slew rate of 1 V/μs. Find the maximum frequency of a 0.1-V-peak sine-wave input that can be amplified without distortion.

Figure 8.16
(Example 8.10)

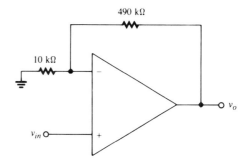

Solution.   The feedback ratio is

$$\beta = \frac{R_1}{R_1 + R_f} = \frac{10 \ \text{k}\Omega}{(10 \ \text{k}\Omega) + (490 \ \text{k}\Omega)} = 0.02$$

By equation 8.34, $\text{BW}_{\text{CL}} = f_t\beta = (1 \ \text{MHz})(0.02) = 20$ kHz. The closed-loop gain for the noninverting configuration is $v_o/v_{in} = 1/\beta = 1/0.02 = 50$. Therefore, the peak value of the output is $K = 50(0.1 \text{ V}) = 5 \text{ V}$. Then $f_S(\text{max}) = S/2\pi K = 10^6/(2\pi)(5 \text{ V}) = 31.83$ kHz. Since we require that f satisfy both $f \leq 20$ kHz and $f \leq 31.83$ kHz, we see that the maximum permissible frequency is 20 kHz. In this case, the bandwidth sets the upper limit.

Example 8.11

1. Derive a design equation that imposes a limit on the value of R_f in Figure 8.17, based on bandwidth and slew-rate limitations of the amplifier. The known values that can be used in the equation are R_1, slew rate S, sinusoidal input frequency f, unity-gain frequency f_t, and peak value of the input, $V_{in}(\text{pk})$.
2. Use the design equation to find the limit on R_f when the input to the amplifier is a 0.5-V-pk sine wave with frequency 5 kHz, $R_1 = 10$ kΩ, $S = 10^6$ V/s, and $f_t = 1$ MHz.

Figure 8.17
(Example 8.11)

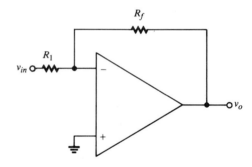

Solution

1. Since the amplifier is in an inverting configuration, the magnitude of the closed-loop gain, G, and the feedback ratio, β, are

$$G = \frac{R_f}{R_1} \qquad \beta = \frac{R_1}{R_1 + R_f}$$

Since the input frequency, f, must be less than the closed-loop bandwidth, we have, from equation 8.34,

$$f < f_t \beta = f_t \left(\frac{R_1}{R_1 + R_f} \right)$$

Solving for R_f leads to

$$R_f < \frac{R_1(f_t - f)}{f}$$

The peak value of the output is

$$K = GV_{in}(\text{pk}) = \frac{R_f}{R_1} V_{in}(\text{pk})$$

By equation 8.40,

$$f < \frac{S}{2\pi(R_f/R_1)V_{in}(\text{pk})}$$

Solving for R_f gives

$$R_f < \frac{R_1 S}{2\pi f V_{in}(\text{pk})}$$

Since R_f must be less than both limits we have found, it must be less than the smaller of the two:

$$R_f < \min\left\{\frac{R_1(f_t - f)}{f}, \frac{R_1 S}{2\pi f V_{in}(\text{pk})}\right\}$$

2. $R_f < \min\left\{\dfrac{10^4 \ \Omega(10^6 \text{ Hz} - 5 \times 10^3 \text{ Hz})}{5 \times 10^3 \text{ Hz}}, \dfrac{(10^4 \ \Omega)(10^6 \text{ V/s})}{2\pi(5 \times 10^3 \text{ Hz})(0.5 \text{ V})}\right\}$

$R_f < \min\{1.99 \text{ M}\Omega, 636.6 \text{ k}\Omega\}$

$R_f < 636.6 \text{ k}\Omega$

If we choose the closest standard-value resistor that is less than 636.6 kΩ, we have $R_f = 620$ kΩ.

Another practical limit that is not considered in the foregoing is the maximum permissible output voltage of the amplifier. If we used $R_f = 620$ kΩ in the present example, then the closed-loop gain would be $R_f/R_1 = 620$ k$\Omega/10$ k$\Omega = 62$, and the peak output voltage would be 62 (0.5 V pk) = 31 V pk, which is too large for many commercially available amplifiers.

We have seen that an amplifier's slew rate affects its ability to track, or follow, a rapidly changing input pulse. When the output voltage must change through ΔV volts, the minimum possible time in which that change can occur is

$$\Delta t = \frac{\Delta V}{S} \text{ seconds} \qquad (8.43)$$

where ΔV is the total *output* voltage change. In terms of input quantities, the minimum time allowed for an input voltage change of ΔV_{in} volts is

$$\Delta t = \frac{(A_{CL})\Delta V_{in}}{S} \text{ seconds} \qquad (8.44)$$

where A_{CL} is the closed-loop gain.

An amplifier's bandwidth also affects the time required for its output to change in response to a pulse input. Recall from Chapter 6 (equation 6.41) that the *rise time* t_r of a single-pole system is

$$t_r = \frac{0.35}{\text{BW}} \text{ seconds} \qquad (8.45)$$

We defined rise time to be the time required for a voltage to change from 10% of its final value to 90% of its final value. Therefore, if we wish the output to follow a pulse through its *entire* variation in Δt seconds, the bandwidth should be larger than that required by equation 8.45. In other words, equation 8.45 requires a bandwidth of $BW = 0.35/t_r$ to achieve a rise time of t_r, but $BW = 0.35/\Delta t$ would not be sufficient to permit a *full* voltage variation in Δt seconds.

We conclude from the foregoing remarks that both slew rate and bandwidth affect the minimum time Δt that an amplifier output can change through ΔV volts. If we now let Δt be the total time through which a given input changes value, then the amplifier must satisfy *both* of the following conditions in order to track that input without distortion:

$$\frac{\Delta V}{S} \le \Delta t \tag{8.46}$$

$$\frac{0.35}{BW_{CL}} << \Delta t \tag{8.47}$$

Note that the inequality in (8.47) is "much less than."

Example 8.12

The operational amplifier shown in Figure 8.18 has a slew rate of 4 V/μs and a unity-gain frequency of 2 MHz. Determine whether the amplifier will distort the input signal shown.

Figure 8.18
(Example 8.12)

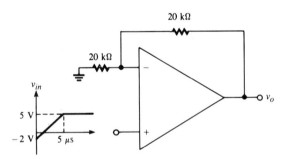

Solution. The closed-loop gain of the amplifier is $v_o/v_{in} = (R_1 + R_f)/R_1 = (40$ k$\Omega)/(20$ k$\Omega) = 2$. Therefore, the output changes from -4 V to 10 V, giving $\Delta V = 14$ V. The voltage change occurs in $\Delta t = 5$ μs, as shown in the figure. Then

$$\frac{\Delta V}{S} = \frac{14 \text{ V}}{4 \times 10^6 \text{ V/s}} = 3.5 \text{ } \mu s < 5 \text{ } \mu s = \Delta t$$

Therefore, condition 8.46 is satisfied. Then, since $\beta = R_1/(R_1 + R_f) = (20$ k$\Omega)/(40$ k$\Omega) = 0.5$, we have $BW_{CL} = \beta f_t = 0.5(2 \text{ MHz}) = 1$ MHz, and

$$\frac{0.35}{BW_{CL}} = \frac{0.35}{10^6} = 0.35 \text{ } \mu s << 5 \text{ } \mu s = \Delta t$$

Therefore, condition 8.47 is also satisfied, and we conclude that no distortion will occur.

8.5   OFFSET CURRENTS AND VOLTAGES

Recall from Chapter 7 that one of the characteristics of an ideal operational amplifier is that it has zero output voltage when both inputs are 0 V (grounded). This characteristic is particularly important in applications where dc or low-frequency signals are involved. If the output is not 0 when the inputs are 0, then the output will not be at its correct dc level when the input is a dc level other than 0.

The actual value of the output voltage when the inputs are 0 is called the *output offset voltage*. Output offset is very much like a dc bias level in the output of a conventional amplifier in that it is added to whatever signal variation occurs there. If an operational amplifier is used only for ac signals, it can be capacitor-coupled if necessary or desirable to block the dc component represented by the offset. However, the capacitors may have to be impractically large if low frequencies and small impedance levels are involved. Also, a dc path must always be present between each input and ground to allow bias currents to flow. Small offsets, on the order of a few millivolts, can often be ignored if the signal variations are large by comparison. On the other hand, a frequent application of operational amplifiers is in precise, high-accuracy signal processing at low levels and low frequencies, and in these situations, very small offsets are crucial.

Manufacturers do not generally specify output offset because, as we shall see, the offset level depends on the closed-loop gain that a user designs through choice of external component values. Instead, *input* offsets are specified, and the designer can use these values to compute the output offset that results in a particular application. Output offset voltages are the result of two distinct input phenomena: input bias currents and input offset voltage. We will use the superposition principle to determine the contribution of each of these input effects to the output offset voltage.

Input Offset Current

In our discussion of differential amplifier circuits in Chapter 7, we ignored base currents because they had negligible effects on the kinds of computations that held our interest then. We know that some dc base current must flow when a transistor is properly biased, and, although small, this current flowing through the external resistors in an amplifier circuit produces a dc input voltage that in turn creates an output offset. To reduce the effect of bias currents, a *compensating resistor R_c* is connected in series with the noninverting (+) terminal of the amplifier. (R_c must provide a dc path to ground, so if a signal is capacitor-coupled to the + input, R_c must be connected between the + input and ground.) We will presently show that proper choice of the value of R_c will minimize the output offset voltage due to bias current. Figure 8.19 shows the bias currents I_B^+ and I_B^- flowing into the + and − terminals of an operational amplifier when the signal inputs are grounded. While the bias currents may actually flow into or out of the terminals, depending on the type of input circuitry, we will, for the sake of convenience, assume that the directions are as shown and that the values are always positive. These assumptions will not affect our ultimate conclusions. The figure also shows the compensating resistor R_c connected in series with the + terminal. Note that this circuit applies to both the inverting and noninverting configurations.

Figure 8.20(a) shows the equivalent circuit of Figure 8.19. Here, the bias currents are represented by current sources having resistances R_1 and R_c. Figure

Figure 8.19
Input bias currents I_B^+ and I_B^- that flow when both signal inputs are grounded. R_c is a compensating resistor used to reduce the effect of bias current on output offset.

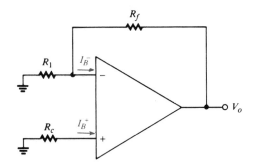

8.20(b) shows the same circuit when the current sources are replaced by their Thevenin equivalent voltage sources.

Using Figure 8.20(b), we can apply the superposition principle to determine the output offset voltage due to each input source acting alone. As illustrated in Figure 8.21(a), the amplifier acts as an inverter when the source connected to the + terminal is shorted to ground, so the output due to $I_B^- R_1$ is

$$V_{o1} = I_B^- R_1 \left(\frac{-R_f}{R_1}\right) = -I_B^- R_f \tag{8.48}$$

When the source connected to the − terminal is shorted to ground, the amplifier is

Figure 8.20
Circuits equivalent to Figure 8.19

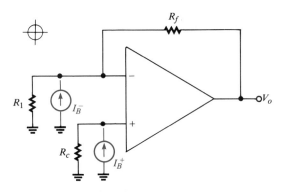

(a) The equivalent circuit of Figure 8.19

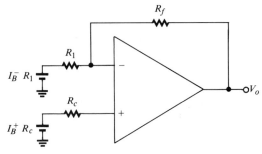

(b) The circuit equivalent to (a) when the current sources are replaced by their Thevenin equivalents

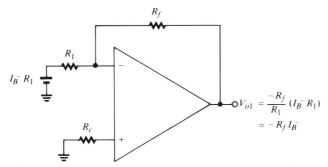

(a) When the noninverting input is grounded, the amplifier inverts and has gain $-R_f/R_1$.

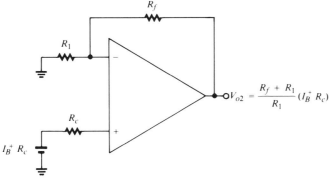

(b) When the inverting input is grounded, the noninverting amplifier has gain $(R_f + R_1)/R_1$.

Figure 8.21
Applying superposition to determine the output offset voltage due to each source in Figure 8.20(b)

in a noninverting configuration, so the output due to $I_B^+ R_c$ is

$$V_{o2} = I_B^+ R_c \left(\frac{R_f + R_1}{R_1}\right) \tag{8.49}$$

Combining (8.48) and (8.49), we obtain the total output offset voltage due to bias current, which we designate by $V_{OS}(I_B)$, as

$$V_{OS}(I_B) = I_B^+ R_c \left(\frac{R_f + R_1}{R_1}\right) - I_B^- R_f \tag{8.50}$$

Depending on which of the terms on the left side of equation 8.50 is greater, $V_{OS}(I_B)$ may be positive or negative. However, the sign of $V_{OS}(I_B)$ is of little interest, since negative offset voltage is just as undesirable as positive offset voltage. Our real interest is in finding a way to minimize the *magnitude* of $V_{OS}(I_B)$. Toward that end, let us make the reasonable assumption that the two inputs are closely matched and that, as a consequence, they have equal bias currents: $I_B^+ = I_B^- = I_{BB}$. Substituting I_{BB} for I_B^- and I_B^+ in equation 8.50 gives

$$V_{OS}(I_B) = I_{BB} \left[R_c \left(\frac{R_f + R_1}{R_1}\right) - R_f \right] \tag{8.51}$$

If the expression enclosed by the brackets in (8.51) were equal to 0, we would have zero offset voltage. To find a value of R_c that accomplishes that goal, we set the bracketed expression equal to 0 and solve for R_c:

$$R_c\left(\frac{R_f + R_1}{R_1}\right) - R_f = 0$$

$$R_c = \frac{R_f}{\dfrac{R_f + R_1}{R_1}} = \frac{R_f R_1}{R_f + R_1} = R_f \parallel R_1 \qquad (8.52)$$

Equation 8.52 reveals the very important result that *output offset due to input bias currents can be minimized by connecting a resistor R_c having value $R_1 \parallel R_f$ in series with the noninverting input.* This method of offset compensation is valid for both inverting and noninverting configurations. Notice that we say the offset can be *minimized* using this remedy, rather than being made exactly 0, because the remedy is based on the assumption that $I_B^+ = I_B^-$, which may not be entirely valid. We can compute the exact value of $V_{OS}(I_B)$ when $R_c = R_1 \parallel R_f$ by substituting this value of R_c back into (8.50), where the assumption is not in force:

$$V_{OS}(I_B) = I_B^+(R_1 \parallel R_f)\left(\frac{R_f + R_1}{R_1}\right) - I_B^- R_f$$

$$= I_B^+\left(\frac{R_1 R_f}{R_1 + R_f}\right)\left(\frac{R_f + R_1}{R_1}\right) - I_B^- R_f \qquad (8.53)$$

$$= (I_B^+ - I_B^-)R_f$$

Equation 8.53 shows that the offset voltage is proportional to the *difference* between I_B^+ and I_B^- when $R_c = R_1 \parallel R_f$. Since the inputs are usually reasonably well matched, the difference between I_B^+ and I_B^- is quite small. The equation confirms the fact that V_{OS} is 0 if I_B^+ exactly equals I_B^-. The quantity $I_B^+ - I_B^-$ is called the *input offset current* and is often quoted in manufacturers' specifications. Remember that it is actually a *difference* current. Letting the input offset current $I_B^+ - I_B^-$ be designated by I_{io}, we have, from equation 8.53,

$$V_{OS}(I_B) = I_{io}R_f \qquad \text{when } R_c = R_1 \parallel R_f \qquad (8.54)$$

$V_{OS}(I_B)$ may be either positive or negative, depending on whether $I_B^+ > I_B^-$ or vice versa. Unless actual measurements are made, we rarely know which current is larger, so a more useful form of (8.54) is

$$|V_{OS}(I_B)| = |I_{io}|R_f \qquad \text{when } R_c = R_1 \parallel R_f \qquad (8.55)$$

Manufacturers' specifications always give a positive value for I_{io}, so it is best interpreted as an absolute value in any case.

Equation 8.55 shows that the output offset is directly proportional to the value of the feedback resistor R_f. For that reason, small resistance values should be used when offset is a critical consideration. However, to achieve large voltage gains when R_f is small may require impractically small values of R_1, to the extent that the amplifier may load the signal source driving it. In any event, large closed-loop gains are detrimental to another aspect of output offset, as we will see in a forthcoming discussion.

Another common manufacturers' specification is called simply *input bias cur-*

rent, I_B. By convention, I_B is the *average* of I_B^+ and I_B^-:

$$I_B = \frac{I_B^+ + I_B^-}{2} \tag{8.56}$$

I_B is typically much larger than I_{io} because I_B is on the same order of magnitude as I_B^+ and I_B^-, while I_{io} is the difference between the two. Given values for I_B and I_{io}, we can find I_B^+ and I_B^-, provided we know which is the larger. If $I_B^+ > I_B^-$, then

$$\left.\begin{aligned} I_B^+ &= I_B + 0.5\,|I_{io}| \\ I_B^- &= I_B - 0.5\,|I_{io}| \end{aligned}\right\} (I_B^+ > I_B^-) \tag{8.57}$$

If $I_B^- > I_B^+$, the + and − signs between terms in (8.57) are interchanged. Proof of these relations is left as an exercise at the end of the chapter.

Example 8.13

The specifications for the operational amplifier in Figure 8.22 state that the input bias current is 80 nA and that the input offset current is 20 nA.

1. Find the optimum value for R_c.
2. Find the magnitude of the output offset voltage due to bias currents when R_c equals its optimum value.
3. Assuming that $I_B^+ > I_B^-$, find the magnitude of the output offset voltage when $R_c = 0$.

Figure 8.22
(Example 8.13)

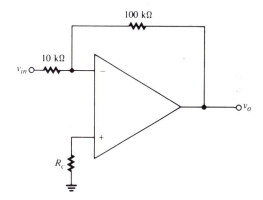

Solution

1. From equation 8.52, $R_c = R_1 \parallel R_f = (10\ \text{k}\Omega) \parallel (100\ \text{k}\Omega) = 9.09\ \text{k}\Omega$.
2. From equation 8.55, $|V_{OS}(I_B)| = |I_{io}|R_f = (20 \times 10^{-9})(100 \times 10^3) = 2\ \text{mV}$.
3. When $R_c = 0$, equation 8.50 becomes $V_{OS}(I_B) = -I_B^- R_f$. From equation 8.57, $I_B^- = I_B - 0.5 I_{io} = (80\ \text{nA}) - 0.5(20\ \text{nA}) = 70\ \text{nA}$. Therefore, the magnitude of the offset voltage when $R_c = 0$ is $|V_{OS}(I_B)| = (70 \times 10^{-9})(100 \times 10^3) = 7\ \text{mV}$. We see that omission of the compensating resistance more than doubles the magnitude of the offset voltage.

Input Offset Voltage

Another input phenomenon that contributes to output offset voltage is an internally generated potential difference that exists because of imperfect matching of the input transistors. This potential may be due, for example, to a difference between the V_{BE} drops of the transistors in the input differential stage of a BJT amplifier. Called *input offset voltage*, the net effect of this potential difference is the same as if a small dc voltage source were connected to one of the inputs. Figure 8.23 shows the equivalent circuit of an amplifier having its signal inputs grounded and its input offset voltage, V_{io}, represented as a dc source in series with the noninverting input. The effect is the same whether it is connected to the inverting or noninverting input. The polarity of the source is arbitrary, because input and output offsets may be either positive or negative. Once again, it is the magnitude of the offset that concerns us.

From Figure 8.23, it is apparent that the output voltage when the input is V_{io} is given by

$$V_{OS}(V_{io}) = V_{io} \frac{(R_1 + R_f)}{R_1} \tag{8.58}$$

where $V_{OS}(V_{io})$ is the output offset voltage due to V_{io}. Note that the compensating resistor R_c is shown in Figure 8.23 for completeness' sake, but it has no effect on the output offset due to V_{io}. Equation 8.58 shows that input offset is magnified at the output by a factor equal to the closed-loop gain of the noninverting amplifier, as we would expect. If the amplifier is operated open-loop, the very large open-loop gain acting on the input offset voltage may well drive the amplifier to one of its output voltage limits. It is therefore important to have an extremely small V_{io} in any application or measurement that requires an open-loop amplifier.

Equation 8.58 is also valid for an amplifier in an inverting configuration. In fact, for a wide variety of amplifier configurations, it is true that

$$V_{OS}(V_{io}) = V_{io}/\beta \tag{8.59}$$

where β is the feedback ratio.

Example 8.14

The specifications for the amplifier in Example 8.13 state that the input offset voltage is 0.8 mV. Find the output offset due to this input offset.

Figure 8.23
The effect of input offset voltage, V_{io}, is the same as if a dc source were connected in series with one of the inputs.

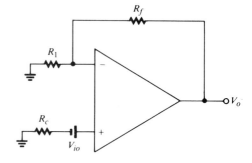

Solution.   From equation 8.58,

$$V_{OS}(V_{io}) = V_{io}\,\frac{(R_1 + R_f)}{R_1} = (0.8 \times 10^{-3}\ \text{V})\,\frac{[(10\ \text{k}\Omega) + (100\ \text{k}\Omega)]}{10\ \text{k}\Omega} = 8.8\ \text{mV}$$

The Total Output Offset Voltage

We have seen that output offset voltage is a function of two distinct input characteristics: input bias currents and input offset voltage. It may be that the polarities
of the offsets caused by these two characteristics are such that they tend to cancel
each other out. Of course, we cannot depend on that happy circumstance, so it is
good design practice to assume a *worst-case* situation, in which the two offsets
have the same polarity and reinforce each other. We can invoke the principle of
superposition and conclude that the total output offset voltage is the sum of the
offsets caused by the individual input phenomena, but for the worst-case situation, we assume that the total offset is the sum of the respective *magnitudes:*

$$|V_{OS}| = |V_{OS}(I_B)| + |V_{OS}(V_{io})| \qquad \text{(worst case)} \qquad \textbf{(8.60)}$$

where V_{OS} is the total output offset voltage.

Example 8.15

The operational amplifier in Figure 8.24 has the following specifications: input bias
current = 100 nA; input offset current = 20 nA; input offset voltage = 0.5 mV.
Find the worst-case output offset voltage. (Consider the two possibilities $I_B^+ > I_B^-$
and vice versa.)

Figure 8.24
(Example 8.15)

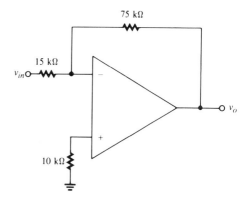

Solution.   We first check to see if the 10-kΩ resistor in series with the noninvert-
ing input has the optimum value of a compensating resistor: $R_1 \| R_f = (15\ \text{k}\Omega) \| (75$
k$\Omega) = 12.5\ \text{k}\Omega$. $R_c = 10\ \text{k}\Omega$ is not optimum, and we will have to use equation 8.50
to find $V_{OS}(I_B)$. Assuming first that $I_B^+ > I_B^-$, we have, from equation 8.57,

$$I_B^+ = I_B + 0.5I_{io} = (100\ \text{nA}) + 0.5(20\ \text{nA}) = 110\ \text{nA}$$
$$I_B^- = I_B - 0.5I_{io} = (100\ \text{nA}) - 0.5(20\ \text{nA}) = 90\ \text{nA}$$

Therefore, by equation 8.50,

$$V_{OS}(I_B) = I_B^+ R_c \left(\frac{R_f + R_1}{R_1}\right) - I_B^- R_f$$

$$= (110 \times 10^{-9})(10 \times 10^3) \left(\frac{75 \times 10^3 + 15 \times 10^3}{15 \times 10^3}\right)$$

$$- (90 \times 10^{-9})(75 \times 10^3)$$

$$= -0.15 \text{ mV}$$

If $I_B^- > I_B^+$, then $I_B^+ = 90$ nA and $I_B^- = 110$ nA. In that case,

$$V_{OS}(I_B) = (90 \times 10^{-9})(10 \times 10^3) \left(\frac{75 \times 10^3 + 15 \times 10^3}{15 \times 10^3}\right)$$

$$- (110 \times 10^{-9})(75 \times 10^3)$$

$$= -2.85 \text{ mV}$$

We see that the worst case occurs for $I_B^- > I_B^+$, and therefore we assume that $|V_{OS}(I_B)| = 2.85$ mV.
By equation 8.58,

$$V_{OS}(V_{io}) = \frac{V_{io}(R_1 + R_f)}{R_1} = (0.5 \text{ mV}) \left[\frac{(15 \text{ k}\Omega) + (75 \text{ k}\Omega)}{15 \text{ k}\Omega}\right] = 3 \text{ mV}$$

Therefore, the worst-case offset is $V_{OS} = |V_{OS}(I_B)| + |V_{OS}(V_{io})| = (2.85 \text{ mV}) + (3 \text{ mV}) = 5.85$ mV. (Note that the "best-case" offset would be 0.15 mV.)

The values we have used for V_{io}, I_B, and I_{io} in the examples of this section are typical for general-purpose, BJT operational amplifiers. Amplifiers with much smaller input offsets are available. The bias currents in amplifiers having FET inputs are in the picoamp range.

Most operational amplifiers have two terminals across which an external potentiometer can be connected to adjust the output to 0 when the inputs are grounded. This operation is called *zeroing,* or *balancing,* the amplifier. However, operational amplifiers are subject to *drift,* wherein characteristics change with time and, particularly, with temperature. For applications in which extremely small offsets are required and in which the effects of drift must be minimized, *chopper-stabilized* amplifiers are available. Internal choppers convert the dc offset to an ac signal, amplify it, and use it to adjust amplifier characteristics so that the output is automatically restored to 0.

8.6   OPERATIONAL AMPLIFIER SPECIFICATIONS

In this section we will examine and interpret a typical set of manufacturer's specifications for an operational amplifier. The specifications shown in Figure 8.25 are those of the 741 amplifier, a popular, inexpensive, general-purpose operational amplifier that has been produced by many different manufacturers for a number of years. Let us first note that, like many integrated circuits, there are several versions of the 741 available. The different versions are identified by letter suffixes (741A, 741C, etc.), and each version has at least one performance specification or operating condition that is different from the others. The Fairchild specifications

A Schlumberger Company

μA741
Operational Amplifier

Linear Products

Description

The μA741 is a high performance Monolithic Operational Amplifier constructed using the Fairchild Planar epitaxial process. It is intended for a wide range of analog applications. High common mode voltage range and absence of latch-up tendencies make the μA741 ideal for use as a voltage follower. The high gain and wide range of operating voltage provides superior performance in integrator, summing amplifier, and general feedback applications.

- **NO FREQUENCY COMPENSATION REQUIRED**
- **SHORT-CIRCUIT PROTECTION**
- **OFFSET VOLTAGE NULL CAPABILITY**
- **LARGE COMMON MODE AND DIFFERENTIAL VOLTAGE RANGES**
- **LOW POWER CONSUMPTION**
- **NO LATCH-UP**

Connection Diagram
8-Pin Metal Package

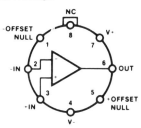

(Top View)

Pin 4 connected to case

Order Information

Type	Package	Code	Part No.
μA741	Metal	5W	μA741HM
μA741A	Metal	5W	μA741AHM
μA741C	Metal	5W	μA741HC
μA741E	Metal	5W	μA741EHC

Connection Diagram
8-Pin DIP

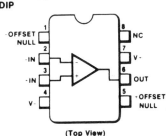

(Top View)

Order Information

Type	Package	Code	Part No.
μA741C	Molded DIP	9T	μA741TC
μA741C	Ceramic DIP	6T	μA741RC

Connection Diagram
10-Pin Flatpak

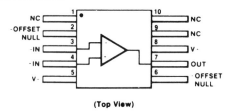

(Top View)

Order Information

Type	Package	Code	Part No.
μA741	Flatpak	3F	μA741FM
μA741A	Flatpak	3F	μA741AFM

Absolute Maximum Ratings

Supply Voltage	
μA741A, μA741, μA741E	± 22 V
μA741C	± 18 V
Internal Power Dissipation (Note 1)	
Metal Package	500 MW
DIP	310 mW
Flatpak	570 mW
Differential Input Voltage	± 30 V
Input Voltage (Note 2)	± 15 V
Storage Temperature Range	
Metal Package and Flatpak	−65°C to +150°C
DIP	−55°C to +125°C

Operating Temperature Range	
Military (μA741A, μA741)	−55°C to +125°C
Commercial (μA741E, μA741C)	0°C to +70°C
Pin Temperature (Soldering 60 s)	
Metal Package, Flatpak, and Ceramic DIP	300°C
Molded DIP (10 s)	260°C
Output Short Circuit Duration (Note 3)	Indefinite

Figure 8.25
Specifications for the μA741 operational amplifier (Courtesy of Fairchild Semiconductor)

Equivalent Circuit

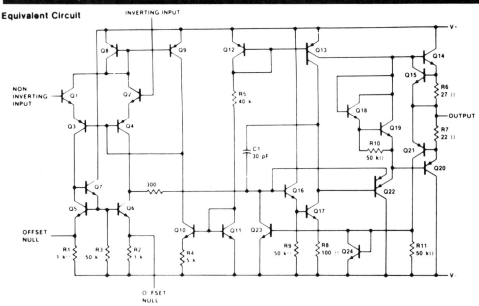

Notes

1. Rating applies to ambient temperatures up to 70 C. Above 70°C ambient derate linearly at 6 3 mW C for the metal package, 7 1 mW C for the flatpak, and 5 6 mW C for the DIP

2. For supply voltages less than + 15 V, the absolute maximum input voltage is equal to the supply voltage

3. Short circuit may be to ground or either supply. Rating applies to +125°C case temperature or 75°C ambient temperature.

μA741 and μA741C
Electrical Characteristics $V_S = \pm 15$ V, $T_A = 25°C$ unless otherwise specified

Characteristic	Condition	μA741			μA741C			Unit
		Min	Typ	Max	Min	Typ	Max	
Input Offset Voltage	$R_S \leq 10$ kΩ		1.0	5.0		2.0	6.0	mV
Input Offset Current			20	200		20	200	nA
Input Bias Current			80	500		80	500	nA
Power Supply Rejection Ratio	$V_S = +10, -20$ $V_S = +20, -10$ V, $R_S = 50$ Ω		30	150		30	150	μV / V
Input Resistance		.3	2.0		.3	2.0		MΩ
Input Capacitance			1.4			1.4		pF
Offset Voltage Adjustment Range			± 15			± 15		mV
Input Voltage Range						± 12	± 13	V
Common Mode Rejection Ratio	$R_S \leq 10$ kΩ					70	90	dB
Output Short Circuit Current			25			25		mA
Large Signal Voltage Gain	$R_L \geq 2$ kΩ, $V_{OUT} = \pm 10$ V	50k	200k		20k	200k		
Output Resistance			75			75		Ω
Output Voltage Swing	$R_L \geq 10$ kΩ					± 12	± 14	V
	$R_L \geq 2$ kΩ					± 10	± 13	V
Supply Current			1.7	2.8		1.7	2.8	mA
Power Consumption			50	85		50	85	mW

Figure 8.25
(Continued)

μA741 and μA741C
Electrical Characteristics (Cont.) $V_S = \pm 15$ V, $T_A = 25°C$ unless otherwise specified

Characteristic	Condition	μA741			μA741C			Unit
		Min	Typ	Max	Min	Typ	Max	
Transient Response (Unity Gain)	Rise Time	$V_{IN} = 20$ mV, $R_L = 2$ kΩ, $C_L \leq 100$ pF	.3			.3		μs
	Overshoot		5.0			5.0		%
Bandwidth (Note 4)			1.0			1.0		MHz
Slew Rate	$R_L \geq 2$ kΩ		.5			.5		V / μs

Notes
4 Calculated value from BW(MHz) = $\dfrac{0.35}{\text{Rise Time (μs)}}$

5 All $V_{CC} = 15$ V for μA741 and μA741C
6 Maximum supply current for all devices
   25°C = 2.8 mA
   125°C = 2.5 mA
   −55°C = 3.3 mA

μA741 and μA741C
Electrical Characteristics (Cont.)   The following specifications apply over the range of $-55°C \leq T_A \leq 125°C$ for μA741, $0°C \leq T_A \leq 70°C$ for μA741C

Characteristic	Condition	μA741			μA741C			Unit
		Min	Typ	Max	Min	Typ	Max	
Input Offset Voltage							7.5	mV
	$R_S \leq 10$ kΩ		1.0	6.0				mV
Input Offset Current							300	nA
	$T_A = +125°C$		7.0	200				nA
	$T_A = -55°C$		85	500				nA
Input Bias Current							800	nA
	$T_A = +125°C$		.03	.5				μA
	$T_A = -55°C$		.3	1.5				μA
Input Voltage Range		± 12	± 13					V
Common Mode Rejection Ratio	$R_S \leq 10$ kΩ	70	90					dB
Adjustment for Input Offset Voltage			± 15			± 15		mV
Supply Voltage Rejection Ratio	$V_S = +10, -20;$ $V_S = +20, -10$ V, $R_S = 50$ Ω		30	150				μV / V
Output Voltage Swing	$R_L \geq 10$ kΩ	± 12	± 14					V
	$R_L \geq 2$ kΩ	± 10	± 13		± 10	± 13		V
Large Signal Voltage Gain	$R_L = 2$ kΩ, $V_{OUT} = \pm 10$ V	25k			15k			
Supply Current	$T_A = +125°C$		1.5	2.5				mA
	$T_A = -55°C$		2.0	3.3				mA
Power Consumption	$T_A = +125°C$		45	75				mW
	$T_A = -55°C$		60	100				mW

Notes
4 Calculated value from BW(MHz) = $\dfrac{0.35}{\text{Rise Time (μs)}}$

5 All $V_{CC} = 15$ V for μA741 and μA741C
6 Maximum supply current for all devices
   25°C = 2.8 mA
   125°C = 2.5 mA
   −55°C = 3.3 mA

Figure 8.25
(Continued)

μA741A and μA741E

Electrical Characteristics $V_S = \pm 15$ V, $T_A = 25°C$ unless otherwise specified.

Characteristic		Condition	μA741A/E			Unit
			Min	Typ	Max	
Input Offset Voltage		$R_S \leq 50\ \Omega$		0.8	3.0	mV
Average Input Offset Voltage Drift					15	μV/°C
Input Offset Current				3.0	30	nA
Average Input Offset Current Drift					0.5	nA/°C
Input Bias Current				30	80	nA
Power Supply Rejection Ratio		$V_S = +10, -20$; $V_S = +20$ V, -10 V, $R_S = 50\ \Omega$		15	50	μV/V
Output Short Circuit Current			10	25	40	mA
Power Consumption		$V_S = \pm 20$ V		80	150	mW
Input Impedance		$V_S = \pm 20$ V	1.0	6.0		MΩ
Large Signal Voltage Gain		$V_S = \pm 20$ V, $R_L = 2$ kΩ, $V_{OUT} = \pm 15$ V	50	200		V/mV
Transient Response (Unity Gain)	Rise Time			0.25	0.8	μs
	Overshoot			6.0	20	%
Bandwidth (Note 4)			.437	1.5		MHz
Slew Rate (Unity Gain)		$V_{IN} = \pm 10$ V	0.3	0.7		V/μs

The following specifications apply over the range of $-55°C \leq T_A \leq 125°C$ for the 741A, and $0°C \leq T_A \leq 70°C$ for the 741E.

Characteristic				Min	Typ	Max	Unit
Input Offset Voltage						4.0	mV
Input Offset Current						70	nA
Input Bias Current						210	nA
Common Mode Rejection Ratio		$V_S = \pm 20$ V, $V_{IN} = \pm 15$ V, $R_S = 50\ \Omega$		80	95		dB
Adjustment For Input Offset Voltage		$V_S = \pm 20$ V		10			mV
Output Short Circuit Current				10		40	mA
Power Consumption		$V_S = \pm 20$ V, μA741A	$-55°C$			165	mW
			$+125°C$			135	mW
		μA741E				150	mW
Input Impedance		$V_S = \pm 20$ V		0.5			MΩ
Output Voltage Swing		$V_S = \pm 20$ V	$R_L = 10$ kΩ	± 16			V
			$R_L = 2$ kΩ	± 15			V
Large Signal Voltage Gain		$V_S = \pm 20$ V, $R_L = 2$ kΩ, $V_{OUT} = \pm 15$ V		32			V/mV
		$V_S = \pm 5$ V, $R_L = 2$ kΩ, $V_{OUT} = \pm 2$ V		10			V/mV

Notes

4. Calculated value from: $BW(MHz) = \dfrac{0.35}{\text{Rise Time }(\mu s)}$

5. All $V_{CC} = 15$ V for μA741 and μA741C.

6. Maximum supply current for all devices
   25°C = 2.8 mA
   125°C = 2.5 mA
   −55°C = 3.3 mA

Figure 8.25
(Continued)

Typical Performance Curves for μA741A and μA741

Open Loop Voltage Gain as a Function of Supply Voltage

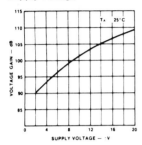

Output Voltage Swing as a Function of Supply Voltage

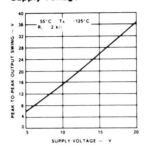

Input Common Mode Voltage as a Function of Supply Voltage

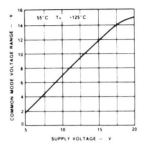

Typical Performance Curves for μA741E and μA741C

Open Loop Voltage Gain as a Function of Supply Voltage

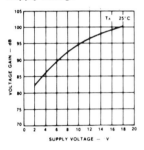

Output Voltage Swing as a Function of Supply Voltage

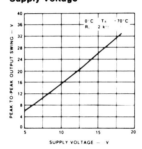

Input Common Mode Voltage Range as a Function of Supply Voltage

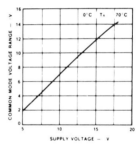

Transient Response

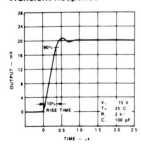

Transient Response Test Circuit

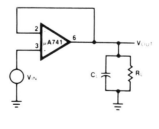

Common Mode Rejection Ratio as a Function of Frequency

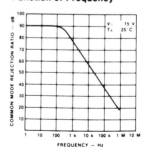

Figure 8.25
(Continued)

Typical Performance Curves for μA741E and μA741C (Cont.)

Frequency Characteristics as a Function of Supply Voltage

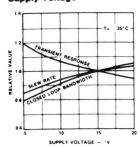

Voltage Offset Null Circuit

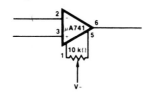

Voltage Follower Large Signal Pulse Response

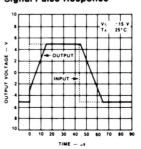

Typical Performance Curves for μA741A, μA741, μA741E and μA741C

Power Consumption as a Function of Supply Voltage

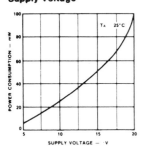

Open Loop Voltage Gain as a Function of Frequency

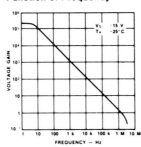

Open Loop Phase Response as a Function of Frequency

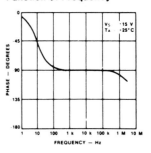

Input Offset Current as a Function of Supply Voltage

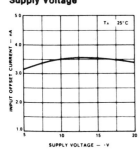

Input Resistance and Input Capacitance as a Function of Frequency

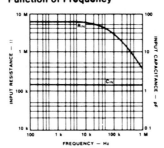

Output Resistance as a Function of Frequency

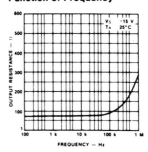

Figure 8.25
(Continued)

Typical Performance Curves for μA741A, μA741, μA741E and μA741C (Cont.)

Output Voltage Swing as a Function of Load Resistance

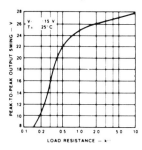

Output Voltage Swing as a Function of Frequency

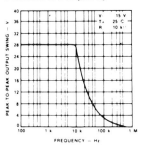

Absolute Maximum Power Dissipation as a* Function of Ambient Temperature

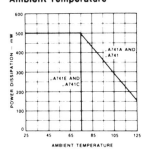

Input Noise Voltage as a Function of Frequency

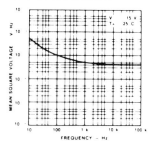

Input Noise Current as a Function of Frequency

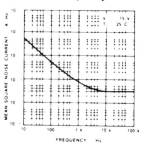

Broadband Noise for Various Bandwidths

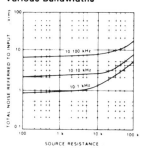

Typical Performance Curves for μA741A and μA741

Input Bias Current as a Function of Ambient Temperature

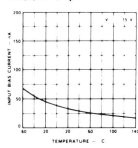

Input Resistance as a Function of Ambient Temperature

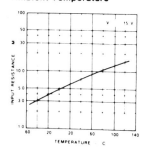

Output Short-Circuit Current as a Function of Ambient Temperature

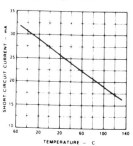

Figure 8.25
(Continued)

Typical Performance Curves for µA741A and µA741 (Cont.)

Input Offset Current as a Function of Ambient Temperature

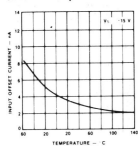

Power Consumption as a Function of Ambient Temperature

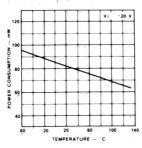

Frequency Characteristics as a Function of Ambient Temperature

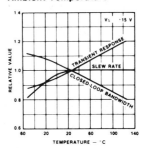

Typical Performance Curves for µA741E and µA741C

Input Bias Current as a Function of Ambient Temperature

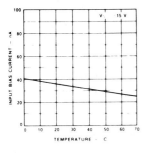

Input Resistance as a Function of Ambient Temperature

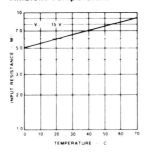

Input Offset Current as a Function of Ambient Temperature

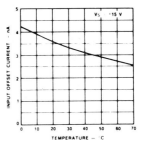

Power Consumption as a Function of Ambient Temperature

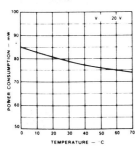

Output Short Circuit Current as a Function of Ambient Temperature

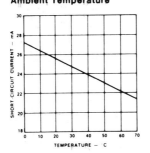

Frequency Characteristics as a Function of Ambient Temperature

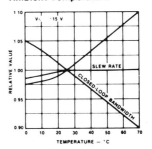

Figure 8.25
(Continued)

shown here use the μA prefix to designate a Fairchild product, and data is given for the μA741, μA741A, μA741C, and μA741E versions. Often the versions differ in respect to the intended market: military or commercial. Specifications for military versions are generally more stringent than their commercial counterparts. For example, we see that the operating temperature range for the military versions (μA741 and μA741A) is $-55°C$ to $+125°C$, while that of the commercial versions (μA741C and μA741E) is $0°C$ to $+70°C$.

Reviewing the specifications, we see that parameter values are comparable to those we have used in the examples of this chapter and are representative of a general-purpose, BJT operational amplifier. Note that most entries show a typical value and a minimum or maximum value. The range of values is the manufacturer's statement of the variation that can be expected among a large number of 741 chips. Those parameters for which a large numerical value is desirable show a minimum value and those for which a small numerical value is desirable show a maximum value. For example, the input offset voltage for the μA741 at 25°C has a typical value of 1 mV but may be as great as 5 mV. Circuit designers who are using the 741, or any other operational amplifier, in the design of a product that will be manufactured in large quantities should use the worst-case specifications.

Note that the specifications include values for the common-mode rejection ratio (CMRR) in dB. The CMRR for an operational amplifier is defined in exactly the same way that we introduced it in connection with differential amplifiers in Chapter 7.

Many of the specifications vary with operating conditions such as frequency, supply voltage, and ambient temperature. Typical variations are shown in the graphs that accompany the value listings. Important examples of which the designer should be aware include the following:

1. The open-loop voltage gain for the μA741 increases from 90 dB to nearly 110 dB as the supply voltage ranges from 2 V to 20 V. Lower supply voltages mean lower values of open-loop gain.
2. Beyond about 100 Hz, the CMRR of the μA741E falls off at the rate of 20 dB/decade, typical behavior for the CMRR of an operational amplifier.
3. The closed-loop bandwidth of the μA741 decreases linearly with increasing ambient temperature, its value at 120°C being about 80% of its value at 20°C.

Example 8.16

Assuming worst-case conditions at 25°C, determine the following, in connection with the μA741 operational-amplifier circuit shown in Figure 8.26:

1. the closed-loop bandwidth
2. the maximum operating frequency when the input is a 0.5 V-peak sine wave
3. the total output offset voltage $|V_{OS}|$

Solution

1. From equation 8.34, $BW_{CL} = f_t\beta$. The graph labeled "Open Loop Voltage Gain as a Function of Frequency" (the open-loop frequency response) in the 741 specifications reveals the unity-gain frequency to be approximately 1 MHz. From Figure 8.26,

$$\beta = \frac{R_1}{R_1 + R_f} = \frac{12 \text{ k}\Omega}{(12 \text{ k}\Omega) + (138 \text{ k}\Omega)} = 0.08$$

Thus, $BW_{CL} = (1 \text{ MHz})(0.08) = 80 \text{ kHz}.$

Figure 8.26
(Example 8.16)

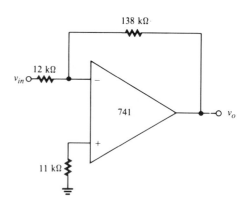

Example 8.18 shows how PSpice can be used to verify this result for a 741 operational amplifier.

2. By equation 8.40, the maximum operating frequency under the slew-rate limitation is

$$f_s(\text{max}) = \frac{S}{2\pi K} \text{ Hz}$$

The specifications show the slew rate to be 0.5 V/μs. From the figure, the closed-loop gain of the inverting configuration is

$$\frac{v_o}{v_{in}} = \frac{-R_f}{R_1} = \frac{-138 \text{ k}\Omega}{12 \text{ k}\Omega} = -11.5$$

Therefore, the magnitude of the peak output voltage is $K = (0.5)(11.5) = 5.75$ V. Thus,

$$f_s(\text{max}) = \frac{0.5 \times 10^6 \text{ V/s}}{2\pi(5.75) \text{ V}} = 13.84 \text{ kHz}$$

Since $f_s(\text{max}) < \text{BW}_{\text{CL}}$, the maximum operating frequency for a 0.5-V-peak input is 13.84 kHz.

3. $R_1 \| R_f = (12 \text{ k}\Omega) \| (138 \text{ k}\Omega) \approx 11 \text{ k}\Omega$. Therefore, the compensating resistor has its optimum value and we can use equation 8.55 to determine the output offset due to bias currents: $|V_{OS}(I_B)| = |I_{io}|R_f$. The μA741 specifications list the maximum value of input offset current to be 200 nA. Therefore, $|V_{OS}(I_B)|_{\text{max}} = (200 \times 10^{-9})(138 \times 10^3) = 27.6$ mV. The specifications list the maximum value of input offset voltage to be 5 mV. Therefore,

$$|V_{OS}(V_{io})| = V_{io}\left(\frac{R_f + R_1}{R_1}\right) = (5 \text{ mV})\left[\frac{(138 \text{ k}\Omega) + (12 \text{ k}\Omega)}{12 \text{ k}\Omega}\right] = 62.5 \text{ mV}$$

Finally, $|V_{OS}|_{\text{worst case}} = (27.6 \text{ mV}) + (62.5 \text{ mV}) = 90.1$ mV.

Notice that the 741 has a pair of pins across which a potentiometer can be connected for offset null (zeroing).

Example 8.17

Use SPICE to find the closed-loop voltage gain, the input resistance seen by v_{in}, and the output resistance of the inverting amplifier in Example 8.3.

Solution. Figure S8.1 shows how we can use a voltage-controlled voltage source (EOP) to model an operational amplifier in SPICE. Notice that the inverting property of the amplifier is realized by connecting the positive terminal (N+) of EOP to ground (node 0). The voltage source is controlled by the voltage between nodes 2 and 0 (NC+ and NC−, respectively). Thus, node 2 corresponds to the inverting input of the amplifier in Figure 8.6. Notice that the simulated amplifier has infinite input impedance, since there is an open circuit between nodes 2 and 0.

Figure S8.1
(Example 8.17)

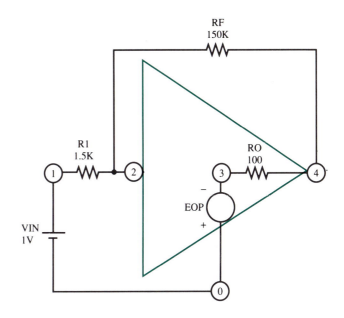

```
EXAMPLE 8.17
VIN  1  0  1V
R1   1  2  1.5K
RF   2  4  150K
EOP  0  3  2  0  2500
RO   3  4  100
.DC  VIN  1  1  1
.PRINT  DC  V(4)  I(VIN)
.END
```

Since VIN = 1 V, the output voltage at node 4, V(4), is numerically equal to the closed-loop voltage gain. The results of a program run reveal that V(4) = −96.11, in close agreement with the gain calculated in Example 8.3. SPICE computes the magnitude of I(VIN) to be 0.641 mA, so the input resistance is

$$r_{in} = \frac{v_{in}}{i_{in}} = \frac{\text{VIN}}{|\text{I(VIN)}|} = \frac{1\ \text{V}}{0.641\ \text{mA}} = 1560\ \Omega$$

To find the output resistance of the amplifier, we must compute the current that flows through a short circuit connected to the output, since $r_o = v_L$(open circuit)$/i_L$(short circuit). A zero-valued dummy voltage source, VDUM, connected between nodes 4 and 0 effectively shorts the output to ground. The current in VDUM is then the short-circuit output current. SPICE computes this value to be I(VDUM) = 24.7 A. Thus,

$$r_o = \frac{\text{V(4) (open circuit)}}{\text{I(VDUM) (short circuit)}} = \frac{96.11\ \text{V}}{24.7\ \text{A}} = 3.89\ \Omega$$

This value is in close agreement with that calculated in Example 8.3.

Example 8.18

Use PSpice and the PSpice library to verify the value of the closed-loop bandwidth calculated in Example 8.16 for the 741 operational amplifier. Assume a 0.1-V-ac input.

Solution.   The PSpice circuit and input circuit file are shown in Figure S8.2. Note that subcircuit call X1 specifies the name UA741 for the subcircuit stored in

Figure S8.2
(Example 8.18)

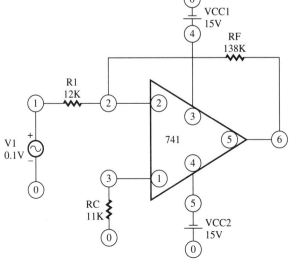

```
EXAMPLE 8.18
V1    1   0   AC  0.1V
VCC1  4   0   15V
VCC2  0   5   15V
R1    1   2   12K
RF    2   6   138K
RC    3   0   11K
X1    3   2   4   5   6   UA741
.LIB
.AC  LIN   1  80KHZ   80KHZ
.PRINT  AC  V(6)
.END
```

the PSpice library (see Appendix Section A.17). (If the Evaluation version of PSpice is used, the .LIB statement must be written .LIB EVAL.LIB.) As shown in Example 8.16, the magnitude of the closed-loop gain is 11.5. Therefore, the midband output voltage is $|A_v|e_{in} = 11.5(0.1\text{ V}) = 1.15$ V. At cutoff, the output voltage should be $0.707(1.15\text{ V}) = 0.813$ V. To verify the calculations of Example 8.16, the frequency of the simulation is set to the theoretical bandwidth, 80 kHz, at which the output voltage should be near 0.813 V. Execution of the program reveals that the output voltage at 80 kHz is 0.822 V, which differs only 1.1% from 0.813 V.

EXERCISES

Section 8.1   Feedback Theory

8.1   An operational amplifier having an open-loop gain of 5000 is used in a noninverting configuration with a feedback resistor of 1 MΩ. For each of the following values of R_1, find the closed-loop gain if the amplifier were ideal, and find the actual closed-loop gain.
   a. $R_1 = 5$ kΩ
   b. $R_1 = 20$ kΩ
   c. $R_1 = 100$ kΩ

8.2   Repeat Exercise 8.1 when the open-loop gain of the amplifier is increased by a factor of 2.

8.3   An operational amplifier is to be used in a noninverting configuration that has an ideal closed-loop gain of 800. What minimum value of open-loop gain should the amplifier have if the actual closed-loop gain must be at least 799?

8.4   The operational amplifier in Exercise 8.1 has a differential input resistance of 40 kΩ and an output resistance of 90 Ω. Find the closed-loop input and output resistance for each of the values of R_1 listed.

8.5   An operational amplifier has open-loop gain 10^4 and output resistance 120 Ω. It is to be used in a noninverting configuration for an application in which its closed-loop output resistance must be no greater than 1 Ω. What is the maximum closed-loop gain that the amplifier can have?

8.6   Derive equation 8.13 from equation 8.12.

8.7   An operational amplifier has an open-loop gain of 5000. It is used in an inverting configuration with a feedback ratio of 0.2. What is its closed-loop gain?

8.8   The operational amplifier in Figure 8.27 has an open-loop gain of 5×10^4. Draw a block diagram of the feedback model for the configuration shown. Label each block with the correct numerical quantity.

8.9   Show that the feedback model shown in Figure 8.28 is equivalent to that shown in Figure 8.5.

8.10   An operational amplifier is to be used in an inverting configuration with feedback resistance 100 kΩ and input resistance 2 kΩ. If the closed-loop gain must be no less than -49.5,
   a. what minimum value of loop gain should it have; and
   b. what minimum value of open-loop gain should it have? (*Hint:* Work with gain *magnitudes*.)

8.11   The operational amplifier shown in Figure

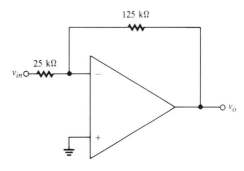

Figure 8.27
(Exercise 8.8)

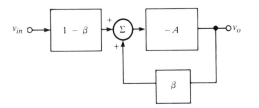

Figure 8.28
(Exercise 8.9)

8.29 has an open-loop gain of 8000 and an open-loop output resistance of 250 Ω. Find the closed-loop
a. input resistance; and
b. output resistance.

Section 8.2   Harmonic Distortion and Feedback

8.12 A square wave having peak value V_P volts and frequency ω rad/s can be expressed as the sum of an infinite number of sine waves as follows: $v(t) = 4V_P/\pi\,[\sin \omega t + (1/3) \sin 3\omega t + (1/5) \sin 5\omega t + \ldots]$. Assuming that the square wave represents a distorted sine wave whose frequency is the same as the fundamental, find its total harmonic distortion. Neglect harmonics beyond the ninth.

8.13 An amplifier has 1% distortion with negative feedback and 10% distortion without feedback. If the feedback ratio is 0.02, what is the open-loop gain of the amplifier?

8.14 An amplifier has a voltage gain of 80 when no feedback is connected. When negative feedback is connected, the amplifier gain is 20 and its harmonic distortion is 10%. What is the harmonic distortion without feedback?

Section 8.3   Frequency Response

8.15 An operational amplifier has gain-bandwidth product equal to 5×10^5 and a dc, open-loop gain of 20,000. At what frequency does the open-loop gain equal 14,142?

8.16 With reference to the operational amplifier in Exercise 8.15,
a. at what frequency is the open-loop gain equal to 0 dB, and
b. what is the open-loop gain at 2.5 kHz?

8.17 The operational amplifier in Figure 8.30 has a unity-gain frequency of 1.2 MHz.
a. What is the closed-loop bandwidth?
b. What is the closed-loop gain at 600 kHz?

8.18 An operational amplifier having a gain-bandwidth product of 8×10^5 is to be used in a noninverting configuration as an audio amplifier (20 Hz–20 kHz). What is the maximum closed-loop gain that can be obtained from the amplifier in this application?

8.19 An operational amplifier has a dc open-loop gain of 25×10^4 and an open-loop cutoff frequency of 40 Hz. It is to be used in an inverting configuration to amplify signals up to 50 kHz. What is the maximum closed-loop gain that can be obtained from the amplifier for this application?

8.20 The operational amplifier in Figure 8.31 has a unity-gain frequency of 2 MHz.
a. What is the closed-loop bandwidth?
b. What is the closed-loop gain at 2 MHz?

8.21 Each of the operational amplifiers in Figure 8.32 has a unity-gain frequency of 750 kHz. What is the approximate upper cutoff frequency of the cascaded system?

Section 8.4   Slew Rate

8.22 What is the rate of change, in volts/second, of a triangular waveform that varies be-

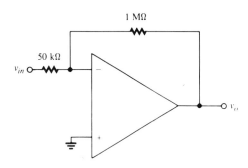

Figure 8.29
(Exercise 8.11)

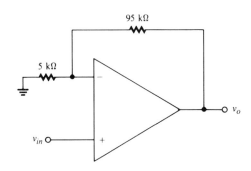

Figure 8.30
(Exercise 8.17)

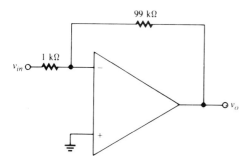

Figure 8.31
(Exercise 8.20)

tween 0 V and 5 V and that has frequency 10 kHz?

8.23 What minimum slew rate is necessary for a unity-gain amplifier that must pass, without distortion, the input waveform shown in Figure 8.33?

8.24 Repeat Exercise 8.23 if the amplifier is in a noninverting configuration with $R_1 = 50$ kΩ and $R_f = 100$ kΩ.

8.25 An inverting amplifier has a slew rate of 2 V/μs. What maximum closed-loop gain can it have without distorting the input waveform shown in Figure 8.33?

8.26 An operational amplifier has a slew rate of 2 V/μs. What maximum peak amplitude can a 1-MHz input signal have without exceeding the slew rate, when the amplifier is used in a voltage-follower circuit?

8.27 What minimum slew rate is required for an operational amplifier whose output must be at least 2 V rms over the audio frequency range?

8.28 The operational amplifier shown in Figure 8.34 has a slew rate of 1.2 V/μs. Determine whether the output will be distorted due to the slew-rate limitation when the input is any one of the following sinusoidal signals: $v_1 = 0.7$ V rms at 30 kHz; $v_2 = 1.0$ V rms at 15 kHz; $v_3 = 0.5$ V rms at 40 kHz; $v_4 = 0.1$ V rms at 20 kHz. If distortion occurs, determine the signals that are responsible.

8.29 If distortion occurs in Exercise 8.28, find remedies other than changing the input signals.

8.30 The operational amplifier shown in Figure 8.35 has a slew rate of 0.5 V/μs and a unity-gain frequency of 1 MHz. Find the maximum frequency of a 0.2-V-rms sine-wave input that can be amplified without distortion.

8.31 To what value could the feedback resistor in Exercise 8.30 be changed if it were necessary that the maximum frequency be 50 kHz?

8.32 If the input to the amplifier in Exercise 8.30 is a triangular wave that varies between -1 V and $+1$ V peak and has frequency 1 kHz, determine whether distortion will occur.

8.33 If the input to the amplifier in Exercise 8.30 is a ramp that rises from -0.5 V to $+1.5$ V in 2.5 μs, determine whether distortion will occur.

Section 8.5   Offset Currents and Voltages

8.34 The operational amplifier shown in Figure 8.36 has $I_B^+ = 100$ nA and $I_B^- = 80$ nA. If $R_1 = R_f = R_c = 20$ kΩ, find the output offset voltage due to the input bias currents.

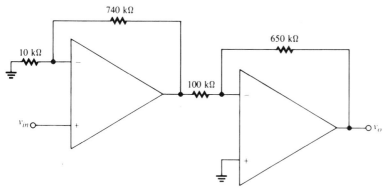

Figure 8.32
(Exercise 8.21)

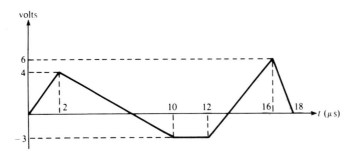

Figure 8.33
(Exercise 8.23)

Figure 8.34
(Exercise 8.28)

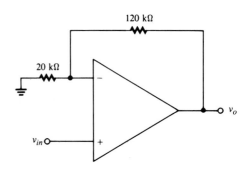

Figure 8.35
(Exercise 8.30)

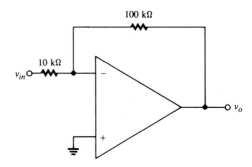

Figure 8.36
(Exercise 8.34)

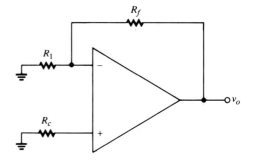

8.35 Find the optimum value for R_c in Exercise 8.34 and repeat the exercise using that value.

8.36 An operational amplifier has an input offset current of 50 nA. It is to be used in an inverting-amplifier application where the output offset voltage due to bias currents cannot exceed 10 mV. If the amplifier must provide an impedance of at least 40 kΩ to the signal source driving it, what is the maximum permissible closed-loop gain of the amplifier?

8.37 An operational amplifier having an input offset current of 80 nA is to be used in a noninverting-amplifier application where the output offset voltage due to bias currents cannot exceed 5 mV. The value of R_1 is 10 kΩ and the amplifier must have the maximum possible closed-loop gain. Design the circuit and find its voltage gain.

8.38 Given $I_B = (I_B^+ + I_B^-)/2$ and $|I_{io}| = |I_B^+ - I_B^-|$, solve these equations simultaneously to show that
   a. when $I_B^+ > I_B^-$, $I_B^+ = I_B + 0.5|I_{io}|$ and $I_B^- = I_B - 0.5|I_{io}|$; and
   b. when $I_B^+ < I_B^-$, $I_B^+ = I_B - 0.5|I_{io}|$ and $I_B^- = I_B + 0.5|I_{io}|$. (*Hint:* When $I_B^+ > I_B^-$, $|I_B^+ - I_B^-| = I_B^+ - I_B^-$.)

8.39 The operational amplifier in Figure 8.37 has $I_B = 100$ nA and $|I_{io}| = 50$ nA. What is the maximum possible value of $|V_{OS}(I_B)|$?

8.40 An operational amplifier has an input offset voltage of 1.2 mV. It is to be used in a noninverting-amplifier application where the output offset due to V_{io} cannot exceed 6 mV. If the feedback resistor is 20 kΩ, what is the minimum permissible value of R_1?

8.41 The input offset voltage of an operational amplifier used in a voltage-follower circuit is 0.5 mV. What is the value of $|V_{OS}(V_{io})|$?

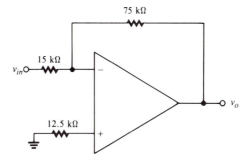

Figure 8.38
(Exercise 8.42)

8.42 The operational amplifier shown in Figure 8.38 has $|I_{io}| = 120$ nA and $|V_{io}| = 1.5$ mV. Find the worst-case output offset voltage.

8.43 The operational amplifier in Figure 8.39 has $I_B = 100$ nA, $|I_{io}| = 30$ nA, and $|v_{io}| = 2$ mV. Find the worst-case value of $|v_{OS}|$.

Section 8.6 Operational Amplifier Specifications

8.44 Using the 741 operational amplifier specifications, determine the following:
   a. the minimum slew rate of the μA741E at 25°C;
   b. the output voltage swing of the μA741C at 70°C when the supply voltage is 15 V;
   c. the input bias current of the μA741A at 60°C ambient temperature; and
   d. the input offset current of the μA741 at 0°C ambient temperature.

8.45 Using the 741 operational amplifier specifications, determine the following:
   a. the maximum input offset voltage of the μA741A at 25°C;
   b. the CMRR of the μA741C at 10 kHz;

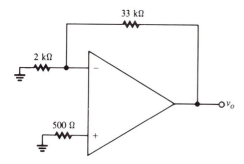

Figure 8.37
(Exercise 8.39)

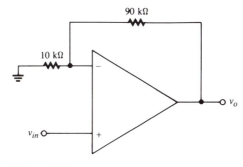

Figure 8.39
(Exercise 8.43)

c. the input offset current of the μA741E at 25°C when the supply voltage is 10 V; and

d. the closed-loop bandwidth of the μA741 at 100°C if its closed-loop bandwidth at 20°C is 40 kHz.

8.46 Using the μA741A, design an inverting operational-amplifier circuit that will meet the following criteria at 25°C:

a. maximum possible closed-loop voltage gain;

b. closed-loop bandwidth of at least 20 kHz;

c. worst-case output offset voltage of 50 mV; and

d. minimum input resistance 10 kΩ.

8.47 Using the μA741C, design a noninverting operational-amplifier circuit that will meet the following criteria at 25°C:

a. maximum possible closed-loop voltage gain;

b. must not distort an input ramp voltage that changes from -1 V to $+1$ V in 150 μs;

c. worst-case output offset voltage of 100 mV; and

d. R_1 must be at least 1 kΩ.

SPICE EXERCISES

8.48 The ideal operational amplifier in Figure 7.17 has $R_1 = 10$ kΩ and $R_f = 25$ kΩ. The input voltage, v_{in}, is $2 \sin(2\pi \times 10^3 t)$ V. Use SPICE to obtain a plot of the output waveform versus time over one full period. Use the results to determine the closed-loop voltage gain and compare that with the theoretical value for an ideal amplifier.

8.49 An operational amplifier is connected in a noninverting configuration with $R_1 = 10$ kΩ and $R_f = 90$ kΩ. The open-loop voltage gain of the amplifier is 1×10^3 and its open-loop output resistance is 150 Ω. Assuming the input impedance is essentially infinite (10^{12} Ω), use SPICE to find the closed-loop voltage gain and output resistance. Compare your results with the theoretical values.

Applications of Operational Amplifiers

9

9.1 VOLTAGE SUMMATION, SUBTRACTION, AND SCALING

Voltage Summation

We have seen that it is possible to *scale* a signal voltage, that is, to multiply it by a fixed constant, through an appropriate choice of external resistors that determine the closed-loop gain of an amplifier circuit. This operation can be accomplished in either an inverting or noninverting configuration. It is also possible to sum several signal voltages in one operational-amplifier circuit and at the same time scale each by a different factor. For example, given inputs v_1, v_2, and v_3, we might wish to generate an output equal to $2v_1 + 0.5v_2 + 4v_3$. The latter sum is called a *linear combination* of v_1, v_2, and v_3, and the circuit that produces it is often called a *linear-combination circuit*.

Figure 9.1 shows an inverting amplifier circuit that can be used to sum and scale three input signals. Note that input signals v_1, v_2, and v_3 are applied through separate resistors R_1, R_2, and R_3 to the summing junction of the amplifier and that there is a single feedback resistor R_f. Resistor R_c is the offset compensation resistor discussed in Chapter 8.

Following the same procedure we used in Chapter 7 to derive the output of an inverting amplifier having a single input, we obtain for the three-input (ideal) amplifier

$$i_1 + i_2 + i_3 = -i_f \tag{9.1}$$

Or, since the voltage at the summing junction is ideally 0,

$$\frac{v_1}{R_1} + \frac{v_2}{R_2} + \frac{v_3}{R_3} = \frac{-v_o}{R_f} \tag{9.2}$$

Solving (9.2) for v_o gives

$$v_o = -\left(\frac{R_f}{R_1} v_1 + \frac{R_f}{R_2} v_2 + \frac{R_f}{R_3} v_3\right) \tag{9.3}$$

Equation 9.3 shows that the output is the inverted sum of the separately scaled inputs, i.e., a *weighted* sum, or linear combination of the inputs. By appropriate

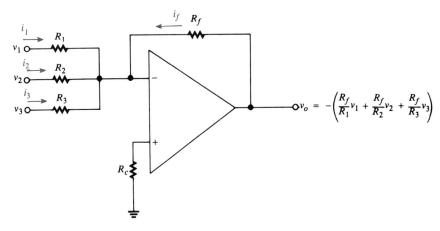

Figure 9.1
An operational-amplifier circuit that produces an output equal to the (inverted) sum of three separately scaled input signals

choice of values for R_1, R_2, and R_3, we can make the scale factors equal to whatever constants we wish, within practical limits. If we choose $R_1 = R_2 = R_3 = R$, then we obtain

$$v_o = \frac{-R_f}{R}(v_1 + v_2 + v_3) \tag{9.4}$$

and, for $R_f = R$,

$$v_o = -(v_1 + v_2 + v_3) \tag{9.5}$$

The theory can be extended in an obvious way to two, four, or any reasonable number of inputs. The feedback ratio for the circuit is

$$\beta = \frac{R_p}{R_p + R_f} \tag{9.6}$$

where $R_p = R_1 \| R_2 \| R_3$. Using this value of β, we can apply the theory developed in Chapter 8 to determine all the performance characteristics that depend on β, including closed-loop bandwidth and output offset $V_{OS}(V_{io})$. The optimum value of the bias-current compensation resistor is

$$R_c = R_f \| R_p = R_f \| R_1 \| R_2 \| R_3 \tag{9.7}$$

Example 9.1

DESIGN

1. Design an operational-amplifier circuit that will produce an output equal to $-(4v_1 + v_2 + 0.1v_3)$.
2. Write an expression for the output and sketch its waveform when $v_1 = 2 \sin \omega t$ V, $v_2 = +5$ V dc, and $v_3 = -100$ V dc.

Solution

1. We arbitrarily choose $R_f = 60$ kΩ. Then

$$\frac{R_f}{R_1} = 4 \Rightarrow R_1 = \frac{60 \text{ k}\Omega}{4} = 15 \text{ k}\Omega$$

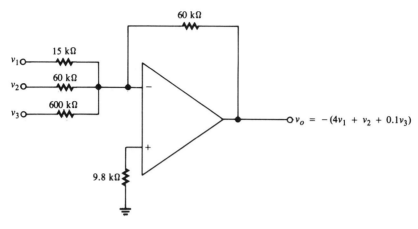

Figure 9.2
(Example 9.1)

Figure 9.3
(Example 9.1)

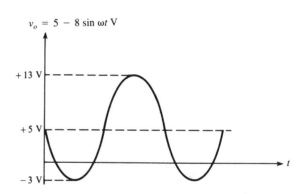

$$\frac{R_f}{R_2} = 1 \Rightarrow R_2 = \frac{60 \text{ k}\Omega}{1} = 60 \text{ k}\Omega$$

$$\frac{R_f}{R_3} = 0.1 \Rightarrow R_3 = \frac{60 \text{ k}\Omega}{0.1} = 600 \text{ k}\Omega$$

By equation 9.7, the optimum value for the compensating resistor is $R_c = R_f \parallel R_1 \parallel R_2 \parallel R_3 = (60 \text{ k}\Omega) \parallel (15 \text{ k}\Omega) \parallel (60 \text{ k}\Omega) \parallel (600 \text{ k}\Omega) = 9.8 \text{ k}\Omega$. The circuit is shown in Figure 9.2.

2. $v_o = -[4(2 \sin \omega t \text{ V}) + 1(5 \text{ V}) + 0.1(-100 \text{ V})] = -8 \sin \omega t \text{ V} - 5 \text{ V} + 10 \text{ V} = 5 - 8 \sin \omega t \text{ V}$. This output is sinusoidal with a 5-V offset and varies between $5 - 8 = -3$ V and $5 + 8 = 13$ V. It is sketched in Figure 9.3.

Figure 9.4 shows a noninverting version of the linear-combination circuit. In this example, only two inputs are connected and it can be shown (Exercise 9.4) that

$$v_o = \frac{R_g + R_f}{R_g} \left(\frac{R_2}{R_1 + R_2} v_1 + \frac{R_1}{R_1 + R_2} v_2 \right) \tag{9.8}$$

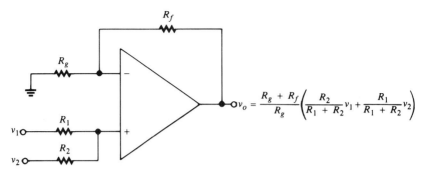

Figure 9.4
A noninverting linear combination circuit

Although this circuit does not invert the scaled sum, it is somewhat more cumbersome than the inverting circuit in terms of selecting resistor values to provide precise scale factors. Also, it is limited to producing outputs of the form $K[av_1 + (1 - a)v_2]$ where K and a are positive constants. Phase inversion is often of no consequence, but in those applications where a noninverted sum is required, it can also be obtained using the inverting circuit of Figure 9.1, followed by a unity-gain inverter.

Voltage Subtraction

Suppose we wish to produce an output voltage that equals the mathematical difference between two input signals. This operation can be performed by using the amplifier in a *differential* mode, where the signals are connected through appropriate resistor networks to the inverting and noninverting terminals. Figure 9.5 shows the configuration. We can use the superposition principle to determine the output of this circuit. First, assume that v_2 is shorted to ground. Then

$$v^+ = \frac{R_2}{R_1 + R_2}\, v_1 \tag{9.9}$$

so

$$v_{o1} = \frac{R_3 + R_4}{R_3}\, v^+ = \left(\frac{R_3 + R_4}{R_3}\right)\left(\frac{R_2}{R_1 + R_2}\right) v_1 \tag{9.10}$$

Assuming now that v_1 is shorted to ground, we have

$$v_{o2} = \frac{-R_4}{R_3}\, v_2 \tag{9.11}$$

Therefore, with both signal inputs present, the output is

$$v_o = v_{o1} + v_{o2} = \left(\frac{R_3 + R_4}{R_3}\right)\left(\frac{R_2}{R_1 + R_2}\right) v_1 - \left(\frac{R_4}{R_3}\right) v_2 \tag{9.12}$$

Equation 9.12 shows that the output is proportional to the difference between scaled multiples of the inputs. To obtain the output

$$v_o = A(v_1 - v_2) \tag{9.13}$$

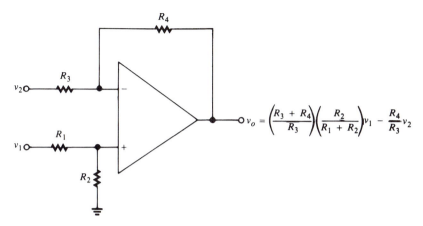

$$v_o = \left(\frac{R_3 + R_4}{R_3}\right)\left(\frac{R_2}{R_1 + R_2}\right)v_1 - \frac{R_4}{R_3}v_2$$

Figure 9.5
Using the amplifier in a differential mode to obtain an output proportional to the difference between two scaled inputs

where A is a fixed constant, select the resistor values in accordance with the following:

$$R_1 = R_3 = R \quad \text{and} \quad R_2 = R_4 = AR \tag{9.14}$$

Substituting these values into (9.12) gives

$$\left(\frac{R + AR}{R}\right)\left(\frac{AR}{R + AR}\right)v_1 - \frac{AR}{R}v_2 = \frac{AR}{R}v_1 - \frac{AR}{R}v_2 = A(v_1 - v_2)$$

as required. When resistor values are chosen in accordance with (9.14), the bias compensation resistance $(R_1 \parallel R_2)$ is automatically the correct value $(R_3 \parallel R_4)$, namely, $R \parallel AR$.

Let the general form of the output of Figure 9.5 be

$$v_o = a_1 v_1 - a_2 v_2 \tag{9.15}$$

where a_1 and a_2 are positive constants. Then, by equation 9.12, we must have

$$a_1 = \left(1 + \frac{R_4}{R_3}\right)\left(\frac{R_2}{R_1 + R_2}\right) \tag{9.16}$$

and

$$a_2 = \frac{R_4}{R_3} \tag{9.17}$$

Substituting (9.17) into (9.16) gives

$$a_1 = (1 + a_2)\frac{R_2}{R_1 + R_2} \tag{9.18}$$

But the quantity $R_2/(R_1 + R_2)$ is always less than 1. Therefore, equation 9.18 shows that in order to use the circuit of Figure 9.5 to produce $v_o = a_1 v_1 - a_2 v_2$, we must have

$$(1 + a_2) > a_1 \qquad\qquad (9.19)$$

This restriction limits the usefulness of the circuit.

Example 9.2

DESIGN

Design an operational-amplifier circuit that will produce the output $v_o = 0.5v_1 - 2v_2$.

Solution.   Note that $a_1 = 0.5$ and $a_2 = 2$, so $(1 + a_2) > a_1$. Therefore, it is possible to construct a circuit in the configuration of Figure 9.5.

Comparing v_o with equation 9.12, we see that we must have

$$\left(1 + \frac{R_4}{R_3}\right)\left(\frac{R_2}{R_1 + R_2}\right) = 0.5$$

and

$$\frac{R_4}{R_3} = 2$$

Let us arbitrarily choose $R_4 = 100 \text{ k}\Omega$. Then $R_3 = R_4/2 = 50 \text{ k}\Omega$. Thus

$$\left(1 + \frac{R_4}{R_3}\right)\left(\frac{R_2}{R_1 + R_2}\right) = \frac{3R_2}{R_1 + R_2} = 0.5$$

Arbitrarily choosing $R_2 = 20 \text{ k}\Omega$, we have

$$\frac{3(20 \text{ k}\Omega)}{R_1 + (20 \text{ k}\Omega)} = 0.5$$

$$60 \text{ k}\Omega = 0.5R_1 + (10 \text{ k}\Omega)$$

$$R_1 = 100 \text{ k}\Omega$$

The completed design is shown in Figure 9.6.

Figure 9.6
(Example 9.2)

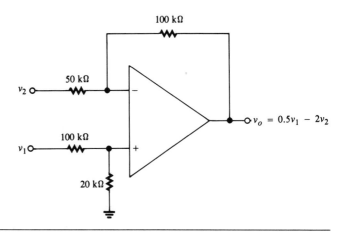

In Example 9.2, we note that the compensation resistance ($R_1 \parallel R_2 = (100 \text{ k}\Omega) \parallel (20 \text{ k}\Omega) = 16.67 \text{ k}\Omega$) is not equal to its optimum value ($R_3 \parallel R_4 = (50 \text{ k}\Omega) \parallel (100 \text{ k}\Omega) = 33.33 \text{ k}\Omega$). With some algebraic complication, we can impose the additional condition $R_1 \parallel R_2 = R_3 \parallel R_4$ and thereby force the compensation resistance to have its optimum value. With $v_o = a_1 v_1 - a_2 v_2$, it can be shown (Exercise 9.8) that the compensation resistance $(R_1 \parallel R_2)$ is optimum when the resistor values are selected in accordance with

$$R_4 = a_1 R_1 = a_2 R_3 = R_2(1 + a_2 - a_1) \tag{9.20}$$

To apply this design criterion, choose R_4 and solve for R_1, R_2, and R_3. In Example 9.2, $a_1 = 0.5$ and $a_2 = 2$. If we choose $R_4 = 100$ kΩ, then $R_1 = (100$ kΩ$)/0.5 = 200$ kΩ, $R_2 = (100$ kΩ$)/2.5 = 40$ kΩ, and $R_3 = (100$ kΩ$)/2 = 50$ kΩ. These choices give $R_1 \parallel R_2 = 33.3$ kΩ $= R_3 \parallel R_4$, as required.

Although the circuit of Figure 9.5 is a useful and economical way to obtain a difference voltage of the form $A(v_1 - v_2)$, our analysis has shown that it has limitations and complications when we want to produce an output of the general form $v_o = a_1 v_1 - a_2 v_2$. An alternate way to obtain a scaled difference between two signal inputs is to use *two* inverting amplifiers, as shown in Figure 9.7. The output of the first amplifier is

$$v_{o1} = \frac{-R_2}{R_1} v_1 \tag{9.21}$$

and the output of the second amplifier is

$$v_{o2} = -\left(\frac{R_5}{R_3} v_{o1} + \frac{R_5}{R_4} v_2\right) = \frac{R_5 R_2}{R_3 R_1} v_1 - \frac{R_5}{R_4} v_2 \tag{9.22}$$

This equation shows that there is a great deal of flexibility in the choice of resistor values necessary to obtain $v_o = a_1 v_1 - a_2 v_2$, since a large number of combinations will satisfy

$$\frac{R_5 R_2}{R_3 R_1} = a_1 \quad \text{and} \quad \frac{R_5}{R_4} = a_2 \tag{9.23}$$

Furthermore, there are no restrictions on the choice of values for a_1 and a_2, nor any complications in setting R_c to its optimum value.

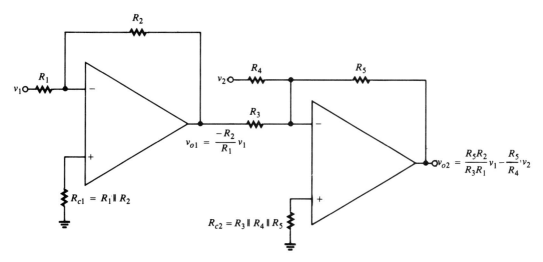

Figure 9.7
Using two inverting amplifiers to obtain the output $v_o = a_1 v_1 - a_2 v_2$

Example 9.3

DESIGN Design an operational-amplifier circuit using two inverting configurations to produce the output $v_o = 20v_1 - 0.2v_2$. (Note that $1 + a_2 = 1.2 < 20 = a_1$, so we cannot use the differential circuit of Figure 9.5.)

Solution.   We have so many choices for resistance values that the best approach is to implement the circuit directly, without bothering to use the algebra of equation 9.20. We can, for example, begin the process by designing the first amplifier to produce $-20v_1$. Choose $R_1 = 10$ kΩ and $R_2 = 200$ kΩ. Then, the second amplifier need only invert $-20v_1$ with unity gain and scale the v_2 input by 0.2.

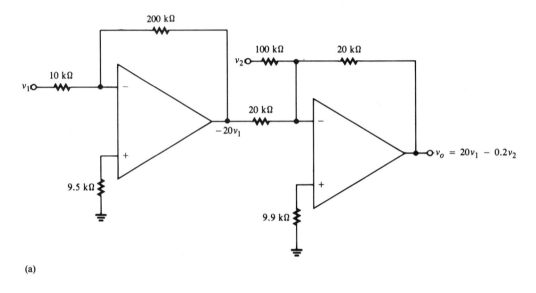

(a)

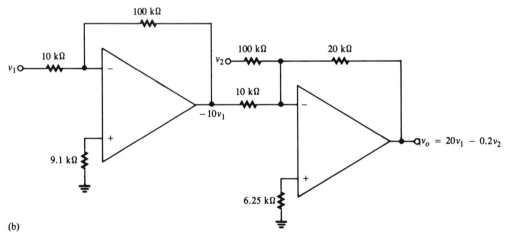

(b)

Figure 9.8
(Example 9.3) Two (of many) equivalent methods for producing $20v_1 - 0.2v_2$ using two inverting amplifiers

Choose $R_5 = 20$ kΩ. Then $R_5/R_3 = 1 \Rightarrow R_3 = 20$ kΩ and $R_5/R_4 = 0.2 \Rightarrow R_4 = 100$ kΩ.

The completed design is shown in Figure 9.8(a). Figure 9.8(b) shows another solution, in which the first amplifier produces $-10v_1$ and the second multiplies that by the constant -2. The compensation resistors have values calculated as shown in Figure 9.7.

Although there are a large number of ways to choose resistor values to satisfy equation 9.21, there may, in practice, be constraints on some of those choices imposed by other performance requirements. For example, R_1 may have to be a certain minimum value to provide adequate input resistance to the v_1 signal source. Recall, also, that the greater the closed-loop gain of a stage, the smaller its bandwidth. Thus it may be necessary to "distribute" gain over two stages, as is done in Figure 9.8(b) to obtain $20v_1$, in order to increase the overall bandwidth. Finally, it may be necessary to minimize the gain of one stage or the other to reduce the effect of its input offset voltage. Note that the input offset voltage of the first stage is amplified by both stages.

The method used to design a subtractor circuit in Example 9.3 can be extended in an obvious way to the design of circuits that produce a linear combination of voltage sums and differences. The most general form of a linear combination is $v_o = \pm a_1v_1 \pm a_2v_2 \pm a_3v_3 \pm \ldots \pm a_nv_n$. Remember that the input signal corresponding to any term that appears in the output with a positive sign must pass through two inverting stages.

Example 9.4

DESIGN

1. Design an operational-amplifier circuit using two inverting configurations to produce the output $v_o = -10v_1 + 5v_2 + 0.5v_3 - 20v_4$.
2. Assuming that the unity-gain frequency of each amplifier is 1 MHz, find the approximate, overall, closed-loop bandwidth of your solution.

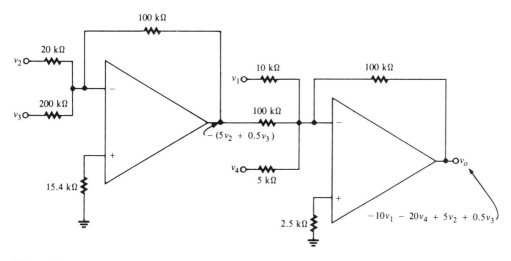

Figure 9.9
(Example 9.4)

Solution

1. Since v_2 and v_3 appear with positive signs in the output, those two inputs must be connected to the first inverting amplifier. We can produce $-(5v_2 + 0.5v_3)$ at the output of the first inverting amplifier and then invert and add it to $-(10v_1 + 20v_4)$ in the second amplifier. One possible solution is shown in Figure 9.9 on page 379.

2. The feedback ratio of the first amplifier is

$$\beta_1 = \frac{(20 \text{ k}\Omega) \| (200 \text{ k}\Omega)}{(20 \text{ k}\Omega) \| (200 \text{ k}\Omega) + (100 \text{ k}\Omega)} = 0.1538$$

Therefore, the closed-loop bandwidth of the first amplifier is $BW_{CL1} = \beta_1 f_t = (0.1538)(1 \text{ MHz}) = 153.8 \text{ kHz}$. Similarly,

$$\beta_2 = \frac{(10 \text{ k}\Omega) \| (100 \text{ k}\Omega) \| (5 \text{ k}\Omega)}{(10 \text{ k}\Omega) \| (100 \text{ k}\Omega) \| (5 \text{ k}\Omega) + (100 \text{ k}\Omega)} = 0.0312$$

and $BW_{CL2} = \beta_2 f_t = (0.0312)(1 \text{ MHz}) = 31.2 \text{ kHz}$. The overall bandwidth is approximately equal to the smaller of BW_{CL1} and BW_{CL2}, or 31.2 kHz.

9.2   CONTROLLED VOLTAGE AND CURRENT SOURCES

Recall that a *controlled* source is one whose output voltage or current is determined by the magnitude of another, independent voltage or current. We have used controlled sources extensively in our study of transistor-circuit models, but those were, in a sense, fictitious devices that served mainly to simplify the circuit analysis. We wish now to explore various techniques that can be used to construct controlled voltage and current sources using operational amplifiers. As we shall see, some of these sources are realized simply by studying already-familiar circuits from a different viewpoint.

Voltage-Controlled Voltage Sources

An ideal, voltage-controlled voltage source (VCVS) is one whose output voltage V_o (1) equals a fixed constant (k) times the value of another, controlling voltage: $V_o = kV_i$; and (2) is independent of the current drawn from it. Notice that the constant k is dimensionless. Both the inverting and noninverting configurations of an ideal operational amplifier meet the two criteria. In each case, the output voltage equals a fixed constant (the closed-loop gain, determined by external resistors) times an input voltage. Also, since the output resistance is (ideally) 0, there is no voltage division at the output and the voltage is independent of load. We have studied these configurations in detail, so we will be content for now with the observation that they do belong to the category of voltage-controlled voltage sources.

Voltage-Controlled Current Sources

An ideal, voltage-controlled current source is one that supplies a current whose magnitude (1) equals a fixed constant (k) times the value of an independent, controlling voltage: $I_o = kV_i$; and (2) is independent of the load to which the current is supplied. Notice that the constant k has the dimensions of conductance (siemens). Since it relates output current to input voltage, it is called the *transcon-ductance, g_m*, of the source.

Figure 9.10 shows two familiar amplifier circuits: the inverting and noninverting configurations of an operational amplifier. Note, however, that we now regard the feedback resistors as *load resistors* and designate each by R_L. We will show that each circuit behaves as a voltage-controlled current source, where the load current is the current I_L in R_L.

In Figure 9.10(a), v^- is virtual ground, so $I_1 = V_{in}/R_1$. Since no current flows into the inverting terminal of the ideal amplifier, $I_L = I_1$, or

$$I_L = \frac{V_{in}}{R_1} \tag{9.24}$$

Equation 9.24 shows that the load current is the constant $1/R_1$ times the control-ling voltage V_{in}. Thus, the transconductance is $g_m = 1/R_1$ siemens. *Note that R_L does not appear in the equation, so the load current is independent of load resistance.* Like any constant-current source, the load voltage (voltage across R_L) will change if R_L is changed, but the current remains the same. The direction of the current through the load is controlled by the polarity of V_{in}. This version of a controlled current source is said to have a *floating load,* because neither side of R_L can be grounded. Thus, it is useful only in applications where the load is not required to have the same ground reference as the controlling voltage, V_{in}.

Figure 9.10
Floating-load, voltage-controlled current sources

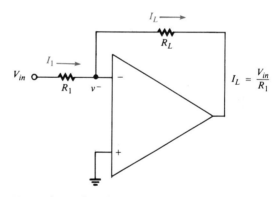

(a) Inverting configuration

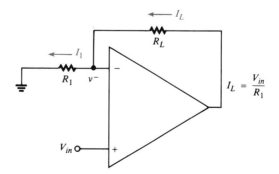

(b) Noninverting configuration

In Figure 9.10(b), $v^- = V_{in}$, so $I_1 = V_{in}/R_1$. Once again, no current flows into the inverting terminal, so $I_L = I_1$. Therefore,

$$I_L = \frac{V_{in}}{R_1} \qquad (9.25)$$

As in the inverting configuration, the load current is independent of R_L and the transconductance is $1/R_1$ siemens. The load is also floating in this version.

Of course, there is a practical limit on the range of load resistance R_L that can be used in each circuit. If R_L is made too large, the output voltage of the amplifier will approach its maximum limit, as determined by the power supply voltages. For successful operation, the load resistance in each circuit must obey

$$R_L < \frac{R_1 |V_{max}|}{V_{in}} \qquad \text{(inverting circuit)} \qquad (9.26)$$

$$R_L < R_1 \left(\frac{|V_{max}|}{V_{in}} - 1 \right) \qquad \text{(noninverting circuit)} \qquad (9.27)$$

where $|V_{max}|$ is the magnitude of the maximum output voltage of the amplifier.

Example 9.5

DESIGN

Design an inverting, voltage-controlled current source that will supply a constant current of 0.2 mA when the controlling voltage is 1 V. What is the maximum load resistance for this supply if the maximum amplifier output voltage is 20 V?

Solution.   The transconductance is $g_m = (0.2 \text{ mA})/(1 \text{ V}) = 0.2 \times 10^{-3}$ S. Therefore, $R_1 = 1/g_m = 5 \text{ k}\Omega$. By (9.26),

$$R_L < \frac{R_1 |V_{max}|}{V_{in}} = \frac{(5 \text{ k}\Omega)(20 \text{ V})}{1 \text{ V}} = 100 \text{ k}\Omega$$

The required circuit is shown in Figure 9.11.

Figure 9.11
(Example 9.5)

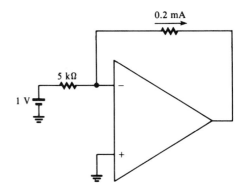

Figure 9.12(a) shows a voltage-controlled current source that can be operated with a grounded load. To understand its behavior as a current source, refer to Figure 9.12(b), which shows the voltages and currents in the circuit. Since there is (ideally) zero current into the + input, Kirchhoff's current law at the node where

Figure 9.12
A voltage-controlled current source with a grounded load

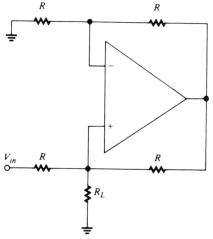

(a) The voltage-controlled current source

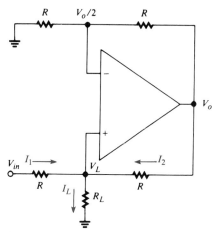

(b) Voltages and currents in the circuit of (a)

R_L is connected to the + input gives

$$I_L = I_1 + I_2 \tag{9.28}$$

or

$$I_L = \frac{V_{in} - V_L}{R} + \frac{V_o - V_L}{R} \tag{9.29}$$

By voltage-divider action,

$$v^- = \left(\frac{R}{R + R}\right) V_o = V_o/2 \tag{9.30}$$

Since $v^- = v^+ = V_L$, we have $V_L = V_o/2$, which, upon substitution in (9.29), gives

$$I_L = \frac{V_{in}}{R} - \frac{V_o}{2R} + \frac{V_o}{R} - \frac{V_o}{2R}$$

or

$$I_L = \frac{V_{in}}{R} \qquad (9.31)$$

This equation shows that the load current is controlled by V_{in} and that it is independent of R_L. These results are valid to the extent that the four resistors labeled R are matched, i.e., truly equal in value. For successful operation, the load resistance must obey

$$R_L < \frac{R|V_{max}|}{2V_{in}} \qquad (9.32)$$

Example 9.6

Find the current through each resistor and the voltage at each node of the voltage-controlled current source in Figure 9.13. What is the transconductance of the source?

Figure 9.13
(Example 9.6)

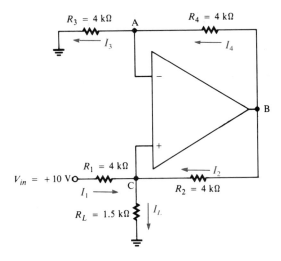

Solution. From equation 9.31, $I_L = V_{in}/R = (10 \text{ V})/(4 \text{ k}\Omega) = 2.5$ mA. Therefore, the voltage at node C (V_L) is $V_C = I_L R_L = (2.5 \text{ mA})(1.5 \text{ k}\Omega) = 3.75$ V. We know that the voltage at node B is twice V_C $(V_o = 2V_L)$: $V_B = 2V_C = 2(3.75 \text{ V}) = 7.5$ V. The voltage at node A is one-half that at node B $(v^- = V_o/2)$: $V_A = (\frac{1}{2})(V_B) = (\frac{1}{2})(7.5 \text{ V}) = 3.75$ V. The currents I_1, I_2, I_3, and I_4 in R_1, R_2, R_3, and R_4 can then be found:

$$I_1 = (V_{in} - V_C)/R_1 = (10 \text{ V} - 3.75 \text{ V})/(4 \times 10^3 \text{ }\Omega) = 1.5625 \text{ mA}$$
$$I_2 = (V_B - V_C)/R_2 = (7.5 \text{ V} - 3.75 \text{ V})/(4 \times 10^3 \text{ }\Omega) = 0.9375 \text{ mA}$$
$$I_3 = V_A/R_3 = 3.75 \text{ V}/(4 \times 10^3 \text{ }\Omega) = 0.9375 \text{ mA}$$
$$I_4 = (V_B - V_A)/R_4 = (7.5 \text{ V} - 3.75 \text{ V})/(4 \times 10^3 \text{ }\Omega) = 0.9375 \text{ mA}$$

The transconductance of the source is $g_m = 1/R = 1/(4 \text{ k}\Omega) = 0.25$ mS.

Current-Controlled Voltage Sources

An ideal current-controlled voltage source has an output voltage that (1) is equal to a constant (k) times the magnitude of an independent current: $v_o = kI_i$, and (2) is independent of the load connected to it. Here, the constant k has the units of ohms. A current-controlled voltage source can be thought of as a *current-to-voltage converter,* since output voltage is proportional to input current. It is useful in applications where current measurements are required, because it is generally more convenient to measure voltages.

Figure 9.14 shows a very simple current-controlled voltage source. Since no current flows into the $-$ input, the controlling current I_{in} is the same as the current in feedback resistor R. Since v^- is virtual ground,

$$V_o = -I_{in}R \qquad (9.33)$$

Once again, the fact that the amplifier has zero output resistance implies that the output voltage will be independent of load. Since output voltage is controlled by input current, R is called the *transresistance* of the source.

Current-Controlled Current Sources

An ideal current-controlled current source is one that supplies a current whose magnitude (1) equals a fixed constant (k) times the value of an independent controlling current: $I_o = kI_i$, and (2) is independent of the load to which the current is supplied. Note that k is dimensionless, since it is the ratio of two currents.

Figure 9.15 shows a current-controlled current source with floating load R_L. Since no current flows into the $-$ input, the current in R_2 must equal I_{in}. Since v^- is at virtual ground, the voltage V_2 is

$$V_2 = -I_{in}R_2 \qquad (9.34)$$

Therefore, the current I_1 in R_1 is

$$I_1 = (0 - V_2)/R_1 = I_{in}R_2/R_1 \qquad (9.35)$$

Writing Kirchhoff's current law at the junction of R_1, R_2, and R_L, we have

$$I_L = I_1 + I_{in} \qquad (9.36)$$

or

$$I_L = \frac{R_2}{R_1} I_{in} + I_{in} = \left(\frac{R_2}{R_1} + 1\right) I_{in} \qquad (9.37)$$

Figure 9.14
A current-controlled voltage source

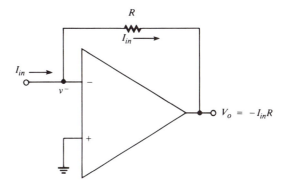

$$V_o = -I_{in}R$$

Figure 9.15
A current-controlled current source with
floating load

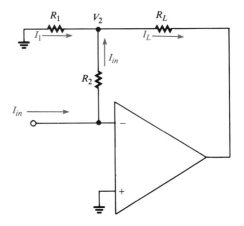

This equation shows that the load current equals the constant $(1 + R_2/R_1)$ times the controlling current and that I_L is independent of R_L. For successful operation, R_L must obey

$$R_L < \left(\frac{|V_{max}|}{I_{in}} - R_2\right)\left(\frac{R_1}{R_1 + R_2}\right) \tag{9.38}$$

Note that the circuit of Figure 9.15 may be regarded as a current *amplifier*, the amplification factor being

$$k = I_L/I_{in} = 1 + R_2/R_1 \tag{9.39}$$

The next example demonstrates the utility of current amplification and illustrates an application where a floating load may be used.

Example 9.7

DESIGN

It is desired to measure a dc current that ranges from 0 to 1 mA using an ammeter whose most sensitive range is 0 to 10 mA. To improve the measurement accuracy, the current to be measured should be amplified by a factor of 10.

1. Design the circuit.
2. Assuming that the meter resistance is 150 Ω and the maximum output voltage of the amplifier is 15 V, verify that the circuit will perform properly.

Solution

1. Figure 9.16 shows the required circuit. I_X is the current to be measured, and the ammeter serves as the load through which the amplified current flows. From equation 9.39, the current amplification is $I_L/I_X = 1 + R_2/R_1 = 10$. Letting $R_1 = 1$ kΩ, we find $R_2 = (10 - 1)1$ kΩ = 9 kΩ.
2. Inequality 9.38 must be satisfied for the smallest possible value of the right-hand side, which occurs when $I_{in} = 1$ mA:

$$R_L < \left[\frac{15 \text{ V}}{1 \text{ mA}} - (9 \text{ k}\Omega)\right]\left[\frac{1 \text{ k}\Omega}{(1 \text{ k}\Omega) + (9 \text{ k}\Omega)}\right] = 600 \ \Omega$$

Since the meter resistance is 150 Ω, the circuit operates satisfactorily.

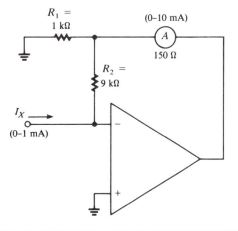

Figure 9.16
(Example 9.7) The current-controlled current source acts as a current amplifier, so a 0–1-mA current can be measured by a 0–10-mA ammeter.

9.3   INTEGRATION

Electronic Integration

An *electronic integrator* is a device that produces an output waveform whose value at any instant of time equals the total *area under the input* waveform up to that point in time. (For those familiar with mathematical integration, the process produces the time-varying function $\int_0^t v_{in}(t)\,dt$.) To illustrate this concept, suppose the input to an electronic integrator is the dc level E volts, which is first connected to the integrator at an instant of time we will call $t = 0$. Refer to Figure 9.17. The plot of the dc "waveform" versus time is simply a horizontal line at level E volts, since the dc voltage is constant. The more time that we allow to pass, the greater the area that accumulates under the dc waveform. At any time-point t, the total area under the input waveform between time 0 and time t is (height) × (width) = Et volts, as illustrated in the figure. For example, if $E = 5$ V dc, then the output will be 5 V at $t = 1$ s, 10 V at $t = 2$ s, 15 V at $t = 3$ s, and so forth. We see that the output is the *ramp* voltage $v_o(t) = Et$.

When the input to a practical integrator is a dc level, the output will rise linearly with time, as shown in Figure 9.17, and will eventually reach the maximum possible output voltage of the amplifier. Of course, the integration process ceases at that time. If the input voltage goes negative for a certain interval of time,

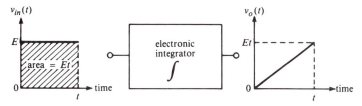

Figure 9.17
The output of the integrator at t seconds is the area, Et, under the input waveform.

the total area during that interval is *negative* and subtracts from whatever positive area had previously accumulated, thus reducing the output voltage. Therefore, the input must periodically go positive and negative to prevent the output of an integrator from reaching its positive or negative limit. We will explore this process in greater detail in a later discussion on waveshaping.

Figure 9.18 shows how an electronic integrator is constructed using an operational amplifier. Note that the component in the feedback path is capacitor C, and that the amplifier is operated in an inverting configuration. Besides the usual ideal-amplifier assumptions, we are assuming *zero input offset,* since any input dc level would be integrated as shown in Figure 9.17 and would eventually cause the amplifier to saturate. Thus, we show an *ideal-integrator* circuit. Using the standard symbol $\int_0^t v \, dt$ to represent integration of the voltage v between time 0 and time t, we can show that the output of this circuit is

$$v_o(t) = \frac{-1}{R_1 C} \int_0^t v_{in} \, dt \tag{9.40}$$

This equation shows that the output is the (inverted) integral of the input, multiplied by the constant $1/R_1 C$. If this circuit were used to integrate the dc waveform shown in Figure 9.17, the output would be a negative-going ramp ($v_o = -Et/R_1 C$).

Readers unfamiliar with calculus can skip the present paragraph, in which we demonstrate why the circuit of Figure 9.18 performs integration. Since the current into the $-$ input is 0, we have, from Kirchhoff's current law,

$$i_1 + i_C = 0 \tag{9.41}$$

where i_1 is the input current through R_1 and i_C is the feedback current through the capacitor. Since $v^- = 0$, the current in the capacitor is

$$i_C = C \frac{dv_o}{dt} \tag{9.42}$$

Thus

$$\frac{v_{in}}{R_1} + C \frac{dv_o}{dt} = 0 \tag{9.43}$$

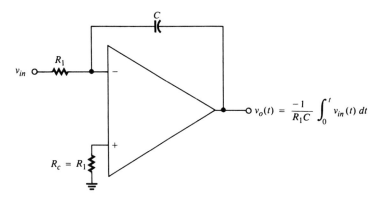

Figure 9.18
An ideal electronic integrator

or

$$\frac{dv_o}{dt} = \frac{-1}{R_1 C} v_{in} \qquad (9.44)$$

Integrating both sides with respect to t, we obtain

$$v_o = \frac{-1}{R_1 C} \int_0^t v_{in} \, dt \qquad (9.45)$$

Hereafter, we will use the abbreviated symbol $\int$ to represent integration. It can be shown, using calculus, that the mathematical integral of the sine wave $A \sin \omega t$ is

$$\int (A \sin \omega t) \, dt = \frac{-A}{\omega} \sin(\omega t + 90°) = \frac{-A}{\omega} \cos(\omega t)$$

Therefore, when the input to the inverting integrator in Figure 9.18 is $v_{in} = A \sin \omega t$, the output is

$$\begin{aligned} v_o &= \frac{-1}{R_1 C} \int (A \sin \omega t) \, dt = \frac{-A}{\omega R_1 C} (-\cos \omega t) \\ &= \frac{A}{\omega R_1 C} \cos \omega t \end{aligned} \qquad (9.46)$$

The most important fact revealed by equation 9.46 is that the output of an integrator with sinusoidal input is a sinusoidal waveform whose *amplitude is inversely proportional to its frequency*. This observation follows from the presence of ω ($= 2\pi f$) in the denominator of (9.46). For example, if a 100-Hz input sine wave produces an output with peak value 10 V, then, all else being equal, a 200-Hz sine wave will produce an output with peak value 5 V. Note also that the output *leads* the input by 90°, regardless of frequency, since $\cos \omega t = \sin(\omega t + 90°)$.

Example 9.8

1. Find the peak value of the output of the ideal integrator shown in Figure 9.19. The input is $v_{in} = 0.5 \sin (100t)$ V.

Figure 9.19
(Example 9.8)

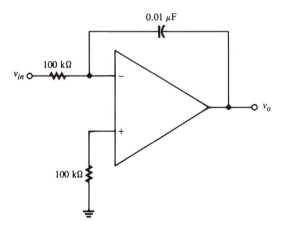

2. Repeat, when $v_{in} = 0.5 \sin (10^3 t)$ V.

Solution

1. From equation 9.46,

$$v_o = \frac{A}{\omega R_1 C} \cos (\omega t) = \frac{0.5}{100(10^5)(10^{-8})} \cos (100t)$$

$$= 5 \cos (100t) \text{ V} \qquad \text{peak value} = 5 \text{ V}$$

2. From equation 9.46,

$$v_o = \frac{0.5}{1000(10^5)(10^{-8})} \cos (1000t)$$

$$= 0.5 \cos (1000t) \text{ V} \qquad \text{peak value} = 0.5 \text{ V}$$

Example 9.8 shows that increasing the frequency by a factor of 10 causes a decrease in output amplitude by a factor of 10. This familiar relationship implies that a Bode plot for the gain of an ideal integrator will have slope -20 dB/decade, or -6 dB/octave. Gain magnitude is the ratio of the peak value of the output to the peak value of the input:

$$\left| \frac{v_o}{v_{in}} \right| = \frac{\left(\dfrac{A}{\omega R_1 C} \right)}{A} = \frac{1}{\omega R_1 C} \qquad (9.47)$$

This equation clearly shows that gain is inversely proportional to frequency. A Bode plot for the case $R_1 C = 0.001$ is shown in Figure 9.20.

Because the integrator's output amplitude decreases with frequency, it is a kind of low-pass filter. It is sometimes called a *smoothing* circuit, because the amplitudes of high-frequency components in a complex waveform are reduced, thus smoothing the jagged appearance of the waveform. This feature is useful for reducing high-frequency noise in a signal. Integrators are also used in *analog computers* to obtain real-time solutions to differential (calculus) equations.

Figure 9.20
Bode plot of the gain of an ideal integrator for the case $R_1C = 0.001$

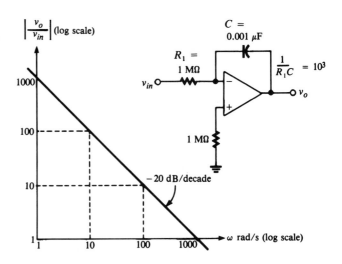

Practical Integrators

Although high-quality, precision integrators are constructed as shown in Figure 9.18 for use in low-frequency applications such as analog computers, these applications require high-quality amplifiers with extremely small offset voltages or chopper stabilization. As mentioned earlier, any input offset is integrated as if it were a dc signal input and will eventually cause the amplifier to saturate. To eliminate this problem in practical integrators using general-purpose amplifiers, a resistor is connected in parallel with the feedback capacitor, as shown in Figure 9.21. Since the capacitor is an open circuit as far as dc is concerned, the integrator responds to dc inputs just as if it were an inverting amplifier. In other words, the *dc closed-loop gain* of the integrator is $-R_f/R_1$. At high frequencies, the impedance of the capacitor is much smaller than R_f, so the parallel combination of C and R_f is essentially the same as C alone, and signals are integrated as usual.

While the feedback resistor in Figure 9.21 prevents integration of dc inputs, it also degrades the integration of low-frequency signals. At frequencies where the capacitive reactance of C is comparable in value to R_f, the net feedback impedance is not predominantly capacitive and true integration does not occur. As a rule of thumb, we can say that satisfactory integration will occur at frequencies much greater than the frequency at which $X_C = R_f$. That is, for integrator action we want

$$|X_C| \ll R_f$$

$$\frac{1}{2\pi f C} \ll R_f$$

or

$$f \gg \frac{1}{2\pi R_f C} \tag{9.48}$$

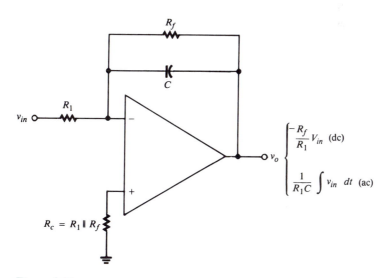

Figure 9.21
A resistor R_f connected in parallel with C causes the practical integrator to behave like an inverting amplifier to dc inputs and like an integrator to high-frequency ac inputs.

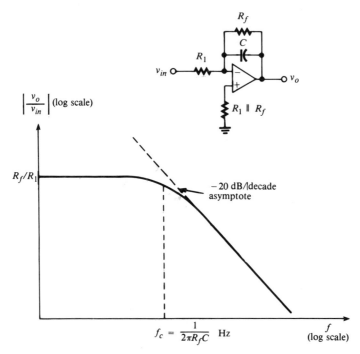

Figure 9.22
Bode plot for a practical integrator, showing that integration occurs at frequencies above $1/(2\pi R_f C)$ hertz

The frequency f_c where $|X_c|$ equals R_f,

$$f_c = \frac{1}{2\pi R_f C} \tag{9.49}$$

defines a *break frequency* in the Bode plot of the practical integrator. As shown in Figure 9.22, at frequencies well above f_c, the gain falls off at the rate of -20 dB/ decade, like that of an ideal integrator, and at frequencies below f_c, the gain approaches its dc value of R_f/R_1.

Example 9.9

DESIGN

Design a practical integrator that

1. integrates signals with frequencies down to 100 Hz; and
2. produces a peak output of 0.1 V when the input is a 10-V-peak sine wave having frequency 10 kHz.

Find the dc component in the output when there is a +50-mV dc input.

Solution.   In order to integrate frequencies down to 100 Hz, we require $f_c \ll 100$ Hz. Let us choose f_c one decade below 100 Hz: $f_c = 10$ Hz. Then, from equation 9.49,

$$f_c = 10 = \frac{1}{2\pi R_f C}$$

Choose $C = 0.01\ \mu$F. Then

$$10 = \frac{1}{2\pi R_f(10^{-8})}$$

or $\qquad$ $R_f = \dfrac{1}{2\pi(10)(10^{-8})} = 1.59\ M\Omega$

To satisfy requirement (2), we must choose R_1 so that the gain at 10 kHz is

$$\left|\frac{v_o}{v_{in}}\right| = \frac{0.1\ V}{10\ V} = 0.01$$

Assuming that we can neglect R_f at this frequency (3 decades above f_c), the gain is the same as that for an ideal integrator, given by equation 9.47:

$$\left|\frac{v_o}{v_{in}}\right| = \frac{1}{\omega R_1 C} = 0.01$$

Thus $\qquad$ $\dfrac{1}{2\pi \times 10^4 R_1(10^{-8})} = 0.01$

or $\qquad$ $R_1 = \dfrac{1}{2\pi \times 10^4 \times (0.01 \times 10^{-8})} = 159\ k\Omega$

The required circuit is shown in Figure 9.23. Note that $R_c = (1.59\ M\Omega)\ \|$ $(159\ k\Omega) = 145\ k\Omega$.

Figure 9.23
(Example 9.9)

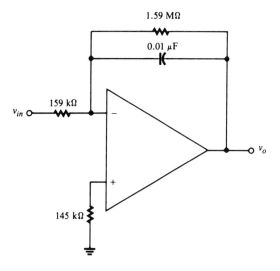

When the input is 50 mV dc, the output is 50 mV times the dc closed-loop gain:

$$v_o = \frac{-R_f}{R_1}(50\ mV) = \frac{-1.59\ M\Omega}{159\ k\Omega}(50\ mV) = -0.5\ V$$

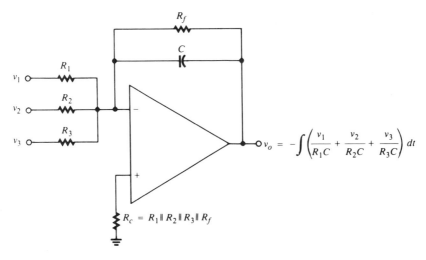

Figure 9.24
A three-input integrator

In closing our discussion of integrators, we should note that it is possible to scale and integrate several input signals simultaneously, using an arrangement similar to the linear combination circuit studied earlier. Figure 9.24 shows a practical, three-input integrator that performs the following operation at frequencies above f_c:

$$v_o = -\int \left(\frac{1}{R_1 C} v_1 + \frac{1}{R_2 C} v_2 + \frac{1}{R_3 C} v_3 \right) dt \qquad (9.50)$$

Equation 9.50 is equivalent to

$$v_o = \frac{-1}{R_1 C} \int v_1 \, dt - \frac{1}{R_2 C} \int v_2 \, dt - \frac{1}{R_3 C} \int v_3 \, dt \qquad (9.51)$$

If $R_1 = R_2 = R_3 = R$, then

$$v_o = \frac{-1}{RC} \int (v_1 + v_2 + v_3) \, dt \qquad (9.52)$$

9.4   DIFFERENTIATION

Electronic Differentiation

An electronic differentiator produces an output waveform whose value at any instant of time is equal to the *rate of change* of the input at that point in time. In many respects, differentiation is just the opposite, or inverse, operation of integration. In fact, integration is also called "antidifferentiation." If we were to pass a signal through an ideal integrator in cascade with an ideal differentiator, the final output would be exactly the same as the original input signal.

Figure 9.25 demonstrates the operation of an ideal electronic differentiator. In this example, the input is the ramp voltage $v_{in} = Et$. The rate of change, or slope, of this ramp is a constant E volts/second. (For every second that passes, the signal

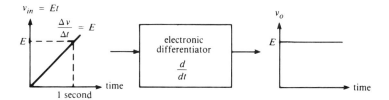

Figure 9.25
The ideal electronic differentiator produces an output equal to the rate of change of the input. Since the rate of change of a ramp is constant, the output in this example is a dc level.

increases by an additional E volts.) Since the rate of change of the input is constant, we see that the output of the differentiator is the constant dc level E volts.

The standard symbol used to represent differentiation of a voltage v is dv/dt. (This should not be interpreted as a fraction. The dt in the denominator simply means that we are finding the rate of change of v with respect to *time, t.*) In the example shown in Figure 9.25, we would write

$$\frac{dv_{in}}{dt} = \frac{d(Et)}{dt} = E$$

Note that the derivative of a constant (dc level) is *zero,* since a constant does not change with time and therefore has zero rate of change.

Figure 9.26 shows how an ideal differentiator is constructed using an operational amplifier. Note that we now have a capacitive input and a resistive feedback—again, just the opposite of an integrator. It can be shown that the output of this differentiator is

$$v_o = -R_f C \frac{dv_{in}}{dt} \qquad (9.53)$$

Thus, the output voltage is the (inverted) derivative of the input, multiplied by the constant $R_f C$. If the ramp voltage in Figure 9.25 were applied to the input of this differentiator, the output would be a negative dc level.

Figure 9.26
An ideal electronic differentiator

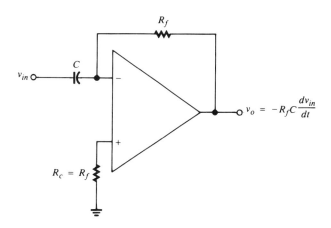

Readers not familiar with calculus can skip the present paragraph, in which we show how the circuit of Figure 9.26 performs differentiation. Since the current into the $-$ terminal is 0, we have, from Kirchhoff's current law,

$$i_C + i_f = 0 \qquad (9.54)$$

Since $v^- = 0$, $v_C = v_{in}$ and

$$i_C = C \frac{dv_{in}}{dt} \qquad (9.55)$$

Also, $i_f = v_o/R_f$, so

$$C \frac{dv_{in}}{dt} + \frac{v_o}{R_f} = 0$$

or

$$v_o = -R_f C \frac{dv_{in}}{dt} \qquad (9.56)$$

It can be shown, using calculus, that

$$\frac{d(A \sin \omega t)}{dt} = A\omega \cos \omega t \qquad (9.57)$$

Therefore, when the input to the differentiator in Figure 9.26 is $v_{in} = A \sin \omega t$, the output is

$$v_o = -R_f C \frac{d(A \sin \omega t)}{dt} = -A\omega R_f C \cos (\omega t) = A\omega R_f C \sin (\omega t - 90°) \quad (9.58)$$

Equation 9.58 shows that when the input is sinusoidal, the *amplitude of the output of a differentiator is directly proportional to frequency* (once again, just the opposite of an integrator). Note also that the output lags the input by 90°, regardless of frequency. The gain of the differentiator is

$$\left| \frac{v_o}{v_{in}} \right| = \frac{A\omega R_f C}{A} = \omega R_f C \qquad (9.59)$$

Practical Differentiators

From a practical standpoint, the principal difficulty with the differentiator is that it effectively amplifies an input in direct proportion to its frequency and therefore increases the level of high-frequency noise in the output. Unlike the integrator, which "smooths" a signal by reducing the amplitude of high-frequency components, the differentiator intensifies the contamination of a signal by high-frequency noise. For this reason, differentiators are rarely used in applications requiring high precision, such as analog computers.

In a practical differentiator, the amplification of signals in direct proportion to their frequencies cannot continue indefinitely as frequency increases, because the amplifier has a finite bandwidth. As we have already discussed in Chapter 8, there is some frequency at which the output amplitude must begin to fall off. Nevertheless, it is often desirable to design a practical differentiator so that it will have a break frequency even lower than that determined by the upper cutoff frequency of the amplifier, that is, to roll off its gain characteristic at some relatively low

Figure 9.27

A practical differentiator. Differentiation occurs at low frequencies, but resistor R_1 prevents high-frequency differentiation.

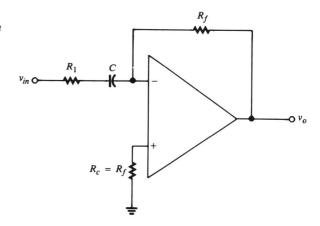

frequency. This action is accomplished in a practical differentiator by connecting a resistor in series with the input capacitor, as shown in Figure 9.27. We can understand how this modification achieves the stated goal by considering the net impedance of the R_1–C combination at low and high frequencies:

$$Z_{in} = R_1 - j/\omega C \qquad (9.60)$$
$$|Z_{in}| = \sqrt{R_1^2 + (1/\omega C)^2} \qquad (9.61)$$

At very small values of ω, Z_{in} is dominated by the capacitive reactance component, so the combination is essentially the same as C alone, and differentiator action occurs. At very high values of ω, $1/\omega C$ is negligible, so Z_{in} is essentially the resistance R_1, and the circuit behaves like an ordinary inverting amplifier (with gain R_f/R_1).

The break frequency f_b beyond which differentiation no longer occurs in Figure 9.27 is the frequency at which the capacitive reactance of C equals the resistance of R_1:

$$\frac{1}{2\pi f_b C} = R_1$$
$$f_b = \frac{1}{2\pi R_1 C} \text{ hertz} \qquad (9.62)$$

In designing a practical differentiator, the break frequency should be set well above the highest frequency at which accurate differentiation is desired:

$$f_b \gg f_h \qquad (9.63)$$

where f_h is the highest differentiation frequency. Figure 9.28 shows Bode plots for the gain of the ideal and practical differentiators. In the low-frequency region where differentiation occurs, note that the gain rises with frequency at the rate of 20 dB/decade. The plot shows that the gain levels off beyond the break frequency f_b and then falls off at -20 dB/decade beyond the amplifier's upper cutoff frequency. Recall that the closed-loop bandwidth, or upper cutoff frequency of the amplifier, is given by

$$f_2 = \beta f_t \qquad (9.64)$$

where β in this case is $R_1/(R_1 + R_f)$.

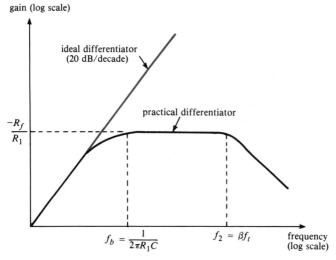

Figure 9.28
Bode plots for the ideal and practical differentiators. f_b is the break frequency due to the input R_1–C combination and f_2 is the upper cutoff frequency of the (closed-loop) amplifier.

In some applications, where very wide bandwidth operational amplifiers are used, it may be necessary or desirable to roll the frequency response off even faster than that shown for the practical differentiator in Figure 9.28. This can be accomplished by connecting a capacitor C_f in parallel with the feedback resistor R_f. This modification will cause the response to roll off at −20 dB/decade beginning at the break frequency

$$f_{b2} = \frac{1}{2\pi R_f C_f} \text{ hertz} \tag{9.65}$$

Obviously, f_{b2} should be set higher than f_b.

Example 9.10

1. Design a practical differentiator that will differentiate signals with frequencies up to 200 Hz. The gain at 10 Hz should be 0.1.
2. If the operational amplifier used in the design has a unity-gain frequency of 1 MHz, what is the upper cutoff frequency of the differentiator?

Solution

1. We must select R_1 and C to produce a break frequency f_b that is well above $f_h = 200$ Hz. Let us choose $f_b = 10\,f_h = 2$ kHz. Letting $C = 0.1\ \mu\text{F}$, we have, from equation 9.62,

$$f_b = 2 \times 10^3 = \frac{1}{2\pi R_1 C}$$

$$R_1 = \frac{1}{2\pi(2 \times 10^3)(10^{-7})} = 796\ \Omega$$

Figure 9.29
(Example 9.10)

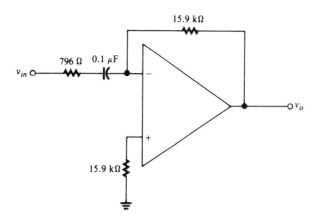

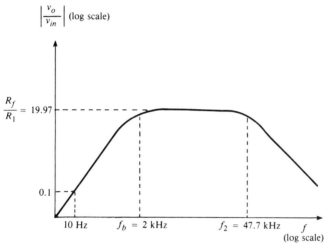

Figure 9.30
(Example 9.10)

In order to achieve a gain of 0.1 at 10 Hz, we have, from equation 9.59,

$$\left| \frac{v_o}{v_{in}} \right| = 0.1 = \omega R_f C = (2\pi \times 10)R_f(10^{-7})$$

$$R_f = \frac{0.1}{2\pi \times 10 \times 10^{-7}} = 15.9 \text{ k}\Omega$$

The completed design is shown in Figure 9.29.

2.

$$\beta = \frac{R_1}{R_1 + R_f} = \frac{796}{796 + 15.9 \times 10^3} = 0.0477$$

From equation 9.64, $f_2 = \beta f_t = (0.0477)(1 \text{ MHz}) = 47.7$ kHz. The Bode plot is sketched in Figure 9.30.

Example 9.11

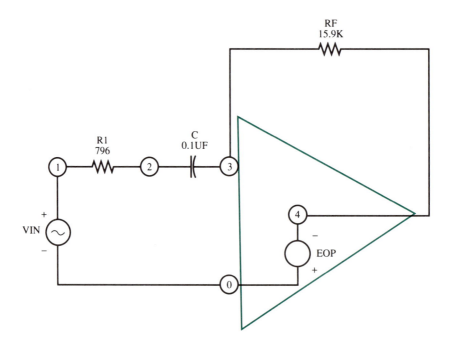

SPICE

Use SPICE to verify the design of the differentiator in Example 9.10. Assume the amplifier is ideal.

Solution.   Figure 9.31(a) shows the SPICE circuit and the input data file. The inverting operational amplifier is modeled like the one in Example 8.17, except that we assume zero output resistance and a (nearly) ideal voltage gain having the very large value 1×10^9.

Figure 9.31(b) shows the frequency response computed by SPICE over the range from 1 Hz to 100 kHz. Since the input signal has default amplitude 1 V, the plot represents values of voltage gain. Notice that the gain at 10 Hz is 0.0999 ≈

```
EXAMPLE 9.11
VIN 1 0 AC
R1 1 2 796
C  2 3 0.1UF
RF 3 4 15.9K
EOP0 4 3 0 1E9
.AC DEC 10 1 100KHZ
.PLOT AC V(4)
.END
```

(a)

Figure 9.31
(Example 9.11)

```
       FREQ        V(4)
                    1.000D-04      1.000D-02      1.000D+00      1.000D+02  1.000D+04
                    - - - - - - - - - - - - - - - - - - - - - - - - - -
  -1.000D+00    9.990D-03 .              *              .              .              .
   1.259D+00    1.258D-02 .             .*              .              .              .
   1.585D+00    1.583D-02 .             .*              .              .              .
   1.995D+00    1.993D-02 .             . *             .              .              .
   2.512D+00    2.509D-02 .             .  *            .              .              .
   3.162D+00    3.159D-02 .             .  *            .              .              .
   3.981D+00    3.977D-02 .             .   *           .              .              .
   5.012D+00    5.007D-02 .             .     *         i              .              .
   6.310D+00    6.303D-02 .             .       *       .              .              .
   7.943D+00    7.935D-02 .             .       *       .              .              .
   1.000D+01    9.990D-02 .             .         *     .              .              .
   1.259D+01    1.258D-01 .             .          *    .              .              .
   1.585D+01    1.583D-01 .             .          *    .              .              .
   1.995D+01    1.993D-01 .             .           *   .              .              .
   2.512D+01    2.509D-01 .             .            *  .              .              .
   3.162D+01    3.159D-01 .             .            *  .              .              .
   3.981D+01    3.976D-01 .             .             * .              .              .
   5.012D+01    5.005D-01 .             .              *.              .              .
   6.310D+01    6.300D-01 .             .              *.              .              .
   7.943D+01    7.929D-01 .             .              *.              .              .
   1.000D+02    9.978D-01 .             .              *              .              .
   1.259D+02    1.255D+00 .             .              .*             .              .
   1.585D+02    1.578D+00 .             .              .*             .              .
   1.995D+02    1.983D+00 .             .              . *            .              .
   2.512D+02    2.490D+00 .             .              .  *           .              .
   3.162D+02    3.120D+00 .             .              .  *           .              .
   3.981D+02    3.901D+00 .             .              .    *         .              .
   5.012D+02    4.857D+00 .             .              .     *        .              .
   6.310D+02    6.011D+00 .             .              .      *       .              .
   7.943D+02    7.375D+00 .             .              .       *      .              .
   1.000D+03    8.935D+00 .             .              .         *    .              .
   1.259D+03    1.064D+01 .             .              .          *   .              .
   1.585D+03    1.241D+01 .             .              .           *  .              .
   1.995D+03    1.411D+01 .             .              .           *  .              .
   2.512D+03    1.563D+01 .             .              .           *  .              .
   3.162D+03    1.688D+01 .             .              .            * .              .
   3.981D+03    1.785D+01 .             .              .            * .              .
   5.012D+03    1.855D+01 .             .              .            * .              .
   6.310D+03    1.904D+01 .             .              .            * .              .
   7.943D+03    1.937D+01 .             .              .            * .              .
   1.000D+04    1.959D+01 .             .              .            * .              .
   1.259D+04    1.973D+01 .             .              .            * .              .
   1.585D+04    1.982D+01 .             .              .            * .              .
   1.995D+04    1.988D+01 .             .              .            * .              .
   2.512D+04    1.991D+01 .             .              .            * .              .
   3.162D+04    1.994D+01 .             .              .            * .              .
   3.981D+04    1.995D+01 .             .              .            * .              .
   5.012D+04    1.996D+01 .             .              .            * .              .
   6.310D+04    1.996D+01 .             .              .            * .              .
   7.943D+04    1.997D+01 .             .              .            * .              .
   1.000D+05    1.997D+01 .             .              .            * .              .
```

(b)

Figure 9.31
(Continued)

0.1, as required. The theoretical gain of the ideal differentiator at 200 Hz is

$$\omega R_f C = (2\pi \times 200)(15.9 \text{ k}\Omega)(0.1 \text{ }\mu\text{F}) = 1.998$$

The plot shows that the gain at a frequency close to 200 Hz (199.5 Hz) is 1.983, so we conclude that satisfactory differentiation occurs up to 200 Hz.

The plot shows that the high-frequency gain (at 100 kHz) is 19.97. At very high frequencies, the reactance of the 0.1-μF capacitor is negligible, so the gain, for all practical purposes, equals the ratio of R_f to R_1: 15.9 kΩ/796 Ω = 19.97. Since the break frequency f_b in the design was selected to be 2 kHz, the gain at 2 kHz should be approximately (0.707)(19.97) = 14.11. The plot shows that the gain at a frequency near 2 kHz (1.995 kHz) is 14.11, and we conclude that the design is valid. Since we assumed an ideal operational amplifier (with infinite bandwidth), the cutoff frequency f_2 does not affect the computed response. (Its presence can be simulated by connecting a 210-pF capacitor between nodes 3 and 4.)

9.5 WAVESHAPING

Waveshaping is the process of altering the shape of a waveform in some pre-scribed manner to produce a new waveform having a desired shape. Examples include altering a triangular wave to produce a square wave, and vice versa, altering a square wave to produce a series of narrow pulses, altering a square or triangular wave to produce a sine wave, and altering a sine wave to produce a square wave. Waveshaping techniques are widely used in function generators, frequency synthesizers, and synchronization circuits that require different wave-forms having precisely the same frequency. In this section we will discuss just two waveshaping techniques, one employing an integrator and one employing a differ-entiator.

Recall that the output of an ideal integrator is a ramp voltage when its input is a positive dc level. Positive area accumulates under the input waveform as time passes, and the output rises, as shown in Figure 9.32. If the dc level is suddenly made negative, and remains negative for an interval of time, then negative area accumulates, so the *net* area decreases and the output voltage begins to fall. At the point in time where the total accumulated negative area equals the previously accumulated positive area, the output reaches 0, as shown in Figure 9.32.

Figure 9.32 shows the basis for generating a triangular waveform using an integrator. When the input is a square wave that continually alternates between positive and negative levels, the output rises and falls in synchronism. Note that the *slope* of the output alternates between $+E$ and $-E$ volts/second. Figure 9.33(a) shows how a triangular wave is generated using a practical integrator. Figure 9.33(b) shows the actual waveform generated. Notice that the figure re-flects the *phase inversion* caused by the amplifier in that the output decreases when the input is positive and increases when the input is negative. The *average*

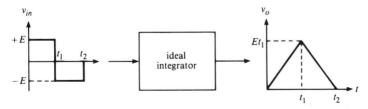

Figure 9.32
The integrator output rises to a maximum of Et_1 while the input is positive and then falls to 0 at time t_2 when the net area under the input is 0.

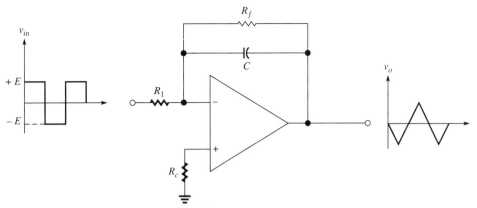

(a) Integrator used to generate a triangular waveform

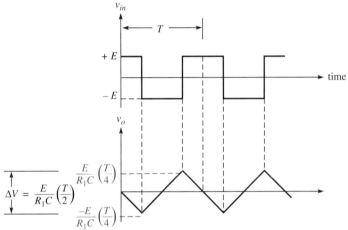

(b) The triangular wave generated by the circuit in (a), showing voltage and time relations. T is the period of the square wave.

Figure 9.33
The practical triangular-waveform generator

level (dc component) of the output is 0, assuming no input offset. Note also that the slopes of the triangular wave are $\pm E/R_1C$ volts/second, since the integrator gain is $1/R_1C$. It is important to be able to predict these slopes to ensure that they will be within the specified slew rate of the amplifier. In the figure, T represents the period of the square wave, and it can be seen that the peak value of the triangular wave is

$$|V_{peak}| = \frac{ET}{4R_1C} \text{ volts} \tag{9.66}$$

Equation 9.66 shows that the amplitude of the triangular wave decreases with increasing frequency, since the period T decreases with frequency. On the other hand, the slopes $\Delta V/\Delta t$ are (contrary to expectation) independent of frequency. This observation is explained by the fact that while an increase in frequency

Figure 9.34
When the frequency of the triangular wave produced by an integrator is doubled, the amplitude is halved, so the slopes remain the same.

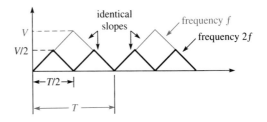

reduces the time Δt over which the voltage changes, it also reduces the total change in voltage, ΔV. The phenomenon is illustrated in Figure 9.34.

Equation 9.66 is a special case of a general relationship that can be used to find the integrator's peak output when the input is any periodic waveform symmetric about the horizontal axis:

$$V_{peak} = \frac{\text{positive area in } \frac{1}{2} \text{ cycle of input}}{2R_1C} \qquad (9.67)$$

For example, note that the positive area in one-half cycle of the square wave in Figure 9.33 is $E(T/2)$, so (9.67) reduces to (9.66) in that case.

Equations 9.66 and 9.67 are based on the assumption that there is *no input offset level or dc level in the input waveform*. When there is an input dc component, there will be an output dc component given by

$$V_o(\text{dc}) = \frac{-R_f}{R_1} V_{in}(\text{dc}) \qquad (9.68)$$

Thus, a triangular output will be shifted up or down by an amount equal to $V_o(\text{dc})$.

Example 9.12

The integrator in Figure 9.35 is to be used to generate a triangular waveform from a 500-Hz square wave connected to its input. Suppose the square wave alternates between ± 12 V.

1. What minimum slew rate should the amplifier have?
2. What maximum output voltage should the amplifier be capable of developing?
3. Repeat (2) if the dc component in the input is -0.2 V. (The square wave alternates between $+11.8$ V and -12.2 V.)

Solution

1. The magnitude of the slope of the triangular waveform is

$$\frac{\Delta V}{\Delta t} = \frac{E}{R_1C} = \frac{12}{400(10^{-6})} = 3 \times 10^4 \text{ V/s}$$

Thus, we must have slew rate $S \geq 3 \times 10^4$ V/s.

2. The period T of the 500-Hz square wave is $T = 1/500 = 2 \times 10^{-3}$ s. By equation 9.66,

$$|V_{peak}| = \frac{ET}{4R_1C} = \frac{12(2 \times 10^{-3})}{4(400)(10^{-6})} = 15 \text{ V}$$

The triangular wave alternates between peak values of $+15$ V and -15 V.

Figure 9.35
(Example 9.12)

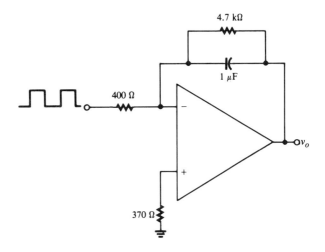

3. By equation 9.68, the dc component in the output is

$$v_o(\text{dc}) = \frac{-R_f}{R_1} V_{in}(\text{dc}) = \frac{-4700}{400}(-0.2\ \text{V}) = 2.35\ \text{V}$$

Therefore, the triangular output is shifted up by 2.35 V and alternates between peak values of 15 + 2.35 = 17.35 V and −15 + 2.35 = −12.65 V. The amplifier must be capable of producing a 17.35-V output to avoid distortion.

If it were possible to generate an *ideal* square wave, it would have zero rise and fall times and would therefore change at an infinite rate every time it switched from one level to the other. If this square wave were the input to an ideal differentiator, the output would be a series of extremely narrow (actually, zero-width) pulses with infinite heights, alternately positive and negative, as illustrated in Figure 9.36. These fictional, zero-width, infinite-height ''spikes'' are called *impulses*. They have zero width because each change in the square wave occurs in zero time, and they have infinite height because the rate of change of the square

Figure 9.36
The output of an ideal differentiator driven by an ideal square wave is a series of infinite-height, zero-width impulses.

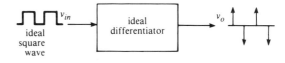

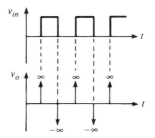

wave is infinite at the points where changes occur. The negative impulses corre-spond to the negative rates of change when the square wave goes from a positive value to a negative value. Between the time points where the square wave changes, the rate of change is 0 and so is the output of the differentiator.

Of course, neither ideal square waves nor ideal differentiators exist. How-ever, many real square waves have small rise and fall times and therefore have large rates of change at points where transitions in level occur. The output of a differentiator driven by such a square wave is therefore a series of narrow, high-amplitude pulses. These are used in *timing*-circuit applications, where it is neces-sary to trigger some circuit action in synchronization with another waveform. Figure 9.37 shows a somewhat idealized square wave in which the nonzero rise and fall times are represented by ramps. The output that would result if this square wave were to drive an ideal differentiator is also shown. Finally, the typical output of a practical differentiator is shown. Here, the output pulse is distorted (although still usable in timing circuits), because of the restricted bandwidth of a practical differentiator (see Figure 9.28) and because of slew-rate limitations. Notice that dc level in the input to a differentiator has no effect on the output, because the rate of change of a dc component is 0. From a circuit viewpoint, the input capacitor blocks dc levels. (The figure does not show the phase inversion caused by the inverting amplifier.)

From studying the ideal output in Figure 9.37, the reader should be able to deduce that an ideal differentiator driven by a triangular wave will produce a square wave. Since the output equals the rate of change of the input, the square wave will alternate between $\pm(R_f C)|\Delta V_o/\Delta t|$ volts, where $\Delta V_o/\Delta t$ is the rate of change of the triangular input. Amplifier slew rate is a serious limitation in these waveshaping applications, because the differentiator output should theoretically have an infinite rate of change whenever the slope of the triangular input changes from positive to negative.

9.6   INSTRUMENTATION AMPLIFIERS

In Section 9.1 we saw how an amplifier can be operated in a differential mode to produce an output voltage proportional to the difference between two input sig-nals (see Figure 9.5). Differential operation is a common requirement in instru-mentation systems and other signal-processing applications where high accuracy is important. The circuit of Figure 9.5 has certain limitations in these applications, including the fact that the signal sources see different input impedances. Also, the circuit does not generally have good common-mode rejection, an important con-

Figure 9.37
Typical outputs from ideal and practical differentiators driven by an imperfect square wave. Phase inversion is not shown.

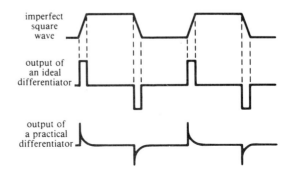

sideration in instrumentation systems, where long signal lines and high electrical noise environments are common. Figure 9.38 shows an improved configuration for producing an output proportional to the difference between two inputs. Notice that the circuit is basically the difference amplifier discussed earlier, with the addition of two input stages. Each input signal is connected directly to the noninverting terminal of an operational amplifier, so each signal source sees a very large input resistance. This circuit arrangement is so commonly used that it is called an *instrumentation amplifier* and is commercially available by that name in single-package units. These devices use closely matched, high-quality amplifiers and have very large common-mode rejection ratios.

In our analysis of the instrumentation amplifier, we will refer to Figure 9.39, which shows current and voltage relations in the circuit. We begin by noting that the usual assumption of ideal amplifiers allows us to equate v_i^+ and v_i^- at each input amplifier ($v_i^+ - v_i^- \approx 0$), with the result that input voltages v_1 and v_2 appear across adjustable resistor R_A in Figure 9.39. For analysis purposes, let us assume that $v_1 > v_2$. Then the current i through R_A is

$$i = \frac{v_1 - v_2}{R_A} \tag{9.69}$$

Since no current flows into either amplifier input terminal, the current i must also flow in each resistor R connected on opposite sides of R_A. Therefore, the voltage

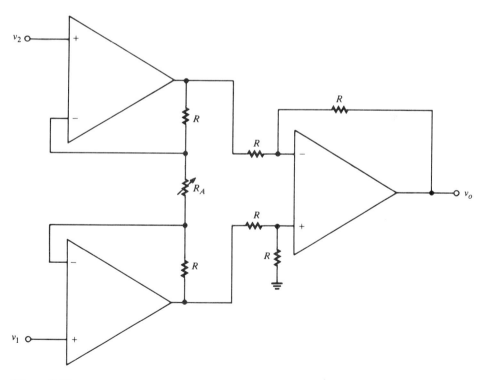

Figure 9.38
An instrumentation amplifier that produces an output proportional to $v_1 - v_2$. Adjustable resistor R_A is used to set the gain.

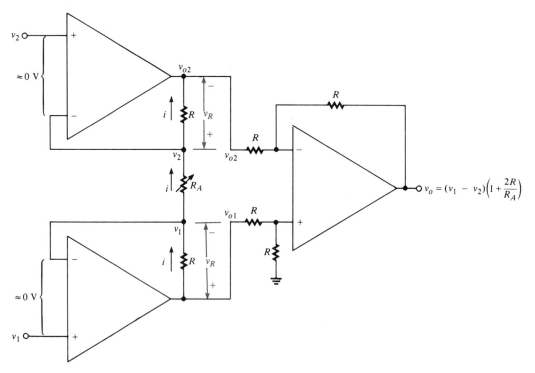

Figure 9.39
Voltage and current relations in the instrumentation amplifier. Note that the overall gain is inversely proportional to the value of adjustable resistor R_A.

drop across each of those resistors is

$$v_R = iR = \frac{(v_1 - v_2)R}{R_A} \tag{9.70}$$

The output voltages v_{o1} and v_{o2} are given by

$$v_{o1} = v_1 + v_R \tag{9.71}$$

and

$$v_{o2} = v_2 - v_R \tag{9.72}$$

Voltages v_{o1} and v_{o2} are the input voltages to the differential stage studied in Section 9.1. Since the external resistors connected to that stage are all equal to R, we recall (from equations 9.13 and 9.14, with $A = 1$) that

$$v_o = v_{o1} - v_{o2} \tag{9.73}$$

Substituting (9.71) and (9.72) into (9.73) gives

$$v_o = (v_1 + v_R) - (v_2 - v_R) = v_1 - v_2 + 2v_R \tag{9.74}$$

Substituting (9.70) into (9.74), we find

$$v_o = v_1 - v_2 + \frac{2(v_1 - v_2)R}{R_A} = (v_1 - v_2)(1 + 2R/R_A) \tag{9.75}$$

Equation 9.75 shows that the output of the instrumentation amplifier is directly proportional to the difference voltage $(v_1 - v_2)$, as required. The overall closed-loop gain is clearly $(1 + 2R/R_A)$. R_A is made adjustable so that gain can be easily adjusted for calibration purposes. Note that the gain is inversely proportional to R_A.

To ensure proper operation of the instrumentation amplifier, all three of the following inequalities must be satisfied at all times:

$$\left|\left(1 + \frac{R}{R_A}\right) v_1 - \frac{R}{R_A} v_2\right| < |V_{max(1)}| \tag{9.76}$$

$$\left|\left(1 + \frac{R}{R_A}\right) v_2 - \frac{R}{R_A} v_1\right| < |V_{max(1)}| \tag{9.77}$$

$$\left(1 + \frac{2R}{R_A}\right) |v_1 - v_2| < |V_{max(2)}| \tag{9.78}$$

where $V_{max(1)}$ is the maximum output voltage of each input stage and $V_{max(2)}$ is the maximum output voltage of the differential stage.

Example 9.13

1. Assuming ideal amplifiers, find the minimum and maximum output voltage V_o of the instrumentation amplifier shown in Figure 9.40 when the 10-kΩ potentiometer R_p is adjusted through its entire range.

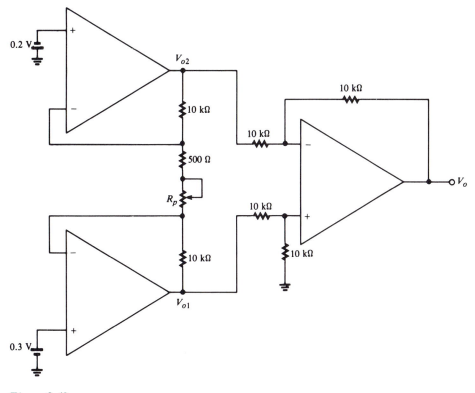

Figure 9.40
(*Example 9.13*)

2. Find V_{o1} and V_{o2} when R_p is set in the middle of its resistance range.

Solution

1. Referring to Figure 9.38, we see that R_A in this example is the sum of R_p and the fixed 500-Ω resistor. Assuming that R_p can be adjusted through a full range from 0 to 10 kΩ, $R_A(\text{min}) = 500\ \Omega$ and $R_A(\text{max}) = (500\ \Omega) + (10\ \text{k}\Omega) = 10.5\ \text{k}\Omega$. Therefore, from equation 9.75, the minimum and maximum values of V_o are

$$v_o(\text{min}) = (V_1 - V_2)\left[1 + \frac{2R}{R_A(\text{max})}\right] = [0.3\ \text{V} - (-0.2\ \text{V})]$$

$$\times \left[1 + \frac{2(10\ \text{k}\Omega)}{10.5\ \text{k}\Omega}\right] = 1.45\ \text{V}$$

$$v_o(\text{max}) = (V_1 - V_2)\left[1 + \frac{2R}{R_A(\text{min})}\right] = [0.3\ \text{V} - (-0.2\ \text{V})]$$

$$\times \left[1 + \frac{2(10\ \text{k}\Omega)}{500\ \Omega}\right] = 20.5\ \text{V}$$

2. When $R_A = 500\ \Omega + (\tfrac{1}{2})(10 \times 10^3\ \Omega) = 5.5\ \text{k}\Omega$, we have, from equation 9.70,

$$V_R = \frac{(V_1 - V_2)R}{R_A} = \frac{[0.3\ \text{V} - (-0.2\ \text{V})]10 \times 10^3\ \Omega}{5.5 \times 10^3\ \Omega} = 0.909\ \text{V}$$

From equations 9.71 and 9.72,

$$V_{o1} = V_1 + V_R = 0.3\ \text{V} + 0.909\ \text{V} = 1.209\ \text{V}$$
$$V_{o2} = V_2 - V_R = -0.2\ \text{V} - 0.909\ \text{V} = -1.109\ \text{V}$$

Example 9.14

DESIGN

The maximum output voltages for all three operational amplifiers in an instrumentation amplifier are ± 15 V. For a particular application, it is known that input signal v_1 may vary from 0 V to 0.8 V and input signal v_2 from 0 V to 1.3 V. Assuming that $R = 2$ kΩ, design the circuit for maximum possible closed-loop gain.

Solution. Since the closed-loop gain is inversely proportional to R_A, we must find the minimum value of R_A that simultaneously satisfies inequalities 9.76 through 9.78. We must consider the worst-case condition for each inequality, that is, the combination of values for v_1 and v_2 that makes the left side of each inequality as large as possible. Thus, for inequality 9.76, we must satisfy both of the following:

$$\left|\left(1 + \frac{R}{R_A}\right)v_1(\text{max}) - \left(\frac{R}{R_A}\right)v_2(\text{min})\right| < |V_{max}|$$

and

$$\left|\left(1 + \frac{R}{R_A}\right)v_1(\text{min}) - \left(\frac{R}{R_A}\right)v_2(\text{max})\right| < |V_{max}|$$

Substituting values gives

$$\left|\left(1 + \frac{2 \times 10^3}{R_A}\right)(0.8) - \left(\frac{2 \times 10^3}{R_A}\right)0\right| < 15$$

and

$$\left|\left(1 + \frac{2 \times 10^3}{R_A}\right)0 - \left(\frac{2 \times 10^3}{R_A}\right)(1.3)\right| < 15$$

The first of these leads to

$$(0.8)\left(\frac{2 \times 10^3}{R_A}\right) < 14.2$$

or $R_A > 112.68 \ \Omega$. The second inequality requires that

$$\frac{2 \times 10^3}{R_A} < 11.54$$

or $R_A > 173.3 \ \Omega$.

Proceeding in a similar manner with inequality 9.77, we require that

$$\left|\left(1 + \frac{2 \times 10^3}{R_A}\right)(1.3) - \left(\frac{2 \times 10^3}{R_A}\right)0\right| < 15$$

and

$$\left|\left(1 + \frac{2 \times 10^3}{R_A}\right)(0) - \left(\frac{2 \times 10^3}{R_A}\right)(0.8)\right| < 15$$

These inequalities lead to $R_A > 189.71 \ \Omega$ and $R_A > 106.67 \ \Omega$.

Finally, inequality 9.78 requires that

$$\left|\left(1 + \frac{4 \times 10^3}{R_A}\right)(0 - 1.3)\right| < 15$$

and

$$\left|\left(1 + \frac{4 \times 10^3}{R_A}\right)(0.8 - 0)\right| < 15$$

which gives $R_A > 379.56 \ \Omega$ and $R_A > 225.35 \ \Omega$.

Summarizing, we require that *all* the following inequalities be satisfied: $R_A > 112.68 \ \Omega$, $R_A > 173.3 \ \Omega$, $R_A > 189.71 \ \Omega$, $R_A > 106.67 \ \Omega$, $R_A > 379.56 \ \Omega$, and $R_A > 225.35 \ \Omega$. Obviously, the only way that all inequalities can be satisfied is for R_A to be larger than the largest of the computed limits: $R_A > 379.56 \ \Omega$. Choosing the closest standard 5% resistor value that is larger than 379.56 Ω, we let $R_A = 390 \ \Omega$. This choice gives us the maximum permissible closed-loop gain, with a small margin for error:

$$\frac{v_o}{v_1 - v_2} = 1 + \frac{2R}{R_A} = 1 + \frac{2(2 \times 10^3 \ \Omega)}{390 \ \Omega} = 11.26$$

9.7   ACTIVE FILTERS

Basic Filter Concepts

A *filter* is a device that allows signals having frequencies in a certain range to pass through it while attenuating all other signals. An *ideal* filter has identical gain at all frequencies in its *passband* and zero gain at all frequencies outside its passband.

Figure 9.41 shows the ideal frequency response for each of several commonly used filter types, along with typical responses for practical filters of the same types. In each case, cutoff frequencies are shown to be those frequencies at which the response is down 3 dB from its maximum value in the passband. Filters are widely used to "extract" desired frequency components from complex signals and/or reject undesired ones, such as noise. *Passive* filters are those constructed with resistors, capacitors, and inductors, while *active* filters employ active components such as transistors and operational amplifiers, along with resistor–capacitor networks. Inductors are rarely used with active filters, because of their bulk, expense, and lack of availability in a wide range of values.

Filters are classified by their *order,* an integer number, also called the number of *poles*. For example, a second-order filter is said to have two poles. In general, the higher the order of a filter, the more closely it approximates an ideal filter and the more complex the circuitry required to construct it. The frequency response outside the passband of a filter of order n has a slope that is asymptotic to $20n$ dB/decade or $6n$ dB/octave. Recall that a simple RC network is called *first-order,* or *single-pole,* filter and its response falls off at 20 dB/decade.

The classification of filters by their order is applicable to all the filter types shown in Figure 9.41. We should note that the response of every high-pass filter must eventually fall off at some high frequency because no physically realizable filter can have infinite bandwidth. Therefore, every high-pass filter is, technically speaking, a bandpass filter. However, it is often the case that a filter behaves like a high-pass for all those frequencies encountered in a particular application, so for

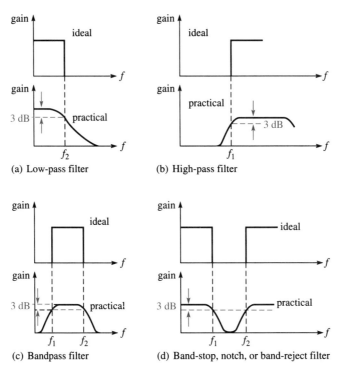

Figure 9.41
Ideal and practical frequency responses of some commonly used filter types

all practical purposes it can be treated as a true high-pass filter. The order of a high-pass filter affects its frequency response in the vicinity of f_1 (Figure 9.41(b)) and is not relevant to the high-frequency region, where it must eventually fall off.

Filters are also classified as belonging to one of several specific design types that, like order, affect their response characteristics within and outside of their passbands. The two categories we will study are called the *Butterworth* and *Chebyshev* types. The gain magnitudes of low- and high-pass Butterworth filters obey the relationships

$$|G| = \frac{M}{\sqrt{1 + (f/f_2)^{2n}}} \qquad \text{(low-pass)} \qquad (9.79)$$

$$|G| = \frac{M}{\sqrt{1 + (f_1/f)^{2n}}} \qquad \text{(high-pass)} \qquad (9.80)$$

where M is a constant, n is the order of the filter, f_1 is the cutoff frequency of the high-pass filter, and f_2 is the cutoff frequency of the low-pass filter. M is the maximum gain approached by signals in the passband. Figure 9.42 shows low- and high-pass Butterworth responses for several different values of the order, n. As frequency f approaches 0, note that the gain of the low-pass filter approaches M. For very large values of f, the gain of the high-pass filter approaches M. The value of M is 1, or less, for passive filters. An active filter is capable of amplifying signals having frequencies in its passband and may therefore have a value of M greater than 1. M is called the "*gain*" of the filter.

The Butterworth filter is also called a *maximally flat* filter, because there is relatively little variation in the gain of signals having different frequencies within its passband. For a given order, the Chebyshev filter has greater variation in the passband than the Butterworth design, but falls off at a faster rate outside the passband. Figure 9.43 shows a typical Chebyshev frequency response. Note that the cutoff frequency for this filter is defined to be the frequency at the point of intersection of the response curve and a line drawn tangent to the lowest gain in the passband. The ripple width (RW) is the total variation in gain within the passband, usually expressed in decibels. A Chebyshev filter can be designed to have a small ripple width, but at the expense of less attenuation outside the passband.

A Chebyshev filter of the same order as a Butterworth filter has a frequency response closer to the ideal filter *outside* the passband, while the Butterworth is closer to ideal within the passband. This fact is illustrated in Figure 9.44, which compares the responses of second-order, low-pass Butterworth and Chebyshev filters having the same cutoff frequency. Note that both filters have responses asymptotic to 40 dB/decade outside the passband, but at any specific frequency beyond cutoff, the Chebyshev shows greater attenuation than the Butterworth.

A bandpass filter is characterized by a quantity called its Q (originating from "quality" factor), which is a measure of how rapidly its gain changes at frequencies outside its passband. Q is defined by

$$Q = \frac{f_o}{\text{BW}} \qquad (9.81)$$

where f_o is the frequency in the passband where the gain is maximum, and BW is the bandwidth measured between the frequencies where the gain is 3 dB down from its maximum value. f_o is often called the *center* frequency because, for many high-Q filters ($Q \geq 10$), it is very nearly in the center of the passband. Figure 9.45

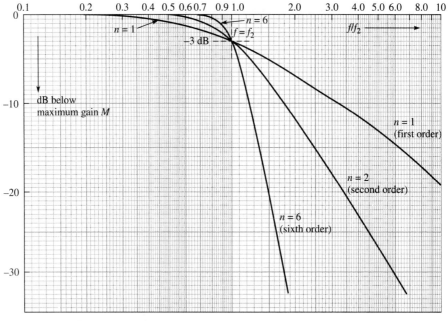

(a) low-pass filters (f_2 = cutoff frequency)

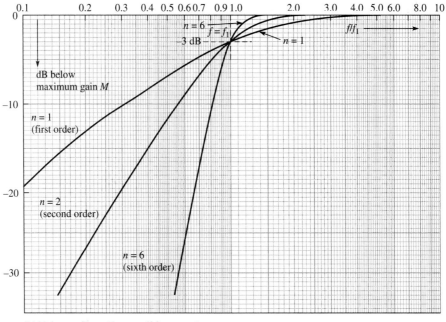

(b) high-pass filters (f_1 = cutoff frequency)

Figure 9.42
Frequency response of low-pass and high-pass Butterworth filters with different orders

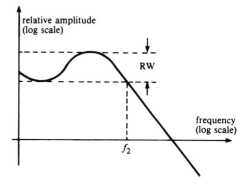

Figure 9.43
Chebyshev low-pass frequency response:
$f_2 =$ *cutoff frequency; RW = ripple width*

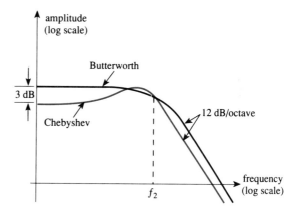

Figure 9.44
Comparison of the frequency responses of second-order, low-pass Butterworth and Chebyshev filters

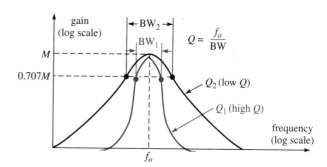

Figure 9.45
Comparison of the frequency responses of low-Q and high-Q bandpass filters

shows the frequency responses of two bandpass filters having the same maximum gain but different values of Q. It is clear that the high-Q filter has a narrower bandwidth and that its response falls off faster on either side of the passband than that of the low-Q filter.

The gain magnitude of any second-order bandpass filter is the following function of frequency f:

$$|G| = \frac{M}{\sqrt{1 + Q^2(f/f_o - f_o/f)^2}} \qquad (9.82)$$

where M is the maximum gain in the band. The center frequency f_o is at the *geometric* center of the band, defined by

$$f_o = \sqrt{f_1 f_2} \qquad (9.83)$$

where f_1 and f_2 are the lower and upper cutoff frequencies. As previously mentioned, f_o is very nearly midway between the cutoff frequencies (the arithmetic mean) in high-Q filters:

$$f_o \approx f_1 + \frac{\text{BW}}{2} \approx f_2 - \frac{\text{BW}}{2} \qquad (9.84)$$

Active Filter Design

The analysis of active filters requires complex mathematical methods that are beyond the scope of this book. We will therefore concentrate on practical design methods that will allow us to construct Butterworth and Chebyshev filters of various types and orders. The discussion that follows is based on design procedures using tables that can be found in *Rapid Practical Designs of Active Filters,* by Johnson and Hilburn (Wiley, 1975). Space limitations prevent us from listing all the tables that are available in that reference, so we will simply illustrate how a few of them are used in filter design. Readers wishing a more comprehensive treatment should consult the reference.

Figure 9.46 shows a general configuration that can be used to construct second-order high-pass and low-pass filters of both the Butterworth and Chebyshev designs. The amplifier is basically operated as a noninverting, voltage-controlled voltage source (see Section 9.2), and the configuration is known as the VCVS design. It is also called a *Sallen-Key* circuit. Each impedance block represents a resistance or capacitance, depending on whether a high-pass or low-pass filter is desired. Table 9.1 shows the impedance type required for each design. We see, for example, that Z_A is a resistor, designated R_1, for a low-pass filter and is a capacitor, designated C, for a high-pass filter. The VCVS design can also be constructed as a bandpass filter using another component arrangement (not shown).

The component values in the VCVS design depend on whether a Butterworth or Chebyshev response is required, the gain required in the passband, and, in the case of the Chebyshev filter, on the tolerable ripple width in the passband. As we

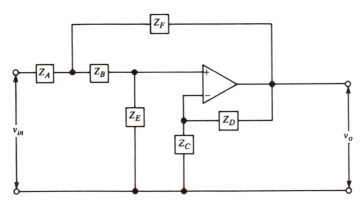

Figure 9.46
Block diagram of a second-order, VCVS low-pass or high-pass filter

shall demonstrate shortly, different "look-up" tables are used to determine component values corresponding to these various options. The design procedure begins with selection of a value for capacitance C. Next, a constant designated K is computed as follows:

$$K = \frac{10^{-4}}{fC} \qquad (9.85)$$

where f is the desired cutoff frequency in Hz and C is the value of capacitance selected, in farads. We then consult an appropriate table to obtain factors by which K is multiplied to give each component value. Tables 9.2 through 9.4 are three such tables. These can be used to design a low-pass Butterworth, a low-pass Chebyshev with 2-dB ripple width, or a high-pass Chebyshev with 2-dB ripple width. The next example illustrates how they are used.

Table 9.1
VCVS Filter Components

	Z_A	Z_B	Z_C	Z_D	Z_E	Z_F
Low-Pass Filter	R_1	R_2	R_3	R_4	C_1	C
High-Pass Filter	C	C	R_3	R_4	R_2	R_1

Table 9.2
Second-Order Low-Pass Butterworth VCVS Filter Designs

	Circuit Element Values[a]					
Gain	**1**	**2**	**4**	**6**	**8**	**10**
R_1	1.422	1.126	0.824	0.617	0.521	0.462
R_2	5.399	2.250	1.537	2.051	2.429	2.742
R_3	Open	6.752	3.148	3.203	3.372	3.560
R_4	0	6.752	9.444	16.012	23.602	32.038
C_1	0.33C	C	2C	2C	2C	2C

[a]Resistances in kilohms for a K parameter of 1.

Table 9.3
Second-Order Low-Pass Chebyshev VCVS Filter Designs (2 dB)

	Circuit Element Values[a]					
Gain	**1**	**2**	**4**	**6**	**8**	**10**
R_1	2.328	1.980	1.141	0.786	0.644	0.561
R_2	13.220	1.555	1.348	1.957	2.388	2.742
R_3	Open	7.069	3.320	3.292	3.466	3.670
R_4	0	7.069	9.959	16.460	24.261	33.031
C_1	0.1C	C	2C	2C	2C	2C

[a]Resistances in kilohms for a K parameter of 1.

Table 9.4
Second-Order High-Pass Chebyshev VCVS Filter Designs (2 dB)

	Circuit Element Values[a]					
Gain	1	2	4	6	8	10
R_1	0.640	1.390	2.117	2.625	3.040	3.399
R_2	3.259	1.500	0.985	0.794	0.686	0.613
R_3	Open	3.000	1.313	0.953	0.784	0.681
R_4	0	3.000	3.939	4.765	5.486	6.133

[a]Resistances in kilohms for a K parameter of 1.

Source: Reprinted from *Rapid Practical Designs of Active Filters,* D. Johnson and J. Hilburn. Copyright © 1975, John Wiley and Sons, Inc., by permission of John Wiley and Sons, Inc.

Example 9.15

DESIGN

Design a second-order, VCVS, low-pass Butterworth filter with cutoff frequency 2.5 kHz. The gain in the passband should be 2.

Solution.   Choose $C = 0.05$ μF. (If this choice results in impractical values for the other components, we can always revise the choice and try again.) From equation 9.85,

$$K = \frac{10^{-4}}{fC} = \frac{10^{-4}}{(2.5 \times 10^3)(0.05 \times 10^{-6})} = 0.8$$

From Table 9.2, in the *Gain = 2* column, we locate the multiplying constants for

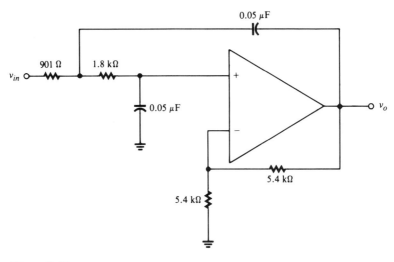

Figure 9.47
(Example 9.15)

each component value and obtain

$R_1 = (1.126)(0.8) \text{ k}\Omega = 901 \ \Omega \qquad R_4 = (6.752)(0.8) \text{ k}\Omega = 5.40 \text{ k}\Omega$

$R_2 = (2.250)(0.8) \text{ k}\Omega = 1.80 \text{ k}\Omega \qquad C_1 = C = 0.05 \ \mu\text{F}$

$R_3 = (6.752)(0.8) \text{ k}\Omega = 5.40 \text{ k}\Omega$

With reference to Table 9.1 and Figure 9.46, we construct the required filter as shown in Figure 9.47. Note that precision, nonstandard component values are required if we demand exact conformity to the original design specifications. Generally speaking, satisfactory performance can be obtained (in low-order filters) by using the closest 5% standard values.

Higher-order filters of the VCVS design can be constructed by cascading VCVS stages. A fourth-order filter requires two stages, a sixth-order filter requires three stages, and so forth. Tables are available in the previously cited reference for designing each stage of VCVS filters with orders up to 8. It should be emphasized that higher-order filters are *not* constructed by cascading identical stages. This precaution applies to all the filters we will discuss in this chapter. Recall from Chapter 6 that a cascade of identical stages has an overall cutoff frequency that is different from that of the individual stages.

Another configuration that can be used to construct low-pass, high-pass, and bandpass filters is called the *infinite-gain multiple-feedback* (IGMF) design. Figure 9.48 shows the component arrangement used to obtain a second-order bandpass filter. One advantage of the IGMF design is that it requires one less component than the VCVS filter. It is also popular because it has good stability and low output impedance. The IGMF filter inverts signals in its passband.

The procedure for designing an IGMF filter is very similar to that for a VCVS filter. We first choose a value of capacitance C and then calculate

$$K = \frac{10^{-4}}{f_o C} \tag{9.86}$$

where f_o is the desired center frequency of the bandpass filter. A table corresponding to the desired Q is then consulted to find the factors that multiply K to determine resistor values R_1, R_2, and R_3 in kΩ. Table 9.5 shows the factors used to

Figure 9.48
The IGMF second-order bandpass filter

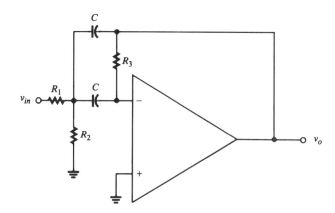

Table 9.5
Second-Order Multiple-Feedback Bandpass Filter Designs (Q = 5)

	Circuit Element Values[a]					
Gain	**1**	**2**	**4**	**6**	**8**	**10**
R_1	7.958	3.979	1.989	1.326	0.995	0.796
R_2	0.162	0.166	0.173	0.181	0.189	0.199
R_3	15.915	15.915	15.915	15.915	15.915	15.915

[a]Resistances in kilohms for a K parameter of 1.
Source: Reprinted from *Rapid Practical Designs of Active Filters*, D. Johnson and J. Hilburn.
Copyright © 1975, John Wiley and Sons, Inc., by permission of John Wiley and Sons, Inc.

design a second-order filter having a Q of 5. The gain headings over the columns
refer to the gain of the filter at its center frequency. Tables are also available in the
previously cited reference for constructing low- and high-pass IGMF filters of
various orders.

Another popular active filter is the *biquad* design, which can be constructed in
low-pass, high-pass, or bandpass configurations. Figure 9.49 shows the low-pass
version. Although this design requires three amplifiers and a greater number of
passive components than the VCVS or IGMF filters, it has the advantage that its
gain and cutoff frequency are easily adjusted. Also, both an inverted and a noninverted output are available, as shown in Figure 9.49. The gain of this filter is given
by

$$G = \frac{R_3}{R_1} \tag{9.87}$$

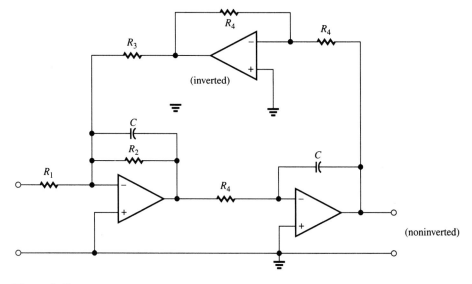

Figure 9.49
Circuit diagram of a second-order low-pass biquad filter

Table 9.6
Second-Order Low-Pass Biquad Filter Designs

		Circuit Element Values[a]				
			Chebyshev			
	Butterworth	**0.1 dB**	**0.5 dB**	**1 dB**	**2 dB**	**3 dB**
R_1	1.592/G	0.480/G	1.050/G	1.444/G	1.934/G	2.248/G
R_2	1.125	0.671	1.116	1.450	1.980	2.468
R_3	1.592	0.480	1.050	1.444	1.934	2.248
R_4	1.592	1.592	1.592	1.592	1.592	1.592

[a]Resistances in kilohms for a K parameter of 1, G = gain.
Source: Reprinted from *Rapid Practical Designs of Active Filters*, D. Johnson and J. Hilburn.
Copyright © 1975, John Wiley and Sons, Inc., by permission of John Wiley and Sons, Inc.

The gain is varied by adjusting R_1, while changing R_3 varies both the gain and the cutoff frequency.

The design procedure for the biquad filter begins with a choice of capacitor C and the calculation of the constant K, as before:

$$K = \frac{10^{-4}}{fC} \tag{9.88}$$

A value of gain G is also chosen. The factors that multiply K to determine the other component values are then taken from a table. Table 9.6 shows the multiplying factors for second-order Butterworth responses and Chebyshev responses having several different ripple widths. Notice that the multiplying factor for R_1 is divided by the user-selected value of gain. Higher-order filters can be designed by cascading stages and using appropriate sets of tables available in the previously cited reference.

Example 9.16

SPICE

Design a low-pass biquad filter with cutoff frequency 1 kHz. The filter should have Chebyshev characteristics with a 3-dB ripple width and a gain of 2. Verify the performance using SPICE.

Solution.   Letting $C = 0.1 \ \mu F$, we have, from equation 9.88,

$$K = \frac{10^{-4}}{fC} = \frac{10^{-4}}{(1 \text{ kHz})(0.1 \ \mu F)} = 1$$

Then, from Table 9.6, $R_1 = (2.248/G) \ k\Omega = (2.248/2) \ k\Omega = 1.124 \ k\Omega$, $R_2 = 2.468$ $k\Omega$, $R_3 = 2.248 \ k\Omega$, and $R_4 = 1.592 \ k\Omega$. Figure 9.50(a) shows the filter circuit in its SPICE format. Note that a subcircuit (SUBCKT) is used to specify the three operational amplifiers. The subcircuit, named OPAMP, models a nearly ideal amplifier having input resistance $1 \times 10^{12} \ \Omega$ and open-loop voltage gain 1×10^9.

Figure 9.50(b) shows the results of an .AC analysis over the frequency range from 10 Hz through 10 kHz. Note the characteristic rise in the response of the Chebyshev filter. The low-frequency gain is seen to be 2, as required by the design specifications. The gain then rises and falls back to $1.999 \approx 2$ at 1 kHz, confirming

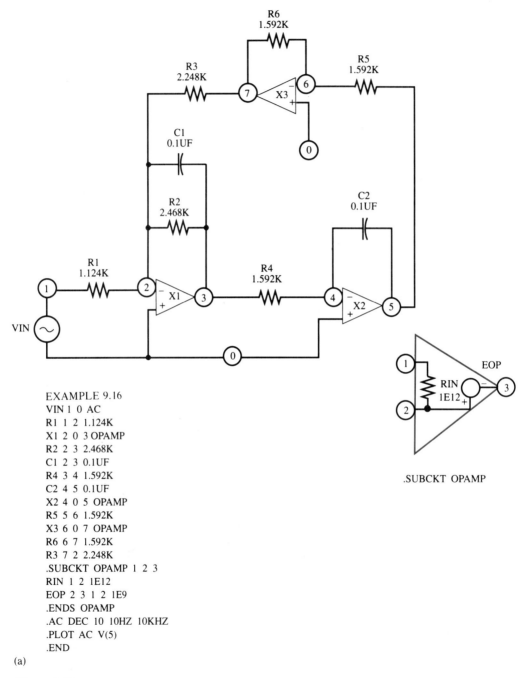

EXAMPLE 9.16
VIN 1 0 AC
R1 1 2 1.124K
X1 2 0 3 OPAMP
R2 2 3 2.468K
C1 2 3 0.1UF
R4 3 4 1.592K
C2 4 5 0.1UF
X2 4 0 5 OPAMP
R5 5 6 1.592K
X3 6 0 7 OPAMP
R6 6 7 1.592K
R3 7 2 2.248K
.SUBCKT OPAMP 1 2 3
RIN 1 2 1E12
EOP 2 3 1 2 1E9
.ENDS OPAMP
.AC DEC 10 10HZ 10KHZ
.PLOT AC V(5)
.END

(a)

Figure 9.50
(Example 9.16)

EXAMPLE 9.16
```
****        AC ANALYSIS                    TEMPERATURE =    27.000 DEG C
*********************************************************************************
     FREQ        V(5)
                   1.000D-02      1.000D-01      1.000D+00      1.000D+01  1.000D+02
              - - - - - - - - - - - - - - - - - - - - - - - - - - - - - -
 1.000D+01   2.000D+00 .               .              .     *            .          .
 1.259D+01   2.000D+00 .               .              .     *            .          .
 1.585D+01   2.001D+00 .               .              .     *            .          .
 1.995D+01   2.001D+00 .               .              .     *            .          .
 2.512D+01   2.001D+00 .               .              .     *            .          .
 3.162D+01   2.002D+00 .               .              .     *            .          .
 3.981D+01   2.003D+00 .               .              .     *            .          .
 5.012D+01   2.005D+00 .               .              .     *            .          .
 6.310D+01   2.008D+00 .               .              .     *            .          .
 7.943D+01   2.013D+00 .               .              .     *            .          .
 1.000D+02   2.020D+00 .               .              .     *            .          .
 1.259D+02   2.032D+00 .               .              .     *            .          .
 1.585D+02   2.051D+00 .               .              .    *             .          .
 1.995D+02   2.081D+00 .               .              .    *             .          .
 2.512D+02   2.130D+00 .               .              .      *           .          .
 3.162D+02   2.208D+00 .               .              .      *           .          .
 3.981D+02   2.335D+00 .               .              .     *            .          .
 5.012D+02   2.530D+00 .               .              .       *          .          .
 6.310D+02   2.768D+00 .               .              .        *         .          .
 7.943D+02   2.733D+00 .               .              .        *         .          .
 1.000D+03   1.999D+00 .               .              .     *            .          .
 1.259D+03   1.184D+00 .               .              .   .*             .          .
 1.585D+03   6.827D-01 .               .          *   .                  .          .
 1.995D+03   4.025D-01 .               .        *     .                  .          .
 2.512D+03   2.428D-01 .               .      *       .                  .          .
 3.162D+03   1.488D-01 .               .   . *        .                  .          .
 3.981D+03   9.218D-02 .               .  *           .                  .          .
 5.012D+03   5.749D-02 .            *   . .            .                  .          .
 6.310D+03   3.601D-02 .         *      .              .                  .          .
 7.943D+03   2.261D-02 .     *          .              .                  .          .
 1.000D+04   1.423D-02 . *              .              .                  .          .
              - - - - - - - - - - - - - - - - - - - - - - - - - - - - - -
```

(b)

Figure 9.50
(*Continued*)

that the Chebyshev cutoff frequency is 1 kHz. The gain at 5.012 kHz is 0.05749 and that at 10 kHz is 0.01423. Thus, the gain changes by

$$20 \log_{10} \left(\frac{0.01423}{0.05749} \right) = -12.12 \text{ dB}$$

over a frequency range close to one octave, confirming that the filter is of second order.

As a final example of an active filter, we will consider a three-amplifier config-uration called the *state-variable* filter. This design simultaneously provides sec-ond-order low-pass, high-pass, and bandpass filtering, each of these functions being taken from a different point in the circuit. It is available in integrated-circuit form as a so-called *universal active filter*. Figure 9.51 shows the configuration. Note that it is essentially two integrators and a summing amplifier.

The following equations define the characteristics of the state-variable filter:

$$\omega_o = 2\pi f_o = \frac{1}{R_1 C} \text{ radians/second} \qquad \textbf{(9.89)}$$

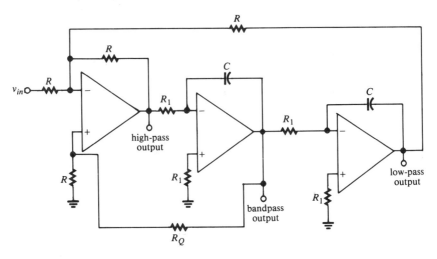

Figure 9.51
A state-variable, or universal, active filter

$$BW = \frac{3R}{R_1C(R + R_Q)} \qquad (9.90)$$

$$Q = \frac{R + R_Q}{3R} \qquad (9.91)$$

where ω_o is the center frequency, BW is the bandwidth, and Q is the Q of the bandpass output. The Q of the circuit determines whether it has Butterworth or Chebyshev characteristics. For a Q of 0.707, it is Butterworth, and for a Q of 0.885, it is Chebyshev. In the case of a Butterworth response, the cutoff frequencies for the low- and high-pass outputs are both equal to ω_o. For the Chebyshev design, the low-pass cutoff frequency is $0.812\omega_o$ rad/s and the high-pass cutoff frequency is $1.23\omega_o$ rad/s.

The resistor labeled R_Q in Figure 9.51 is used to set the Q of the filter. Solving equation 9.91 for R_Q, we find

$$R_Q = R(3Q - 1) \qquad (9.92)$$

Resistors R_1 or capacitors C can be adjusted to change the values of ω_o. Note that *both* resistors or *both* capacitors must be adjusted simultaneously. A band-*reject* filter (see Figure 9.41(d)) can be obtained by summing the low-pass and high-pass outputs of the state-variable filter in a fourth amplifier. The summation should be performed with equal gain for each signal.

Example 9.21

DESIGN   Design a state-variable filter with Butterworth characteristics and center frequency 1.59 kHz. What are the cutoff frequencies of the low-pass and high-pass outputs? What is the bandwidth of the bandpass output?

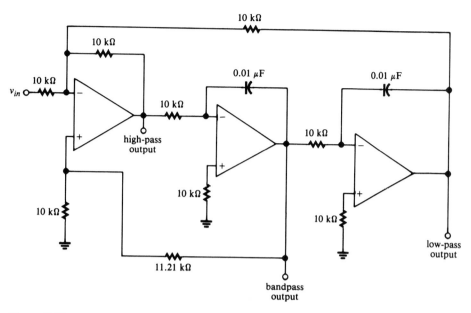

Figure 9.52
(Example 9.17)

Solution. Choose $C = 0.01$ μF. Then, from equation 9.89,

$$2\pi \times (1.59 \times 10^3) = \frac{1}{R_1(10^{-8})}$$

$$R_1 = \frac{1}{2\pi(1.59 \times 10^3)(10^{-8})} = 10 \text{ k}\Omega$$

For a Butterworth response, we must have $Q = 0.707$. Let $R = 10$ kΩ. Then, from equation 9.92, $R_Q = R(3Q - 1) = 10^4[3(0.707) - 1] = 11.21$ kΩ. The completed design is shown in Figure 9.52.

Since the filter has Butterworth characteristics, the cutoff frequencies of both the low- and high-pass outputs are equal to the center frequency of the bandpass output: 1.59 kHz.

EXERCISES

Section 9.1 Voltage Summation, Subtraction, and Scaling

9.1
  a. Write an expression for the output of the amplifier in Figure 9.53 in terms of v_1, v_2, v_3, and v_4.
  b. What is the output when $v_1 = 5 \sin \omega t$, $v_2 = -3$ V dc, $v_3 = -\sin \omega t$, and $v_4 = 2$ V dc?
  c. What value should R_c have?

9.2
  a. Design an operational-amplifier circuit that will produce the output $v_o = -10v_1 - 50v_2 + 10$. Use only one amplifier. (*Hint:* One of the inputs is a dc source.)
  b. Sketch the output waveform when $v_1 = v_2 = -0.1 \sin \omega t$ volts.

9.3
  The operational amplifier in Exercise 9.1 has unity-gain frequency 1 MHz and input offset voltage 3 mV. Find

Figure 9.53
(*Exercise 9.1*)

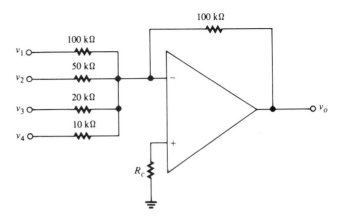

a. the closed-loop bandwidth of the config-uration; and
b. the magnitude of the output offset volt-age due to V_{io}.

9.4 Derive equation 9.8 for the output of the circuit shown in Figure 9.4. (*Hint:* Using superposition, write an expression for v^+. Then use the fact that $v^- = v^+$ and $v_o = v^- + v^- R_f/R_g$.)

9.5 **a.** Write an expression, in terms of v_1 and v_2, for the output of the amplifier shown in Figure 9.54.
b. Write an expression for the output in the special case in which v_1 and v_2 are equal-magnitude, out-of-phase signals.

9.6 Design a noninverting circuit using a single operational amplifier that will produce the output $v_o = 4v_1 + 6v_2$.

9.7 Using a single operational amplifier in each case, design circuits that will produce the following outputs:
a. $v_o = 0.1v_1 - 5v_2$
b. $v_o = 10(v_1 - v_2)$
Design the circuits so that the compensa-tion resistance has an optimum value.

9.8 If the resistor values in Figure 9.5 are cho-sen in accordance with $R_4 = a_1 R_1 = a_2 R_3 = R_2(1 + a_2 - a_1)$, then, assuming that $1 + a_2 > a_1$, show that
a. $v_o = a_1 v_1 - a_2 v_2$; and
b. the compensation resistance $(R_1 \parallel R_2)$ has its optimum value $(R_3 \parallel R_4)$.

9.9 Design operational-amplifier circuits to pro-duce each of the following outputs:
a. $v_o = 0.4v_2 - 10v_1$
b. $v_o = v_1 + v_2 - 20v_3$

9.10 Assuming that the unity-gain frequency of each amplifier used in Exercise 9.9 is 750 kHz, find the approximate, overall band-width of each circuit.

Section 9.2 Controlled Voltage and Current Sources

9.11 **a.** Design an inverting voltage-controlled current source that will supply a current of 1 mA to a floating load when the con-trolling voltage is 2 V.
b. If the source designed in (a) must supply its current to loads of up to 20 kΩ, what

Figure 9.54
(*Exercise 9.5*)

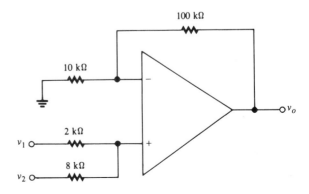

maximum output voltage should the amplifier have?

9.12 **a.** Design a voltage-controlled current source that will supply a current of 2 mA to a floating load when the controlling voltage is 10 V. The input resistance seen by the controlling voltage source would have to be greater than 10 kΩ.

b. If the maximum output voltage of the amplifier is 15 V, what is the maximum load resistance for which your design will operate properly?

9.13 **a.** Design a voltage-controlled current source that will supply a current of 0.5 mA to a grounded load when the controlling voltage is 5 V.

b. What will be the value of the amplifier output voltage if the load resistance is 12 kΩ?

9.14 The voltage-controlled current source in Figure 9.12 is to be used to supply current to a grounded 10-kΩ load when the controlling voltage is 5 V. If the maximum output voltage of the amplifier is 20 V, what is the maximum current that can be supplied to the load?

9.15 A certain temperature-measuring device generates current in direct proportion to temperature, in accordance with the relation $I = 2.5T$ μA, where T is in degrees Celsius. It is desired to construct a current-to-voltage converter for usc with this device so that an output of 20 mV/°C can be obtained. Design the circuit.

9.16 The circuit shown in Figure 9.14 is used with the temperature-measuring device described in Exercise 9.15. If $R = 10$ kΩ, what is the output voltage when the temperature is 75°C?

9.17 Find the currents I_1, I_2, and I_3, and the voltages V_A and V_B, in the circuit of Figure 9.55. Assume an ideal operational amplifier.

9.18 **a.** Design a current amplifier that will produce, in a 1-kΩ load, five times the current supplied to it.

b. If the input current supplied to the amplifier is 2 mA, what should be the magnitude of the maximum output voltage of the amplifier?

Section 9.3   Integration

9.19 The input to an ideal electronic integrator is 0.25 V dc. Assume that the integrator inverts and multiplies by the constant 20.

a. What is the output 2 s after the input is connected?

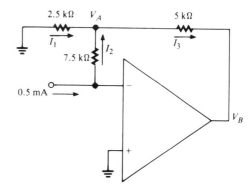

Figure 9.55
(Exercise 9.17)

b. How long would it take the output to reach −15 V?

9.20 Using an ideal operational amplifier, design an ideal integrator whose output will reach 5 V 200 ms after a −0.1 V-dc input is connected.

9.21 The input to the circuit shown in Figure 9.56 is $v_{in} = 6 \sin (500t − 30°)$ V. Write a mathematical expression for the output voltage.

9.22 The input to the integrator in Exercise 9.21 is a 100-Hz sine wave with peak value 5 V.

a. What is the peak value of the output?

b. What is the peak value of the output if the capacitance in the feedback is halved?

c. What is the peak value of the output

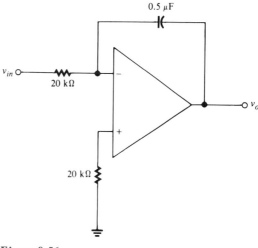

Figure 9.56
(Exercise 9.21)

(with the original capacitance) if the frequency is halved?

d. What is the peak value of the output (with original capacitance and original frequency) if the input resistance is halved?

e. Name three ways to double the closed-loop gain of an integrator.

9.23 Using an ideal operational amplifier, design an ideal integrator that will produce the output $v_o = 0.04 \cos (2 \times 10^3 t)$ when the input is $v_{in} = 8 \sin (2 \times 10^3 t)$.

9.24 What is the closed-loop gain, in decibels, of the integrator designed in Exercise 9.23 when the angular frequency is 500 rad/s?

9.25 Design a practical integrator that will integrate signals with frequencies down to 500 Hz and that will provide unity gain to dc inputs.

9.26 **a.** Design a practical integrator that will integrate signals with frequencies down to 500 Hz and that will provide a closed-loop gain of 0.005 at $f = 20$ kHz.

b. What will be the output of the integrator when the input is -10 mV dc?

9.27 **a.** Write an expression for the output v_o of the system shown in Figure 9.57.

b. Show how the same output could be obtained using a single amplifier.

Section 9.4   Differentiation

9.28 An ideal differentiator has a closed-loop gain of 2 when the input is a 50-Hz signal. What is its closed-loop gain when the signal frequency is changed to 2 kHz?

9.29 The differentiator shown in Figure 9.27 has $R_1 = 2.2$ kΩ, $C = 0.015$ μF, and $R_f = 27$ kΩ.

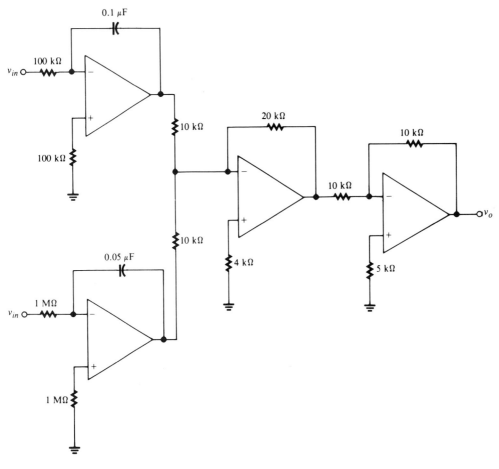

Figure 9.57
(Exercise 9.27)

a. Assuming that the circuit performs satisfactory differentiation at frequencies up to 1 decade below its first break frequency, find the frequency range of satisfactory differentiation.

b. Find the magnitude of the voltage gain at 120 Hz.

c. Sketch a Bode plot of the gain, as it would appear if plotted on log-log graph paper. Label all break frequencies. Assume that the operational amplifier has a gain–bandwidth product of 2×10^6.

9.30 Design a practical differentiator that will perform satisfactory differentiation of signals up to 1 kHz. The *maximum* closed-loop gain of the differentiator (at any frequency) should be 60. (You may assume an ideal operational amplifier that has a wide bandwidth.)

9.31 Sketch the Bode plot for the gain of the circuit shown in Figure 9.58 the way it would appear if plotted on log-log graph paper. Assume that the amplifier has a unity-gain frequency of 1 MHz. On your sketch, label

a. all break frequencies, in Hz;

b. the slopes of all gain asymptotes, in dB/decade; and

c. the value of the maximum closed-loop gain.

Section 9.5   Waveshaping

9.32 An integrator having $R_1 = 10$ kΩ and $C = 0.02$ μF is to be used to generate a triangu-

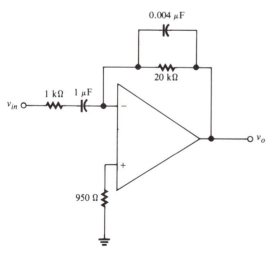

Figure 9.58
(Exercise 9.31)

lar wave from a square-wave input. The operational amplifier has a slew rate of 10^5 V/s. Assume that the input has zero dc component.

a. What maximum positive level can the square wave have when its frequency is 500 Hz?

b. Repeat when the square-wave frequency is 5 kHz.

9.33 Sketch the output of the integrator shown in Figure 9.59. Label the peak positive and peak negative output voltages.

9.34 Design a triangular-waveform generator whose output alternates between +9 V peak and −9 V peak when the input is a 10-Hz square wave that alternates between +1.5 V and −1.5 V. The input resistance to the generator must be at least 15 kΩ. You may assume that there is no dc component or offset in the input.

9.35 Sketch the output of an ideal differentiator having $R_f = 40$ kΩ and $C = 0.5$ μF when the input is each of the waveforms shown in Figure 9.60. Label maximum and minimum values on your sketches.

9.36 Sketch the output of an ideal differentiator having $R_f = 60$ kΩ and $C = 0.5$ μF when the input is each of the waveforms shown in Figure 9.61. Label minimum and maximum values on your sketches.

Section 9.6   Instrumentation Amplifiers

9.37 **a.** Assuming that the amplifiers shown in Figure 9.62 are ideal, find V_{o1}, V_{o2}, I_1, I_2, I_3, and V_o.

b. Repeat when the inputs V_1 and V_2 are interchanged.

9.38 **a.** The instrumentation amplifier in Figure 9.38 is required to have gain $|v_o/(v_1 - v_2)|$ equal to 25. Assuming that $R = 5$ kΩ, what should be the value of R_A?

b. Verify that the amplifier will perform satisfactorily if the maximum permissible output voltage of each operational amplifier is 23 V. v_1 varies from 0.3 V to 1.2 V and v_2 varies from 0.5 V to 0.8 V.

Section 9.7   Active Filters

9.39 A low-pass Butterworth filter having a maximum low-frequency gain of 1 and a cutoff frequency of 4 kHz is to be designed so that its gain is no less than 0.92 at 3 kHz. What minimum order must the filter have? (*Hint:* Use logarithms.)

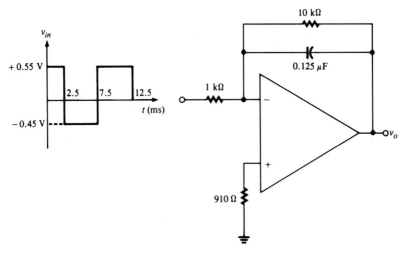

Figure 9.59
(Exercise 9.33)

Figure 9.60
(Exercise 9.35)

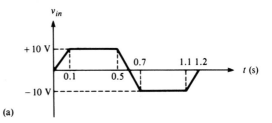

(a)

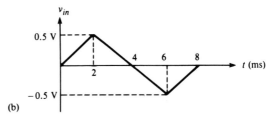

(b)

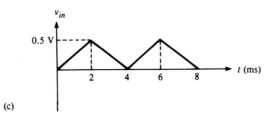

(c)

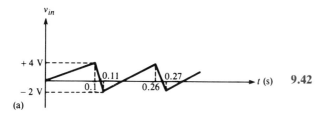

(a)

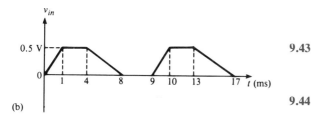

(b)

Figure 9.61
(Exercise 9.36)

9.40 A low-pass Butterworth filter must have an attenuation of at least −20 dB at 1 octave above its cutoff frequency. What minimum order must the filter have?

9.41 A high-pass Butterworth filter having a maximum high-frequency gain of 5 and a cutoff frequency of 500 Hz must have a gain no less than 4.9 at 2 kHz and no greater than 0.1 at 100 Hz. What minimum order must the filter have?

9.42 The gain of a bandpass filter at its upper cutoff frequency is 42. The lower cutoff frequency is 10.8 kHz and the Q is 50. Assume that the center frequency is midway between the cutoff frequencies.
a. What is the gain at the center frequency?
b. What is the center frequency?
c. What is the bandwidth?

9.43 Design a second-order, high-pass, VCVS Chebyshev filter having a cutoff frequency of 8 kHz and a ripple width of 2 dB. The filter should have a gain of 4.

9.44 Design a second-order, low-pass, VCVS Butterworth filter with gain 4 and cutoff frequency 1 kHz. If the input to this filter has a 500-Hz component with amplitude 1.2 V and a 4-kHz component with amplitude 5 V, what are the amplitudes of these components in the output?

9.45 Design a second-order, IGMF bandpass filter with a center frequency of 20 kHz and a bandwidth of 4 kHz. The gain should be 2 at 20 kHz.

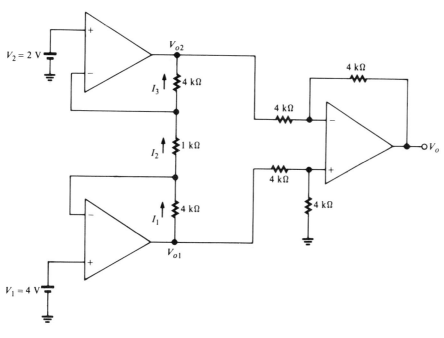

Figure 9.62
(Exercise 9.37)

9.46   The input to the filter in Exercise 9.45 has the following components:
a. v_1: 6 V rms at 2 kHz
b. v_2: 2 V rms at 20 kHz
c. v_3: 10 V rms at 100 kHz
What are the rms values of these components in the filter's output?

9.47   Design a second-order, low-pass, Butterworth filter using a biquad design having a cutoff frequency 400 Hz. The gain of the filter should be 1.5.

9.48   Repeat Exercise 9.47 with the additional requirement that the input resistance to the filter must be at least 20 kΩ.

9.49   Design a state-variable filter with Chebyshev characteristics and a center frequency of 40 kHz. What is the bandwidth of the bandpass output?

9.50   If the low-pass and high-pass outputs of the filter designed in Exercise 9.49 were summed in a fourth amplifier, with equal gain for each signal, what would be the *approximate* bandwidth of the resulting band-reject filter? Why is your answer approximate? Is the actual bandwidth greater or smaller than your approximation? Why?

SPICE EXERCISES

9.51   Use SPICE to obtain a plot of the output of the amplifier designed in Example 9.1 when the inputs are $v_1 = -5 \sin(2\pi \times 1000t)$ V, $v_2 = 10 \sin(2\pi \times 1000t)$ V, and $v_3 = -50$ V dc. Compare these results with theoretically calculated values.

9.52   Use SPICE to verify the theoretical computations of the voltages and currents in the voltage-controlled current source of Example 9.6 (Figure 9.13). Assume an ideal operational amplifier.

9.53   Design an operational-amplifier integrator that integrates signals with frequencies down to 200 Hz. The magnitude of the gain at 10 kHz should be 0.02. Verify your design using SPICE. Obtain a plot of the frequency response and compare the gain at 20 Hz, 200 Hz, and 10 kHz with theoretical values.

9.54   Design an IGMF bandpass filter having center frequency 1 kHz, $Q = 5$, and gain 1. Use SPICE to verify your design. Obtain a plot of the frequency response extending from 2 decades below the center frequency to 1 decade above the center frequency. Using SPICE to determine the bandwidth of the filter, verify that its Q is 5.

Oscillators 10

10.1 INTRODUCTION

An *oscillator* is a device that generates a periodic, ac output signal without any form of input signal required. The term is generally used in the context of a sine-wave signal generator, while a square-wave generator is usually called a *multivibrator*. A *function generator* is a laboratory instrument that a user can set to produce sine, square, or triangular waves, with amplitudes and frequencies that can be adjusted at will. Desirable features of a sine-wave oscillator include the ability to produce a low distortion ("pure") sinusoidal waveform, and, in many applications, the capability of being easily adjusted so that the user can vary the frequency over some reasonable range.

Oscillation is a form of instability caused by feedback that *regenerates,* or reinforces, a signal that would otherwise die out due to energy losses. In order for the feedback to be regenerative, it must satisfy certain amplitude and phase relations that we will discuss shortly. Oscillation often plagues designers and users of high-gain amplifiers because of unintentional feedback paths that create signal regeneration at one or more frequencies. By contrast, an oscillator is designed to have a feedback path with known characteristics, so that a predictable oscillation will occur at a predetermined frequency.

10.2 THE BARKHAUSEN CRITERION

We have stated that an oscillator has no input *per se,* so the reader may wonder what we mean by "feedback"—feedback to where? In reality, it makes no difference where, because we have a closed loop with no summing junction at which any external input is added. Thus, we could start anywhere in the loop and call that point both the "input" and the "output"; in other words, we could think of the "feedback" path as the entire path through which signal flows in going completely around the loop. However, it is customary and convenient to take the output of an amplifier as a reference point and to regard the feedback path as that portion of the loop that lies between amplifier output and amplifier input. This viewpoint is illustrated in Figure 10.1, where we show an amplifier having gain A

and a feedback path having gain β. β is the usual feedback ratio that specifies the portion of amplifier output voltage fed back to amplifier input. Every oscillator must have an amplifier, or equivalent device, that supplies energy (from the dc supply) to replenish resistive losses and thus sustain oscillation.

In order for the system shown in Figure 10.1 to oscillate, the loop gain $A\beta$ must satisfy the *Barkhausen criterion,* namely,

$$A\beta = 1 \qquad\qquad (10.1)$$

Imagine a small variation in signal level occurring at the input to the amplifier, perhaps due to noise. The essence of the Barkhausen criterion is that this variation will be reinforced and signal regeneration will occur only if the net gain around the loop, beginning and ending at the point where the variation occurred, is unity. It is important to realize that unity gain means not only a gain magnitude of 1, but also an *in-phase* signal reinforcement. Negative feedback causes signal cancellation because the feedback voltage is out of phase. By contrast, the unity loop-gain criterion for oscillation is often called *positive feedback.*

To understand and apply the Barkhausen criterion, we must regard both the gain and the phase shift of $A\beta$ as *functions of frequency.* Reactive elements, capacitance in particular, contained in the amplifier and/or feedback cause the gain magnitude and phase shift to change with frequency. In general, there will be only one frequency at which the gain magnitude is unity and at which, simultaneously, the total phase shift is equivalent to 0 degrees (in phase—a multiple of 360°). *The system will oscillate at the frequency that satisfies those conditions.* Designing an oscillator amounts to selecting reactive components and incorporating them into circuitry in such a way that the conditions will be satisfied at a predetermined frequency.

To show the dependence of the loop gain $A\beta$ on frequency, we write $A\beta(j\omega)$, a complex phasor that can be expressed in both polar and rectangular form:

$$A\beta(j\omega) = |A\beta|\underline{/\theta} = |A\beta|\cos\theta + j|A\beta|\sin\theta \qquad\qquad (10.2)$$

where $|A\beta|$ is the gain magnitude, a function of frequency, and θ is the phase shift, also a function of frequency. The Barkhausen criterion requires that

$$|A\beta| = 1 \qquad\qquad (10.3)$$

and

$$\theta = \pm360°n \qquad\qquad (10.4)$$

Figure 10.1
Block diagram of an oscillator

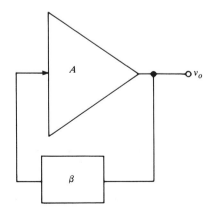

where n is any integer, including 0. In polar and rectangular forms, the Barkhausen criterion is expressed as

$$A\beta(j\omega) = 1\underline{/\pm 360°n} = 1 + j0 \qquad (10.5)$$

Example 10.1

The gain of a certain amplifier as a function of frequency is $A(j\omega) = -16 \times 10^6/j\omega$. A feedback path connected around it has $\beta(j\omega) = 10^3/(2 \times 10^3 + j\omega)^2$. Will the system oscillate? If so, at what frequency?

Solution.   The loop gain is

$$A\beta = \left(\frac{-16 \times 10^6}{j\omega}\right)\left[\frac{10^3}{(2 \times 10^3 + j\omega)^2}\right] = \frac{-16 \times 10^9}{j\omega(2 \times 10^3 + j\omega)^2}$$

To determine if the system will oscillate, we will first determine the frequency, if any, at which the phase angle of $A\beta$ $(\theta = \underline{/A\beta})$ equals 0 or a multiple of 360°. Using phasor algebra, we have

$$\theta = \underline{/A\beta} = \left|\frac{-16 \times 10^9}{\underline{/j\omega(2 \times 10^3 + j\omega)^2}}\right| = \underline{/-16 \times 10^9} + \underline{/1/j\omega} + \left|\frac{1}{\underline{/(2 \times 10^3 + j\omega)^2}}\right|$$

$$= -180° - 90° - 2\arctan(\omega/2 \times 10^3)$$

This expression will equal $-360°$ if $2\arctan(\omega/2 \times 10^3) = 90°$, or

$$\arctan(\omega/2 \times 10^3) = 45°$$
$$\omega/2 \times 10^3 = 1$$
$$\omega = 2 \times 10^3 \text{ rad/s}$$

Thus, the phase shift around the loop is $-360°$ at $\omega = 2000$ rad/s. We must now check to see if the gain magnitude $|A\beta|$ equals 1 at $\omega = 2 \times 10^3$. The gain magnitude is

$$|A\beta| = \left|\frac{-16 \times 10^9}{j\omega(2 \times 10^3 + j\omega)^2}\right| = \frac{|-16 \times 10^9|}{|j\omega| |(2 \times 10^3 + j\omega)|^2}$$

$$= \frac{16 \times 10^9}{\omega[(2 \times 10^3)^2 + \omega^2]}$$

Substituting $\omega = 2 \times 10^3$, we find

$$|A\beta| = \frac{16 \times 10^9}{2 \times 10^3(4 \times 10^6 + 4 \times 10^6)} = 1$$

Thus, the Barkhausen criterion is satisfied at $\omega = 2 \times 10^3$ rad/s and oscillation occurs at that frequency $(2 \times 10^3/2\pi = 318.3$ Hz).

Example 10.1 illustrated an application of the *polar* form of the Barkhausen criterion, since we solved for $\underline{/A\beta}$ and then determined the frequency at which that angle equals $-360°$. It is instructive to demonstrate how the same result can be obtained using the *rectangular* form of the criterion: $A\beta = 1 + j0$. Toward that end, we first expand the denominator:

$$A\beta = \frac{-16 \times 10^9}{j\omega(2 \times 10^3 + j\omega)^2} = \frac{-16 \times 10^9}{j\omega(4 \times 10^6 + j4 \times 10^3\omega - \omega^2)}$$

$$= \frac{-16 \times 10^9}{j\omega[(4 \times 10^6 - \omega^2) + j4 \times 10^3\omega]} = \frac{16 \times 10^9}{4 \times 10^3\omega^2 - j\omega(4 \times 10^6 - \omega^2)}$$

To satisfy the Barkhausen criterion, this expression for $A\beta$ must equal 1. We therefore set it equal to 1 and simplify:

$$1 = \frac{16 \times 10^9}{4 \times 10^3\omega^2 - j\omega(4 \times 10^6 - \omega^2)}$$

$$4 \times 10^3\omega^2 - j\omega(4 \times 10^6 - \omega^2) = 16 \times 10^9$$

$$(4 \times 10^3\omega^2 - 16 \times 10^9) - j\omega(4 \times 10^6 - \omega^2) = 0$$

In order for this expression to equal 0, *both the real and imaginary parts must equal 0*. Setting either part equal to 0 and solving for ω will give us the same result we obtained before:

$$4 \times 10^3\omega^2 - 16 \times 10^9 = 0 \Rightarrow \omega = 2 \times 10^3$$

$$4 \times 10^6 - \omega^2 = 0 \Rightarrow \omega = 2 \times 10^3$$

This result was obtained with somewhat more algebraic effort than previously. In some applications, it is easier to work with the polar form than the rectangular form, and in others, the reverse is true.

10.3   THE RC PHASE-SHIFT OSCILLATOR

One of the simplest kinds of oscillators incorporating an amplifier can be constructed as shown in Figure 10.2. Here we see that an operational amplifier is connected in an inverting configuration and drives three cascaded (high-pass) RC

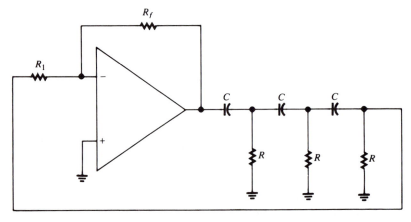

Figure 10.2
An RC phase-shift oscillator

sections. The arrangement is called an *RC phase-shift oscillator*. The inverting amplifier causes a 180° phase shift in the signal passing through it, and the purpose of the cascaded RC sections is to introduce an additional 180° at some frequency. Recall that the output of a single, high-pass RC network leads its input by a phase angle that depends on the signal frequency. When the signal passes through all three RC sections, there will be some frequency at which the cumulative phase shift is 180°. When the signal having that frequency is fed back to the inverting amplifier, as shown in the figure, the total phase shift around the loop will equal 180° + 180° = 360° (or, equivalently, −180° + 180° = 0°) and oscillation will occur at that frequency, provided the loop gain is 1. The gain necessary to overcome the loss in the RC cascade and bring the loop gain up to 1 is supplied by the amplifier ($v_o/v_{in} = -R_f/R_1$).

With considerable algebraic effort, it can be shown (Exercise 10.4) that the feedback ratio determined by the RC cascade (with the feedback connection to R_1 opened) is

$$\beta = \frac{R^3}{(R^3 - 5RX_C^2) + j(X_C^3 - 6R^2X_C)} \tag{10.6}$$

In order for oscillation to occur, the cascade must shift the phase of the signal by 180°, which means the angle of β must be 180°. When the angle of β is 180°, β is a purely real number. In that case, the imaginary part of the denominator of equation 10.6 is 0. Therefore, we can find the oscillation frequency by finding the value of ω that makes the imaginary part equal 0. Setting it equal to 0 and solving for ω, we find

$$X_C^3 - 6R^2X_C = 0$$
$$X_C^3 = 6R^2X_C$$
$$X_C^2 = 6R^2$$
$$\frac{1}{(\omega C)^2} = 6R^2$$
$$\omega = \frac{1}{\sqrt{6}RC} \text{ rad/s}$$

or

$$f = \frac{1}{2\pi\sqrt{6}RC} \text{ Hz} \tag{10.7}$$

Notice that resistor R_1 in Figure 10.2 is effectively in parallel with the rightmost resistor R in the RC cascade, because the inverting input of the amplifier is at virtual ground. Therefore, when the feedback loop is closed by connecting the cascade to R_1, the frequency satisfying the phase criterion will be somewhat different than that predicted by equation 10.7. If $R_1 \gg R$, so that $R_1 \parallel R \approx R$, then equation 10.7 will closely predict the oscillation frequency.

We can find the gain that the amplifier must supply by finding the reduction in gain caused by the RC cascade. This we find by evaluating the magnitude of β at the oscillation frequency: $\omega = 1/(\sqrt{6}RC)$. At that frequency, the imaginary term in equation 10.6 is 0 and β is the real number

$$|\beta| = \frac{R^3}{R^3 - 5RX_C^2} = \frac{R^3}{R^3 - 5R\left(\dfrac{\sqrt{6}RC}{C}\right)^2}$$

$$= \frac{R^3}{R^3 - 30R^3}$$

$$= -1/29$$

(10.8)

The minus sign confirms that the cascade inverts the feedback at the oscillation frequency. We see that the amplifier must supply a gain of -29 to make the loop gain $A\beta = 1$. Thus, we require

$$\frac{R_f}{R_1} = 29$$

(10.9)

In practice, the feedback resistor is made adjustable to allow for small differences in component values and for the loading caused by R_1.

Example 10.2

DESIGN

Design an RC phase-shift oscillator that will oscillate at 100 Hz.

Solution.   From equation 10.7,

$$f = 100 \text{ Hz} = \frac{1}{2\pi\sqrt{6}RC}$$

Let $C = 0.5\mu\text{F}$. Then

$$R = \frac{1}{100(2\pi)\sqrt{6}(0.5 \times 10^{-6})]} = 1300 \ \Omega$$

To prevent R_1 from loading this value of R, we choose $R_1 = 20 \text{ k}\Omega$. (For greater

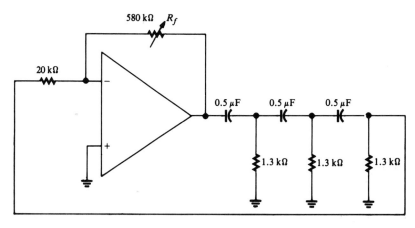

Figure 10.3
(Example 10.2)

precision, we could choose the last R in the cascade and the value R_1 so that $R \parallel R_1 = 1300 \ \Omega$.) From equation 10.9, $R_f = 29R_1 = 29(20 \ \text{k}\Omega) = 580 \ \text{k}\Omega$. The completed circuit is shown in Figure 10.3. R_f is made adjustable so the loop gain can be set precisely to 1.

10.4   THE WIEN-BRIDGE OSCILLATOR

Figure 10.4 shows a widely used type of oscillator called a *Wien bridge*. The operational amplifier is used in a noninverting configuration, and the impedance blocks labeled Z_1 and Z_2 form a voltage divider that determines the feedback ratio. Note that a portion of the output voltage is fed back through this impedance divider to the + input of the amplifier. Resistors R_g and R_f determine the amplifier gain and are selected to make the magnitude of the loop gain equal to 1. If the feedback impedances are chosen properly, there will be some frequency at which there is zero phase shift in the signal fed back to the amplifier input (v^+). Since the amplifier is noninverting, it also contributes zero phase shift, so the total phase shift around the loop is 0 at that frequency, as required for oscillation.

In the most common version of the Wien-bridge oscillator, Z_1 is a series RC combination and Z_2 is a parallel RC combination, as shown in Figure 10.5. For this configuration,

$$Z_1 = R_1 - jX_{C_1}$$

and
$$Z_2 = R_2 \parallel -jX_{C_2} = \frac{-jR_2X_{C_2}}{R_2 - jX_{C_2}}$$

The feedback ratio is then

$$\beta = \frac{v^+}{v_o} = \frac{Z_2}{Z_1 + Z_2} = \frac{-jR_2X_{C_2}/(R_2 - jX_{C_2})}{R_1 - jX_{C_1} - jR_2X_{C_2}/(R_2 - jX_{C_2})} \qquad (10.10)$$

Figure 10.4

The Wien-bridge oscillator. Z_1 and Z_2 determine the feedback ratio to the non-inverting input. R_f and R_g control the magnitude of the loop gain.

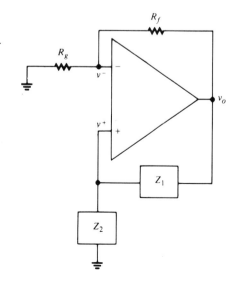

Figure 10.5
The Wien-bridge oscillator showing the
RC networks that form Z_1 and Z_2

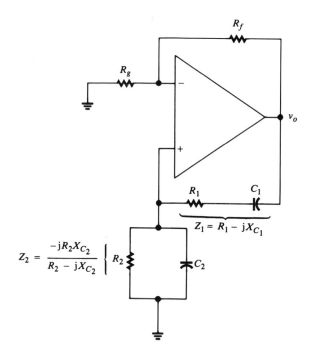

which, upon simplification, becomes

$$\frac{v^+}{v_o} = \frac{R_2 X_{C_2}}{(R_1 X_{C_2} + R_2 X_{C_1} + R_2 X_{C_2}) + j(R_1 R_2 - X_{C_1} X_{C_2})} \qquad (10.11)$$

In order for v^+ to have the same phase as v_o, this ratio must be a purely real number. Therefore, the imaginary term in (10.11) must be 0. Setting the imaginary term equal to 0 and solving for ω gives us the oscillation frequency:

$$R_1 R_2 - X_{C_1} X_{C_2} = 0$$

$$R_1 R_2 = \left(\frac{1}{\omega C_1}\right)\left(\frac{1}{\omega C_2}\right)$$

$$\omega^2 = \frac{1}{R_1 R_2 C_1 C_2}$$

$$\omega = \frac{1}{\sqrt{R_1 R_2 C_1 C_2}} \text{ rad/s} \qquad (10.12)$$

In most applications, the resistors are made equal and so are the capacitors: $R_1 = R_2 = R$ and $C_1 = C_2 = C$. In this case, the oscillation frequency becomes

$$\omega = \frac{1}{\sqrt{R^2 C^2}} = \frac{1}{RC} \text{ rad/s} \qquad (10.13)$$

or

$$f = \frac{1}{2\pi RC} \text{ Hz} \qquad (10.14)$$

When $R_1 = R_2 = R$ and $C_1 = C_2 = C$, the capacitive reactance of each capacitor at the oscillation frequency is

$$X_{C_1} = X_{C_2} = \frac{1}{\omega C} = \frac{1}{\left(\dfrac{1}{RC}\right)C} = R$$

Substituting $X_{C_1} = X_{C_2} = R = R_1 = R_2$ in equation 10.11, we find that the feedback ratio at the oscillation frequency is

$$\frac{v^+}{v_o} = \frac{R^2}{3R^2 + j0} = \frac{1}{3}$$

Therefore, the amplifier must provide a gain of 3 to make the magnitude of the loop gain unity and sustain oscillation. Since the amplifier gain is $(R_g + R_f)/R_g$, we require

$$\frac{R_g + R_f}{R_g} = 3, \text{ or}$$

$$1 + \frac{R_f}{R_g} = 3 \Rightarrow \frac{R_f}{R_g} = 2 \qquad\qquad \textbf{(10.15)}$$

(Another way of reaching this same conclusion is to recognize that the operational amplifier maintains $v^+ \approx v^-$. But, from Figure 10.5, we see that $v^- = v_o R_g/(R_g + R_f)$, so $v^-/v_o = (R_g + R_f)/R_g = v^+/v_o = 1/3$.)

Example 10.3

DESIGN

Design a Wien-bridge oscillator that oscillates at 25 kHz.

Solution.   Let $C_1 = C_2 = 0.001\ \mu\text{F}$. Then, from equation 10.14,

$$f = 25 \times 10^3 \text{ Hz} = \frac{1}{2\pi R(10^{-9}\ \text{F})}$$

$$R = \frac{1}{2\pi(25 \times 10^3 \text{ Hz})(10^{-9}\ \text{F})} = 6366\ \Omega$$

Let $R_g = 10\ \text{k}\Omega$. Then, from equation 10.15,

$$\frac{R_f}{R_g} = 2 \Rightarrow R_f = 20\ \text{k}\Omega$$

In practical Wien-bridge oscillators, R_f is not equal to exactly $2R_g$ because component tolerances prevent R_1 from being exactly equal to R_2 and C_1 from being exactly equal to C_2. Furthermore, the amplifier is not ideal, so v^- is not exactly equal to v^+. In a circuit constructed for laboratory experimentation, R_f should be made adjustable so that the loop gain can be set as necessary to sustain oscillation. Practical oscillators incorporate a nonlinear device in the R_f–R_g feedback loop to provide automatic adjustment of the loop gain, as necessary to sustain oscillation. This arrangement is a form of *automatic gain control* (AGC), whereby a reduction in signal level changes the resistance of the nonlinear device in a way that restores gain.

10.5   THE COLPITTS OSCILLATOR

In the Colpitts oscillator, the impedance in the feedback circuit is a resonant LC network. See Figure 10.6(a). The frequency of oscillation is the resonant frequency of the network, which is the frequency at which the phase shift through the network is 180°. At that frequency, the impedance is a real number. Figure 10.6(b) shows the feedback network. The impedance looking into the network from the amplifier output is the parallel combination of $-jX_{C_1}$ and $(jX_L - jX_{C_2})$:

$$Z = \frac{(-jX_{C_1})(jX_L - jX_{C_2})}{-jX_{C_1} + jX_L - jX_{C_2}} = \frac{X_L C_1 - X_{C_1}X_{C_2}}{j(X_L - X_{C_1} - X_{C_2})}$$

In order for Z to equal a real number, the imaginary term must equal 0. Thus, at

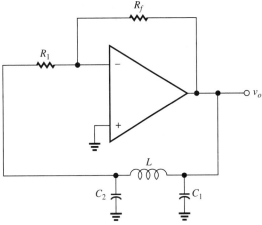

(a) The Colpitts oscillator

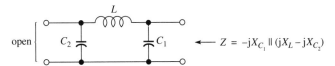

(b) The feedback network, whose impedance is derived under the assumption that the feedback path is open (valid for $R_1 \gg X_{C_2}$).

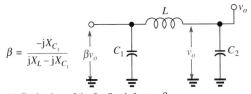

(c) Derivation of the feedback factor β.

Figure 10.6
The Colpitts oscillator

resonance,

$$j(X_L - X_{C_1} - X_{C_2}) = 0 \Rightarrow X_L = X_{C_1} + X_{C_2}$$

or

$$\omega L = \frac{1}{\omega C_1} + \frac{1}{\omega C_2}$$

Solving for ω gives the oscillation frequency:

$$\omega = \frac{1}{\sqrt{LC_T}} \text{ radians/second} \quad \text{or} \quad f = \frac{1}{2\pi\sqrt{LC_T}} \text{ hertz} \qquad \textbf{(10.16)}$$

where
$$C_T = \frac{C_1 C_2}{C_1 + C_2}$$

Note that this computation is based on the assumption that the left-hand side of the network is open, as can be seen in Figure 10.6(b). In reality, it is loaded by input resistor R_1 (Figure 10.6(a)). Therefore, equation 10.16 is an approximation, but it is valid when $R_1 \gg X_{C_2}$.

Since v_0 appears across the voltage divider consisting of L and C_2 (see Figure 10.6(c)), the feedback factor (again neglecting R_1) is

$$\beta = \frac{-jX_{C_2}}{jX_L - jX_{C_2}} \qquad \textbf{(10.17)}$$

Since $X_L = X_{C_1} + X_{C_2}$ at resonance, the feedback factor at resonance is

$$\beta = \frac{-jX_{C_2}}{j(X_{C_1} + X_{C_2}) - jX_{C_2}} = \frac{-X_{C_2}}{X_{C_1}} = \frac{-C_1}{C_2} \qquad \textbf{(10.18)}$$

Note that β is real and has angle 180°, as required. In order for the loop gain to equal 1, we require

$$|A_v \beta| = 1 \Rightarrow |A_v| = \frac{1}{|\beta|} = \frac{C_2}{C_1} \qquad \textbf{(10.19)}$$

Thus, the closed-loop gain of the inverting amplifier must be at least C_2/C_1. In practice, $|A_v|$ is adjusted to be greater than C_2/C_1, but not so much greater that distortion results.

Example 10.4

DESIGN

Design a Colpitts oscillator that will oscillate at 100 kHz.

Solution.   Let us choose $C_1 = C_2 = 0.01 \ \mu F$. Then

$$C_T = \frac{C_1 C_2}{C_1 + C_2} = \frac{(0.01 \ \mu F)(0.01 \ \mu F)}{0.01 \ \mu F + 0.01 \ \mu F} = 0.005 \ \mu F$$

From equation 10.16,

$$L = \frac{1}{(2\pi)^2 f^2 C_T} = \frac{1}{(2\pi)^2 (100 \times 10^3 \text{ Hz})^2 (0.005 \times 10^{-6} \text{ F})} \approx 0.5 \text{ mH}$$

We wish to make $R_1 \gg X_{C_2}$ at resonance:

$$X_{C_2} = \frac{1}{\omega C_2} = \frac{1}{2\pi(100 \text{ kHz})(0.01 \ \mu\text{F})} = 159 \ \Omega$$

Choose $R_1 = 10 \text{ k}\Omega$. Then, since $|A_v| = R_f/R_1$ and we require that $|A_v| \geq C_2/C_1 = 1$, we must choose $R_f \geq 10 \text{ k}\Omega$.

10.6   THE HARTLEY OSCILLATOR

The impedance in the feedback of a Hartley oscillator is a resonant LC network consisting of two inductors and one capacitor. See Figure 10.7. As in the Colpitts oscillator, oscillation occurs at the resonant frequency of the network, the frequency at which its impedance is real. Following the same procedure used to derive the oscillation frequency and feedback factor of the Colpitts oscillator, we can show (Exercise 10.9) that

$$f = \frac{1}{2\pi\sqrt{CL_T}} \text{ hertz} \qquad (10.20)$$

where $L_T = L_1 + L_2$ and

$$\beta = \frac{-L_2}{L_1} \qquad \text{(at resonance)} \qquad (10.21)$$

If L_1 and L_2 are wound on the same core and therefore have mutual inductance M, then L_T in equation 10.20 is $L_T = L_1 + L_2 + 2M$. In order to satisfy $|A_v\beta| \geq 1$ in the Hartley oscillator, we require, from equation 10.21,

$$|A_v| = \frac{R_f}{R_1} \geq \frac{L_1}{L_2} \qquad (10.22)$$

Figure 10.7
The Hartley oscillator

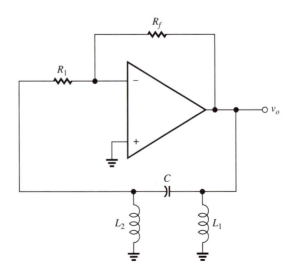

10.7   RELAXATION OSCILLATORS

A *relaxation* oscillator is an oscillator that produces a waveform whose frequency is determined by the charging and discharging of a capacitor (or inductor) through a resistor. The waveform is typically rectangular or sawtooth (or a somewhat distorted version of one of these), rather than sinusoidal. The general principle governing the operation of a relaxation oscillator is that the reactive component, such as a capacitor, charges until its voltage is large enough to dramatically alter the characteristics of an electronic device to which it is connected. The capacitor voltage is said to *trigger* the electronic device into an alternate state. The altered state of the electronic device allows the capacitor to *discharge*. As a result, the capacitor voltage falls until the voltage is small enough to trigger the electronic device back into its original state, at which time the capacitor begins to recharge. The continually alternating charge and discharge cycles determine the period, and hence the frequency, of the oscillation.

Figure 10.8 illustrates the concept of one type of relaxation oscillator. In this case, the input to the electronic device is initially an open circuit (or large resistance) that allows the capacitor to charge through resistance R. When the capacitor voltage becomes sufficiently large, the electronic device is triggered into a state where its input becomes conductive, thus allowing the capacitor to discharge. The capacitor voltage falls until the electronic device reverts to its high-resistance input state, once again allowing the capacitor to charge. Each time the electronic device changes state, its output switches from a low voltage to a high voltage or vice versa. Consequently, the output waveform is rectangular and has a period equal to the sum of the capacitor charge and discharge times.

An example of an electronic device that can be used in the construction of a relaxation oscillator is a *voltage comparator*. Voltage comparators are discussed in Chapter 11, and an example of a relaxation oscillator is given there.

The 8038 Integrated-Circuit Function Generator

Function generators capable of producing sinusoidal, triangular, and rectangular waveforms are commercially available in integrated-circuit form. An example is the 8038, manufactured by Intersil as the ICL8038 and also available from other

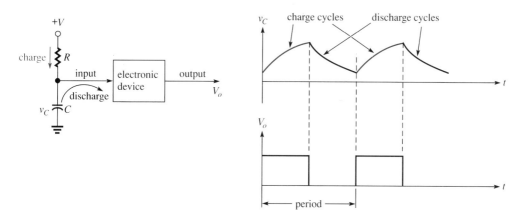

Figure 10.8
Illustration of the concept of a relaxation oscillator

manufacturers. This versatile circuit is capable of generating the aforementioned waveforms simultaneously (at three separate output terminals) over a frequency range from 0.0001 Hz to 1 MHz. Frequency is determined by externally connected resistor–capacitor combinations and can also be controlled by an external voltage. In the latter mode of operation, the generator serves as a *voltage-controlled oscillator* (VCO). (A VCO is also called a *voltage-to-frequency converter*.)

The 8038 is basically a relaxation oscillator that generates a triangular waveform. The triangular waveform is converted internally to a rectangular waveform using voltage comparators and a digital-type storage device (flip-flop). Sixteen internal transistors are used to shape the triangular wave into an approximation of a sine wave. Under ideal conditions (using external circuit connections specified in manufacturer's literature), the total harmonic distortion of the sine-wave output can be reduced to less than 1%.

Figure 10.9(a) shows a pin diagram of the 8038 and identifies the function of each pin. Pins 1 and 12, "sine wave adjust," are used for connecting external resistors to minimize sine-wave distortion, as previously mentioned. Pins 4 and 5, "duty cycle and frequency adjust," are used for connecting exteral resistors that, in conjunction with an external capacitor connected to pin 10, "timing capacitor," determine the *duty cycle* and frequency of the outputs. The duty cycle of an

Figure 10.9
The 8038 function generator

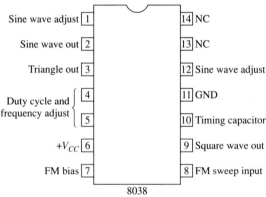

(a) Pin diagram (NC = no connection)

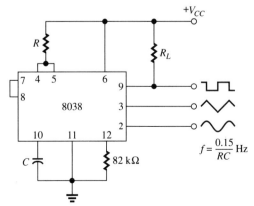

$$f = \frac{0.15}{RC} \text{ Hz}$$

(b) Connection for operation as a 50%–duty-cycle, fixed-frequency function generator

output is the time it is high expressed as a percent of the period of oscillation. For example, a square wave, which is high for one-half of a period and low for the other half, has a 50% duty cycle. Pin 8, "FM sweep input," is the input to which an external voltage may be connected to adjust the frequency of oscillation when the 8038 is operated as a VCO. (FM refers to *frequency modulation*: As the input voltage is adjusted, the output frequency changes in direct proportion to the change in voltage.)

Figure 10.9(b) shows the simplest possible circuit connections for using the 8038 as a fixed-frequency, 50%–duty-cycle function generator. In the configuration shown, the frequency of oscillation is

$$f = \frac{0.15}{RC} \text{ Hz} \tag{10.23}$$

Manufacturer's product literature can be consulted for external circuit connections required when the 8038 is used as a VCO and for duty cycles other than 50%.

EXERCISES

Section 10.2   The Barkhausen Criterion

10.1   The gain of an amplifier as a function of frequency is $A(j\omega) = -2.56 \times 10^8/j\omega$ and the feedback factor is $\beta(j\omega) = 500/(4 \times 10^3 + j\omega)^2$. Determine if the amplifier will oscillate and, if so, the frequency of oscillation.

10.2   The gain of an amplifier as a function of frequency is $-K/j\omega$ and the feedback factor is $10^3/(10^3 + j\omega)^2$. What should be the value of K if the amplifier is to oscillate at $\omega = 10^3$ rad/s?

Section 10.3 The RC Phase-shift Oscillator

10.3   Design an RC phase-shift oscillator that will oscillate at 1.5 kHz.

10.4   Figure 10.10 shows the cascaded RC sections that form the feedback network for the RC phase-shift oscillator. Show that the feedback ratio is

$$\beta = \frac{v_o}{v_{in}} = \frac{R^3}{(R^3 - 5RX_C^2) + j(X_C^3 - 6R^2X_C)}$$

(*Hint:* Write three loop equations and solve for i_3, the current in the rightmost loop. Then $v_o = i_3R$. Solve this equation for v_o/v_{in}.)

Section 10.4   The Wien-bridge Oscillator

10.5   Design a Wien-bridge oscillator that oscillates at 180 kHz.

10.6   The Wien-bridge oscillator shown in Figure 10.5 has $C_1 = C_2 = 0.001 \ \mu F$ and $R_1 =$

10 kΩ. It is desired to make the oscillation frequency variable over the range from 10 kHz to 50 kHz by making R_2 adjustable. Through what range of resistance values should it be possible to adjust R_2?

Section 10.5   The Colpitts Oscillator

10.7   A Colpitts oscillator has $C_1 = 1000$ pF, $C_2 = 2200$ pF, and $L = 2.5$ mH.
  a. At what frequency does it oscillate?
  b. If $R_1 = 10$ kΩ, what should be the minimum value of R_f?

10.8   Design a Colpitts oscillator that oscillates at 15 kHz.

Section 10.6   The Hartley Oscillator

10.9   Derive equations 10.20 and 10.21 for the oscillation frequency and feedback factor of the Hartley oscillator.

10.10   Design a Hartley oscillator that oscillates at

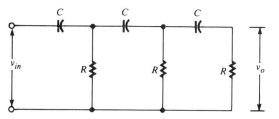

Figure 10.10
(*Exercise 10.4*)

20 kHz. Assume the inductors are wound on separate cores and have no mutual inductance.

Section 10.7   Relaxation Oscillators

10.11   The output of the electronic device in Figure 10.8 switches from low to high when its input voltage rises to 6 V and switches from high to low when its input falls to 2 V. If the capacitor charges from 2 V to 6 V in 0.4 ms and discharges from 6 V to 2 V in 0.1 ms, what is the frequency of the waveform produced by the oscillator?

10.12   A 0.1-μF timing capacitor is to be used with an 8038 function generator to provide output frequencies in the range from 200 Hz to 10 kHz. Through what range should external timing resistor R be adjustable?

Special-Purpose Circuits

11.1   VOLTAGE COMPARATORS

As the name implies, a *voltage comparator* is a device used to compare two voltage levels. The output of the comparator reveals which of its two inputs is larger, so it is basically a switching device, producing a high output when one input is the larger, and switching to a low output if the other input becomes larger. An operational amplifier is used as a voltage comparator by operating it open-loop (no feedback) and by connecting the two voltages to be compared to the inverting and noninverting inputs. Because it has a very large open-loop gain, the amplifier's output is driven all the way to one of its output voltage limits when there is a very small difference between the input levels. For example, if the + input voltage is slightly greater than the − input voltage, the amplifier quickly switches to its maximum positive output, and when the − input voltage is slightly greater than the + input voltage, the amplifier switches to its maximum negative output.* This behavior is illustrated in Figure 11.1.

Note that Figure 11.1(b) is a transfer characteristic showing output voltage versus *differential* input voltage, $v^+ - v^-$. It can be seen that the output switches when $v^+ - v^-$ passes through 0.

To further clarify this behavior of the comparator, Figure 11.2 shows the output waveform when the noninverting input is a 10-V-peak sine wave and a +6-V-dc source is connected to the inverting input. The comparator output is assumed to switch between ±15 V. Notice that the output switches to +15 V each time the sine wave rises through +6 V, because $v^+ - v^- = (6\ \text{V}) - (6\ \text{V}) = 0\ \text{V}$ at those points in time. The output remains high so long as $v^+ - v^- > 0$, i.e., $v^+ > 6$ V, and when v^+ falls below 6 V, the comparator output switches to −15 V. As an exercise, plot the output when the sine wave is connected to the inverting input and the +6-V-dc source is connected to the noninverting input.

---

*Many authors use the term *saturation voltage* to describe the minimum or maximum output voltage of the amplifier. We are avoiding this usage only to prevent confusion with the saturation voltage of a transistor, which, as we have seen, is near 0.

Figure 11.1
The operational amplifier used as a voltage comparator

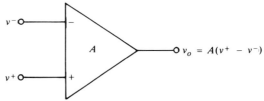

(a) Open-loop operation as a voltage comparator

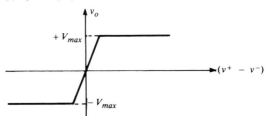

(b) Transfer characteristic of the voltage comparator. When $v+ > v-$, the output is at its maximum positive limit, and when $v+ < v-$, the output switches to its maximum negative limit.

In some applications, either the inverting or noninverting input is grounded, so the comparator is effectively a zero-crossing detector. It switches output states when the ungrounded input passes through 0. For example, if the inverting input is grounded, the output switches to its maximum positive voltage when v^+ is slightly positive and to its maximum negative voltage when v^+ is slightly negative. The reverse action occurs if the noninverting input is grounded. The transfer characteristics for these two cases are shown in Figure 11.3.

In the context of a voltage comparator, the *input offset voltage* is defined to be the *minimum* differential input voltage that will cause the output to switch from

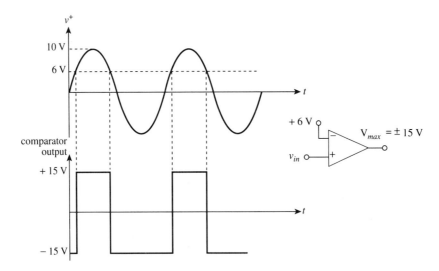

Figure 11.2
The comparator output switches to $+V_{max}$ when $v^+ - v^- > 0$ V, which corresponds to the time points where v^+ rises through $+6$ V. The output remains high so long as $v^+ - v^- > 0$, or $v^+ > 6$ V.

Figure 11.3
Operation of the voltage comparator as a zero-level detector

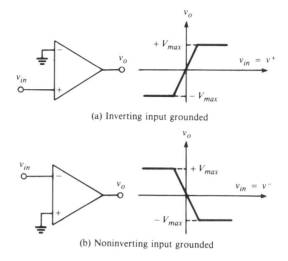

(a) Inverting input grounded

(b) Noninverting input grounded

one state to the other. The smaller the input offset voltage, the more accurate the voltage comparator in terms of its ability to detect the equality of two input levels. Clearly, the greater the open-loop gain, the smaller the input offset voltage. For example, a gain of 20,000 will cause the output to switch from -10 V to $+10$ V when $v^+ - v^-$ is (20 V)/20,000 = 1 mV.

Two other important characteristics of a voltage comparator are its *response time* and *rise time* illustrated in Figure 11.4. The response time is the delay between the time a step input is applied and the time the output begins to change state. It is measured from the edge of the step input to the time point where the output reaches a fixed percentage of its final value, such as 10% of $+V_{max}$. (For clarity, Figure 11.4 shows the output switching from 0 V toward $+V_{max}$; in many

Figure 11.4
Response time and rise time of a voltage comparator

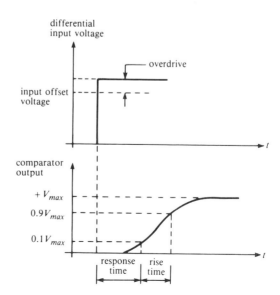

applications, one of the output levels actually is 0 V.) Response time is strongly dependent on the amount of *overdrive* in the input: the voltage in excess of that required to cause switching to occur. The greater the overdrive, the shorter the response time. Rise time is defined in the usual way: the time required for the output to change from 10% of its final value to 90% of its final value. Recall that rise time is inversely proportional to amplifier bandwidth: $t_r \approx 0.35/\text{BW}$. Note this important point: A large voltage gain improves the input offset voltage (reduces it), but lengthens the rise time, because large gains mean smaller bandwidths, the gain–bandwidth product being constant.

Although general-purpose operational amplifiers can be, and are, used as voltage comparators in the way we have described, there are also more elaborate, specially designed operational amplifiers manufactured and marketed specifically for voltage-comparator applications. One feature of some comparators is their ability to switch between output levels that are not necessarily related to the amplifier supply voltages. These are useful in digital systems where *level shifting* is necessary to interface logic circuitry of different types. For example, one part of a digital system may be designed to operate with logic levels of "one" (high) = +5 V and "zero" (low) = 0 V, while another part of the system uses "one" = 0 V and "zero" = −10 V. A level-shifting voltage comparator can be used to make these components compatible.

Hysteresis and Schmitt Triggers

In its most general sense, *hysteresis* is a property that means a device behaves differently when its input is increasing from the way it behaves when its input is decreasing. In the context of a voltage comparator, hysteresis means that the output will switch when the input increases to one level but will not switch back until the input falls below a *different* level. In some applications, hysteresis is a desirable characteristic because it prevents the comparator from switching back and forth in response to random noise fluctuations in the input. For example, if $v^+ - v^-$ is near 0 V, and if the input offset voltage is 1 mV, then noise voltages on the order of 1 mV will cause random switching of the comparator output. On the other hand, if the output will switch to one state only when the input rises past −1 V, and will switch to the other state only when the input falls below +1 V, then only a very large (2-V) noise voltage will cause it to switch states when the input is in the vicinity of one of these "trigger" points.

Figure 11.5(a) shows how hysteresis can be introduced into comparator operation. In this case, the input is connected to the inverting terminal and a voltage divider is connected across the noninverting terminal between v_o and a fixed reference voltage V_{REF} (which may be 0). Figure 11.5(b) shows the resulting transfer characteristic (called a *hysteresis loop*). This characteristic shows that the output switches to $+V_{max}$ when v_{in} *falls* below a lower trigger level (LTL), but will not switch to $-V_{max}$ unless v_{in} *rises* past an upper trigger level (UTL). The arrows indicate the portions of the characteristic followed when the input is increasing (upper line) and when it is decreasing (lower line). A comparator having this characteristic is called a *Schmitt trigger*.

We can derive expressions for UTL and LTL using the superposition principle. Suppose first that the comparator output is shorted to ground. Then

$$v^+ = \frac{R_2}{R_1 + R_2} V_{REF} \qquad (v_o = 0) \qquad (11.1)$$

Figure 11.5
A voltage comparator with hysteresis (Schmitt trigger)

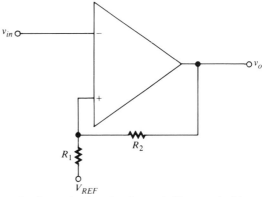

(a) Circuitry used to introduce hysteresis (V_{REF} may be 0.)

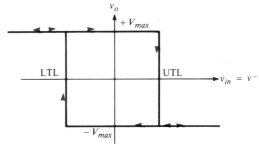

(b) Transfer characteristic of (a). The arrowheads show the portions of the characteristic followed when v_{in} is increasing (arrows pointing right) and when v_{in} is decreasing (arrows pointing left). Double-headed arrows mean that the output remains on that portion of the characteristic whether v_{in} is increasing or decreasing.

When V_{REF} is 0, we find

$$v^+ = \frac{R_1}{R_1 + R_2} v_o \qquad (V_{REF} = 0) \tag{11.2}$$

Therefore, when the output is at its negative limit ($v_o = -V_{max}$),

$$v^+ = \frac{R_2}{R_1 + R_2} V_{REF} + \frac{R_1}{R_1 + R_2} (-V_{max}) \tag{11.3}$$

As can be seen in Figure 11.5(b), v^- must *fall* to this value of v^+ before the comparator switches to $+V_{max}$. Therefore,

$$\text{LTL} = \frac{R_2}{R_1 + R_2} V_{REF} + \frac{R_1}{R_1 + R_2} (-V_{max}) \tag{11.4}$$

Similarly, when $v_o = +V_{max}$, v_{in} must *rise* to

$$\text{UTL} = \frac{R_2}{R_1 + R_2} V_{REF} + \frac{R_1}{R_1 + R_2} (+V_{max}) \tag{11.5}$$

In these equations, $+V_{max}$ is the maximum positive output voltage (a positive number) and $-V_{max}$ is the maximum negative output voltage (a negative number).

The magnitudes of these quantities may be different; for example, $+V_{max} = +10$ V and $-V_{max} = -5$ V.

Quantitatively, the hysteresis of a Schmitt trigger is defined to be the difference between the input trigger levels. From equations 11.4 and 11.5,

$$\text{hysteresis} = \text{UTL} - \text{LTL}$$

$$= \left(\frac{R_1}{R_1 + R_2}\right)(+V_{max}) - \left(\frac{R_1}{R_1 + R_2}\right)(-V_{max}) \qquad (11.6)$$

If the magnitudes of the maximum output voltages are equal, we have

$$\text{hysteresis} = \frac{2R_1 V_{max}}{R_1 + R_2} \qquad (11.7)$$

Example 11.1

1. Find the upper and lower trigger levels and the hysteresis of the Schmitt trigger shown in Figure 11.6. Sketch the hysteresis loop. The output switches between ± 15 V.
2. Repeat (1) if $V_{REF} = 0$ V.
3. Repeat (1) if $V_{REF} = 0$ V and the output switches between 0 V and $+15$ V.

Solution

1. From equations 11.4 and 11.5

$$\text{LTL} = \left[\frac{10\ k\Omega}{(5\ k\Omega) + (10\ k\Omega)}\right](6\ V) + \left[\frac{5\ k\Omega}{(5\ k\Omega) + (10\ k\Omega)}\right](-15\ V) = -1\ V$$

$$\text{UTL} = \left[\frac{10\ k\Omega}{(5\ k\Omega) + (10\ k\Omega)}\right](6\ V) + \left[\frac{5\ k\Omega}{(5\ k\Omega) + (10\ k\Omega)}\right](15\ V) = +9\ V$$

$$\text{hysteresis} = (9\ V) - (-1\ V) = 10\ V$$

(Note also, from equation 11.7, hysteresis = 2(5 kΩ)(15 V)/(15 kΩ) = 10 V.)

Figure 11.6
(Example 11.1)

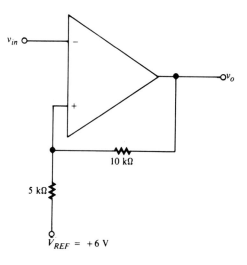

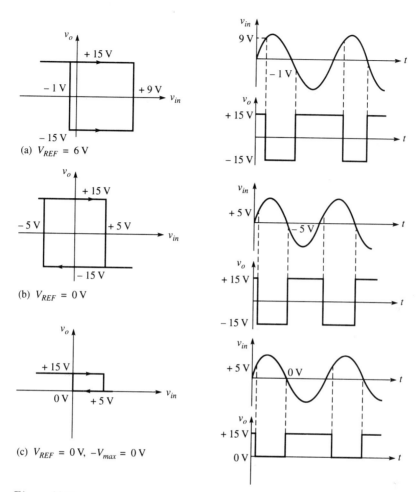

(a) $V_{REF} = 6$ V

(b) $V_{REF} = 0$ V

(c) $V_{REF} = 0$ V, $-V_{max} = 0$ V

Figure 11.7
(Example 11.1)

2. Since $V_{REF} = 0$,

$$\text{LTL} = \left(\frac{R_1}{R_1 + R_2}\right)(-V_{max}) = \left[\frac{5 \text{ k}\Omega}{(5 \text{ k}\Omega) + (10 \text{ k}\Omega)}\right](-15 \text{ V}) = -5 \text{ V}$$

$$\text{UTL} = \left(\frac{R_1}{R_1 + R_2}\right)(+V_{max}) = \left[\frac{5 \text{ k}\Omega}{(5 \text{ k}\Omega) + (10 \text{ k}\Omega)}\right](15 \text{ V}) = +5 \text{ V}$$

$$\text{hysteresis} = (5 \text{ V}) - (-5 \text{ V}) = 10 \text{ V}$$

(Again, equation 11.7 may be used.)

3. Since the output switches between 0 V and +15 V, we must use 0 in place of $-V_{max}$ in the trigger-level equations:

$$\text{LTL} = \left(\frac{R_1}{R_1 + R_2}\right)(-V_{max}) = \left[\frac{5 \text{ k}\Omega}{(5 \text{ k}\Omega) + (10 \text{ k}\Omega)}\right]0 = 0 \text{ V}$$

$$\text{UTL} = \left(\frac{R_1}{R_1 + R_2}\right)(+V_{max}) = +5 \text{ V}$$

$$\text{hysteresis} = (5 \text{ V}) - (0 \text{ V}) = 5 \text{ V}$$

(Note that equation 11.7 is not applicable in this case.)

Figure 11.7 shows the hysteresis loops for these cases, along with the output waveforms that result when v_{in} is a 10-V-peak sine wave.

The comparator we have discussed is called an *inverting* Schmitt trigger because the output is high when the input is low, and vice versa, as can be seen in Figure 11.7. Figure 11.8 shows a noninverting Schmitt trigger. For this circuit, the lower and upper trigger levels are

$$\text{LTL} = \frac{-R_1}{R_2}(+V_{max}) \tag{11.8}$$

$$\text{UTL} = \frac{R_1}{R_2}\left|-V_{max}\right| \tag{11.9}$$

Notice that these equations permit the magnitudes of $+V_{max}$ and $-V_{max}$ to be different values. For example, if $R_1 = 10 \text{ k}\Omega$ and $R_2 = 20 \text{ k}\Omega$, and if the output switches between $+10$ V and -5 V, then LTL $= -(0.5)(10 \text{ V}) = -5$ V and UTL $= 0.5 \left|-5 \text{ V}\right| = +2.5$ V. The derivation of equations 11.8 and 11.9 is an exercise at the end of this chapter.

An Astable Multivibrator

The word *astable* means "unstable," and, like other unstable devices, an astable multivibrator is a (square-wave) oscillator. (A *bi*stable multivibrator, also called a *flip-flop,* is a digital device with two stable stages; a *mono*stable multivibrator has one stable state, and an astable multivibrator has zero stable states.) An astable multivibrator can be constructed by using an operational amplifier as a voltage comparator in a circuit like that shown in Figure 11.9. This circuit is an example of a relaxation oscillator, as discussed in Chapter 10.

For analysis purposes, let us assume that the output voltages of the comparator are equal in magnitude and opposite in polarity: $\pm V_{max}$. Figure 11.10 shows the voltage across capacitor C and the output waveform produced by the comparator. We begin by assuming that the output is at $+V_{max}$. Then, the voltage fed back to

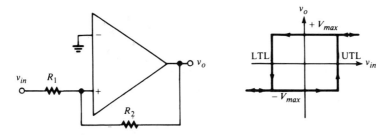

Figure 11.8
The noninverting Schmitt trigger

Figure 11.9
An astable multivibrator

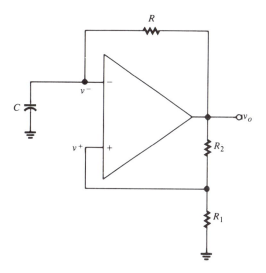

the noninverting input is

$$v^+ = \frac{R_1}{R_1 + R_2}V_{max} = +\beta V_{max} \qquad (11.10)$$

Notice that v^- equals the voltage across the capacitor. The capacitor begins to charge through R *toward* a final voltage of $+V_{max}$. However, as soon as the capacitor voltage reaches a voltage equal to v^+, the comparator switches state. In other words, switching occurs at the point in time where $v^- = v^+ = +\beta V_{max}$. After the comparator switches state, its output is $-V_{max}$, and the voltage fed back to the

Figure 11.10
The waveforms on the capacitor and at the output of the astable multivibrator shown in Figure 11.9

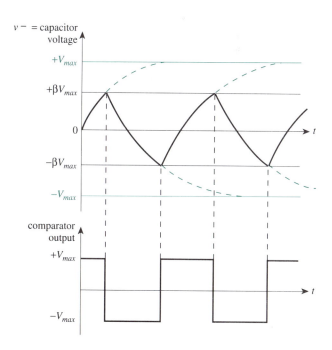

noninverting input becomes

$$v^+ = \frac{R_1}{R_1 + R_2}(-V_{max}) = -\beta V_{max} \qquad (11.11)$$

Since the comparator output is now negative, the capacitor begins to discharge through R toward $-V_{max}$. But, when that voltage falls to $-\beta V_{max}$, we once again have $v^+ = v^-$, and the comparator switches back to $+V_{max}$. This cycle repeats continuously, as shown in Figure 11.10, with the result that the output is a square wave that alternates between $\pm V_{max}$ volts.

It can be shown that the period of the multivibrator oscillation is

$$T = 2RC \ln\left(\frac{1 + \beta}{1 - \beta}\right) \text{ seconds} \qquad (11.12)$$

11.2   CLIPPING AND RECTIFYING CIRCUITS

Clipping Circuits

In Chapter 5, we referred to *clipping* as the undesirable result of overdriving an amplifier. We have seen that any attempt to push an output voltage beyond the limits through which it can "swing" causes the tops and/or bottoms of a waveform to be "clipped" off, resulting in distortion. However, in numerous practical applications, including waveshaping and nonlinear function generation, waveforms are *intentionally* clipped.

Figure 11.11 shows how the transfer characteristic of a device is modified to reflect the fact that its output is clipped at certain levels. In each of the examples shown, note that the characteristic becomes horizontal at the output level where clipping occurs. The horizontal line means that the output remains constant regardless of the input level in that region. Outside of the clipping region, the transfer characteristic is simply a line whose slope equals the gain of the device. This is the region of normal, *linear* operation. In these examples, the devices are assumed to have unity gain, so the slope of each line in the linear region is 1.

Figure 11.12 illustrates a somewhat different kind of clipping action. Instead of the positive or negative peaks being "chopped off," the output follows the input when the signal is above or below a certain level. The transfer characteristics show that linear operation occurs only when certain signal levels are reached and that the output remains constant below those levels. This form of clipping can also be thought of as a special case of that shown in Figure 11.11. Imagine, for example, that the clipping level in Figure 11.11(b) is raised to a positive value; then the result is the same as Figure 11.12(a).

Clipping can be accomplished using *biased diodes,* a technique that is more efficient than overdriving an amplifier. Clipping circuits rely on the fact that diodes have very low impedances when they are forward biased and are essentially open circuits when reverse biased. If a certain point in a circuit, such as the output of an amplifier, is connected through a very small impedance to a *constant* voltage, then the voltage at the circuit point cannot differ significantly from the constant voltage. We say in this case that the point is *clamped* to the fixed voltage. (However, we will reserve the term *clamping circuit* for a special application to be discussed later.) An ideal, forward-biased diode is like a closed switch, so if it is connected between a point in a circuit and a fixed voltage source, the diode very effectively holds the point to the fixed voltage. Diodes can be connected in operational-amplifier circuits, as well as other circuits, in such a way that they become

Figure 11.11
Waveforms and transfer characteristics
of clipping circuits

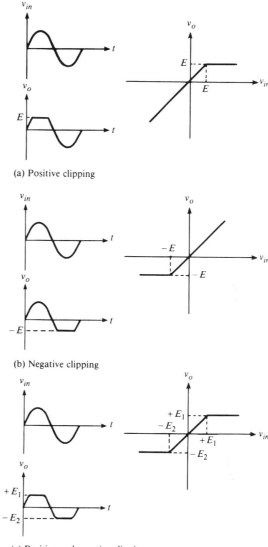

(a) Positive clipping

(b) Negative clipping

(c) Positive and negative clipping

forward biased when a signal reaches a certain voltage. When the forward-biasing level is reached, the diode serves to hold the output to a fixed voltage and thereby establishes a clipping level.

A biased diode is simply a diode connected to a fixed voltage source. The value and polarity of the voltage source determine what value of total voltage across the combination is necessary to forward bias the diode. Figure 11.13 shows several examples. (In practice, a series resistor would be connected in each circuit to limit current flow when the diode is forward biased.) In each part of the figure, we can write Kirchhoff's voltage law around the loop to determine the value of input voltage v_i that is necessary to forward bias the diode. Assuming that the diodes are ideal (neglecting their forward voltage drops), we determine the value of v_i necessary to forward bias each diode by determining the value of v_i necessary

Figure 11.12
*Another form of clipping. Compare with
Figure 11.11.*

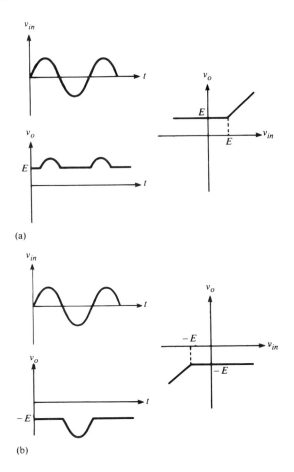

to make $V_D > 0$. Whenever v_i reaches the voltage necessary to make $V_D > 0$, the diode becomes forward biased and the signal source is forced to, or held at, the dc source voltage. If the forward voltage drop across the diode is not neglected, the clipping level is found by determining the value of v_i necessary to make V_D greater than that forward drop (e.g., $V_D > 0.7$ V for a silicon diode). Although these conditions can be determined through formal application of Kirchhoff's voltage law, as shown in the figure, the reader is urged to develop a mental image of circuit behavior in each case. For example, in (a), think of the diode as being reverse biased by 6 V, so the input must "overcome" that reverse bias by reaching +6 V to forward bias the diode.

Figure 11.14 shows three examples of clipping circuits using ideal biased diodes and the waveforms that result when each is driven by a sine-wave input. In each case, note that the output equals the dc source voltage when the input reaches the value necessary to forward bias the diode. Note also that the type of clipping we showed in Figure 11.11 occurs when the fixed bias voltage tends to *reverse* bias the diode, and the type shown in Figure 11.12 occurs when the fixed voltage tends to *forward* bias the diode. When the diode is reverse biased by the input signal, it is like an open circuit that disconnects the dc source, and the output follows the input. These circuits are called *parallel* clippers because the biased diode is in parallel with the output. Although the circuits behave the

Figure 11.13
Examples of biased diodes and the signal voltages v_i required to forward bias them. (Ideal diodes are assumed.) In each case, we solve for the value of v_i that is necessary to make $V_D > 0$.

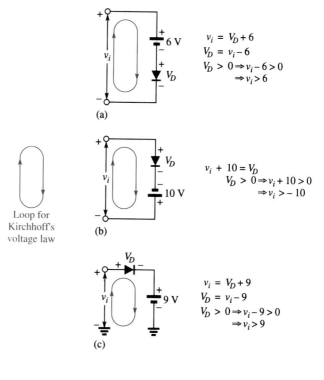

$v_i = V_D + 6$
$V_D = v_i - 6$
$V_D > 0 \Rightarrow v_i - 6 > 0$
$\Rightarrow v_i > 6$

(a)

Loop for
Kirchhoff's
voltage law

$v_i + 10 = V_D$
$V_D > 0 \Rightarrow v_i + 10 > 0$
$\Rightarrow v_i > -10$

(b)

$v_i = V_D + 9$
$V_D = v_i - 9$
$V_D > 0 \Rightarrow v_i - 9 > 0$
$\Rightarrow v_i > 9$

(c)

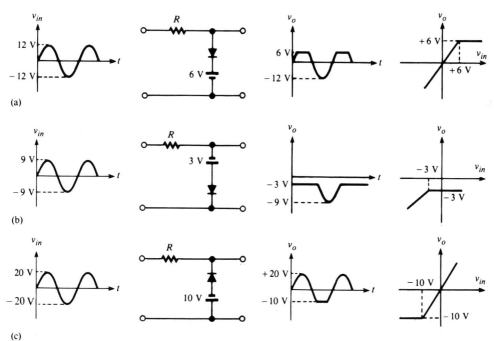

(a)

(b)

(c)

Figure 11.14
Examples of parallel clipping circuits

same way whether or not one side of the dc voltage source is connected to the common (low) side of the input and output, the connections shown in Figures 11.14(a) and (c) are preferred to that in (b), because the latter uses a floating source.

Figure 11.15(a) shows a biased diode connected in the feedback path of an inverting operational amplifier. The diode is in parallel with the feedback resistor and forms a parallel clipping circuit like that shown in Figure 11.14(a). Since v^- is at virtual ground, the voltage across R_f is the same as the output voltage v_o. Therefore, when the output voltage reaches the bias voltage E, the output is held at E volts. Figure 11.15(b) illustrates this fact for a sinusoidal input. So long as the diode is reverse biased, it acts like an open circuit and the amplifier behaves like a conventional inverting amplifier. Notice that output clipping occurs at *input* voltage $-(R_1/R_f)E$, since the amplifier inverts and has closed-loop gain magnitude

Figure 11.15
An operational-amplifier limiting circuit

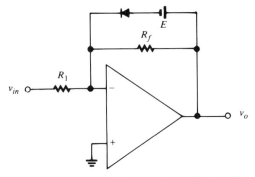

(a) The biased diode in the feedback path provides (parallel) clipping of the output at E volts.

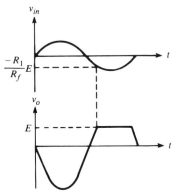

(b) The output clamps at E volts when the input reaches $\dfrac{-R_1}{R_f}E$ volts.

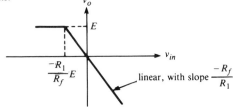

(c) Transfer characteristic

R_f/R_1. The resulting transfer characteristic is shown in Figure 11.15(c). This circuit is often called a *limiting* circuit because it limits the output to the dc level clamped by the diode. (In this and future circuits in this section, we are omitting the bias compensation resistor, R_c, for clarity; it should normally be included, following the guidelines of Chapter 8.)

In practice, the biased diode shown in the feedback of Figure 11.15(a) is often replaced by a *zener* diode in series with a conventional diode. This arrangement eliminates the need for a floating voltage source. We will study zener diodes in more detail in Chapter 13 (Section 13.7) and will learn that in many respects they are equivalent to biased diodes. Figure 11.16 shows two operational-amplifier clipping circuits using zener diodes. The zener diode conducts like a conventional diode when it is forward biased, so it is necessary to connect a reversed diode in series with it to prevent shorting of R_f. When the reverse voltage across the zener diode reaches V_Z, the diode breaks down and conducts heavily, while maintaining an essentially constant voltage, V_Z, across it. Under those conditions, the *total* voltage across R_f, i.e., v_o, equals V_Z plus the forward drop, V_D, across the conventional diode.

Figure 11.17 shows *double-ended* limiting circuits, in which both positive and negative peaks of the output waveform are clipped. Figure 11.17(a) shows the conventional parallel clipping circuit and (b) shows how double-ended limiting is accomplished in an operational-amplifier circuit. In each circuit, note that no more

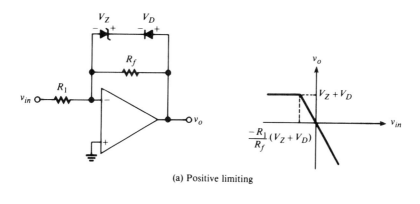

(a) Positive limiting

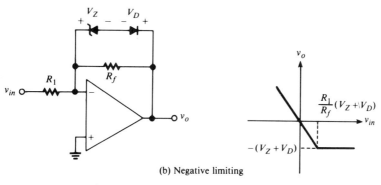

(b) Negative limiting

Figure 11.16
Operational-amplifier limiting circuits using zener diodes

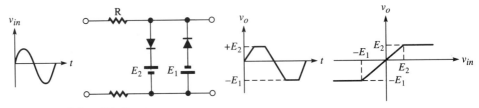

(a) Double-ended parallel clipper

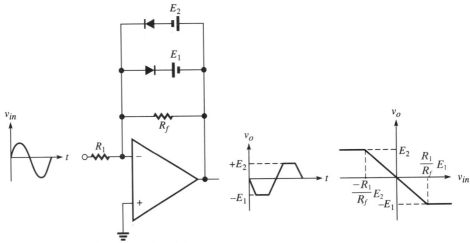

(b) Operational amplifier with double-ended clipping

Figure 11.17
Double-ended clipping, or limiting

than one diode is forward biased at any given time, and that both diodes are reverse biased for $-E_1 < v_o < E_2$, the linear region.

Figure 11.18 shows a double-ended limiting circuit using *back-to-back* zener diodes. Operation is similar to that shown in Figure 11.16, but no conventional diode is required. Note that diode D_1 is conducting in a forward direction when D_2 conducts in its reverse breakdown (zener) region, while D_2 is forward biased when D_1 is conducting in its reverse breakdown region. Neither diode conducts when $-(V_{Z2} + 0.7) < v_o < (V_{Z1} + 0.7)$, which is the region of linear amplifier operation.

Precision Rectifying Circuits

Recall that a *rectifier* is a device that allows current to pass through it in one direction only (see Figure 2.19). A diode can serve as a rectifier because it permits generous current flow in only one direction—the direction of forward bias. Rectification is the same as clipping at the 0-V level: all of the waveform below (or above) the zero-axis is eliminated. Recall, however, that a diode rectifier has certain intervals of nonconduction and produces resulting "gaps" at the zero-crossing points of the output voltage, due to the fact that the input must overcome the diode drop (0.7 V for silicon) before conduction begins. See Figures 2.21 and 2.22. In power-supply applications, where input voltages are quite large, these gaps are of no concern. However, in many other applications, especially in instru-

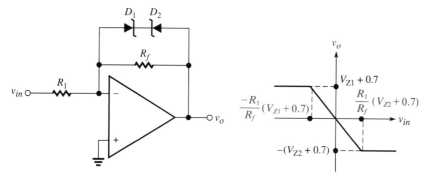

Figure 11.18
A double-ended limiting circuit using zener diodes

mentation, the 0.7-V drop can be a significant portion of the total input voltage swing and can seriously affect circuit performance. For example, most ac instruments rectify ac inputs so they can be measured by a device that responds to dc levels. It is obvious that small ac signals could not be measured if it were always necessary for them to reach 0.7 V before rectification could begin. For these applications, *precision* rectifiers are necessary.

Figure 11.19 shows one way to obtain precision rectification using an operational amplifier and a diode. The circuit is essentially a noninverting voltage follower when the diode is forward biased. When v_{in} is positive, the output of the amplifier, v_o, is positive, the diode is forward biased, and a low-resistance path is established between v_o and v^-, as necessary for a voltage follower. The load voltage, v_L, then follows the positive variations of $v_{in} = v^+$. Note that even a very small positive value of v_{in} will cause this result, because of the large differential gain of the amplifier. That is, the large gain and the action of the feedback cause the usual result that $v^+ \approx v^-$. Note also that the drop across the diode does not appear in v_L.

When the input goes negative, v_o becomes negative, and the diode is reverse biased. This effectively opens the feedback loop, so v_L no longer follows v_{in}. The amplifier itself, now operating open-loop, is quickly driven to its maximum negative output, thus holding the diode well into reverse bias.

Another precision rectifier circuit is shown in Figure 11.20. In this circuit, the load voltage is an amplified and inverted version of the *negative* variations in the

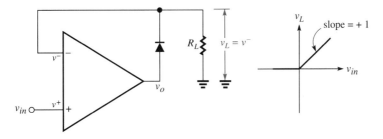

Figure 11.19
A precision rectifier. When v_{in} is positive, the diode is forward biased, and the amplifier behaves like a voltage follower, maintaining $v^+ \approx v^- = v_L$.

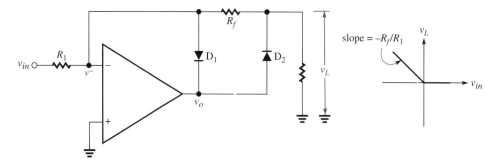

Figure 11.20
A precision rectifier circuit that amplifies and inverts the negative variations in the input voltage

input signal, and is 0 when the input is positive. Also in contrast with the previous circuit, the amplifier in this rectifier is not driven to one of its output extremes. When v_{in} is negative, the amplifier output, v_o, is positive, so diode D_1 is reverse biased and diode D_2 is forward biased. D_1 is open and D_2 connects the amplifier output through R_f to v^-. Thus, the circuit behaves like an ordinary inverting amplifier with gain $-R_f/R_1$. The load voltage is an amplified and inverted (positive) version of the negative variations in v_{in}. When v_{in} becomes positive, v_o is negative, D_1 is forward biased, and D_2 is reverse biased. D_1 shorts the output v_o to v^-, which is held at virtual ground, so v_L is 0. It is an exercise at the end of the chapter to analyze this circuit when the diodes are turned around.

11.3   CLAMPING CIRCUITS

Clamping circuits are used to shift an ac waveform up or down by adding a dc level equal to the positive or negative peak value of the ac signal. In the author's opinion, "clamping" is not a particularly good term for this operation: *Level shifting* is more descriptive. Clamping circuits are also called dc *level restorers,* because they are used in systems (television, for example) where the original dc level is lost in capacitor-coupled amplifier stages. It is important to recognize that the amount of dc-level shift required in these applications *varies* as the peak value of the ac signal varies over a period of time. In other words, it is not possible to simply add a fixed dc level to the ac signal using a summing amplifier. To illustrate, Figure 11.21 shows the outputs required from a clamping circuit for two different inputs. Note in both cases that the *peak-to-peak* value of the output is the same as the peak-to-peak value of the input, and that the output is shifted up (in this case) by an amount equal to the negative peak of the input.

Figure 11.22(a) shows a clamping circuit constructed from passive components. When the input first goes negative, the diode is forward biased, and the capacitor charges rapidly to the peak negative input voltage, V_1. The charging time-constant is very small because the forward resistance of the diode is small. The capacitor voltage, V_1, then has the polarity shown in the figure. Assuming that the capacitor does not discharge appreciably through R_L, the total load volt-

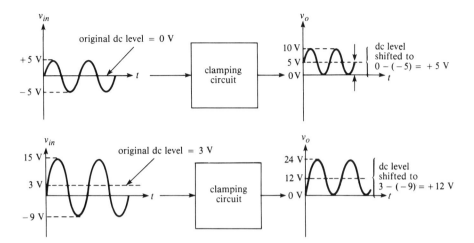

Figure 11.21
A clamping circuit that shifts a waveform up by an amount equal to the negative peak value

age as v_{in} begins to increase is

$$v_L = V_1 + v_{in} \qquad (11.13)$$

(Verify this equation by writing Kirchhoff's voltage law around the loop.) Notice that the polarity of the capacitor voltage keeps the diode reverse biased, so it is like an open circuit during this time and does not discharge the capacitor. Equa-

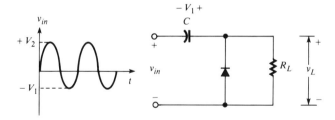

(a) The capacitor charges to V_1 volts and holds that voltage, so $v_L \cong v_{in} + V_1$.

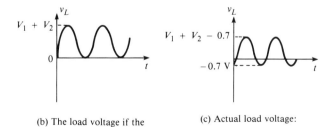

(b) The load voltage if the diode were ideal

(c) Actual load voltage:
$v_L = v_{in} + V_1 - 0.7$.

Figure 11.22
A clamping circuit consisting of a diode and a capacitor

tion 11.13 shows that the load voltage equals the input voltage shifted up by an amount equal to V_1, as required. When the input again reaches its negative peak, the capacitor may have to recharge slightly to make up for any decay that occurred during the cycle. For proper circuit performance, the discharge time-constant, $R_L C$, must be much greater than the period of the input. If the diode connections are reversed, the waveform is shifted down by an amount equal to the positive peak voltage, V_2. If the diode is biased by a fixed voltage, the waveform can be shifted up or down by an amount equal to a peak value plus or minus the bias voltage. Examples are given in the exercises at the end of the chapter.

Figure 11.22(b) shows the load voltage that results if the diode is assumed to have zero voltage drop. In reality, since the capacitor charges through the diode, the voltage across the capacitor reaches only V_1 minus the diode drop. Consequently,

$$v_L = v_{in} + V_1 - 0.7 \qquad (11.14)$$

The waveform that results is shown in Figure 11.22(c). If the input voltage is large, this offset due to an imperfect diode can be neglected, as is the case in many practical circuits.

If precision clamping is required, the operational-amplifier circuit shown in Figure 11.23 can be used. When v_{in} in Figure 11.23 first goes negative, the amplifier output, v_o, is positive and the diode is forward biased. The capacitor quickly charges to V_1, with the polarity shown. Notice that $v_L = v_{in} + V_1$ and that the drop across the diode does not appear in v_L. With the capacitor voltage having the polarity shown, v^- becomes positive, and remains positive throughout the cycle, so the amplifier output is negative. Therefore, the diode is reverse biased and the feedback loop is opened. The amplifier is driven to its maximum negative output level and the diode remains reverse biased. During one cycle of the input, the capacitor may discharge somewhat into the load, causing its voltage to fall below V_1. If so, then when v_{in} once again reaches its maximum negative voltage, v^- will once again be negative, v_o will be positive, and the capacitor will be allowed to recharge to V_1 volts, as before.

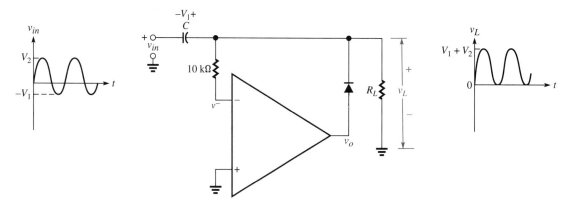

Figure 11.23
A precision clamping circuit

11.4 LOGARITHMIC AND ANTILOGARITHMIC AMPLIFIERS

A logarithmic (log) amplifier produces an output that is proportional to the logarithm of its input. Since the log function is nonlinear, it is clear that a log amplifier is not linear in the sense discussed in Chapter 1. A logarithmic transfer characteristic is shown in Figure 11.24. We see that the slope of the characteristic, $\Delta V_o / \Delta V_{in}$, and hence the voltage gain, is small for large values of V_{in} and large for small values of V_{in}. Since the gain decreases with increasing input signal level, the amplifier is said to *compress* signals.

One important application of log amplifiers is in the amplification of signals having a wide *dynamic range*: signals that may be very small as well as very large. Suppose, for example, that a temperature sensor generates a few millivolts at very low temperatures and a few volts at very high temperatures. To obtain good resolution, we would like the small signals to undergo significant amplification. The same amplification applied to the large signals, as would occur in a linear amplifier, would overdrive the amplifier and create clipping and distortion. The log amplifier eliminates this problem. Of course, the nonlinear characteristic of the log amplifier creates output waveforms that are distorted versions of input waveforms. If necessary for a particular application, the distortion can be removed by an *antilogarithmic* (inverse log, or exponential) amplifier, which has a transfer characteristic that is exactly the opposite of the log amplifier. On the other hand, in some applications the antilog operation is not necessary, as, for example, when it is desired to create a display of signal magnitudes on a logarithmic scale. An example is a *spectrum analyzer*, in which the frequency content of a complex signal is displayed as a plot of decibel voltage levels versus frequency. Another application of log amplifiers is in *analog computation*, where signal voltages must be multiplied or divided. For example, if we wished to generate the product voltage $v_1 v_2$, we could sum the outputs of two log amplifiers to obtain $\log v_1 + \log v_2 = \log v_1 v_2$. The output of an antilog amplifier whose input is $\log v_1 v_2$ would then be a voltage proportional to $v_1 v_2$.

The logarithmic characteristic of a log amplifier stems from the relationship between the collector current and base-to-emitter voltage of a BJT, which is similar to the diode equation (equation 2.1):

$$I_C = I_s(e^{V_{be}/V_T} - 1) \qquad (11.15)$$

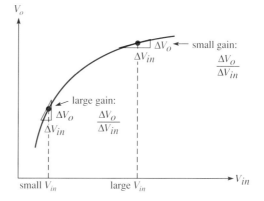

Figure 11.24
The transfer characteristic of a log amplifier shows that its voltage gain decreases with increasing values of V_{in}.

where I_s is the reverse saturation current of the base–emitter diode and V_T is the thermal voltage, $V_T = q/kT$ (as defined in equation 2.1). When V_{be} is a few tenths of a volt, $e^{V_{be}/V_T} \gg 1$, and (11.15) becomes

$$I_C = I_s e^{V_{be}/V_T}$$

Figure 11.25 shows the basic configuration of a log amplifier, in which a BJT is connected in the feedback path of an inverting operational amplifier. Taking the logarithm of both sides of the above and solving for V_{be}, we find

$$V_{be} = V_t \ln \left(\frac{I_C}{I_s}\right) \tag{11.16}$$

In Figure 11.25, note that the collector is at virtual ground (≈ 0 V), so

$$I_C = \frac{V_{in}}{R_1} \tag{11.17}$$

Also note that $V_{be} = -V_o$. Substituting these relationships in (11.16), we obtain

$$V_o = -V_T \ln \left(\frac{V_{in}}{I_s R_1}\right) \tag{11.18}$$

We see that V_o is a logarithmic function of V_{in}. Since the common logarithm (base 10) is related to the natural logarithm by $\ln x = 2.303 \log_{10} x$, equation 11.18 can also be written in terms of $\log_{10}$ as

$$V_o = -2.303 \, V_T \log_{10} \left(\frac{V_{in}}{I_s R_1}\right) \tag{11.19}$$

At room temperature, $V_T \approx 0.0257$ V, for which equations 11.18 and 11.19 become

$$V_o = -0.0257 \ln \left(\frac{V_{in}}{I_s R_1}\right) = -0.0592 \log_{10} \left(\frac{V_{in}}{I_s R_1}\right) \tag{11.20}$$

The practical difficulty of the configuration we have described is that the value of I_s cannot usually be predicted accurately and is, in any event, highly sensitive to temperaure variations. To overcome this problem, a practical log amplifier is constructed as shown in Figure 11.26. The two transistors are closely matched in an integrated circuit, so their values of I_s are essentially equal and change equally with temperature. Following the same procedure we used to obtain equation

Figure 11.25
Basic configuration of a log amplifier

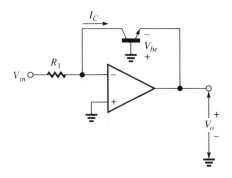

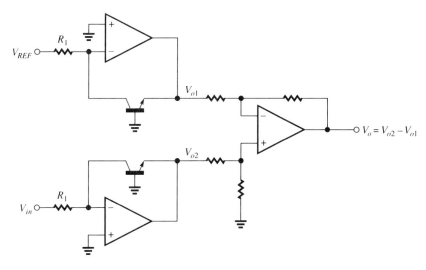

Figure 11.26
A practical log amplifier designed to compensate for the variability of I_s

11.20, we find

$$V_{01} = -0.0257 \ln \left(\frac{V_{REF}}{I_s R_1} \right) \tag{11.21}$$

and

$$V_{02} = -0.0257 \ln \left(\frac{V_{in}}{I_s R_1} \right) \tag{11.22}$$

As discussed in Section 9.1, the output amplifier forms the difference voltage $V_{o2} - V_{o1}$:

$$V_o = V_{o2} - V_{o1} = -0.0257 \left(\ln \frac{V_{in}}{I_s R_1} - \ln \frac{V_{REF}}{I_s R_1} \right)$$

$$= -0.0257 \ln \left[\frac{(V_{in}/I_s R_1)}{(V_{REF}/I_s R_1)} \right] \tag{11.23}$$

$$= -0.0257 \ln \left(\frac{V_{in}}{V_{REF}} \right)$$

Here we see that the output is no longer dependent on the value of I_s. Also, the external voltage V_{REF} can be adjusted to control the overall sensitivity of the amplifier (as can the gain of the output difference amplifier).

Example 11.2

When V_{in} in Figure 11.26 is 1 V, it is desired that $V_o = 50$ mV. What value of V_{REF} should be used? Assume the difference amplifier has unity gain.

Solution.   From equation 11.23,

$$50 \text{ mV} = -0.0257 \ln \left(\frac{1 \text{ V}}{V_{REF}} \right)$$

$$\frac{50 \times 10^{-3}}{-0.0257} = -1.9456 = \ln \left(\frac{1}{V_{REF}} \right)$$

Taking the inverse logarithm of both sides,

$$0.143 = \frac{1}{V_{REF}}$$

$$V_{REF} = 7.0 \text{ V}$$

Figure 11.27 shows the basic configuration of an antilogarithmic (inverse log) amplifier. Note that the transistor in this case is connected to the input of an inverting operational amplifier. As before,

$$I_C = I_s e^{V_{be}/V_T} \tag{11.24}$$

Since I_C flows through R_f and the collector is at virtual ground, we have

$$V_o = R_f I_C = R_f I_s e^{V_{be}/V_T} \tag{11.25}$$

Since $V_{be} = -V_{in}$, we obtain

$$V_o = R_f I_s e^{V_{in}/V_T} \tag{11.26}$$

Equation 11.26 is equivalent to

$$V_o = R_f I_s \text{ antilog} \left(-V_{in}/V_T \right) \tag{11.27}$$

Note that values of V_{in} must be negative, so that $-V_{in}/V_T$ is always positive. As in practical log amplifiers, practical antilog amplifiers use two transistors and two operational amplifiers to compensate for the variability of I_s.

Examples of integrated-circuit log and antilog amplifiers are the ICL 8048 (log) and ICL 8049 (antilog) amplifiers, manufactured by Intersil. These devices can be used over a 60-dB (1000 to 1) dynamic voltage range, with output voltages up to 14 V. The 8048 does not incorporate a difference amplifier to compensate for I_s, but arranges two matched transistors and two operational amplifiers in such a way that the output of the second amplifier is proportional to the logarithm of the input difference voltage, which accomplishes the same goal.

Figure 11.27
Basic configuration of an antilog amplifier

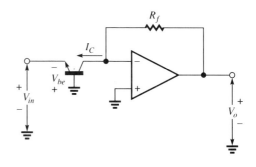

11.5   TRANSCONDUCTANCE AMPLIFIERS

A *transconductance amplifier* is basically a voltage-controlled current source. As shown in Figure 11.28, the amplifier typically has a differential input and single-ended output. Recall that transconductance is defined to be output current divided by input voltage. The transconductance of the amplifier in Figure 11.28 is

$$g_m = \frac{i_o}{v_i^+ - v_i^-} \text{ siemens} \tag{11.28}$$

Since the transconductance amplifier supplies a constant current to a load, the voltage gain of the amplifier depends directly on its load resistance. For example, if $g_m = 10$ mS and $v_i^+ - v_i^- = 50$ mV, then

$$i_o = g_m(v_i^+ - v_i^-) = (10 \times 10^{-3} \text{ S})(50 \text{ mV}) = 0.5 \text{ mA}$$

If $R_L = 1$ kΩ, the voltage gain is

$$\frac{v_L}{v_i^+ - v_i^-} = \frac{i_o R_L}{v_i^+ - v_i^-} = \frac{(0.5 \text{ mA}) 1 \text{ k}\Omega}{50 \text{ mV}} = 10$$

and if R_L is changed to 5 kΩ,

$$\frac{v_L}{v_i^+ - v_i^-} = \frac{(0.5 \text{ mA}) 5 \text{ k}\Omega}{50 \text{ mV}} = 50$$

It is an exercise at the end of the chapter to show that the voltage gain of a transconductance amplifier is

$$\frac{v_L}{v_i^+ - v_i^-} = g_m R_L \tag{11.29}$$

A *programmable* transconductance amplifier is one whose value of g_m is determined by the value of an externally supplied control current or voltage. In the programmable transconductance amplifier shown in Figure 11.29, the value of

Figure 11.28
The transconductance amplifier represented as a voltage-controlled current source

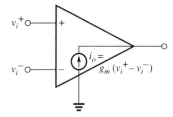

Figure 11.29
A programmable transconductance amplifier. $g_m = kI_{ABC}$.

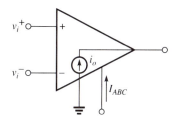

g_m is controlled by the value of *amplifier bias current, I_{ABC}*. A programmable transconductance amplifier can be used to construct a programmable voltage amplifier, that is, an amplifier whose voltage gain can be controlled electronically. The next example illustrates this application.

Example 11.3

The transconductance of the programmable transconductance amplifier in Figure 11.30 depends on I_{ABC} according to $g_m = 20I_{ABC}$.

1. Determine the voltage gain of the amplifier as a function of I_{ABC}.
2. For what value of I_{ABC} will the voltage gain of the amplifier equal 15?

Figure 11.30
(Example 11.3)

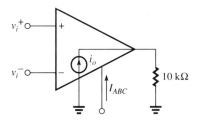

Solution

1.
$$i_o = g_m(v_i^+ - v_i^-) = 20I_{ABC}(v_i^+ - v_i^-)$$
$$v_L = i_o R_L = 20\ I_{ABC}R_L(v_i^+ - v_i^-)$$

$$\frac{v_L}{v_i^+ - v_i^-} = 20\ I_{ABC}R_L = 20 \times 10\ \text{k}\Omega\ I_{ABC} = 2 \times 10^5\ I_{ABC}$$

2.
$$15 = 2 \times 10^5\ I_{ABC}$$

$$I_{ABC} = \frac{15}{2 \times 10^5} = 75\mu\text{A}$$

Another application of a programmable transconductance amplifier is its use as a *programmable resistor*: a resistor whose resistance is determined by an external control voltage or current. Figure 11.31 illustrates this perspective. Suppose that the relationship between I_{ABC} and g_m is

$$g_m = k\ I_{ABC} \tag{11.30}$$

Figure 11.31
Using the programmable transconduct-ance amplifier as a programmable resis-tor

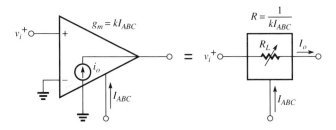

Figure 11.32
*Pin diagram for the 3080 programmable
transconductance amplifier*

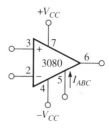

Then, since the inverting input is grounded,

$$i_o = g_m v_i^+ = k\, I_{ABC}\, v_i^+ \tag{11.31}$$

The effective resistance of the amplifier is

$$\frac{v_i^+}{i_o} = \frac{1}{k\, I_{ABC}} \tag{11.32}$$

We see that the programmable resistance is inversely proportional to I_{ABC}.

The 3080 Programmable Transconductance Amplifier

An example of an integrated-circuit programmable transconductance amplifier is
the 3080, available from RCA as the CA3080 and National Semiconductor as the
LM 3080. Figure 11.32 shows a pin diagram of the amplifier. Manufacturer's
specifications show that a typical room-temperature value for g_m is 9.6 mS when
$I_{ABC} = 500\ \mu A$. Thus, a typical value for k (in equation 11.30) is

$$k = \frac{g_m}{I_{ABC}} = \frac{9.6\ \text{mS}}{500\ \mu A} = 19.2$$

However, considerable variation is possible, since g_m has a specified range of 6.7
mS to 13 mS when $I_{ABC} = 500\ \mu A$. I_{ABC} can be supplied using any one of the three
methods shown in Figure 11.33. If resistor R_{ABC} is connected between pin 5 and
ground, as shown in Figure 11.33(a), then

$$I_{ABC} = \frac{V_{CC} - 0.6\ \text{V}}{R_{ABC}} \tag{11.33}$$

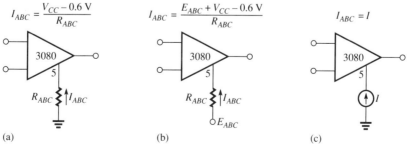

(a)   (b)   (c)

Figure 11.33
Three methods for supplying control current I_{ABC} to the 3080

I_{ABC} can also be controlled by an external control voltage, E_{ABC}, as shown in Figure 11.33(b). In this case,

$$I_{ABC} = \frac{E_{ABC} + V_{CC} - 0.6 \text{ V}}{R_{ABC}} \tag{11.34}$$

With this method of control, the previously described applications of programmable gain and programmable resistance can be implemented using control voltage E_{ABC}. Finally, I_{ABC} can be supplied directly from an external constant-current source, as shown in Figure 11.33(c).

Example 11.4

Find the range of voltage gains of the amplifier in Figure 11.34 when E_{ABC} is adjusted from 5 V to 15 V. Assume k in equation 11.30 equals 19.2.

Figure 11.34
(Example 11.4)

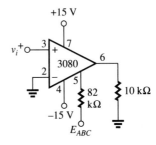

Solution

$$v_L = i_o R_L = g_m v_i^+ R_L$$

Substituting $g_m = 19.2\, I_{ABC}$,

$$v_L = 19.2\, I_{ABC}\, v_i^+ R_L$$

Substituting I_{ABC} from equation 11.34,

$$v_L = 19.2 \left(\frac{E_{ABC} + V_{CC} - 0.6 \text{ V}}{R_{ABC}}\right) v_i^+ R_L$$

$$\frac{v_L}{v_i^+} = 19.2 \left(\frac{E_{ABC} + 15 \text{ V} - 0.6 \text{ V}}{82 \text{ k}\Omega}\right) 10 \text{ k}\Omega$$

$$= 2.34 \, (E_{ABC} + 14.4)$$

When $E_{ABC} = 5$ V,

$$\frac{v_L}{v_i^+} = 2.34(5 + 14.4) = 45.4$$

When $E_{ABC} = 15$ V,

$$\frac{v_L}{v_i^-} = 2.34(15 + 14.4) = 68.8$$

11.6   PHASE-LOCKED LOOPS

A *phase-locked loop* (PLL) is a set of components that through the use of feedback generate a signal whose frequency tracks that of another externally connected input signal. The term *phase-locked* is derived from the fact that the (apparent) phase difference between two signals of different frequencies is used to control, or maintain, the frequency of the output, as will be discussed in more detail presently. Entire books have been written on the many applications of PLLs, particularly in signal-processing and communications equipment, but we will discuss just two of the most common: FM demodulation and frequency synthesis.

Figure 11.35 shows a block diagram of a phase-locked loop. Note the voltage-controlled oscillator (VCO), which, as discussed in Section 10.7, produces a signal whose frequency is proportional to input voltage. The output of the VCO is connected to a *phase comparator,* which generates an output voltage proportional to the difference in phase between the VCO signal and the externally connected input signal, v_{in}. When the frequency of the input signal fluctuates, the output of the phase comparator is a fluctuating voltage that is, in effect, an *error* voltage proportional to the difference in *frequency* between the two inputs to the comparator. The fluctuating error voltage is smoothed by the low-pass filter and applied to the input of the VCO. The system is designed so that the error voltage causes the VCO to adjust its frequency to match the frequency of v_{in}. When the VCO frequency and the frequency of v_{in} are equal, the loop is "locked" at that frequency.

Strictly speaking, phase difference is not defined for two signals having different frequencies. However, the phase comparator "sees" an apparent phase difference when the frequency of one of its input changes, as illustrated in Figure 11.36. We see that a small frequency difference creates a small (apparent) phase difference and a larger frequency difference creates a larger apparent phase difference. By considering how these diagrams would appear if one frequency were very much larger than the other, we can understand why a practical PLL may not achieve lock in such a situation without some design modifications.

FM Demodulation

Recall that frequency modulation (FM) is the process in which the frequency of one signal is controlled by the magnitude of another signal. A VCO generates an FM signal, as illustrated in Figure 11.37. In this example, the input to the VCO (the *modulating* signal) is a ramp voltage whose amplitude increases with time. Consequently, the output of the VCO (the *modulated* signal) is a signal whose

Figure 11.35
Block diagram of a phase-locked loop

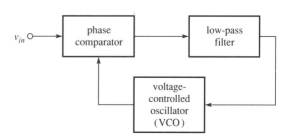

Figure 11.36
Apparent phase difference between sig-
nals of different frequencies

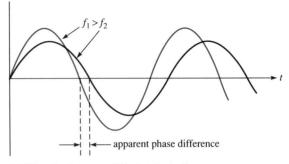

(a) When the frequency difference is small,
the apparent phase difference is small.

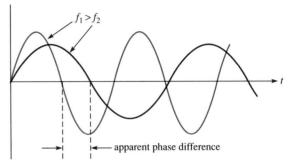

(b) When the frequency difference is larger,
the apparent phase difference is larger.

frequency increases with time. In a communications system, the modulated signal is transmitted to a receiver, where it must be *demodulated;* that is, the original modulating signal must be recovered from the modulated signal. As shown in the example of Figure 11.37, the FM demodulator performs this function. The input to the demodulator is the FM signal and the output is the ramp voltage that created it. An FM demodulator is also called a frequency-to-voltage converter.

A phase-locked loop can serve as an FM demodulator by connecting the FM signal to the input and taking the output from the low-pass filter. As the input frequency changes the output of the phase comparator and filter changes the same way. For example, an increasing input frequency (originally produced as the result of an increasing modulating signal) will create an increasing error voltage because the frequency of the VCO must be increased to track the input. Thus, the error voltage duplicates the original modulating signal.

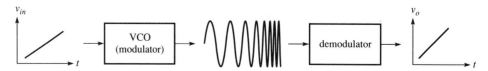

Figure 11.37
Frequency modulation and demodulation

Frequency Synthesizers

A frequency synthesizer is simply a signal generator whose frequency can be readily adjusted. A phase-locked loop can be used as a frequency synthesizer by inserting a frequency divider (counter) in the feedback path to the phase comparator and taking the output from the VCO. This arrangement is shown in Figure 11.38. Note that a low-pass filter is not required. The counter is a digital-type device that produces a signal having frequency f/N, where N is an integer, when the input to the counter has frequency f. As can be seen in the figure, a signal with fixed reference frequency, f_{REF}, is compared in the phase comparator to the output of the counter. Consider what happens when this arrangement is first connected: The signal fed back to the phase comparator has a frequency that is initially smaller (by the factor $1/N$) than f_{REF}. Consequently, the phase comparator produces an output that causes the VCO to increase its frequency, in the comparator's usual attempt to bring the two frequencies to equality. The VCO increases its frequency until the signal fed back to the comparator has frequency f_{REF} and lock is achieved. But, in order for the output of the counter to have frequency f_{REF}, the VCO must have reached frequency Nf_{REF}, since the counter divides its input frequency by N. Thus, the output of the phase-locked loop (the output of the VCO) is a signal whose frequency is the multiple Nf_{REF} of the reference frequency.

One advantage of a frequency synthesizer using a PLL is that the signal providing the reference frequency can be very stable, as, for example, from a crystal-controlled oscillator. The higher frequency, Nf_{REF}, will then also be very stable. In some high-frequency applications, it is not possible to achieve such crystal-controlled stability at high frequencies without using such a scheme. We should note that a frequency-dividing counter can also be connected btween the f_{REF} input and the phase comparator. This connection has just the opposite effect of connecting a divider in the feedback: The frequency of the VCO is *divided* by the same factor that the input divider provides.

The 565 Integrated-Circuit PLL

An example of a commercially available, integrated-circuit PLL is the LM565, manufactured by National Semiconductor. The inputs and outputs of the phase comparator and VCO are accessible at separate pins so that external components, such as filters and frequency dividers, can be inserted as required for different applications. Some important specifications of a PLL, and their values for the LM565, are the maximum VCO frequency (500 kHz), the demodulated output voltage (300 mV for a 10% change in input frequency), and the phase comparator sensitivity (0.68 V/radian).

Figure 11.38
Block diagram of a PLL used as a fre-
quency multiplier (synthesizer)

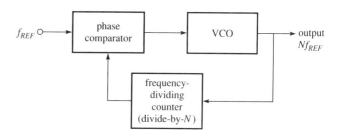

EXERCISES

Section 11.1 Voltage Comparators

11.1   The maximum output voltages of each of the voltage comparators shown in Figure 11.39 are ±15 V. Sketch the output waveforms for each when v_{in} is a 10-V-pk sine wave. (In each case, show v_{in} and v_o on your sketch, and label voltage levels where switching occurs.)

11.2   Repeat Exercise 11.1 when each of the v^+ and v^- inputs are interchanged.

11.3   The output of the comparator shown in Figure 11.40 switches between +10 V and −10 V.
   a. Find the lower and upper trigger levels.
   b. Find the hysteresis.
   c. Sketch the hysteresis loop.
   d. Sketch the output when v_{in} is a 10-V-pk sine wave.

11.4   Repeat Exercise 11.3 when $V_{REF} = +4$ V, under the assumption that the comparator output switches between +10 V and −5 V.

11.5   Design a Schmitt trigger circuit whose output switches to a high level when the input falls to −1 V and switches to a low level when the input rises to +1 V. Assume that the output switches between ±10 V.

11.6   Derive equations 11.8 and 11.9 for the lower and upper trigger levels of a noninverting Schmitt trigger.

11.7   Design a Schmitt trigger that switches to a high level when the input rises through +2 V and switches to a low level when the input falls through −1 V. You may choose the high and low output levels of the comparator. Sketch the output when the input is a 5-V-pk sine wave.

11.8   The output of the voltage comparator in Figure 11.41 switches between ±10 V.
   a. What maximum and minimum values does the capacitor voltage reach?
   b. What is the frequency of the output oscillation?

11.9   Design an astable multivibrator that produces a 1-kHz square wave.

Section 11.2 Clipping and Rectifying Circuits

11.10   Assume that the input to each device whose transfer characteristic is shown in Figure 11.42 is a 10-V-pk sine wave. Sketch input and output waveforms for each device. Label voltage values on your sketches.

11.11   Sketch the output waveform for the input shown to each circuit in Figure 11.43. Label voltage values on your sketches. Assume ideal diodes.

11.12   Design an operational-amplifier clipping circuit whose output is that shown in Figure 11.44(b) when its input is as shown in 11.44(a). Sketch the transfer characteristic.

11.13   Design an operational amplifier having the following characteristics:
   a. When the input is +0.4 V, the output is −2 V.

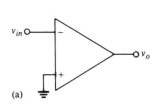

(a)

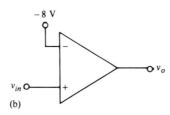

−8 V

(b)

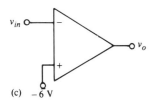

(c) −6 V

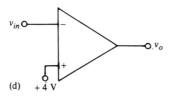

(d) +4 V

Figure 11.39
(Exercise 11.1)

Figure 11.40
(Exercise 11.3)

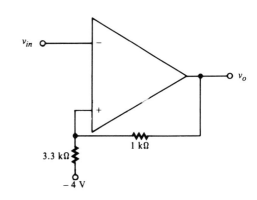

Figure 11.41
(Exercise 11.8)

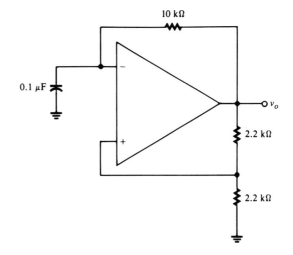

Figure 11.42
(Exercise 11.10)

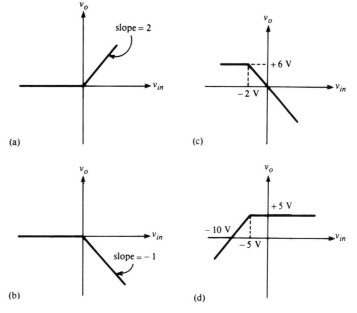

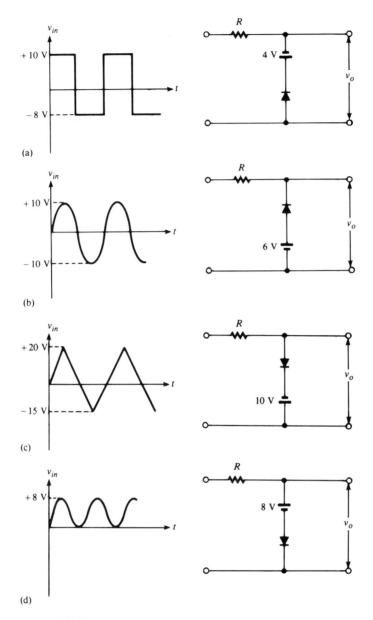

(a)

(b)

(c)

(d)

Figure 11.43
(Exercise 11.11)

Figure 11.44
(Exercise 11.12)

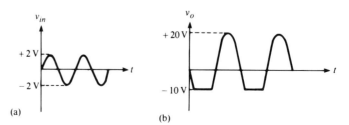

(a)

(b)

Figure 11.45
(Exercise 11.15)

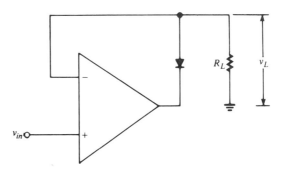

b. When the input is a 6-V-pk sine wave, the output is clipped at +10 V and at −5 V.

c. The input resistance is at least 10 kΩ. Sketch the transfer characteristic.

11.14 Using zener diodes, design an operational-amplifier clipping circuit that clips its output at +8.3 V and −4.8 V. (Specify the zener voltages in your design.) The gain in the linear region should be −4, and the input resistance to the circuit must be at least 25 kΩ. Sketch the transfer characteristic.

11.15 Sketch v_L in the circuit shown in Figure 11.45 when the input is a sine wave that alternates between +10 V peak and −6 V peak. Sketch the transfer characteristic of the circuit.

11.16 Repeat Exercise 11.15 for the circuit shown in Figure 11.46.

Section 11.3 Clamping Circuits

11.17 Sketch the load-voltage waveforms in each of the circuits shown in Figure 11.47. Assume ideal diodes, and label voltage values on your sketches.

11.18 Sketch the load-voltage waveforms in each of the circuits shown in Figure 11.48. Assume ideal diodes, and label voltage values on your sketches.

11.19 Sketch the load-voltage waveform in the circuit shown in Figure 11.49. Label voltage values on your sketch.

Section 11.4 Logarithmic and Antilogarithmic Amplifiers

11.20 The log amplifier in Figure 11.26 has $V_{REF} = 5$ V. It is necessary for the output voltage to be 0.25 V when the input is 0.5 V. What should be the voltage gain of the output difference amplifier?

11.21 Assuming the output difference amplifier in Figure 11.26 has unit voltage gain, what is the change in the output voltage of the log amplifier when V_{in} doubles in value?

Section 11.5 Transconductance Amplifiers

11.22 The voltage gain of a transconductance amplifier is 32 when its load resistance is 4 kΩ. What is the value of its transconductance?

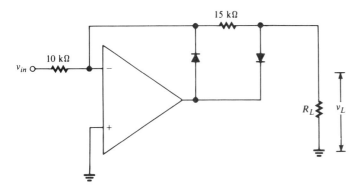

Figure 11.46
(Exercise 11.16)

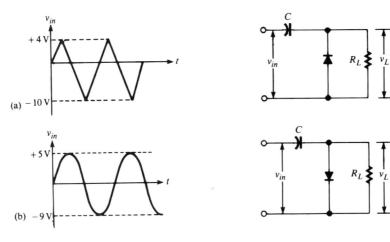

Figure 11.47
(Exercise 11.17)

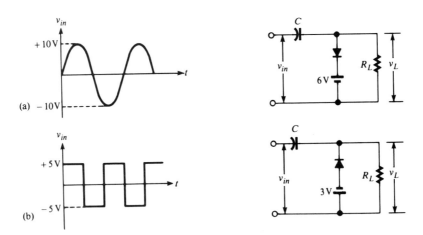

Figure 11.48
(Exercise 11.18)

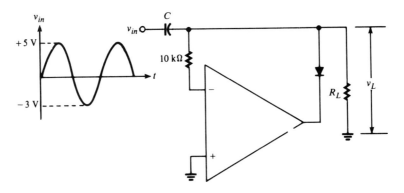

Figure 11.49
(Exercise 11.19)

11.23 Derive equation 11.29 for the voltage gain of a transconductance amplifier.

11.24 A programmable transconductance amplifier has $g_m = 16I_{ABC}$ and $R_L = 22$ kΩ. The manufacturer specifies that the permissible range of I_{ABC} is 10 μA $\leq I_{ABC} \leq 400$ μA. What range of programmable voltage gains is possible?

11.25 A programmable transconductance amplifier having $g_m = 20I_{ABC}$ is to be used as a programmable resistor having resistance 2 kΩ. What should be the value of I_{ABC}?

11.26 A 3080 transconductance amplifier is programmed as shown in Figure 11.33(a). If $V_{CC} = 15$ V and $g_m = 19.2I_{ABC}$, what should

be the value of R_{ABC} to obtain a voltage gain of 20 when $R_L = 4$ kΩ?

Section 11.6 Phase-locked Loops

11.27 A phase-locked loop used as a frequency synthesizer is to generate an output frequency of 500 kHz. If the counter divides frequency by 10, what should be the input reference frequency to the PLL?

11.28 **a.** What phase difference (in radians) will cause the phase comparator of the 565 PLL to generate an output of 0.5 V?
    b. What percent change in the frequency of the input to the 565 PLL will create a demodulated output voltage of 100 mV?

Digital-to-Analog and Analog-to-Digital Converters

12

OVERVIEW

Analog and Digital Voltages

As mentioned in Chapter 1, an *analog* voltage is one that may vary continuously throughout some range. For example, the output of an audio amplifier is an analog voltage that may have any value (of an infinite number of values) between its minimum and maximum voltage limits. Most of the devices and circuitry we have studied in this book are analog in nature. In contrast, a digital voltage has only two useful values: a "low" and a "high" voltage, such as 0 V and +5 V. Digital voltages are used to represent numerical values in the binary number system, which has just the two digits 0 and 1. For example, four digital voltages having the values 0 V, +5 V, 0 V, +5 V would represent the binary number 0101, which equals 5 in decimal. Digital computers perform computations using binary numbers represented by digital voltages.

Most physical variables in our environment are analog in nature. Quantities such as temperature, pressure, velocity, weight, etc. can have any of an infinite number of values. However, because of the high speed and accuracy of digital systems, such as computers, it is frequently the case that the transmission of data representing such quantities and computations performed on them are operations that are best accomplished using digital equivalents of analog values. Two modern examples of functions that have traditionally been analog in nature and that are now performed digitally include communications systems using fiber-optic cables and music reproduction using compact disks. Converting an analog value to a digital equivalent (binary number) is called *digitizing* the value. An analog-to-digital converter (ADC, or A/D converter) performs this function. Figure 12.1(a) illustrates the operation of an ADC. After a digital system has transmitted, analyzed, or otherwise processed digital data, it is often necessary to convert the results of such operations back to analog values. This function is performed by a digital-to-analog converter (DAC, or D/A converter). Figure 12.1(b) illustrates the operation of a DAC. Figure 12.1(c) shows a practical example of a system that incorporates both an ADC and a DAC. In this example, an accelerometer mounted on a vibration table (used to test components for vibration damage)

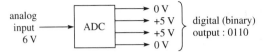

(a) The ADC converts a 6-V analog input to an equivalent digital output. (The binary number 0110 is equivalent to decimal 6.)

(b) The DAC converts 0110 to 6 V.

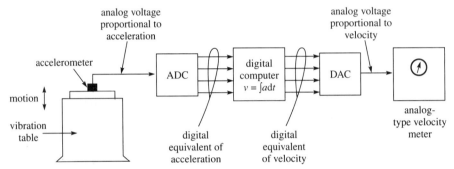

(c) Example of an instrumentation system that uses an analog-to-digital converter and a digital-to-analog converter.

Figure 12.1
The analog-to-digital converter (ADC) and digital-to-analog converter (DAC)

produces an analog voltage proportional to the instantaneous acceleration of the table. The analog voltage is converted to digital form and transmitted to a computer, which calculates the instantaneous velocity of the table (by mathematical integration: $v = \int a \, dt$). The digital quantity representing the computed velocity is then converted by a DAC to an analog voltage, which becomes the input to an analog-type velocity meter.

Converting Binary Numbers to Decimal Equivalents

As an aid in understanding subsequent discussions on digital-to-analog and analog-to-digital converters, we present here for review and reference purposes a brief summary of the mathematics used to convert binary numbers to their decimal equivalents.

As in the decimal number system, the position of each digit in a binary number carries a certain *weight*. In the decimal number system, the weights are powers of 10 (the units position: $10^0 = 1$, the tens position: $10^1 = 10$, the hundreds position: $10^2 = 100$, and so forth). In the binary number system, each position carries a weight equal to a power of 2: $2^0 = 1$, $2^1 = 2$, $2^2 = 4$, $2^3 = 8$, and so forth. A binary digit is called a *bit,* so a binary number consists of a sequence of bits, each of which can be either 0 or 1. The rightmost bit in a binary number carries the least weight and is called the *least significant bit* (LSB). If the binary number is an integer (whole number), the LSB carries weight $2^0 = 1$, and each bit to the left of the LSB has a weight equal to a power of 2 that is one greater than the weight of the bit to its right. To convert a binary number to its decimal equivalent, we compute the sum of all the weights of the positions where binary 1s occur. The following is an example:

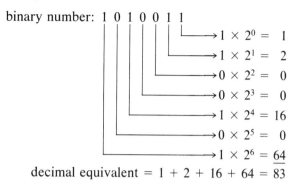

$$\text{binary number: } 1\ 0\ 1\ 0\ 0\ 1\ 1$$

$$1 \times 2^0 = 1$$
$$1 \times 2^1 = 2$$
$$0 \times 2^2 = 0$$
$$0 \times 2^3 = 0$$
$$1 \times 2^4 = 16$$
$$0 \times 2^5 = 0$$
$$1 \times 2^6 = 64$$

$$\text{decimal equivalent} = 1 + 2 + 16 + 64 = 83$$

The leftmost bit in a binary number is called the *most significant bit* (MSB). An easy way to convert a binary number to its decimal equivalent is to write the values of the powers of 2 (1, 2, 4, 8, . . .) above each bit, beginning with the LSB and proceeding through the MSB. Then we simply add those powers where 1s occur. For example:

$$\text{power of 2: } 16\ 8\ 4\ 2\ 1$$
$$\text{binary number: } \underline{1\ 0\ 1\ 1\ 0}$$
$$16 + 4 + 2 = 22 = \text{decimal equivalent}$$

If a binary number has a fractional part, then a *binary point* (like a decimal point) separates the integer part from the fractional part. The powers of 2 become negative and descend as we move right past the binary point. The following is an example:

$$
\begin{array}{cc}
2^{-1} & 2^{-2} \\
\| & \|
\end{array}
$$
$$\text{power of 2: } 8\ 4\ 2\ 1\quad 0.5\ 0.25$$
$$\text{binary number: } \underline{1\ 0\ 0\ 1.\quad\ 0\ 1}$$
$$8 + 1 + 0.25 = 9.25$$

Some Digital Terminology

Binary 0 and 1 are called logical 0 and logical 1 to distinguish them from the voltages used to represent them in a digital system. The voltages are called *logic levels*. For example, a common set of logic levels is 0 V (ground) for logical 0 and +5 V for logical 1.

   A *binary counter* is a device whose binary output is numerically equal to the number of pulses that have occurred at its input. For example, the output of a 4-bit binary counter would be 0110 after 6 pulses had occurred at its input. The input pulses often occur at a fixed frequency from a signal called the system *clock*. The largest binary number that a 4-bit counter can contain is 1111, or decimal 15. (The counter *resets* to 0000 after the sixteenth pulse.) Thus, the total number of binary numbers that a 4-bit counter can produce (counting 0000) is 16, or 2^4. The total number of binary numbers that can be represented by n bits is 2^n. The counter we have just described is called an *up-counter* because its binary output increases by

1 (increments) each time a new clock pulse occurs. In a *down-counter*, the binary output decreases by 1 (decrements) each time a new clock pulse occurs. Thus, the sequence of outputs from a 4-bit down-counter would be 1111, 1110, 1101, . . . , 0000, 1111, . . .

A *latch* is a digital device that stores the value (0 or 1) of a binary input. It is especially useful when a binary input is changing and we want to "latch onto" its value at a particular instant of time—as, for example, when performing a digital-to-analog conversion of a digital quantity that is continuously changing in value. A *register* is a set of latches used to store all the bits of a digital quantity, one latch for each bit.

Some analog-to-digital converters produce outputs that are in 8-4-2-1 *binary-coded–decimal* (BCD) form, rather than true binary. In 8-4-2-1 BCD, *each* decimal digit is represented by 4 bits. For example, if the input to an ADC having this type of output were 14 V, then the output would be

analog input: 14
8-4-2-1 output: 0001   0100

Note that this type of output is quite different from true binary. (The number 14 in binary is 1110.)

Resolution

A digital-to-analog converter having a 4-bit binary input produces only $2^4 = 16$ different analog output voltages, corresponding to the 16 different values that can be represented by the 4-bit input. The output of the converter is, therefore, not truly analog, in the sense that it cannot have an infinite number of values. If the input were a 5-bit binary number, the output could have $2^5 = 32$ different values. In short, the greater the number of input bits, the greater the number of output values and the closer the output resembles a true analog quantity. *Resolution* is a measure of this property. The greater the resolution of the DAC, the finer the increments between output voltage levels.

Similarly, an analog-to-digital converter having a 4-bit binary output produces only 16 different binary outputs, so it can convert only 16 different analog inputs to digital form. Since an analog input has an infinite number of values, the ADC does not truly convert (every) analog input to equivalent digital form. The greater the number of output bits, the greater the number of analog inputs the ADC can convert and the greater the resolution of the device. Resolution is clearly an important performance specification for both DACs and ADCs, and we will examine quantitative measures for it in later discussions.

12.2   THE R-2R LADDER DAC

The most popular method for converting a digital input to an analog output incorporates a ladder network containing series-parallel combinations of two resistor values: R and $2R$. Figure 12.2 shows an example of an R-$2R$ ladder having a 4-bit digital input.

To understand the operation of the R-$2R$ ladder, let us first determine the output voltage in Figure 12.2 when the input is 1000. We will assume that a logical-1 input is E volts and that logical 0 is 0 V (ground). Figure 12.3(a) shows the circuit with input D_3 connected to E and all other inputs connected to ground, corre-

Figure 12.2
A 4-bit R-2R ladder network used for digital-to-analog conversion

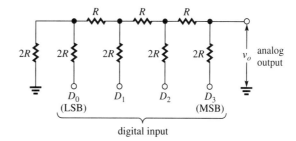

sponding to the binary input 1000. We wish to find the total equivalent resistance, R_{eq}, looking to the left from node A. At the left end of the ladder, we see that $2R$ is in parallel with $2R$, so that combination is equivalent to R. That R is in series with another R, giving $2R$. That $2R$ is in parallel with another $2R$, which is equivalent to R once again. Continuing in this manner, we ultimately find that $R_{eq} = 2R$. In fact, we see that the equivalent resistance looking to the left from every node is $2R$. Figure 12.3(b) shows the circuit when it is redrawn with R_{eq} replacing all of the network to the left of node A. Figure 12.3(c) shows an equivalent way to draw the circuit in (b), and it is now readily apparent that $v_o = E/2$.

Figure 12.3
Calculating the output of the R-2R ladder when the input is 1000

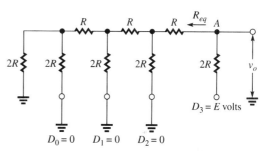

(a) When the input is 1000, D_0, D_1, and D_2 are grounded (0 V) and D_3 is E volts.

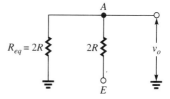

(b) The circuit equivalent to (a) when the network to the left of node A is replaced by its equivalent resistance, R_{eq}.

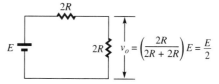

(c) Calculation of v_o using the voltage-divider rule. (Note that v_o in (b) is the voltage across $R_{eq} = 2R$.)

Let us now find v_o when the input is 0100. Figure 12.4(a) shows the circuit, with D_2 connected to E and all other inputs grounded. As demonstrated earlier, the equivalent resistance looking to the left from node B is $2R$. The equivalent circuit with R_{eq} replacing the network to the left of B is shown in Figure 12.4(b). We proceed with the analysis by finding the (Thevenin) equivalent circuit to the left of node A, as indicated by the bracketed arrows. The Thevenin equivalent resistance (with E shorted to ground) is $R_{TH} = R + 2R \parallel 2R = R + R = 2R$. The Thevenin equivalent voltage is $[2R/(2R + 2R)]E = E/2$. The Thevenin equivalent circuit is shown in Figure 12.4(c). Figure 12.4(d) shows the ladder with the Thevenin equivalent circuit replacing everything to the left of node A. It is now apparent that $v_o = E/4$.

Figure 12.4
Calculating the output of the R-2R ladder when the input is 0100

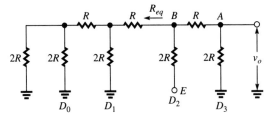

(a) When the input is 0100, D_0, D_1, and D_3 are grounded and D_2 is E volts.

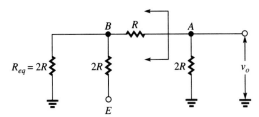

(b) The circuit equivalent to (a) when the network to the left of node B is replaced by its equivalent resistance ($2R$). The Thevenin equivalent circuit to the left of the bracketed arrows is shown in (c).

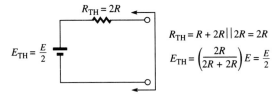

$$R_{TH} = R + 2R \parallel 2R = 2R$$

$$E_{TH} = \left(\frac{2R}{2R + 2R}\right)E = \frac{E}{2}$$

(c) The Thevenin equivalent circuit to the left of node A.

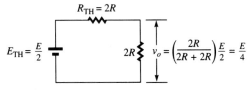

$$v_o = \left(\frac{2R}{2R + 2R}\right)\frac{E}{2} = \frac{E}{4}$$

(d) Calculation of v_o using the voltage-divider rule.

By an analysis similar to the foregoing, we find that the output of the ladder is $E/8$ when the input is 0010 and that the output is $E/16$ when the input is 0001. In general, when the D_n input is 1 and all other inputs are 0, the output is

$$v_o = \frac{E}{2^{N-n}} \qquad\qquad\qquad \textbf{(12.1)}$$

where N is the total number of binary inputs. For example, if $E = 5$ V, the output voltage when the input is 0010 is $(5 \text{ V})/(2^{4-1}) = (5 \text{ V})/8 = 0.625$ V.

To find the output voltage corresponding to *any* input combination, we can invoke the principle of superposition and simply add the voltages produced by the inputs where 1s are applied. For example, when the input is 1100, the output is $E/2 + E/4 = 3E/4$. The next example illustrates these computations.

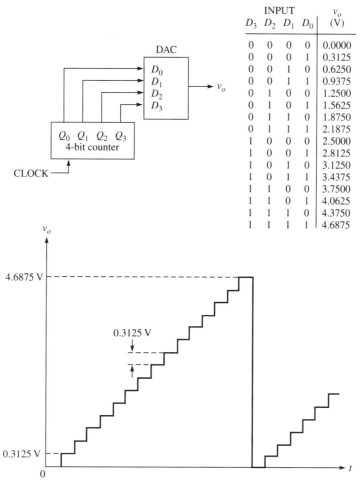

INPUT				v_o
D_3	D_2	D_1	D_0	(V)
0	0	0	0	0.0000
0	0	0	1	0.3125
0	0	1	0	0.6250
0	0	1	1	0.9375
0	1	0	0	1.2500
0	1	0	1	1.5625
0	1	1	0	1.8750
0	1	1	1	2.1875
1	0	0	0	2.5000
1	0	0	1	2.8125
1	0	1	0	3.1250
1	0	1	1	3.4375
1	1	0	0	3.7500
1	1	0	1	4.0625
1	1	1	0	4.3750
1	1	1	1	4.6875

Figure 12.5
(Example 12.1)

Example 12.1

The logic levels used in a 4-bit R-$2R$ ladder DAC are $1 = +5$ V and $0 = 0$ V.

1. Find the output voltage when the input is 0001 and when it is 1010.

2. Sketch the output when the inputs are driven from a 4-bit binary up-counter.

Solution

1. By equation 12.1, the output when the input is 0001 is

$$v_o = \frac{5 \text{ V}}{2^{4-0}} = \frac{5 \text{ V}}{16} = 0.3125 \text{ V}$$

When the input is 1000, the output is $(5 \text{ V})/2 = 2.5$ V, and when the input is 0010, the output is $(5 \text{ V})/8 = 0.625$ V. Therefore, when the input is 1010, the output is $2.5 \text{ V} + 0.625 \text{ V} = 3.125$ V.

2. Figure 12.5 shows a table of the output voltages corresponding to every input combination, calculated using the method illustrated in part (1) of this example. As the counter counts up, the output voltage increases by 0.3125 V at each new count. Thus, the output is the *staircase* waveform shown in the figure. The voltage steps from 0 V to 4.6875 V each time the counter counts from 0000 to 1111.

Typical values for R and $2R$ are 10 kΩ and 20 kΩ. For accurate conversion, the output voltage from the R-$2R$ ladder should be connected to a high impedance to prevent loading. Figure 12.6 shows how an operational amplifier can be used for that purpose. The output of the ladder is connected to a unity-gain voltage follower, whose input impedance is extremely large and whose output voltage is the same as its input voltage.

12.3   A WEIGHTED-RESISTOR DAC

Figure 12.7 illustrates another approach to digital-to-analog conversion. The operational amplifier is used to produce a *weighted sum* of the digital inputs, where the weights are proportional to the weights of the bit positions of the inputs. Recall from Chapter 9 that each input is amplified by a factor equal to the ratio of the

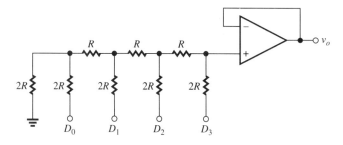

Figure 12.6
Using a unity-gain voltage follower to provide a high impedance for the R-2R ladder

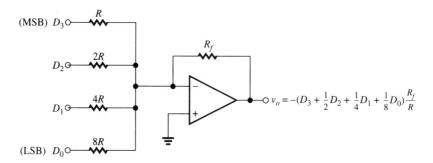

Figure 12.7
A weighted-resistor DAC using an inverting operational amplifier

feedback resistance divided by the input resistance to which it is connected. Thus, D_3, the most significant bit, is amplified by R_f/R, D_2 by $R_f/2R = 1/2(R_f/R)$, D_1 by $R_f/4R = 1/4(R_f/R)$, and D_o by $R_f/8R = 1/8(R_f/R)$. Since the amplifier sums and inverts, the output is

$$v_o = -\left(D_3 + \frac{1}{2}D_2 + \frac{1}{4}D_1 + \frac{1}{8}D_o\right)\frac{R_f}{R} \qquad (12.2)$$

The principal disadvantage of this type of converter is that a different-valued precision resistor must be used for each digital input. In contrast, the R-$2R$ ladder network uses only two values of resistance.

Example 12.2

Design a 4-bit, weighted-resistor DAC whose full-scale output voltage is -10 V. Logic levels are $1 = +5$ V and $0 = 0$ V. What is the output voltage when the input is 1010?

Solution.   The full-scale output voltage is the output voltage when the input is maximum: 1111. In that case, from equation 12.2, we require

$$\left(5\text{ V} + \frac{5\text{ V}}{2} + \frac{5\text{ V}}{4} + \frac{5\text{ V}}{8}\right)\frac{R_f}{R} = 10\text{ V}$$

or

$$9.375\,\frac{R_f}{R} = 10$$

Let us choose $R_f = 10$ kΩ. Then

$$R = \frac{9.375(10\text{ k}\Omega)}{10} = 9.375\text{ k}\Omega$$

$$2R = 18.75\text{ k}\Omega$$
$$4R = 37.50\text{ k}\Omega$$
$$8R = 75\text{ k}\Omega$$

When the input is 1010, the output voltage is

$$V_o = -\left(5\text{ V} + \frac{0\text{ V}}{2} + \frac{5\text{ V}}{4} + \frac{0\text{ V}}{8}\right) \frac{10\text{ k}\Omega}{9.375\text{ k}\Omega} = -6.667\text{ V}$$

12.4   THE SWITCHED CURRENT-SOURCE DAC

The D/A converters we have discussed so far can be regarded as switched voltage-source converters: When a binary input goes high, the high voltage is effectively switched into the circuit, where it is summed with other input voltages. Because of the technology used to construct integrated-circuit DACs, currents can be switched in and out of a circuit faster than voltages can. For that reason, most integrated-circuit DACs utilize some form of current switching, where the binary inputs are used to open and close switches that connect and disconnect internally generated currents. The currents are weighted according to the bit positions they represent and are summed in an operational amplifier. Figure 12.8 shows an example. Note that an R-$2R$ ladder is connected to a voltage source identified as E_{REF}. The current that flows in each $2R$ resistor is

$$I_n = \left(\frac{E_{REF}}{R}\right) \frac{1}{2^{N-n}} \tag{12.3}$$

when $n = 0, 1, \ldots, N - 1$ is the subscript for the current created by input D_n and N is the total number of inputs. Thus, each current is weighted according to the bit position it represents. For example, the current in the $2R$ resistor at the D_1 input in the figure ($n = 1$ and $N = 4$) is $(E_{REF}/R)(1/2)^3$. The binary inputs control switches that connect the currents either to ground or to the input of the amplifier. The amplifier sums all currents whose corresponding binary inputs are high. The amplifier also serves as a current-to-voltage converter. It is connected in an inverting configuration and its output is

$$v_o = -I_T R \tag{12.4}$$

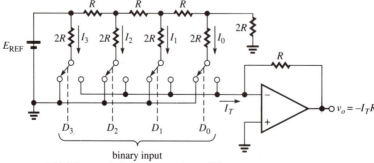

binary input
A high input switches current to the amplifier
input, and a low input switches current to ground.

Figure 12.8
A 4-bit switched current-source DAC

where I_T is the sum of the currents that have been switched to its input. For example, if the input is 1001, then

$$I_T = \frac{E_{REF}}{R}\left(\frac{1}{2}\right) + \frac{E_{REF}}{R}\left(\frac{1}{16}\right) = \frac{E_{REF}}{R}\left(\frac{7}{16}\right)$$

and

$$v_o = -\left(\frac{E_{REF}}{R}\right)\left(\frac{7}{16}\right) R = -\frac{7}{16} E_{REF}$$

Example 12.3

The switched current-source DAC in Figure 12.8 has $R = 10$ kΩ and $E_{REF} = 10$ V. Find the total current delivered to the amplifier and the output voltage when the binary input is 1010.

Solution.   From equation (12.3),

$$I_3 = \left(\frac{10 \text{ V}}{10 \text{ k}\Omega}\right) \frac{1}{2^{4-3}} = 0.5 \text{ mA}$$

and

$$I_1 = \left(\frac{10 \text{ V}}{10 \text{ k}\Omega}\right) \frac{1}{2^{4-1}} = 0.125 \text{ mA}$$

Therefore, $I_T = I_3 + I_1 = 0.5$ mA $+ 0.125$ mA $= 0.625$ mA. From equation (12.4),

$$v_o = -I_T R = -(0.625 \text{ mA})(10 \text{ k}\Omega) = -6.25 \text{ V}$$

The reference voltage E_{REF} in Figure 12.8 may be fixed—as, for example, when it is generated internally in an integrated circuit—or it may be externally variable. When it is externally variable, the output of the DAC is proportional to the *product* of the variable E_{REF} and the variable binary input. In that case, the circuit is called a *multiplying* D/A converter, and the output represents the product of an analog input (E_{REF}) and a digital input.

In most integrated-circuit DACs utilizing current-source switching, the output is the total current I_T produced in the R-$2R$ ladder. The user may then connect a variety of operational amplifier configurations at the output to perform magnitude scaling and/or phase inversion. If E_{REF} can be both positive and negative and if the binary input always represents a positive number, then the output of the DAC is both positive and negative. In that case, the DAC is called a *two-quadrant multiplier*. A DAC whose output can be both positive and negative is said to be *bipolar*, and one whose output is only positive or only negative is *unipolar*. A four-quadrant multiplier is one in which both inputs can be either negative or positive and in which the output (product) has the correct sign for every case. Negative binary inputs can be represented using the *offset binary code*, the 4-bit version of which is shown in Table 12.1. Note that the numbers +7 through −8 are represented by 0000 through 1111, respectively, with 0)$_{10}$ represented by 1000. The table also shows the analog outputs that are produced by a multiplying D/A converter. In general, for an N-bit DAC, the offset binary code represents decimal numbers

Table 12.1
The offset binary code used to represent positive and negative binary inputs to a D/A converter

Decimal Number	Offset Binary Code	Analog Output
+7	1111	$+(7/8)E_{REF}$
+6	1110	$+(6/8)E_{REF}$
+5	1101	$+(5/8)E_{REF}$
+4	1100	$+(4/8)E_{REF}$
+3	1011	$+(3/8)E_{REF}$
+2	1010	$+(2/8)E_{REF}$
+1	1001	$+(1/8)E_{REF}$
0	1000	0
−1	0111	$-(1/8)E_{REF}$
−2	0110	$-(2/8)E_{REF}$
−3	0101	$-(3/8)E_{REF}$
−4	0100	$-(4/8)E_{REF}$
−5	0011	$-(5/8)E_{REF}$
−6	0010	$-(6/8)E_{REF}$
−7	0001	$-(7/8)E_{REF}$
−8	0000	$-(8/8)E_{REF}$

from $+2^{N-1} - 1$ through -2^{N-1} and the analog outputs range from

$$+ \left(\frac{2^{N-1} - 1}{2^N - 1}\right) E_{REF} \quad \text{through} \quad -E_{REF}$$

in steps of size $(1/2^{N-1})E_{REF}$.

12.5 SWITCHED-CAPACITOR DACS

The newest technology used to construct D/A converters employs weighted capacitors instead of resistors. In this method, charged capacitors form a capacitive voltage divider whose output is proportional to the sum of the binary inputs.

As an aid in understanding the method, let us review the theory of capacitive voltage dividers. Figure 12.9 shows a two-capacitor example. The total equivalent capacitance of the two series-connected capacitors is

$$C_T = \frac{C_1 C_2}{C_1 + C_2} \tag{12.5}$$

Therefore, the total charge delivered to the circuit, which is the same as the charge on both C_1 and C_2, is

$$Q_1 = Q_2 = Q_T = C_T E = \left(\frac{C_1 C_2}{C_1 + C_2}\right) E \tag{12.6}$$

Figure 12.9
The capacitive voltage divider

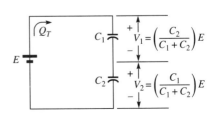

The voltage across C_2 is

$$V_2 = \frac{Q_2}{C_2} = \frac{Q_T}{C_2} = \frac{\left(\dfrac{C_1 C_2}{C_1 + C_2}\right) E}{C_2} = \left(\frac{C_1}{C_1 + C_2}\right) E \qquad (12.7)$$

Similarly,

$$V_1 = \left(\frac{C_2}{C_1 + C_2}\right) E \qquad (12.8)$$

Figure 12.10(a) shows an example of a 4-bit switched-capacitor DAC. Note that the capacitance values have binary weights. A *two-phase* clock is used to control switching of the capacitors. The two-phase clock consists of clock signals ϕ_1 and ϕ_2; ϕ_1 goes high while ϕ_2 is low, and ϕ_2 goes high while ϕ_1 is low. When ϕ_1 goes high, all capacitors are switched to ground and discharged. When ϕ_2 goes

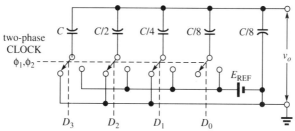

(a) All capacitors are switched to ground by ϕ_1. Those capacitors whose digital inputs are high are switched to E_{REF} by ϕ_2.

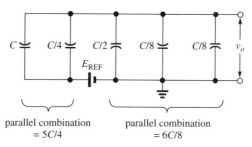

(b) Equivalent circuit when the input is 1010. The capacitors switched to E_{REF} are in parallel as are the ones connected to ground.

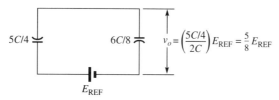

(c) Circuit equivalent to (b). The output is determined by a capacitive voltage divider.

Figure 12.10
The switched-capacitor D/A converter

high, those capacitors where the digital input is high are switched to E_{REF}, whereas those whose inputs are low remain grounded. Figure 12.10(b) shows the equivalent circuit when ϕ_2 is high and the digital input is 1010. We see that the two capacitors whose digital inputs are 1 are in parallel, as are the two capacitors whose digital inputs are 0. The circuit is redrawn in Figure 12.10(c) with the parallel capacitors replaced by their equivalents (sums). The output of the capacitive voltage divider is

$$v_o = \left(\frac{5C/4}{5C/4 + 6C/8}\right) E_{REF} = \left(\frac{5C/4}{2C}\right) E_{REF} = \frac{5}{8} E_{REF} \qquad (12.9)$$

The denominator in (12.9) will always be $2C$, the sum of all the capacitance values in the circuit. From the foregoing analysis, we see that the output of the circuit in the general case is

$$v_o = \left(\frac{C_{eq}}{2C}\right) E_{REF} \qquad (12.10)$$

where C_{eq} is the equivalent (sum) of all the capacitors whose digital inputs are high. Table 12.2 shows the outputs for every possible input combination, and it is apparent that the analog output is proportional to the digital input. For the case where the input is 0000, note that the positive terminal of E_{REF} in Figure 12.10 is effectively open-circuited, so the output is 0 V.

Switched-capacitor technology evolved as a means for implementing analog functions in integrated circuits, particularly MOS circuits. It has been used to construct filters, amplifiers, and many other special devices. The principal advantage of the technology is that small capacitors, on the order of a few picofarads, can be constructed in the integrated circuits to perform the function of the much larger capacitors that are normally needed in low-frequency analog circuits. When capacitors are switched at a high enough frequency, they can be effectively "transformed" into other components, including resistors. The transformations

Table 12.2
Output voltages produced by a 4-bit switched-capacitor DAC

Binary Input $D_3D_2D_1D_0$	V_0
0000	0
0001	$(1/16)E_{REF}$
0010	$(1/8)E_{REF}$
0011	$(3/16)E_{REF}$
0100	$(1/4)E_{REF}$
0101	$(5/16)E_{REF}$
0110	$(3/8)E_{REF}$
0111	$(7/16)E_{REF}$
1000	$(1/2)E_{REF}$
1001	$(9/16)E_{REF}$
1010	$(5/8)E_{REF}$
1011	$(11/16)E_{REF}$
1100	$(3/4)E_{REF}$
1101	$(13/16)E_{REF}$
1110	$(7/8)E_{REF}$
1111	$(15/16)E_{REF}$

are studied from the standpoint of sampled-data theory, which is beyond the scope of this book.

12.6   DAC PERFORMANCE SPECIFICATIONS

As discussed in Section 12.1, the resolution of a D/A converter is a measure of the fineness of the increments between output values. Given a fixed output voltage range, say, 0 to 10 V, a DAC that divides that range into 1024 distinct output values has greater resolution than one that divides it into 512 values. Since the output increment is directly dependent on the number of input bits, the resolution is often quoted as simply that total number of bits. The most commonly available integrated-circuit converters have resolutions of 8, 10, 12, or 16 bits. Resolution is also expressed as the reciprocal of the total number of output voltages, often in terms of a percentage. For example, the resolution of an 8-bit DAC may be specified as

$$\left(\frac{1}{2^8}\right) \times 100\% = 0.39\%$$

Some DAC specifications are quoted with reference to one or to one-half LSB (least significant bit). In this context, an LSB is simply the increment between successive output voltages. Since an n-bit converter has $2^n - 1$ such increments,

$$\text{LSB} = \frac{\text{FSR}}{2^n - 1} \qquad\qquad\qquad \textbf{(12.11)}$$

where FSR is the full-scale range of the output voltage.

When the output of a DAC changes from one value to another, it typically overshoots the new value and may oscillate briefly around that new value before it settles to a constant voltage. The *settling time* of a D/A converter is the total time between the instant that the digital input changes and the time that the output enters a specified error band for the last time, usually $\pm 1/2$ LSB around the final value. Figure 12.11 illustrates the specification. Settling times of typical integrated-circuit converters range from 50 ns to several microseconds. Settling time may depend on the magnitude of the change at the input and is often specified for a prescribed input change.

Linearity error is the maximum deviation of the analog output from the ideal output. Since the output is ideally in direct proportion to the input, the ideal output is a straight line drawn from 0 V. Linearity error may be specified as a percentage of the full-scale range or in terms of an LSB.

Differential linearity error is the difference between the ideal output increment (1 LSB) and the actual increment. For example, each output increment of an 8-bit DAC whose full-scale range is 10 V should be $(10 \text{ V})/(2^8 - 1) = 39.22 \text{ mV}$. If any one increment between two successive values is, say, 30 mV, then there is a differential linearity error of 9.22 mV. This error is also specified as a percentage of the full-scale range or in terms of an LSB. If the differential linearity error is greater than 1 LSB, it is possible for the output voltage to *decrease* when there is an increase in the value of the digital input or to increase when the input decreases. Such behavior is said to be *nonmonotonic*. In a monotonic DAC, the output always increases when the input increases and decreases when the input decreases.

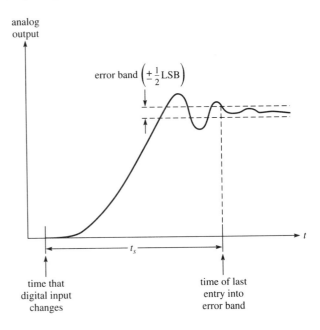

Figure 12.11
The settling time, t_s, of a D/A converter

The input of a DAC is said to undergo a *major change* when every input bit changes, as, for example, from 01111111 to 10000000. If the switches in Figure 12.8 open faster than they close or vice versa, the output of the DAC will momentarily go to 0 or to full scale when a major change occurs, thus creating an output *glitch*. *Glitch area* is the total area of an output voltage glitch in volt-seconds or of an output current glitch in ampere-seconds. Commercial units are often equipped with *deglitchers* to minimize glitch area.

An Integrated-Circuit DAC

The AD7524 CMOS integrated-circuit DAC, manufactured by Analog Devices and available from other manufacturers, is an example of an 8-bit, multiplying D/A converter. Figure 12.12 shows a functional block diagram. Note that the digital input is latched under the control of $\overline{CS}$ and $\overline{WR}$. When both of these control inputs are low, the output of the DAC responds directly to the digital inputs, with no latching occurring. If either control input goes high, the digital input is latched and the analog output remains at the level corresponding to the latched data, independent of any changes in the digital input. The device is then said to be in a HOLD mode, with the data bus *locked out*. The OUT2 output is normally grounded. Maximum settling time to a ±1/2 LSB error band for the Texas Instruments version is 100 ns and maximum linearity error is ±0.2% of the full-scale range. The device can be used as a 2- or 4-quadrant multiplier and E_{REF} can vary ±25 V.

12.7   THE COUNTER-TYPE ADC

The simplest type of A/D converter employs a binary counter, a voltage comparator, and a D/A converter, as shown in Figure 12.13(a). Recall from Section 11.1 that the output of the voltage comparator is high as long as its v^+ input is greater

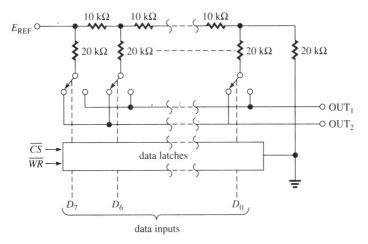

Figure 12.12
Functional block diagram of the AD7524 D/A converter

than its v^- input. Notice that the analog input is the v^+ input to the comparator. As long as it is greater than the v^- input, the AND gate is enabled and clock pulses are passed to the counter. The digital output of the counter is converted to an analog voltage by the DAC, and that voltage is the other input to the comparator. Thus, the counter counts up until its output has a value equal to the analog input. At that time, the comparator switches low, inhibiting the clock pulses, and counting ceases. The count it reaches is the digital output proportional to the analog input. Control circuitry, not shown, is used to latch the output and reset the counter. The cycle is repeated, with the counter reaching a new count proportional to whatever new value the analog input has acquired. Figure 12.13(b) illustrates the output of a 4-bit DAC in an ADC over several counting cycles when the analog input is a slowly varying voltage. The principal disadvantage of this type of converter is that the conversion time depends on the magnitude of the analog input: the larger the input, the more clock pulses must pass to reach the proper count. An 8-bit converter could require as many as 255 clock pulses to perform a conversion, so the counter-type ADC is considered quite slow in comparison to other types we will study.

Tracking A/D Converter

To reduce conversion times of the counter-type ADC, the up counter can be replaced by an up/down counter, as illustrated in Figure 12.14. In this design, the counter is not reset after each conversion. Instead, it counts up or down from its last count to its new count. Thus, the total number of clock pulses required to perform a conversion is proportional to the *change* in the analog input between counts rather than to its magnitude. Since the count more or less keeps up with the changing analog input, the type of ADC is called a *tracking converter*. A disadvantage of the design is that the count may oscillate up and down from a fixed count when the analog input is constant.

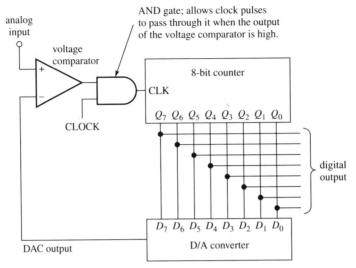

(a) Block diagram of an 8-bit ADC.

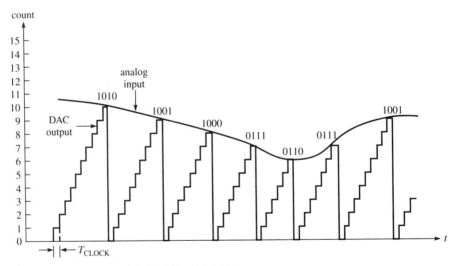

(b) Example of the output of the DAC in a 4-bit ADC.

Figure 12.13
The counter-type A/D converter

12.8   FLASH A/D CONVERTERS

The fastest type of A/D converter is called the *flash* (or simultaneous, or parallel) type. As shown for the 3-bit example in Figure 12.15, a reference voltage is connected to a voltage divider that divides it into 7 ($2^n - 1$) equal-increment levels. Each level is compared to the analog input by a voltage comparator. For any given analog input, one comparator and all those below it will have a high output. All comparator outputs are connected to a *priority encoder*. A priority

Figure 12.14
The tracking counter-type A/D converter.
The counter counts up or down from its
last count to reach its next count rather
than resetting to 0 between counts.

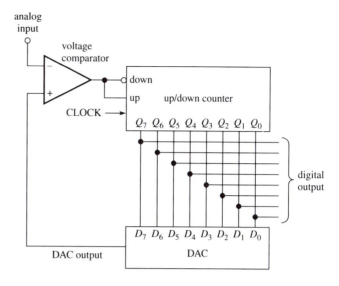

Figure 12.15
A 3-bit flash A/D converter

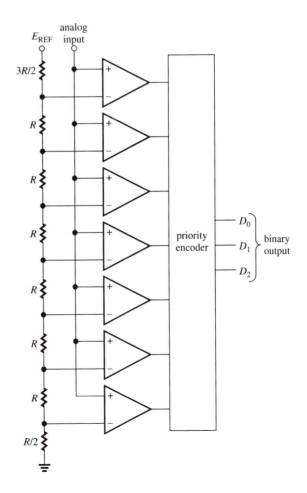

encoder produces a binary output corresponding to the input having the highest priority, in this case, the one representing the largest voltage level equal to or less than the analog input. Thus, the binary output represents the voltage that is closest in value to the analog input.

The voltage applied to the v^- input of the uppermost comparator in Figure 12.15 is, by voltage-divider action,

$$\left(\frac{6R + R/2}{6R + R/2 + 3R/2}\right) E_{\text{REF}} = \frac{13R/2}{16R/2} E_{\text{REF}} = \frac{13}{16} E_{\text{REF}} \qquad \textbf{(12.12)}$$

Similarly, the voltage applied to the v^- input of the second comparator is $(11/16)E_{\text{REF}}$, that applied to the third is $(9/16)E_{\text{REF}}$, and so forth. The increment between voltages is seen to be $(2/16)E_{\text{REF}}$, or $(1/8)E_{\text{REF}}$. An n-bit flash comparator has $(2^n - 2)$ R-valued resistors, and the increment between voltages is

$$\Delta V = \frac{1}{2^n} E_{\text{REF}} \qquad \textbf{(12.13)}$$

The voltage levels range from

$$\left(\frac{2^{n+1} - 3}{2^{n+1}}\right) E_{\text{REF}} \quad \text{through} \quad \left(\frac{1}{2^{n+1}}\right) E_{\text{REF}}$$

The flash converter is fast because the only delays in the conversion are in the comparators and the priority encoder. Under the control of a clock, a new conversion can be performed very soon after one conversion is complete. The principal disadvantage of the flash converter is the need for a large number of voltage comparators $(2^n - 1)$. For example, an 8-bit flash ADC requires 255 comparators.

Figure 12.16 shows a block diagram of a modified flash technique that uses 30 comparators instead of 255 to perform an 8-bit A/D conversion. One 4-bit flash converter is used to produce the 4 most significant bits. Those 4 bits are converted back to an analog voltage by a D/A converter and the voltage is subtracted from the analog input. The difference between the analog input and the analog voltage corresponding to the 4 most significant bits is an analog voltage corresponding to

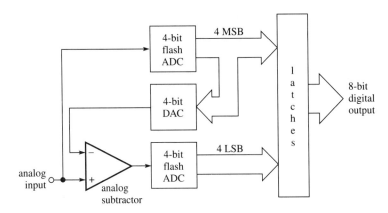

Figure 12.16
Modified flash converter that uses 30 comparators instead of 255 to produce an 8-bit output

the 4 least significant bits. Therefore, that voltage is converted to the 4 least significant bits by another 4-bit flash converter.

12.9   THE DUAL-SLOPE (INTEGRATING) ADC

A dual-slope ADC uses an operational amplifier to integrate the analog input. Recall from Chapter 9 that the output of an integrator is a ramp when the input is a fixed level (Figure 9.33). The slope of the ramp is $\pm E_{in}/R_1 C$, where E_{in} is the input voltage that is integrated and R_1 and C are the fixed components of the integrating operational amplifier. Since R_1 and C are fixed, the slope of the ramp is directly dependent on the value of E_{in}. If the ramp is allowed to continue for a fixed time, the voltage it reaches in that time depends on the slope of the ramp and hence on the value of E_{in}. The basic principle of the integrating ADC is that the voltage reached by the ramp controls the length of time that a binary counter is allowed to count. Thus, a binary number proportional to the value of E_{in} is obtained. In the dual-slope ADC, two integrations are performed, as described next.

Figure 12.17(a) shows a functional block diagram of the dual slope ADC. Recall that the integrating operational amplifier inverts, so a positive input generates a negative-going ramp and vice versa. A conversion begins with the switch connected to the analog input. Assume that the input is negative, so a positive-going ramp is generated by the integrator. As discussed earlier, the ramp is allowed to continue for a fixed time, and the voltage it reaches in that time is directly dependent on the analog input. The fixed time is controlled by sensing the time when the counter reaches a specific count. At that time, the counter is reset and control circuitry causes the switch to be connected to a reference voltage having a polarity *opposite* to that of the analog input—in this case, a positive voltage. As a conscquence, the output of the integrator becomes a negative-going ramp, beginning from the positive value it reached during the first integration. Since the reference voltage is fixed, so is the slope of the negative-going ramp. When the negative-going ramp reaches 0 V, the voltage comparator switches, the clock pulses are inhibited, and the counter ceases to count. The count it contains at that time is proportional to the time required for the negative-going ramp to reach 0 V, which is proportional to the positive voltage reached in the first integration. Thus, the binary count is proportional to the value of the analog input. Figure 12.17(b) shows examples of the ramp waveforms generated by a small analog input and by a large analog input. Note that the slope of the positive-going ramp is variable (depending on E_{in}), and the slope of the negative-going ramp is fixed. The origin of the name *dual-slope* is now apparent.

One advantage of the dual-slope converter is that its accuracy depends neither on the values of the integrator components R_1 and C nor upon any long-term changes that may occur in them. This fact is demonstrated by examining the equations governing the times required for the two integrations. Since the slope of the positive-going ramp is $E_{in}/R_1 C$, the maximum voltage V_M reached by the ramp in time t_1 is

$$V_M = \frac{|E_{in}|t_1}{R_1 C} \qquad \text{(12.14)}$$

The magnitude of the slope of the negative-going ramp is $|E_{REF}|/R_1 C$, so

$$V_M = \frac{|E_{REF}|}{R_1 C}(t_2 - t_1) \qquad \text{(12.15)}$$

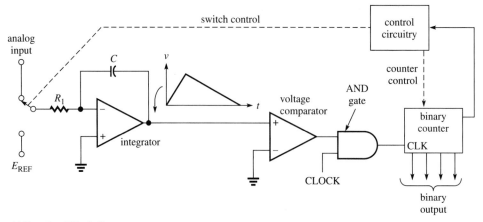

(a) Functional block diagram.

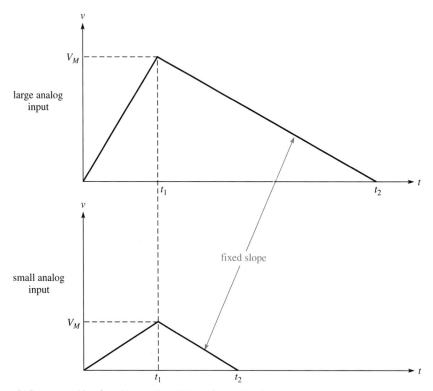

(b) Ramps resulting from large and small (negative) analog inputs.

Figure 12.17
The dual-slope integrating A/D converter

Equating (12.14) and (12.15),

$$\frac{|E_{in}|}{R_1 C} t_1 = \frac{|E_{REF}|}{R_1 C} (t_2 - t_1) \tag{12.16}$$

Cancelling $R_1 C$ on both sides and solving for $t_2 - t_1$ gives

$$t_2 - t_1 = \frac{|E_{\text{in}}|}{|E_{\text{REF}}|} t_1 \qquad \textbf{(12.17)}$$

Since the counter contains a count proportional to $t_2 - t_1$ (the time required for the negative-going ramp to reach 0 V) and t_1 is fixed, equation 12.17 shows that the count is directly proportional to E_{in}, the analog input. Note that this expression does not contain R_1 or C, since those quantities cancelled out in (12.16). Thus, accuracy does not depend on their values. Furthermore, accuracy does not depend on the frequency of the clock. Equation 12.17 shows that accuracy does depend on E_{REF}, so the reference voltage should be very precise.

An important advantage of the dual-slope A/D converter is that the integrator suppresses noise. Recall from equation 9.46 that the output of an integrator has amplitude inversely proportional to frequency. Thus, high-frequency noise components in the analog input are attenuated. This property makes it useful for instrumentation systems, and it is widely used for applications such as digital voltmeters. However, the integrating ADC is not particularly fast, so its use is restricted to signals having low to medium frequencies.

12.10   THE SUCCESSIVE-APPROXIMATION ADC

The method called successive approximation is the most popular technique used to construct A/D converters, and, with the exception of flash converters, successive-approximation converters are the fastest of those we have discussed. Figure 12.18(a) shows a block diagram of a 4-bit version. The method is best explained by way of an example. For simplifying purposes, let us assume that the output of the D/A converter ranges from 0 V through 15 V as its binary input ranges from 0000 through 1111, with 0000 producing 0 V, 0001 producing 1 V, and so forth. Suppose the "unknown" analog input is 13 V. On the first clock pulse, the output register is loaded with 1000, which is converted by the DAC to 8 V. The voltage comparator determines that 8 V is less than the analog input (13 V), so on the next clock pulse, the control circuitry causes the output register to be loaded with 1100. The output of the DAC is now 12 V, which the comparator again determines is less than the analog input. Consequently, the register is loaded with 1110 on the next clock pulse. The output of the DAC is 14 V, which the comparator now determines is larger than the 13-V analog input. Therefore, the last 1 that was loaded into the register is replaced with a 0, and a 1 is loaded into the LSB. This time, the output of the DAC is 13 V, which equals the analog input, so the conversion is complete. The output register contains 1101. We see that the method of successive approximation amounts to testing a sequence of trial values, each of which is adjusted to produce a number closer in value to the input than the previous value. This "homing-in" on the correct value is illustrated in part (b) of the figure. Note that an n-bit conversion requires n clock pulses. As another example, the following is the sequence of binary numbers that would appear in the output register of an 8-bit successive-approximation converter when the analog input is a voltage that is ultimately converted to 01101001:

10000000
01000000
01100000
01110000

01101000
01101100
01101010
01101001

Some modern successive-approximation converters have been constructed using the switched-capacitor technology discussed in connection with DACs.

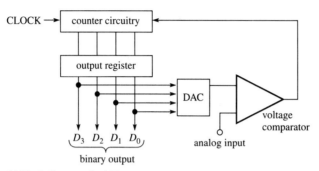

(a) Block diagram of a 4-bit converter.

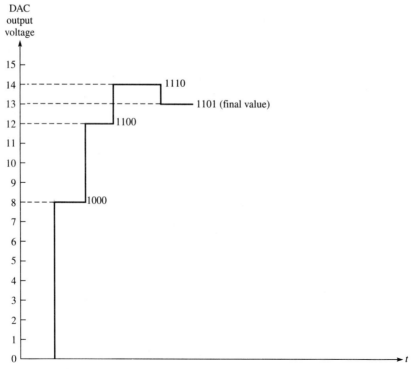

(b) Output of the DAC and contents of the output register, showing successive approximations when a 13-V analog input is converted.

Figure 12.18
The successive-approximation A/D converter

The primary component affecting the accuracy of a successive-approximation converter is the D/A converter. Consequently, the reference voltage connected to it and its ladder network must be very precise for accurate conversions. Also, the analog input should remain fixed during the conversion time. Some units employ a *sample-and-hold* circuit to ensure that the input voltage being compared at the voltage comparator does not vary during the conversion. A sample-and-hold is the analog counterpart of a digital data latch. It is constructed using electronic switches, an operational amplifier, and a capacitor that charges to and holds a particular voltage level.

12.11   ADC PERFORMANCE SPECIFICATIONS

The resolution of an A/D converter is the smallest change that can be distinguished in the analog input. As in DACs, resolution depends directly on the number of bits, so it is often quoted as simply the total number of output bits. The actual value depends on the full-scale range (FSR) of the analog input:

$$\text{resolution} = \frac{\text{FSR}}{2^n} \tag{12.18}$$

Some A/D converters have 8-4-2-1 BCD outputs rather than straight binary. This is especially true of dual-slope types designed for use in digital instruments. The BCD outputs facilitate driving numerical displays. The resolution of a BCD converter is quoted as the number of (decimal) digits available at the output, where each digit is represented by 4 bits. In this context, the term $\frac{1}{2}$ *digit* is used to refer to a single binary output. For example, a $1\frac{1}{2}$-digit output is represented by 5 bits. The $\frac{1}{2}$ digit is used as the most significant bit, and its presence doubles the number of decimal values that can be represented by the 4 BCD bits. Some BCD converters employ multiplexers to expand the number of output digits. An example is Texas Instruments' TLC135C $4\frac{1}{2}$-digit A/D converter. Expressed as a voltage, the resolution of a BCD A/D converter is $\text{FSR}/10^d$, where d is the number of output digits.

The time required to convert a single analog input to a digital output is called the *conversion time* of an A/D converter. Conversion time may be quoted as including any other delays, such as access time, associated with acquiring and converting an analog input. In that case, the total number of conversions that can be performed each second is the reciprocal of the conversion time. The reason this specification is important is that it imposes a limit on the rate at which the analog input can be allowed to change. In effect, the A/D converter *samples* the changing analog input when it performs a sequence of conversions. The *Shannon sampling theorem* states that sampled data can be used to faithfully reproduce a time-varying signal provided that the sampling rate is at least *twice* the frequency of the highest-frequency component in the signal. Thus, in order for the sequence of digital outputs to be a valid representation of the analog input, the A/D converter must perform conversions at a rate equal to at least twice the frequency of the highest-frequency component of the input.

Example 12.4

What maximum conversion time can an A/D converter have if it is to be used to convert *audio* input signals? (The audio frequency range is considered to be 20 Hz to 20 kHz.)

Solution.   Since the highest frequency in the input may be 20 kHz, conversions should be performed at a rate of at least 40×10^3 conversions/s. The maximum allowable conversion time is therefore equal to

$$\frac{1}{40 \times 10^3} = 25 \ \mu s$$

One LSB for an A/D converter is defined in the same way it is for a D/A converter:

$$\text{LSB} = \frac{\text{FSR}}{2^{n-1}} \tag{12.19}$$

where FSR is the full-scale range of the analog input. Other A/D converter specifications may be quoted in terms of an LSB or as a percentage of FSR.

Integrated-Circuit A/D Converters

A wide variety of A/D converters of all the types we have discussed are available in integrated circuits. Most have additional features, such as latched *three-state outputs* (0, 1, and open circuit), that make them compatible with microprocessor systems. Some have *differential* inputs, wherein the voltage converted to a digital output is the difference between two analog inputs. Differential inputs are valuable in reducing the effects of noise, because any noise signal common to both inputs is "differenced out." Recall that this property is called common-mode rejection. Differential inputs also allow the user to add or subtract a fixed voltage to the analog input, thereby offsetting the values converted. Conventional (single-ended) inputs can be accommodated simply by grounding one of the differential inputs.

The advances in MOS technology that have made high-density memory circuits possible have also made it possible to incorporate many special functions into a single-integrated circuit containing an ADC. Examples include sample-and-hold circuitry and analog multiplexers. With these functions, and versatile control circuitry, complete microprocessor-compatible *data-acquisition* systems have become available in a single-integrated circuit. The microprocessor can be programmed to control the multiplexer so that analog data from many different sources (up to 19 in some versions) can be sampled in a desired sequence. The analog data from an instrumentation system, for example, is sampled, converted, and transmitted directly to the microprocessor for storing in memory or further processing.

EXERCISES

Section 12.2   The *R-2R* Ladder DAC

12.1   The logic levels used in an 8-bit *R-2R* ladder DAC are 1 = 5 V and 0 = 0 V. Find the output voltage for each input:
  a. 00100000
  b. 10100100

12.2   The logic levels used in a 6-bit *R-2R* ladder DAC are 1 = +5 V and 0 = 0 V. What is the binary input when the analog output is 3.28125 V?

Section 12.3   A Weighted-Resistor DAC

12.3   Design a 5-bit weighted-resistor DAC whose full-scale output voltage is −15 V.

Logic levels are $1 = +5$ V and $0 = 0$ V. What is the output voltage when the input is 01010?

12.4 Design an 8-bit weighted-resistor DAC using operational amplifiers and two *identical* 4-bit DACs of the design shown in Figure 12.7. The output should be $+10$ V when the input is 11110000. (*Hint:* Sum the outputs of two amplifiers in a third amplifier, using appropriate voltage gains in the summation.) What is the full-scale output of your design?

Section 12.4 The Switched Current-Source DAC

12.5 In the switched current-source DAC shown in Figure 12.8, $R = 10$ kΩ and $E_{REF} = 20$ V. Find the current in each 20-kΩ resistor.

12.6 An 8-bit switched current-source DAC of the design shown in Figure 12.8 has $R = 10$ kΩ and $E_{REF} = 15$ V. Find the total current I_T delivered to the amplifier and the output voltage when the input is 01101100.

12.7 An 8-bit switched current-source DAC is operated as a two-quadrant multiplier. The binary input is positive and the reference voltage can range from -10 V to $+10$ V. If $R = 10$ kΩ, what is the total range of the output voltage?

12.8 A 4-bit switched current-source DAC is operated as a four-quadrant multiplier. The input is offset binary code.
Find the output voltage in each case:
   a. The reference voltage is -10 V and the input is 0011.
   b. The reference voltage is $+5$ V and the input is 1010.
   c. The reference voltage is $+10$ V and the input is 0001.

Section 12.5 Switched-Capacitor DACs

12.9 The binary input to a 4-bit switched-capacitor DAC having $C = 2$ pF is 0101.
   a. What is the total capacitance connected to E_{REF} when the ϕ_2 clock is high?
   b. What is the total capacitance connected to ground when the ϕ_2 clock is high?
   c. If $E_{REF} = 8$ V, what is the output voltage?

12.10 Draw a schematic diagram of an 8-bit switched capacitor DAC. Label all capacitor values in terms of capacitance C. What is the total capacitance when all capacitors are in parallel?

Section 12.6 DAC Performance Specifications

12.11 A 12-bit D/A converter has a full-scale range of 15 V. Its maximum differential linearity error is specified to be $\pm(\frac{1}{2})$LSB.
   a. What is its percentage resolution?
   b. What are the minimum and maximum possible values of the increment in its output voltage?

12.12 The LSB of a 10-bit D/A converter is 20 mV.
   a. What is its percentage resolution?
   b. What is its full-scale range?
   c. A differential linearity error greater than what percentage of FSR could make its output nonmonotonic?

Section 12.7 The Counter-Type ADC

12.13 The A/D converter in Figure 12.13(a) is clocked at 1 MHz. What is the maximum possible time that could be required to perform a conversion?

12.14 The minimum conversion time of a tracking type A/D converter is 400 ns. At what frequency is it clocked?

Section 12.8 Flash A/D Converters

12.15 A flash-type 5-bit A/D converter has a reference voltage of 10 V.
   a. How many voltage comparators does it have?
   b. What is the increment between the fixed voltages applied to the comparators?

12.16 The largest fixed voltage applied to a comparator in a flash-type A/D converter is 14.824218 V when the reference voltage is 15 V. What is the number of bits in the digital output?

Section 12.9 The Dual-Slope (Integrating) ADC

12.17 In an 8-bit, dual-slope A/D converter, $R_1 = 20$ kΩ and $C = 0.001$ μF. An analog input of -0.25 V is integrated for $t_1 = 160$ μs.
   a. What is the maximum voltage reached in the integration?
   b. If the input to the integrator is then switched to a reference voltage of $+5$ V, how long does it take the output to reach 0 V?
   c. If the counter is clocked at 3.125 MHz, what is the digital output after the conversion?

12.18 An 8-bit dual-slope A/D converter integrates analog inputs for $t_1 = 50$ μs. What should be the magnitude of the reference voltage if an input of 25 V is to produce a binary output of 11111111 when the clock frequency is 1 MHz?

Section 12.10 The Successive-Approximation ADC

12.19 List the sequence of binary numbers that would appear in the output register of a 4-bit successive-approximation A/D converter when the analog input has a value that is ultimately converted to 1011.

12.20 Sketch the output of the DAC in an 8-bit successive-approximation A/D converter when the analog input is a voltage that is ultimately converted to 10101011. Label each step of the DAC output with the decimal number corresponding to the binary value it represents.

Section 12.11 ADC Performance Specifications

12.21 List four types of A/D converters in descending order of speed (fastest converter type first).

12.22 List the principal advantage of each of four types of A/D converters.

12.23 The analog input of an 8-bit A/D converter can range from 0 V to 10 V. Find its resolution in volts and as a percentage of full-scale range.

12.24 The resolution of a 12-bit A/D converter is 7 mV. What is its full scale range?

12.25 The frequency components of the analog input to an A/D converter range from 50 Hz to 10 kHz. What maximum total conversion time should the converter have?

12.26 The analog input to an A/D converter consists of a 500-Hz fundamental waveform and its harmonics. If the converter has a total conversion time of 40 μs, what is the highest-order harmonic that should be in the input?

Power Supplies and Voltage Regulators

13.1   INTRODUCTION

The term *power supply* generally refers to a source of dc power that is itself operated from a source of ac power, such as a 120-V, 60-Hz line. A power supply of this type can therefore be regarded as an *ac-to-dc converter*. A supply that produces constant-frequency, constant-amplitude, *ac* power from a dc source is called an *inverter*. Some power supplies are designed to operate from a dc source and to produce power at a different dc level; these are called *dc-to-dc converters*.

A dc power supply operated from an ac source consists of one or more of the following fundamental components:

1. A *rectifier* that converts an ac voltage to a pulsating dc voltage and permits current to flow in one direction only;
2. A low-pass *filter* that suppresses the pulsations in the rectified waveform and passes its dc (average value) component; and
3. A *voltage regulator* that maintains a substantially constant output voltage under variations in the load current drawn from the supply and under variations in line voltage.

The extent to which each of the foregoing components is required in a given supply, and the complexity of its design, depend entirely on the application for which the supply is designed. For example, a dc supply designed exclusively for service as a battery charger can consist simply of a half-wave rectifier. No filter or regulator is required, since it is necessary to supply only pulsating, unidirectional current to recharge a battery. Special-purpose, fixed-voltage supplies that are subjected to little or no variation in current demand can be constructed with a rectifier and a large capacitive filter, and require no regulator. On the other hand, devices such as operational amplifiers and output power amplifiers require elaborate, well-regulated supplies.

Power supplies are classified as regulated or unregulated and as adjustable or fixed. Adjustable supplies are generally well regulated and are used in laboratory applications for general-purpose experimental and developmental work. The most demanding requirement imposed on these supplies is that they maintain an extremely constant output voltage over a wide range of loads and over a continu-ously adjustable range of voltages. Some fixed-voltage supplies are also designed so that their output can be adjusted, but over a relatively small range, to permit periodic recalibration.

13.2   RECTIFIERS

Half-Wave Rectifiers

In Chapter 2 (Section 2.5), we showed how a single diode performs half-wave rectification. Figure 13.1 shows the circuit and the half-wave–rectified waveform developed across load R_L. The average (dc) and rms values of a half-wave–rectified *sine-wave* voltage are

$$\left. \begin{aligned} V_{avg} &= \frac{V_{PR}}{\pi} \\ V_{rms} &= \frac{V_{PR}}{2} \end{aligned} \right\} \quad \text{(half-wave)} \qquad \text{(13.1)}$$

where V_{PR} is the peak value of the rectified voltage. Notice how the forward voltage drop across the diode affects the peak value of the rectified waveform shown in Figure 13.1. The peak value V_{PR} of v_L equals $V_P - 0.7$ V (for a silicon diode). The 0.7-V drop can be neglected in many power-supply applications be-cause of the large ac voltages present.

Assuming that the ac voltage is a symmetrical sine wave, the diode in Figure 13.1 must be capable of withstanding a peak reverse voltage of V_P volts, since that is the maximum reverse biasing voltage that occurs during one complete cycle. Recall that the *peak-inverse-voltage* (PIV) rating of a diode determines its maxi-mum permissible reverse bias without breakdown.

Full-Wave Rectifiers

Figure 13.2 shows the waveform that results when a sine wave is *full-wave recti-fied*. In effect, the negative half-cycles of the sine wave are inverted to create a

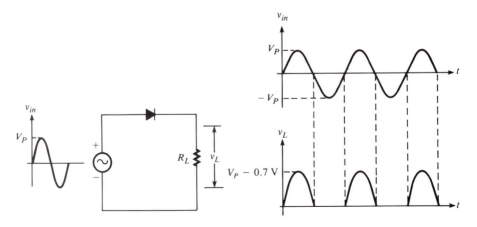

Figure 13.1
A half-wave rectifier

Figure 13.2
A full-wave–rectified sine wave

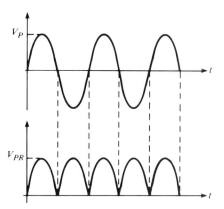

continuous series of positive half-cycles. Comparing Figure 13.1 with the full-wave–rectified waveform in Figure 13.2, it is apparent that the latter has twice the average value of the former. Thus, a full-wave–rectified voltage with peak value V_{PR} has average value

$$V_{avg} = \frac{2V_{PR}}{\pi} \qquad \text{(full-wave)} \qquad (13.2)$$

The rms value of a full-wave–rectified waveform is the same as that of a sine wave: $V_{PR}/\sqrt{2}$.

Figure 13.3(a) shows how a center-tapped transformer can be connected in a circuit with two diodes to perform full-wave rectification. Assume that the transformer is wound so that terminal A on the secondary is positive with respect to terminal B at an instant of time when v_{in} is positive, as signified by the polarity symbols (dot convention) shown in the figure. Then, with the center tap as reference (ground), v_A is positive with respect to ground and v_B is negative with respect to ground. Similarly, when v_{in} is negative, v_A is negative with respect to ground and v_B is positive with respect to ground. (For a detailed discussion of the voltages developed across the secondary of a center-tapped transformer, see Section 14.6.)

Figure 13.3(b) shows that when v_{in} is positive, v_A forward biases diode D_1. As a consequence, current flows in a clockwise loop through R_L. Figure 13.3(c) shows that when v_{in} is negative, D_1 is reverse biased, D_2 is forward biased, and current flows through R_L in a counterclockwise loop. Notice that the voltage developed across R_L has the same polarity in either case. Therefore, positive voltage pulses are developed across R_L during both the positive and negative half-cycles of v_{in}, and a full-wave–rectified waveform is created.

The peak rectified voltage is the secondary voltage in the transformer, between center tap and one side, less the diode drop:

$$V_{PR} = (N_s/N_p)V_P - 0.7 \text{ V} \qquad (13.3)$$

where V_P is the peak primary voltage, N_p is the total number of turns on the primary, and N_s is the number of turns between the center tap and either end of the secondary.

To determine the maximum reverse bias to which each diode is subjected, refer to the circuit in Figure 13.4. Here, we show the voltage drops in the rectifier when diode D_1 is forward biased and diode D_2 is reverse biased. Neglecting the

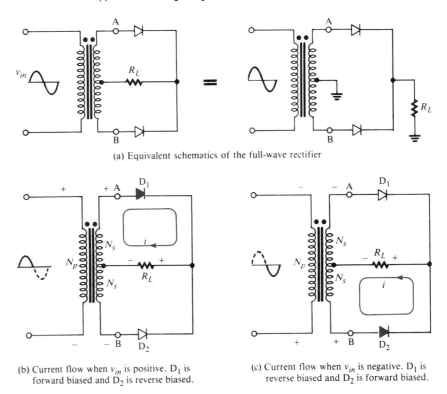

(a) Equivalent schematics of the full-wave rectifier

(b) Current flow when v_{in} is positive. D_1 is forward biased and D_2 is reverse biased.

(c) Current flow when v_{in} is negative. D_1 is reverse biased and D_2 is forward biased.

Figure 13.3
A full-wave rectifier employing a center-tapped transformer and two diodes

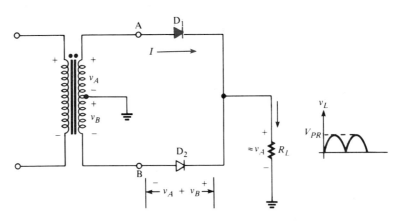

Figure 13.4
Diode D_2 is reverse biased by $v_A + v_B$ volts, which has a maximum value of $2V_{PR}$ volts.

0.7-V drop across D_1, the voltage across R_L is v_A volts. Thus the cathode-to-ground voltage of D_2 is v_A volts. Now, the anode-to-ground voltage of D_2 is $-v_B$ volts, as shown in the figure. Therefore, the total reverse bias across D_2 is $v_A + v_B$ volts, as shown. When v_A is at its positive peak, v_B is at its negative peak, so the maximum reverse bias equals *twice* the peak value of either. We conclude that the PIV rating of each diode must be equal to at least twice the peak value of the rectified voltage:

$$\text{PIV} \geq 2V_{PR} \tag{13.4}$$

Example 13.1

The primary voltage in the circuit shown in Figure 13.5 is 120 V rms, and the transformer has $N_p : N_s = 4:1$. Find

1. the average value of the voltage across R_L;
2. the (approximate) average power dissipated by R_L; and
3. the minimum PIV rating required for each diode.

Figure 13.5
(Example 13.1)

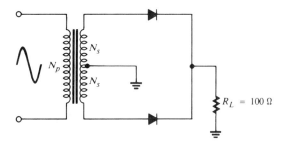

Solution

1. $V_P = \sqrt{2}(120) = 169.7$ V

   From equation 13.3, $V_{PR} = (N_s/N_p)V_P - 0.7 \text{ V} = (1/4)(169.7) - 0.7 \text{ V} = 41.7 \text{ V}.$

   Although equation 13.2 does not strictly apply to a rectified waveform with 0.7-V nonconducting gaps, it is a good approximation when the peak value is so much greater than 0.7 V:

   $$V_{avg} \approx \frac{2V_{PR}}{\pi} = \frac{2(41.7 \text{ V})}{\pi} = 26.5 \text{ V}$$

2. $$V_{rms} = \frac{V_{PR}}{\sqrt{2}} = \frac{41.7 \text{ V}}{\sqrt{2}} = 29.5 \text{ V rms}$$

   $$P_{avg} = \frac{V_{rms}^2}{R_L} = \frac{(29.5 \text{ V})^2}{100 \text{ }\Omega} = 8.7 \text{ W}$$

3. PIV $\geq 2V_{PR} = 2(41.7) = 83.4$ V

Full-Wave Bridge Rectifiers

Figure 13.6 shows another circuit used to perform full-wave rectification. One advantage of this circuit is that it does not require a transformer (although a transformer is often used to isolate the ac power line from the rest of the power

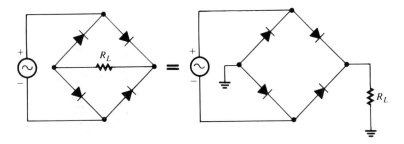

Figure 13.6
A full-wave bridge rectifier

supply). The diodes are arranged in the form of a bridge, and the circuit is called a
full-wave bridge rectifier.

Figure 13.7 demonstrates that current flows through R_L in the same direction
when v_{in} is positive as it does when v_{in} is negative. In Figure 13.7(a), v_{in} is positive,
so D_2 and D_3 are forward biased, while D_1 and D_4 are reverse biased. Current
therefore flows through R_L from right to left, as shown. In (b), v_{in} is negative, so D_1

Figure 13.7
*Current flow in the full-wave bridge
rectifier*

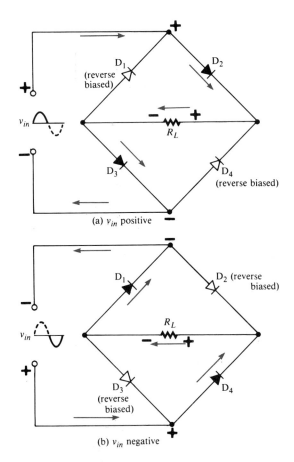

(a) v_{in} positive

(b) v_{in} negative

Diode Bridges

Molded diode bridges are available from 1 to 100 Amps in a variety of space-saving, easy-to-use packages.

Single ended bridges are available in three popular case styles for 1 to 6 Amps PC board-mount applications.

The 10 to 35 Amp range is covered by the JB and MB series which are isolated base, **U.L. recognized bridges** with very high surge current capability.

To make a bridge circuit into a center tap circuit, connect to terminals as shown.

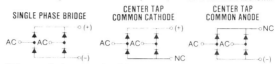

Ratings per diode are same for all three applications. NC = No Connection.

SINGLE PHASE DIODE MOLDED BRIDGES — 1.0 TO 100 AMPS (I_O)

I_O (A) Output Current	1.0	1.2	1.8	1.9	2.0	3.0	6.0	10	25	35	80	100
@ T_C (°C)	25	45	50	45	50	50	50	65	65	55	85	80
I_{FSM} (A) (Surge)	30	52	52	52	60	50	125	130	350	420	800	1000
Notes	(1)	(1)	(1)	(1) (3)	(1)	(1)	(1)	(2)	(2)	(2)	(2)	(2)
Case Style	D-43	D-38	D-2	D-37	D-44	D-45	D-46	D-34	D-34	D-34	D-20-6	D-20-6

PART NUMBERS

V_{RRM}												
50 Volts	—	—	18DB05A	—	2KBP005	KBPC1005	KBPC6005	100JB05L	250JB05L	35MB5A	800HB05U	1000HB05U
100 Volts	1DMB10	1KAB10E	18DB1A	2KBB10		KBPC105		100JB1L	250JB1L	35MB10A	800HB1U	1000HB1U
200 Volts	1DMB20	1KAB20E	18DB2A	2KBB20	2KBP02	KBPC102	KBPC602	100JB2L	250JB2L	35MB20A	800HB2U	1000HB2U
300 Volts	—	—	—	—							800HB3U	1000HB3U
400 Volts	1DMB40	1KAB40E	18DB4A	2KBB40	2KBP04	KBPC104	KBPC604	100JB4L	250JB4L	35MB40A	800HB4U	1000HB4U
600 Volts	—	1KAB60E	18DB6A	2KBB60	2KBP06	KBPC106	KBPC606	100JB6L	250JB6L	35MB60A	800HB6U	1000HB6U
800 Volts	—	1KAB80E	18DB8A	2KBB80	2KBP08	KBPC108	KBPC608	100JB8L	250JB8L	35MB80A	—	—
1000 Volts	—	1KAB100E	18DB10A	2KBB100	2KBP10	KBPC110	KBPC610	100JB10L	250JB10L	35MB100A	—	—
1200 Volts	—	—	—	—				100JB12L	250JB12L	35MB120A	—	—

NOTES: (1) Ambient temperature (T_A). (2) Must be used with suitable heatsink. (3) For lead configuration - ~ ~ + add 'R' to part number.

\# For detailed specifications, contact your local IR Field Office or IR Distributor.

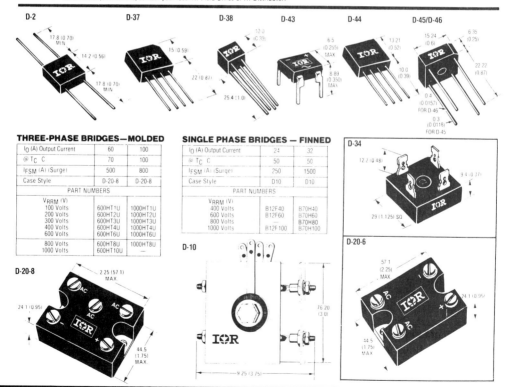

THREE-PHASE BRIDGES—MOLDED

I_O (A) Output Current	60	100
@ T_C C	70	100
I_{FSM} (A) (Surge)	500	800
Case Style	D-20-8	D-20-8

PART NUMBERS

V_{RRM} (V)		
100 Volts	600HT1U	1000HT1U
200 Volts	600HT2U	1000HT2U
300 Volts	600HT3U	1000HT3U
400 Volts	600HT4U	1000HT4U
600 Volts	600HT6U	1000HT6U
800 Volts	600HT8U	1000HT8U
1000 Volts	600HT10U	

SINGLE PHASE BRIDGES — FINNED

I_O (A) Output Current	24	32
@ T_C C	50	50
I_{FSM} (A) (Surge)	250	1500
Case Style	D10	D10

PART NUMBERS

V_{RRM} (V)		
400 Volts	B12F40	B70H40
600 Volts	B12F60	B70H60
800 Volts		B70H80
1000 Volts	B12F100	B70H100

Dimensions in Millimeters and (Inches)

Figure 13.8
Diode bridge specifications (Courtesy of International Rectifier)

and D_4 are forward biased, and D_2 and D_3 are reverse biased. Notice that current still flows through R_L from right to left. Therefore, the polarity of the voltage across R_L is always the same, confirming that full-wave rectification occurs. Note that the common side of the ac source *must* be isolated from the common side of the load voltage. Thus, the negative terminal of R_L in Figure 13.7(a) cannot be the same as the negative terminal of v_{in}. As previously mentioned, a transformer is often used to provide this isolation.

Since load current flows through two forward-biased diodes during each half-cycle of v_{in}, the peak rectified voltage across R_L is the peak input voltage reduced by $2(0.7 \text{ V}) = 1.4 \text{ V}$:

$$V_{PR} = V_P - 1.4 \text{ V} \qquad (13.5)$$

Another advantage of the bridge rectifier is that the reverse bias across each diode never exceeds V_{PR}. Thus, the PIV rating of the diodes is half that required in the rectifier using a center-tapped transformer.

Diode bridges are now commonly available in single-package units. These packages have a pair of terminals to which the ac input is connected and another pair at which the full-wave–rectified output is taken. Figure 13.8 shows a typical manufacturer's specification sheet for a line of single-package bridges with current ratings from 1 to 100 A. The V_{RRM} specifications refer to the maximum repetitive reverse voltage ratings of each, i.e., the peak inverse voltage ratings when operated with repetitive inputs, such as sinusoidal voltages. These ratings range from 50 to 1200 V. Note the I_{FSM} specifications, which refer to maximum forward surge current. As we will learn in the next section, a filter is often connected to the output of a bridge, and the nature of the filter determines the amount of current that surges through the rectifier when ac power is first applied. For example, if a large capacitor is connected directly across the bridge output, then the rectifier is essentially shorted out when power is first applied. The filter must be designed so that this initial surge current does not exceed the bridge rating. The specifications show that bridges are available with I_{FSM} ratings from 30 to 1000 A.

Example 13.2

DESIGN

Figure 13.9 shows a bridge rectifier isolated from the 120-V-rms power line by a transformer.

1. What turns ratio should the transformer have in order to produce an average current of 1 A in R_L?

Figure 13.9
(Example 13.2)

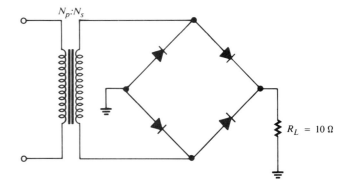

$N_p{:}N_s$

$R_L = 10 \text{ }\Omega$

2. What is the average current in each diode under the conditions of (1)?
3. What minimum PIV rating should each diode have?
4. How much power is dissipated by each diode?

Solution

1. $I_{avg} = 2I_{PR}/\pi$, where I_{PR} is the peak value of the rectified current in R_L. Thus,

$$I_{PR} = \frac{\pi I_{avg}}{2} = \frac{\pi(1\ A)}{2} = 1.57\ A\ pk$$

$$V_{PR} = (I_{PR})R_L = (1.57)(10) = 15.7\ V$$

From equation 13.5, the peak value of the voltage applied across the bridge is $V_P = V_{PR} + 1.4 = 15.7 + 1.4 = 17.1$ V. Since the peak value of the 120-V-rms primary is $(\sqrt{2})(120\ V\ rms) = 169.7$ V-pk,

$$\frac{N_p}{N_s} = \frac{169.7}{17.1} = 9.92:1$$

2. Each diode conducts during one half-cycle only, so the average current in each is the same as that of a half-wave–rectified sine wave:

$$I_{avg} = \frac{I_{PR}}{\pi} = \frac{1.57}{\pi} = 0.5\ A\ dc$$

Note that this result is one-half the total average current in R_L, as expected.

3. PIV $\geq V_{PR} = 15.7$ V.

4. Assuming that negligible current flows through each diode when it is reverse biased, power is dissipated by a diode only when it is conducting forward current. Assuming that the forward voltage is 0.7 V during the full half-cycle that a diode conducts (a conservative, worst-case assumption), the rms diode voltage over one full cycle is*

$$V_{rms} = \frac{0.7\ V}{\sqrt{2}} = 0.5\ V\ rms$$

The rms value of one-half cycle of the sinusoidal current having peak value $I_{PR} = 1.57$ A is

$$I_{rms} = \frac{I_{PR}}{2} = \frac{1.57\ A}{2} = 0.785\ A\ rms$$

Therefore,

$$P_{avg} = V_{rms}I_{rms} = (0.5\ V)(0.785\ A) = 0.39\ W$$

13.3   **CAPACITOR FILTERS**

The frequency of the *fundamental* component of a half-wave–rectified waveform is the same as the frequency of its original (unrectified) ac waveform. Since the ac power source used in most dc supplies has frequency 60 Hz, a half-wave–rectified

---

*The rms value of a waveform that equals V for one half-cycle and equals 0 for the other half-cycle is $V/\sqrt{2}$.

waveform contains a 60-Hz fundamental, plus *harmonic* components that are integer multiples of 60 Hz, plus a dc, or average value, component. A full-wave–rectified waveform has one-half the period of a half-wave–rectified waveform, so it has twice the frequency. It therefore contains a 120-Hz fundamental, plus harmonics of 120 Hz, plus a dc component that is twice that of the half-wave–rectified waveform, as we have seen. A low-pass filter is connected across the output of a rectifier to suppress the ac components and to pass the dc component.

A rudimentary low-pass filter used in power supplies consists simply of a capacitor connected across the rectifier output, that is, in parallel with the load, as illustrated in Figure 13.10. Here, we show a simple half-wave rectifier with capacitor C connected in parallel with R_L. The forward resistance of the diode is small in comparison to R_L, so during positive half-cycles, the capacitor charges to the peak value of the ac input.

It is not convenient to analyze the circuit in Figure 13.10 in terms of filter theory, because the nonlinear operation of the diode effectively changes the circuit resistance, as the diode is alternately forward and reverse biased. Instead, we analyze it from the standpoint of the transient voltage across the capacitor.

Figure 13.11(a) shows how the capacitor charges and discharges during a full cycle of the ac input. Notice that the capacitor charges and its voltage rises with the input voltage when the input is large enough to forward bias the diode. The capacitor discharges through R_L when the input falls to a level below which the diode is reverse biased. The smaller the R_LC time constant, the further the capacitor voltage decays before another positive pulse arrives and recharges the capacitor. Figure 13.11(b) compares the output waveforms that result when large and small time constants are used. The voltage fluctuation in the filtered waveform is called the *ripple voltage,* which in most applications should be kept as small as possible. Figure 13.11(b) shows that a heavy load (small R_L) will result in an undesirably large ripple voltage.

Figure 13.12 shows a full-wave bridge rectifier with a capacitor filter. Also shown is the filtered waveform. In this case, positive pulses are present during every half-cycle of input, so the capacitor voltage does not decay as far as it does in the half-wave circuit before another pulse is available to recharge it. As a consequence, the peak-to-peak ripple voltage is smaller, for a given R_LC time constant, than it is in the half-wave circuit.

Percent Ripple

The percent ripple in a rectified waveform (also called the *ripple factor*) is defined by

$$r = \frac{\text{rms ripple voltage}}{\text{dc (average) voltage}} \times 100\% \qquad (13.6)$$

Figure 13.10
A half-wave rectifier with capacitor filter

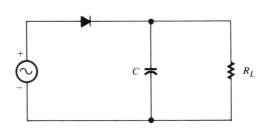

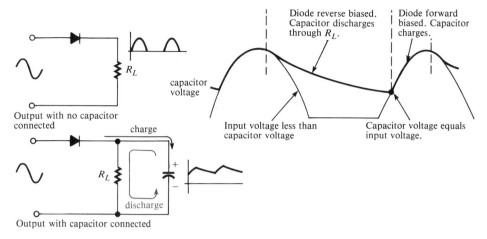

(a) The filter capacitor charges when the diode is forward biased and discharges when the diode is reverse biased.

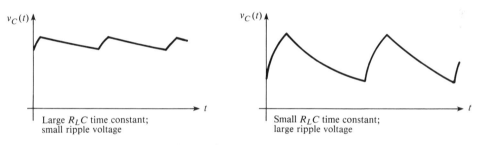

(b) The effect of the $R_L C$ time constant on the filtered waveform

Figure 13.11
Waveforms produced by a half-wave rectifier with capacitor filter

Note carefully that the numerator is the rms value of just the ripple (ac components) in the rectified waveform, not the rms value of the total waveform.

Let us first compute the percent ripple in *unfiltered* half- and full-wave–rectified waveforms. Using calculus, it can be shown that the rms values of the *ac components* of half- and full-wave–rectified voltages are

$$V(\text{rms}) = 0.385 V_{PR} \qquad \text{(half-wave)} \qquad \textbf{(13.7)}$$

$$V(\text{rms}) = 0.308 V_{PR} \qquad \text{(full-wave)} \qquad \textbf{(13.8)}$$

where V_{PR} is the peak value of the rectified waveform in each case. (Note carefully that equations 13.7 and 13.8 give the rms values of the ac components *only*, *not* the rms values of the waveforms themselves.) Then

$$r(\text{half-wave, unfiltered}) = \frac{0.385 V_{PR}}{V_{PR}/\pi} \times 100\% = 121\% \qquad \textbf{(13.9)}$$

and

$$r(\text{full-wave, unfiltered}) = \frac{0.308 V_{PR}}{2 V_{PR}/\pi} \times 100\% = 48.4\% \qquad \textbf{(13.10)}$$

These ripple values are quite large and demonstrate the need for a filter in most power-supply applications. If a filter is not used, or if it is improperly designed, the ac components of the ripple are superimposed on the signal lines in the device that receives power from the supply. Power-supply ripple is therefore a source of *noise* in an electronic system. It is one of the principal causes of 60-Hz (or 120-Hz) *hum* in an audio amplifier.

We will now derive an expression for the ripple in the output of a rectifier having a capacitor filter and load resistance R_L (Figures 13.11 and 13.12). Since a knowledge of ripple magnitude is important in only those applications where ripple affects the performance of a system using the power supply, we assume that the filter is well designed and that, as a consequence, the ripple voltage is small compared to the dc component. In other words, we assume that the capacitor voltage does not decay significantly from its peak value between the occurrences of the rectified pulses that recharge the capacitor. This circumstance is called *light loading,* because the charge supplied to R_L by the capacitor is small compared to the total charge stored on the capacitor. To simplify the computations, we can assume that the ripple voltage in a lightly loaded filter is a sawtooth wave, as illustrated in Figure 13.13.

Figure 13.13(a) shows the waveform in a lightly loaded capacitor filter across the output of a full-wave rectifier. The derivation that follows is applicable to both half-wave and full-wave rectifiers, provided the assumption of light loading is valid for both cases. Note that the period T shown in the figure is the fundamental period of the *rectified* waveform, typically 1/60 s for half-wave rectifiers and 1/120 s for full-wave rectifiers. The light-loading assumption in the half-wave case is more restrictive than in the full-wave case, because there is a longer time between pulses, during which the capacitor voltage can decay.

Figure 13.13(b) and (c) show the ripple waveform approximated by a sawtooth voltage with peak-to-peak value V_{PP}. This approximation is equivalent to assum-

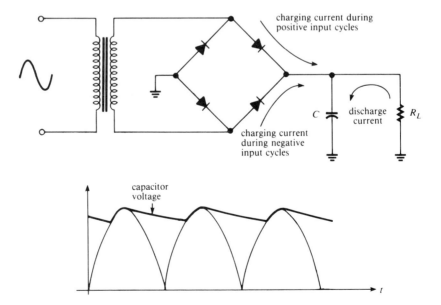

Figure 13.12
A full-wave bridge rectifier with capacitor filter

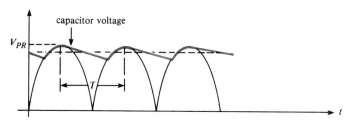

(a) In the lightly loaded filter, the capacitor voltage does not decay significantly from V_{PR} between charging pulses.

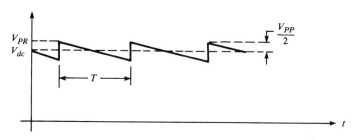

(b) The capacitor voltage in (a) approximated by a sawtooth waveform

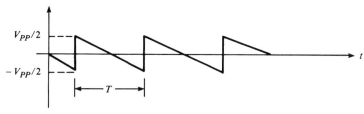

(c) The ac component (ripple voltage) in (b)

Figure 13.13
Approximating the ripple voltage in a lightly loaded capacitor filter

ing that the capacitor charges instantaneously and that its voltage decays linearly, instead of exponentially. Assuming that the voltage decays linearly is equivalent to assuming that the discharge current is constant and equal to V_{dc}/R_L, where V_{dc} is the dc value of the filtered waveform. As shown in (c), the total change in capacitor voltage is V_{PP} volts, and this change occurs over the period of time T. Therefore, since $\Delta Q = I\Delta t$,

$$V_{PP} = \frac{\Delta Q}{C} = \frac{(V_{dc}/R_L)T}{C} \qquad (13.11)$$

Since $T = 1/f_r$, where f_r is the frequency of the fundamental component of the ripple (typically 60 Hz or 120 Hz), equation 13.11 can be written

$$V_{PP} = \frac{V_{dc}}{f_r R_L C} \qquad (13.12)$$

or

$$V_{dc} = V_{PP} f_r R_L C \qquad (13.13)$$

From Figure 13.13(b), it is apparent that

$$V_{dc} = V_{PR} - V_{PP}/2 \qquad\qquad (13.14)$$

Substituting from (13.12), we obtain

$$V_{dc} = V_{PR} - \frac{V_{dc}}{2f_r R_L C} \qquad\qquad (13.15)$$

Solving for V_{dc}, we obtain an expression for V_{dc} in terms of the peak rectifier voltage:

$$V_{dc} = \frac{V_{PR}}{1 + \dfrac{1}{2f_r R_L C}} \qquad\qquad (13.16)$$

It is clear that the dc value cannot exceed V_{PR} volts, and that it equals V_{PR} when $R_L = \infty$ (i.e., when the load is an open circuit).

The rms value of a sawtooth waveform having peak-to-peak value V_{PP} is known to be

$$V(\text{rms}) = \frac{V_{PP}}{2\sqrt{3}} \qquad\qquad (13.17)$$

Therefore, from (13.13) and (13.17), the percent ripple is

$$r = \frac{V(\text{rms})}{V_{dc}} \times 100\% = \frac{V_{PP}/2\sqrt{3}}{V_{PP} f_r R_L C} \times 100\%$$

$$= \frac{1}{2\sqrt{3} f_r R_L C} \times 100\% \qquad\qquad (13.18)$$

Equation 13.18 confirms our previous analysis of the capacitor filter (Figure 13.11): A large $R_L C$ time constant results in a small ripple voltage, and vice versa. The light-load assumption on which our derivation is based is generally valid for percent ripple less than about 6.5%. From a design standpoint, the values of f_r and R_L are usually fixed, and the designer's task is to select a value of C that keeps the ripple below a prescribed value.

Example 13.3

DESIGN

A full-wave rectifier is operated from a 60-Hz line and has a filter capacitor connected across its output. What minimum value of capacitance is required if the load is 200 Ω and the ripple must be no greater than 4%?

Solution.   Using the decimal form of r ($r = 0.04$), we find, from equation 13.18,

$$\frac{1}{C} = 2\sqrt{3} f_r R_L r$$

or

$$C = \frac{1}{2\sqrt{3}(120)(200)(0.04)} \approx 300 \ \mu\text{F}$$

Example 13.4

The rectifier shown in Figure 13.10 is operated from a 60-Hz, 120-V-rms line. It has a 100-μF filter capacitor and a 1-kΩ load.

1. What is the percent ripple?
2. What is the average current in R_L?

Solution

1. We must first assume that the filter is lightly loaded, then perform the computation based on that assumption, and then verify from the result that the assumption is valid. From equation 13.18,

$$r = \frac{1}{2\sqrt{3}f_rR_LC} \times 100\% = \frac{100\%}{2\sqrt{3}(60)(10^3)(10^{-4})} = 4.8\%$$

   We see that $r < 6.5\%$ and that our assumption of a light load is therefore appropriate.

2. Neglecting the drop across the diode,

$$V_{PR} = \sqrt{2}(120) = 169.7 \text{ V}$$

   From equation 13.16,

$$V_{dc} = \frac{V_{PR}}{1 + \dfrac{1}{2f_rR_LC}} = \frac{169.7}{1 + \dfrac{1}{2(60)(10^3)(10^{-4})}} = \frac{169.7}{1.083} = 157 \text{ V}$$

   Therefore, $I_{dc} = V_{dc}/R_L = 157/1000 = 157$ mA.

Example 2.9 at the end of Chapter 2 shows how SPICE can be used to determine the ripple voltage on a capacitor filter that is *not* lightly loaded.

Repetitive Surge Currents

We have already discussed the initial surge current that flows through rectifier diodes when power is first applied and the uncharged filter capacitor behaves like a momentary short circuit. Figure 13.14 shows that current also surges through the diodes during the short time intervals that they are forward biased in normal operation. When the capacitor voltage is below the rectifier voltage, the total amount of charge that flows through the forward-biased diode(s) to recharge the capacitor must equal the amount of charge lost from the capacitor during the time it discharged into R_L. For a lightly loaded filter, the capacitor voltage does not decay significantly, so the conduction interval T_1 shown in the figure is small compared to the period T between rectifier pulses. If we approximate the current pulse as a rectangle with height I_P and width T_1, and the discharge current as a rectangle with height I_{dc} and width T, we have

$$Q(\text{discharge}) = Q(\text{charge})$$
$$I_{dc}T = I_PT_1$$

or   (13.19)

$$I_P = I_{dc}\frac{T}{T_1}$$

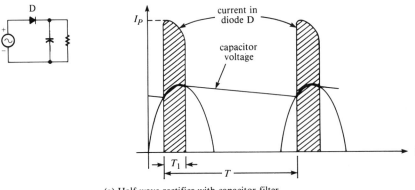

(a) Half-wave rectifier with capacitor filter

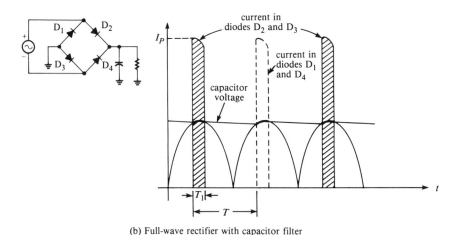

(b) Full-wave rectifier with capacitor filter

Figure 13.14
Repetitive surge currents in rectifier diodes

Since $T \gg T_1$, equation 13.19 shows that the peak current through the diode may be many times larger than the average current supplied to the load. The diodes must be capable of supplying this repetitive surge current.

Increasing the value of filter capacitance C decreases the ripple (equation 13.18) and increases the dc voltage (equation 13.16), both of which are generally desirable outcomes. However, increased capacitance also shortens the conduction interval T_1 in Figure 13.14, because the capacitor voltage decays less when the $R_L C$ time constant is large. Consequently, by equation 13.19, large capacitance increases the peak diode current, and the permissible repetitive surge current imposes a practical limitation on the size of the filter capacitor that can be used. To limit surge current, especially that which occurs at initial turn-on, a small resistance is sometimes connected between the rectifier and the filter capacitor, as shown in Figure 13.15. One disadvantage of this resistance is that it reduces the dc voltage level at the load, since it forms a voltage divider with R_L.

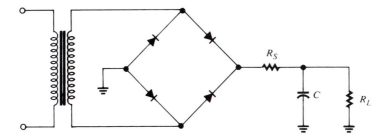

Figure 13.15
Use of series resistance R_S to limit diode surge currents

13.4   RC AND LC FILTERS

RC Filters

To obtain a greater reduction in the ripple of a capacitor-filtered waveform, a low-pass RC filter section can be connected across the capacitor, as shown in Figure 13.16. The capacitor–resistor–capacitor combination is often called an $RC\,\pi$ (PI) filter. The figure shows that the low-pass RC section further attenuates the ac components represented by the ripple and passes the dc value, although there is some reduction in dc level.

An approximate analysis of the filter shown in Figure 13.16 can be performed by assuming that the impedance looking back into node 1 is small compared to the impedance of the rest of the filter network connected across node 1. That assumption, which is valid for large C_1 and small diode on-resistance, allows us to treat node 1 as a voltage source that drives the rest of the filter. We can then apply the superposition principle to determine the dc and ac voltage components at node 2 due to the dc and ac components at node 1. Figure 13.17 shows the dc and ac equivalent circuits of the filter with node 1 replaced by dc and ac voltage sources.

Figure 13.16
Adding a low-pass RC filter section (RC_2) reduces the ripple voltage and the dc value of the filtered waveform.

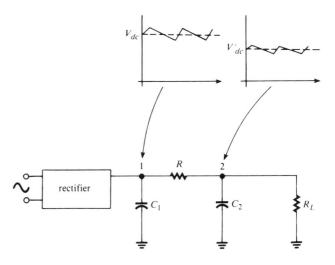

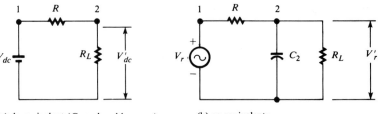

(a) dc equivalent (C_2 replaced by open)

(b) ac equivalent;
V_r = rms ripple voltage at node 1.

Figure 13.17
DC and ac equivalent circuits of Figure 13.16

It is apparent from Figure 13.17(a) that the dc load voltage is, by voltage-divider action,

$$V'_{dc} = \frac{R_L}{R + R_L} V_{dc} \tag{13.20}$$

Equation 13.20 shows that the use of an RC filter causes a reduction in the dc voltage available at the load. To minimize this effect, R is usually made much smaller than R_L.

From Figure 13.17(b),

$$V'_r = \frac{Z}{R + Z} V_r \tag{13.21}$$

where

$$Z = R_L \parallel (-jX_{C_2}) = \frac{-jR_L/\omega C_2}{R_L - jR/\omega C_2}$$

We are interested only in the magnitude of V'_r, so

$$|V'_r| = \frac{|Z|}{|R + Z|} |V_r| \tag{13.22}$$

where

$$|Z| = \frac{|-jR_L/\omega C_2|}{|R_L - j/\omega C_2|} = \frac{R_L}{\sqrt{1 + (\omega R_L C_2)^2}} \tag{13.23}$$

and

$$|R + Z| = \left| R - \frac{jR_L/\omega C_2}{R_L - j/\omega C_2} \right| = \sqrt{\frac{(R + R_L)^2 + (\omega R R_L C_2)^2}{1 + (\omega R_L C_2)^2}} \tag{13.24}$$

Substituting (13.23) and (13.24) into (13.22), we find

$$|V'_r| = \frac{R_L}{\sqrt{(R + R_L)^2 + (\omega R R_L C_2)^2}} |V_r| \tag{13.25}$$

Although the ripple voltage contains harmonics, ac calculations are usually performed using only the fundamental frequency of the ripple: $\omega = 2\pi(60)$ rad/s for a half-wave rectifier, and $\omega = 2\pi(120)$ rad/s for a full-wave rectifier.

Equation 13.25 can be written

$$|V'_r| = \frac{|V_r|}{\sqrt{\left(\frac{R}{R_L} + 1\right)^2 + (\omega R C_2)^2}} \tag{13.26}$$

Since R is usually much smaller than R_L, as noted earlier, it follows that $R/R_L + 1 \approx 1$, and (13.26) can be approximated as

$$|V_r'| \approx \frac{|V_r|}{\sqrt{1 + (\omega R C_2)^2}} \qquad (13.27)$$

Example 13.5

1. Find the average value and the percent ripple in the load voltage in Figure 13.18(a).
2. Repeat when an RC filter is inserted between the capacitor and the load, as shown in Figure 13.18(b).

Solution

1. Since the rectifier is full-wave, $f_r = 120$ Hz. $V_{PR} = \sqrt{2}(30 \text{ V rms}) - 1.4 \text{ V} = 41 \text{ V}$. From equation 13.16,

$$V_{dc} = \frac{V_{PR}}{1 + \dfrac{1}{2 f_r R_L C}} = \frac{41 \text{ V}}{1 + \dfrac{1}{2(120)(200)(10^{-3})}} = 40.2 \text{ V}$$

From equation 13.12, the peak-to-peak ripple voltage is

$$V_{PP} = \frac{V_{dc}}{f_r R_L C} = \frac{40.2 \text{ V}}{(120)(200)(10^{-3})} = 1.68 \text{ V}$$

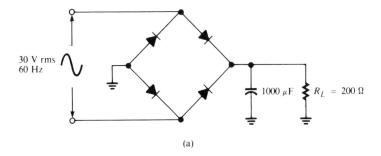

(a)

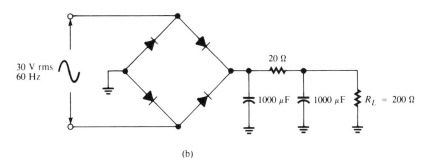

(b)

Figure 13.18
(Example 13.5)

From equation 13.17, the rms ripple voltage is

$$V_r = \frac{V_{PP}}{2\sqrt{3}} = \frac{1.68}{2\sqrt{3}} = 0.48 \text{ V rms}$$

Therefore,

$$r = \frac{V_r}{V_{dc}} \times 100\% = \frac{0.48}{40.2} \times 100\% = 1.2\%$$

2. We will assume that V_{dc} and V_r have the same values calculated in 1. In reality, the presence of the new RC filter section affects the values of V_{dc} and V_r somewhat. In practical circuits, the effect is small and the computational error resulting from the assumption can be neglected. From equation 13.20,

$$V'_{dc} = \frac{R_L}{R + R_L} V_{dc} = \left(\frac{200}{20 + 200}\right)(40.2 \text{ V}) = 36.5 \text{ V}$$

From equation 13.33,

$$|V'_r| = \frac{|V_r|}{\sqrt{\left(\frac{R}{R_L} + 1\right)^2 + (\omega R C_2)^2}} = \frac{0.48 \text{ V}}{\sqrt{\left(\frac{20}{200} + 1\right)^2 + [2\pi(120)(20)(10^{-3})]^2}}$$

$$= 0.032 \text{ V rms}$$

Therefore,

$$r' = \frac{V'_r}{V'_{dc}} \times 100\% = \frac{0.032}{36.5} \times 100\% = 0.087\%$$

We see that the RC filter section has reduced the percent ripple from 1.2% to 0.087%, i.e., by a factor of more than 13. The penalty paid for this improvement is a 9.2% decrease in the dc voltage and a reduction in the ability of the supply to *regulate* the output (hold a constant voltage, independent of load), as we shall discuss presently.

If approximation 13.27 is used to calculate the ripple voltage, we find

$$|V'_r| \approx \frac{0.48 \text{ V}}{\sqrt{1 + [2\pi(120)(20)(10^{-3})]^2}} = 0.032 \text{ V rms}$$

This approximation equals the previously calculated value out to three decimal places.

LC π Filters

The series resistance in the RC filter causes a voltage drop that reduces load voltage, as we have seen. Furthermore, the voltage drop increases as load current increases, so the output voltage varies with load. This variation is undesirable in a power supply, which ideally should maintain a constant output voltage, independent of load current. (Series resistance always degrades the *voltage regulation* of a supply. We will discuss percent voltage regulation in detail in a subsequent discussion.) By connecting a low-resistance inductor in place of the series resistor, as shown in Figure 13.19, we can reduce the dc voltage variation with load and at the same time attenuate ripple by virtue of the inductive reactance presented to the ac components. The filter is called an LC π, and while it is an improvement over the RC π, it does require a large and usually expensive inductor.

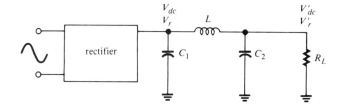

Figure 13.19
The LC π filter. The inductor presents a low resistance to the dc component and a large impedance to the ac (ripple) components.

By an analysis similar to that used for the RC filter, the following relationships can be derived:

$$V'_{dc} = \frac{R_L}{R' + R_L} V_{dc} \tag{13.28}$$

where R' is the resistance of the inductor; and

$$|V'_r| \approx \frac{|V_r|}{\omega^2 L C_2} \tag{13.29}$$

As in the case of the RC π, V_{dc} is the dc voltage across C_1 and V_r is the rms ripple voltage across C_1. Approximation 13.29 is valid when $R' \ll \omega L$, $R' \ll R_L$, $|X_{C_2} \parallel R_L| \approx |X_{C_2}|$, and $|X_L + (X_{C_2} \parallel R_L)| \approx |X_L|$, which is the usual situation, as illustrated in the next example.

Example 13.6

Assuming that $V_{dc} = 40.2$ V and $V_r = 0.48$ V rms in Figure 13.19 (the same values found in Example 13.5), find V'_{dc} and V'_r when $C_1 = C_2 = 1000$ μF, $R_L = 200$ Ω, and L is a 6-H inductor with resistance 2 Ω.

Solution. By equation 13.28,

$$V'_{dc} = \frac{R_L}{R' + R_L} V_{dc} = \left(\frac{200}{2 + 200}\right) 40.2 \text{ V} = 39.8 \text{ V}$$

To verify that the conditions under which approximation 13.29 can be used are applicable in this example, we compute

$$\omega L = 2\pi(120)(6) = 4524 \ \Omega$$
$$X_{C_2} = 1/(2\pi)(120)(10^{-3}) = 1.33 \ \Omega$$

Thus,

$$R' = 2 \ \Omega \ll 4524 \ \Omega = \omega L$$
$$R' = 2 \ \Omega \ll 200 \ \Omega = R_L$$
$$|X_{C_2} \parallel R_L| = \frac{R_L}{\sqrt{1 + (\omega R_L C_2)^2}} \quad \text{(equation 13.23)}$$
$$= \frac{200}{\sqrt{1 + [2\pi(120)(200)(10^{-3})]^2}} = 1.32 \ \Omega \approx |X_{C_2}|$$
$$|X_L + (X_{C_2} \parallel R_L)| \approx \sqrt{(4524)^2 + (1.32)^2} \approx 4524 \ \Omega = |X_L|$$

The approximating conditions are therefore valid, and we have, from (13.29),

$$|V_r'| \approx \frac{|V_r|}{\omega^2 L C_2} = \frac{0.48}{(2\pi \times 120)^2 (6)(10^{-3})} = 0.141 \text{ mV rms}$$

$$r' = \frac{|V_r'|}{V_{dc}'} \times 100\% = \frac{0.141 \times 10^{-3}}{39.8} \times 100\% = 3.54 \times 10^{-4}\%$$

This very substantial reduction in ripple is attributable to the series impedance presented by the inductor to the ripple voltage and to the two capacitors that shunt ripple current to ground.

13.5   VOLTAGE MULTIPLIERS

Diodes and capacitors can be connected in various configurations to produce filtered, rectified voltages that are integer multiples of the peak value of an input sine wave. The principle of operation of these circuits is similar to that of the *clamping* circuits discussed in Chapter 11. By using a transformer to change the amplitude of an ac voltage before it is applied to a voltage multiplier, a wide range of dc levels can be produced using this technique. One advantage of a voltage multiplier is that high voltages can be obtained without using a high-voltage transformer.

Half-Wave Voltage Doubler

Figure 13.20(a) shows a half-wave voltage *doubler*. When v_{in} first goes positive, diode D_1 is forward biased and diode D_2 is reverse biased. Because the forward resistance of D_1 is quite small, C_1 charges rapidly to V_P (neglecting the diode drop), as shown in (b). During the ensuing negative half-cycle of v_{in}, D_1 is reverse biased and D_2 is forward biased, as shown in (c). Consequently, C_2 charges rapidly, with polarity shown. Neglecting the drop across D_2, we can write Kirchhoff's voltage law around the loop at the instant v_{in} reaches its negative peak, and obtain

$$V_P = -V_P + V_{C_2}$$

or

$$V_{C_2} = 2V_P \qquad\qquad (13.30)$$

During the next positive half-cycle of v_{in}, D_2 is again reverse biased and the voltage across the output terminals remains at $V_{C_2} = 2V_P$ volts. Note carefully the polarity of the output. If a load resistor is connected across C_2, then C_2 will discharge into the load during positive half-cycles of v_{in}, and will recharge to $2V_P$ volts during negative half-cycles, creating the usual ripple waveform. The PIV rating of each diode must be at least $2V_P$ volts.

Full-Wave Voltage Doubler

Figure 13.21(a) shows a full-wave voltage doubler. This circuit is the same as the full-wave bridge rectifier shown in Figure 13.6, with two of the diodes replaced by

Figure 13.20
A half-wave voltage doubler

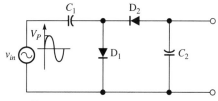

(a) Half-wave voltage-doubler circuit

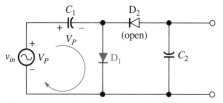

(b) C_1 charges to V_P during the positive half-cycle of v_{in}.

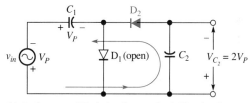

(c) C_2 charges to $2V_P$ during the negative half-cycle of v_{in}.

capacitors. When v_{in} is positive, D_1 conducts and C_1 charges to V_P volts, as shown in (b). When v_{in} is negative, D_2 conducts and C_2 charges to V_P volts, with the polarity shown in (c). It is clear that the output voltage is then $V_{C_1} + V_{C_2} = 2V_P$ volts. Since one or the other of the capacitors is charging during every half-cycle, the output is the same as that of a capacitor-filtered, full-wave rectifier. Note, however, that the effective filter capacitance is that of C_1 and C_2 in series, which is less than either C_1 or C_2. The PIV rating of each diode must be at least $2V_P$ volts.

Voltage Tripler and Quadrupler

By connecting additional diode–capacitor sections across the half-wave voltage doubler, output voltages equal to three and four times the input peak voltage can be obtained. The circuit is shown in Figure 13.22. When v_{in} first goes positive, C_1 charges to V_P through forward-biased diode D_1. On the ensuing negative half-cycle, C_2 charges through D_2 and, as demonstrated earlier, the voltage across C_2 equals $2V_P$. During the next positive half-cycle, D_3 is forward biased and C_3 charges to the same voltage attained by C_2: $2V_P$ volts. On the next negative half-cycle, D_2 and D_4 are forward biased and C_4 charges to $2V_P$ volts. As shown in the figure, the voltage across the combination of C_1 and C_3 is $3V_P$ volts, and that across C_2 and C_4 is $4V_P$ volts. Additional stages can be added in an obvious way to obtain even greater multiples of V_P. The PIV rating of each diode in the circuit must be at least $2V_P$ volts.

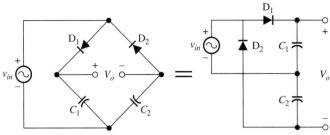

(a) The full-wave voltage-doubler circuit

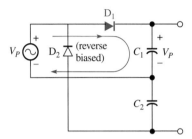

(b) C_1 charges to V_P during the positive half-cycle of V_{in}.

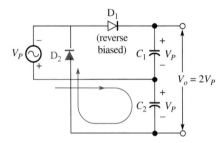

(c) C_2 charges to V_P during the negative half-cycle of V_{in}.

Figure 13.21
A full-wave voltage doubler

Figure 13.22
Voltage tripler and quadrupler

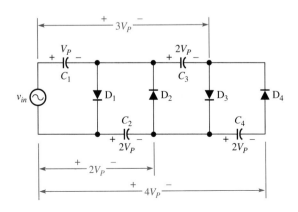

13.6 VOLTAGE REGULATION

An *ideal* power supply maintains a constant voltage at its output terminals, no matter what current is drawn from it. The output voltage of a practical power supply changes with load current, generally dropping as load current increases. Power-supply specifications include a *full-load current* (I_{FL}) rating, which is the maximum current that can be drawn from the supply. The terminal voltage when full-load current is drawn is called the *full-load voltage* (V_{FL}). The *no-load voltage* (V_{NL}) is the terminal voltage when zero current is drawn from the supply, that is, the open-circuit terminal voltage. Figure 13.23 illustrates these terms.

One measure of power-supply performance, in terms of how well the power supply is able to maintain a constant voltage between no-load and full-load conditions, is called its *percent voltage regulation:*

$$\text{VR} = \frac{V_{NL} - V_{FL}}{V_{FL}} \times 100\% \qquad (13.31)$$

More precisely, equation 13.31 defines the percent *output,* or *load,* voltage regulation, since it is based on changes that occur due to changes in load conditions, all other factors (including input voltage) remaining constant. It is clear that the numerator of equation 13.31 is the total change in output voltage between no-load and full-load, and that the ideal supply therefore has *zero* percent voltage regulation.

Figure 13.24 shows the Thevenin equivalent circuit of a power supply. The Thevenin voltage is the no-load voltage V_{NL}, and the Thevenin equivalent resistance is called the *output resistance, R_o,* of the supply. Many power-supply manufacturers specify output resistance rather than percent voltage regulation. We will show that one can be obtained from the other, if the full-load–voltage and full-load–current ratings are known. Let the full-load resistance be designated

$$R_{FL} = \frac{V_{FL}}{I_{FL}} \qquad (13.32)$$

When R_L in Figure 13.24 equals R_{FL}, we have, by voltage-divider action,

$$V_{FL} = \left(\frac{R_{FL}}{R_{FL} + R_o}\right) V_{NL} \qquad (13.33)$$

Substituting (13.33) into (13.31) (and omitting the percent factor) gives

$$\text{VR} = \frac{V_{NL} - \left(\dfrac{R_{FL}}{R_{FL} + R_o}\right) V_{NL}}{\left(\dfrac{R_{FL}}{R_{FL} + R_o}\right) V_{NL}} = \frac{R_o}{R_{FL}} \qquad (13.34)$$

Figure 13.23
No-load and full-load conditions in a power supply

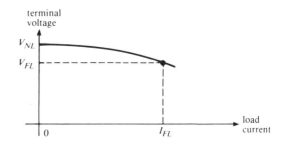

Figure 13.24
Thevenin equivalent circuit of a power supply

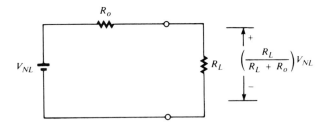

or

$$\text{VR} = R_o \left(\frac{I_{FL}}{V_{FL}} \right) \tag{13.35}$$

It is clear that the ideal supply has zero output resistance, corresponding to zero percent voltage regulation. Like the output resistance we have studied in earlier chapters, R_o can also be determined as the slope of a plot of load voltage versus load current: $R_o = \Delta V_L / \Delta I_L$.

Example 13.7

A power supply having output resistance 1.5 Ω supplies a full-load current of 500 mA to a 50-Ω load.

1. What is the percent voltage regulation of the supply?
2. What is the no-load output voltage of the supply?

Solution

1. $V_{FL} = I_{FL}R_{FL} = (500 \text{ mA})(50 \ \Omega) = 25$ V. From equation 13.35,

$$\text{VR(\%)} = R_o \left(\frac{I_{FL}}{V_{FL}} \right) \times 100\% = 1.5 \left(\frac{0.5 \text{ A}}{25 \text{ V}} \right) \times 100\% = 3.0\%$$

(Note also that VR $= R_o/R_{FL} = 1.5 \ \Omega/50 \ \Omega = 0.03$.)

2. From equation 13.33,

$$V_{NL} = \frac{V_{FL}(R_{FL} + R_o)}{R_{FL}} = \frac{25(50 + 1.5)}{50} = 25.75 \text{ V}$$

Example 13.8

Assuming that the transformer and forward-biased–diode resistances in Figure 13.25 are negligible, find the percent voltage regulation of the power supply. The full-load current is 2 A at a full-load voltage of 15 V.

Solution.   Since the transformer and diode resistances can be neglected, the dc output resistance looking back into the load terminals is $R_o = 8 \ \Omega$. Therefore, by equation 13.35,

$$\text{VR(\%)} = R_o \left(\frac{I_{FL}}{V_{FL}} \right) \times 100\% = 8 \left(\frac{2 \text{ A}}{15 \text{ V}} \right) \times 100\% = 106.6\%$$

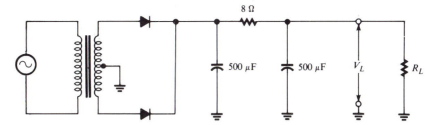

Figure 13.25
(Example 13.8)

This example illustrates that the series resistance used in an RC π filter to reduce ripple can seriously degrade voltage regulation. Note that the no-load voltage in this example is, from equation 13.31, $V_{NL} = \text{VR}(V_{FL}) + V_{FL} = (1.066)15 + 15 = 31$ V, so there is a 16-V change in output voltage between no-load and full-load! This is an example of an *unregulated* supply, since there is no circuitry designed to correct, or compensate for, the effect of load variations. It would be unacceptable for most applications in which the load could vary over such a wide range.

Line Regulation

Percent *line regulation* is another measure of the ability of a power supply to maintain a constant output voltage. In this case, it is a measure of how sensitive the output is to changes in *input*, or line, voltage rather than to changes in load. The specification is usually expressed as the percent change in output voltage that occurs per volt change in input voltage, with the load R_L assumed constant. For example, a line regulation of 1%/V means that the output voltage will change 1% for each one-volt change in input voltage. If the input voltage to a 20-V supply having that specification were to change by 5 V, then the output could be expected to change by (5 V)(1%/V) = 5%, or 0.05(20) = 1 V.

$$\% \text{ line regulation} = \frac{(\Delta V_o / V_o) \times 100\%}{\Delta V_{in}} \qquad (13.36)$$

13.7   ZENER-DIODE CIRCUITS

Zener Diodes

Figure 13.26 shows a typical *I–V* characteristic for a zener diode. The forward-biased characteristic is identical to that of a forward-biased silicon diode and obeys the same diode equation that we developed in Chapter 2 (equation 2.1). The zener diode is normally operated in its reverse-biased breakdown region, where the voltage across the device remains substantially constant as the reverse current varies over a large range. Like a fixed-voltage source, this ability to maintain a constant voltage across its terminals, independent of current, is what makes the device useful as a voltage reference. The fixed breakdown voltage is called the *zener voltage*, V_Z, as illustrated in the figure.

Figure 13.26
The I–V characteristic of a zener diode

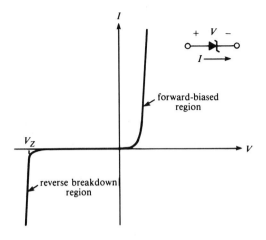

The Zener-Diode Voltage Regulator

To demonstrate how a zener diode can serve as a constant voltage reference, Figure 13.27 shows a simple but widely used configuration that maintains a constant voltage across a load resistor. The circuit is an elementary *voltage regulator* (Section 13.8) that holds the load voltage near V_Z volts as R_L and/or V_{in} undergo changes. In order that the voltage across the parallel combination of the zener and R_L remain at V_Z volts, the reverse current I_Z through the diode must at all times be large enough to keep the device in its breakdown region, as shown in Figure 13.26. The value selected for R_S is critical in that respect. As we shall presently demonstrate, R_S must be small enough to permit adequate zener current, yet large enough to prevent the zener current and power dissipation from exceeding permissible limits.

It is apparent in Figure 13.27 that

$$I_S = I_Z + I_L \qquad\qquad (13.37)$$

Also, I_S is the voltage difference across R_S divided by R_S:

$$I_S = \frac{V_{in} - V_Z}{R_S} \qquad\qquad (13.38)$$

The power dissipated in the zener diode is

$$P_Z = V_Z I_Z \qquad\qquad (13.39)$$

Figure 13.27
A simple voltage regulator using a zener diode

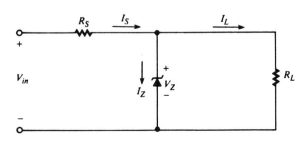

Example 13.9

In the circuit of Figure 13.27, $R_S = 20\ \Omega$, $V_Z = 18$ V, and $R_L = 200\ \Omega$. If V_{in} can vary from 20 V to 30 V, find

1. the minimum and maximum currents in the zener diode;
2. the minimum and maximum power dissipated in the diode; and
3. the minimum rated power dissipation that R_S should have.

Solution

1. Assuming that the zener diode remains in breakdown, then the load voltage remains constant at $V_Z = 18$ V, and the load current therefore remains constant at

$$I_L = \frac{V_Z}{R_L} = \frac{18\ \text{V}}{200\ \Omega} = 90\ \text{mA}$$

From equation 13.38, when $V_{in} = 20$ V,

$$I_S = \frac{(20\ \text{V}) - (18\ \text{V})}{20\ \Omega} = 100\ \text{mA}$$

Therefore, $I_Z = I_S - I_L = (100\ \text{mA}) - (90\ \text{mA}) = 10\ \text{mA}$. When $V_{in} = 30$ V,

$$I_S = \frac{(30\ \text{V}) - (18\ \text{V})}{20\ \Omega} = 600\ \text{mA}$$

and $I_Z = I_S - I_L = (600\ \text{mA}) - (90\ \text{mA}) = 510\ \text{mA}$.

2. $P_Z(\text{min}) = V_Z I_Z(\text{min}) = (18\ \text{V})(10\ \text{mA}) = 180\ \text{mW}$
   $P_Z(\text{max}) = V_Z I_Z(\text{max}) = (18\ \text{V})(510\ \text{mA}) = 9.18\ \text{W}$
3. $P_{R_S}(\text{max}) = I_S^2(\text{max})R_S = (0.6)^2(20) = 7.2\ \text{W}$

Solving equation 13.38 for R_S, we find

$$R_S = \frac{V_{in} - V_Z}{I_S} \tag{13.40}$$

Substituting (13.37) into (13.40) gives

$$R_S = \frac{V_{in} - V_Z}{I_Z + I_L} \tag{13.41}$$

Let $I_Z(\text{min})$ denote the minimum zener current necessary to ensure that the diode is in its breakdown region. As mentioned earlier, R_S must be small enough to ensure that $I_Z(\text{min})$ flows under worst-case conditions, namely, when V_{in} falls to its smallest possible value, $V_{in}(\text{min})$, and I_L is its largest possible value, $I_L(\text{max})$. Thus, from (13.41), we require

$$R_S \leq \frac{V_{in}(\text{min}) - V_Z}{I_Z(\text{min}) + I_L(\text{max})} \tag{13.42}$$

Maximum load current flows when R_L has its minimum possible value, $R_L(\text{min})$. Substituting $I_L(\text{max}) = V_Z/R_L(\text{min})$ in (13.42) gives

$$R_S \leq \frac{V_{in}(\text{min}) - V_Z}{I_Z(\text{min}) + V_Z/R_L(\text{min})} \tag{13.43}$$

The value of R_S must also be large enough to ensure that the zener current does not exceed the manufacturer's specified maximum or cause the zener power dissipation to exceed the rated maximum at the operating temperature. Let $I_Z(\text{max})$ denote the minimum of those two limits, i.e., the smaller of the rated current and $P_d(\text{max})/V_Z$. Then we require that R_S be large enough to ensure that I_Z does not exceed $I_Z(\text{max})$ under the worst-case conditions where $V_{in} = V_{in}(\text{max})$ and $I_L = I_L(\text{min})$, or $R_L = R_L(\text{max})$. Thus, from equation 13.41, it is necessary that

$$R_S \geq \frac{V_{in}(\text{max}) - V_Z}{I_Z(\text{max}) + I_L(\text{min})} = \frac{V_{in}(\text{max}) - V_Z}{I_Z(\text{max}) + V_Z/R_L(\text{max})} \qquad \textbf{(13.44)}$$

It is important to note that, as far as zener dissipation is concerned, the worst-case load condition in some applications may correspond to an open output, that is, $R_L = \infty$ and $I_L = 0$. In that case, all of the current through R_S flows in the zener diode.

Example 13.10

The reverse current in a certain 10-V, 2-W zener diode must be at least 5 mA to ensure that the diode remains in breakdown. The diode is to be used in the regulator circuit shown in Figure 13.28, where V_{in} can vary from 15 V to 20 V. Note that the load can be switched out of the regulator circuit in this application. Find a value for R_S. What power dissipation rating should R_S have?

Solution

$$V_{in}(\text{min}) = 15 \text{ V}$$
$$I_Z(\text{min}) = 5 \text{ mA}$$
$$R_L(\text{min}) = R_L = 500 \text{ } \Omega \quad \text{(when the switch is closed)}$$

Therefore, from inequality 13.43,

$$R_S \leq \frac{V_{in}(\text{min}) - V_Z}{I_Z(\text{min}) + V_Z/R_L(\text{min})} = \frac{(15 - 10) \text{ V}}{(5 \text{ mA}) + (10 \text{ V})/(500 \text{ } \Omega)} = 200 \text{ } \Omega$$

$$V_{in}(\text{max}) = 20 \text{ V}$$

$$I_Z(\text{max}) = \frac{P_Z(\text{max})}{V_Z} = \frac{2 \text{ W}}{10 \text{ V}} = 200 \text{ mA}$$

$$I_L(\text{min}) = 0 \quad \text{(when the switch is open)}$$

From inequality 13.44,

$$R_S \geq \frac{V_{in}(\text{max}) - V_Z}{I_Z(\text{max}) + I_L(\text{min})} = \frac{(20 - 10) \text{ V}}{(200 \text{ mA}) + (10 \text{ A})} = 50 \text{ } \Omega$$

Figure 13.28
(Example 13.10)

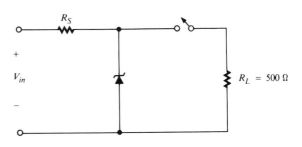

Thus, we require $50 \le R_S \le 200$. Choosing $R_S = 100 \ \Omega$, the maximum current in R_S is

$$I_S(\text{max}) = \frac{V_{in}(\text{max}) - V_Z}{R_S} = \frac{(20 - 10) \ \text{V}}{100 \ \Omega} = 0.1 \ \text{A}$$

Therefore, the maximum power dissipated in R_S is $P(\text{max}) = I_S^2(\text{max})R_S = (0.1)^2(100) = 1 \ \text{W}$.

Temperature Effects

The breakdown voltage of a zener diode is a function of the width of its depletion region, which is controlled during manufacturing by the degree of impurity doping. Heavy doping increases conductivity, which narrows the depletion region, and therefore decreases the voltage at which breakdown occurs. Zener diodes are available with breakdown voltages ranging from 2.4 V to 200 V. The mechanism by which breakdown occurs depends on the breakdown voltage itself. When V_Z is less than about 5 V, the high electric field intensity across the narrow depletion region (around 3×10^7 V/m) strips carriers directly from their bonds, a phenomenon usually called *zener breakdown*. For V_Z greater than about 8 V, breakdown occurs as a result of collisions between high-energy carriers, the mechanism called *avalanching*. Between 5 V and 8 V, both the avalanching and zener mechanisms contribute to breakdown. The practical significance of these facts is that the breakdown mechanism determines how temperature variations affect the value of V_Z. Low-voltage zener diodes that break down by the zener mechanism have negative temperature coefficients (V_Z decreases with increasing temperature) and higher-voltage avalanche zeners have positive temperature coefficients. When V_Z is between about 3 V and 8 V, the temperature coefficient is also strongly influenced by the current in the diode: the coefficient may be positive or negative, depending on current, becoming more positive as current increases.

The temperature coefficient of a zener diode is defined to be its change in breakdown voltage per degree Celsius increase in temperature. For example, a temperature coefficient of $+8$ mV/°C means that V_Z will increase 8 mV for each degree Celsius increase in temperature. *Temperature stability* is the ratio of the temperature coefficient to the breakdown voltage. Expressed as a percent,

$$S(\%) = \frac{\text{T.C.}}{V_Z} \times 100\% \tag{13.45}$$

where T.C. is the temperature coefficient. Clearly, small values of S are desirable.

In applications requiring a zener diode to serve as a highly stable voltage reference, means must be taken to *temperature compensate* the diode. A technique that is used frequently is to connect the zener in series with one or more semiconductor devices whose voltage drops change with temperature in the opposite way that V_Z changes, i.e., devices having the opposite kind of temperature coefficient. If a temperature change causes V_Z to increase, then the voltage across the other components decreases, so the total voltage across the series combination is (ideally) unchanged. For example, the temperature coefficient of a forward-biased silicon diode is negative, so one or more of these can be connected in series with a zener diode having a positive temperature coefficient, as illustrated in Figure 13.29. The next example illustrates that several forward-biased diodes, which have relatively small temperature coefficients, may be required to compensate a single zener diode.

Figure 13.29
Temperature compensating a zener di-
ode by connecting it in series with for-
ward-biased diodes having opposite
temperature coefficients

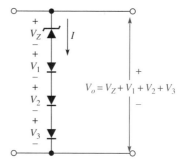

$$V_o = V_Z + V_1 + V_2 + V_3$$

Example 13.11

A zener diode having a breakdown voltage of 10 V at 25°C has a temperature coefficient of +5.5 mV/°C. It is to be temperature compensated by connecting it in series with three forward-biased diodes, as shown in Figure 13.29. Each compensating diode has a forward drop of 0.65 V at 25°C and a temperature coefficient of −2 mV/°C.

1. What is the temperature stability of the uncompensated zener diode?
2. What is the breakdown voltage of the uncompensated zener diode at 100°C?
3. What is the voltage across the compensated network at 25°C? At 100°C?
4. What is the temperature stability of the compensated network?

Solution

1. From equation 13.45,
$$S = \frac{\text{T.C.} \times 100\%}{V_Z} = \frac{5.5 \times 10^{-3}}{10 \text{ V}} \times 100\% = 0.055\%$$

2. $V_Z = (10 \text{ V}) + \Delta T(\text{T.C.}) = (10 \text{ V}) + (100°C - 25°C)(5.5 \text{ mV}/°C) = 10.4125 \text{ V}$

3. As shown in Figure 13.29, $V_o = V_Z + V_1 + V_2 + V_3$. At 25°C, $V_o = 10 + 3(0.65) = 11.95$ V. At 100°C, the drop V_d across each forward-biased diode is $V_d = (0.65 \text{ V}) + (100°C - 25°C)(-2 \text{ mV}/°C) = 0.5$ V. Therefore, at 100°C, $V_o = (10.4125 \text{ V}) + 3(0.5 \text{ V}) = 10.5625$ V.

4. The temperature coefficient of the compensated network is T.C. = (+5.5 mV/°C) + 3(−2 mV/°C) = (+5.5 mV/°C) − (6 mV/°C) = −0.5 mV/°C. The voltage drop across the network (at 25°C) was found to be 11.95 V, so
$$S = \frac{-0.5 \text{ mV}/°C}{11.95 \text{ V}} \times 100\% = -0.00418\%$$

We see that compensation has improved the stability by a factor greater than 10.

Temperature-compensated zener diodes are available from manufacturers in single-package units called *reference diodes*. These units contain specially fabricated junctions that closely track and oppose variations in V_Z with temperature. Although it is possible to obtain an extremely stable reference this way, it may be necessary to maintain the zener current at a manufacturer's specified value in order to realize the specified stability. Figure 13.45 shows a temperature-stabilized zener diode whose current is maintained by a constant-current source to ensure its stability as reference for the 723 integrated-circuit voltage regulator.

Zener-Diode Impedance

The breakdown characteristic of an *ideal* zener diode is a perfectly vertical line, signifying zero change in voltage for any change in current. Thus, the ideal diode has zero impedance (or ac resistance) in its breakdown region. A practical zener diode has nonzero impedance, which can be computed in the usual way:

$$Z_Z = \frac{\Delta V_Z}{\Delta I_Z} \; \Omega \tag{13.46}$$

Z_Z is the reciprocal of the slope of the breakdown characteristic on an I_Z–V_Z plot. The slope is not constant, so the value of Z_Z depends on the point along the characteristic where the measurement is made. The impedance decreases as I_Z increases; that is, the breakdown characteristic becomes steeper at points further down the line, corresponding to greater reverse currents. For this reason, the diode should be operated with as much reverse current as possible, consistent with rating limitations.

Manufacturers' specifications for zener impedances are usually given for a specified ΔI_Z that covers a range from a small I_Z near the onset of breakdown to some percentage of the maximum rated I_Z. The values may range from a few ohms to several hundred ohms. There is also a variation in the impedance of zener diodes among those having different values of V_Z. Diodes with breakdown voltages near 7 V have the smallest impedances.

Example 13.12

A zener diode has impedance 40 Ω in the range from $I_Z = 1$ mA to $I_Z = 10$ mA. The voltage at $I_Z = 1$ mA is 9.1 V. Assuming that the impedance is constant over the given range, what minimum and maximum zener voltages can be expected if the diode is used in an application where the zener current changes from 2 mA to 8 mA?

Solution. From equation 13.46, the voltage change between $I_Z = 1$ mA and $I_Z = 2$ mA is $\Delta V_Z = \Delta I_Z Z_Z = [(2 \text{ mA}) - (1 \text{ mA})](40 \; \Omega) = 0.04$ V. Therefore, the minimum voltage is $V_Z(\min) = (9.1 \text{ V}) + \Delta V_Z = (9.1 \text{ V}) + (0.04 \text{ V}) = 9.14$ V. The voltage change between $I_Z = 2$ mA and $I_Z = 8$ mA is $\Delta V_Z = [(8 \text{ mA}) - (2 \text{ mA})](40 \; \Omega) = 0.16$ V. Therefore, the maximum voltage is $V_Z(\max) = V_Z(\min) + \Delta V_Z = (9.14 \text{ V}) + (0.16 \text{ V}) = 9.3$ V.

13.8   SERIES AND SHUNT VOLTAGE REGULATORS

A voltage regulator is a device, or combination of devices, designed to maintain the output voltage of a power supply as nearly constant as possible. It can be regarded as a *closed-loop control system* because it monitors output voltage and generates *feedback* that automatically increases or decreases the supply voltage, as necessary to compensate for any tendency of the output to change. Thus, the purpose of a regulator is to eliminate any output voltage variation that might otherwise occur because of changes in load, changes in line (input) voltage, or changes in temperature. Regulators are constructed in a very large variety of circuit configurations, ranging from one or two discrete components to elaborate integrated circuits.

Series Regulators

Voltage regulators may be classified as being of either a *series* or a *shunt* design. Figure 13.30 is a *functional* block diagram of the series-type regulator. While the individual components shown in the diagram may not be readily identifiable in every series regulator circuit, the functional block diagram serves as a useful model for understanding the underlying principles of series regulation. V_{in} is an unregulated dc input, such as might be obtained from a rectifier with a capacitor filter, and V_o is the regulated output voltage. Notice that the *control element*, which is a device whose operating state adjusts as necessary to maintain a constant V_o, is in a series path between V_{in} and V_o. A *sampling circuit* produces a feedback voltage proportional to V_o and this voltage is compared to a reference voltage. The output of the comparator circuit is the control signal that adjusts the operating state of the control element. If V_o decreases, due, for example, to an increased load, then the comparator produces an output that causes the control element to increase V_o. In other words, V_o is automatically raised until the comparator circuit no longer detects any difference between the reference and the feedback. Similarly, any tendency of V_o to increase results in a signal that causes the control element to reduce V_o.

Figure 13.31 shows a simple transistor voltage regulator of the series type. Here, the control element is the NPN transistor, often called the *pass transistor* because it conducts, or passes, all the load current through the regulator. It is usually a power transistor, and it may be mounted on a heat sink in a heavy-duty power supply that delivers substantial current. The zener diode provides the voltage reference, and the base-to-emitter voltage of the transistor is the control voltage. In this case there is no identifiable sampling circuit, since the entire output voltage level V_o is used for feedback.

In Figure 13.31, notice that the zener diode is reverse biased and that reverse current is furnished to it through resistor R. Although V_{in} is unregulated, it must remain sufficiently large, and R must be sufficiently small, to keep the zener in its reverse breakdown region, as shown in Figure 13.26. Thus, as the unregulated input voltage varies, V_Z remains essentially constant. Writing Kirchhoff's voltage law around the output loop, we find

$$V_{BE} = V_Z - V_o \tag{13.47}$$

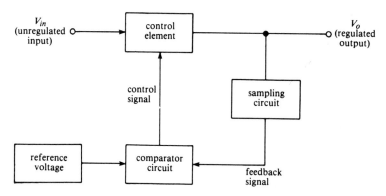

Figure 13.30
Block diagram of a series voltage regulator

Figure 13.31
A transistor series voltage regulator

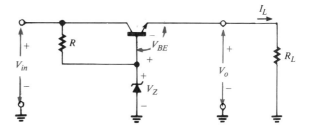

We now treat V_Z as perfectly constant. Since (13.47) is valid at all times, any change in V_o must cause a change in V_{BE}, in order to maintain the equality. For example, if V_o decreases, V_{BE} must increase, since V_Z is constant. Similarly, if V_o increases, V_{BE} must decrease.

The behavior we have just described accounts for the ability of the circuit to provide voltage regulation. When V_o decreases, V_{BE} increases, which causes the NPN transistor to conduct more heavily and to produce more load current. The increase in load current causes an increase in V_o, since $V_o = I_L R_L$. For example, suppose the regulator supplies 1 A to a 10-Ω load, so $V_o = (1 \text{ A})(10 \text{ } \Omega) = 10$ V. Now suppose the load resistance is reduced to 8 Ω. The output voltage would then fall to $(1 \text{ A})(8 \text{ } \Omega) = 8$ V. However, this reduction in V_o causes the transistor to conduct more heavily, and the load current increases to 1.25 A. Thus, the output voltage is restored to $V_o = (1.25 \text{ A})(8 \text{ } \Omega) = 10$ V. The regulating action is similar when V_o increases: V_{BE} decreases, transistor conduction is reduced, load current decreases, and V_o is reduced.

Notice that the transistor is used essentially as an emitter follower. The load is connected to the emitter and the emitter follows the base, which is the constant zener voltage.

Example 13.13

In Figure 13.31, $V_{in} = 20$ V, $R = 200 \text{ } \Omega$, and $V_Z = 12$ V. If $V_{BE} = 0.65$ V, find

1. V_o;
2. the collector-to-emitter voltage of the pass transistor; and
3. the current in the 200-Ω resistor.

Solution

1. From equation 13.37, $V_o = V_Z - V_{BE} = 12$ V $- 0.65$ V $= 11.35$ V.
2. By writing Kirchhoff's voltage law around the loop that includes V_{in}, V_{CE}, and V_o, we find $V_{CE} = V_{in} - V_o = 20$ V $- 11.35$ V $= 8.65$ V.
3. The voltage drop across the 200-Ω resistor is $V_{in} - V_Z = 20$ V $- 12$ V $= 8$ V. Therefore, the current in the resistor is $I = (8 \text{ V})/(200 \text{ } \Omega) = 0.04$ A.

For successful regulator operation, the pass transistor must remain in its active region, V_{in} must not drop to a level so small that the zener diode is no longer in its breakdown region, and the zener voltage V_Z should be highly independent of both current and temperature. Also, the feedback *loop gain*, from output back through the control element, should be very large, so that small changes in output voltage can be detected and rapidly corrected. Figure 13.32 shows an improved

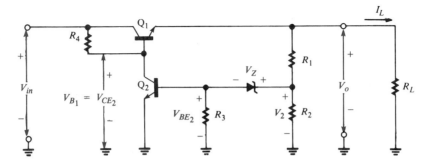

Figure 13.32
An improved series regulator that incorporates transistor Q_2 to increase sensitivity

series regulator that incorporates an additional transistor in the feedback path to increase gain.

Note that resistors R_1 and R_2 form a voltage divider across V_o and serve as the sampling circuit that produces voltage V_2 proportional to V_o. Assuming that the divider is designed so that the zener-diode current does not load it appreciably,

$$V_2 = \frac{R_2}{R_1 + R_2} V_o \qquad \text{(13.48)}$$

Writing Kirchhoff's voltage law around the loop containing R_2, R_3, and the zener diode, we have

$$V_{BE_2} + V_Z = V_2 \qquad \text{(13.49)}$$

Any decrease in V_o will cause a decrease in V_2 and, assuming that V_Z remains constant, V_{BE_2} must therefore decrease to maintain equality in (13.49). A decrease in V_{BE_2} causes V_{CE_2} to increase, since Q_2 is an inverting common-emitter stage. Since the base voltage V_{B_1} on Q_1 experiences this same increase, Q_1 conducts more heavily, produces more load current, and increases V_o, as previously described. Another way of viewing this feedback action is to recognize that a decrease in V_{BE_2} causes Q_2 to conduct less, and therefore allows more base current to flow into Q_1. Of course, any *increase* in V_o causes actions that are just the opposite of those we have described and results in less load current.

Example 13.14

In the regulator shown in Figure 13.32, $R_1 = 50$ kΩ, $R_2 = 43.75$ kΩ, and $V_Z = 6.3$ V. If the 15-V output drops 0.1 V, find the change in V_{BE_2} that results.

Solution.   From equation 13.48, when $V_o = 15$ V,

$$V_2 = \frac{R_2}{R_1 + R_2} V_o = \left[\frac{43.75 \text{ k}\Omega}{(50 \text{ k}\Omega) + (43.75 \text{ k}\Omega)} \right] (15 \text{ V}) = 7 \text{ V}$$

Therefore, by (13.49), $V_{BE_2} = V_2 - V_Z = 7$ V $- 6.3$ V $= 0.7$ V. When $V_o = 15$ V $- 0.1$ V $= 14.9$ V, V_2 becomes

$$V_2' = \left[\frac{43.75 \text{ k}\Omega}{(50 \text{ k}\Omega) + (43.75 \text{ k}\Omega)} \right] (14.9 \text{ V}) = 6.953 \text{ V}$$

Therefore,

$$V'_{BE_2} = V'_2 - V_Z = 6.953 \text{ V} - 6.3 \text{ V} = 0.653 \text{ V}$$
$$\Delta V_{BE_2} = V_{BE_2} - V'_{BE_2} = 0.7 \text{ V} - 0.653 \text{ V} = 0.047 \text{ V}$$

Figure 13.33 shows how an operational amplifier is used in a series voltage regulator. Resistors R_1 and R_2 form a voltage divider that feeds a voltage proportional to V_o back to the inverting input: $V^- = V_o R_2/(R_1 + R_2)$. The zener voltage V_Z on the noninverting input is greater than V^-, so the amplifier output is positive and is proportional to $V_Z - V^-$. If V_o decreases, V^- decreases, and the amplifier output increases. The greater voltage applied to the base of the pass transistor causes it to conduct more heavily, and V_o increases. Similarly, an increase in V_o causes V^- to increase, $V_Z - V^-$ to decrease, and the amplifier output to decrease. The pass transistor conducts less heavily and V_o decreases.

The operational amplifier in Figure 13.33 can be regarded as a noninverting configuration with input V_Z and gain

$$A = 1 + \frac{R_1}{R_2} \tag{13.50}$$

Therefore, neglecting the base-to-emitter drop of the pass transistor, the regulated output voltage is

$$V_o = \left(1 + \frac{R_1}{R_2}\right) V_Z \tag{13.51}$$

R_3 is chosen to ensure that sufficient reverse current flows through the zener diode to keep it in breakdown.

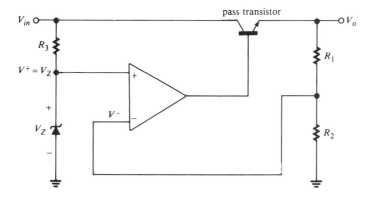

Figure 13.33
An operational amplifier used in a series voltage regulator

Example 13.15

DESIGN

Design a series voltage regulator using an operational amplifier and a 6-V zener diode to maintain a regulated output of 18 V. Assume that the unregulated input varies between 20 V and 30 V and that the current through the zener diode must be at least 20 mA to keep it in its breakdown region.

Solution.   From equation 13.51, 18 V = $(1 + R_1/R_2)$6 V. Thus, $(1 + R_1/R_2) = 3$. Let $R_1 = 20$ kΩ. Then

$$\left(1 + \frac{20 \text{ k}\Omega}{R_2}\right) = 3$$

$$R_2 = 10 \text{ k}\Omega$$

The current through R_3 is

$$I = \frac{V_{in} - V_Z}{R_3} \qquad (13.52)$$

Since the current into the noninverting input is negligibly small, the current in R_3 is the same as the current in the zener diode. This current must be at least 20 mA, so in equation 13.52, I must be 20 mA for the smallest possible value of V_{in} (20 V):

$$20 \text{ mA} = \frac{(20 \text{ V}) - (6 \text{ V})}{R_3}$$

$$R_3 = \frac{14 \text{ V}}{20 \text{ mA}} = 700 \text{ } \Omega$$

Current Limiting

Many general-purpose power supplies are equipped with short-circuit or overload protection. One form of protection is called *current limiting,* whereby specially designed circuitry limits the current that can be drawn from the supply to a certain specific maximum, even if the output terminals are short-circuited. Figure 13.34 shows a popular current-limiting circuit incorporated into the operational-amplifier regulator. As load current increases, the voltage drop across resistor R_{SC} increases. Notice that R_{SC} is in parallel with the base-to-emitter junction of transistor Q_2. If the load current becomes great enough to create a drop of about 0.7 V across R_{SC}, then Q_2 begins to conduct a substantial collector current. As a consequence, current that would otherwise enter the base of Q_1 is diverted through Q_2. In this way, the pass transistor is prevented from supplying additional load cur-

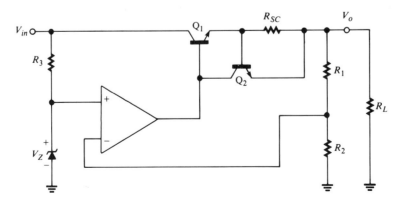

Figure 13.34
Q_2 and R_{SC} are used to limit the maximum current that can be drawn from the supply.

rent. The maximum (short-circuit) current that can be drawn from the supply is the current necessary to create the 0.7-V drop across R_{SC}:

$$I_L(\text{max}) = \frac{0.7 \text{ V}}{R_{SC}} \qquad (13.53)$$

Example 13.16

The voltage regulator in Figure 13.34 maintains an output voltage of 25 V.

1. What value of R_{SC} should be used if it is desired to limit the maximum current to 0.5 A?
2. With the value of R_{SC} found in (1), what will be the output voltage when $R_L = 100 \ \Omega$? When $R_L = 10 \ \Omega$?

Solution

1. From equation 13.53,

$$R_{SC} = \frac{0.7 \text{ V}}{I_L(\text{max})} = \frac{0.7 \text{ V}}{0.5 \text{ A}} = 1.4 \ \Omega$$

2. We must determine whether a 100-Ω load would attempt to draw more than the current limit of 0.5 A at the regulated output voltage of 25 V:

$$I_L = \frac{25 \text{ V}}{100 \ \Omega} = 0.25 \text{ A}$$

Since $0.25 \text{ A} < I_L(\text{max}) = 0.5 \text{ A}$, the output voltage will remain at 25 V. If $R_L = 10 \ \Omega$, the (unlimited) current would be

$$I_L = \frac{25 \text{ V}}{10 \ \Omega} = 2.5 \text{ A}$$

Since $2.5 \text{ A} > I_L(\text{max}) = 0.5 \text{ A}$, current limiting will occur, and the output voltage will be $V_o = I_L(\text{max})R_L = (0.5 \text{ A})(10 \ \Omega) = 5 \text{ V}$.

Example 13.16 demonstrates that the output voltage of a current-limited regulator decreases if the load resistance is made smaller than that which would draw maximum current at the regulated output voltage. Figure 13.35 shows a typical load-voltage–load-current characteristic for a current-limited regulator. The characteristic shows that load current may increase slightly beyond $I_L(\text{max})$ as the output approaches a short-circuit condition ($V_L = 0$). Note that Q_2 in Figure 13.34

Figure 13.35
A typical load-voltage–load-current characteristic for a current-limited regulator

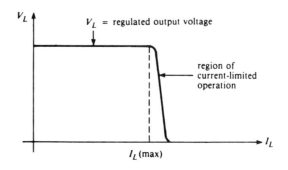

supplies a small amount of additional current to the load once current limiting takes place.

Foldback Limiting

Another form of overcurrent protection, called *foldback limiting*, is used to reduce both the output current and the output voltage if the load resistance is made smaller than that which would draw a specified maximum current, I_L(max). We have already seen that the output voltage of a current-limited regulator decreases as the load resistance is made smaller (Example 13.16). In foldback limiting, this decrease in output voltage is sensed and is used to further decrease the amount of current that can flow to the load. Thus, as load resistance decreases beyond a certain minimum, both load voltage and load current decrease. If the load becomes a short circuit, output voltage and output current both approach 0. The foldback characteristic is shown in Figure 13.36. The principal purpose of foldback is to protect a load from overcurrent, as well as to protect the regulator itself.

Figure 13.37 shows one method used to add foldback limiting to the basic current-limited regulator. Notice the similarity of this circuit to the current-limited regulator circuit shown in Figure 13.34. The only difference is that the base of Q_2 is now connected to the R_3–R_4 voltage divider. Writing Kirchhoff's voltage law around the loop, we find

$$V_{BE} = V_{R_{SC}} - V_{R_3} \qquad (13.54)$$

Notice that V_{R_3} will increase or decrease if the load voltage increases or decreases. When the load current increases to its maximum permissible value, $V_{R_{SC}}$ becomes large enough to make V_{BE} approximately 0.7 V; that is, $V_{R_{SC}}$ becomes large enough to exceed the drop across V_{R_3} by about 0.7 V: $0.7 \approx V_{BE} = V_{R_{SC}} - V_{R_3}$. At this point, current limiting occurs, just as it does in the current-limited regulator. If the load resistance is now made smaller, the load voltage will drop, as we have seen previously. But when the load voltage drops, V_{R_3} drops. Consequently, by equation 13.54, a smaller value of $V_{R_{SC}}$ is required to maintain $V_{BE} = 0.7$ V. Since V_{BE} remains essentially constant at 0.7 V, a smaller load current must flow to produce the smaller drop across R_3. Further decreases in load resistance produce further drops in load voltage and a further reduction in load current, so foldback limiting occurs. If the load resistance is restored to its normal operating value, the circuit resumes normal regulator action.

Figure 13.36
Foldback limiting (Compare with Figure 13.35.)

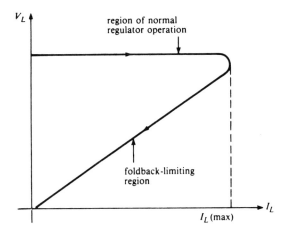

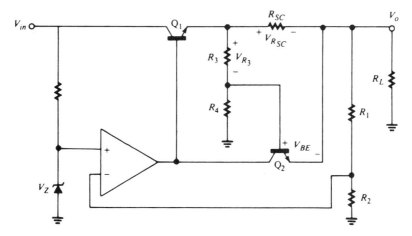

Figure 13.37
A current-limited regulator modified to provide foldback limiting

Example 13.17

The circuit in Figure 13.37 maintains a regulated output of 6 V. If $R_3 = 1$ kΩ and $R_4 = 9$ kΩ, what should be the value of R_{SC} to impose a maximum current limit of 1 A?

Solution. At the onset of current limiting,

$$V_{R_3} = \left(\frac{R_3}{R_3 + R_4}\right) V_o = \left[\frac{1 \text{ k}\Omega}{(1 \text{ k}\Omega) + (9 \text{ k}\Omega)}\right] (6 \text{ V}) = 0.6 \text{ V}$$

From equation 13.54,

$$V_{BE} = 0.7 \text{ V} = V_{R_{SC}} - 0.6 \text{ V}$$
$$V_{R_{SC}} = 1.3 \text{ V}$$

Thus,

$$R_{SC} = \frac{V_{R_{SC}}}{I_L(\text{max})} = \frac{1.3 \text{ V}}{1 \text{ A}} = 1.3 \text{ }\Omega$$

Shunt Regulators

Figure 13.38 is a functional block diagram of the shunt-type regulator. Each of the components shown in the figure performs the same function as its counterpart in the series regulator (Figure 13.30), but notice in this case that the control element is in parallel with the load. The control element maintains a constant load voltage by shunting more or less current from the load.

It is convenient to think of the control element in Figure 13.38 as a variable resistance. When the load voltage decreases, the resistance of the control element is made to increase, so less current is diverted from the load, and the load voltage rises. Conversely, when the load voltage increases, the resistance of the control element decreases, and more current is shunted away from the load. From an-

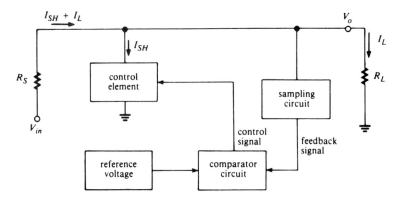

Figure 13.38
Functional block diagram of a shunt-type voltage regulator

other viewpoint, the source resistance R_S on the unregulated side of Figure 13.38 forms a voltage divider with the parallel combination of the control element and R_L. Thus, when the resistance of the control element increases, the resistance of the parallel combination increases, and, by voltage-divider action, the load voltage increases.

Figure 13.39 shows a discrete shunt regulator in which transistor Q_1 serves as the shunt control element. Since V_Z is constant, any change in output voltage creates a proportional change in the voltage across R_1. Thus, if V_o decreases, the voltage across R_1 decreases, as does the base voltage of Q_2. Q_2 conducts less heavily and the current into the base of Q_1 is reduced. Q_1 then conducts less heavily and shunts less current from the load, allowing the load voltage to rise. Conversely, an increase in V_o causes both Q_1 and Q_2 to conduct more heavily, and more current is diverted from the load.

Figure 13.40 shows a shunt regulator incorporating an operational amplifier. Resistors R_1 and R_2 form a voltage divider that feeds a voltage proportional to V_o back to the noninverting input. This voltage is greater than the reference voltage V_Z applied to the inverting input, so the output of the amplifier is a positive voltage proportional to $V_o - V_Z$. If V_o decreases, the amplifier output decreases and Q_1

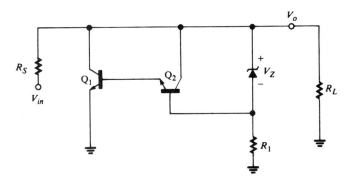

Figure 13.39
A discrete shunt regulator

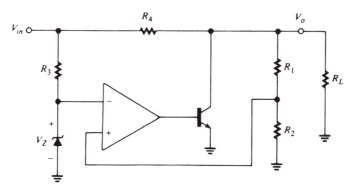

Figure 13.40
A shunt voltage regulator incorporating an operational amplifier

conducts less heavily. Thus, less current is diverted from the load and the output voltage rises.

One advantage of the shunt-regulator circuit shown in Figure 13.40 is that it has inherent current limiting. It is clear that the load current cannot exceed V_{in}/R_4, which is the current that would flow through R_4 if the output were short-circuited. Since load current must flow through R_4, the power dissipation in the resistor may be quite large, particularly under short-circuit conditions.

13.9   SWITCHING REGULATORS

A principal disadvantage of the series and shunt regulators discussed in Section 13.8 is the fact that the control element in each type must dissipate a large amount of power. As a consequence, they have poor *efficiency;* that is, the ratio of load power to total input power is relatively small. Since the control element, such as a pass transistor, is operated in its active region, power dissipation can be quite large, particularly when there is a large voltage difference between V_{in} and V_o and when substantial load current flows. When efficiency is a major concern, the switching-type regulator is often used. In this design, the control element is switched on and off at a rapid rate and is therefore either in saturation or cutoff most of the time. As is the case with other digital switching devices, this mode of operation results in very low power consumption.

The fundamental component of a switching regulator is a *pulse-width modula-tor,* illustrated in Figure 13.41. This device produces a train of rectangular pulses having widths that are proportional to the device's input. The figure illustrates the pulse output corresponding to a small dc input, the output corresponding to a large dc input, and that corresponding to a ramp-type input. In order to produce a periodic sequence of pulses, the pulse-width modulator must employ some form of oscillator. Practical modulators used in switching regulators usually have inherent oscillation capabilities that are incorporated into their circuit designs. Figure 14.38 shows how a pulse-width modulator (used in power amplifiers) can be constructed using a voltage comparator driven by a sawtooth waveform.

Recall that the *duty cycle* of a pulse train is the proportion of the period during which the pulse is present, i.e., the fractional part of the period during which the

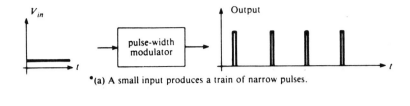

*(a) A small input produces a train of narrow pulses.

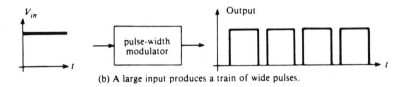

(b) A large input produces a train of wide pulses.

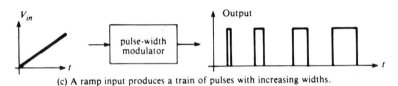

(c) A ramp input produces a train of pulses with increasing widths.

Figure 13.41
Operation of a pulse-width modulator

output is high. For example, a square wave, which is high for one half-cycle and low for the other half-cycle, has a duty cycle of 0.5, or 50%. By definition

$$\text{duty cycle} = T_{HI}/T \qquad (13.55)$$

where T_{HI} is the total time during period T that the waveform is high. As illustrated in Figure 13.42, the dc, or average, value of a pulse train is directly proportional to its duty cycle:

$$V_{dc} = V_{HI}\left(\frac{T_{HI}}{T}\right) \qquad (13.56)$$

where V_{HI} is the high, or peak, value of the pulse. Equation 13.56 is valid for pulses that alternate between 0 volts (low) and V_{HI} volts.

A switching regulator uses a pulse-width modulator to produce a pulse train whose duty cycle is automatically adjusted as necessary to increase or decrease the dc value of the train. The basic circuit is shown in Figure 13.43. Note that the output of the pulse-width modulator drives pass transistor Q_1. When the drive pulse is high, Q_1 is saturated and maximum current flows through it. When the drive pulse is low, Q_1 is cut off and no current flows. Thus, the current through Q_1 is a series of pulses having the same duty cycle as the modulator output. As in the regulator circuits discussed earlier, the zener diode provides a reference voltage that is compared in the operational amplifier to the feedback voltage obtained from the R_1–R_2 voltage divider. If the load voltage V_o tends to fall, then the output of the operational amplifier increases and a larger voltage is applied to the pulse-width modulator. It therefore produces a pulse train having a larger duty cycle. The pulse train switches Q_1 on and off with a greater duty cycle, so the dc value of

Figure 13.42
The dc value of a pulse train is directly proportional to its duty cycle.

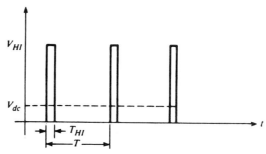

(a) A low–duty-cycle pulse train has a small dc value.

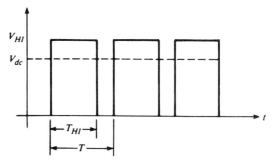

(b) A high–duty-cycle pulse train has a large dc value.

the current through Q_1 is increased. The increased current raises the load voltage to compensate for its initial decrease.

Series inductor L and shunt capacitor C in Figure 13.43 form a low-pass filter that recovers the dc value of the pulse waveform supplied to it by Q_1. The fundamental frequency of the pulse train and its harmonics are suppressed by the filter. Diode D_1 is used to suppress the negative voltage transient generated by the inductor when Q_1 turns off. It is called a *catcher* diode, or *free-wheeling* diode. Modern switching regulators are operated at frequencies in the kilohertz range, usually less than 100 kHz. High frequencies are desirable because the LC filter components can then be smaller, less bulky, and less expensive. However, high-

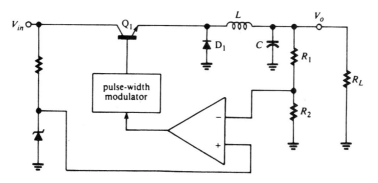

Figure 13.43
A basic switching-type regulator circuit

frequency switching of heavy currents creates large magnetic fields that induce noise voltages in surrounding conductors. This generation of *electromagnetic interference* (EMI) is a principal limitation of switching regulators, and they require careful shielding. Also, switching frequencies are limited by the inability of the power (pass) transistor to turn on and off at high speeds. Many modern regulators use power MOSFETs of the VMOS design, instead of bipolars, because of their superior ability to switch heavy loads at high frequencies.

Example 13.18

The switching regulator in Figure 13.43 is designed to maintain a 12-V-dc output when the unregulated input voltage varies from 15 V to 24 V. When pass transistor Q_1 is conducting, its collector-to-emitter saturation voltage is 0.5 V. Assuming that the load is constant and the LC filter is ideal (so its output is the dc value of its input), find the minimum and maximum duty cycles of the pulse-width modulator.

Solution.   When V_{in} has its minimum value of 15 V, the high voltage of the pulse train at the input to the filter is $15 - V_{CE(sat)} = 15 - 0.5 = 14.5$ V. The duty cycle of this pulse train must be sufficient to produce a dc value of 12 V. From equation 13.56, $T_{HI}/T = V_{dc}/V_{HI} = 12/14.5 = 0.828$. Thus, the *maximum* duty cycle, corresponding to the smallest value of V_{in}, is 0.828, or 82.8%. When V_{in} has its maximum value of 24 V, we find the minimum duty cycle to be $T_{HI}/T = 12/(24 - 0.5) = 0.51$.

13.10   THREE-TERMINAL INTEGRATED-CIRCUIT REGULATORS

A three-terminal regulator is a compact, easy-to-use, fixed-voltage regulator packaged in a single integrated circuit. To use the regulator, it is necessary only to make external connections to the three terminals: V_{in}, V_o, and ground. These devices are widely used to provide *local regulation* in electronic systems that may require several different supply voltages. For example, a 5-V regulator could be used to regulate the power supplied to all the chips mounted on one printed-circuit board, and a 12-V regulator could be used for a similar purpose on a different board. The regulators might well use the same unregulated input voltage, say, 20 V.

A popular series of three-terminal regulators is the 7800/7900 series, available from several manufacturers with a variety of output voltage ratings. Figure 13.44 shows National Semiconductor specifications for their 7800-series regulators, which carry the company's standard LM prefix and which are available with regulated outputs of +5 V, +12 V, and +15 V. The last two digits of the 7800 number designate the rated output voltage. For example, the 7805 is a +5-V regulator and the 7815 is a +15-V regulator. The 7900-series regulators provide negative output voltages. Notice that the integrated circuitry shown in the schematic diagram is considerably more complex than that of the simple discrete circuits discussed earlier. It can be seen that the circuit incorporates a zener diode as an internal voltage reference. The 7800/7900 series also has internal current-limiting circuitry.

 **National Semiconductor**

Voltage Regulators

LM78XX Series

LM78XX Series Voltage Regulators

General Description

The LM78XX series of three terminal regulators is available with several fixed output voltages making them useful in a wide range of applications. One of these is local on card regulation, eliminating the distribution problems associated with single point regulation. The voltages available allow these regulators to be used in logic systems, instrumentation, HiFi, and other solid state electronic equipment. Although designed primarily as fixed voltage regulators these devices can be used with external components to obtain adjustable voltages and currents.

The LM78XX series is available in an aluminum TO-3 package which will allow over 1.0A load current if adequate heat sinking is provided. Current limiting is included to limit the peak output current to a safe value. Safe area protection for the output transistor is provided to limit internal power dissipation. If internal power dissipation becomes too high for the heat sinking provided, the thermal shutdown circuit takes over preventing the IC from overheating.

Considerable effort was expended to make the LM78XX series of regulators easy to use and minimize the number of external components. It is not necessary to bypass the output, although this does improve transient response. Input bypassing is needed only if the regulator is located far from the filter capacitor of the power supply.

For output voltage other than 5V, 12V and 15V the LM117 series provides an output voltage range from 1.2V to 57V.

Features

- Output current in excess of 1A
- Internal thermal overload protection
- No external components required
- Output transistor safe area protection
- Internal short circuit current limit
- Available in the aluminum TO-3 package

Voltage Range

LM7805C	5V
LM7812C	12V
LM7815C	15V

Schematic and Connection Diagrams

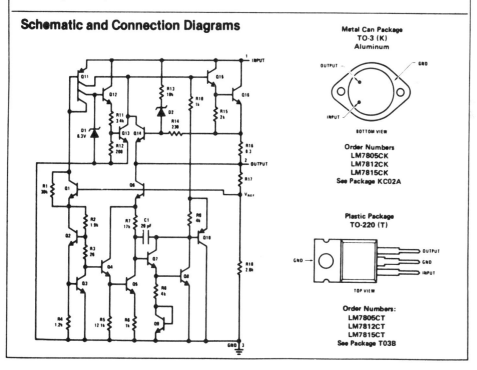

Figure 13.44
7800-series voltage-regulator specifications (Courtesy of National Semiconductor)

LM78XX Series

Absolute Maximum Ratings

Input Voltage (V_O = 5V, 12V and 15V)	35V
Internal Power Dissipation (Note 1)	Internally Limited
Operating Temperature Range (T_A)	
	0°C to +70°C
Maximum Junction Temperature	
(K Package)	150°C
(T Package)	125°C
Storage Temperature Range	−65°C to +150°C
Lead Temperature (Soldering, 10 seconds)	
TO-3 Package K	300°C
TO-220 Package T	230°C

Electrical Characteristics LM78XXC (Note 2) 0°C ≤ Tj ≤ 125°C unless otherwise noted.

OUTPUT VOLTAGE			5V			12V			15V			
INPUT VOLTAGE (unless otherwise noted)			10V			19V			23V			UNITS
PARAMETER		CONDITIONS	MIN	TYP	MAX	MIN	TYP	MAX	MIN	TYP	MAX	
V_O	Output Voltage	T_j = 25°C, 5 mA ≤ I_O ≤ 1A	4.8	5	5.2	11.5	12	12.5	14.4	15	15.6	V
		P_D ≤ 15W, 5 mA ≤ I_O ≤ 1A	4.75		5.25	11.4		12.6	14.25		15.75	V
		V_{MIN} ≤ V_{IN} ≤ V_{MAX}	(7 ≤ V_{IN} ≤ 20)			(14.5 ≤ V_{IN} ≤ 27)			(17.5 ≤ V_{IN} ≤ 30)			V
ΔV_O	Line Regulation	T_j = 25°C ΔV_{IN}		3	50		4	120		4	150	mV
			(7 ≤ V_{IN} ≤ 25)			(14.5 ≤ V_{IN} ≤ 30)			(17.5 ≤ V_{IN} ≤ 30)			V
		I_O = 500 mA 0°C ≤ T_j ≤ +125°C ΔV_{IN}			50			120			150	mV
			(8 ≤ V_{IN} ≤ 20)			(15 ≤ V_{IN} ≤ 27)			(18.5 ≤ V_{IN} ≤ 30)			V
		T_j = 25°C ΔV_{IN}			50			120			150	mV
			(7.3 ≤ V_{IN} ≤ 20)			(14.6 ≤ V_{IN} ≤ 27)			(17.7 ≤ V_{IN} ≤ 30)			V
		I_O ≤ 1A 0°≤ T_j ≤ +125°C ΔV_{IN}			25			60			75	mV
			(8 ≤ I_{IN} ≤ 12)			(16 ≤ V_{IN} ≤ 22)			(20 ≤ V_{IN} ≤ 26)			V
ΔV_O	Load Regulation	T_j = 25°C 5 mA ≤ I_O ≤ 1.5A		10	50		12	120		12	150	mV
		250 mA ≤ I_O ≤ 750 mA			25			60			75	mV
		5 mA ≤ I_O ≤ 1A, 0°C ≤ T_j ≤ +125°C			50			120			150	mV
I_Q	Quiescent Current	I_O ≤ 1A T_j = 25°C			8			8			8	mA
		0°C ≤ T_j ≤ +125°C			8.5			8.5			8.5	mA
ΔI_Q	Quiescent Current Change	5 mA ≤ I_O ≤ 1A			0.5			0.5			0.5	mA
		T_j = 25°C, I_O ≤ 1A V_{MIN} ≤ V_{IN} ≤ V_{MAX}			1.0			1.0			1.0	mA
			(7.5 ≤ V_{IN} ≤ 20)			(14.8 ≤ V_{IN} ≤ 27)			(17.9 ≤ V_{IN} ≤ 30)			V
		I_O ≤ 500 mA, 0°C ≤ T_j ≤ +125°C V_{MIN} ≤ V_{IN} ≤ V_{MAX}			1.0			1.0			1.0	mA
			(7 ≤ V_{IN} ≤ 25)			(14.5 ≤ V_{IN} ≤ 30)			(17.5 ≤ V_{IN} ≤ 30)			V
V_N	Output Noise Voltage	T_A = 25°C, 10 Hz ≤ f ≤ 100 kHz		40			75			90		μV
$\dfrac{\Delta V_{IN}}{\Delta V_{OUT}}$ Ripple Rejection		f = 120 Hz { I_O ≤ 1A, T_j = 25°C or I_O ≤ 500 mA 0°C ≤ T_j ≤ +125°C	62 62	80		55 55	72		54 54	70		dB dB
		V_{MIN} ≤ V_{IN} ≤ V_{MAX}	(8 ≤ V_{IN} ≤ 18)			(15 ≤ V_{IN} ≤ 25)			(18.5 ≤ V_{IN} ≤ 28.5)			V
R_O	Dropout Voltage	T_j = 25°C, I_{OUT} = 1A		2.0			2.0			2.0		V
	Output Resistance	f = 1 kHz		8			18			19		mΩ
	Short-Circuit Current	T_j = 25°C		2.1			1.5			1.2		A
	Peak Output Current	T_j = 25°C		2.4			2.4			2.4		A
	Average TC of V_{OUT}	0°C ≤ T_j ≤ +125°C, I_O = 5 mA		0.6			1.5			1.8		mV/°C
V_{IN}	Input Voltage Required to Maintain Line Regulation	T_j = 25°C, I_O ≤ 1A	7.3			14.6			17.7			V

NOTE 1: Thermal resistance of the TO-3 package (K, KC) is typically 4°C/W junction to case and 35°C/W case to ambient. Thermal resistance of the TO-220 package (T) is typically 4°C/W junction to case and 50°C/W case to ambient.

NOTE 2: All characteristics are measured with capacitor across the inut of 0.22 μF, and a capacitor across the output of 0.1 μF. All characteristics except noise voltage and ripple rejection ratio are measured using pulse techniques (t_W ≤ 10 ms, duty cycle ≤ 5%). Output voltage changes due to changes in internal temperature must be taken into account separately.

Figure 13.44
(Continued)

Important points to note in the 7800-series specifications include the following:

1. The output voltage of an arbitrarily chosen device might not *exactly* equal its nominal value. For example, with a 23-V input, the 7815 output may be anywhere from 14.4 V to 15.6 V. This specification does not mean that the output voltage of a single device will vary over that range, but that one 7815 chosen at random from a large number will hold its output constant at some voltage within that range.
2. The input voltage cannot exceed 35 V and must not fall below a certain minimum value, depending on type number, if output regulation is to be maintained. The minimum specified inputs are 7.3, 14.6, and 17.7 V for the 7805, 7812, and 7815, respectively.
3. Load regulation is specified as a certain output voltage change (ΔV_o) as the load current (I_o) is changed over a certain range. For example, the output of the 7805 will change a maximum of 50 mV as load current changes from 5 mA to 1.5 A, and will change a maximum of 25 mV as load current changes from 250 mA to 750 mA.

Example 13.19

A 7815 regulator is to be used with a full-wave–rectified (120-Hz), capacitor-filtered input whose dc value may have long-term variations between 19 V and 23 V.

1. What is the maximum peak-to-peak ripple voltage that can be tolerated on the input?
2. What maximum peak-to-peak ripple voltage could appear in the output of the regulator if the input has the ripple found in (1)? (Assume that the dc value of the input is 23 V and the load current is 0.5 A.)

Solution

1. The ripple voltage must not cause the input to fall below the minimum required to maintain regulation, which is 17.7 V for the 7815. Under worst-case conditions, when the dc value of the input is 19 V, the decrease in V_{in} due to ripple cannot therefore exceed $19 - 17.7 = 1.3$ V. Thus, the maximum tolerable peak-to-peak ripple is 2(1.3 V) = 2.6 V p–p.
2. The 7815 specifications show that ripple rejection under the given operating conditions is a minimum of 54 dB. Thus,

$$20 \log_{10}\left(\frac{\Delta V_{in}}{\Delta V_o}\right) = 54$$

$$V_o = \frac{\Delta V_{in}}{\text{antilog}(54/20)} = \frac{2.6 \text{ V}}{501.19} = 5.18 \text{ mV p–p}$$

13.11 ADJUSTABLE INTEGRATED-CIRCUIT REGULATORS

As the name implies, an *adjustable* voltage regulator is one that can be set to maintain any output voltage that is within some prescribed range. Unlike three-terminal regulators, adjustable IC regulators must have external components connected to them in order to perform voltage regulation.

The 723 integrated-circuit regulator is an example of a popular and very versatile adjustable regulator. It can be connected to produce positive or negative outputs over the range from 2 to 37 V, can provide either current limiting or foldback limiting, can be used with an external pass transistor to handle load currents up to 10 A, and can be used as a switching regulator. Figure 13.45 is a block diagram of the regulator. Note the terminal labeled V_{REF} at the output of the voltage reference amplifier. This is an internally generated voltage of approximately 7 V that is available at an external pin. To set a desired regulator output voltage, the user connects this 7-V output, or an externally divided-down portion of it, to one of the inputs of the error amplifier. The error amplifier is a comparator that compares the externally connected reference to a voltage proportional to V_o. Depending on whether the reference is connected to the noninverting or the inverting input, the regulated output is either positive or negative. For normal positive voltage regulation, the unregulated input is connected to the terminals labeled V^+ and V_C, and V^- is connected to ground.

Note the transistor labeled "current limiter" in Figure 13.45. By making external resistor connections to the CL and CS terminals, either the current-limiting circuit of Figure 13.34 or the foldback circuit of Figure 13.37 can be implemented. The current limiter performs the function of Q_2 in each of those figures. Of course, the terminals can be left open if no limiting is desired.

Figure 13.46 shows manufacturer's specifications for the 723 regulator, along with some typical applications.

Figure 13.47 shows the 723 regulator connected to maintain its output at any voltage between +2 V and +7 V. The output is determined from

$$V_o = V_{REF} \left(\frac{R_2}{R_1 + R_2} \right) \tag{13.57}$$

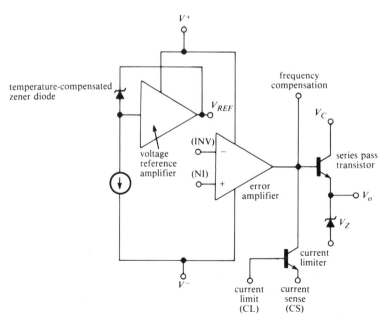

Figure 13.45
Block diagram of the 723 adjustable voltage regulator

 **National Semiconductor**

Voltage Regulators

LM723/LM723C Voltage Regulator

General Description

The LM723/LM723C is a voltage regulator designed primarily for series regulator applications. By
itself, it will supply output currents up to 150 mA,
but external transistors can be added to provide
any desired load current. The circuit features extremely low standby current drain, and provision
is made for either linear or foldback current limiting. Important characteristics are:

- 150 mA output current without external pass
  transistor

- Output currents in excess of 10A possible by
  adding external transistors

- Input voltage 40V max

- Output voltage adjustable from 2V to 37V

- Can be used as either a linear or a switching
  regulator.

The LM723/LM723C is also useful in a wide range
of other applications such as a shunt regulator, a
current regulator or a temperature controller.

The LM723C is identical to the LM723 except
that the LM723C has its performance guaranteed
over a 0°C to 70°C temperature range, instead of
−55°C to +125°C

Schematic and Connection Diagrams *

Equivalent Circuit *

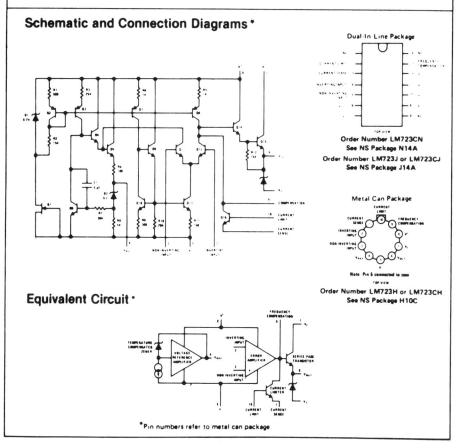

*Pin numbers refer to metal can package.

Figure 13.46
723–voltage-regulator specifications (Courtesy of National Semiconductor)

LM723/LM723C

Absolute Maximum Ratings

Pulse Voltage from V^+ to V^- (50 ms)	50V
Continuous Voltage from V^+ to V^-	40V
Input Output Voltage Differential	40V
Maximum Amplifier Input Voltage (Either Input)	7.5V
Maximum Amplifier Input Voltage (Differential)	5V
Current from V_Z	25 mA
Current from V_{REF}	15 mA
Internal Power Dissipation Metal Can (Note 1)	800 mW
Cavity DIP (Note 1)	900 mW
Molded DIP (Note 1)	660 mW
Operating Temperature Range LM723	-55°C to +125°C
LM723C	0°C to +70°C
Storage Temperature Range Metal Can	-65°C to +150°C
DIP	-55°C to +125°C
Lead Temperature (Soldering, 10 sec)	300°C

Electrical Characteristics (Note 2)

PARAMETER	CONDITIONS	LM723			LM723C			UNITS
		MIN	TYP	MAX	MIN	TYP	MAX	
Line Regulation	$V_{IN} = 12V$ to $V_{IN} = 15V$		01	01		01	01	% V_{OUT}
	-55°C < T_A < +125°C			03				% V_{OUT}
	0°C < T_A < +70°C						03	% V_{OUT}
	$V_{IN} = 12V$ to $V_{IN} = 40V$		02	02		01	05	% V_{OUT}
Load Regulation	$I_L = 1$ mA to $I_L = 50$ mA		03	0.15		03	02	% V_{OUT}
	-55°C < T_A < +125°C			06				% V_{OUT}
	0°C < T_A < +70°C						06	% V_{OUT}
Ripple Rejection	f = 50 Hz to 10 kHz, $C_{REF} = 0$		74			74		dB
	f = 50 Hz to 10 kHz, $C_{REF} = 5 \mu F$		86			86		dB
Average Temperature Coefficient of Output Voltage	-55°C < T_A < +125°C		002	015				%/°C
	0°C < T_A < +70°C					003	015	%/°C
Short Circuit Current Limit	$R_{SC} = 10\Omega$, $V_{OUT} = 0$		65			65		mA
Reference Voltage		6.95	7.15	7.35	6.80	7.15	7.50	V
Output Noise Voltage	BW = 100 Hz to 10 kHz, $C_{REF} = 0$		20			20		μVrms
	BW = 100 Hz to 10 kHz, $C_{REF} = 5 \mu F$		25			25		μVrms
Long Term Stability			01			01		%/1000 hrs
Standby Current Drain	$I_L = 0$, $V_{IN} = 30V$		13	35		13	40	mA
Input Voltage Range		95		40	95		40	V
Output Voltage Range		20		37	20		37	V
Input Output Voltage Differential		30		38	30		38	V

Note 1: See derating curves for maximum power rating above 25°C.

Note 2: Unless otherwise specified, T_A = 25°C, V_{IN} = V^+ = V_C = 12V, V^- = 0, V_{OUT} = 5V, I_L = 1 mA, R_{SC} = 0, C_1 = 100 pF, C_{REF} = 0 and divider impedance as seen by error amplifier $\leq$ 10 kΩ connected as shown in Figure 1. Line and load regulation specifications are given for the condition of constant chip temperature. Temperature drifts must be taken into account separately for high dissipation conditions.

Note 3: L_1 is 40 turns of No. 20 enameled copper wire wound on Ferroxcube P36/22-3B7 pot core or equivalent with 0.009 in. air gap.

Note 4: Figures in parentheses may be used if R1/R2 divider is placed on opposite input of error amp.

Note 5: Replace R1/R2 in figures with divider shown in Figure 13.

Note 6: V^+ must be connected to a +3V or greater supply.

Note 7: For metal can applications where V_Z is required, an external 6.2 volt zener diode should be connected in series with V_{OUT}.

Figure 13.46
(Continued)

LM723/LM723C

TABLE I   RESISTOR VALUES (kΩ) FOR STANDARD OUTPUT VOLTAGE

POSITIVE OUTPUT VOLTAGE	APPLICABLE FIGURES	FIXED OUTPUT ±5%		OUTPUT ADJUSTABLE ±10% (Note 5)			NEGATIVE OUTPUT VOLTAGE	APPLICABLE FIGURES	FIXED OUTPUT ±5%		5% OUTPUT ADJUSTABLE ±10%		
(Note 4)		R1	R2	R1	P1	R2			R1	R2	R1	P1	R2
+3.0	1,5,6,9,12 (4)	4.12	3.01	1.8	0.5	1.2	-100	7	3.57	102	2.2	10	91
+3.6	1,5,6,9,12 (4)	3.57	3.65	1.5	0.5	1.5	-250	7	3.57	255	2.2	10	240
+5.0	1,5,6,9,12 (4)	2.15	4.99	.75	0.5	2.7	-6 (Note 6)	3,(10)	3.57	2.43	1.2	0.5	75
+6.0	1,5,6,9,12 (4)	1.15	6.04	0.5	0.5	2.7	-9	3,10	3.48	5.36	1.2	0.5	2.0
+9.0	2,4,(5,6,12,9)	1.87	7.15	.75	1.0	2.7	-12	3,10	3.57	8.45	1.2	0.5	3.3
+12	2,4,(5,6,9,12)	4.87	7.15	2.0	1.0	3.0	-15	3,10	3.65	11.5	1.2	0.5	4.3
+15	2,4,(5,6,9,12)	7.87	7.15	3.3	1.0	3.0	-28	3,10	3.57	24.3	1.2	0.5	10
+28	2,4,(5,6,9,12)	21.0	7.15	5.6	1.0	2.0	-45	8	3.57	41.2	2.2	10	33
+45	7	3.57	48.7	2.2	10	34	-100	8	3.57	97.6	2.2	10	91
+75	7	3.57	78.7	2.2	10	68	-250	8	3.57	249	2.2	10	240

TABLE II FORMULAE FOR INTERMEDIATE OUTPUT VOLTAGES

Outputs from +2 to +7 volts
[Figures 1, 5, 6, 9, 12, (4)]

$$V_{OUT} = V_{REF} \left[\frac{R2}{R1+R2} \right]$$

Outputs from +4 to +250 volts
(Figure 7)

$$V_{OUT} = \frac{V_{REF}}{2} \left[\frac{R2 - R1}{R1} \right] ; \quad R3 = R4$$

Current Limiting

$$I_{LIMIT} = \frac{V_{SENSE}}{R_{SC}}$$

Outputs from +7 to +37 volts
[Figures 2, 4, (5,6,9,12)]

$$V_{OUT} = V_{REF} \left[\frac{R1+R2}{R2} \right]$$

Outputs from -6 to -250 volts
[Figures 3, 8, 10]

$$V_{OUT} = \frac{V_{REF}}{2} \left[\frac{R1+R2}{R1} \right] ; \quad R3 = R4$$

Foldback Current Limiting

$$I_{KNEE} = \frac{V_{OUT}}{R_{SC}} \frac{R3}{R4} + \frac{V_{SENSE}}{R_{SC}} \frac{(R3+R4)}{R4}$$

$$I_{SHORT CKT} = \frac{V_{SENSE}}{R_{SC}} \times \frac{R3+R4}{R4}$$

Typical Applications

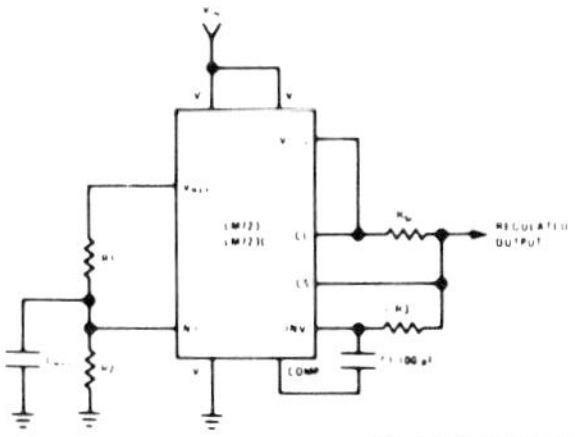

Note: $R3 = \frac{R1\,R2}{R1+R2}$ for minimum temperature drift.

TYPICAL PERFORMANCE
Regulated Output Voltage   5V
Line Regulation (ΔV_IN = 3V)   0.5 mV
Load Regulation (ΔI_L = 50 mA)   1.5 mV

FIGURE 1. Basic Low Voltage Regulator
(V_OUT = 2 to 7 Volts)

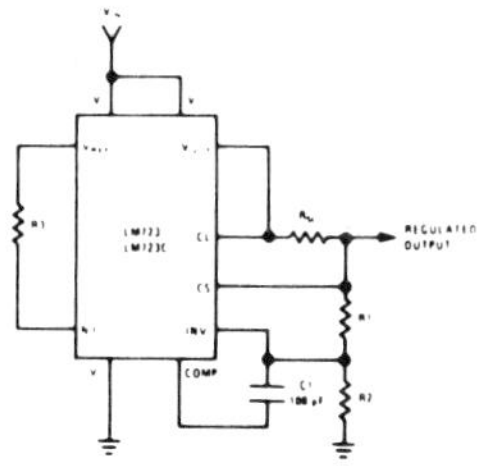

Note: $R3 = \frac{R1\,R2}{R1+R2}$ for minimum temperature drift.
R3 may be eliminated for minimum component count.

TYPICAL PERFORMANCE
Regulated Output Voltage   15V
Line Regulation (ΔV_IN = 3V)   1.5 mV
Load Regulation (ΔI_L = 50 mA)   4.5 mV

FIGURE 2. Basic High Voltage Regulator
(V_OUT = 7 to 37 Volts)

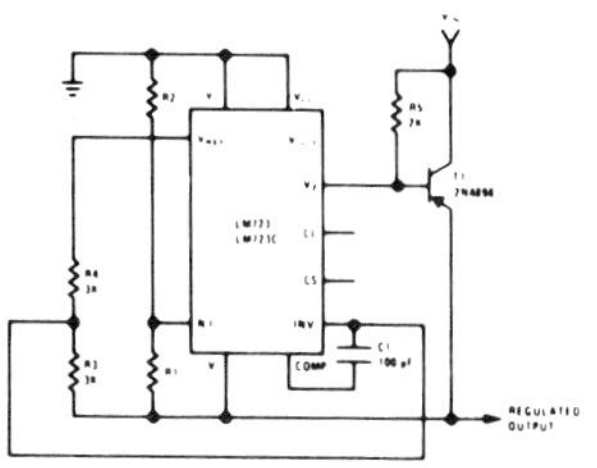

TYPICAL PERFORMANCE
Regulated Output Voltage   -15V
Line Regulation (ΔV_IN = 3V)   1 mV
Load Regulation (ΔI_L = 100 mA)   2 mV

FIGURE 3. Negative Voltage Regulator

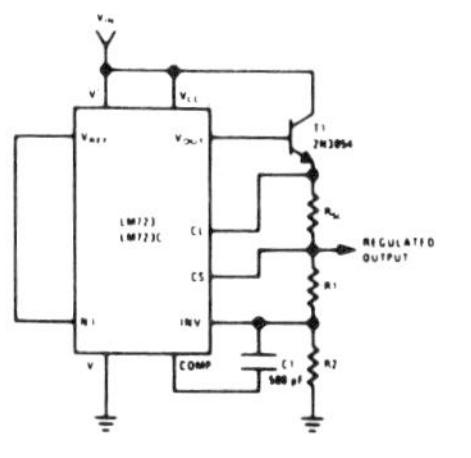

TYPICAL PERFORMANCE
Regulated Output Voltage   +15V
Line Regulation (ΔV_IN = 3V)   1.5 mV
Load Regulation (ΔI_L = 1A)   15 mV

FIGURE 4. Positive Voltage Regulator
(External NPN Pass Transistor)

Figure 13.46
(Continued)

Figure 13.47
The 723 regulator connected to provide
output voltages between +2 V and +7 V

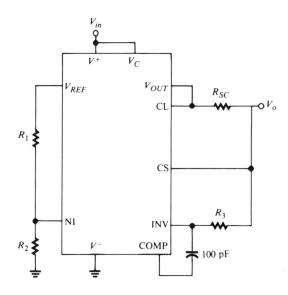

From the specifications in Figure 13.46, we see that V_{REF} may be between 6.95 V and 7.35 V. Therefore, the actual value produced by a given device should be measured before selecting values for R_1 and R_2, if a very accurate output voltage is required. Notice that the full (undivided) output voltage V_o is fed back to the inverting input (INV) through R_3. For maximum thermal stability, R_3 should be set equal to $R_1 \parallel R_2$.

The circuit shown in Figure 13.47 is connected to provide current limiting, where

$$I_L(\text{max}) \approx \frac{0.7 \text{ V}}{R_{SC}} \tag{13.58}$$

The 100-pF capacitor shown in the figure is used to ensure circuit stability. When the circuit is connected to provide foldback limiting, a voltage divider (identified as resistors R_3 and R_4 in Figure 13.37) is connected across V_{OUT} in Figure 13.47. The CL terminal on the 723 regulator is then connected to the middle of the divider, instead of to V_{OUT}.

Example 13.20

DESIGN

Design a 723-regulator circuit that will maintain an output voltage of +5 V and that will provide current limiting at 0.1 A. Assume that $V_{REF} = 7.0$ V.

Solution.   We arbitrarily choose $R_2 = 1$ kΩ. Then, from equation 13.57,

$$V_o = 5 \text{ V} = (7 \text{ V}) \left[\frac{1 \text{ k}\Omega}{R_1 + (1 \text{ k}\Omega)} \right]$$

Solving for R_1, we find $R_1 = 400$ Ω. The optimum value of R_3 is

$$R_3 = R_1 \parallel R_2 = \frac{400(1000)}{1400} = 285.7 \text{ }\Omega$$

From equation 13.58,

$$I_L(\text{max}) = 0.1 \text{ A} = \frac{0.7 \text{ V}}{R_{SC}}$$

$$R_{SC} = 7 \text{ } \Omega$$

EXERCISES

Section 13.2 Rectifiers

13.1 What peak-to-peak sinusoidal voltage must be connected to a half-wave rectifier if the rectified waveform is to have a dc value of 6 V? Assume that the forward drop across the diode is 0.7 V.

13.2 What should be the rms voltage of a sinusoidal wave connected to a full-wave rectifier if the rectified waveform is to have a dc value of 50 V? Neglect diode voltage drops.

13.3 The primary voltage on the transformer shown in Figure 13.48 is 120 V rms and $R_L = 10 \text{ } \Omega$. Neglecting the forward voltage drops across the diodes, find
   a. the turns ratio $N_p : N_s$, if the average current in the resistor must be 1.5 A;
   b. the average power dissipated in the resistor, under the conditions of (a); and
   c. the maximum PIV rating required for the diodes, under the conditions of (a).

13.4 The primary voltage on the transformer in Figure 13.48 is 120 V rms and $N_P : N_s = 15 : 1$. Diode voltage drops are 0.7 V.
   a. What should be the value of R_L if the average current in R_L must be 0.5 A?
   b. What power is dissipated in R_L under the conditions of (a)?
   c. What minimum PIV rating is required for the diodes under the conditions of (a)?

13.5 The primary voltage on the transformer in Figure 13.49 is 110 V rms at 50 Hz. The diode voltage drops are 0.7 V. Sketch the waveforms of the voltage across and current through the 20-Ω resistor. Label peak values and the time points where the waveforms go to 0.

13.6 Each of the diodes in Figure 13.50 has a forward voltage drop of 0.7 V. Find
   a. the average voltage across R_L;
   b. the average power dissipated in the 1-Ω resistor; and
   c. the minimum PIV rating required for the diodes.

13.7 Repeat Exercise 13.6 if R_L is changed to 5 Ω and the transformer turns ratio is changed to 1 : 1.5.

13.8 Sketch the waveform of the voltage v_L in the circuit shown in Figure 13.51. Include the ripple and show the value of its period on the sketch. Also show the value of V_{PR}. Neglect the forward drop across the diode.

13.9 What is the percent ripple of a full-wave–rectified waveform having peak value 75 V and frequency 120 Hz? What is the percent ripple if the peak value is doubled? If the frequency is halved?

13.10 A full-wave bridge is to be connected to a 240-V-rms power line. The output will be filtered and will supply an average voltage of 150 V to a 50-Ω load. The worst-case current that will flow through the bridge when power is first applied is twenty times the average load current. Using the International Rectifier specifications in Figure 13.8, select a bridge (give its part number) that can be used for this application.

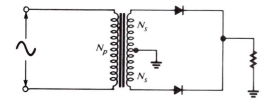

Figure 13.48
(Exercises 13.3 and 13.4)

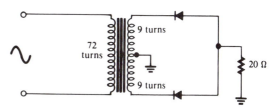

Figure 13.49
(Exercise 13.5)

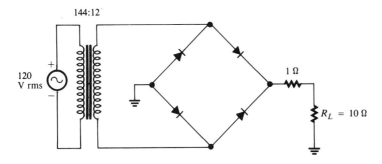

Figure 13.50
(Exercise 13.6)

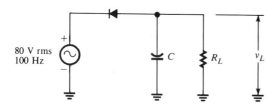

Figure 13.51
(Exercise 13.8)

Section 13.3   Capacitor Filters

13.11 A half-wave rectifier is operated from a 60-Hz line and has a 1000-μF filter capacitance connected across it. What is the minimum value of load resistance that can be connected across the capacitor if the percent ripple cannot exceed 5%?

13.12 A full-wave rectifier is operated from a 60-Hz, 50-V-rms source. It has a 500-μF filter capacitor and a 750-Ω load. Find
a. the average value of the load voltage;
b. the peak-to-peak ripple voltage; and
c. the percent ripple.

13.13 A half-wave rectifier has a 1000-μF filter capacitor and a 500-Ω load. It is operated from a 60-Hz, 120-V-rms source. It takes 1 ms for the capacitor to recharge during each input cycle. For what minimum value of repetitive surge current should the diode be rated?

13.14 Repeat Exercise 13.13 if the rectifier is full-wave and the capacitor takes 0.5 ms to recharge.

Section 13.4   RC and LC Filters

13.15 **a.** Find the rms ripple voltage across C_1 in Figure 13.52(a). (The filter is lightly loaded.)
b. Assuming that $R \ll R_L$, find the value of RC_2 in Figure 13.52(b) that will make the ripple voltage across C_2 equal to one-tenth of that found in (a).
c. Under the conditions of (b), find values for R and C_2 if it is required that the dc voltage across R_L in Figure 13.52(b) be not less than nine-tenths of the dc voltage across R_L in 13.52(a).

13.16 **a.** Find the percent ripple in Exercise 13.15(a) (Figure 13.52(a)).

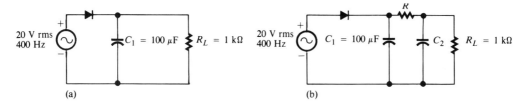

(a)   (b)

Figure 13.52
(Exercise 13.15)

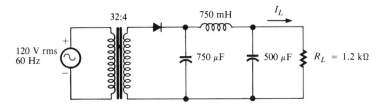

Figure 13.53
(Exercise 13.17)

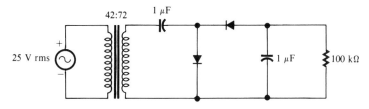

Figure 13.54
(Exercise 13.18)

b. Find the percent ripple in the circuit of Figure 13.52(b), using the values for R and C_2 found in Exercise 13.15(c).

c. Explain why the values of R and C_2 reduce the rms ripple voltage by a factor of 10 but do not reduce the percent ripple by the same amount.

13.17 The inductor in Figure 13.53 has resistance 4 Ω. Find the average load current I_L and the percent ripple in the load voltage. Be certain to confirm the validity of any approximating assumptions made in your computations.

Section 13.5 Voltage Multipliers

13.18 Assuming negligible ripple, find the average current in the 100-kΩ resistor in Figure 13.54.

13.19 The transformer shown in Figure 13.55 has a *tapped secondary*, each portion of the secondary winding having the number of turns shown. Assuming that the primary voltage is that shown in the figure, design two separate circuits that can be used with the transformer to obtain an (unloaded) dc voltage of 1200 V. It is not necessary to specify capacitor sizes. What minimum PIV ratings should the diodes in each design have?

Section 13.6 Voltage Regulation

13.20 A power supply has 4% voltage regulation and an open-circuit output voltage of 48 V dc.

a. What is the full-load voltage of the supply?

b. If a 120-Ω resistor draws full-load current from the supply, what is its output resistance?

13.21 One way to determine the output resistance of a power supply is to vary its load resistance while measuring load voltage. The output resistance of the supply equals the value of load resistance that is found to make the load voltage equal to one-half of the open-circuit voltage of the supply. Using an appropriate equation, explain why this method is valid.

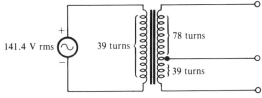

Figure 13.55
(Exercise 13.19)

13.22 A 50-V power supply has line regulation 0.2%/V. How large would the 75-V input voltage to the supply have to become in order for the output voltage to rise to 52 V?

13.23 The specifications for a certain 24-V power supply state that the output voltage increases 18 mV per volt increase in the input voltage. What is the percent line regulation of the supply?

Section 13.7   Zener-Diode Circuits

13.24 In the circuit shown in Figure 13.56, the zener diode has a reverse breakdown voltage of 10 V. If $R_S = 200\ \Omega$,
  a. find V_{in} when $I_Z = 15$ mA and $I_L = 50$ mA; and
  b. find I_Z when $V_{in} = 30$ V.

13.25 In the circuit shown in Figure 13.56, the zener diode has a reverse breakdown voltage of 12 V. $R_S = 50\ \Omega$, $V_{in} = 20$ V, and R_L can vary from 100 Ω to 200 Ω. Assuming that the zener diode remains in breakdown, find
  a. the minimum and maximum current in the zener diode;
  b. the minimum and maximum power dissipated in the diode; and
  c. the minimum rated power dissipation that R_S should have.

13.26 Repeat Exercise 13.25 if, in addition to the variation in R_L, V_{in} can vary from 19 V to 30 V.

13.27 The 6-V zener diode in Figure 13.57 has a maximum rated power dissipation of 0.5 W. Its reverse current must be at least 5 mA to keep it in breakdown. Find a suitable value for R_S if V_{in} can vary from 8 V to 12 V and R_L can vary from 500 Ω to 1 kΩ.

13.28 a. If R_S in Exercise 13.27 is set to its maximum permissible value, what is the maximum permissible value of V_{in}?

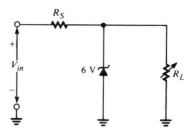

Figure 13.57
(Exercise 13.27)

  b. If R_S in Exercise 13.27 is set equal to its minimum permissible value, what is the minimum permissible value of R_L?

13.29 A zener diode has a breakdown voltage of 12 V at 25°C and a temperature coefficient of +0.5 mV/°C.
  a. Design a temperature-stabilizing circuit using silicon diodes that have temperature coefficients of −0.21 mV/°C. The forward drop across each diode at 25°C is 0.68 V.
  b. Find the voltage across the stabilized network at 25°C and at 75°C.
  c. Find the temperature stability of the stabilized network.

13.30 A zener diode has a breakdown voltage of 15.1 V at 25°C. It has a temperature coefficient of +0.78 mV/°C and is to be operated between 25°C and 100°C. It is to be temperature stabilized in such a way that the voltage across the network is never less than its value at 25°C.
  a. Design a temperature-stabilizing network using silicon diodes whose temperature coefficients are −0.2 mV/°C. The forward drop across each diode at 25°C is 0.65 V.
  b. What is the maximum voltage across the stabilized network?

13.31 Following is a set of measurements that were made on the voltage across and current through a zener diode:

I_Z(mA)	V_Z(volts)
0.5	30.1
1.0	30.15
2.0	30.25
3.5	30.37
6	30.56
8	30.68
10	30.80
30	31.90
40	32.40
90	34.00

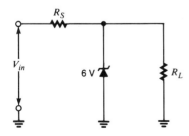

Figure 13.56
(Exercises 13.24 and 13.25)

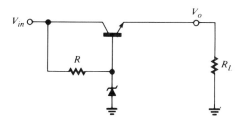

Figure 13.58
(Exercises 13.33 and 13.34)

a. Find the approximate zener impedance over the range from $I_Z = 3.5$ mA to $I_Z = 10$ mA.
b. Show that the zener impedance decreases with increasing current.

13.32 The breakdown voltage of a zener diode when it is conducting 2.5 mA is 7.5 V. If the voltage must not increase more than 10% when the current increases 50%, what maximum impedance can the diode have?

Section 13.8 Series and Shunt Voltage Regulators

13.33 The base-to-emitter voltage of the transistor in Figure 13.58 is 0.7 V. V_{in} can vary from 12 V to 24 V.
a. What breakdown voltage should the zener diode have if the load voltage is to be maintained at 9 V?
b. If the zener diode must conduct 10 mA of reverse current to remain in breakdown, what maximum value should R have?
c. With the value of R found in (b), what is the maximum power dissipated in the zener diode?

13.34 The base-to-emitter voltage of the transistor shown in Figure 13.58 is 0.7 V. V_{in} can vary from 18 V to 30 V. If $V_Z = 10$ V, $R = 500$ Ω, and $R_L = 1$ kΩ, find
a. the minimum collector-to-emitter voltage of the transistor;
b. the maximum power dissipated in the transistor; and
c. the maximum current supplied by V_{in}.

13.35 Draw a schematic diagram of a series voltage regulator that employs one transistor and one zener diode and that can be operated from an unregulated negative voltage to produce a regulated negative voltage. It is not necessary to show component values.

13.36 In the regulator circuit shown in Figure 13.31, V_{BE_2} is 0.68 V when V_o is maintained at 24 V. The zener diode has $V_Z = 9$ V. Find values for R_1 and R_2.

13.37 In the regulator circuit shown in Figure 13.33, $R_1 = 4$ kΩ, $R_2 = 1$ kΩ, and $R_3 = 2$ kΩ. V_{in} can vary from 30 V to 50 V.
a. What breakdown voltage should the zener diode have in order that the regulated output voltage be 25 V?
b. Using the V_Z found in (a), what is the maximum current in the zener diode?

13.38 The potentiometer in Figure 13.59 has a total resistance of 10 kΩ.
a. What range of regulated output voltages can be obtained by adjusting the potentiometer through its entire range?
b. What power is dissipated in the pass transistor when the potentiometer is set for maximum resistance?

13.39 In the circuit shown in Figure 13.34, $R_1 = R_2 = 5$ kΩ, $V_Z = 6$ V, and $R_{SC} = 2$ Ω. Find V_o
a. when $R_L = 100$ Ω; and
b. when $R_L = 30$ Ω.

13.40 Design a current-limited, series voltage regulator that includes an operational amplifier

Figure 13.59
(Exercise 13.38)

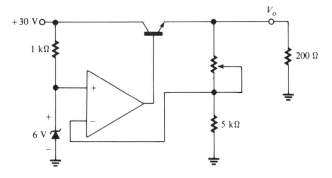

and a 10-V zener diode. The regulated output voltage should be 36 V and the output current should be limited to 700 mA. The zener current should be 10 mA when V_{in} = 40 V.

13.41 In Figure 13.37, R_1 = 1.5 kΩ, R_2 = 1 kΩ, R_3 = 820 Ω, R_4 = 8.2 kΩ, R_{SC} = 2 Ω, and V_Z = 6.3 V.

    a. Find the regulated output voltage.

    b. Find the maximum load current.

13.42 When the regulated output voltage in Figure 13.39 is 18 V, V_{in} supplies a total current of 0.5 A. If R_L = 100 Ω, V_Z = 6 V, R_1 = 200 Ω, and the collector current in Q_2 is negligible, find the power dissipated by Q_1.

13.43 In Figure 13.40, the unregulated input voltage varies from 6 to 9 volts and R_4 = 15 Ω.

    a. What is the maximum (short-circuit) current that can be drawn from the regulator?

    b. What is the minimum power-dissipation rating that R_4 should have?

Section 13.9   Switching Regulators

13.44 What should be the duty cycle of a train of pulses that alternate between 0 V and +5 V if the dc value of the train must be 1.2 V?

13.45 **a.** In Figure 13.60(a), T_1 = 2 ms, T = 5 ms, V_1 = 0.5 V, and V_2 = 6 V. Find the dc value.

    b. In Figure 13.60(b), T_1 = 2 ms, T = 5 ms, V_1 = −0.5 V, and V_2 = 6 V. Find the dc value.

13.46 Derive one general expression involving V_1, V_2, T_1, and T that can be used to determine the dc value of either of the pulse trains in Figure 13.60. (Note that V_1 would be entered in the expression as a positive number for Figure 13.60(a) and as a negative number for Figure 13.60(b).) Use your expression to solve Exercise 13.45.

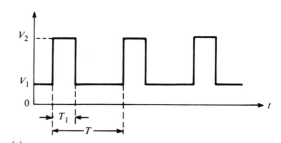

(a)

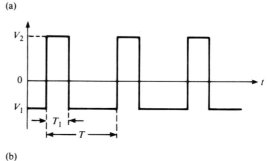

(b)

Figure 13.60
(Exercise 13.45)

13.47 The pulse train produced by the pulse-width modulator in Figure 13.43 has a 50% duty cycle when V_{in} = 20 V. What is its duty cycle when V_{in} = 30 V? When V_{in} = 15 V?

Section 13.10   Three-Terminal Integrated-Circuit Regulators

13.48 A 7815 regulator is to be operated with an unregulated input whose dc value may range from 23 V to 33 V.

    a. What is the maximum peak-to-peak ripple that can be tolerated on the input?

b. With the peak-to-peak ripple found in (a), what maximum percent ripple could the output have? (Assume that the ripple waveform is triangular and that the 7815 is operated at 23 V with load current 0.5 A.)

13.49 A 7812 voltage regulator is operated at 25°C with a 24-Ω load and a 19-V input.

a. What is the worst-case percent line regulation that could be expected?

b. What maximum change in output voltage could be expected if the load resistance were changed over the range from 16 Ω to 48 Ω?

Section 13.11 Adjustable Integrated-Circuit Regulators

13.50 In Figure 13.47, $R_1 = 330$ Ω, $R_2 = 4.7$ kΩ, and $R_{SC} = 3$ Ω. Assuming that $V_{REF} = 7$ V,

a. find the value of the regulated output voltage;

b. find the optimum value of R_3; and

c. find the maximum load current.

13.51 Design a 723-regulator circuit whose output voltage can be varied from 3 V to 6 V by adjusting a potentiometer connected between R_1 and R_2 in Figure 13.47. Assume that $V_{REF} = 7$ V.

13.52 Using the reference material given in the 723-regulator specifications (Figure 13.46), design a +20-V regulator circuit. Assume that $V_{REF} = 7$ V.

SPICE EXERCISES

13.53 The filter capacitor in Figure 13.10 is 75 μF and the input is a 50-V peak sine wave hav-

ing frequency 60 Hz. By trial-and-error runs of SPICE programs, determine the smallest value that R_L can have without the peak-to-peak ripple exceeding 8.7 V ± 0.05 V.

13.54 Use SPICE to find the peak-to-peak ripple voltage of the filtered output in each of the circuits shown in Figure 13.18. Also, use the SPICE results to estimate the dc value of each filtered output. (Because of the long initial transient in each circuit, it will be necessary to examine the output after the input has been applied for about one second. Use the *TSTART* specification in a .TRAN statement to plot only the last 10 ms of a 1-s transient. The iteration limit in SPICE must be overridden using an .OPTIONS statement and a specification such as ITL5 = 20000. Considerable computer time will be required to execute this program.)

13.55 Use SPICE to determine the values of $2V_p$, $3V_p$, and $4V_p$ in Figure 13.22 when v_{in} is a 20-V peak sine wave with frequency 400 Hz. Each capacitor is 1 μF. To allow initial transients time to settle, observe only the last 10 ms of a 100-ms transient. Why are the computed values not exactly twice, three times, and four times the peak value of the input?

13.56 The voltage regulator in Figure 13.32 has $R_1 = 50$ kΩ, $R_2 = 43.75$ kΩ, $R_3 = 47$ kΩ, and $R_4 = 1$ kΩ. The zener diode has a breakdown voltage of 6.3 V and the input voltage to the regulator is 20 V dc. The regulator supplies full-load current when the load resistance is 200 Ω. Use SPICE to find the percent voltage regulation of the regulator. Model the zener diode by specifying a conventional diode having a reverse breakdown voltage of 6.3 V.

Power Amplifiers

14

14.1   DEFINITIONS, APPLICATIONS, AND TYPES OF POWER AMPLIFIERS

As the name implies, a *power amplifier* is one that is designed to deliver a large amount of power to a load. To perform this function, a power amplifier must itself be capable of *dissipating* large amounts of power; that is, it must be designed so that the heat generated when it is operated at high current and voltage levels is released into the surroundings at a rate fast enough to prevent destructive temperature build-up. Consequently, power amplifiers typically contain bulky components having large surface areas to enhance heat transfer to the environment. A *power transistor* is a discrete device with a large surface area and a metal case, characteristics that make it suitable for incorporation into a power amplifier.

A power amplifier is often the last, or *output,* stage of an amplifier system. The preceding stages may be designed to provide voltage amplification, to provide buffering to a high-impedance signal source, or to modify signal characteristics in some predictable way, functions that are collectively referred to as *signal conditioning*. The output of the signal-conditioning stages drives the power amplifier, which in turn drives the load. Some amplifiers are constructed with signal-conditioning stages and the output power stage all in one integrated circuit. Others, especially those designed to deliver very large amounts of power, have a *hybrid* structure, in the sense that the signal-conditioning stages are integrated and the power stage is discrete.

Power amplifiers are widely used in audio components—radio and television receivers, phonographs and tape players, stereo and high-fidelity systems, recording-studio equipment, and public address systems. The load in these applications is most often a loudspeaker ("speaker"), which requires considerable power to convert electrical signals to sound waves. Power amplifiers are also used in electromechanical *control systems* to drive electric motors. Examples include computer disk and tape drives, robotic manipulators, autopilots, antenna rotators, pumps and motorized valves, and manufacturing and process controllers of all kinds.

Large-Signal Operation

Because a power amplifier is required to produce large voltage and current variations in a load, it is designed so that at least one of its semiconductor components, typically a power transistor, can be operated over substantially the *entire* range of its output characteristics, from saturation to cutoff. This mode of operation is called *large-signal operation*. Recall that small-signal operation occurs when the range of current and voltage variation is small enough that there is no appreciable change in device parameters, such as β and r_e. By contrast, the parameters of a large-signal amplifier at one output voltage may be considerably different from those at another output voltage. There are two important consequences of this fact:

1. Signal distortion occurs because of the change in amplifier characteristics with signal level. *Harmonic distortion* always results from such *nonlinear* behavior of an amplifier. Compensating techniques, such as negative feedback, must be incorporated into a power amplifier if low-distortion, high-level outputs are required.
2. Many of the equations we have developed for small-signal analysis of amplifiers are no longer valid. Those equations were based on the assumption that device parameters did not change, contrary to fact in large-signal amplifiers. Rough approximations of large-signal–amplifier behavior can be obtained by using average parameter values and applying small-signal analysis techniques. However, graphical methods are used more frequently in large-signal–amplifier design.

As a final note on terminology, we should mention that the term *large-signal operation* is also applied to devices used in digital switching circuits. In these applications, the output level switches between "high" and "low" (cutoff and saturation), but remains in those states *most of the time*. Power dissipation is therefore not a problem. Either the output voltage or the output current is near 0 when a digital device is in an ON or an OFF state, so power, which is the product of voltage and current, is near 0 except during the short time when the device switches from one state to the other. On the other hand, the variations in the output level of a power amplifier occur in the active region, *between* the two extremes of saturation and cutoff, so a substantial amount of power is dissipated.

14.2   TRANSISTOR POWER DISSIPATION

Recall that power is, by definition, the *rate* at which energy is consumed or dissipated (1 W = 1 J/s). If the rate at which heat energy is dissipated in a device is less than the rate at which it is generated, the temperature of the device must rise. In electronic devices, electrical energy is converted to heat energy at a rate given by $P = VI$ watts, and temperature rises when this heat energy is not removed at a comparable rate. Since semiconductor material is irreversibly damaged when subjected to temperatures beyond a certain limit, temperature is the parameter that ultimately limits the amount of power a semiconductor device can handle.

Transistor manufacturers specify the maximum permissible junction temperature and the maximum permissible power dissipation that a transistor can withstand. As we shall presently see, the maximum permissible power dissipation is specified as a function of *ambient temperature* (temperature of the surroundings), because the rate at which heat can be liberated from the device depends on the temperature of the region to which the heat must be transferred. Most of the conversion of electrical energy to heat energy in a bipolar junction transistor occurs at the junctions. Since power is the product of voltage and current, and since collector and emitter currents are approximately equal, the greatest power dissipation occurs at the junction where the voltage is greatest. In normal transistor operation, the collector–base junction is reverse biased and has, on average, a large voltage across it, while the base–emitter junction has a small forward-biasing voltage. Consequently, most of the heat generated in a transistor is produced at the collector–base junction. The total power dissipated at the junctions is

$$P_d = V_{CB}I_C + V_{BE}I_E \approx (V_{CB} + V_{BE})I_C = V_{CE}I_C \qquad \textbf{(14.1)}$$

Equation 14.1 gives, for all practical purposes, the total power dissipation of the transistor.

For a fixed value of P_d, the graph of equation 14.1 is a *hyperbola* when plotted on I_C–V_{CE}-axes. Larger values of P_d correspond to hyperbolas that move outward from the axes, as illustrated in Figure 14.1. Each hyperbola represents all possible combinations of V_{CE} and I_C that give a product equal to the same value of P_d. The figure shows sample coordinate values (I_C, V_{CE}) at points on each hyperbola. Note that the product of the coordinate values is the same for points on the same hyperbola, and that the product equals the power dissipation corresponding to the hyperbola.

Figure 14.2 shows a simple common-emitter amplifier and its dc load line plotted on I_C–V_{CE}-axes. Also shown is a set of hyperbolas corresponding to different values of power dissipation. Recall that the load line represents all possible combinations of I_C and V_{CE} corresponding to a particular value of R_C. As the amplifier output changes in response to an input signal, the collector current and voltage undergo variations along the load line and intersect different hyperbolas of

Figure 14.1
A family of graphs (hyperbolas) corresponding to $P_d = V_{CE}I_C$ for different values of P_d. Each hyperbola represents all combinations of collector voltage and collector current that result in a specific power dissipation (value of P_d).

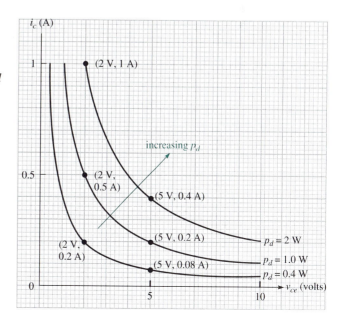

Figure 14.2
Load lines and hyperbolas of constant
power dissipation

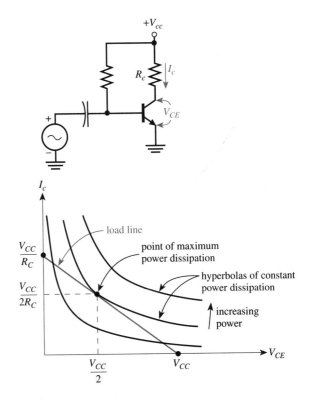

power dissipation. It is clear that the power dissipation changes as the amplifier output changes. For safe operation, the load line must lie below and to the left of the hyperbola corresponding to the maximum permissible power dissipation. When the load line meets this requirement, there is no possible combination of collector voltage and current that results in a power dissipation exceeding the rated maximum.

It can be shown that the point of maximum power dissipation occurs at the *center* of the load line, where $V_{CE} = V_{CC}/2$ and $I_C = V_{CC}/2R_C$ (see Figure 14.2). Therefore, the maximum power dissipation is

$$P_d(\text{max}) = \left(\frac{V_{CC}}{2}\right)\left(\frac{V_{CC}}{2R_C}\right) = \frac{V_{CC}^2}{4R_C} \tag{14.2}$$

To ensure that the load line lies below the hyperbola of maximum dissipation, we therefore require that

$$\frac{V_{CC}^2}{4R_C} < P_d(\text{max}) \tag{14.3}$$

where $P_d(\text{max})$ is the manufacturer's specified maximum dissipation at a specified ambient temperature. Inequality 14.3 enables us to find the minimum permissible value of R_C for a given V_{CC} and a given $P_d(\text{max})$:

$$R_C > \frac{V_{CC}^2}{4P_d(\text{max})} \tag{14.4}$$

Example 14.1

The amplifier shown in Figure 14.2 is to be operated with $V_{CC} = 20$ V and $R_C = 1$ kΩ.

1. What maximum power dissipation rating should the transistor have?
2. If an increase in ambient temperature reduces the minimum rating found in (1) by a factor of 2, what new value of R_C should be used to ensure safe operation?

Solution

1. From inequality 14.3,

$$P_d(\text{max}) > \frac{V_{CC}^2}{4R_C} = \frac{(20 \text{ V})^2}{4(10^3 \text{ }\Omega)} = 0.1 \text{ W}$$

2. When the dissipation rating is decreased to 0.05 W, the minimum permissible value of R_C is, from (14.4),

$$R_C > \frac{V_{CC}^2}{4P_d(\text{max})} = \frac{(20 \text{ V})^2}{4(0.05 \text{ W})} = 2 \text{ k}\Omega$$

14.3 HEAT TRANSFER IN SEMICONDUCTOR DEVICES

Conduction, Radiation, and Convection

Heat transfer takes place by one or more of three fundamental mechanisms: *conduction, radiation,* and *convection,* as illustrated in Figure 14.3. Conduction occurs when the energy that the atoms of a material acquire through heating is transferred to adjacent, less energetic atoms in a cooler region of the material. For example, conduction is the process by which heat flows from the heated end of a metal rod towards the cooler end, and the process by which heat is transferred from the heating element of a stove to a pan, and from the pan to the water it contains. Heat radiation is similar to other forms of radiation in that no physical medium is required between the heat source and its destination. For example,

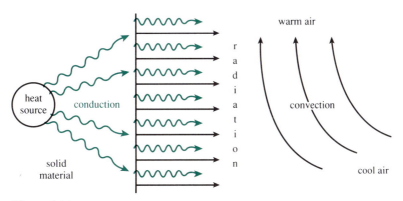

Figure 14.3
Examples illustrating heat transfer by conduction, radiation, and convection

heat is transferred by radiation through space from the sun to the earth. Convection occurs when a physical medium that has been heated by conduction or radiation *moves* away from the source of heat, i.e., is displaced by a cooler medium, so more conduction or radiation can occur. Examples include *forced* convection, where a fan is used to push air past a heated surface, and *natural* convection, where heated air rises and is replaced by cooler air from below.

Semiconductor devices are cooled by all three heat-transfer mechanisms. Heat is conducted from a junction, through the semiconductor material, through the case, and into the surrounding air. It is radiated from the surface of the case. Heating of the surrounding air creates air flow around the device, so convection cooling occurs. Various measures are taken to enhance heat transfer by each mechanism. To improve conduction, good physical contact is made between a junction and a device's metal case, or enclosure. In fact, in many power transistors the collector is in direct contact with the case, so the two are electrically the same as well as being at approximately the same temperature. To improve the conduction and radiation of heat from the case to the surrounding air, power devices are often equipped with *heat sinks*. These are attached to the device to be cooled and conduct heat outward to metal *fins* that increase the total surface area from which conduction and radiation into the air can take place. See Figure 14.4. Finally, convection cooling is enhanced by the use of fans that blow cool air past the surface of the case and/or the heat sink.

Thermal Resistance

Heat flow by conduction is very much like the conduction of electrical charge, i.e., current. Recall that current is the *rate* of flow of charge and is proportional to the difference in voltage across a resistance. Similarly, power is the rate of flow of heat energy and is proportional to the difference in *temperature* across the region through which heat is conducted. We can regard any impediment to heat flow as *thermal resistance*, θ. Thus, the rate of flow of heat (i.e., power) is directly

Figure 14.4
Typical heat sinks (Courtesy of EG & G Wakefield Engineering)

proportional to temperature difference and inversely proportional to thermal resistance:

$$P = \frac{\Delta T}{\theta} = \frac{T_2 - T_1}{\theta} \text{ joules/second, or watts} \tag{14.5}$$

Insulating materials, like wool, have large thermal resistances and metals have small thermal resistances. Note the similarity of equation 14.5 to Ohm's law: $I = (V_2 - V_1)/R$. Solving for θ in (14.5), we obtain the following equivalent relation that allows us to determine the units of thermal resistance:

$$\theta = \frac{T_2 - T_1}{P} \text{ degrees/watt} \tag{14.6}$$

In a power amplifier, heat is transferred through different types of materials and across boundaries of dissimilar materials, each of which presents different values of thermal resistance. These are usually treated like series electrical circuits, so thermal resistances are added when computing the total heat flow through the system or the temperatures at various points in the system.

Example 14.2

The collector–base junction of a certain transistor dissipates 2 W. The thermal resistance from junction to case is 8°C/W and the thermal resistance from case to air is 20°C/W. The free-air temperature (ambient temperature) is 25°C.

1. What is the junction temperature?
2. What is the case temperature?

Solution

1. The total thermal resistance between junction and ambient is the sum of that from junction to case and that from case to ambient: $\theta_T = \theta_{JC} + \theta_{CA} = 8°C/W + 20°C/W = 28°C/W$. Therefore, by equation 14.5, with $T_2 = T_J =$ junction temperature and $T_1 = T_A =$ ambient temperature,

$$T_J - T_A = P\theta_T$$
$$T_J = P\theta_T + T_A = (2 \text{ W})(28°C/W) + 25°C = 81°C$$

2. The rate of heat flow, P, is constant through the series configuration, so

$$T_J - T_C = P\theta_{JC}$$
$$T_C = T_J - P\theta_{JC} = 81°C - (2 \text{ W})(8°C/W) = 65°C$$

To reduce thermal resistance and improve heat conductivity, a special silicone grease is often used between contact surfaces, such as between case and heat sink. Mica washers are used to isolate case and heat sink *electrically* when the case is electrically common to one of the device terminals, such as a collector. The washer creates additional thermal resistance in the heat flow path between case and heat sink and may have to be taken into account in heat-flow computations.

The values of the thermal resistance of various types of heat sinks, washers, and metals can be determined from published manufacturers' data. The data are often given in terms of thermal *resistivity* ρ (°C · in./W or °C · m/W), and thermal resistance is computed by

$$\theta = \frac{\rho t}{A} \text{ degrees Celsius/watt} \qquad (14.7)$$

where

$$\rho = \text{resistivity of material}$$
$$t = \text{thickness of material}$$
$$A = \text{area of material}$$

For example, a typical value of ρ for mica is 66°C · in/W, so a 0.002-in.-thick mica washer having a surface area of 0.6 in.² will have a thermal resistance of

$$\theta = \frac{(66°C \cdot \text{in./W})(2 \times 10^{-3} \text{ in.})}{0.6 \text{ in.}^2} = 0.22°C/W$$

Example 14.3

The maximum permissible junction temperature of a certain power transistor is 150°C. It is desired to operate the transistor with a power dissipation of 15 W in an ambient temperature of 40°C. The thermal resistances are as follows:

$$\theta_{JC} = 0.5°C/W \qquad \text{(junction to case)}$$
$$\theta_{CA} = 10°C/W \qquad \text{(case to ambient)}$$

1. Determine whether a heat sink is required for this application.
2. If a heat sink is required, determine the maximum thermal resistance it can have. Assume that a mica washer having thermal resistance $\theta_W = 0.5°C/W$ must be used between case and heat sink.

Solution

1.
$$\theta_T = \theta_{JC} + \theta_{CA} = 0.5°C/W + 10°C/W = 10.5°C/W$$
$$T_J - T_A = \theta_T P = (10.5°C/W)(15 \text{ W}) = 157.5°C$$
$$T_J = 157.5°C + T_A = 157.5°C + 40°C = 197.5°C$$

Since the junction temperature, 197.5°C, will exceed the maximum permissible value of 150°C under the given conditions, a heat sink must be used to reduce the total thermal resistance.

2. Setting T_J equal to its maximum permissible value of 150°C, we can solve for the maximum total thermal resistance:

$$T_J - T_A = \theta_T P$$
$$150 - 40 = (\theta_T)(15)$$
$$\theta_T = \frac{110°C}{15 \text{ W}} = 7.33°C/W$$

The thermal resistance of the heat sink, θ_H, will replace the thermal resistance θ_{CA} between case and ambient that was present in (1). Thus, taking into account the thermal resistance of the washer, we have $\theta_T = 7.33°C/W = \theta_{JC} + \theta_W + \theta_H = 0.5°C/W + 0.5°C/W + \theta_H$, or $\theta_H = (7.33 - 1.0)°C/W = 6.33°C/W$.

Derating

The maximum permissible power dissipation of a semiconductor device is specified by the manufacturer at a certain temperature, either case temperature or ambient temperature. For example, the maximum power dissipation of a power transistor may be given as 10 W at 25°C ambient, and that of another device as 20 W at 50°C case temperature. These ratings mean that each device can dissipate the specified power at temperatures up to and including the given temperature, but not at greater temperatures. The maximum permissible dissipation *decreases* as a function of temperature when temperature increases beyond the given value. The decrease in permissible power dissipation at elevated temperatures is called *derating*, and the manufacturer specifies a *derating factor*, in W/°C, that is used to find the *decrease* in dissipation rating beyond a certain temperature.

To illustrate derating, suppose a certain device has a maximum rated dissipation of 20 W at 25°C ambient and a derating factor of 100 mW/°C at temperatures above 25°C. Then the maximum permissible dissipation at 100°C is

$$P_d(\text{max}) = (\text{rated dissipation at } T_1) - (T_2 - T_1)(\text{derating factor})$$
$$= (20 \text{ W}) - (100°C - 25°C)(100 \text{ mW/}°C) \qquad (14.8)$$
$$= (20 \text{ W}) - (75°C)(100 \text{ mW/}°C) = (20 \text{ W}) - (7.5 \text{ W}) = 12.5 \text{ W}$$

Example 14.4

A certain semiconductor device has a maximum rated dissipation of 5 W at 50°C case temperature and must be derated above 50°C case temperature. The thermal resistance from case to ambient is 5°C/W.

1. Can the device be operated at 5 W of dissipation without auxiliary cooling (heat sink or fan) when the ambient temperature is 40°C?
2. If not, what is the maximum permissible dissipation, with no auxiliary cooling, at 40°C ambient?
3. What derating factor should be applied to this device, in watts per ambient degree?

Solution

1. We determine the case temperature when the ambient temperature is 40°C.

$$T_C - T_A = \theta_{CA}P$$
$$T_C - 40°C = (5°C/W)(5 \text{ W})$$
$$T_C = 40 + 25 = 65°C$$

Since the case temperature exceeds 50°C, the full 5 W of power dissipation is not permitted when the ambient temperature is 40°C.

2. The derated dissipation at $T_A = 40°C$ is

$$P_d(\text{max}) = (T_C - T_A)/\theta = \frac{(50 - 40)°C}{5°C/W} = 2 \text{ W}$$

3. We must first find the maximum ambient temperature at which the device can dissipate 5 W when the case temperature is 50°C:

$$5 \text{ W} = \frac{(50°C - T_A)}{5°C/W}$$
$$T_A = 50 - 25 = 25°C$$

Thus, derating will be necessary when the ambient temperature exceeds 25°C. Let D be the derating factor. Then $P_d(\text{max}) = (5 \text{ W}) - (T_A - 25)D$. Substituting the results of (2), we find

$$2 \text{ W} = (5 \text{ W}) - (40 - 25)D$$
$$D = 0.2 \text{ W/}°\text{C}$$

Power Dissipation in Integrated Circuits

All the heat transfer concepts we have described in connection with power devices apply equally to integrated circuits, including derating. Of course, a maximum junction temperature is not specified for an integrated circuit, but a maximum device temperature or case temperature is usually given. Many integrated
circuits are available in a variety of case types, and the rated power dissipation
may depend on the case used. For example, the specifications given in Chapter 8
for the μA741 operational amplifier show that the maximum dissipation is 500 mW
for the metal package, 310 mW for the DIP, and 570 mW for the flatpak. Note that
these specifications require the use of derating factors above 70°C ambient, and
that the factors differ, depending on case type.

The total power dissipation of an integrated circuit can be calculated by measuring the current drawn from each supply voltage used in a particular application
and forming the product of each current with the respective voltage. For example,
if an operational amplifier employing ± 15-V-dc power supplies draws 10 mA from
the positive supply and 8 mA from the negative supply, the total dissipation is
$P_d = (15 \text{ V})(10 \text{ mA}) + (15 \text{ V})(8 \text{ mA}) = 270 \text{ mW}$.

14.4   AMPLIFIER CLASSES AND EFFICIENCY

Class-A Amplifiers

All the small-signal amplifiers we have studied in this book have been designed so
that output voltage can vary in response to both positive and negative inputs; that
is, the amplifiers are *biased* so that under normal operation the output never
saturates or cuts off. An amplifier that has that property is called a *class-A* amplifier. More precisely, an amplifier is class A if its output remains in the active
region during a complete cycle (one full period) of a sine-wave input signal.

Figure 14.5 shows a typical class-A amplifier and its input and output waveforms. In this case, the transistor is biased at $V_{CE} = V_{CC}/2$, which is midway
between saturation and cutoff, and which permits maximum output voltage swing.
Note that the output can vary through (approximately) a full V_{CC} volts, peak-to-
peak. The output is in the transistor's active region during a full cycle (360°) of the
input sine wave.

Figure 14.5
The output of a class-A ampli-
fier remains in the active region
during a full period (360°) of the
input sine wave. In this exam-
ple, the transistor output is
biased midway between satura-
tion and cutoff.

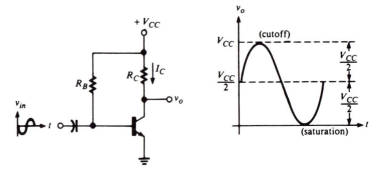

Efficiency

The efficiency of a power amplifier is defined to be

$$\eta = \frac{\text{average signal power delivered to load}}{\text{average power drawn from dc source(s)}} \qquad (14.9)$$

Note that the numerator of (14.9) is average *signal* power, that is, average *ac* power, excluding any dc or bias components in the load. Recall that when voltages and currents are sinusoidal, average ac power can be calculated using any of the following relations:

$$P = V_{rms}I_{rms} = \frac{V_P I_P}{2} = \frac{V_{PP}I_{PP}}{8} \qquad (14.10)$$

$$P = I_{rms}^2 R = \frac{I_P^2 R}{2} = \frac{I_{PP}^2 R}{8} \qquad (14.11)$$

$$P = \frac{V_{rms}^2}{R} = \frac{V_P^2}{2R} = \frac{V_{PP}^2}{8R} \qquad (14.12)$$

where the subscripts P and PP refer to peak and peak-to-peak, respectively.

As a consequence of the definition of efficiency (equation 14.9), the efficiency of a class-A amplifier is 0 when no signal is present. (The amplifier is said to be in *standby* when no signal is applied to its input.) We will now derive a general expression for the efficiency of the class-A amplifier shown in Figure 14.5. In doing so, we will not consider the small power consumed in the base-biasing circuit, i.e., the power at the input side: $I_B^2 R_B + v_{be}i_b$. Figure 14.6 shows the voltages and currents used in our analysis. Notice that resistance R is considered to be the load. We will refer to this configuration as a *series-fed* class-A amplifier, and we will consider capacitor- and transformer-coupled loads in a later discussion.

The instantaneous power from the dc supply is

$$p_S(t) = V_{CC}i = V_{CC}(I_Q + I_P\sin \omega t) = V_{CC}I_Q + V_{CC}I_P\sin \omega t \qquad (14.13)$$

Since the average value of the sine term is 0, the average power from the dc supply is

$$P_S = V_{CC}I_Q \text{ watts} \qquad (14.14)$$

Figure 14.6
Voltages and currents used in the derivation of an expression for the efficiency of a series-fed, class-A amplifier

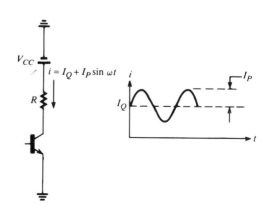

The average *signal* power in load resistor R is, from equation 14.11,

$$P_R = \frac{I_P^2 R}{2} \text{ watts} \qquad (14.15)$$

Therefore, by equation 14.9,

$$\eta = \frac{P_R}{P_S} = \frac{I_P^2 R}{2 V_{CC} I_Q} \qquad (14.16)$$

We see again that the efficiency is 0 under no-signal conditions ($I_P = 0$) and that efficiency rises as the peak signal level I_P increases. The maximum possible efficiency occurs when I_P has its maximum possible value without distortion. When the bias point is at the center of the load line, as shown in Figure 14.2, the quiescent current is one-half the saturation current, and the output current can swing through the full range from 0 to V_{CC}/R amperes without distorting (clipping). Thus, the maximum undistorted peak current is also one-half the saturation current:

$$I_Q = I_P(\text{max}) = \frac{V_{CC}}{2R} \qquad (14.17)$$

Substituting (14.17) into (14.16), we find the maximum possible efficiency of the series-fed, class-A amplifier:

$$\eta(\text{max}) = \frac{(V_{CC}/2R)^2 R}{2 V_{CC}(V_{CC}/2R)} = 0.25 \qquad (14.18)$$

This result shows that the best possible efficiency of a series-fed, class-A amplifier is undesirably small: only ¼ of the total power consumed by the circuit is delivered to the load, under optimum conditions. For that reason, this type of amplifier is not widely used in heavy power applications. The principal advantage of the class-A amplifier is that it generally produces less signal distortion than some of the other, more efficient classes that we will consider later.

Another type of efficiency used to characterize power amplifiers relates signal power to total power dissipated at the collector. Called *collector efficiency,* its practical significance stems from the fact that a major part of the cost and bulk of a power amplifier is invested in the output device itself and the means used to cool it. Therefore, it is desirable to maximize the ratio of signal power in the load to power consumed by the device. Collector efficiency η_c is defined by

$$\eta_c = \frac{\text{average signal power delivered to load}}{\text{average power dissipated at collector}} \qquad (14.19)$$

The average power P_C dissipated at the collector of the class-A amplifier in Figure 14.6 is the product of the dc (quiescent) voltage and current:

$$P_C = V_Q I_Q = (V_{CC} - I_Q R) I_Q \qquad (14.20)$$

Therefore,

$$\eta_c = \frac{I_P^2 R/2}{(V_{CC} - I_Q R) I_Q} \qquad (14.21)$$

The maximum value of η_c occurs when I_P is maximum, $I_P = I_Q = V_{CC}/2R$, as previously discussed. Substituting these values into (14.21) gives

$$\eta_c(\text{max}) = \frac{(V_{CC}/2R)^2 \ R/2}{(V_{CC} - V_{CC}/2)V_{CC}/2R} = 0.5 \qquad (14.22)$$

Figure 14.7 shows the output side of a class-A amplifier with capacitor-coupled load R_L. Also shown are the dc and ac load lines that result. In this case, the average power delivered to the load is

$$P_L = I_{PL}^2 R_L/2 \qquad (14.23)$$

where I_{PL} is the peak ac load current. The average power from the dc source is computed in the same way as for the series-fed amplifier: $P_S = V_{CC}I_Q$, so the efficiency is

$$\eta = \frac{I_{PL}^2 R_L}{2V_{CC}I_Q} \qquad (14.24)$$

As in the case of the series-fed amplifier, the efficiency is 0 under no-signal (standby) conditions and increases with load current I_{PL}.

Maximum output swing can be achieved by setting the Q-point in the center of the ac load line at

$$I_Q = \frac{V_{CC}}{R_C + r_L} \qquad (14.25)$$

The peak collector current under those circumstances is $V_{CC}/(R_C + r_L)$. Neglecting the transistor output resistance, the portion of the collector current that flows

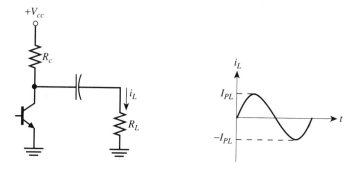

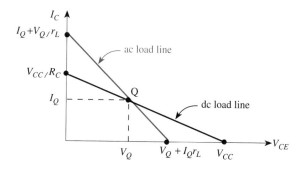

Figure 14.7
A class-A amplifier with capacitor-coupled load R_L

in R_L is, by the current-divider rule,

$$I_{PL} = \left(\frac{V_{CC}}{R_C + r_L}\right)\left(\frac{R_C}{R_C + R_L}\right) \tag{14.26}$$

The average ac power in the load resistance R_L is then

$$P_L = \frac{I_{PL}^2 R_L}{2} = \left[\left(\frac{V_{CC}}{R_C + r_L}\right)\left(\frac{R_C}{R_C + R_L}\right)\right]^2 (R_L/2) \tag{14.27}$$

The average power supplied from the dc source is

$$P_S = V_{CC} I_Q = \frac{V_{CC}^2}{R_C + r_L} \tag{14.28}$$

Therefore, the efficiency under the conditions of maximum possible undistorted output is

$$\eta = \frac{P_L}{P_S} = \frac{\left[\left(\dfrac{V_{CC}}{R_C + r_L}\right)\left(\dfrac{R_C}{R_C + R_L}\right)\right]^2 R_L}{2\left(\dfrac{V_{CC}^2}{R_C + r_L}\right)} \tag{14.29}$$

Algebraic simplification of (14.29) leads to

$$\eta = \frac{R_C R_L}{2(R_C + 2R_L)(R_C + R_L)} = \frac{r_L}{2(R_C + 2R_L)} \tag{14.30}$$

Equation 14.30 shows that the efficiency depends on both R_C and R_L. In practice, R_L is a fixed and known value of load resistance, while the value of R_C is selected by the designer. If R_C is fixed and R_L can be selected, it can be shown using calculus (differentiating equation 14.30 with respect to R_L) that η is maximized by setting

$$R_L = \frac{R_C}{\sqrt{2}} \tag{14.31}$$

With this value of R_L, the maximum efficiency is

$$\eta(\text{max}) = 0.0858 \tag{14.32}$$

Another criterion for choosing R_L is to select its value so that maximum power is transferred to the load. Under the constraint of maintaining maximum output swing, it can be shown that maximum power transfer occurs when $R_L = R_C/2$. Under that circumstance, by substituting into (14.30), we find the maximum possible efficiency with maximum power transfer to be

$$\eta(\text{max}) = 0.0833 \quad (\text{max power transfer}) \tag{14.33}$$

It is interesting to note that the efficiency under maximum power transfer (0.0833) is somewhat less than that which can be achieved (0.0858) without regard to power transfer. In either case, the maximum efficiency is substantially less than that attainable in the series-fed class-A amplifier.

Example 14.5

The class-A amplifier shown in Figure 14.8 is biased at $V_{CE} = 12$ V. The output voltage is the maximum possible without distortion. Find

1. the average power from the dc supply;
2. the average power delivered to the load;
3. the efficiency; and
4. the collector efficiency.

Solution

1. $I_Q = (V_{CC} - V_Q)/R_C = (24 - 12)$ V/50 $\Omega = 0.24$ A
   $P_S = V_{CC}I_Q = (24$ V$)(0.24$ A$) = 5.76$ W

2. As can be seen in Figure 14.7, the maximum value of the peak output voltage is the smaller of V_Q and $I_Q r_L$. In this case, $V_Q = 12$ V and $I_Q r_L = (0.24$ A$)$ $(50 \, \Omega \,\|\, 50 \, \Omega) = 6$ V. Thus, the peak undistorted output voltage is 6 V and the ac power delivered to the 50-Ω load is

$$P_L = \frac{V_{PL}^2}{2R_L} = \frac{6^2}{100} = 0.36 \text{ W}$$

3. $$\eta = \frac{P_L}{P_S} = \frac{0.36 \text{ W}}{5.76 \text{ W}} = 0.0625$$

The efficiency is less than the theoretical maximum because the bias point permits only a ± 6-V swing. As an exercise, find the quiescent value of V_{CE} that maximizes the swing and calculate the efficiency under that condition.

4. The average power dissipated at the collector is $P_C = V_Q I_Q = (12$ V$)(0.24$ A$) = 2.88$ W. Therefore,

$$\eta_c = \frac{P_L}{P_C} = \frac{0.36 \text{ W}}{2.88 \text{ W}} = 0.125$$

Figure 14.8
(Example 14.5)

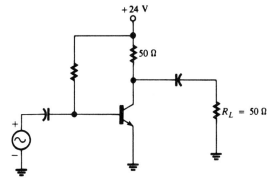

Transformer-Coupled Class-A Amplifiers

In Chapter 4, we reviewed basic transformer theory and discussed transformer-coupled amplifiers. Transformers are also used to couple power amplifiers to their loads and, in those applications, are called *output transformers*. As in other coupling applications, the advantages of a transformer are that it provides an opportunity to achieve impedance matching for maximum power transfer and that it blocks the flow of dc current in a load.

Figure 14.9 shows a transformer used to couple the output of a transistor to load R_L. Also shown are the dc and ac load lines for the amplifier. Here we assume that the dc resistance of the primary winding is negligibly small, so the dc load line is vertical (slope $= -1/R_{dc} = -\infty$). Recall that the ac resistance r_L reflected to the primary side is

$$r_L = (N_p/N_s)^2 R_L \tag{14.34}$$

where N_p and N_s are the numbers of turns on the primary and secondary windings, respectively. As shown in the figure, the slope of the ac load line is $-1/r_L$.

Since we are assuming that there is negligible resistance in the primary winding, there is no dc voltage drop across the winding, and the quiescent collector voltage is therefore V_{CC} volts, as shown in Figure 14.9. Conventional base-bias circuitry (not shown in the figure) is used to set the quiescent collector current I_Q. The Q-point is the point on the dc load line at which the collector current equals I_Q.

Since V_{CE} cannot be negative, the maximum permissible decrease in V_{CE}

Figure 14.9
Transformer-coupled, class-A amplifier and load lines. The amplifier is biased for maximum peak-to-peak variation in V_{CE}.

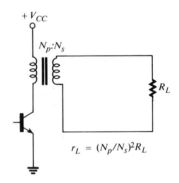

$$r_L = (N_p/N_s)^2 R_L$$

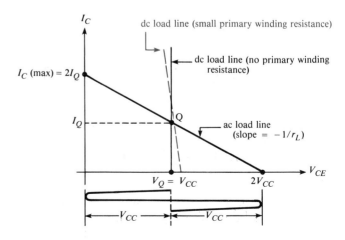

below its quiescent value is $V_Q = V_{CC}$ volts. Thus, the maximum possible peak value of V_{CE} is V_{CC} volts. To achieve maximum peak-to-peak output variation, the intercept of the ac load line on the V_{CE}-axis should therefore be $2V_{CC}$ volts, as shown in Figure 14.9. The quiescent current I_Q is selected so that the ac load line, a line having slope $-1/r_L$, intersects the V_{CE}-axis at $2V_{CC}$ volts.

The ac load line intersects the I_C-axis in Figure 14.9 at $I_C(\text{max})$. Note that there is no theoretical limit to the value that I_C may have, since there is no limiting resistance in the collector circuit. However, in practice, I_C must not exceed the maximum permissible collector current for the transistor and it must not be so great that the magnetic flux of the transformer saturates. When the transformer saturates, it can no longer induce current in the secondary winding and signal distortion results. When I_Q is set for maximum signal swing (so that $V_{CE}(\text{max}) = 2V_{CC}$), I_Q is one-half $I_C(\text{max})$; that is, $I_C(\text{max}) = 2I_Q$, as shown in Figure 14.9. Thus, the maximum values of the peak primary voltage and peak primary current are V_{CC} and I_Q, respectively. Since the ac output can vary through this range, below and above the quiescent point, the amplifier is of the class-A type.

Note that, unlike the case of the capacitor-coupled or series-fed amplifier, the collector voltage can exceed the supply voltage. A transistor having a collector breakdown voltage equal to at least twice the supply voltage should be used in this application.

The ac power delivered to load resistance R_L in Figure 14.9 is

$$P_L = \frac{V_s^2}{2R_L} = \frac{V_{PL}^2}{2R_L} \text{ watts} \tag{14.35}$$

where $V_s = V_{PL}$ is the peak value of the secondary, or load, voltage. The average power from the dc supply is

$$P_S = V_{CC}I_Q \text{ watts} \tag{14.36}$$

Therefore, the efficiency is

$$\eta = \frac{P_L}{P_S} = \frac{V_{PL}^2}{2R_L V_{CC}I_Q} \tag{14.37}$$

Under maximum signal conditions, the peak primary voltage is V_{CC} volts, so the peak load voltage is

$$V_{PL} = (N_s/N_p)V_{CC} \tag{14.38}$$

Also, since the slope of the ac load line is $-1/r_L$, we have

$$\frac{|\Delta I_C|}{|\Delta V_{CE}|} = \frac{1}{r_L} = \frac{I_Q}{V_Q}$$

or

$$I_Q = \frac{V_Q}{r_L} = \frac{V_{CC}}{(N_p/N_s)^2 R_L} \tag{14.39}$$

Substituting (14.38) and (14.39) into (14.37), we find the maximum possible efficiency of the class-A, transformer-coupled amplifier:

$$\eta(\text{max}) = \frac{(N_s/N_p)^2 V_{CC}^2}{(2R_L V_{CC})\dfrac{V_{CC}}{(N_p/N_s)^2 R_L}} = 0.5 \tag{14.40}$$

We see that the maximum efficiency is twice that of the series-fed class-A ampli-fier and six times that of the capacitor-coupled class-A amplifier. This improvement in efficiency is attributable to the absence of external collector resistance that would otherwise consume dc power. Note that the collector efficiency of the transformer-coupled class-A amplifier is the same as the overall amplifier efficiency, because the average power from the dc supply is the same as the collector dissipation:

$$P_S = V_{CC}I_Q = V_Q I_Q = P_C \qquad (14.41)$$

In practice, a full output voltage swing of $2V_{CC}$ volts cannot be achieved in a power transistor. The device is prevented from cutting off entirely by virtue of a relatively large leakage current, and it cannot be driven all the way into saturation $(I_C = I_C(\text{max}))$ without creating excessive distortion. These points are illustrated in the next example.

Example 14.6

The transistor in the power amplifier shown in Figure 14.10 has the output characteristics shown in Figure 14.11. Assume that the transformer has zero resistance.

1. Construct the (ideal) dc and ac load lines necessary to achieve maximum output voltage swing. What quiescent values of collector and base current are necessary to realize the ac load line?
2. What is the smallest value of $I_C(\text{max})$ for which the transistor should be rated?
3. What is the maximum peak-to-peak collector voltage, and what peak-to-peak base current is required to achieve it? Assume that the base current cannot go negative and that, to minimize distortion, the collector should not be driven below 2.5 V in the saturation region.
4. Find the average power delivered to the load under the maximum signal conditions of (3).
5. Find the power dissipated in the transistor under no-signal conditions (standby).
6. Find the efficiency.

Solution

1. The vertical dc load line intersects the V_{CE}-axis at $V_{CC} = 15$ V, as shown in Figure 14.11. To find the slope of the ac load line, we must find r_L. From

Figure 14.10
(Example 14.6)

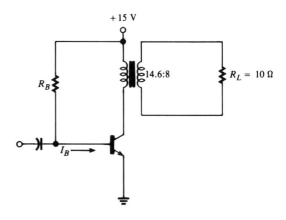

Figure 14.11
(Example 14.6)

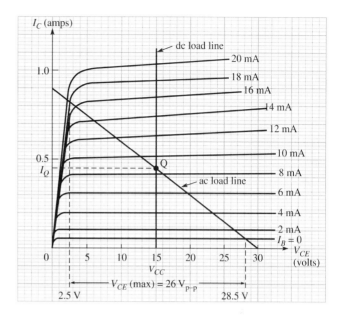

equation 14.34, $r_L = (N_p/N_s)^2 R_L = (14.6/8)^2(10) = 33.3\ \Omega$. Thus, the slope of the ac load line is $-1/r_L = -1/33.3 = -0.03$. To achieve the ideal maximum output swing, we want the ac load line to intercept the V_{CE}-axis at $2V_{CC} = 30$ V. Since the slope of that line has magnitude 0.03, it will intercept the I_C-axis at $I_C = (0.03\ \text{A/V})(30\ \text{V}) = 0.9$ A. The ac load line is then drawn between the two intercepts (0 A, 30 V) and (0.9 A, 0 V), as shown in Figure 14.11.

The ac load line intersects the dc load line at the Q-point. The quiescent collector current at that point is seen to be $I_Q = 0.45$ A. The corresponding base current is approximately halfway between $I_B = 8$ mA and $I_B = 10$ mA, so the quiescent base current must be 9 mA.

2. The maximum collector current is $I_C(\text{max}) = 0.9$ A, at the intercept of the ac load line on the I_C-axis. Actually, we will not operate the transistor that far into saturation, since we do not allow V_{CE} to fall below 2.5 V. However, a maximum rating of 0.9 A (or 1 A) will provide us with a margin of safety.

3. The maximum value of V_{CE} occurs on the ac load line at the point where $I_B = 0$. As shown in Figure 14.11, this value is 28.5 V. Since the minimum permissible value of V_{CE} is 2.5 V, the maximum peak-to-peak voltage swing is $28.5 - 2.5 = 26$ V p-p. As can be seen on the characteristic curves, the base current must vary from $I_B = 0$ to $I_B = 18$ mA, or 18 mA peak-to-peak, to achieve that voltage swing.

4. The peak primary voltage in the transformer is $(1/2)(26\ \text{V}) = 13$ V. Therefore, the peak secondary, or load, voltage is $V_{PL} = (N_s/N_p)(13\ \text{V}) = (8/14.6)13 = 7.12$ V. The average load power is then

$$P_L = \frac{V_{PL}^2}{2R_L} = \frac{(7.12)^2}{20} = 2.53\ \text{W}$$

5. The standby power dissipation is $P_d = V_Q I_Q = V_{CC} I_Q = (15\ \text{V})(0.45\ \text{A}) = 6.75$ W.

6. The standby power dissipation found in (5) is the same as the average power supplied from the dc source, so

$$\eta = \frac{2.53 \text{ W}}{6.75 \text{ W}} = 0.375$$

(Why is this value less than the theoretical maximum of 0.5?)

Class-B Amplifiers

Transistor operation is said to be *class B* when output current varies during only one half-cycle of a sine-wave input. In other words, the transistor is in its active region, responding to signal input, only during a positive half-cycle or only during a negative half-cycle of the input. This operation is illustrated in Figure 14.12.

It is clear that class-B operation produces an output waveform that is severely clipped (half-wave rectified) and therefore highly distorted. The waveform *by itself* is not suitable for audio applications. However, in practical amplifiers, *two* transistors are operated class B: one to amplify positive signal variations and the other to amplify negative signal variations. The amplifier output is the composite waveform obtained by combining the waveforms produced by each class-B transistor. We will study these amplifiers in detail in the next section, for they are more efficient and more widely used in power applications than are class-A amplifiers. An amplifier utilizing transistors that are operated class B is called a *class-B* amplifier.

14.5   PUSH-PULL AMPLIFIER PRINCIPLES

A *push-pull* amplifier uses two output devices to drive a load. The name is derived from the fact that one device is primarily (or entirely) responsible for driving current through the load in one direction (pushing), while the other device drives current through the load in the opposite direction (pulling). The output devices are typically two transistors, each operated class B, one of which conducts only when the input is positive, and the other of which conducts only when the input is negative. This arrangement is called a *class-B, push-pull* amplifier and its principle is illustrated in Figure 14.13.

Figure 14.12
In class-B operation, output current variations occur during only each positive or each negative half-cycle of input. In the example shown, output current flows during positive half-cycles of the input, and the amplifier is cut off during negative half-cycles.

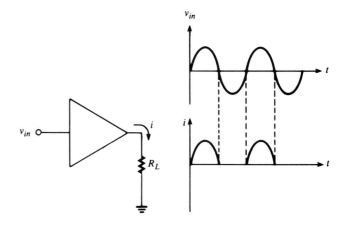

Figure 14.13
The principle of class-B, push-pull oper-
ation. Output amplifying device 1 drives
current i_L through the load in one direc-
tion while 2 is cut off, and device 2
drives current in the opposite direction
while 1 is cut off.

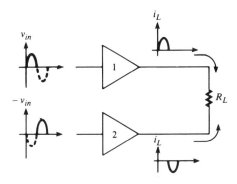

Note in Figure 14.13 that amplifying devices 1 and 2 are driven by equal-amplitude, *out-of-phase* input signals. The signals are identical except for phase. Here we assume that each device conducts only when its input is positive and is cut off when its input is negative. The net effect is that device 1 produces load current when the input is positive and device 2 produces load current, in the opposite direction, when the input is negative. An example of a device that has the property that it produces output (collector) current only when its input (base-to-emitter) voltage is positive is an NPN transistor having no base-biasing circuitry, that is, one that is biased at cutoff. As we shall see, NPN transistors can be used as the output amplifying devices in push-pull amplifiers. However, the circuitry must be somewhat more elaborate than that diagrammed in Figure 14.13, since we must make provisions for load current to flow through a *complete circuit,* regardless of current direction. Obviously, when amplifying device 1 in Figure 14.13 is cut off, it cannot conduct current produced by device 2, and vice versa.

Push-Pull Amplifiers with Output Transformers

Figure 14.14 shows a push-pull arrangement that permits current to flow in both directions through a load even though one or the other of the amplifying devices (NPN transistors) is always cut off. The *output transformer* shown in the figure is the key component. Note that the primary winding is connected between the transistor collectors and that its *center tap* is connected to the dc supply, V_{CC}. The center tap is simply an electrical connection made at the center of the winding, so

Figure 14.14
A center-tapped output transformer used
in a push-pull amplifier

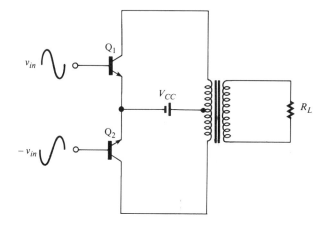

there are an equal number of turns between each end of the winding and the center tap. The figure does not show the push-pull *driver* circuitry, which must produce out-of-phase signals on the bases of Q_1 and Q_2. We will discuss that circuitry later.

Figure 14.15 shows how current flows through the amplifier during a positive half-cycle of input and during a negative half-cycle of input. In 14.15(a), the input to Q_1 is the positive half-cycle of the signal, and since the input to Q_2 is out-of-phase with that to Q_1, Q_2 is driven by a negative half-cycle. Notice that neither class-B transistor is biased. Consequently, the positive base voltage on Q_1 causes it to turn on and conduct current in the counterclockwise path shown. The negative base voltage on Q_2 keeps that transistor cut off. Current flowing in the upper half of the transformer's primary induces current in the secondary, and current flows through the load. In Figure 14.15(b), the input signal on the base of Q_1 has gone negative, so its inverse on the base of Q_2 is positive. Therefore, Q_2 conducts current in the clockwise path shown, and Q_1 is cut off. Current induced in the secondary winding is in the direction opposite that shown in Figure 14.15(a). The

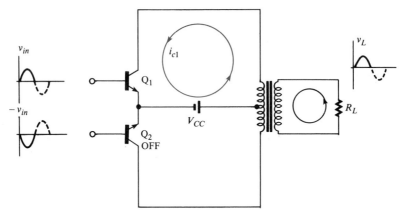

(a) When v_{in} is positive, Q_1 conducts and Q_2 is cut off. A counterclockwise current is induced in the load.

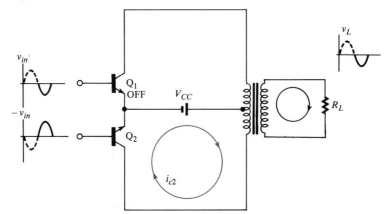

(b) When v_{in} is negative, Q_2 conducts and Q_1 is cut off. A clockwise current is induced in the load.

Figure 14.15
Current flow in a push-pull amplifier with output transformer. (Currents are those that flow during the portion of the input shown as a solid line.)

upshot is that current flows through the load in one direction when the input signal is positive and in the opposite direction when the input signal is negative, just as it should in an ac amplifier.

Figure 14.16 displays current flow in the push-pull amplifier in the form of a *timing diagram*. Here the complete current waveforms are shown over two full cycles of input. For purposes of this illustration, counterclockwise current (in Figure 14.15) is arbitrarily assumed to be positive and clockwise current is therefore negative. Note that, as far as the load is concerned, current flows during the full 360° of input signal. Figure 14.16(e) shows that the current i_S from the power supply varies from 0 to the peak value I_P every half-cycle. Because the current variation is so large, the power supply used in a push-pull amplifier must be particularly well regulated—that is, it must maintain a constant voltage, independent of current demand.

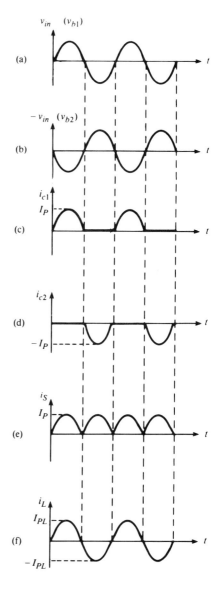

Figure 14.16

A timing diagram showing currents in the push-pull amplifier of Figure 14.15 over two full cycles of input

Class-B Efficiency

The principal advantage of using a class-B power amplifier is that it is possible to achieve an efficiency greater than that attainable in a class-A amplifier. The improvement in efficiency stems from the fact that no power is dissipated in a transistor during the time intervals that it is cut off. Also, like the transformer-coupled class-A amplifier, there is no external collector resistance that would otherwise consume power.

We will derive an expression for the maximum efficiency of a class-B push-pull amplifier assuming ideal conditions: perfectly matched transistors and zero resistance in the transformer windings. The current supplied by each transistor is a half-wave–rectified waveform, as shown in Figure 14.16. Let I_p represent the peak value of each. Then the peak value of the current in the secondary winding, which is the same as the peak load current, is

$$I_{PL} = (N_p/N_s)I_P \tag{14.42}$$

where N_p/N_s is the turns ratio between one-half the primary winding and the secondary winding. (Note that only those primary turns between one end of the winding and its center tap are used to induce current in the secondary.) Similarly, the peak value of the load voltage is

$$V_{PL} = (N_s/N_p)V_P \tag{14.43}$$

where V_P is the peak value of the primary (collector) voltage. Since the load voltage and load current are sinusoidal, the average power delivered to the load is, from equation 14.10,

$$P_L = \frac{V_{PL}I_{PL}}{2} = \frac{(N_s/N_p)V_P(N_p/N_s)I_P}{2} = \frac{V_PI_P}{2} \tag{14.44}$$

As shown in Figure 14.16(e), the power-supply current is a full-wave–rectified waveform having peak value I_P. The dc, or average, value of such a waveform is known to be $2I_P/\pi$. Therefore, the average power delivered to the circuit by the dc supply is

$$P_S = \frac{2I_PV_{CC}}{\pi} \tag{14.45}$$

The efficiency is then

$$\eta = \frac{P_L}{P_S} = \frac{V_PI_P/2}{2I_PV_{CC}/\pi} = \frac{\pi V_P}{4V_{CC}} \tag{14.46}$$

Under maximum signal conditions, $V_P = V_{CC}$, and (14.46) becomes

$$\eta(\text{max}) = \frac{\pi}{4} = 0.785 \tag{14.47}$$

Equation 14.47 shows that a class-B push-pull amplifier can be operated with a much higher efficiency than the class-A amplifiers studied earlier. Furthermore, unlike the case of class-A amplifiers, the power dissipated in the transistors is 0 under standby (zero-signal) conditions, because both transistors are cut off. A general expression for the total power dissipated in the transistors can be obtained by realizing that it equals the difference between the total power supplied by the dc source and the total power delivered to the load:

$$P_d = \frac{2I_P V_{CC}}{\pi} - \frac{V_P I_P}{2} \qquad\qquad (14.48)$$

Using calculus, it can be shown that P_d is maximum when $V_P = 2V_{CC}/\pi = 0.636V_{CC}$. We conclude that maximum transistor dissipation does *not* occur when maximum load power is delivered ($V_P = V_{CC}$), but at the intermediate level $V_P = 0.636V_{CC}$.

Example 14.7

The push-pull amplifier in Figure 14.14 has $V_{CC} = 20$ V and $R_L = 10\ \Omega$. The *total* number of turns on the primary winding is 100 and the secondary winding has 50 turns. Assume that the transformer has zero resistance.

1. Find the maximum power that can be delivered to the load.
2. Find the power dissipated in each transistor when maximum power is delivered to the load.
3. Find the power delivered to the load and the power dissipated in each transistor when transistor power dissipation is maximum.

Solution

1. The turns ratio between each half of the primary and the secondary is $N_p/N_s = (100/2):50 = 50:50$. Therefore, the peak values of primary and secondary voltages are equal, as are the peak values of primary and secondary current.

$$V_P(\text{max}) = V_{PL}(\text{max}) = V_{CC} = 20 \text{ V}$$

$$I_P(\text{max}) = I_{PL}(\text{max}) = \frac{V_{CC}}{R_L} = \frac{20 \text{ V}}{10\ \Omega} = 2 \text{ A}$$

Therefore,

$$P_L(\text{max}) = \frac{V_P(\text{max})I_P(\text{max})}{2} = \frac{(20 \text{ V})(2 \text{ A})}{2} = 20 \text{ W}$$

2. From equation 14.48,

$$P_d = \frac{2I_P V_{CC}}{\pi} - \frac{V_P I_P}{2} = \frac{2(2 \text{ A})(20 \text{ V})}{\pi} - 20 \text{ W} = 5.46 \text{ W}$$

Since 5.46 W is the total power dissipated by both transistors, each dissipates one-half that amount, or 2.73 W.

3. Transistor power dissipation is maximum when $V_P = 0.636V_{CC} = (0.636)(20) = 12.72$ V. Then

$$I_P = I_{PL} = \frac{12.72 \text{ V}}{10\ \Omega} = 1.272 \text{ A}$$

and

$$P_L = \frac{V_P I_P}{2} = \frac{(12.72 \text{ V})(1.272 \text{ A})}{2} = 8.09 \text{ W}$$

From equation 14.48,

$$P_d(\text{max}) = \frac{2I_P V_{CC}}{\pi} - \frac{V_P I_P}{2} = \frac{2(1.272 \text{ A})(20 \text{ V})}{\pi} - 8.09 \text{ W} = 8.09 \text{ W}$$

The maximum power dissipation in each transistor is then $8.09/2 = 4.05$ W.

The preceding example demonstrates some results that are true in general for push-pull amplifiers: (1) When transistor power dissipation is maximum, its value equals the power delivered to the load; and (2) the maximum total transistor power dissipation equals approximately 40% of the maximum power that can be delivered to the load.

14.6   PUSH-PULL DRIVERS

We have seen that the push-pull amplifier described earlier must be driven by out-of-phase input signals. Figure 14.17 shows how a transformer can be used to provide the required drive signals. Here, the *secondary* winding has a grounded center tap that effectively splits the secondary voltage into two out-of-phase signals, each having one-half the peak value of the total secondary voltage. The input signal is applied across the primary winding and a voltage is developed across secondary terminals A–B, in the usual transformer fashion. To understand the phase-splitting action, consider the instant at which the voltage across A–B is +6 V, as shown in the figure. Then, since the center point is at ground, the voltage from A to ground must be +3 V and that from B to ground must be −3 V: $V_{AB} = V_A - V_B = 3 - (-3) = +6$ V. The same logic applied at every instant throughout a complete cycle shows that v_B with respect to ground is always the negative of v_A with respect to ground; in other words, v_A and v_B are equal-amplitude, out-of-phase driver signals, as required.

A specially designed amplifier, called a *phase-splitter,* can be used instead of a driver transformer to produce equal amplitude, out-of-phase drive signals. Figure 14.18 shows two possible designs. Figure 14.18(a) is a conventional amplifier circuit with outputs taken at the collector and at the emitter. The collector output is out of phase with the input and the emitter output is in phase with the input, so the two outputs are out of phase with each other. With no load connected to either output, the output signals will have approximately equal amplitudes if $R_C = R_E$. However, the output impedance at the collector is significantly greater than that at the emitter, so when loads are connected, each output will be affected differently. As a consequence, the signal amplitudes will no longer be equal, and gain adjustments will be required. Furthermore, the nonlinear nature of the large-signal load

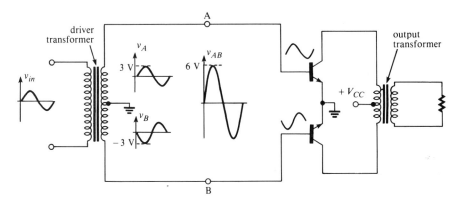

Figure 14.17
Using a driver transformer to create equal-amplitude, out-of-phase drive signals (v_A and v_B) for a push-pull amplifier

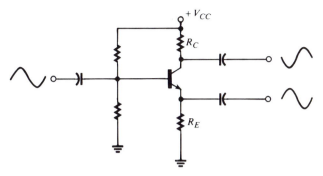

(a) A phase-splitting driver

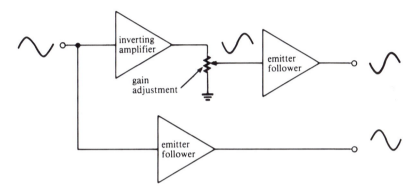

Figure 14.18
Phase-splitting methods for obtaining push-pull drive signals

(the output transistors) means that the gain of the collector output may vary appreciably with signal level. Intolerable distortion can be created as a result. A better way to drive the output transistors is from two low-impedance signal sources, as shown in Figure 14.18(b). Here, the output from an inverting amplifier is buffered by an emitter-follower stage, as is the original signal.

14.7 DISTORTION IN PUSH-PULL AMPLIFIERS

Cancellation of Even Harmonics

Recall that push-pull operation effectively produces in a load a waveform proportional to the *difference* between two input signals. Under normal operation, the signals are out of phase, so their waveform is reproduced in the load. If the signals were in phase, cancellation would occur. It is instructive to view a push-pull output as the difference between two distorted (half-wave rectified) sine waves that are out of phase with each other. This viewpoint is illustrated in Figure 14.19. It can be shown that a half-wave–rectified sine wave contains only the fundamental and all *even* harmonics. Figure 14.19(a) shows the two out-of-phase, half-wave–rectified sine waves that drive the load, and 14.19(b) shows the fundamental and second-harmonic components of each. Notice that the fundamental compo-

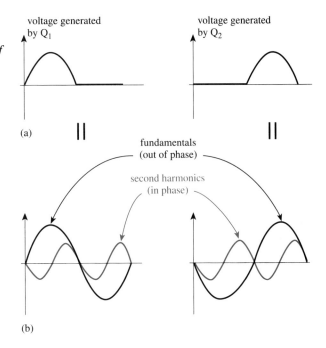

Figure 14.19
The fundamental components of the half-wave–rectified waveforms are out of phase, but the second harmonics are in phase. Therefore, second-harmonic distortion is cancelled in push-pull operation.

nents are out of phase. Therefore, the fundamental component is reproduced in the load, as we have already seen (Figure 14.16). However, the second-harmonic components are in phase, and therefore cancel in the load. Although not shown in this figure, the fourth and all other even harmonics are also in phase and therefore also cancel. Our conclusion is an important property of push-pull amplifiers: *even harmonics are cancelled in push-pull operation.*

The cancellation of even harmonics is an important factor in reducing distortion in push-pull amplifiers. However, perfect cancellation would occur only if the two sides were perfectly matched and perfectly balanced: identical transistors, identical drivers, and a perfectly center-tapped transformer. Of course, this is not the case in practice, but even imperfect push-pull operation reduces even harmonic distortion. Odd harmonics are out of phase, so cancellation of those components does not occur.

Crossover Distortion

Recall that a forward-biasing voltage applied across a PN junction must be raised to a certain level (about 0.7 V for silicon) before the junction will conduct any significant current. Similarly, the voltage across the base–emitter junction of a transistor must reach that level before any appreciable base current, and hence collector current, can flow. As a consequence, the drive signal applied to a class-B transistor must reach a certain minimum level before its collector current is properly in the active region. This fact is the principal source of distortion in a class-B, push-pull amplifier, as illustrated in Figure 14.20. Figure 14.20(a) shows that the initial rise of collector current in a class-B transistor lags the initial rise of input voltage, for the reason we have described. Also, collector current prematurely drops to 0 when the input voltage approaches 0. Figure 14.20(b) shows the voltage waveform that is produced in the load of a push-pull amplifier when the distortion generated during each half-cycle by each class-B transistor is combined. This

Figure 14.20
Crossover distortion

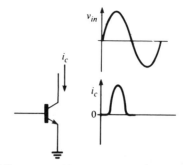

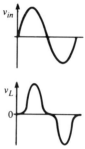

(a) Collector current in a class-B transistor does not follow
input voltage in the regions near 0 (crossovers)

(b) The load voltage in a class-B push-pull amplifier, showing
the combined effects of the distortion generated during each
half-cycle of input

distortion is called *crossover* distortion, because it occurs where the composite waveform crosses the zero-voltage axis. Clearly, the effect of crossover distortion becomes more serious as the signal level becomes smaller.

Class-AB Operation

Recall from Section 8.2 that negative feedback can be used to reduce distortion, and this remedy is often used in power amplifiers. Crossover distortion can also be reduced or eliminated in a push-pull amplifier by biasing each transistor slightly into conduction. When a small forward-biasing voltage is applied across each base–emitter junction, and a small base current flows under no-signal conditions, it is not necessary for the base drive signal to overcome the built-in junction potential before active operation can occur. A simple voltage-divider bias network can be connected across each base for this purpose, as shown in Figure 14.21. Figure 14.21(a) shows how two resistors can provide bias for both transistors when a driver transformer is used. Figure 14.21(b) shows the use of two voltage dividers when the drive signals are capacitor-coupled. Typically, the base–emitter junctions are biased to about 0.5 V for silicon transistors, or so that the collector current under no-signal conditions is about 1% of its peak signal value.

When a transistor is biased slightly into conduction, output current will flow during more than one-half cycle of a sine-wave input, as illustrated in Figure 14.22. As can be seen in the figure, conduction occurs for more than one-half but

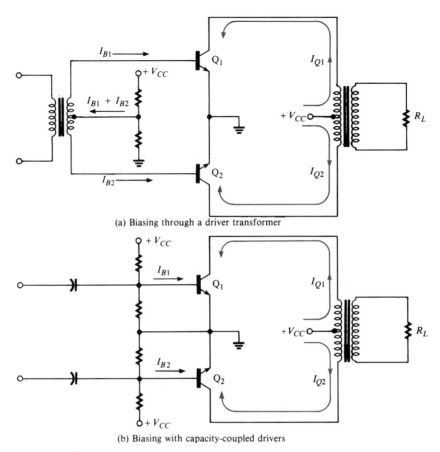

(a) Biasing through a driver transformer

(b) Biasing with capacity-coupled drivers

Figure 14.21
Methods for providing a slight forward bias for push-pull transistors to reduce cross-over distortion

Figure 14.22
Class-AB operation. Output current i_{out} flows during more than one-half but less than a full cycle of input.

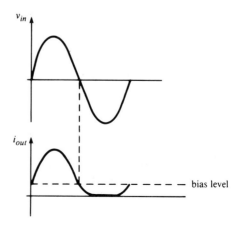

less than a full cycle of input. This operation, which is neither class A nor class B, is called *class-AB* operation.

While class-AB operation reduces crossover distortion in a push-pull amplifier, it has the disadvantage of reducing amplifier efficiency. The fact that bias current is always present means that there is continuous power dissipation in both transistors, including the time intervals during which one of the transistors would be cut off if the operation were class B. The extent to which efficiency is reduced depends directly on how heavily the transistors are biased, and the maximum achievable efficiency is somewhere between that which can be obtained in class-A operation (0.5) and that attainable in class-B operation (0.785).

In Figure 14.21, notice that the quiescent collector currents I_{Q1} and I_{Q2} flow in opposite directions through the primary of the transformer. Thus, the magnetic flux created in the transformer by one dc current opposes that created by the other, and the net flux is 0. This is an advantageous situation, in comparison with the class-A transformer-coupled amplifier, because it means that transformer current can swing positive and negative through a maximum range. If the transformer flux had a bias component, the signal swing would be limited in one direction by the onset of magnetic saturation.

14.8   TRANSFORMERLESS PUSH-PULL AMPLIFIERS

Complementary Push-Pull Amplifiers

The principal disadvantage of the push-pull amplifier circuits we have discussed so far is the cost and bulk of their output transformers. High-power amplifiers in particular are encumbered by the need for very large transformers capable of conducting large currents without saturating. Figure 14.23(a) shows a popular design using *complementary* (PNP and NPN) output transistors to eliminate the need for an output transformer in push-pull operation. This design also eliminates the need for a driver transformer or any other drive circuitry producing out-of-phase signals.

Figure 14.23(b) shows that current flows in a counterclockwise path through the load when the input signal on the base of NPN transistor Q_1 is positive. The positive input simultaneously appears on the base of PNP transistor Q_2 and keeps it cut off. When the input is negative, Q_1 is cut off and Q_2 conducts current through the load in the opposite direction, as shown in Figure 14.23(c).

Note that each transistor in Figure 14.23 drives the load in an emitter-follower configuration. The advantageous consequence is that low-impedance loads can be driven from a low-impedance source. Also, the large negative feedback that is inherent in emitter-follower operation reduces the problem of output distortion. However, as is the case in all emitter followers, voltage gains greater than unity cannot be realized. The maximum positive voltage swing is V_{CC1} and the maximum negative swing is V_{CC2}. Normally, $|V_{CC1}| = |V_{CC2}| = V_{CC}$, so the maximum peak-to-peak swing is $2V_{CC}$ volts. Since the voltage gain is near unity, the input must also swing through $2V_{CC}$ volts to realize maximum output swing. Notice that under conditions of maximum swing, the cut-off transistor experiences a maximum reverse-biasing collector-to-base voltage of $2V_{CC}$ volts. For example, when Q_1 is off, its collector is at $+V_{CC}$ and its base voltage (the input signal) swings to $-V_{CC}$. Thus, each transistor must have a rated breakdown voltage of at least $2V_{CC}$.

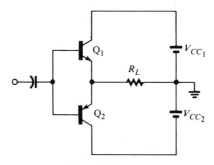

(a) Complementary push-pull amplifier

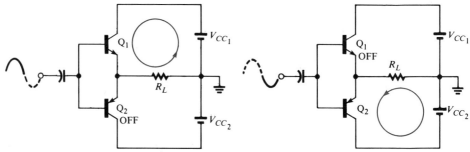

(b) When the input is positive, Q_1 conducts and Q_2 is cut off.

(c) When the input is negative, Q_2 conducts and Q_1 is cut off.

Figure 14.23
Push-pull amplification using complementary transistors. Note that the direction of current through R_L alternates each half-cycle, as required.

Example 14.8

The amplifier in Figure 14.23 must deliver 30 W to a 15-Ω load under maximum drive.

1. What is the minimum value required for each supply voltage?
2. What minimum collector-to-base breakdown-voltage rating should each transistor have?

Solution

1.
$$P_L(\text{max}) = \frac{V_P^2(\text{max})}{2R_L}$$

$V_p(\text{max}) = \sqrt{2R_L P_L(\text{max})} = \sqrt{2(15)(30)} = 30$ V

Therefore, for maximum swing, $V_{CC} = V_P(\text{max}) = 30$ V; i.e., $V_{CC1} = +30$ V and $V_{CC2} = -30$ V.

2. The minimum rated breakdown is $2V_{CC} = 60$ V.

Example 14.9

The push-pull amplifier in Figure 14.23 has $|V_{CC1}| = |V_{CC2}| = 30$ V and $R_L = 15$ Ω. The input coupling capacitor is 100 μF. Assuming the transistors have their default parameters, use SPICE to obtain a plot of one full cycle of output when the input is

1. a 1-kHz sine wave with peak value 30 V;
2. a 1-kHz sine wave with peak value 2 V.

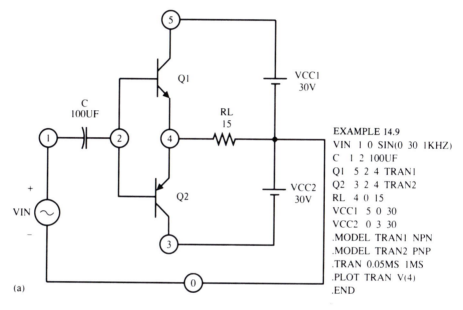

```
EXAMPLE 14.9
VIN  1 0 SIN(0 30 1KHZ)
C    1 2 100UF
Q1   5 2 4 TRAN1
Q2   3 2 4 TRAN2
RL   4 0 15
VCC1 5 0 30
VCC2 0 3 30
.MODEL TRAN1 NPN
.MODEL TRAN2 PNP
.TRAN 0.05MS 1MS
.PLOT TRAN V(4)
.END
```

(a)

```
 TIME         V(4)
              -4.000D+01     -2.000D+01      0.000D+00      2.000D+01  4.000D+01
              - - - - - - - - - - - - - - - - - - - - - - - - - - - - - -
0.000D+00  -5.841D-29  .                              *                .
5.000D-05   8.326D+00  .                  .           .        *       .
1.000D-04   1.664D+01  .                  .           .           *    .
1.500D-04   2.328D+01  .                  .           .           .  *
2.000D-04   2.749D+01  .                  .           .           .      *
2.500D-04   2.898D+01  .                  .           .           .       *
3.000D-04   2.747D+01  .                  .           .           .      *
3.500D-04   2.324D+01  .                  .           .           .  *
4.000D-04   1.659D+01  .                  .           .           *  .
4.500D-04   8.270D+00  .                  .           .        *      .
5.000D-04  -6.253D-14  .                  .           *               .
5.500D-04  -8.374D+00  .                  .        *  .               .
6.000D-04  -1.673D+01  .                  .  *        .               .
6.500D-04  -2.331D+01  .              *    .           .               .
7.000D-04  -2.759D+01  .          *        .           .               .
7.500D-04  -2.900D+01  .        *          .           .               .
8.000D-04  -2.757D+01  .          *        .           .               .
8.500D-04  -2.327D+01  .              *    .           .               .
9.000D-04  -1.668D+01  .                  . *          .               .
9.500D-04  -8.317D+00  .                  .        *   .               .
1.000D-03   3.754D-12  .                  .           *               .
              - - - - - - - - - - - - - - - - - - - - - - - - - - - - - -
```

(b)

Figure 14.24
(Example 14.9)

```
     TIME         V(4)
                  -2.000D+00      -1.000D+00      0.000D+00      1.000D+00    2.000D+00
                  - - - - - - - - - - - - - - - - - - - - - - - - - - - - - - - - - - - -
 -0.000D+00  -5.841D-29  .                            .            *            .                  .
  5.000D-05   1.359D-03  .                            .            *            .                  .
  1.000D-04   3.212D-01  .                            .            .  *         .                  .
  1.500D-04   7.411D-01  .                            .            .          * .                  .
  2.000D-04   1.015D+00  .                            .            .            .*                 .
  2.500D-04   1.111D+00  .                            .            .            . *                .
  3.000D-04   1.015D+00  .                            .            .            .*                 .
  3.500D-04   7.397D-01  .                            .            .          * .                  .
  4.000D-04   3.200D-01  .                            .            .  *         .                  .
  4.500D-04   4.242D-03  .                            .            *            .                  .
  5.000D-04  -4.141D-13  .                            .            *            .                  .
  5.500D-04  -1.922D-04  .                            .            *            .                  .
  6.000D-04  -3.227D-01  .                            .        *   .            .                  .
  6.500D-04  -7.428D-01  .                            . *          .            .                  .
  7.000D-04  -1.017D+00  .                         *  .            .            .                  .
  7.500D-04  -1.113D+00  .                       *  . .            .            .                  .
  8.000D-04  -1.016D+00  .                         *  .            .            .                  .
  8.500D-04  -7.414D-01  .                          .*            .            .                  .
  9.000D-04  -3.216D-01  .                            .       *    .            .                  .
  9.500D-04  -1.290D-04  .                            .            *            .                  .
  1.000D-03   1.776D-18  .                            .            *            .                  .
                  - - - - - - - - - - - - - - - - - - - - - - - - - - - - - - - - - - - -
(c)
```

Figure 14.24
(Continued)

Solution. Figure 14.24(a) shows the SPICE circuit and input data file for the case where the input signal has peak value 30 V. The resulting plot is shown in Figure 14.24(b). We see that the computed output swings from -29 to $+28.98$ V. The output that results when the peak value of the input is changed to 2 V is shown in Figure 14.24(c). Crossover distortion is now clearly apparent in this low-amplitude case.

Figure 14.25 shows two variations on the basic complementary, push-pull amplifier. In (a), the transistors are replaced by emitter-follower *Darlington pairs*. Since power transistors tend to have low betas, particularly at high current levels, the Darlington pair improves the drive capabilities and the current gain of the amplifier. These devices are available in matched complementary sets with current ratings up to 20 A. Figure 14.25(b) shows how emitter-follower transistors can be operated in parallel to increase the overall current-handling capability of the amplifier. In this variation, the parallel transistors must be closely matched to prevent "current hogging," wherein one device carries most of the load, thus subverting the intention of load sharing. Small emitter resistors R_E shown in the figure introduce negative feedback and help prevent current hogging, at the expense of efficiency. Amplifiers capable of dissipating several hundred watts have been constructed using this arrangement.

One disadvantage of the complementary push-pull amplifier is the need for two power supplies. Also, like the transformer-coupled push-pull amplifier, the complementary class-B amplifier produces crossover distortion in its output. Figure 14.26(a) shows another version of the complementary amplifier that eliminates these problems and that incorporates some additional features.

The complementary amplifier in Figure 14.26 can be operated with a single power supply because the output v_o is biased at half the supply voltage and is capacitor-coupled to the load. The resistor–diode network connected across the

Figure 14.25
Variations on the basic complementary
push-pull amplifier

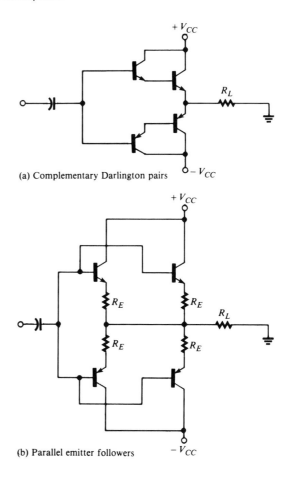

(a) Complementary Darlington pairs

(b) Parallel emitter followers

transistor bases is used to bias each transistor near the threshold of conduction. Crossover distortion can be reduced or eliminated by inserting another resistor (not shown in the figure) in series with the diodes to bias the transistors further into AB operation. Assuming that all components are perfectly matched, the supply voltage will divide equally across each half of the amplifier, as shown in Figure 14.26(b). (In practice, one of the resistors R can be made adjustable for balance purposes.) Resistors R_E provide bias stability to prevent *thermal runaway* (where increased heat causes increased dissipation, which causes increased heat, and so on), but are made as small as possible since they adversely affect efficiency. Since each half of the amplifier has $V_{CC}/2$ volts across it, the forward-biased diode drops appear across the base–emitter junctions with the proper polarity to bias each transistor toward conduction. The diodes are selected so that their characteristics track the base–emitter junctions under temperature changes and thus ensure bias stability. The diodes are typically mounted on the same heat sinks as the transistors so that both change temperature in the same way.

In ac operation, when input v_1 is positive and Q_1 is conducting, current is drawn from the power supply and flows through Q_1 to the load. When Q_1 is cut off by a negative input, no current can flow from the supply. At those times, Q_2 is conducting and capacitor C_c discharges through that transistor. Thus, current flows from the load, through C_c, and through Q_2 to ground whenever the input is

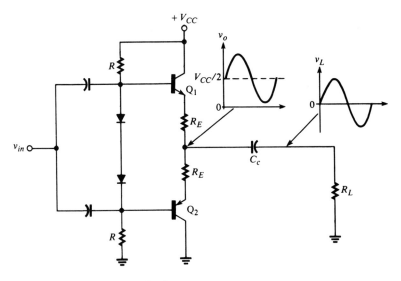

(a) The output is biased at $V_{CC}/2$ volts.

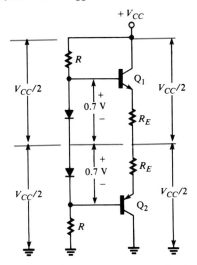

(b) DC bias voltages (All components are assumed to be perfectly matched.)

Figure 14.26
Complementary push-pull amplifier using a single power supply

negative. The $R_L C_c$ time constant must be much greater than the period of the lowest signal frequency. The lower cutoff frequency due to C_c is given by

$$f_1(C_c) = \frac{1}{2\pi(R_L + R_E)C_c} \text{ Hz} \qquad (14.49)$$

The peak load current is the peak input voltage V_P divided by $R_L + R_E$:

$$I_{PL} = \frac{V_P}{R_L + R_E} \qquad (14.50)$$

Therefore, the average ac power delivered to the load is

$$P_L = \frac{I_{PL}^2 R_L}{2} = \frac{V_P^2 R_L}{2(R_L + R_E)^2} \qquad (14.51)$$

Since current is drawn from the power supply only during positive half-cycles of input, the supply-current waveform is half-wave rectified, with peak value $V_P/(R_L + R_E)$ amperes. Therefore, the average value of the supply current is

$$I_S(\text{avg}) = \frac{V_P}{\pi(R_L + R_E)} \qquad (14.52)$$

and the average power from the supply is

$$P_S = V_{CC} I_S(\text{avg}) = \frac{V_{CC} V_P}{\pi(R_L + R_E)} \qquad (14.53)$$

Dividing (14.51) by (14.53), we find the efficiency to be

$$\eta = \frac{P_L}{P_S} = \frac{\pi}{2}\left(\frac{R_L}{R_L + R_E}\right)\left(\frac{V_P}{V_{CC}}\right) \qquad (14.54)$$

The efficiency is maximum when the peak voltage V_P has its maximum possible value, $V_{CC}/2$. In that case,

$$\eta(\text{max}) = \frac{\pi}{2}\left(\frac{R_L}{R_L + R_E}\right)\left(\frac{V_{CC}/2}{V_{CC}}\right) = \frac{\pi}{4}\left(\frac{R_L}{R_L + R_E}\right) \qquad (14.55)$$

Equations 14.54 and 14.55 show that efficiency decreases, as expected, when R_E is increased. If $R_E = 0$, then the maximum possible efficiency becomes $\eta(\text{max}) = \pi/4 = 0.785$, the theoretical maximum for a class-B amplifier.

Example 14.10

Assuming that all components in Figure 14.27 are perfectly matched, find

1. the base-to-ground voltages V_{B1} and V_{B2} of each transistor;
2. the power delivered to the load under maximum signal conditions;
3. the efficiency under maximum signal conditions; and
4. the value of capacitor C_c if the amplifier is to be used at signal frequencies down to 20 Hz.

Solution

1. Since all components are matched, the supply voltage divides equally across the resistor–diode network, as shown in Figure 14.28. Then, as can be seen from the figure, $V_{B1} = 10 + 0.7 = 10.7$ V and $V_{B2} = V_{B1} - 1.4 = 9.3$ V.
2. From equation 14.51, with $V_P(\text{max}) = V_{CC}/2 = 10$ V,

$$P_L(\text{max}) = \frac{V_P^2(\text{max})R_L}{2(R_L + R_E)^2} = \frac{10^2(10)}{2(10 + 1)^2} = 4.132 \text{ W}$$

3. From equation 14.55,

$$\eta(\text{max}) = \frac{\pi}{4}\left(\frac{R_L}{R_L + R_E}\right) = \frac{\pi}{4}\left(\frac{10}{11}\right) = 0.714$$

Figure 14.27
(Example 14.10)

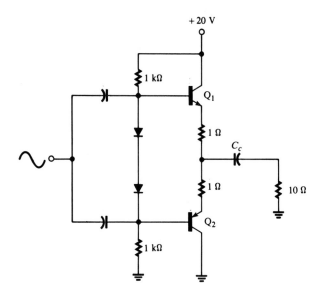

Figure 14.27
(Example 14.10)

Figure 14.28
(Example 14.10)

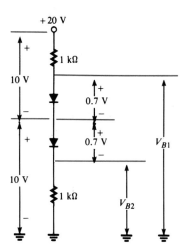

4. From equation 14.49,

$$C_c = \frac{1}{2\pi(R_L + R_E)f_1(C_c)} = \frac{1}{2\pi(11)(20)} = 723 \ \mu\text{F}$$

This result demonstrates a principal disadvantage of the complementary, single-supply, push-pull amplifier: the coupling capacitor must be quite large to drive a low-impedance load at a low frequency. These operating requirements are typical for audio power amplifiers.

Quasi-Complementary Push-Pull Amplifiers

As has been discussed in previous chapters, modern semiconductor technology is such that NPN transistors are generally superior to PNP types. Figure 14.29 shows a popular push-pull amplifier design that uses NPN transistors for both output devices. This design is called *quasi-complementary* because transistors Q_3 and Q_4 together perform the same function as the PNP transistor in a complementary push-pull amplifier. When the input signal is positive, PNP transistor Q_3 is cut off, so NPN transistor Q_4 receives no base current and is also cut off. When the input is negative, Q_3 conducts and supplies base current to Q_4, which can then conduct load current. Transistors Q_1 and Q_2 form an NPN Darlington pair, so Q_1 and Q_2 provide emitter-follower action to the load when the input is positive, and Q_3 and Q_4 perform that function when the input is negative. The entire configuration thus performs push-pull operation in the same manner as the complementary push-pull amplifier. The current gain of the Q_3Q_4 combination is

$$\frac{I_E}{I_B} = \beta_3\beta_4 + (1 + \beta_3) \tag{14.56}$$

This gain is very nearly that of a Darlington pair. Transistors Q_1 and Q_2 are connected as a Darlington pair to ensure that both sides of the amplifier have similar gain.

Integrated-Circuit Power Amplifiers

Low- to medium-power audio amplifiers (in the 1- to 20-W range) are available in integrated-circuit form. Many have differential inputs and quasi-complementary outputs. Figure 14.30 shows an example, the LM380 2.5-W audio amplifier manufactured by National Semiconductor. Transistors Q_1 and Q_2 are input emitter followers that drive the differential pair consisting of Q_3 and Q_4. Transistors Q_5 and Q_6 serve as *active loads*. Transistors Q_{10} and Q_{11} form a current mirror

Figure 14.29
A quasi-complementary push-pull ampli-fier. Transistors Q_3 and Q_4 conduct dur-ing negative half-cycles of input.

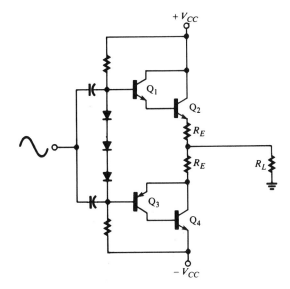

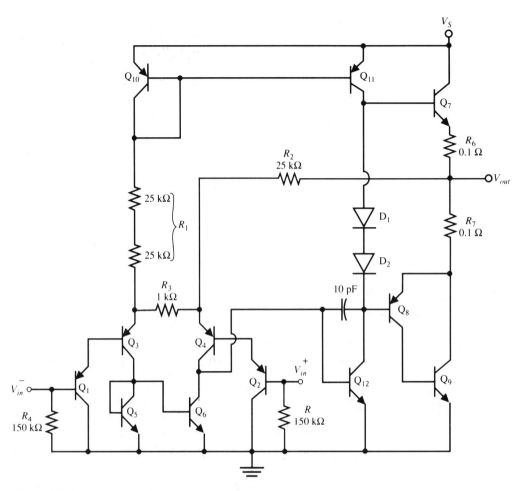

Figure 14.30
Simplified schematic diagram of the LM380 integrated-circuit amplifier

supplying bias current (Figure 5.12). The output of the differential stage, at the collector of Q_4, drives a common-emitter stage (Q_{12}), which in turn drives a quasi-complementary output stage. The 10-pF capacitor provides internal frequency compensation.

One feature of the LM380 is that internal resistor networks (R_1 and R_2) are used to set the output bias level automatically at one-half the supply voltage, V_S. The output is normally capacitor-coupled to the load. The voltage gain of the amplifier is fixed at 50 (34 dB). Manufacturer's specifications state that the bandwidth is 100 kHz and that the total harmonic distortion is typically 0.2%. The supply voltage can be set from 10 to 22 V.

14.9   CLASS-C AMPLIFIERS

A class-C amplifier is one whose output conducts load current during *less* than one-half cycle of an input sine wave. Figure 14.31 shows a typical class-C current

Figure 14.31
Output current in a class-C amplifier

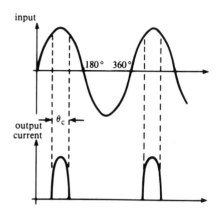

waveform, and it is apparent that the total angle during which current flows is less than 180°. This angle is called the *conduction angle, θ_c*.

Of course, the output of a class-C amplifier is a highly distorted version of its input. It could not be used in an application requiring high fidelity, such as an audio amplifier. Class-C amplifiers are used primarily in high-power, high-frequency applications, such as radio-frequency transmitters. In these applications, the high-frequency pulses handled by the amplifier are not themselves the signal, but constitute what is called the *carrier* for the signal. The signal is transmitted by varying the amplitude of the carrier, using the process called *amplitude modulation* (AM). The signal is ultimately recovered in a *receiver* by filtering out the carrier frequency. The principal advantage of a class-C amplifier is that it has a very high efficiency, as we shall presently demonstrate.

Figure 14.32 shows a simple class-C amplifier with a resistive load. Note that the base of the NPN transistor is biased by a *negative* voltage, $-V_{BB}$, connected through a coil labeled *RFC*. The RFC is a *radio-frequency choke* whose inductance presents a high impedance to the high-frequency input and thereby prevents the dc source from shorting the ac input. In order for the transistor to begin

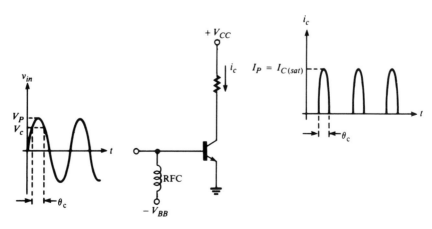

Figure 14.32
A class-C amplifier with resistive load. The transistor conducts when $v_{in} \geq |V_{BB}| + 0.7$.

conducting, the input must reach a level sufficient to overcome both the negative bias and the V_{BE} drop of about 0.7 V:

$$V_c = |V_{BB}| + 0.7 \text{ V} \tag{14.57}$$

where V_c is the input voltage at which the transistor begins to conduct. As shown in the figure, the transistor is cut off until v_{in} reaches V_c, then it conducts, and then it cuts off again when v_{in} falls below V_c. Clearly, the more negative the value of V_{BB}, the shorter the conduction interval. In most class-C applications, the amplifier is designed so that the peak value of the input, V_P, is just sufficient to drive the transistor into saturation, as shown in the figure.

The conduction angle θ_c in Figure 14.32 can be found from

$$\theta_c = 2 \arccos \left(\frac{V_c}{V_P} \right) \tag{14.58}$$

where V_P is the peak input voltage that drives the transistor to saturation. If the peak input only just reaches V_c, then $\theta_c = 2 \arccos(1) = 0°$. At the other extreme, if $V_{BB} = 0$, then $V_c = 0.7$, $(V_c/V_P) \approx 0$, and $\theta_c = 2 \arccos(0) = 180°$, which corresponds to class-B operation.

Figure 14.33 shows the class-C amplifier as it is normally operated, with an LC *tank* network in the collector circuit. Recall that the tank is a *resonant* network whose center frequency, assuming small coil resistance, is closely approximated by

$$f_o \approx \frac{1}{2\pi\sqrt{LC}} \tag{14.59}$$

The purpose of the tank is to produce the fundamental component of the pulsed, class-C waveform, which has the same frequency as v_{in}. The configuration is called a *tuned* amplifier, and the center frequency of the tank is set equal to (tuned to) the input frequency. There are several ways to view its behavior as an aid in understanding how it recovers the fundamental frequency. We may regard the tank as a highly selective (high-Q) filter that suppresses the harmonics in the class-C waveform and passes its fundamental. We may also recall that the voltage

Figure 14.33
A tuned class-C amplifier with an LC
tank circuit as load

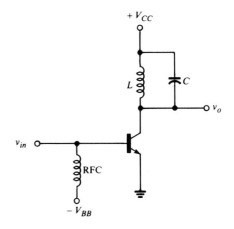

gain of the transistor equals the impedance in the collector circuit divided by the emitter resistance. Since the impedance of the tank is very large at its center frequency, the gain is correspondingly large at that frequency, while the impedance and the gain at harmonic frequencies are much smaller.

The amplitude of the fundamental component of a class-C waveform depends on the conduction angle θ_c. The greater the conduction angle, the greater the ratio of the amplitude of the fundamental component to the amplitude of the total waveform. Let r_1 be the ratio of the peak value of the fundamental component to the peak value of the class-C waveform. The value of r_1 is closely approximated by

$$r_1 \approx (-3.54 + 4.1\theta_c - 0.0072\theta_c^2) \times 10^{-3} \tag{14.60}$$

where $0° \leq \theta_c \leq 180°$. The values of r_1 vary from 0 to 0.5 as θ_c varies from 0° to 180°.

Let r_0 be the ratio of the dc value of the class-C waveform to its peak value. The value of r_0 can be found from

$$r_0 = \frac{\text{dc value}}{\text{peak value}} = \frac{\theta_c}{\pi(180)} \tag{14.61}$$

where $0° \leq \theta_c \leq 180°$. The values of r_0 vary from 0 to $1/\pi$ as θ_c varies from 0° to 180°.

The efficiency of a class-C amplifier is large because very little power is dissipated when the transistor is cut off, and it is cut off during most of every full cycle of input. The output power at the fundamental frequency under maximum drive conditions is

$$P_o = \frac{(r_1 I_P) V_{CC}}{2} \tag{14.62}$$

where I_P is the peak output (collector) current. The average power supplied by the dc source is V_{CC} times the average current drawn from the source. Since current flows only when the transistor is conducting, this current waveform is the same as the class-C collector-current waveform having peak value I_P. Therefore,

$$P_S = (r_0 I_P) V_{CC} \tag{14.63}$$

The efficiency is then

$$\eta = \frac{P_o}{P_S} = \frac{r_1 I_P V_{CC}}{2 r_0 I_P V_{CC}} = \frac{r_1}{2 r_0} \tag{14.64}$$

Example 14.11

A class-C amplifier has a base bias voltage of -5 V and $V_{CC} = 30$ V. It is determined that a peak input voltage of 9.8 V at 1 MHz is required to drive the transistor to its saturation current of 1.8 A.

1. Find the conduction angle.
2. Find the output power at 1 MHz.
3. Find the efficiency.
4. If an LC tank having $C = 200$ pF is connected in the collector circuit, find the inductance necessary to tune the amplifier.

Solution

1. From equation 14.57, $V_c = |V_{BB}| + 0.7\text{ V} = 5 + 0.7 = 5.7\text{ V}$. From equation 14.58,

$$\theta_c = 2 \arccos\left(\frac{V_c}{V_P}\right) = 2 \arccos\left(\frac{5.7}{9.8}\right) = 108.9°$$

2. From equation 14.60, $r_1 \approx [-3.54 + 4.1(108.9) - 0.0072(108.9)^2] \times 10^{-3} = 0.357$. From equation 14.62,

$$P_o = \frac{(r_1 I_P)V_{CC}}{2} = \frac{(0.357)(1.8\text{ A})(30\text{ V})}{2} = 9.64\text{ W}$$

3. From equation 14.61,

$$r_0 = \frac{\theta_c}{\pi(180)} = \frac{108.9}{\pi(180)} = 0.193$$

From equation 14.64,

$$\eta = \frac{r_1}{2r_0} = \frac{0.357}{2(0.193)} = 0.925 \quad (\text{or } 92.5\%)$$

This result demonstrates that a very high efficiency can be achieved in a class-C amplifier.

4. From equation 14.59,

$$L = \frac{1}{(2\pi f_o)^2 C} = \frac{1}{(2\pi \times 10^6)^2(200 \times 10^{-12})} = 0.127\text{ mH}$$

Amplitude Modulation

We have mentioned that amplitude modulation is a means used to transmit signals by varying the amplitude of a high-frequency carrier. In a typical application, the signal is a low-frequency audio waveform and the amplitude of a high-frequency (radio-frequency, or rf) carrier is made to increase and decrease as the audio signal increases and decreases. Figure 14.34 shows the waveforms that are generated by an amplitude modulator when the signal input is a small dc value, a large dc value, and a low-frequency sinusoidal wave. Notice that the carrier input to the modulator is a constant-amplitude, constant-frequency rf sine wave. It can be seen that the audio waveform is reproduced in the variations of the positive and negative peaks of the output. The audio waveform superimposed on the high-frequency peaks is called the *envelope* of the modulated wave.

It is important to note that amplitude modulation is achieved by *multiplying* two waveforms (signal × carrier), which is a *nonlinear* process. As such, the modulated output contains frequency components that are not present in either the signal or the carrier. Amplitude modulation cannot be achieved by simply adding two waveforms. Unfortunately, the term *mixing* is used in the broadcast industry to mean both the summation of signals and the multiplication of signals (some modulators are called *mixers*), but these are two very distinct processes.

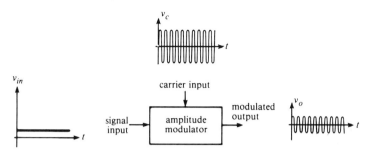

(a) The output of the modulator is a low-amplitude, high-frequency waveform when the signal input is a small dc value.

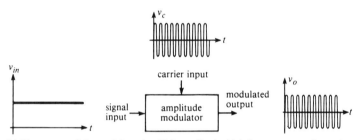

(b) The output of the modulator is a high-amplitude, high-frequency waveform when the signal input is a large dc value.

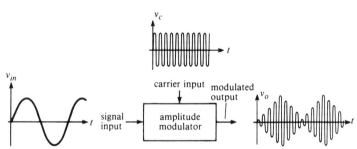

(c) The amplitude of the output varies in the same way that the signal input varies.

Figure 14.34
Amplitude-modulated (AM) waveforms

We will now demonstrate that amplitude modulation creates new frequency components. Let the input signal be a pure sine wave designated

$$v_s(t) = A_s \sin \omega_s t \qquad (14.65)$$

and let the carrier input be designated

$$v_c(t) = A_c \sin \omega_c t \qquad (14.66)$$

where A_s and A_c are the peak values of the signal and the carrier, respectively.

The signal and carrier frequencies are $\omega_s = 2\pi f_s$ rad/s and $\omega_c = 2\pi f_c$ rad/s, respectively. The envelope of the modulated output is the time-varying signal voltage added to the peak carrier voltage:

$$\text{envelope} = e(t) = v_s(t) + A_c = A_s \sin \omega_s t + A_c \qquad (14.67)$$

The modulated AM waveform has frequency ω_c and has a (time-varying) peak value equal to the envelope:

$$
\begin{aligned}
v_o(t) &= \underbrace{e(t)}_{\text{envelope = peak value}} \sin \omega_c t \\
&= (A_s \sin \omega_s t + A_c)\sin \omega_c t \\
&= A_s(\sin \omega_c t)(\sin \omega_s t) + A_c \sin \omega_c t
\end{aligned}
\qquad (14.68)
$$

Equation 14.68 makes it apparent now that amplitude modulation involves the multiplication of two sine waves. Using a trigonometric identity, equation 14.68 may be expressed as (Exercise 14.31)

$$v_o(t) = A_c \sin \omega_c t + \tfrac{1}{2}A_s \cos(\omega_c - \omega_s)t - \tfrac{1}{2}A_s \cos(\omega_c + \omega_s)t \qquad (14.69)$$

Equation 14.69 shows that the modulated waveform contains the new frequency components $\omega_c + \omega_s$ and $\omega_c - \omega_s$, called the *sum and difference frequencies,* as well as a component at the carrier frequency, ω_c. Note that there is *no* frequency component equal to ω_s. In practice, the input signal will consist of a complex waveform containing many different frequencies. Therefore, the AM output will contain many different sum and difference frequencies. The band of difference frequencies is called the *lower sideband* because its frequencies are all less than the carrier frequency, and the band of sum frequencies is called the *upper sideband,* each of its frequencies being greater than the carrier frequency. The significance of this result is that it allows us to determine the bandwidth that an amplifier must have to pass an AM waveform. The bandwidth must extend from the smallest difference frequency to the largest sum frequency.

Example 14.12

An amplitude modulator is driven by a 580-kHz carrier and has an audio signal input containing frequency components between 200 Hz and 9.5 kHz.

1. What range of frequencies must be included in the passband of an amplifier that will be used to amplify the modulated signal?
2. What frequency components are in the lower sideband of the AM output? In the upper sideband?

Solution

1. The smallest difference frequency is $2\pi(580 \times 10^3) - 2\pi(9.5 \times 10^3)$ rad/s, or (580 kHz) − (9.5 kHz) = 570.5 kHz, and the largest sum frequency is (580 kHz) + (9.5 kHz) = 589.5 kHz. Therefore, the amplifier must pass frequencies in the range from 570.5 kHz to 589.5 kHz. An amplifier having only this range would be considered a *narrowband* amplifier.
2. The lowest sideband extends from the smallest difference frequency to the largest difference frequency: [(580 kHz) − (9.5 kHz)] to [(580 kHz) − (200 Hz)], or 570.5 kHz to 579.8 kHz. The upper sideband extends from the smallest

sum frequency to the largest sum frequency: [(580 kHz) + (200 Hz)] to [(580 kHz) + (9.5 kHz)], or 580.2 kHz to 589.5 kHz.

Figure 14.35 shows how a class-C amplifier can be used to produce amplitude modulation. The circuit is similar to Figure 14.33, except that the coil in the collector circuit has been replaced by the secondary winding of a transformer. The low-frequency signal input is connected to the transformer primary. The voltage at the collector is then the sum of V_{CC} and a signal proportional to v_s. As v_s increases and decreases, so does the collector voltage. In effect, we are varying the supply voltage on the collector. When the carrier signal on the base drives the transistor to saturation, the peak collector current $I_{c(sat)}$ that flows depends directly on the supply voltage: the greater the supply voltage, the greater the peak current. Since the supply voltage varies with the signal input, so does the peak collector current. As a consequence, the collector current is amplitude modulated by v_s, and the tank circuit produces an AM output voltage, as shown.

14.10 MOSFET AND CLASS-D POWER AMPLIFIERS

MOSFET Amplifiers

MOSFET devices of the VMOS design (in which the channel is a V-shaped region) are becoming a popular choice for power amplifiers, particularly those that are designed to switch large currents on and off. Examples include line drivers for digital switching circuits, switching-mode voltage regulators (discussed in Chapter 13), and class-D amplifiers, which we will discuss presently. One advantage of a MOSFET switch is the fact that turn-off time is not delayed by minority-carrier storage, as it is in a bipolar switch that has been driven deeply into saturation. Recall that current in field-effect transistors is due to the flow of majority carriers only. Also, MOSFETs are not susceptible to thermal runaway like bipolar transistors. Finally, a MOSFET has a very large input impedance, which simplifies the design of driver circuits.

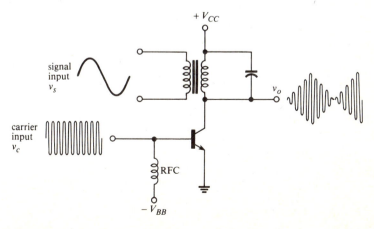

Figure 14.35
A class-C amplitude modulator

Figure 14.36 shows a simple, class-A audio amplifier that uses a VN64GA VMOS transistor to drive an output transformer. Notice that the driver for the output stage is a J108 JFET transistor connected in a common-source configuration. Also notice the feedback path between the transformer secondary and the input to the JFET driver. This negative feedback is incorporated to reduce distortion, as discussed earlier. The amplifier will reportedly deliver between 3 and 4 W to an 8-Ω speaker with less than 2% distortion up to 15 kHz.

MOSFET power amplifiers are also constructed for class-B and class-AB operation. Like their bipolar counterparts, these amplifiers can be designed without the need for an output transformer, since power MOSFETs are available in complementary pairs (N-channel and P-channel). Figure 14.37 shows a typical push-pull MOSFET amplifier using complementary output devices.

Class-D Amplifiers

A class-D amplifier is one whose output is switched on and off, that is, one whose output is in its linear range for essentially *zero* time during each cycle of an input sine wave. As we progress through the letters designating the various classes of operation, A, B, C, and D, we see that linear operation occurs for shorter and shorter intervals of time, and in class D we reach the limiting case where no linear operation occurs at all. The only time that a class-D output device is in its linear region is during that short interval required to switch from saturation to cutoff, or vice versa. In other words, the output device is a digital power switch, an application ideally suited for VMOS transistors.

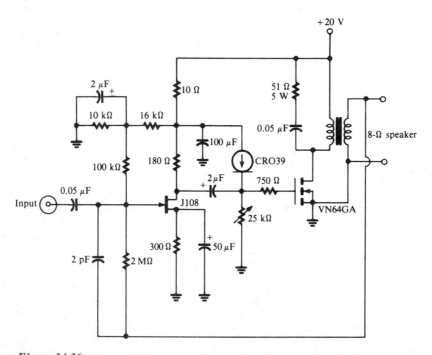

Figure 14.36
A two-stage, class-A audio amplifier with a VMOS output transistor (Courtesy of Siliconix Incorporated)

Figure 14.37
A complementary MOSFET amplifier

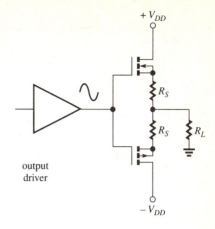

A fundamental component of a class-D amplifier is a *pulse-width modulator,* which produces a train of pulses having widths that are proportional to the level of the amplifier's input signal. When the signal level is small, a series of narrow pulses is generated, and when the input level is large, a series of wide pulses is generated. (See Figure 13.41, which shows some typical outputs of a pulse-width modulator used in a voltage-regulator application.) As the input signal increases and decreases, the pulse widths increase and decrease in direct proportion.

Figure 14.38 shows how a pulse-width modulator can be constructed using a sawtooth generator and a voltage comparator. A sawtooth waveform is one that rises linearly and then quickly switches back to its low level, to begin another linear rise, as illustrated in the figure. When the sawtooth voltage is greater than v_{in}, the output of the comparator is low, and when the sawtooth falls below v_{in}, the comparator switches to its high output. Notice that the comparator must switch high each time the sawtooth makes its vertical descent. These time points mark the beginning of each new pulse. The comparator output remains high until the sawtooth rises back to the value of v_{in}, at which time the comparator output switches low. Thus the width of the high pulse is directly proportional to the length of time it takes the sawtooth to rise to v_{in}, which is directly proportional to the level of v_{in}. As shown in the figure, the result is a series of pulses whose widths are proportional to the level of v_{in}. Notice that the peak-to-peak voltage of the sawtooth must exceed the largest peak-to-peak input voltage for successful operation. Also, the frequency of the sawtooth should be at least ten times as great as the highest frequency component of v_{in}.

The pulse-width modulator drives the output stage of the class-D amplifier, causing it to switch on and off as the pulses switch between high and low. Figure 14.39 shows a popular switching circuit, called a *totem pole,* that is used to drive heavy loads. The MOSFET version of the totem pole is shown in the figure, because MOSFETs are generally used in class-D amplifiers. Bipolar versions of the totem pole are also widely used in digital logic circuits, particularly in the integrated-circuit family called *TTL* (transistor-transistor logic). The totem pole shown in the figure *inverts,* in the sense that the output is low when the input is high, and vice versa.

When the input in Figure 14.39 is high, Q_1 and Q_3 are on, so the output is low (R_L is effectively connected to ground through Q_3). Since Q_1 is on, the gate of Q_2 is low and Q_2 is held off. Thus, when the input is high, Q_2 is like an open switch and

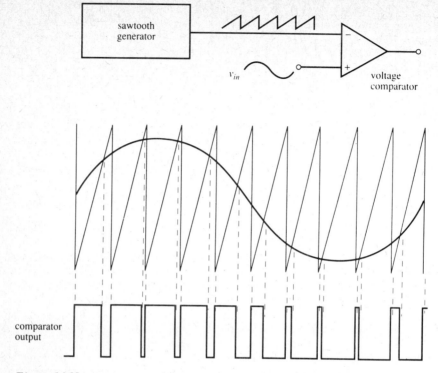

Figure 14.38
Construction of a pulse-width modulator using a sawtooth generator and a voltage comparator

Q_3 is like a closed switch. When the input is low, Q_1 is off and a high voltage (V_{DD}) is applied through R to the gate of Q_2, turning it on. Thus, R_L is connected through Q_2 to the high level V_{DD}. Since Q_3 is also off, a low input makes Q_3 an open switch and Q_2 a closed switch. The advantage of this arrangement is that the load is driven from a low-impedance signal source, a turned-on MOSFET, both when the output is low and when the output is high.

Like the class-C amplifier, a class-D amplifier must have a *filter* to extract, or recover, the signal from the pulsed waveform. However, in this case the signal may have many frequency components, so a tank circuit, which resonates at a single frequency, cannot be used. Instead, a low-pass filter having a cutoff frequency near the highest signal frequency is used. The low-pass filter suppresses the high-frequency components of the pulse train and, in effect, recovers the *average value* of the pulse train. Since the average value of the pulses depends on the pulse widths, the output of the filter is a waveform that increases and decreases as the pulse widths increase and decrease, that is, a waveform that duplicates the input signal, v_{in}. Figure 14.40 shows a block diagram of a complete class-D amplifier, including negative feedback to reduce distortion.

The principal advantage of a class-D amplifier is that it may have a very high efficiency, approaching 100%. Like that of a class-C amplifier, the high efficiency is due to the fact that the output device spends very little time in its active region, so power dissipation is minimal. The principal disadvantages are the need for a very good low-pass filter and the fact that high-speed switching of heavy currents

Figure 14.39
A MOSFET totem pole, used to switch heavy load currents

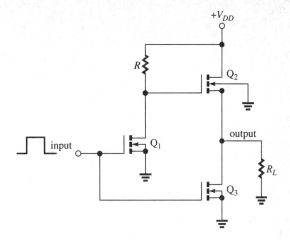

generates noise through electromagnetic coupling, called *electromagnetic interference,* or *EMI.*

Figure 14.41 shows a 100-W class-D amplifier using a MOSFET *totem pole* with a bipolar driver.

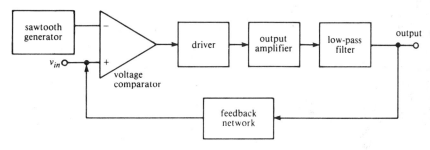

Figure 14.40
Block diagram of a class-D amplifier

Example 14.13

A class-D audio amplifier is to be driven by a signal that varies between ± 5 V. The output is a MOSFET totem pole with $V_{DD} = 30$ V.

1. What minimum frequency should the sawtooth waveform have?
2. What minimum peak-to-peak voltage should the sawtooth waveform have, assuming that the input is connected directly to the voltage comparator?
3. Assume that the amplifier is 100% efficient and is operated at the frequency found in (1). What average current is delivered to an 8-Ω load when the pulse-width modulator produces pulses that are high for 2.5 μs?

Solution

1. The nominal audio-frequency range is 20 Hz to 20 kHz, and the sawtooth waveform should have a frequency equal to at least 10 times the highest input frequency, i.e., 10(20 kHz) = 200 kHz.

2. The peak-to-peak sawtooth voltage should at least equal the maximum peak-to-peak input voltage, which is 10 V.

3. The period of the 200-kHz sawtooth is

$$T = \frac{1}{200 \times 10^3} = 5 \ \mu s$$

Therefore, the pulse train is high for $(2.5 \ \mu s)/(5 \ \mu s)$ = one-half the period. The totem-pole output therefore switches between 0 V and 30 V with pulse widths equal to one-half the period. Thus, the average voltage is

$$V_{avg} = \frac{1}{2}(30) = 15 \ V$$

and the average current is

$$I_{avg} = \frac{V_{avg}}{R_L} = \frac{15 \ V}{8 \ \Omega} = 1.875 \ A$$

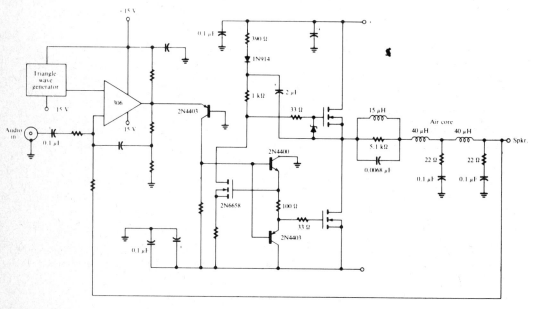

Figure 14.41
A 100-W class-D amplifier (Reprinted with permission from Power FETs and Their Applications *by Edwin S. Oxner, Prentice Hall, 1982)*

EXERCISES

Section 14.2   Transistor Power Dissipation

14.1. The transistor in Figure 14.42 is biased at $V_{CE} = 12$ V. If an increase in temperature causes the quiescent collector current to in-

crease, will the power dissipation at the Q-point increase or decrease? Explain. What is the power dissipation at the point of maximum power dissipation?

14.2. Repeat Exercise 14.1 if the supply voltage is

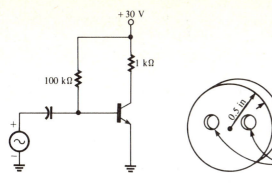

Figure 14.42
(Exercise 14.1)

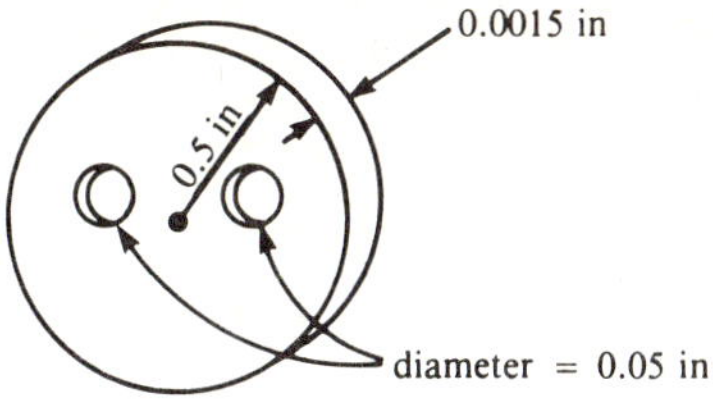

Figure 14.44
(Exercise 14.9)

changed to 20 V and the Q-point is set at $V_{CE} = 13$ V.

14.3. The amplifier in Figure 14.43 is to be operated with a 40-V-dc supply. If the maximum permissible power dissipation in the transistor is 1 W, what is the minimum permissible value of R_C?

14.4. If the amplifier in Exercise 14.3 has $R_C = 100$ Ω, what is the maximum permissible value of the supply voltage that can be used?

Section 14.3 Heat Transfer in Semiconductor Devices

14.5. The collector–base junction of a transistor conducts 0.4 A when it is reverse biased by 21 V. At what rate, in joules/second, must heat be removed from the junction to prevent its temperature from rising?

14.6. A semiconductor junction at temperature 64°C releases heat into a 25°C ambient at the rate of 7.8 J/s. What is the thermal resistance between junction and ambient?

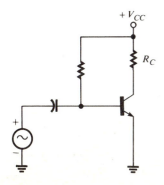

Figure 14.43
(Exercise 14.3)

14.7. The thermal resistance between a semiconductor device and its case is 0.8°C/W. It is used with a heat sink whose thermal resistance to ambient is 0.5°C/W. If the device dissipates 10 W and the ambient temperature is 47°C, what is the maximum permissible thermal resistance between case and heat sink if the device temperature cannot exceed 70°C?

14.8. A semiconductor dissipates 25 W through its case and its heat sink to the surrounding air. The thermal resistances are the following: device-to-case, 1°C/W; case-to-heat sink, 1.2°C/W; and heat sink-to-air, 0.7°C/W. In what maximum air temperature can the device be operated if its temperature cannot exceed 100°C?

14.9. The mica washer shown in Figure 14.44 has a thermal resistivity of 66°C·in./W. Find its thermal resistance.

14.10. A transistor has a maximum dissipation rating of 10 W at ambient temperatures up to 40°C and is derated at 100 mW/°C at higher temperatures. At what ambient temperature is the dissipation rating one-half its low-temperature value?

14.11. The temperature of a certain semiconductor device is 100°C when the device dissipates 1.2 W. The device temperature cannot exceed 100°C. If the total thermal resistance from device to ambient is 50°C/W, above what ambient temperature should the dissipation be derated?

Section 14.4 Amplifier Classes and Efficiency

14.12. The amplifier shown in Figure 14.45 is biased at $I_Q = 20$ mA. Find
  a. the ac power in the load resistance when

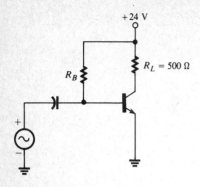

Figure 14.45
(Exercise 14.12)

   the voltage swing is the maximum possible without distortion;
  b. the amplifier efficiency under the conditions of (a); and
  c. the collector efficiency under the conditions of (a).

14.13. a. Find the value of quiescent current in the amplifier of Exercise 14.12 that will maximize the efficiency under maximum output voltage swing.
  b. Find the power delivered to the load resistance under the conditions of (a).

14.14. The amplifier shown in Figure 14.46 is biased at $I_Q = 0.2$ A. The maximum possible voltage swing without distortion is 8 V pk-pk. Find
  a. the ac power in the load resistance when the voltage swing is the maximum possible without distortion;
  b. the amplifier efficiency under the conditions of (a); and

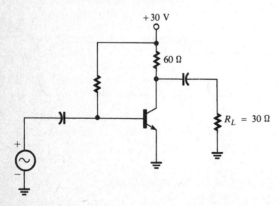

Figure 14.46
(Exercise 14.14)

  c. the collector efficiency under the conditions of (a).

14.15. a. The efficiency of the amplifier in Exercise 14.14 is maximum when $I_Q = 0.375$ A, in which case V_{PL} (max) $= 7.5$ V pk. Find the maximum efficiency.
  b. Why is the efficiency less than the theoretical maximum (0.0858)?

14.16. The class-A amplifier in Figure 14.47 is biased at $I_C = 0.2$ A. The transformer resistance is negligible.
  a. What is the slope of the ac load line?
  b. At what value does the ac load line intersect the V_{CE}–axis?
  c. What is the maximum peak value of the collector voltage without distortion?
  d. What is the maximum power delivered to the load under the conditions of (c)?
  e. What is the amplifier efficiency under the conditions of (c)?

14.17. The transistor in Exercise 14.16 has a beta of 40.
  a. Find the value of base current necessary to set a new Q-point that will permit maximum possible peak-to-peak output voltage.
  b. What power can be delivered to the load under the conditions of (a)?
  c. What are the amplifier and collector efficiencies under the conditions of (a)?

Section 14.5   Push-Pull Amplifier Principles

14.18. Draw a schematic diagram of a class-B, push-pull, transformer-coupled amplifier using PNP transistors. Draw a loop showing current direction through the primary when the input is positive and another showing current when the input is negative.

14.19. The peak collector current and voltage in each transistor in Figure 14.48 are 4 A and 12 V, respectively. Assuming that the transformer has negligible resistance, find
  a. the average power delivered to the load;
  b. the average power supplied by the dc source;
  c. the average power dissipated by each transistor; and
  d. the efficiency.

14.20. Repeat Exercise 14.19 when the drive signals are increased so that the peak collector voltage in each transistor is the maximum possible. Assume that the peak current increases proportionally.

14.21. A push-pull amplifier utilizes a transformer whose primary has a total of 160 turns and

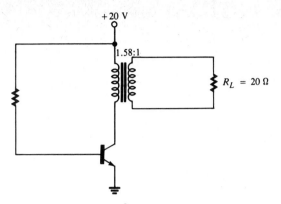

Figure 14.47
(Exercise 14.16)

whose secondary has 40 turns. It must be capable of delivering 40 W to an 8-Ω load under maximum power conditions. What is the minimum possible value of V_{CC}?

Section 14.7   Distortion in Push-Pull Amplifiers

14.22. Draw schematic diagrams of class-AB amplifiers using PNP transistors and an output transformer and having
a. an input coupling transformer; and
b. capacitor-coupled inputs.
It is not necessary to show component values, but label the polarities of all power supplies used. In each diagram, draw arrows showing the directions of the quiescent base currents and the dc currents in the primary of the output transformer.

Section 14.8   Transformerless Push-Pull Amplifiers

14.23. a. What is the maximum permissible peak value of input v_{in} in the amplifier shown in Figure 14.49?
b. What is the maximum power that can be delivered to the load?
c. What should be the breakdown voltage rating of each transistor?
14.24. If the amplifier in Exercise 14.23 must be redesigned so that it can deliver 36% more power to the load, by what percentage must the supply voltages be increased? What should be the minimum breakdown voltage of the transistors used in the new design?
14.25. Assuming that all the components in Figure 14.50 are perfectly matched, find
a. the current in each 680-Ω resistor;
b. the base-to-ground voltage of each transistor;
c. the efficiency when the peak voltage V_P is 10 V; and
d. the lower cutoff frequency.
14.26. a. What new value of load resistance is required in the amplifier of Exercise 14.25 if the lower cutoff frequency must be 15 Hz?
b. What is the maximum efficiency with the new load resistance determined in (a)?

Section 14.9   Class-C Amplifiers

14.27. The peak input voltage required to saturate the transistor in Figure 14.32 is 5.5 V. What should be the value of V_{BB} in order to achieve a conduction angle of 90°?

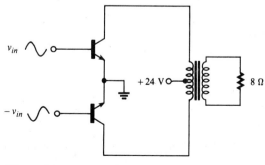

Figure 14.48
(Exercise 14.19)

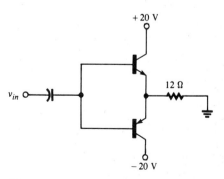

Figure 14.49
(Exercise 14.23)

Figure 14.50
(Exercise 14.25)

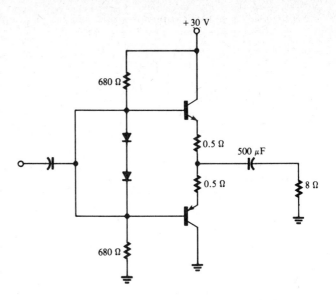

14.28. A peak input of 8 V is required to drive the transistor in Figure 14.32 to its 1.2-A saturation current. It is desired to produce a collector current that has a dc (average) value of 0.3 A. What should be the value of V_{BB}?

14.29. A peak input of 9 V is required to drive the transistor in Figure 14.32 to its 2.0-A saturation current. It is desired to produce a collector current whose fundamental has a peak value of 0.5 A. What should be the value of V_{BB}?

14.30. The input to a class-C amplifier must be 6.8 V to drive the output to its saturation current of 1.0 A. The amplifier begins to conduct when the input voltage reaches 2.6 V.
  a. What should be the value of V_{CC} if it is desired to obtain 5 W of output power at the fundamental frequency?
  b. What is the efficiency of the amplifier under the conditions of (a)?

14.31. Derive equation 14.69 from equation 14.68. (*Hint:* Use the trigonometric identity for $(\sin A)(\sin B)$.)

14.32. An amplitude-modulated waveform has a carrier frequency 810 kHz and an upper sideband that extends from 812.5 kHz to 829 kHz. What is the frequency range of the lower sideband?

14.33. An amplitude modulator is driven by an 8-V-rms, 10-kHz carrier and has a 2-V-rms, 10-kHz signal input. Write the mathematical expression for the modulated output.

Section 14.10 MOSFET and Class-D Power Amplifiers

14.34. The pulse-width modulator in Figure 14.38 is driven by a 100-kHz sawtooth. When v_{in} is +2 V, the modulator output is a series of pulses having widths equal to 1 μs. What are the pulse widths when $v_{in} = +3.5$ V?

14.35. The voltage comparator in Figure 14.38 switches between 0 V and +10 V. A 1-kHz sawtooth that varies between 0 V and +5 V is applied to the inverting input. A ramp voltage that rises from 0 V to 5 V in 5 ms is applied to the noninverting input. Sketch the input and output waveforms. Label the time points where the comparator output switches from low to high.

14.36. Repeat Exercise 14.35 if the inputs to the comparator are reversed. Label time points where the comparator switches from high to low instead of low to high. What would be the purpose of this input reversal, assuming that the comparator output was connected directly to a MOSFET totem-pole output stage?

14.37. Draw a schematic diagram of a MOSFET totem pole using P-channel devices. Describe an input signal that would be appropriate for driving this totem pole. Describe the operation of the totem pole, including its output levels, for each input level.

14.38. A class-D amplifier is to be designed to deliver a maximum average current of 2 A to a

12-Ω load. A 50-kHz sawtooth is to be used for the pulse-width modulator. Under maximum drive, the totem-pole output produces pulses that are 15 μs wide.

a. What should be the maximum input signal frequency?

b. Assuming 100% efficiency, what should be the value of V_{DD} used in the totem pole?

SPICE EXERCISES

14.39. Use SPICE to obtain a plot of the load voltage in the class-B amplifier shown in Figure 14.25(a). The supply voltages are ± 15 V dc, the coupling capacitor is 100 μF, and the load resistance is 100 Ω. The input signal is a sine wave having frequency 1 kHz. Obtain a plot of one full cycle of output when

a. the peak input is 12.5 V. What is the (approximate) power delivered to the load, based on the SPICE results?

b. the peak input is 2.5 V. Why is the crossover distortion so much more severe than in Example 14.9?

14.40. The amplifier in Figure 14.25(b) has $R_E = 1$ Ω, $V_{CC} = \pm 50$ V, $R_L = 20$ Ω, and a 100-μF coupling capacitor. Use SPICE to obtain a plot of one full cycle of load voltage when the input is a 40-V peak sine wave having frequency 1 kHz. Also obtain plots of the collector current in each of transistors Q_3 and Q_4. Use the SPICE results to determine the power delivered to the load.

14.41. The input coupling capacitors in the amplifier in Figure 14.26 are each 100 μF and the output coupling capacitor, C_c, is 723 μF. The input signal is a 200-Hz sine wave having peak value 10 V. Use SPICE to obtain a plot of the load voltage over one full cycle and a plot of the current drawn from the power supply over one full cycle. Use the SPICE results to answer the following questions:

a. What is the dc base-to-ground voltage of each transistor? Compare with values calculated in Example 14.10.

b. What is the power delivered to the load?

c. What is the voltage gain of the amplifier?

d. What is the peak current drawn from the power supply? Explain the appearance of the current waveform. (Why is it not sinusoidal?)

14.42. The quasi-complementary amplifier in Figure 14.29 has $R_E = 1$ Ω, $V_{CC} = \pm 20$ V, $R_L = 10$ Ω and 100-μF-input coupling capacitors. The input signal is a 12-V-pk sine wave having frequency 1 kHz. Use SPICE to obtain a plot of the load voltage over one full cycle. Using the results, determine the peak-to-peak load current, the voltage gain of the amplifier, and the power delivered to the load.

SPICE and PSpice

A.1   INTRODUCTION

SPICE—Simulation Program with Integrated Circuit Emphasis—was developed at the University of California, Berkeley, as a computer aid for designing integrated circuits. However, it is readily used to analyze discrete circuits as well and can, in fact, analyze circuits containing no semiconductor devices at all. In addition to this versatility, SPICE owes its current popularity to the ease with which a circuit model can be constructed and the wide range of output options (analysis types) available to the user.

As a brief note on terminology, observe that SPICE is a computer *program,* stored in computer memory, which we do not normally inspect or alter. As users, we merely supply the program with data, in the form of an *input data file,* which describes the circuit we wish to analyze and the type of output we desire. This input data file is usually supplied to SPICE by way of a keyboard. SPICE then executes a program run, using the data we have supplied, and displays or prints the results on a video terminal or printer. The mechanisms, or *commands,* that must be used to supply the input data to SPICE and to cause it to execute a program run vary widely with the computer system used, so we cannot describe that procedure here. Furthermore, there are numerous versions of SPICE, some designed for use with microcomputers and others capable of analyzing more complex circuits at greater speed, designed for use on large mainframe computers. Depending on the version used, minor variations in *syntax* (the format for specifying the input data) may be encountered. The *User's Guide,* supplied with most versions, should be consulted if any difficulty is experienced with any of the programs in this book. All programs here have been run successfully using SPICE version 2G.6 on a Honeywell DPS 90 computer.

PSPICE

One of the most widely used versions of SPICE designed for operation on microcomputers is PSpice (a registered trademark of the MicroSim Corporation). It has numerous features and options that make it more versatile and somewhat easier to use than the original Berkeley version of SPICE. However, with a few minor exceptions, any input data file written to run on Berkeley SPICE will also run on PSpice. In the discussions that follow, features of PSpice are highlighted immediately after the corresponding capabilities or requirements of Berkeley SPICE are described.

A.2   DESCRIBING A CIRCUIT FOR A SPICE INPUT FILE

The input data file consists of successive lines, which we will hereafter refer to as *statements,* each of which serves a specific purpose, such as identifying one component in the circuit. The statements do not have to be numbered and, except for the first and last, can appear in any order.

The Title

The first statement in every input file must be a *title*. Subject only to the number of characters permitted by a particular version, the title can be anything we wish. Examples are:

AMPLIFIER
EXERCISE 2.25
A DIODE (1N54) TEST

Nodes and Component Descriptions

The first step in preparing the circuit description is to identify and number all the *nodes* in the circuit. It is good practice to draw a schematic diagram with nodes shown by circles containing the node numbers. Node numbers can be any positive integers, and one of them must be 0. (The zero node is usually, but not necessarily, the circuit common, or ground.) Node numbers can be assigned in any sequence, such as 0, 1, 2, 3, . . . or 0, 2, 4, 6, ...

Once the node numbers have been assigned, each component in the circuit is identified by a separate statement that specifies the type of component it is and the node numbers between which it is connected. The first letter appearing in the statement identifies the component type. Passive components (resistors, capacitors, and inductors) are identified by the letters R, C, and L. Any other characters can follow the first letter, but each component must have a unique designation. For example, the resistors in a circuit might be identified by R1, R2, RB, and REQUIV.

The node numbers between which the component is connected appear next in the statement, separated by one or more spaces. Except in some special cases, it does not matter which node number appears first. The component value, in ohms, farads, or henries, appears next. Resistors cannot have value 0. Following are some examples:

R25 6 0 100   (Resistor R25 is connected between nodes 6 and 0 and has value 100 Ω.)
R25 0 6 1E2   (Interpreted by SPICE the same as in the first case.)
CIN 3 5 22E−6   (Capacitor CIN is connected between nodes 3 and 5 and has value 22 μF.)
LSHUNT 12 20   0.01   (Inductor LSHUNT is connected between nodes 12 and 20 and has value 0.01H.)

PSPICE

In the foregoing examples, all letters used in the input data files are capitalized, a requirement of Berkeley SPICE. In PSpice, either lowercase or capital letters (or both) may be used. For example, resistor R1 can be listed as r1 in one line and referred to again as R1 in another line, and PSpice will recognize both as representing the same component.

Node numbers in PSpice do not have to be integers; they can be identified by any set of alphanumeric characters (letters or numbers) up to 31 in length. However, one node must be node 0.

Specifying Numerical Values

Suffixes representing powers of 10 can be appended to value specifications (with no spaces in between). Following are the SPICE suffix designations:

T	(tera: 10^{12})	U	(micro: 10^{-6})
G	(giga: 10^9)	N	(nano: 10^{-9})
MEG	(mega: 10^6)	P	(pico: 10^{-12})
K	(kilo: 10^3)	F	(femto: 10^{-15})
M	(milli: 10^{-3})		

Note that M represents *milli* (10^{-3}) and that MEG is used for 10^6. Following are some examples of equivalent ways of representing values, all of which are interpreted by SPICE in the same way:

$$0.002 = 2M = 2E-3 = 2000U$$
$$5000E-12 = 5000P = 5N = .005U = 0.005E-6$$
$$0.15MEG = 150K = 150E3 = 150E+3 = .15E+6$$

Any characters can follow a value specification, and unless the characters are one of the powers-of-10 suffixes just given, SPICE simply ignores them. Characters are often added to designate units. Following are some examples, all of which are interpreted by SPICE in the same way:

$$100UF = 100E-6F = 100U = 100UFARADS$$
$$56N = 56NSEC = 0.056US = 56E-9SECONDS$$
$$2.2K = 2200OHMS = 0.0022MEGOHMS = 2.2E3$$
$$0.05MV = 50UVOLTS = 50E-6V = 0.05MILLIV$$

Be careful not to use a unit that begins with one of the power-of-10 suffixes. For example, a 0.0001-F capacitor specified as 0.0001F would be interpreted by SPICE as 0.0001×10^{-15} F.

DC Voltage Sources

The first letter designating a voltage source, dc or ac, is V. As with passive components, any characters can follow the V, and each voltage source must have a unique designation. Examples are V1, VIN, and VSIGNAL. Following is the format for representing a dc source:

```
V******* N+ N- ⟨DC⟩ value
```

where ******* are arbitrary characters, N+ is the number of the node to which the positive terminal of the source is connected, and N− is the number of the node to which the negative terminal is connected. The symbol ⟨ ⟩ enclosing DC means that the specification DC is optional: If a source is not designated DC, SPICE will automatically assume that it is DC. *Value* is the source voltage, in volts. *Value* can be negative, which is equivalent to reversing the N+ and N− node numbers. The

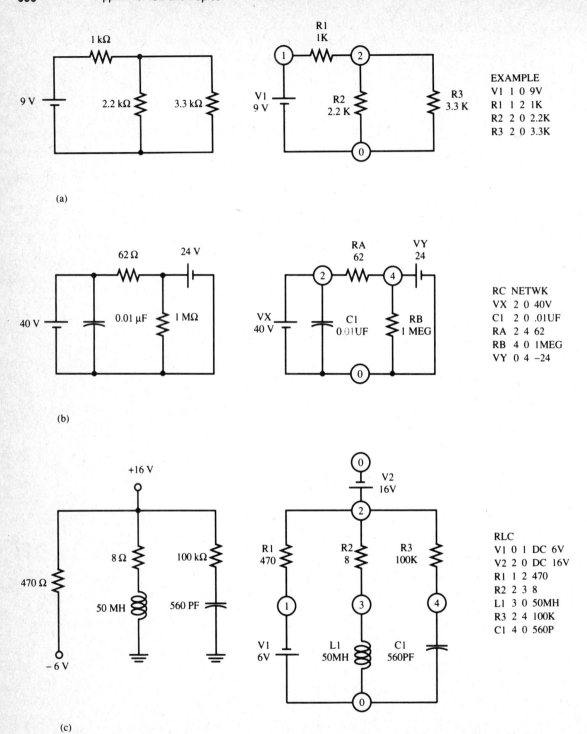

(a)

(b)

(c)

Figure A.1
Examples of circuit descriptions for a SPICE input data file

Figure A.2
Equivalent ways of specifying a dc current source. Note that the "negative" terminal is the one to which current is delivered (N− = 2).

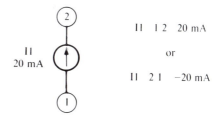

following are examples:

```
VIN 5 0 24VOLTS
VIN 0 5 DC -24
```

Both of these statements specify a 24-V-dc source designated VIN whose positive terminal is connected to node 5 and whose negative terminal is connected to node 0. Both statements are treated the same by SPICE.

Figure A.1 shows some examples of circuits and the statements that could be used to describe them in a SPICE input file. (These examples are not complete input files because we have not yet discussed *control* statements, used to specify the type of analysis and the output desired.)

DC Current Sources

The format for specifying a dc current source is

```
I******* N+ N− ⟨DC⟩ value
```

where *value* is the value of the source in amperes and DC is optional. Note the following *unconventional* definition of N+ and N−: *N− is the number of the node to which the source delivers current*. To illustrate, Figure A.2 shows two equivalent ways of specifying a 20-mA dc current source.

A.3   THE .DC AND .PRINT CONTROL STATEMENTS

A control statement is one that specifies the type of analysis to be performed or the type of output desired. Every control statement begins with a period, followed immediately by a group of characters that identifies the type of control statement it is.

The .DC Control Statement

The .DC control statement tells SPICE that a dc analysis is to be performed. This type of analysis is necessary when the user wishes to determine dc voltages and/ or currents in a circuit. Although ac sources can be present in the circuit, there must be at least one dc (voltage or current) source present if a dc analysis is to be performed. The format for the .DC control statement is

```
.DC name1 start stop incr. ⟨name2 start2 stop 2 incr.2⟩
```

where *name1* is the name of one dc voltage or current source in the circuit and *name2* is (optionally) the name of another. The .DC control statement can be used to *step* a source through a sequence of values, a useful feature when plotting characteristic curves. For that use, *start* is the first value of voltage or current in the sequence, *stop* is the last value, and *incr.* is the value of the increment, or step, in the sequence. A second source, *name2*, can also be stepped. If analysis is desired at a *single* dc value, we set *start* and *stop* both equal to that value and arbitrarily set *incr.* equal to 1. In cases where the circuit contains several fixed-value sources, one of them (any one) *must* be specified in the .DC control statement. Following are some examples:

.DC	V1	24	24	1	V1 is a fixed 24-V-dc voltage source.
.DC	IB	50U	50U	1	IB is a fixed 50-μA-dc current source.
.DC	VCE	0	50	10	VCE is a dc voltage source that is stepped from 0 to 50 V in 10-V increments.

When two sources are stepped in a .DC control statement, the first source is stepped over its entire range for each value of the second source. The following is an example:

```
.DC VCC 0 25 5 IBB 0 20U 2U
```

In this example, the dc voltage source named VCC is stepped from 0 to 25 V for *each* value of the dc current source named IBB.

Identifying Output Voltages and Currents

To tell SPICE the voltages whose values we wish to determine (the *output* voltages we desire), we must identify them in one of the following formats:

$$V(N+, N-) \quad \text{or} \quad V(N+)$$

In the first case, $V(N+, N-)$ refers to the voltage at node $N+$ with respect to node $N-$. In the second case, $V(N+)$ is the voltage at node $N+$ with respect to node 0. The following are examples:

V(5, 1)   The voltage at node 5 with respect to node 1.
V(3)   The voltage at node 3 with respect to node 0.

The only way to obtain the value of a current in Berkeley SPICE is to request the value of the current in a *voltage* source. Thus, a voltage source must be in the circuit at any point where we wish to know the value of the current. The current is identified by I(*Vname*), where *Vname* is the name of the voltage source. *We can insert zero-valued voltage sources (dummy sources) anywhere in a circuit for the purpose of obtaining a current.* These dummy voltage sources serve as ammeters for SPICE and do not in any way affect circuit behavior. Note carefully the following unconventional way that SPICE assigns polarity to the current through a voltage source: Positive current flows *into* the positive terminal (N+) of the voltage source. Figure A.3 shows an example. Here, conventional positive current flows in a clockwise direction, but a SPICE output would show I(V1) equal to -2 A. On the other hand, SPICE would show the current in the zero-valued dummy source I(VDUM), to be $+2$ A.

The .PRINT Control Statement

The .PRINT statement tells SPICE to print the values of voltages and/or currents resulting from an analysis. The format is

```
.PRINT type out1 (out2 out3 ...)
```

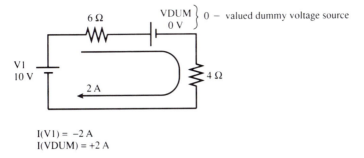

$I(V1) = -2$ A
$I(VDUM) = +2$ A

Figure A.3
SPICE treats conventional current flowing out of a positive terminal as negative current and current flowing into a positive terminal as positive current.

where *type* is the type of analysis and *out1, out2, . . .* , identify the output voltages and/or currents whose values we desire. So far, DC is the only analysis type we have discussed. For example, the statement

```
.PRINT DC V(1,2) V(3) I(VDUM)
```

tells SPICE to print the values of the voltages $V(1, 2)$ and $V(3)$ and the value of the current I(VDUM) resulting from a dc analysis. If the .DC control statement specifies a stepped source, the .PRINT statement will print all the output values resulting from all the stepped values. The number of output variables whose values can be requested by a single .PRINT statement may vary with the version of SPICE used (up to 8 can be requested in version 2G.6). Any number of .PRINT statements can appear in a SPICE input file.

PSPICE   In PSpice, the current through any component can be requested directly, without using a dummy voltage source. For example, I(R1) is the current through resistor R1. The reference polarity of the current is from the first node number of R1 to the second node number. In other words, positive current is assumed to flow from the first node given in the description of R1 to the second node. For example, if 5 A flows from node 1 to node 2 in a circuit whose input data file contains

```
R1 1 2 10
.PRINT DC I(R1)
```

then the .PRINT statement will produce 5 A. On the other hand, if the description of the same R1 were changed to R1 2 1 10, then the same .PRINT statement would produce −5 A.

Also, voltages across components can be requested directly, as, for example, V(RX), the voltage across resistor RX. The reference polarity is from the first node listed in the description of RX to the second node. For example, if the voltage across RX from node 5 to node 6 is 12 V in a circuit whose input data file contains

```
RX 5 6 1K
.PRINT DC V(RX)
```

then the .PRINT statement will produce 12 V. On the other hand, if the description of the same resistor were changed to RX 6 5 1K, the same .PRINT statement would produce −12 V.

The .END Statement

The last statement in every SPICE input file must be .END. We have now discussed enough statements to construct a complete SPICE input file, as demonstrated in the next example.

Example A.1

Use SPICE to determine the voltage drop across and the current through every resistor in Figure A.4(a).

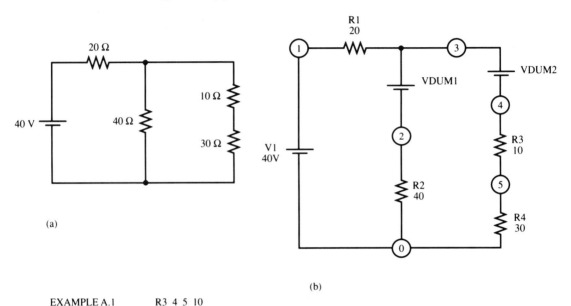

(a)

(b)

```
EXAMPLE A.1        R3 4 5 10
V1 1 0             R4 5 0 30
VDUM1 3 2          .DC V1 40 40 1
VDUM2 3 4          .PRINT DC I(V1) I(VDUM1) I(VDUM2) V(1,3) V(2) V(4,5) V(5)
R1 1 3 20          .END
R2 2 0 40
```

(c)

```
EXAMPLE A.1
****      DC TRANSFER CURVES                    TEMPERATURE =    27.000 DEG C
****************************************************************************
   V1          I(V1)         I(VDUM1)      I(VDUM2)      V(1,3)        V(2)
 4.000E+01    -1.000E+00     5.000E-01     5.000E-01     2.000E+01     2.000E+01


****************************************************************************
   V1          V(4,5)        V(5)
 4.000E+01     5.000E+00     1.500E+01
```

(d)

Figure A.4
(Example A.1)

Solution.   Figure A.4(b) shows the circuit when redrawn and labeled for analysis by SPICE. Note that two dummy voltage sources are inserted to obtain currents in two branches. The polarities of these sources are such that positive currents will be computed. In PSpice, we could simply request I(R2) and I(R3) in the PRINT statement. The current in R1 is the same as the current in V1, and we must simply remember that SPICE will print a negative value for that current.

Figure A.4(c) shows the SPICE input file. Note that it is not necessary to specify a voltage value in the statement defining V1, since that value is given in the .DC statement. Figure A.4(d) shows the results of a program run. The outputs appear under the heading ''DC TRANSFER CURVES,'' which refers to the type of output obtained when the source(s) are stepped. In our case, the heading is irrelevant. Note that the analysis is performed under the (default) assumption that the circuit temperature is 27°C (80.6°F). We will see later that we can specify different temperatures. Referring to the circuit nodes in Figure A.4(b), we see that the printed results give the following voltages and currents:

	I	V
R_1	1 A	20 V
R_2	0.5 A	20 V
R_3	0.5 A	5 V
R_4	0.5 A	15 V

Circuit Restrictions

Every node in a circuit defined for SPICE must have a *dc path to ground* (node 0). A dc path to ground can be through a resistor, inductor, or voltage source but not through a capacitor or current source. Figure A.5(a) shows two examples of nodes that do not have dc paths to ground and that cannot, therefore, appear in a SPICE

Figure A.5
Examples of circuits that cannot be simulated in SPICE (for the reasons cited). To force a simulation, a very large resistance can be connected in parallel with a capacitor in (a), and a very small resistance can be connected in series with either V1 or L in (b).

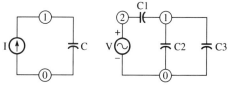

(a) Neither circuit has a dc path from node 1 to ground (node 0).

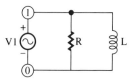

(b) Voltage source V1 and inductance L appear in a closed loop.

circuit. However, we can connect a very large resistance in parallel with a capacitor or current source to provide a dc path to ground. The resistance should be very large in comparison to other impedances in the circuit so that it will have a negligible effect on the computations. For example, if the capacitive reactances in the circuits of Figure A.5(a) are less than 1 MΩ, we can specify a 10^{12}-Ω resistor, RDUM (1 million megohms), between nodes 1 and 0:

```
RDUM 1 0 1E12
```

SPICE does not permit any loop (closed circuit path) to consist exclusively of inductance(s) and voltage source(s). Figure A.5(b) shows an example. Here, inductance L appears in a closed loop with voltage source V1. To circumvent this problem, we can insert a very small resistance in such a loop. The resistance should be much smaller than the impedances of other elements in the circuit in order to have a negligible effect on the computations. For example, if the impedances of R and L in Figure A.5(b) are greater than 1 Ω, we could insert a 1-pΩ resistor, RDUM, (10^{-12} Ω) in series with either L or V1.

A.4   THE .TRAN AND .PLOT CONTROL STATEMENTS

The .TRAN Control Statement

The .TRAN control statement (derived from *transient*) is used when we want to obtain values of voltages or currents versus *time* (whether they are transients in the traditional sense or not). We must specify the total time interval over which we wish to obtain the time-varying values and the increment of time between each using the format

```
.TRAN STEP TSTOP ⟨TSTART⟩
```

where *STEP* is the time increment and *TSTOP* is the largest value of time at which values will be computed. Unless we optionally specify the start time, *TSTART,* SPICE assumes it to be 0. If we do specify *TSTART,* computations still begin at $t = 0$, but only those in the interval from $t = TSTART$ through *TSTOP* are provided as output. To illustrate, the statement

```
.TRAN 5M 100M
```

will cause SPICE to produce values of the output(s) at the 21 time points 0, 5 ms, 10 ms, . . . , 100 ms. When TRAN is used as the analysis type in a .PRINT statement, 21 values of each output variable specified in the .PRINT statement will be printed. We will show an example of a .TRAN analysis (Example A.2) after discussing a few more statement types.

The .PLOT Control Statement

The .PLOT control statement can be used to obtain many different kinds of plots, depending on the analysis type specified. The format is

```
.PLOT type out1 ⟨out2 out3 ...⟩
```

where *type* is the analysis type and *out1, out2,* . . . are the outputs whose values are to be plotted. If the analysis type is TRAN, then the output variables are plotted versus time, with time increasing downward along the vertical axis. If more than one output is specified in the .PLOT statement, all will be plotted, using different symbols, on the same axes. Although it is possible to specify the scale desired, it is easier to let SPICE automatically determine the scale (using the minimum and maximum values it computes). When more than one output is plotted, SPICE automatically determines and displays all scales needed for all outputs. It also prints the time increments and the values of the points plotted. If more than one output is plotted, the values of the first output specified in the .PLOT statement are the only ones printed. Separate .PLOT statements can be used to obtain separate plots and value printouts if desired.

.DC Plots

When the analysis type is .DC, a .PLOT statement causes SPICE to plot the output(s) specified in that statement versus the values of a stepped source. The stepped source values are printed downward along the vertical axis. An example is shown in Figure 2.33. Here, the statement .PLOT DC I(VDUM) causes the computed values of I(VDUM) to be plotted versus the stepped values of V1: .DC V1 0.6 0.7 5MV. Note that the stepped values of V1 are printed down the left column, along with the computed values of I(VDUM).

When there are two stepped sources, the values of the stepped source appearing first in the .DC statement are printed downward along the vertical axis. These sets of values and the plots are repeated for each value of the stepped source appearing second in the .DC statement. Figure 3.51 is an example. Here, the combination

```
.DC VCE 0 50 5 IB 0 40U 10U
.PLOT DC I(VDUM)
```

causes SPICE to

1. plot values of I(VDUM) versus the 11 values of VCE when IB = 0;
2. plot values of I(VDUM) versus the 11 values of VCE when IB = 10 μA;

.
.
.

5. plot values of I(VDUM) versus the 11 values of VCE when IB = 40 μA.

LIMPTS and the .OPTIONS Control Statement

Normally, SPICE will not print or plot more than 201 values. We can override this limit by specifying a different limit on the number of points (LIMPTS) using the .OPTIONS control statement. The .OPTIONS statement can also be used to change many other operating characteristics and limits that are normally imposed by SPICE, most of them related to the mathematical techniques used in the computations. The majority of these will not concern us. The format for changing the limit on the number of points plotted or printed is

```
.OPTIONS LIMPTS = n
```

where n is the number of points.

A.5 THE SIN AND PULSE SOURCES

If we wish to obtain a printout or plot of an output versus time, as in a .TRAN analysis, at least one source in the circuit should itself be a time-varying voltage or current. In other words, we will not be able to observe an output *waveform* unless an input waveform is defined. It is not sufficient to indicate in a component definition that a particular source is AC instead of DC. A source designated AC is used by SPICE in an .AC analysis, to be discussed presently, and that analysis type does *not* cause SPICE to display time-varying outputs. For a .TRAN analysis, we use a different format to define the time-varying sources. The two sources that are most widely used for that purpose are the SIN (sinusoidal ac) and PULSE sources.

The SIN Source

The format for specifying a sinusoidal voltage source is

```
V******* N+ N- SIN(VO VP FREQ TD θ)
```

where ******* are arbitrary characters, $N+$ and $N-$ are the node numbers of the positive and negative terminals, *VO* is the *offset* (dc, or bias level), *VP* is the peak value in volts, and *FREQ* is the frequency in hertz. *TD* and θ are special parameters related to time delay and damping, both of which are set to 0 to obtain a conventional sine wave. A sinusoidal current source is defined by using I instead of V as the first character. Note that it is not possible to specify a phase shift (other than 180°, by reversing $N+$ and $N-$). Figure A.6 shows an example of a SIN source definition. The zero values for TD and θ can be omitted in the specification, and SPICE will assume they are zero by default.

In PSpice, a phase angle can be specified for a SIN source. The format (for a sinusoidal voltage) is

```
V******* N+ N- SIN(VO VP FREQ TD θ PH)
```

where *PH* is phase angle in degrees and all other parameters are the same as in the SIN source description in Berkeley SPICE.

The PULSE Source

The PULSE source can be used to simulate a dc source that is switched into a circuit at a particular instant of time (a *step* input) or to generate a sequence of square, trapezoidal, or triangular pulses. Figure A.7(a) shows the parameters used to define a voltage pulse or pulse-type waveform: one having time delay (*TD*) that elapses before the voltage begins to rise linearly with time, a *rise time* (*TR*) that represents the time required for the voltage to change from *V1* volts to *V2* volts, a

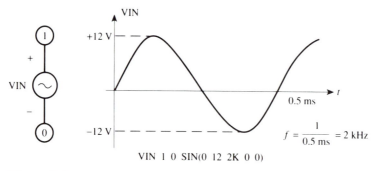

VIN 1 0 SIN(0 12 2K 0 0)

Figure A.6
An example of the specification of a sinusoidal voltage source.

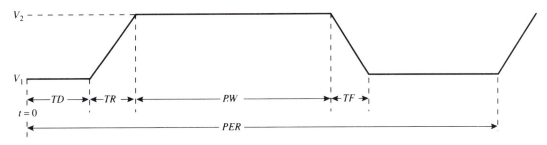

$V \ast\ast\ast\ast\ast\ast\ast N+ \ N- \ \text{PULSE}(V_1 \ V_2 < TD \ TR \ TF \ PW \ PER >)$
Default values: $TD = 0, TR = STEP, TF = STEP, PW = TSTOP,$
$\qquad\qquad PER = TSTOP.$

(a)

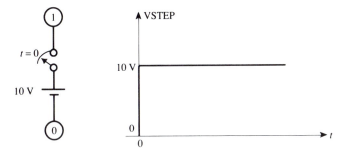

VSTEP 1 0 PULSE(0 10 0 0 0 0)

(b) Using the PULSE source to simulate a 10-V dc source
switched into a circuit at $t = 0$. Even though TR and TF
are set to 0, SPICE assigns each the default value of
$STEP$ specified in a .TRAN statement. PW and PER both
default to the value of $TSTOP$.

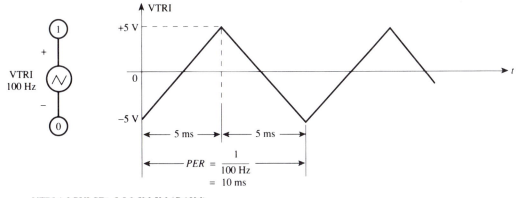

VTRI 1 0 PULSE(−5 5 0 5M 5M 1P 10M)

(c) Using the PULSE source to generate a 100-Hz
triangular waveform. Note that $PW = 1$ ps ≈ 0.

Figure A.7
The PULSE source

pulse width (*PW*), and a *fall time* (*TF*), during which the voltage falls from *V2* volts to *V1* volts. If the pulse is repetitive, a value for the period of the waveform (*PER*) is also specified. If *PER* is not specified, its default value is the value of *TSTOP* in a .TRAN analysis; that is, the pulse is assumed to remain at *V2* volts for the duration of the analysis, simulating a step input. The default value for the rise and fall times is the *STEP* time specified in a .TRAN analysis. The figure shows the format for identifying a pulsed voltage source. Current pulses can be obtained by using I instead of V as the first character. Figure A.7(b) shows an example of how the PULSE source is used to simulate a step input caused by switching a 10-V-dc source into a circuit at $t = 0$. Figure A.7(c) shows an example of how the PULSE source is used to define a triangular waveform that alternates between ±5 V with a frequency of 100 Hz. SPICE does not accept a zero pulse width, as would be necessary to define an ideal triangular waveform, but *PW* can be made negligibly small. In this example, we set $PW = 10^{-12}$ s $= 1$ ps, which makes the period 10^{10} times as great as the pulse width.

A.6   THE INITIAL TRANSIENT SOLUTION

The next example demonstrates the use of the PULSE source to generate a square wave and contains some important discussion on how SPICE performs a dc analysis in conjunction with every transient analysis.

Example A.2

Use SPICE to obtain a plot of the capacitor voltage versus time in Figure A.8(a). The plot should cover two full periods of the square-wave input.

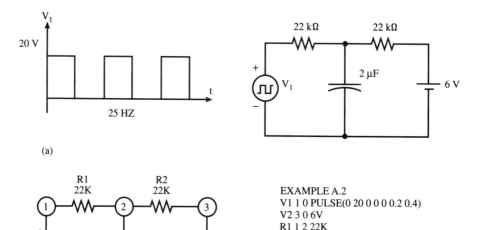

(a)

(b)

EXAMPLE A.2
V1 1 0 PULSE(0 20 0 0 0 0.2 0.4)
V2 3 0 6V
R1 1 2 22K
R2 2 3 22K
C 2 0 2UF
.TRAN 0.02 0.8
.PLOT TRAN V(2)
.END

Figure A.8
(Example A.2)

```
EXAMPLE A.2
****      INITIAL TRANSIENT SOLUTION        TEMPERATURE =   27.000 DEG C
*********************************************************************************
  NODE     VOLTAGE      NODE    VOLTAGE      NODE    VOLTAGE
(   1)      .0000     (   2)    3.0000     (   3)    6.0000
     VOLTAGE SOURCE CURRENTS
     NAME        CURRENT
     V1        1.364D-04
     V2       -1.364D-04
     TOTAL POWER DISSIPATION    8.18D-04   WATTS


EXAMPLE A.2
****      TRANSIENT ANALYSIS             TEMPERATURE =   27.000 DEG C
*********************************************************************************
     TIME       V(2)
                     0.000D+00     5.000D+00     1.000D+01     1.500D+01   2.000D+01
                 - - - - - - - - - - - - - - - - - - - - - - - - - - - - - - - - - -
 0.000D+00   3.000D+00 .        *         .             .             .           .
 2.000D-02   6.394D+00 .             .    *    .             .             .           .
 4.000D-02   1.023D+01 .             .         .      .*             .           .
 6.000D-02   1.191D+01 .             .         .             *             .           .
 8.000D-02   1.259D+01 .             .         .             .   *         .           .
 1.000D-01   1.285D+01 .             .         .             .    *        .           .
 1.200D-01   1.294D+01 .             .         .             .    *        .           .
 1.400D-01   1.298D+01 .             .         .             .    *        .           .
 1.600D-01   1.299D+01 .             .         .             .    *        .           .
 1.800D-01   1.300D+01 .             .         .             .    *        .           .
 2.000D-01   1.300D+01 .             .         .             .    *        .           .
 2.200D-01   1.300D+01 .             .         .             .    *        .           .
 2.400D-01   9.591D+00 .             .         .         *.             .           .
 2.600D-01   5.658D+00 .             .    *    .             .             .           .
 2.800D-01   4.071D+00 .         *    .         .             .             .           .
 3.000D-01   3.420D+00 .        *     .         .             .             .           .
 3.200D-01   3.159D+00 .       *      .         .             .             .           .
 3.400D-01   3.059D+00 .       *      .         .             .             .           .
 3.600D-01   3.024D+00 .      *       .         .             .             .           .
 3.800D-01   3.009D+00 .      *       .         .             .             .           .
 4.000D-01   3.003D+00 .      *       .         .             .             .           .
 4.200D-01   6.410D+00 .             .    *    .             .             .           .
 4.400D-01   1.034D+01 .             .         .     .*             .           .
 4.600D-01   1.193D+01 .             .         .             *             .           .
 4.800D-01   1.258D+01 .             .         .             .  *          .           .
 5.000D-01   1.284D+01 .             .         .             .   *         .           .
 5.200D-01   1.294D+01 .             .         .             .    *        .           .
 5.400D-01   1.298D+01 .             .         .             .    *        .           .
 5.600D-01   1.299D+01 .             .         .             .    *        .           .
 5.800D-01   1.300D+01 .             .         .             .    *        .           .
 6.000D-01   1.300D+01 .             .         .             .    *        .           .
 6.200D-01   1.300D+01 .             .         .             .    *        .           .
 6.400D-01   9.591D+00 .             .         .         *.             .           .
 6.600D-01   5.658D+00 .             .    *    .             .             .           .
 6.800D-01   4.071D+00 .         *    .         .             .             .           .
 7.000D-01   3.420D+00 .        *     .         .             .             .           .
 7.200D-01   3.159D+00 .       *      .         .             .             .           .
 7.400D-01   3.059D+00 .       *      .         .             .             .           .
 7.600D-01   3.024D+00 .      *       .         .             .             .           .
 7.800D-01   3.009D+00 .      *       .         .             .             .           .
 8.000D-01   3.003D+00 .      *       .         .             .             .           .
                 - - - - - - - - - - - - - - - - - - - - - - - - - - - - - - - - - -
```

Figure A.9
(Example A.2)

Solution.   Figure A.8(b) shows the circuit when redrawn for analysis by SPICE and the corresponding input data file. The period of the input is T = 1/(2.5 Hz) = 0.4 s, so *TSTOP* in the .TRAN statement is set to 0.8 s to obtain a plot covering two full periods. Note that *PW* in the definition of V1 is set to 0.2 s. The square wave is idealized by setting the rise and fall times, *TR* and *TF*, to 0, so the actual value assigned by SPICE to *TR* and *TF* is the value of *STEP*: 0.02 s.

Figure A.9 shows the results of a program run. When SPICE performs a .TRAN analysis, it first obtains an "initial transient solution." This solution is obtained from a dc analysis with all time-varying sources set to zero. Thus, the initial solution represents the *quiescent,* or dc operating conditions in the circuit, useful information for determining the bias point(s) in circuits containing transistors. The actual time-varying outputs are computed with the initial voltages and currents as the starting points, so the outputs do not reflect initial transients associated with the charging of capacitors, such as coupling capacitors, in the circuit.

In our example, the initial transient solution gives the dc voltages and currents when V1, the square-wave generator, is set to 0. The figure shows that the dc voltages at all nodes are printed, as are the dc currents in all voltage sources and the total dc power dissipation in the circuit. (Note that dc current flows *into* V1 when it is set to 0 V.) Since the capacitor is charged to +3 VDC, the time-varying plot shows its voltage to begin at +3 V. The actual transient that would occur (beginning at $t = 0$) while the capacitor charged to 3 V does not appear in the output.

A.7   DIODE MODELS

When there is a semiconductor device in a circuit to be analyzed by SPICE, the SPICE input file must contain two new types of statements: one that identifies the device by name and gives its node numbers in the circuit and another, called a .MODEL statement, that specifies the values of the device *parameters* (saturation current and the like).

All diode names must begin with D. The format for identifying a diode in a circuit is

```
D******* NA NC Mname
```

where *NA* and *NC* are the numbers of the nodes to which the anode and cathode are connected, respectively, and *Mname* is the *model name*. The model name associates the diode with a particular .MODEL statement that specifies the parameter values of the diode:

```
.MODEL Mname D (Pval1 = n1 Pval2 = n2 ...)
```

where D, signifying diode, *must* appear as shown, and *Pval1 = n1, . . .* specify parameter values, to be described shortly. Note that several diodes, having different names, can all be associated with the same .MODEL statement, and other diodes can be associated with a different .MODEL statement. Figure A.10 shows an example: a diode bridge in which diodes D1 and D3 are modeled by MODA and diodes D2 and D4 are modeled by MODB. MODA specifies a diode having saturation current (IS) 0.5 pA and MODB specifies a diode having saturation current 0.1 pA.

Table A.1 lists the diode parameters whose values can be specified in a .MODEL statement and the default values of each. The default values are typical,

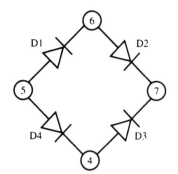

```
D1 5 6 MODA
D2 6 7 MODB
D3 4 7 MODA
D4 5 4 MODB
.MODEL MODA D IS = 0.5P
.MODEL MODB D IS = 0.1P
```

Figure A.10
The parameter values of diodes D1 and D3 are given in the model whose name is MODA, and the parameter values of D2 and D4 are given in the model whose name is MODB.

or average, values and are acceptable for most electronic circuit analysis at our level of study. In some examples in the book, we specify parameter values different from the default values to illustrate certain points, but average diode behavior is adequate for most of our purposes and can be realized without knowledge of specific values. In practice, some of these diode parameters are very difficult to obtain or measure and are necessary only when a highly accurate model is essential. For example, if we were designing a new diode to have specific low-noise characteristics, we would want to know the noise parameters, KF and AF, very accurately. Examples 2.7 and 2.8 illustrate the .DC and .TRAN analysis of circuits containing a diode.

Table A.1
Diode parameters

Parameter	Identification	Units	Default Value
Saturation current	IS	A	1×10^{-14}
Ohmic resistance	RS	Ω	0
Emission coefficient	N	—	1
Transit time	TT	s	0
Zero-bias junction capacitance	CJO	F	0
Junction potential	VJ	V	1
Grading coefficient	M	—	0.5
Activation energy	EG	eV	1.11
Saturation current temperature exponent	XTI	—	3
Flicker-noise coefficient	KF	—	0
Flicker-noise exponent	AF	—	0
Coefficient for forward-bias depletion capacitance equation	FC	—	0.5
Reverse breakdown voltage	BV	V	Infinite
Current at breakdown voltage	IBV	A	1×10^{-3}

PSPICE

In PSpice, the current through and/or the voltage across a diode can be printed or plotted by specifying output variables I(D*******) and V(D*******), where D******* is the diode name. For example, the following statements request the value of the dc current through diode D1 and a plot of the ac voltage across diode D2:

```
.PRINT DC I(D1)
.PLOT AC V(D2)
```

The reference polarity in each case is from the anode node to the cathode node.

A.8   BJT MODELS

The format for identifying a bipolar junction transistor in a circuit is

```
Q******* NC NB NE Mname
```

where *NC*, *NB*, and *NE* are the numbers of the nodes to which the collector, base, and emitter are connected and *Mname* is the name of the model that specifies the transistor parameters. The format of the .MODEL statement for a BJT is

```
.MODEL Mname type ⟨Pval1 = n1 Pval2 = n2 •••⟩
```

where *type* is either NPN or PNP. Figure A.11 shows an example of how an NPN transistor and a PNP transistor in a circuit are identified and modeled. The NPN model specifies that the "ideal maximum forward beta" (BF) of the transistor is 150, and the PNP model allows that parameter to have its default value of 100.

As shown in Table A.2, we can specify up to 40 different parameter values for a BJT. The mathematical model used by SPICE to simulate a BJT is very complex but very accurate, provided that all parameter values are accurately known. However, to an even greater extent than in the diode model, many of the parameters are very difficult to measure, estimate, or deduce from other characteristics. Once again, for routine circuit analysis, it is generally sufficient to allow most of the parameters to have their default values. This approach is further justified by the fact that there is usually a wide variation in parameter values among transistors of the same type. Therefore, it is unrealistic and unnecessary in many practical applications to seek a highly accurate analysis of a circuit containing transistors. On the other hand, there are situations, such as the design and development of a

Figure A.11
An example of how an NPN transistor and a PNP transistor are identified and modeled

```
Q1 4 3 2 TYPE1
Q2 0 1 2 TYPE2
.MODEL TYPE1 NPN BF=150
.MODEL TYPE2 PNP
```

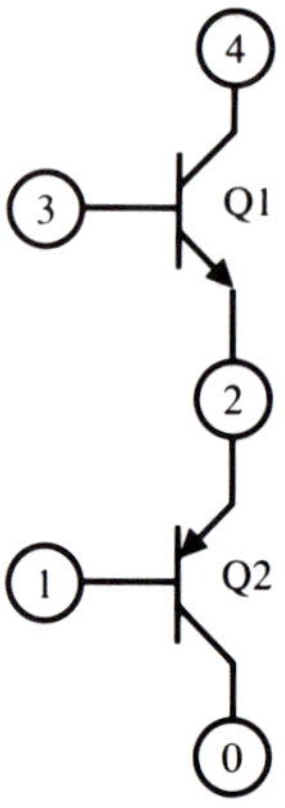

Table A.2
BJT parameters

Parameter	Identification	Units	Default Value
Transport saturation current	IS	A	10^{-16}
Ideal maximum forward beta	BF	—	100
Forward current emission coefficient	NF	—	1
Forward Early voltage	VAF	V	Infinity
Corner for forward beta high-current roll-off	IKF	A	Infinity
Base-emitter leakage saturation current	ISE	A	0
Base-emitter leakage emission coefficient	NE	—	1.5
Ideal maximum reverse beta	BR	—	1
Reverse current emission coefficient	NR	—	1
Reverse Early voltage	VAR	V	Infinity
Corner for reverse beta high-current roll-off	IKR	A	Infinity
Base-collector–leakage saturation current	ISC	A	0
Base-collector–leakage emission coefficient	NC	—	2
Zero-bias base resistance	RB	Ω	0
Current where base resistance falls to half of its minimum value	IRB	A	Infinity
Minimum base resistance at high currents	RBM	Ω	RB
Emitter resistance	RE	Ω	0
Collector resistance	RC	Ω	0
Base-emitter zero-bias depletion capacitance	CJE	F	0
Base-emitter built-in potential	VJE	V	0.75
Base-emitter–junction exponential factor	MJE	—	0.33
Ideal forward transit time	TF	s	0
Coefficient for bias dependence of TF	XTF	—	0
Voltage describing V_{BC}-dependence of TF	VTF	V	Infinity
High-current parameter for effect on TF	ITF	A	0
Excess phase at $f = 1/(2\pi\,TF)$ Hz	PTF	Degrees	0
Base-collector zero-bias depletion capacitance	CJC	F	0
Base-collector built-in potential	VJC	V	0.75
Base-collector–junction exponential factor	MJC	—	0.33
Fraction of base-collector depletion capacitance to internal base node	XCJC	—	1
Ideal reverse transit time	TR	s	0
Zero-bias collector-substrate capacitance	CJS	F	0
Substrate-junction built-in potential	VJS	V	0.75
Substrate-junction exponential factor	MJS	—	0
Forward and reverse beta temperature exponent	XTB	—	0
Energy gap for temperature effect on IS	EG	eV	1.11
Temperature exponent for effect on IS	XTI	—	3
Flicker-noise coefficient	KF	—	0
Flicker-noise exponent	AF	—	1
Coefficient for forward-bias depletion capacitance formula	FC	—	0.5

new integrated circuit required to have certain properties, where an accurate determination of parameter values is warranted.

PSPICE

PSpice has a *library* option that allows access to model statements for many commonly used diodes and transistors. These model statements specify values for the many device parameters that we could not ordinarily determine from a manufacturer's specifications. The PSpice library is discussed in Section A.17.

In PSpice, the voltage across any pair of BJT terminals and/or the current into any BJT terminal can be printed or plotted using the letters B, E, and C to represent the base, emitter, and collector terminals. Following are some examples of output variables that could be specified with a .PRINT or .PLOT statement:

`VBE(Q1)`	The base-to-emitter voltage of transistor Q1
`IC(Q2)`	The collector current of transistor Q2
`VC(QA)`	The collector-to-ground voltage of transistor QA

A.9   THE .TEMP STATEMENT

As noted earlier, all SPICE computations are performed under the assumption that the temperature of the circuit is 27°C, unless a different temperature is specified. The .TEMP statement is used to request an analysis at one or more different temperatures:

`.TEMP T1 ⟨T2 T3 ...⟩`

where *T1* is the temperature in degrees Celsius at which an analysis is desired. *T2*, *T3*, . . . are optional additional temperatures at which SPICE will repeat the analysis, once for each temperature. Temperature is a particularly important parameter in circuits containing semiconductor devices, since their characteristics are temperature-sensitive. However, SPICE will not adjust all device characteristics for temperature unless the parameters in the .MODEL statement that relate to temperature sensitivity are given specific values. A case in point is β in a BJT. The parameter BF is called the "ideal maximum forward beta." The actual value of β used in the computations depends on other factors, including the forward Early voltage (VAF) and the temperature. However, no temperature variation in the value of BF will occur unless the parameter XTB (forward and reverse beta temperature exponent) is specified to have a value other than 0 (its default value). When SPICE performs an analysis at a temperature other than 27°C, it prints a list of temperature-adjusted values of device parameters that are temperature-sensitive.

The values of resistors in a circuit are not adjusted for temperature unless either first- or second-order temperature coefficients of resistance, tc_1 and tc_2 (or both) are given values in the statements defining resistors. The format is

`R******** N1 N2 value TC = tc`$_1$`, tc`$_2$

The temperature-adjusted value is then computed by

$$R_T = R_{27}[1 + tc_1(T - 27°) + 1 + tc_2(T - 27°)]$$

where R_T is the resistance at temperature T and R_{27} is the resistance at temperature 27°C. The default values of tc_1 and tc_2 are 0.

A.10   AC SOURCES AND THE .AC CONTROL STATEMENT

The format for identifying an ac voltage source in a circuit is

```
V******* N+ N- AC (mag) (phase)
```

where ******* is an arbitrary sequence of characters, $N+$ and $N-$ are the numbers of the nodes to which the positive and negative terminals are connected, *mag* is the magnitude of the voltage in volts, and *phase* is its phase angle in degrees. All ac sources are assumed to be sinusoidal. If *mag* is omitted, its default value is 1 V, and if *phase* is omitted, its default value is 0°. An ac current source is identified by making the first character I instead of V. Note that *mag* may be regarded as either a peak or an rms value, since SPICE output from an ac analysis does not consist of instantaneous (time-varying) values.

The .AC Control Statement

The .AC control statement is used to compute ac voltages and currents in a circuit *versus frequency*. A single frequency or a range of frequencies can be specified. The circuit must contain at least one source designated AC, and all AC sources are assumed to be sinusoidal and to have identical frequencies or to undergo the same frequency variation, if any. The format is

```
.AC vartype N fstart fstop
```

where *vartype* specifies the way frequency is to be varied in the range from *fstart* through *fstop*. N is a number related to the number of frequencies at which computations are to be performed, as will be discussed shortly. The *vartype* is one of DEC, OCT, or LIN (decade, octave, or linear). If analysis at a single frequency is desired, we can use any *vartype*, set *fstart* equal to *fstop*, and let $N = 1$.

When *vartype* is LIN, the frequencies at which analysis is performed vary linearly from *fstart* through *fstop*. In that case, N is the total number of frequencies at which the analysis is performed (counting *fstart* and *fstop*). Thus, the interval between frequencies is

$$\Delta f = \frac{fstop - fstart}{N - 1}$$

For example, the statement

```
.AC LIN 21 100 1K
```

will tell SPICE to perform an ac analysis at 21 frequencies from 100 Hz through 1 kHz. The frequency interval will be $(1000 - 100)/20 = 45$ Hz, so computations will be performed at 100 Hz, 145 Hz, 190 Hz, . . . , 1 kHz.

When the *vartype* is DEC or OCT, the analysis is performed at logarithmically spaced intervals and N is the total number of frequencies *per decade or per octave*. For example, the statement

```
.AC DEC 10 100 10K
```

will cause SPICE to analyze the circuit at ten frequencies in each of the decades 100 Hz to 1 kHz and 1 kHz through 10 kHz. The frequencies will be at one-tenth–decade intervals, so each interval will be different. The frequencies at one-tenth–

decade intervals are 10^x, $10^{x+0.1}$, $10^{x+0.2}$, . . . , where $x = \log_{10}(fstart)$. In general, the frequencies at which SPICE performs an ac analysis using the DEC *vartype* are 10^x, $10^{x+1/N}$, $10^{x+2/N}$, . . . , where $x = \log_{10}(fstart)$. The last frequency in this sequence is not necessarily *fstop*, but SPICE will compute at frequencies in the sequence up through the first frequency that are equal to or greater than *fstop*. The frequencies at which SPICE performs an ac analysis when the *vartype* is OCT are 2^x, $2^{x+1/N}$, $2^{x+2/N}$, . . . , where $x = \log_2(fstart)$.

AC Outputs

AC voltages and currents whose values are desired from an .AC analysis are specified the same way as dc voltages and currents, using V(*N1, N2*), V(*N1*), or I(*Vname*) in a .PRINT or .PLOT statement. The values of the magnitudes of these quantities are printed or plotted. In addition, we can request certain other values, as indicated in the following list of voltage characteristics:

VR   real part
VI   imaginary part
VM   magnitude, $|V|$
VP   phase, degrees
VDB $20 \log_{10}|V|$

The same values for ac currents can be obtained by substituting I for V. To illustrate, the statement

```
.PRINT AC V(1) VLP(2,3) II(VDUM) IR(VX)
```

causes SPICE to print the magnitude of the ac voltage between nodes 1 and 0, the phase angle of the ac voltage VL between nodes 2 and 3, the imaginary part of the current in VDUM, and the real part of the current in VX. Note that ac must be listed as the analysis type.

AC Plots

When ac voltages or currents are specified in a .PLOT statement, we obtain a linear, semilog, or log-log plot, depending on the type of voltage or current output requested and the *vartype*. Table A.3 summarizes the types of plots produced for

Table A.3

Vartype	Output Requested	Frequency (Vertical) Axis	Output (Horizontal) Axis
LIN	Magnitude	Linear	Log
	Phase	Linear	Linear
	Imaginary part	Linear	Linear
	Real part	Linear	Linear
	Decibels	Linear	Linear
DEC or OCT	Magnitude	Log	Log
	Phase	Log	Linear
	Imaginary part	Log	Linear
	Real part	Log	Linear
	Decibels	Log	Linear

each combination. ''Log'' in the table means that the scale supplied by SPICE has logarithmically spaced values.

Small-Signal Analysis and Distortion

When a circuit contains active devices such as transistors, an .AC analysis by SPICE is assumed to be a small-signal analysis. That is, ac variations are assumed to be small enough that the values of device parameters do not change. As in a .TRAN analysis, SPICE performs an initial dc analysis to determine quiescent voltages and currents so that the values of those device parameters affected by dc levels can be computed. In the ac analysis, the device parameters are assumed to retain those initial values, regardless of the actual magnitudes of the ac variations. *Thus, in an .AC analysis, SPICE does not take into account any distortion, even clipping, that would actually occur if we were to severely overdrive a transistor by specifying very large ac inputs.* For example, if the output swing of an actual transistor circuit were limited to 10 V, this fact would not be ''known'' to SPICE, and by overdriving the computer-simulated circuit, we could obtain outputs of several hundred volts from an .AC analysis.

To observe the effects of distortion, such as clipping, it is necessary to perform a .TRAN analysis and obtain a plot of the output waveform versus time. There is also a .DISTO (distortion) control statement that can be used to obtain limited information on harmonic distortion, but we will not have occasion to use that statement.

Example 5.11 illustrates an .AC analysis at a single frequency. The next example illustrates an .AC analysis over a range of frequencies.

Example A.3

Use SPICE to perform an ac analysis of the transistor amplifier in Figure A.12(a) over the frequency range from 100 Hz through 100 kHz. Obtain a plot of the

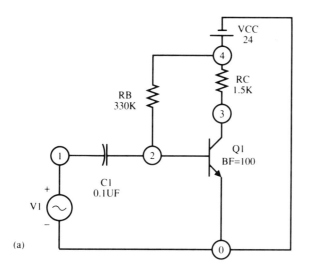

```
EXAMPLE A.3
VCC 4 0 24
V1 1 0 AC
RC 4 3 1.5K
RB 4 2 330K
C1 1 2 0.1UF
Q1 3 2 0 TRANS
.MODEL TRANS NPN BF=100
.AC DEC 10 100 100K
.PLOT AC V(3) VP(3)
.END
```

Figure A.12
(Example A.3)

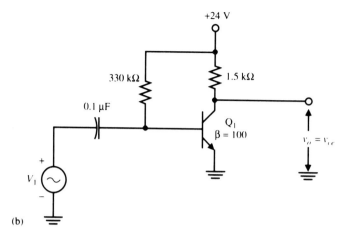

Figure A.12
(Example A.3)

magnitude and phase angle of the collector-to-emitter voltage over the frequency range, with 10 frequencies per decade.

Solution.   Figure A.12(b) shows the circuit redrawn for analysis by SPICE and the corresponding input data file. Note that we allow the magnitude and phase of the ac source (V1) to have the default values 1 V and 0°, respectively.

Figure A.13 shows the results of a program run. Note the following points in connection with an .AC analysis of a circuit containing a transistor:

1. SPICE prints a list of the values of the transistor model parameters, which in this case are default values. (Since the default value of BF is 100, we could have omitted that specification in the .MODEL statement.)
2. SPICE obtains a "small-signal bias solution" to determine the dc voltages and

```
EXAMPLE A.3
****       BJT MODEL PARAMETERS                    TEMPERATURE =    27.000 DEG C
****************************************************************************************
            TRANS
TYPE        NPN
IS          1.00D-16
BF          100.000
NF          1.000
BR          1.000
NR          1.000
******02/22/89 ********   SPICE 2G.6    3/15/83 ********13:15:17*****
EXAMPLE A.3
****       SMALL SIGNAL BIAS SOLUTION              TEMPERATURE =    27.000 DEG C
****************************************************************************************
  NODE    VOLTAGE      NODE    VOLTAGE      NODE    VOLTAGE      NODE    VOLTAGE
(   1)      .0000    (   2)     .8246    (   3)    13.4657    (   4)    24.0000
    VOLTAGE SOURCE CURRENTS
    NAME         CURRENT
    VCC        -7.093D-03
    V1          0.000D+00
    TOTAL POWER DISSIPATION    1.70D-01   WATTS
*
```

Figure A.13
(Example A.3)

```
EXAMPLE A.3
****        OPERATING POINT INFORMATION       TEMPERATURE =   27.000 DEG C
************************************************************************************
**** BIPOLAR JUNCTION TRANSISTORS
             Q1
MODEL        TRANS
IB           7.02E-05
IC           7.02E-03
VBE          .825
VBC          -12.641
VCE          13.466
BETADC       100.000
GM           2.72E-01
RPI          3.68E+02
RX           0.00E+00
RO           1.00E+12
CPI          0.00E+00
CMU          0.00E+00
CBX          0.00E+00
CCS          0.00E+00
BETAAC       100.000
FT           4.32E+18
*


EXAMPLE A.3
****        AC ANALYSIS                        TEMPERATURE =   27.000 DEG C
************************************************************************************
LEGEND:
*:  V(3)
+:  VP(3)
     FREQ        V(3)
*)------------- 1.000D+00    1.000D+01    1.000D+02    1.000D+03 1.000D+04
                - - - - - - - - - - - - - - - - - - - - - - - -
+)------------- -2.000D+02   -1.500D+02   -1.000D+02   -5.000D+01 0.000D+00
                - - - - - - - - - - - - - - - - - - - - - - - -

1.000D+02  9.412D+00 .              *            . +                .           .
1.259D+02  1.185D+01 .            .*             . +                .           .
1.585D+02  1.491D+01 .            .  *           . +                .           .
1.995D+02  1.876D+01 .                 *         . +                .           .
2.512D+02  2.361D+01 .                  *        . +                .           .
3.162D+02  2.969D+01 .                    *      . +                .           .
3.981D+02  3.732D+01 .                     *     .+                 .           .
5.012D+02  4.687D+01 .                      *    .+                 .           .
6.310D+02  5.878D+01 .                     *  +                     .           .
7.943D+02  7.355D+01 .                     *  +                     .           .
1.000D+03  9.172D+01 .                        X.                    .           .
1.259D+03  1.138D+02 .                      + .*                    .           .
1.585D+03  1.401D+02 .                    +   . *                   .           .
1.995D+03  1.706D+02 .                  +     .   *                 .           .
2.512D+03  2.045D+02 .                +       .      *              .           .
3.162D+03  2.403D+02 .              +         .         *           .           .
3.981D+03  2.758D+02 .           +            .            *        .           .
5.012D+03  3.083D+02 .        +               .               *     .           .
6.310D+03  3.359D+02 .      .+                .                 *   .           .
7.943D+03  3.577D+02 .     +                  .                  *  .           .
1.000D+04  3.738D+02 .         +  .           .                  * .           .
1.259D+04  3.852D+02 .        +   .           .                  * .           .
1.585D+04  3.929D+02 .       +    .           .                  * .           .
1.995D+04  3.980D+02 .      +     .           .                  * .           .
2.512D+04  4.014D+02 .    +       .           .                  * .           .
3.162D+04  4.035D+02 .    +       .           .                  * .           .
3.981D+04  4.049D+02 .   +        .           .                 *  .           .
5.012D+04  4.058D+02 .   +        .           .                 *  .           .
6.310D+04  4.063D+02 .    +       .           .                 *  .           .
7.943D+04  4.067D+02 .  +         .           .                 *  .           .
1.000D+05  4.069D+02 .   +        .           .                 *  .           .
                      - - - - - - - - - - - - - - - - - - - - - - - -
```

Figure A–13
(Continued)

currents in the circuit with the ac source set to 0. This is similar to the "initial transient solution" obtained in a .TRAN analysis.

3. Using the dc values obtained from the small-signal bias solution, SPICE computes "operating point information." This information is provided in the form of a list of important bias-dependent parameter values.

4. The two outputs are plotted using * to represent V(3), the magnitude of V_{CE}, on a log scale and + to represent VP(3), the phase angle of V_{CE} on a linear scale. The values of V(3) are printed along the vertical scale, along with the logarithmically spaced frequencies. Thus, the plot is a log-log plot of voltage magnitude and a semilog plot of phase angle.

5. The plots intersect at $f = 1$ kHz, and SPICE prints an X where that occurs.

6. The maximum output voltage (at 100 kHz) is 406.9 V, which is clearly impossible in the actual circuit. As noted earlier, SPICE does not take output voltage limits (clipping) into account during an .AC analysis. Since the input signal magnitude is 1 V, the output magnitude also represents the voltage gain at each frequency.

7. At high frequencies, the phase angle approaches $-180°$, confirming the phase inversion caused by a common-emitter amplifier.

A.11   JUNCTION FIELD-EFFECT TRANSISTORS (JFETs)

An N-channel or P-channel JFET in a circuit to be analyzed by SPICE must be given a name that begins with J, such as JX or JFET1. The format for specifying both P-channel and N-channel devices is

`J******* ND NG NS Mname`

where *ND* is the node number of the drain, *NG* the node number of the gate, *NS* the node number of the source, and *Mname* is the model name. The model name appears in the MODEL statement for a JFET using the format

`.MODEL Mname type ⟨Pval1 = n1 Pval2 = n2 ···⟩`

where *type* is NJF for an N-channel JFET and PJF for a P-channel JFET. Table A.4 shows the JFET parameters that can be specified in a .MODEL statement and their default values.

The two principal parameters related to the dc characteristics of a JFET are VTO (the pinch-off voltage, V_p) and BETA (the transconductance parameter, β). In the pinch-off region, these are related by

$$\beta = \frac{I_D}{(V_{GS} - V_p)^2} \quad \text{A/V}^2$$

Since $I_D = I_{DSS}$ when $V_{GS} = 0$, we have

$$\beta = \frac{I_{DSS}}{V_p^2}$$

When modeling a JFET for which values of I_{DSS} and V_p are known, we must calculate β using this equation, so that the parameter BETA can be specified in the .MODEL statement. *Note this important point: The value entered for the parameter VTO is always negative, whether the JFET is an N-channel or a P-channel device.*

Table A.4
JFET parameters

Parameter	Identification	Units	Default Value
Threshold voltage	VTO	V	-2
Transconductance parameter	BETA	A/V^2	10^{-4}
Channel-length-modulation parameter	LAMBDA	$1/V$	0
Drain ohmic resistance	RD	Ω	0
Source ohmic resistance	RS	Ω	0
Zero-bias gate-to-source–junction capacitance	CGS	F	0
Zero-bias gate-to-drain–junction capacitance	CGD	F	0
Gate-junction saturation current	IS	A	10^{-14}
Gate-junction potential	PB	V	1
Flicker-noise coefficient	KF	—	0
Flicker-noise exponent	AF	—	1
Coefficient for forward-bias depletion capacitance formula	FC	—	0.5

The parameter LAMBDA (λ, the channel-length-modulation factor) controls the extent to which the characteristic curves rise for increasing values of V_{DS} in the pinch-off region. If λ is 0 (the default value), the characteristic curves are perfectly flat. A typical value for λ is 10^{-4}. Example 3.13 illustrates a SPICE simulation of a JFET.

A.12   MOS FIELD-EFFECT TRANSISTORS (MOSFETs)

An N-channel or P-channel MOSFET must be identified in a SPICE data file by a name beginning with the letter M, using the format:

```
M******* ND NG NS NSS Mname
```

where *ND, NG, NS,* and *NSS* are the node numbers of the drain, gate, source, and substrate, respectively, and *Mname* is the model name. The values of certain geometric parameters, such as the length and width of the channel, can be optionally specified with the MOSFET identification, but these will not concern us and we can allow all of them to default.

The MOSFET model is very complex. Like other semiconductor models in SPICE, it involves many parameters whose values are difficult to determine and many that are beyond the scope of our treatment. There are actually three built-in models, referred to as *levels* 1, 2, and 3. LEVEL is one of the parameters that can be specified in a MOSFET .MODEL statement, and its value prescribes the particular model to be used. The default level is 1, which we can assume for all purposes in this book. The format of the MOSFET .MODEL statement is

```
.MODEL Mname type ⟨Pval1 = n1 Pval2 = n2 ···⟩
```

where *type* is NMOS for an N-channel MOSFET and PMOS for a P-channel MOSFET. A MOSFET can be of either the depletion-mode type or of the enhancement-mode type, as is discussed shortly.

Since the MOSFET parameters in the three-level SPICE model are so numerous and complex, we will not present a table showing their identifications, units, and default values. (Such information should be available in a user's guide furnished with the SPICE software used.) In any case, there is a significant variation among the several versions of SPICE in the number and type of MOSFET parameters that can be specified. However, there are two fundamental parameters whose values should probably be specified in every MOSFET simulation: β and V_T. In SPICE, β is called the *intrinsic transconductance parameter* and is identified in a .MODEL statement by KP. Its value is always positive. V_T is called the zero-bias threshold voltage and is identified by VTO. Following is an example of a .MODEL statement for an N-channel MOSFET having $\beta = 0.5 \times 10^{-3}$ A/V^2 and $V_T = 2$ V:

```
.MODEL M1 NMOS KP = 0.5E-3 VTO = 2
```

The threshold voltage, VTO, is positive or negative, according to the following table:

Mode	Channel	Sign of VTO
Depletion	N	−
Depletion	P	+
Enhancement	N	+
Enhancement	P	−

For example, the value specified for the VTO of an N-channel, enhancement-mode MOSFET should be positive.

Example 5.23 illustrates a SPICE simulation of a circuit containing a MOSFET.

PSPICE

In PSpice, the voltage across any pair of JFET or MOSFET terminals and/or the current into any JFET or MOSFET terminal can be printed or plotted using the letters D, G, and S to represent the drain, gate, and source terminals. The following are some examples of output variables that could be specified with a .PRINT or .PLOT statement:

VGS(J1)   gate-to-source voltage of the JFET named J1
ID(MXY)   drain current of the MOSFET named MXY
VD(M25)   drain-to-ground voltage of the MOSFET named M25

A.13   CONTROLLED (DEPENDENT) SOURCES

A controlled voltage source is one whose output voltage is controlled by (depends on) the value of a voltage or current elsewhere in the circuit. The simplest and most familiar example is a voltage amplifier: It is a voltage-controlled voltage source because its output voltage *depends* on its input voltage. The output voltage equals the input voltage multiplied by the gain. A current-controlled voltage source obeys the relation $v_o = ki$, where i is the controlling current and k is a constant having the units of resistance: $k = v_o/i$ volts per ampere, or ohms. In the context of a current-controlled voltage source, k is called a *transresistance*.

Similarly, a controlled current source produces a current whose value depends on a voltage or a current elsewhere in the circuit. A voltage-controlled current source obeys the relation $i_o = kv$, where v is the controlling voltage and k

has the units of conductance: $k = i_o/v$ amperes per volt, or siemens. In the context of a controlled source, k is called a *transconductance*.

The four types of controlled sources, voltage-controlled voltage sources, current-controlled voltage sources, voltage-controlled current sources, and current-controlled current sources, can be modeled in SPICE. Figure A.14 shows the format used to model each type. Note that the controlling voltage in voltage-controlled sources (the voltage between NC+ and NC− in (a) and (b) of the figure) can be the voltage between any two nodes; it is not necessary that a component be connected between those nodes. Also note that the controlling current in controlled current sources (parts (c) and (d) of the figure) is always the current in a voltage source. Thus, it may be necessary to insert a dummy voltage source in a circuit in order to specify a controlling current at a desired point in the circuit. Observe how the "plus" and "minus" nodes are defined in the figure, in connection with the polarity assumptions made by SPICE for the currents and current sources. Example 8.17 illustrates the use of a voltage-controlled voltage source (EOP) in a SPICE program.

A.14   TRANSFORMERS

A transformer can be modeled in SPICE using three statements: one to specify the nodes and inductance of the primary winding, one to specify the nodes and inductance of the secondary winding, and one to specify the *coefficient of coupling*

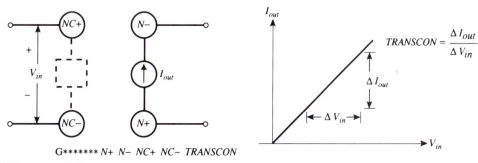

G******* N+ N− NC+ NC− TRANSCON

(a) Voltage-controlled current source

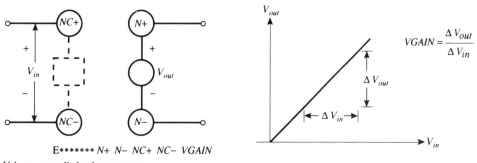

E******* N+ N− NC+ NC− VGAIN

(b) Voltage-controlled voltage source

Figure A.14
Specification of controlled sources in SPICE (Continued on page 664)

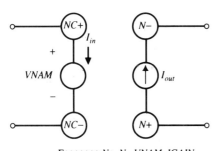 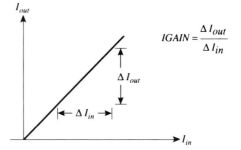

F******* N+ N− VNAM IGAIN

(c) Current-controlled current source

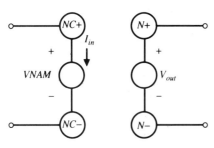 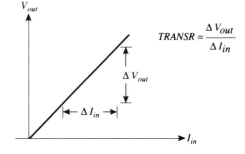

H******* N+ N− VNAM TRANSR

(d) Current-controlled voltage source

Figure A–14
(Continued)

between the windings. The primary and secondary windings are specified exactly as ordinary inductors are, using an L prefix, node numbers, and the inductance (in henries). The coefficient of coupling is specified in a statement that must begin with K:

```
K******* LNAM1 LNAM2 k
```

where *LNAM1* and *LNAM2* are the inductors comprising the primary and secondary windings and *k* is the value of the coefficient of coupling ($0 < k < 1$). For an ideal transformer, in which all the magnetic flux in the primary is coupled to the secondary, the coefficient of coupling equals 1. (In PSpice, *k* cannot be set equal to 1, but can be made arbitrarily close to it, such as 0.999.) To simulate an ideal iron-core transformer in SPICE, it is necessary to specify inductance values so that the reactance of the primary winding is much greater than the impedance of the signal source driving it and so that the reactance of the secondary winding is much greater than the load impedance. (The primary and secondary windings of an ideal transformer have infinite inductance.) Thus, it may be necessary to specify unrealistically large values for the inductances of the primary and secondary windings. (Such will be the case in all examples found in this book.) The following is an example showing the specification of a transformer (KXFRMR) having primary and secondary windings named LPRIM and LSEC:

```
LPRIM 5 0 20H
LSEC  6 0 0.2H
KXFRMR LPRIM LSEC 1
```

If operation of this transformer is to be simulated at 1 kHz, the source impedance should be much smaller than $2\pi f(\text{LPRIM}) = 2\pi(10^3 \text{ Hz})(20 \text{ H}) = 125.7 \text{ k}\Omega$, and the load impedance should be much smaller than $2\pi f(\text{LSEC}) = 2\pi(10^3 \text{ Hz})$ $(0.2 \text{ H}) = 1.257 \text{ k}\Omega$. In SPICE, an inductor cannot appear in a loop isolated from ground, so it is *not* possible to isolate the primary and secondary windings from each other, as is done in a real transformer.

The turns ratio of an ideal transformer is determined by the inductance values of the primary and secondary windings, according to

$$\frac{N_p}{N_s} = \sqrt{\frac{L_p}{L_s}}$$

where N_p/N_s is the turns ratio and L_p and L_s are the primary and secondary inductances, respectively. For example, the turns ratio of the transformer in the foregoing example is

$$\frac{N_p}{N_s} = \sqrt{\frac{20 \text{ H}}{0.2 \text{ H}}} = \sqrt{100} = 10$$

Thus, the transformer is a step-down type whose secondary voltage is one-tenth of its primary voltage. Example 4.16 demonstrates the use of a transformer in a SPICE program.

A.15 SUBCIRCUITS

A complex electronic circuit will often contain several components, or subsections of circuitry, that are identical to each other. Examples include active filters containing several identical operational amplifiers, multistage amplifiers consisting of identical amplifier stages, and digital logic systems containing numerous identical logic gates. When modeling such circuits in SPICE, it is a tedious and time-consuming task to write numerous sets of identical statements describing identical circuitry. Furthermore, the input data file for such a system may become so long and cumbersome that it is difficult to interpret or modify. To alleviate those kinds of problems, SPICE allows a user to create a *subcircuit:* circuitry that can be defined just once and then, effectively, inserted into a larger system (which we will call the *main* circuit) at as many places as desired. The concept is similar to that of a *subroutine* in conventional computer programming. In SPICE, it is possible to define several different subcircuits in one data file, and, in fact, one subcircuit can contain other subcircuits.

The first statement of a subcircuit is

`.SUBCKT NAME N1 (N2 N3 ...)`

where *NAME* is any name chosen to identify the subcircuit and *N1, N2, . . .* are the numbers of the nodes in the subcircuit that will be joined to other nodes in the main circuit. None of these can be node 0. Components in a subcircuit are defined in exactly the same way they are in any SPICE data file, using successive statements following the .SUBCKT statement. A subcircuit can contain .MODEL statements, but it cannot contain any control statements, such as .DC, .TRAN, .PRINT, or .PLOT. The node numbers in a subcircuit do not have to be different from those in the main program. SPICE treats a subcircuit as a completely separate (isolated) entity, and a node having the same number in a subcircuit as another node in the main circuit will still be treated as a different node. The exception is node 0: If node 0 appears in a subcircuit, it is treated as the same node 0 as in the main circuit. (In the language of computer science, subcircuit nodes are said to be *local,* except node 0, which is *global.*) The last statement in a subcircuit must be

```
.ENDS ⟨NAME⟩
```

If a subcircuit itself contains subcircuits, the *NAME* must be given in the .ENDS statement to specify which subcircuit definition has been ended.

In order to "insert" a subcircuit into the main circuit, we must write a subcircuit *call* statement in the main program. A different call statement is required for each location where the subcircuit is to be inserted. The format of a call statement is

```
X******* N1 ⟨N2 N3 ...⟩ NAME
```

where ******* are arbitrary characters that must be different for each call; *N1, N2, . . . ,* are the node numbers in the main circuit that are to be joined to the subcircuit nodes specified in the .SUBCKT statement; and *NAME* is the name of the subcircuit to be inserted. The node numbers in the call statement will be joined to the nodes in the .SUBCKT statement in exactly the same order as they both appear. That is, *N1* in the subcircuit will be joined to *N1* in the main circuit, and so forth. The following is an example:

```
.SUBCKT OPAMP 1 2 3
  -  ⎫
  -  ⎬ Statements describing components in the subcircuit
  -  ⎭
.ENDS
X1 8 4 12 OPAMP
X2 1 4 3  OPAMP
  -
  -
```

In this example, the subcircuit named OPAMP is inserted at two locations in the main program. The first call statement (X1) connects subcircuit nodes 1, 2, and 3 to main-circuit nodes 8, 4, and 12, respectively. The X2 call statement connects subcircuit nodes 1, 2, and 3 to main-circuit nodes 1, 4, and 3, respectively. (In this case some of the subcircuit and main-circuit node numbers are the same.)

The next example illustrates the use of a subcircuit to model an RC filter containing three identical stages. Although this circuit could not be considered complex enough to warrant the use of a subcircuit, it does serve to demonstrate the syntax we have described. Another example can be found in Example 9.16.

Example A.4

Using a SPICE subcircuit, determine the magnitude and phase angle of the output of the RC filter in Figure A.15(a) when the input is a 1-kHz sine wave with peak value 10 V.

Solution. Figure A.15(b) shows the RC subcircuit, named STAGE, and the way that it is inserted into the main circuit in three locations. Its locations are identified by rectangles labeled X1, X2, and X3. The node numbers inside the rectangles are the subcircuit node numbers and those outside are the main-circuit node numbers.

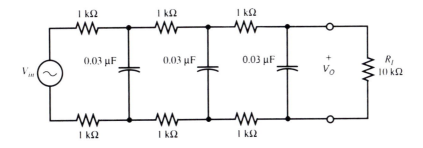

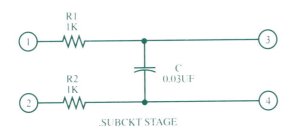

.SUBCKT STAGE

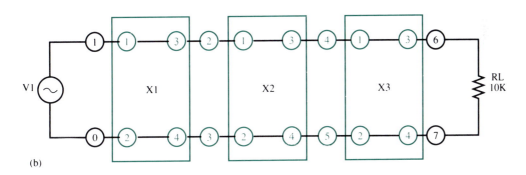

(b)

```
EXAMPLE A.4
V1 1 0 AC 10V
.SUBCKT STAGE 1 2 3 4
R1 1 3 1K
R2 2 4 1K
C 3 4 0.03UF
.ENDS
X1 1 0 2 3 STAGE
X2 2 3 4 5 STAGE
X3 4 5 6 7 STAGE
RL 6 7 10K
.AC LIN 1 1K 1K
.PRINT AC VM(6,7) VP(6,7)
.END
```

(c)

```
EXAMPLE A.4
****      AC ANALYSIS                              TEMPERATURE =   27.000 DEG C
**********************************************************************************:
    FREQ        VM(6,7)       VP(6,7)
 1.000E+03      3.769E+00    -7.107E+01
```

(d)

Figure A.15
(Example A.4)

In the input data file, shown in part (c) of the figure, note that the node numbers in the three call statements are the main-circuit nodes that are connected to subcircuit nodes 1, 2, 3, and 4, in that order. The results of a program run are shown in Figure A.15(d). We see that the output voltage has peak value 3.769 V and lags the input by 71.07°.

A.16   PROBE AND CONTROL SHELL

PSPICE

Probe

Probe is a PSpice option that makes it possible to obtain output plots having greater resolution than those produced by .PLOT statements. Instead of plotting a sequence of asterisks, Probe produces a virtually solid line on a high-resolution monitor. Also, Probe has other features that make it very useful for analyzing output data. It behaves very much like a high-quality oscilloscope that allows the user to position a cursor on a trace and to obtain a direct readout of the value of the variable displayed at the position of the cursor.

If the statement

```
.PROBE
```

appears anywhere in an input data file that requests a .DC, .AC, or .TRAN analysis, then Probe automatically generates plotting data for the voltage (with respect to ground) at every node in the circuit and for the current entering every device in the circuit. Probe stores the data in an *output data file* named PROBE.DAT. To initiate a Probe run, the user enters the command PROBE directly from the keyboard. If more than one analysis type appears in the input data file (called the *circuit file* in PSpice), a "start-up menu" appears, and the user selects a single analysis type for the Probe run. If only one analysis type appeared in the circuit file, then a set of axes and a menu are displayed immediately after Probe is entered. Selecting the option ADD TRACE from this menu prompts the user to enter the variable to be plotted, using the same format used to specify outputs in PSpice (such as V(2), I(R1), etc.). The plot is scaled automatically and displayed on the monitor. Additional variables can be plotted simultaneously by repeatedly selecting the ADD TRACE option. Also, plots can be deleted by selecting the DELETE TRACE option.

Selecting the CURSOR option from the Probe menu creates two sets of cross hairs, identified as cursor 1 (C1) and cursor 2 (C2), on the display. These cross hairs can be moved along the plot using direction keys (← and →) on the keyboard. Using the direction keys alone moves C1 and using the direction keys with SHIFT depressed moves C2. The numerical values of the variable at the positions of the cross hairs on the plot are displayed in a window, along with the difference in the values. Figure A.16 shows an example.

If desired, Probe plots can be limited to specific variables by giving their names in the .PROBE statement in the circuit file. For example, the statement

```
.PROBE V(2)
```

causes Probe to create an output data file containing only plotting data for the

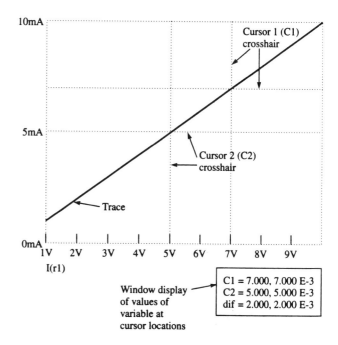

Figure A.16
Example of a PSpice Probe display

voltage at node 2. This option is useful when memory capacity is too small to store all voltages and currents.

Another feature of Probe is that it can create plots of mathematical functions of the variables. For example, after selecting ADD TRACE, entering the expression

```
I(R1)*I(R1)*1K
```

creates a plot of the power (I^2R) dissipated by 1-kΩ resistor R1. Numerous mathematical functions, including trigonometric functions, logarithmic functions, average values, square roots, and absolute values, of the variables can also be plotted. One difference between Probe and PSpice is that suffixes in expressions written for Probe must use lowercase m to represent milli and capital M to represent meg (as opposed to M for milli and MEG for meg).

It should be noted that the high-resolution plots created by Probe are *not* the result of Probe increasing the number of output values computed in a circuit simulation. Rather, Probe *interpolates* values between the data points that a circuit file specifies are to be computed. Thus, if the output variable has a sudden change in value (such as a sharp peak) and the circuit file does not specify a fine-enough increment between data points to detect that change, Probe cannot be expected to detect it either.

Control Shell

Control Shell is a PSpice option that coordinates and simplifies writing, editing, and running input circuit files and using the various options and features available in PSpice. For example, Control Shell can be used to make Probe run automatically after every circuit simulation. Options are selected by the user from various menus displayed by Control Shell.

In a system equipped with Control Shell, the first (main) menu displayed contains, among others, selections entitled Files, Analysis, and Probe. These are the principal choices that will be used in creating and running most circuit simulations. When the user selects Files, another menu is displayed with options that include Edit, Browse Output, and Current File. Edit and Current File allow the user to edit an existing circuit file or to create and name a new circuit file. Selecting Browse Output allows the user to scroll through the entire output created after a circuit simulation has been run.

To perform a program run of an input circuit file that has already been created, the option Analysis is selected from the main menu. This selection creates another menu that allows the user to specify the analysis type (AC, DC, or transient). Since at least one analysis type is (presumably) already specified in the input circuit file, an arrow in this menu points to one analysis type and the user can simply press the ENTER key to initiate the run. Alternatively, another analysis type specified in the circuit file can be selected, and the user is prompted to enter (or change) analysis parameters (such as start and stop voltages and step size in a .DC analysis). It should be noted that if there is an error in the input circuit file, Control Shell does not allow the user to select the Analysis option.

Selecting Probe from the main menu creates another menu that allows the user to request automatic running of Probe after a PSpice simulation run. In this option, it is not necessary to include a .PROBE statement in the input circuit file.

A.17 THE PSPICE LIBRARY

PSPICE

The MicroSim Corporation has created models and subcircuits for over 3500 standard analog devices, including transistors, diodes, and operational amplifiers. These can be stored in a computer system as a *library* that can be accessed by users who wish to specify any one or more of the devices in an input circuit file. It is not necessary to write the entire model or subcircuit defining a particular device in the input file; it is necessary only to refer to it by name. PSpice automatically retrieves the description from the library and uses it in the circuit simulation.

Transistors and diodes are stored in the library as model statements. If we wish to incorporate a particular transistor in an input circuit file, we specify the device name in place of the model name in the statement defining the transistor in our circuit. The statement .LIB is used to inform PSpice that we are accessing the library. The following is an example:

```
Q1 3 5 8 Q2N2222A
.LIB
```

These statements cause PSpice to use the model statement in the library for the 2N2222A transistor as the model statement for Q1. That model statement is as follows:

```
.model Q2N2222A NPN (Is=14.34f Xti=3 Eg=1.11 Vaf=74.03 Bf=255.9 Ne=1.307
      Ise=14.34f Ikf=.2847 Xtb=1.5 Br=6.092 Nc=2 Isc=0 Ikr=0 Rc=1
      Cjc=7.306p Mjc=.3416 Vjc=.75 Fc=.5 Cje=22.01p Mje=.377 Vje=.75
      Tr=46.91n Tf=411.1p Itf=.6 Vtf=1.7 Xtf=3 Rb=10)
```

We should note that the PSpice library actually consists of a number of library files, each having a name and containing devices of a certain type. For example, DIODE.LIB is the name of a library file that contains model statements for diodes, and LINEAR.LIB contains subcircuits for operational amplifiers. PSpice can be directed to a particular library file by including the name of the file in the .LIB statement. For example,

```
.LIB LINEAR.LIB
```

directs PSpice to the linear library file. In particular, in the *evaluation version* (student version) of PSpice, the library file is called EVAL.LIB, and this must be included in the .LIB statement. Linear devices in EVAL.LIB include the following:

Q2N2222A	NPN transistor
Q2N2907A	PNP transistor
Q2N3904	NPN transistor
Q2N3906	PNP transistor
D1N750	zener diode
MV2201	voltage variable-capacitance diode
D1N4148	switching diode
J2N3819	N-channel JFET
J2N4398	N-channel JFET
LM324	operational amplifier
UA741	operational amplifier
LM111	voltage comparator

Example A.5

Use the PSpice library to obtain a plot of v_{ce} versus time in the circuit shown in Figure A.17(a). The plot should cover two full cycles of output. Also find the quiescent values of V_{CE} and I_C.

Solution.   To obtain a plot of v_{ce} versus time, we must use a SIN source and perform a .TRAN analysis. In order for the plot to cover two full cycles, we set *TSTOP* in the .TRAN statement to two times the period of the sine wave: 2(1/10 kHz) = 0.2 ms. The circuit is redrawn in a PSpice format in Figure A.17(b). Also shown is the PSpice input circuit file. Note how the PSpice library is used to access the model statement for the 2N2222A transistor. Also note that the Q-point can be obtained in a PSpice .PRINT statement by requesting outputs IC(Q1) (the dc collector current of Q1) and VCE(Q1) (the dc collector-to-emitter voltage of Q1).

Execution of the program reveals that IC(Q1) = 1.125 mA and VCE(Q1) = 13.31 V. Since V_{CC} = 15 V, this quiescent point is close to the cutoff region of the transistor. We might therefore expect positive clipping to occur if the input to the amplifier is sufficiently large. Figure A.17(c) shows a plot produced by the PSpice Probe option, and we see that clipping does indeed occur. The Probe cursors are set at the minimum and maximum values of v_{ce}, C1 = 14.403 V and C2 = 6.6941 V. Thus, positive clipping occurs at v_{ce} = 14.403 V.

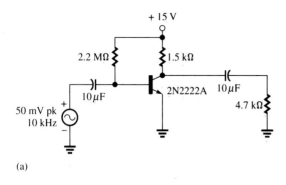

(a)

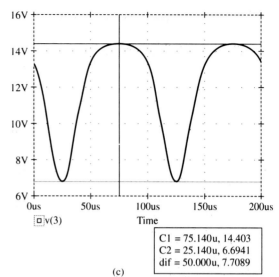

C1 = 75.140u, 14.403
C2 = 25.140u, 6.6941
dif = 50.000u, 7.7089

(c)

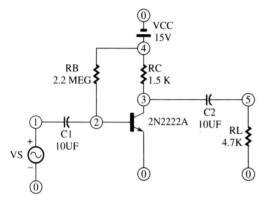

EXAMPLE A.5
VS 1 0 SIN(0 50MV 10KHZ 0 0)
VCC 4 0 15V
C1 1 2 10UF
C2 3 5 10UF
RB 2 4 2.2MEG
RC 3 4 1.5K
RL 5 0 4.7K
Q1 3 2 0 Q2N2222A
.LIB EVAL.LIB
.TRAN 500NS 0.2MS
.DC VCC 15V 15V 1
.PRINT DC IC(Q1) VCE(Q1)
.PLOT TRAN V(3)
.END

(EVAL.LIB for Evaluation version of PSpice only)

(b)

Figure A.17
(Example A.5)

Since operational amplifiers are stored in the library as subcircuits, they must be accessed in the input circuit file by subcircuit *calls* (statements beginning with X; see Section A.15). Figure A.18 shows an example of an input circuit file containing a call for the 741 operational amplifier. Note that the subcircuit nodes are

1   noninverting input
2   inverting input
3   positive supply voltage
4   negative supply voltage
5   output

In this example, the amplifier is connected in an inverting configuration with voltage gain $-R_f/R_1 = -10\text{ k}\Omega/1\text{ k}\Omega = -10$. The input is a 1-V, 1-kHz sine wave. Execution of the program reveals that $v_o = V(30) = 9.999$ V, $\underline{/v_o} = VP(30) = 179.4°$, and $i_{in} = I(RIN) = 0.9999$ mA.

The subcircuits for operational amplifiers are the *functional equivalents* of the devices they represent. That is, the circuits do not contain the transistors, diodes,

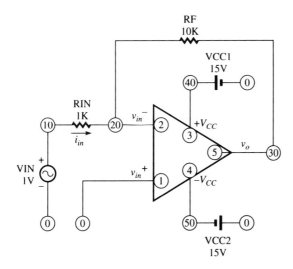

```
OPAMP EXAMPLE
VIN 10 0 AC 1V
RIN 10 20 1K
RF 20 30 10K
VCC1 40 0 15V
VCC2 0 50 15V
X1 0 20 40 50 30 UA741
.LIB EVAL.LIB
.AC DEC 1 1KHZ 1KHZ
.PRINT AC V(30) VP(30) I(RIN)
.END
```

(EVAL.LIB for Evaluation version of PSpice only)

Figure A.18
Example of a call for an operational-amplifier subcircuit from the PSpice library

resistors, etc., that are in the actual amplifier circuits. These functionally equivalent circuits are designed to exhibit nominal amplifier characteristics (such as bandwidth), not the worst-case values given in manufacturers' specifications. These characteristics do *not* change with temperature, as they do in actual amplifiers. Finally, we should note that component names used in library subcircuits are all *local*. In other words, PSpice will recognize that R1, for example, used in a library subcircuit is different from another resistor named R1 in the input circuit file.

Derivation of Frequency-Response Equations

B.1 CUTOFF FREQUENCY OF TWO ISOLATED HIGH-PASS RC FILTERS IN CASCADE

Let

$$f_A = \text{cutoff frequency of first filter}$$
$$f_B = \text{cutoff frequency of second filter}$$

Since the filters are assumed to be isolated (by an ideal, unity-gain amplifier), the overall voltage gain of the cascade at frequency f is

$$\frac{|v_o|}{|v_{in}|} = \frac{1}{\sqrt{1 + (f_A/f)^2}} \frac{1}{\sqrt{1 + (f_B/f)^2}}$$

At cutoff ($f = f_1$),

$$\frac{1}{\sqrt{1 + (f_A/f_1)^2}} \frac{1}{\sqrt{1 + (f_B/f_1)^2}} = \frac{1}{\sqrt{2}}$$

Let

$$x = \frac{1}{f_1^2}, \, a = f_A^2, \, b = f_B^2$$

Then

$$\frac{1}{\sqrt{1 + ax} \sqrt{1 + bx}} = \frac{1}{\sqrt{2}}$$

$$\frac{1}{(1 + ax)(1 + bx)} = \frac{1}{2}$$

$$(1 + ax)(1 + bx) = 2$$

$$abx^2 + (a + b)x - 1 = 0$$

$$x = \frac{-(a + b) \pm \sqrt{(a + b)^2 + 4ab}}{2ab}$$

$$f_1 = \frac{1}{\sqrt{x}} = \sqrt{\frac{2ab}{-(a + b) + \sqrt{(a + b)^2 + 4ab}}}$$

B.2   CUTOFF FREQUENCY OF TWO ISOLATED LOW-PASS RC FILTERS IN CASCADE

Let

$$f_A = \text{cutoff frequency of first filter}$$
$$f_B = \text{cutoff frequency of second filter}$$

Since the filters are assumed to be isolated (by an ideal, unity-gain amplifier), the overall voltage gain of the cascade at frequency f is

$$\frac{|v_o|}{|v_{in}|} = \frac{1}{\sqrt{1 + (f/f_A)^2}} \frac{1}{\sqrt{1 + (f/f_B)^2}}$$

At cutoff ($f = f_2$)

$$\frac{1}{\sqrt{1 + (f_2/f_A)^2}} \frac{1}{\sqrt{1 + (f_2/f_B)^2}} = \frac{1}{\sqrt{2}}$$

Let

$$x = f_2^2, \ a = 1/f_A^2, \ b = 1/f_B^2$$

Then

$$\frac{1}{\sqrt{1 + ax} \sqrt{1 + bx}} = \frac{1}{\sqrt{2}}$$
$$\frac{1}{(1 + ax)(1 + bx)} = \frac{1}{2}$$
$$(1 + ax)(1 + bx) = 2$$
$$abx^2 + (a + b)x - 1 = 0$$
$$x = \frac{-(a + b) \pm \sqrt{(a + b)^2 + 4ab}}{2ab}$$
$$f_2 = \sqrt{x} = \sqrt{\frac{-(a + b) + \sqrt{(a + b)^2 + 4ab}}{2ab}}$$

Standard Values of 5% and 10% Resistors

Resistors with 5% tolerance are available in all values shown. Resistors with 10% tolerance are available only in the boldfaced values.

Ohms (Ω)					Kilohms (kΩ)		Megohms (MΩ)	
0.10	**1.0**	**10**	**100**	**1000**	**10**	**100**	**1.0**	**10.0**
0.11	1.1	11	110	1100	11	110	1.1	11.0
0.12	**1.2**	**12**	**120**	**1200**	**12**	**120**	**1.2**	**12.0**
0.13	1.3	13	130	1300	13	130	1.3	13.0
0.15	**1.5**	**15**	**150**	**1500**	**15**	**150**	**1.5**	**15.0**
0.16	1.6	16	160	1600	16	160	1.6	16.0
0.18	**1.8**	**18**	**180**	**1800**	**18**	**180**	**1.8**	**18.0**
0.20	2.0	20	200	2000	20	200	2.0	20.0
0.22	**2.2**	**22**	**220**	**2200**	**22**	**220**	**2.2**	**22.0**
0.24	2.4	24	240	2400	24	240	2.4	
0.27	**2.7**	**27**	**270**	**2700**	**27**	**270**	**2.7**	
0.30	3.0	30	300	3000	30	300	3.0	
0.33	**3.3**	**33**	**330**	**3300**	**33**	**330**	**3.3**	
0.36	3.6	36	360	3600	36	360	3.6	
0.39	**3.9**	**39**	**390**	**3900**	**39**	**390**	**3.9**	
0.43	4.3	43	430	4300	43	430	4.3	
0.47	**4.7**	**47**	**470**	**4700**	**47**	**470**	**4.7**	
0.51	5.1	51	510	5100	51	510	5.1	
0.56	**5.6**	**56**	**560**	**5600**	**56**	**560**	**5.6**	
0.62	6.2	62	620	6200	62	620	6.2	
0.68	**6.8**	**68**	**680**	**6800**	**68**	**680**	**6.8**	
0.75	7.5	75	750	7500	75	750	7.5	
0.82	**8.2**	**82**	**820**	**8200**	**82**	**820**	**8.2**	
0.91	9.1	91	910	9100	91	910	9.1	

Answers to Odd-Numbered Exercises

CHAPTER 1

1.1   2 μS
1.3   0.637 mS
1.5   **(a)** 250 kΩ; **(b)** 250 kΩ; **(c)** 4 μS
1.7   **(a)** 60; **(b)** 60; **(c)** 0 V
1.9   **(a)** 10 V pk-pk; **(b)** $5 + 5 \sin \omega t$ V

CHAPTER 2

2.1   Silicon and germanium; the ability to form crystals having special electrical properties.
2.3   Positively; free electron falling into a hole.
2.5   Doping; donors and acceptors.
2.7   Barrier voltage; from N to P.
2.9   **(a)** -1.97 pA; **(b)** -2.4 pA; **(c)** 3.404 mA; **(d)** 337.9 mA
2.11   12 pA
2.13   **(a)** 275 Ω; **(b)** 10 Ω
2.15   10.4 Ω
2.17   0.067 V rms
2.19 $I_{max} = 0.591$ mA; $V_R(\max) = 1.3$ V
2.21   **(a)** 165.3 V; **(b)** 7.35 V pk-pk
2.23   \$3.00
2.25   1N4004
2.27   Aluminum and N-type silicon; silicon.
2.29   Aluminum and P-type silicon; aluminum and N$^+$ silicon.
2.31   3.2 V pk
2.33   250 μF

CHAPTER 3

3.1   **(a)** electron; **(b)** hole
3.3   **(a)** positive; **(b)** negative; **(c)** positive
3.5   See Figure 3.5.
3.7   **(a)** 0.7 V; **(b)** -0.3 V
3.9   **(a)** 0.01 mA; **(b)** See Figure 3.5(a).
3.11   Reverse current from collector to base when the emitter is open.

3.13   0.99476
3.15   99
3.17 $\beta = \dfrac{\alpha}{1 - \alpha} \Rightarrow \beta(1 - \alpha) = \alpha \Rightarrow \beta - \alpha\beta = \alpha$
        $\Rightarrow \alpha + \alpha\beta = \beta \Rightarrow \alpha(\beta + 1) = \beta \Rightarrow \alpha = \dfrac{\beta}{\beta + 1}$
3.19   0.99%
3.21   **(a)** 10 mA; **(b)** 213
3.23   **(a)** 228 (approx.); **(b)** 240 (approx.); **(c)** 50 V (approx.)
3.25   1. Breakdown region; 2. saturation region; 3. cutoff region; 4. linear region
3.27 $I_B = 25$ μA; $I_C = 3.75$ mA
3.29   Current in an FET is due to one type of charge carrier only; field.
3.31   Electron; from source to drain.
3.33   **(a)** negative; **(b)** positive
3.35 I_{DSS}
3.37   5.1 mA
3.39   **(a)** 3.8 mA; **(b)** -4 V; 15 mA
3.41 $I_D = 15 \times 10^{-3} \left(1 - \dfrac{V_{GS}}{-4}\right)^2$ A
3.43   **(a)** 2.57 mA; **(b)** (i) 0 V; (ii) -1.32 V; (iii) -4.5 V
3.45   **(a)** 10 mA; -3.5 V; **(b)** 1.8 mA;
        (c) $I_D = 10 \times 10^{-3} \left(1 - \dfrac{V_{GS}}{-3.5}\right)^2$ A
3.47   **(a)** P-channel; **(b)** 13 mA; 4.5 V; **(c)** 2 V;
        (d) $I_D = 13 \times 10^{-3} \left(1 - \dfrac{V_{GS}}{4.5}\right)^2$ A
3.49   **(a)** 4.2 mS; **(b)** 4.1 mS; **(c)** 1.23 mA
3.51   From equation 3.11, $\left(1 - \dfrac{V_{GS}}{V_p}\right) = \sqrt{\dfrac{I_D}{I_{DSS}}}$;
        Substituting into equation 3.13 gives equation 3.14.
3.53   Gate is insulated from the channel (no PN junction); IGFET
3.55   By making the gate positive (in an N-channel MOSFET), electrons are drawn into the

channel, thus increasing the conductivity of (enhancing) the channel, without the consequences of forward biasing a PN junction.

3.57 (a) 8.07 mA; (b) 15 mA; (c) 19 mA

3.59 The channel of the N-channel depletion-type MOSFET is N material and is P material in the enhancement type.

3.61 12 mA

3.63 (a) 2.7 mS (approx.); (b) 2.5 mS

3.65 ICs are economical in terms of cooling, wiring, and packaging costs and they are more reliable than discrete circuits. Short signal paths in ICs reduce delays and phase shifts. Discrete circuitry is more appropriate for power handling and in applications where few components are required as, for example, a relay or lamp driver, a protective diode, and an output power amplifier.

3.67 An ingot is a cylinder of silicon crystal from which wafers are sliced. Numerous identical chips are fabricated on and cut from a single wafer. A chip is also called a die.

3.69 Silicon dioxide (SiO_2) blocks impurity diffusion and is applied by heating silicon in the presence of oxygen.

3.71 (a) oxygen; (b) hydrofluoric acid; (c) boron; (d) phosphorous; (e) aluminum

3.73 An epitaxial layer is a thin crystalline layer grown on the surface of a wafer. It can be formed by chemical vapor deposition, in which gaseous chemicals are deposited on the wafer surface. A buried layer is an N^+ region diffused into the crystal beneath the epitaxial layer to reduce resistance in the collector path.

3.75 1.7 kΩ

3.77 Ion implantation; a very thin impurity layer (small value of t) can be achieved.

3.79 0.422 pF

CHAPTER 4

4.1 (a) 10 V; (b) 10 V pk-pk; (c) 5 V, 15 V; (d) 1.59 kHz; (e) 10 V; (f) 0 V;

(g)

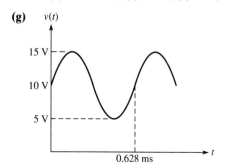

4.3 32.5 + 12.5 sin(3141t) mA

4.5 (a) 5 V; (b) 9 V; (c) 0.01 s

4.7 (a) 3 V; (b) 1.5 sin ωt V; (c) 3 + 1.5 sin ωt V

4.9 23.53 + 5 sin ωt mA

4.11 (a) 150; (b) 1.59 V rms

4.13 (a) 7.5 kΩ; (b) 17.5 kΩ

4.15 1 + 0.4 sin ωt mA

4.17 200 Ω

4.19 0.806 V pk-pk

4.21 3 kΩ

4.23 4.92 V rms

4.25 36.9 V pk-pk

4.27 (a) $-3 \sin(2\pi \times 2 \times 10^5 t)$ V or $3 \sin(2\pi \times 2 \times 10^5 t \pm 180°)$ V;

(b)

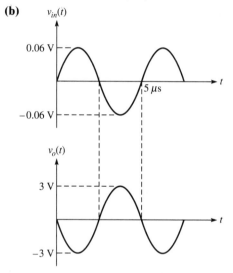

4.29 (a)

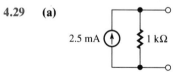

(b) 250

4.31 60

4.33 136

4.35 4.91 kΩ

4.37 675

4.39 $A_p = \dfrac{V_o^2(\text{rms})/r_o}{I_{in}^2(\text{rms})\, r_{in}} = \dfrac{[V_o(\text{rms})/I_{in}(\text{rms})]^2}{r_o r_{in}}$ (The same result for pk or pk-pk values.)

4.41 (a) 0.247 W; (b) 49.92 W

4.43 (a) 7 V; (b) 14.5 μF (for $|X_C| = 22$ Ω)

4.45 2.5 kΩ

4.47 $P_1 = 0.8$ mW, $P_L = 4$ mW; not matched.

4.49 (a) 10 : 1; (b) $P_1 = P_L = 0.144$ W

4.51 1 μF

4.53 400

CHAPTER 5

5.1

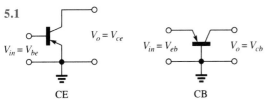

CE   CB

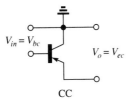

CC

5.3 (a) 2.74 mA, 6.52 V; (b) $V_{CB} = 5.82$ V
5.5 680 kΩ
5.7 (a) $V_{CE} = 12 - I_C(2 \times 10^3)$; (b) 2 mA
5.9 (a) $I_C = 2$ mA, $V_{CE} = 8$ V; (b) $I_C = 2$ mA, $V_{CE} = 8$ V
5.11 5.1 + sin ωt mA
5.13 $I_C = 5.16$ mA, $V_{CE} = 9.2$ V
5.15 $I_C = 4.78$ mA, $V_{CE} = 10.23$ V
5.17 (a) $V_{CE1} = 3.32$ V, $V_{CE2} = 2.04$ V;
(b) $V_{CE1} = 3.37$ V, $V_{CE2} = 2.13$ V
5.19 770 Ω
5.21 (a) See Figure 5.17(b); $R_B = 330$ kΩ, $(\beta + 1)r_e = 770$ Ω, $\beta i_b = 80\ i_b$, $r_c/\beta = 100$ kΩ, $R_C = 2$ kΩ;
(b) −210.5
5.23 (a) See Figure 5.21(b); $r_S = 1$ kΩ, $R_1 = 39$ kΩ, $R_2 = 10$ kΩ, $(\beta + 1)r_e = 1322$ Ω, $\beta i_b = 150\ i_b$, $r_c/\beta = 120$ kΩ, $R_C = 2.7$ kΩ, $R_L = 3.9$ kΩ;
(b) −97.2
(c) −1.4
5.25 $I_E = 3.55$ mA, $V_E = 3.55$ V, $V_{CE} = 8.45$ V
5.27 $V_{in} = (\beta + 1)r_e i_b + i_e(R_E \| R_L)$
$= (\beta + 1)r_e i_b + (\beta + 1)i_b r_L$
$= (\beta + 1)i_b(r_e + r_L)$
5.29 (a) 0.94; (b) 16.4 Ω
5.31 (a) $I_C = 0.928$ mA, $V_{CB} = 7.79$ V;
(b) 28 Ω
(c) 76.9
5.33 (a) 2.5 mA, 5.3 V (approx.);
(b) 4.1 mA, 1 V (approx.);
(a) is in pinch-off; (b) is not.
5.35 (a) 1.098 mA, 13.02 V; valid;
(b) 2.24 mA, 10.97 V; valid;
(c) 6.09 mA, 4.04 V; not valid.
5.37 $I_D \approx 2.3$ mA, $V_{GS} = -1.15$ V; $V_{DS} = 4.59$ V
5.39 (a) $I_D \approx 6.6$ mA, $V_{Gs} \approx 4$ V; $V_{DS} = -11.52$ V;
(b) $I_D = 6.52$ mA, $V_{GS} = 3.91$ V, $V_{DS} = -11.74$ V

5.41 $I_D \approx 9$ mA, $V_{DS} = -6.12$ V
5.43 (a) $I_D \approx 6.1$ mA, $V_{DS} = 6.7$ V;
(b) $I_D = 6.09$ mA, $V_{DS} = 6.73$ V
5.45 75 kΩ
5.47 (a) $I_D = 3.05$ mA, $V_{DS} = 9.51$ V; (b) 4.36 mS; (c) See Figure 5.46; $R_G = 1$ MΩ, $g_m v_{gs} = 4.36 \times 10^{-3}\ v_{gs}$, $r_d = 120$ kΩ, $R_D = 1.8$ kΩ; (d) −7.85
5.49 (a) −12.33; (b) 90.9 kΩ
5.51 −9.04
5.53 (a) 0.922; (b) 90.9 kΩ
5.55 (a) 5.41 mA; (b) 10.15 V
5.57 $I_D \approx 1.8$ mA, $V_{GS} \approx 4.5$ V; $V_{DS} = 14.82$ V
5.59 (a) 2.79 MΩ; (b) 833 Ω; (c) 12.72 V; (d) Yes.
5.61 $V_{DS} = V_{DD} - I_D R_D$ (1); $I_D = 0.5\beta(V_{DS} - V_T)^2$ (2); Substitute (1) into (2) and solve for I_D using the quadratic formula.
5.63 No change.
5.65 (a) 468 kΩ; (b) −1.62
5.67 (a) $R_B = 1.5$ MΩ, $R_C = 10$ kΩ;
(b) I_C(min) = 0.2724 mA, V_{CE}(max) = 12.41 V; I_C(max) = 1.51 mA > I_C(sat) ⇒ not valid.
5.69 See Figure 5.21(a); $R_C = 8.2$ kΩ, $R_E = 1.5$ kΩ, $R_1 = 100$ kΩ, $R_2 = 15$ kΩ; $V_{CE} = 9.22$ V; $A_v = -631$
5.71 See Figure 5.24; (a) $R_E = 1.8$ kΩ, $R_b = 110$ kΩ; (b) 9.89 V
5.73 (One solution) See Figure 5.41(a).
(a) $R_S = 1.46$ kΩ, $R_D = 2.87$ kΩ, $R_1 = 1.43$ MΩ, $R_2 = 330$ kΩ; (b) I_D: 2.65 mA to 3.24 mA; V_{DS}: 3.43 V to 6.08 V; (c) Yes.

CHAPTER 6

6.1 (a) 0.88 MHz; (b) 0.6 V rms; (c) 0.8 W; (d) 0.4 W
6.3 0.125 V rms, 0.25 ms
6.5 1 V rms
6.7 (a) 32 dB; (b) 30.46 dB
6.9 (a) 0.856 V rms; (b) 18 dB
6.11 (a) 3.6 V rms; (b) 72 V rms; (c) 0.18 V rms; (d) 360 V rms
6.13 4 cycles by 4 cycles (4 × 4)
6.15 (a) $f_1 \approx 500$ Hz, $f_2 \approx 15$ kHz; (b) 0.5 V rms (approx.); (c) 63° (approx.); (d) 5 (approx.)
6.17 (a) 159.15 Hz; (b) 70.67; (c) 84.29°; (d) 52.63 Hz
6.19 (a) 72.34 Hz; (b) 19.38 Hz; (c) 97.01
6.21 (a) 491.2 Hz; (b) 170.1; (c) 1.94 μF
6.23 (a) 120.3 Hz; (b) 37 dB; (c) 34 dB; (d) 126° (approx.); (e) 13.2 Hz (approx.)
6.25 (a) 3.04 MHz; (b) 157.5; (c) −73.1°
6.27 3383 Ω
6.29 (a) 271 kHz; (b) 105.47;

(c)

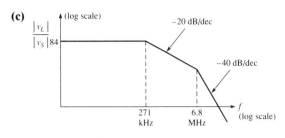

6.31 59.2 kHz
6.33 **(a)** 6029 Hz; **(b)** 58 μs; **(c)** decreases (with both R and C); **(d)** increases (with both R and C)
6.35 **(a)** -8568; **(b)** 11.6
6.37 $A_1 = 10$, $A_2 = 10$, $A_3 = 5$
6.39 $f_1 = 34.5$ Hz, $f_2 = 12.9$ kHz
6.41 207.1 kHz
6.43 135 Hz
6.45 307 Hz
6.47 260 kHz
6.49 1092 pF
6.51 18.7 Hz
6.53 30.6 MHz
6.55 $C_E = 40$ μF, $C_1 = 4$ μF, $C_2 = 0.4$ μF
6.57 **(a)** $C_1 = 1.5$ μF, $C_2 = 10$ μF; **(b)** 30 Hz; **(c)** 31 Hz

CHAPTER 7

7.1 **(a)** $4 \sin \omega t - 2$ V; **(b)** $4 \sin \omega t$ V; **(c)** $0.04 \sin \omega t$ V; **(d)** $0.02 \sin \omega t$ V
7.3 **(a)** -80, $+80$; **(b)** -160
7.5 **(a)** -132.7, $+132.7$; **(b)** -265.4
7.7 **(a)** 5 V, 5V; **(b)** -269.5; **(c)** 7.49 kΩ
7.9 **(a)** 6 V, 6 V; **(b)** -15.48; **(c)** 1.55 V rms
7.11 0.104 mV rms
7.13 **(a)** -0.312 V dc; **(b)** $-1.3 \sin \omega t$ V; **(c)** 6.5 V dc; **(d)** $-10.4 + 2.6 \sin \omega t$ V; **(e)** $-2.08 \sin(\omega t + 75°)$ V
7.15 **(a)** 8.06 μA dc; **(b)** $-0.067 \sin \omega t$ μA; **(c)** $21.3 - 133.3 \sin \omega t$ μA; **(d)** $26.6 \sin(\omega t - 30°)$ μA
7.17 (One solution) See Figure 7.17. $R_1 = 10$ kΩ, $R_f = 200$ kΩ
7.19 **(a)** 4.5 V rms; **(b)** 0.6068 V rms; **(c)** 5.5 V rms; **(d)** 0.55 V rms

CHAPTER 8

8.1 **(a)** Ideal: 201, Actual: 193.23; **(b)** Ideal: 51, Actual: 50.485; **(c)** Ideal: 11, Actual: 10.976
8.3 639,200
8.5 84
8.7 -3.996

8.9 $1 - \beta = 1 - \dfrac{R_1}{R_1 + R_f} = \dfrac{R_f}{R_1 + R_f}$
8.11 **(a)** 50.125 kΩ; **(b)** 0.654 Ω
8.13 450
8.15 25 Hz
8.17 **(a)** 60 kHz; **(b)** 2
8.19 199
8.21 10 kHz
8.23 3 V/μs
8.25 0.667
8.27 355,376 V/s
8.29 **(1)** new slew rate of 1.31×10^6 V/s **(2)** change amplifier gain to 6.4
8.31 56.2 kΩ
8.33 Distortion occurs.
8.35 10 kΩ; 0.4 mV
8.37 See Figure 8.19. $R_1 = 10$ kΩ, $R_f = 62.5$ kΩ, $R_c = 8.62$ kΩ; $V_o/V_{in} = 7.25$
8.39 3.469 mV
8.41 0.5 mV
8.43 30.35 mV
8.45 **(a)** 3 mV; **(b)** 58 dB (approx.); **(c)** 3.5 nA; **(d)** 34 kHz (approx.)
8.47 See Figure 8.19. $R_1 = 1$ kΩ, $R_f = 18.27$ kΩ, $R_c = 948$ Ω

CHAPTER 9

9.1 **(a)** $-(v_1 + 2v_2 + 5v_3 + 10v_4)$; **(b)** -14 v dc; **(c)** 5.26 kΩ
9.3 **(a)** 53 kHz; **(b)** 57 mV
9.5 **(a)** $8.8 v_1 + 2.2 v_2$; **(b)** $6.6 v_1$ or $-6.6 v_2$
9.7 See Figure 9.5. **(a)** (One solution) $R_1 = 1$ MΩ, $R_2 = 16.95$ kΩ, $R_3 = 20$ kΩ, $R_4 = 100$ kΩ; **(b)** (One solution) $R_1 = 10$ kΩ, $R_2 = 100$ kΩ, $R_3 = 10$ kΩ, $R_4 = 100$ kΩ
9.9 See Figure 9.7. **(a)** (One solution) $R_1 = 100$ kΩ, $R_2 = 40$ kΩ, $R_{c1} = 28.6$ kΩ, $R_3 = 100$ kΩ, $R_4 = 10$ kΩ, $R_5 = 100$ kΩ, $R_{c2} = 8.3$ kΩ; **(b)** (One solution) Input resistors for v_1 and v_2 in first amplifier: 100 kΩ each, R_2 (feedback) $= 100$ kΩ, $R_{c1} = 33.3$ kΩ, $R_3 = 100$ kΩ, $R_4 = 5$ kΩ, $R_5 = 100$ kΩ, $R_{c2} = 4.5$ kΩ
9.11 See Figure 9.10 (a). **(a)** $R_1 = 2$ kΩ; **(b)** 20 V
9.13 See Figure 9.12(a). **(a)** $R = 10$ kΩ; **(b)** 12 V
9.15 See Figure 9.14. $R = 8$ kΩ
9.17 $I_1 = 1.5$ mA, $I_2 = 0.5$ mA, $I_3 = 2$ mA, $V_A = -3.75$ V, $V_B = -13.75$ V
9.19 **(a)** -10 V; **(b)** 3 s
9.21 $1.2 \cos(500t - 30°)$ or $1.2 \sin(500t + 60°)$ V
9.23 See Figure 9.18. (One solution) $R_1 = 100$ kΩ, $C = 1$ μF, $R_c = 100$ kΩ
9.25 See Figure 9.21. (One solution) $R_1 = 318$ kΩ, $R_f = 318$ kΩ, $R_c = 159$ kΩ, $C = 0.01$ μF

9.27 **(a)** $-\int(200\,v_1 + 40\,v_2)\,dt$; **(b)** See Figure 9.24 (with 2 inputs). (One solution) $R_1 = 50$ kΩ, $R_2 = 250$ kΩ, $R_c = 41.67$ kΩ, $C = 0.1$ μF

9.29 **(a)** $0 - 482$ Hz; **(b)** 0.305; **(c)** See Figure 9.28. $R_f/R_1 = 12.27$, $f_b = 4.82$ kHz, $f_2 = 150.7$ kHz

9.31

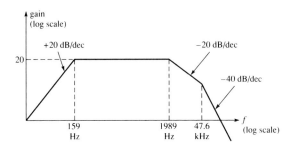

9.33

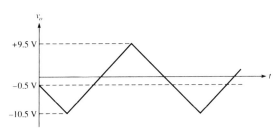

9.35 (Phase inversions not shown)

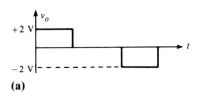

(a)

(b)

(c) Same as (b)

9.37 **(a)** $V_{o1} = 12$ V, $V_{o2} = -6$ V, $I = I_1 = I_2 = I_3 = 2$ mA, $V_o = 18$ V; **(b)** $V_{o1} = -6$ V, V_{o2} $= 12$ V, $I = I_1 = I_2 = I_3 = -2$ mA, $V_o = -18$ V

9.39 3

9.41 3

9.43 See Figure 9.46. (One design): $C = 0.01$ μF, $R_1 = 2.65$ kΩ, $R_2 = 1.23$ kΩ, $R_3 = 1.64$ kΩ, $R_4 = 4.92$ kΩ

9.45 See Figure 9.48. (One design): $C = 0.01$ μF, $R_1 = 1.99$ kΩ, $R_2 = 83$ Ω, $R_3 = 7.96$ kΩ

9.47 See Figure 9.49. (One design): $C = 0.1$ μF, $R_1 = 2.65$ kΩ, $R_2 = 2.81$ kΩ, $R_3 = 3.98$ kΩ, $R_4 = 3.98$ kΩ

9.49 See Figure 9.49. (One design): $C = 0.001$ μF, $R_1 = 3.98$ kΩ, $R_Q = 16.55$ kΩ, BW $= 45.2$ kHz

CHAPTER 10

10.1 Oscillates at 4×10^3 rad/s

10.3 See Figure 10.2. (One solution) $C = 0.1$ μF, $R = 433$ Ω, $R_1 = 5$ kΩ

10.5 See Figure 10.4. (One solution) $R_1 = R_2 = 8842$ Ω, $C_1 = C_2 = 100$ pF, $R_f = 20$ kΩ, $R_g = 10$ kΩ

10.7 **(a)** 121.4 kHz; **(b)** 22 kΩ

10.9 Z (feedback) $= jX_{L1} \parallel (j\omega L_2 - jX_C)$

$$= \frac{-X_{L1}X_{L2} + X_{L1}X_C}{j(X_{L1} + X_{L2} - X_C)}$$

$$X_{L1} + X_{L2} - X_C = 0 \Rightarrow \omega L_1 + \omega L_2 = \frac{1}{\omega C} \Rightarrow$$

$$\omega = \frac{1}{\sqrt{L_T C}} \text{ where } L_T = L_1 + L_2;$$

$$\beta = \frac{jX_{L2}}{jX_{L2} - j(X_{L1} + X_{L2})} \text{ (at resonance)}$$

$$= \frac{X_{L2}}{-X_{L1}} = \frac{-L_2}{L_1}$$

10.11 2 kHz

CHAPTER 11

11.1

(a)

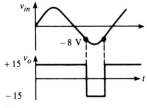

(b)

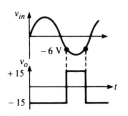

(c)

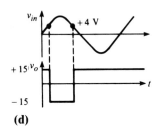

(d)

11.3 **(a)** LTL = -8.6 V, UTL = $+6.74$ V; **(b)** 15.34 V;

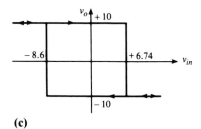

(c)

11.5 (One design):

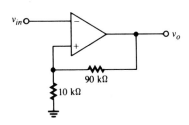

11.7 (One design);

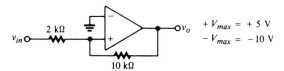

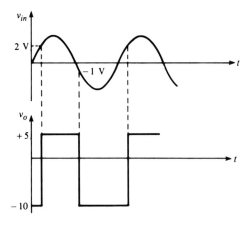

11.9 See Figure 11.5. (One design): $C = 0.1$ μF, $R = 4.55$ kΩ, $R_1 = R_2 = 10$ kΩ

11.11

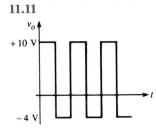

(a)

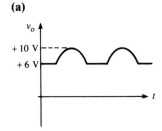

(b)

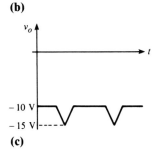

(c)

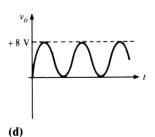

(d)

11.13

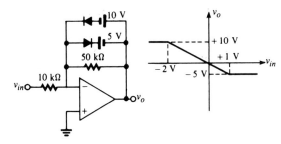

11.15

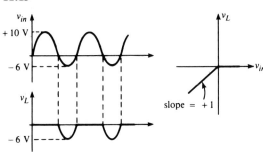

11.17

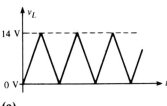

(a)

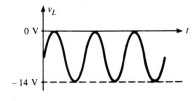

(b)

11.19

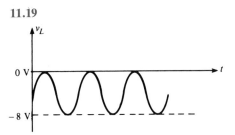

11.21   17.8 mV

11.23 $v_L = i_o R_L = g_m(v_i^+ - v_i^-)R_L \Rightarrow \dfrac{v_L}{v_i^+ - v_i^-} = g_m R_L$

11.25   25 μA

11.27   50 kHz

CHAPTER 12

12.1   **(a)** 0.625 V; **(b)** 3.203125 V

12.3   One solution (see Figure 12.7): $R_f = 10$ kΩ, $R = 6.4583$ kΩ, $2R = 12.917$ kΩ, $4R = 25.833$ kΩ, $8R = 51.667$ kΩ, $16R = 103.33$ kΩ; output = -4.8387 V

12.5 $I_o = 0.125$ mA; $I_1 = 0.250$ mA; $I_2 = 0.500$ mA; $I_3 = 1.00$ mA

12.7 -9.9609375 V to $+9.9609375$ V

12.9   **(a)** 1.25 pF; **(b)** 2.75 pF; **(c)** 2.5 V

12.11   **(a)** 0.0244%; **(b)** minimum = 1.8315 mV; maximum = 5.4945 mV

12.13   255 μs

12.15   **(a)** 31; **(b)** 0.3125 V

12.17   **(a)** 2 V; **(b)** 8 μs; **(c)** 00011001

12.19   1000, 1100, 1010, 1011

12.21   **(1)** flash; **(2)** successive approximation; **(3)** dual-slope; **(4)** tracking counter; **(5)** counter

12.23   39.06 mV; 0.3906%

12.25   50 μs

CHAPTER 13

13.1   39.1 V p-p

13.3   **(a)** 120/16.66 = 7.2 : 1; **(b)** 27.76 W; **(c)** 47.12 V

13.5

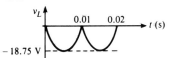

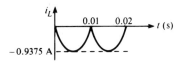

13.7 **(a)** 134.3 V; **(b)** 890 W; **(c)** 253.15 V
13.9 48.4% in every case
13.11 96 Ω
13.13 5.54 A
13.15 **(a)** 0.202 V rms; **(b)** 3.96 × 10⁻³; **(c)** 111.11 Ω, 35.6 µF
13.17 17.5 mA, 0.01%
13.19

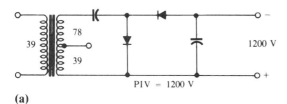

(a)

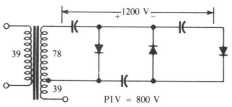

(b)

13.21 $V_L = [R_L/(R_L + R_o)]V_{NL}$; $R_L = R_o \Rightarrow V_L = 0.5 V_{NL}$
13.23 0.075%
13.25 **(a)** 40 mA, 100 mA; **(b)** 0.48 W, 1.2 W; **(c)** 1.28 W
13.27 67.2 Ω ≤ R_S ≤ 117.7 Ω
13.29

(a)
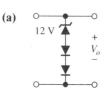

(b) At 25°C, 13.36 V; At 75°C, 13.364 V; **(c)** 5.998 × 10⁻⁴%
13.31 **(a)** 66.15 Ω; **(b)** $\Delta V_Z/\Delta I_Z$ decreases from 100 Ω to 32 Ω as I_Z increases from 0.5 mA to 90 mA.
13.33 **(a)** 9.7 V; **(b)** 230 Ω; **(c)** 0.603 W
13.35

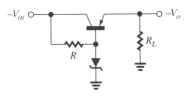

13.37 **(a)** 5 V; **(b)** 22.5 mA
13.39 **(a)** 12 V; **(b)** 10.5 V
13.41 **(a)** 15.75 V; **(b)** 1.065 A
13.43 **(a)** 0.6 A; **(b)** 5.4 W
13.45 **(a)** 2.7 V; **(b)** 2.1 V
13.47 33.3%; 66.7%
13.49 **(a)** 0.0645%; **(b)** 60 mV
13.51

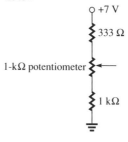

CHAPTER 14

14.1 decrease; 225 mW
14.3 400 Ω
14.5 8.4 J/s
14.7 1°C/W
14.9 0.127°C/W
14.11 40°C
14.13 **(a)** 24 mA; **(b)** 0.144 W
14.15 **(a)** 0.08333; **(b)** $R_C \neq \sqrt{2}R_L$
14.17 **(a)** 0.01 A; **(b)** 4 W; **(c)** $\eta = \eta_C = 0.5$
14.19 **(a)** 24 W; **(b)** 61.12 W; **(c)** 18.56 W; **(d)** 0.3927
14.21 50.6 V
14.23 **(a)** 20 V; **(b)** 16.66 W; **(c)** 40 V
14.25 **(a)** 21 mA; **(b)** $V_{B1} = 15.7$ V, $V_{B2} = 14.3$ V; **(c)** 0.493; **(d)** 37.5 Hz
14.27 3.19 V
14.29 6.65 V
14.31 $\sin A \sin B = \frac{1}{2}\cos(A - B) - \frac{1}{2}\cos(A + B)$. Let $A = \omega_c t$ and $B = \omega_s t$ in 14.68; then 14.69 follows immediately.
14.33 $11.314 \sin(200\pi \times 10^3 t) + 1.414 \cos(180\pi \times 10^3 t) - 1.414 \cos(220\pi \times 10^3 t)$
14.35

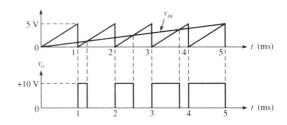

14.37

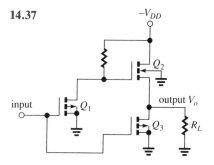

The input must be $-V_{DD}$(low) and 0 V (high). When the input is $-V_{DD}$, Q_1 and Q_3 are ON, and Q_2 is OFF. V_o is therefore 0 V (high). When the input is 0 V, Q_1 and Q_3 are OFF and Q_2 is ON. V_o is therefore $-V_{DD}$(low).

Index

C